S0-BZS-908

Wood

TECHNOLOGY & PROCESSES

5th Edition

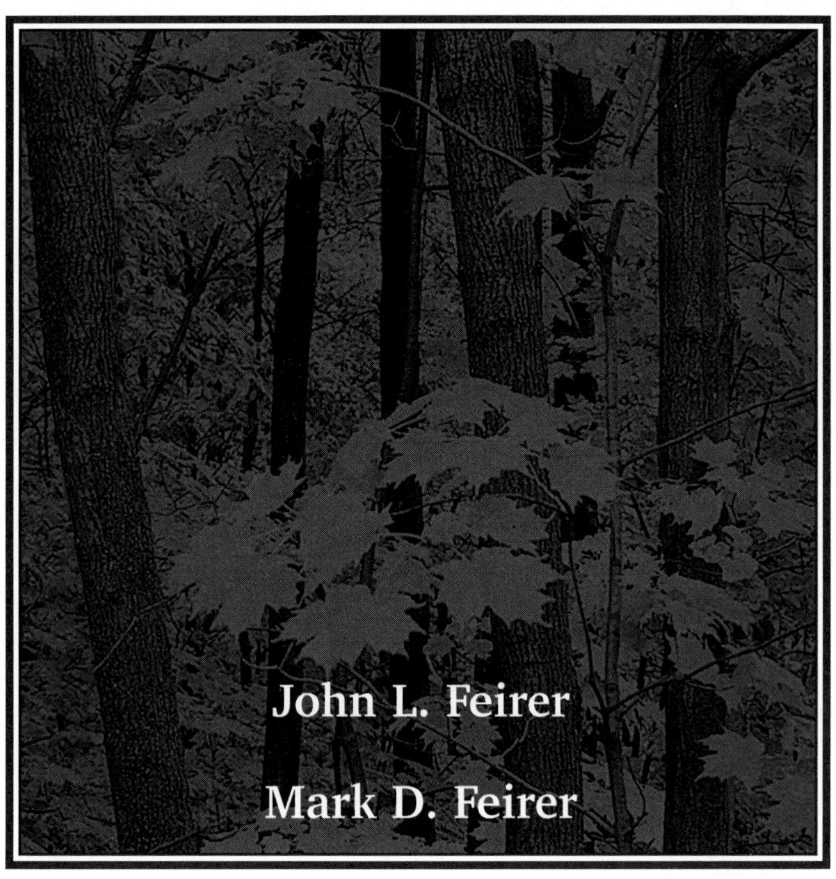

New York, New York   Columbus, Ohio   Woodland Hills, California   Peoria, Illinois

NOTICE
Publisher does not make any representation, warranty, guarantee or endorsement of any of the products, their use or the methods or techniques described in this book. Publisher has not, and expressly disclaims any obligation to independently test any products or to verify facts, information, methods or techniques described in this book.

The reader is expressly advised to consider and use all safety precautions described in this book or that might also be indicated by undertaking the activities described herein. In addition, common sense should be exercised to help avoid all potential hazards.

Publisher assumes no responsibility for the activities of the reader or for the subject matter experts who prepared this book. Publisher makes no representation or warranties of any kind, including but not limited to, the warranties of fitness for particular purpose or merchantability, nor for any implied warranties related thereto, or otherwise. Publisher will not be liable for damages of any type, including any consequential, special or exemplary damages resulting, in whole or in part, from reader's use or reliance upon the information, instructions, warnings or other matter contained in this book.

NOTICE
Publisher does not necessarily recommend or endorse any particular company or brand name product that may be discussed or pictured in this text. Brand name products are used because they are readily available, likely to be known to the reader, and their use may aid in the understanding of the text. Publisher recognizes other brand name or generic products may be substituted and work as well or better than those featured in the text.

Glencoe/McGraw-Hill
A Division of The ***McGraw-Hill*** *Companies*

Copyright © 2002 by Glencoe/McGraw-Hill. Previous copyrights 1994 by Glencoe/McGraw-Hill and 1987, 1980, and 1975 by John L. Feirer. All rights reserved. Except as permitted under the United States Copyright Act, no part of this publication may be reproduced or distributed in any form or by any means, or stored in a database or retrieval system, without prior written permission of the publisher.

Send all inquiries to:
Glencoe/McGraw-Hill
3008 W. Willow Knolls Drive
Peoria, IL 61614

ISBN 0-07-822411-X (Student text)
ISBN 0-07-822412-8 (Teacher's Resource Guide)
ISBN 0-07-822413-6 (Student Workbook)

Printed in the United States of America.

4 5 6 7 8 9 10 071/071 04 03 02

ACKNOWLEDGMENTS

Contributors and Reviewers

The following people helped in the revision of this text either by reviewing the text or by offering their woodworking expertise as consultants.

Gerald L. Andrew
Oologah-Talala Public Schools
Oologah, OK

Steve E. Burns
R-S Central High School
Rutherfordton, NC

Roger Buss and Phil Fifield
Woodcraft Store #522
Peoria, IL

Dick Coers
Coers Custom Woodworking
Peoria, IL

Ron Foy
Woodworking Instructor
Pflugerville, TX

Stan M. Hoer
Nevada Union High School
Grass Valley, CA

Darren Howlett
Howlett's Industries
Tremont, IL

David Keller
Keller and Company
Petaluma, CA

Rev. Lawrence Kruzan
Woodworker and Instructor
Pekin, IL

Jim Myers
Taylor High School
Kokomo, IN

Jennifer Wofford Pratt
Houston High School
Houston, MS

Jason Snyder
Woodworkers Shop
Pekin, IL

Laze Stewart
Booker T. Washington High School
Memphis, TN

James Ulmer
Elloree High School
Elloree, SC

Kevin Wagner and Richard Jones
Limestone High School
Bartonville, IL

George D. Wertz
Deming High School
Deming, NM

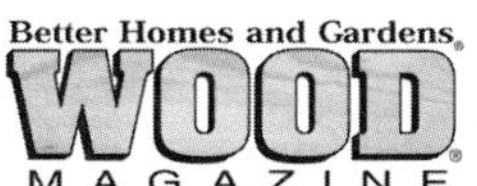

Woodworking Projects

The woodworking projects beginning on page 513 are adapted from projects published in *Better Homes and Gardens®WOOD®* magazine.

CONTENTS IN BRIEF

Using Machines

Finishing

Construction

TABLE OF CONTENTS

Basic Tools and Operations

Joinery and Assembly

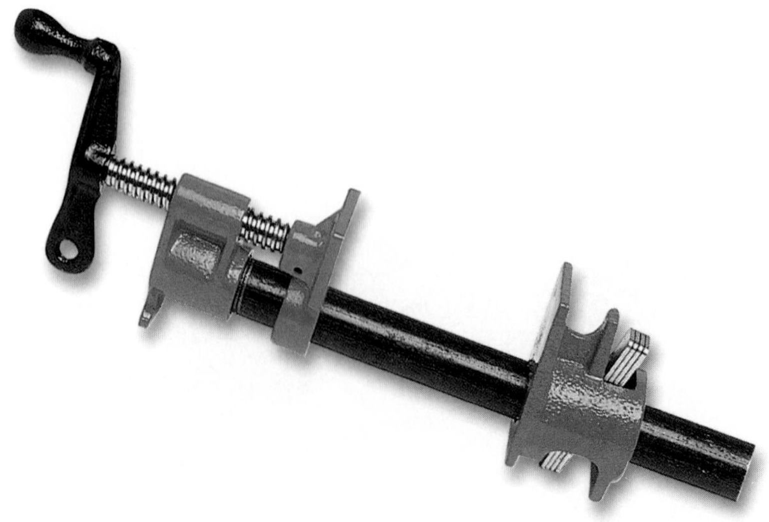

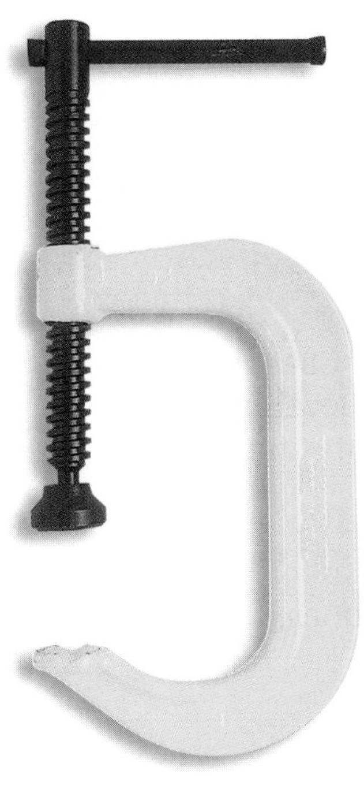

Using Machines

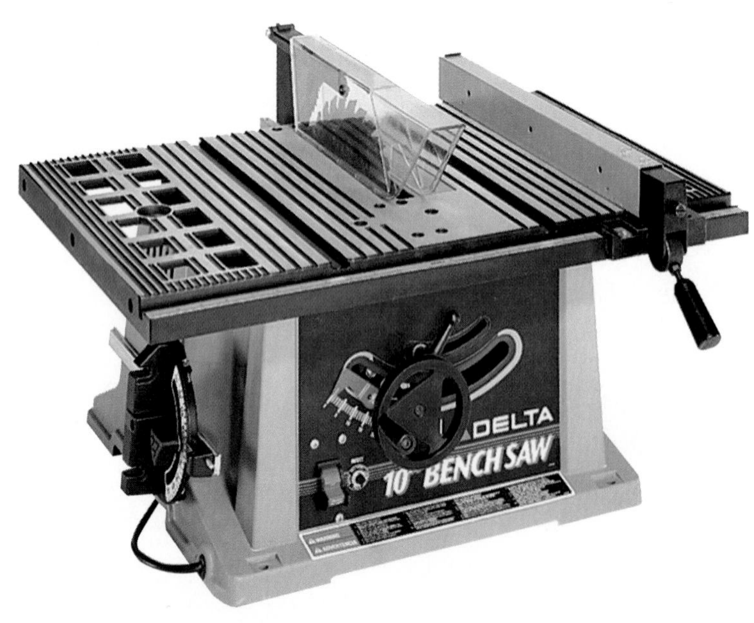

PORTER+CABLE
SPEEDMATIC

Finishing

Construction

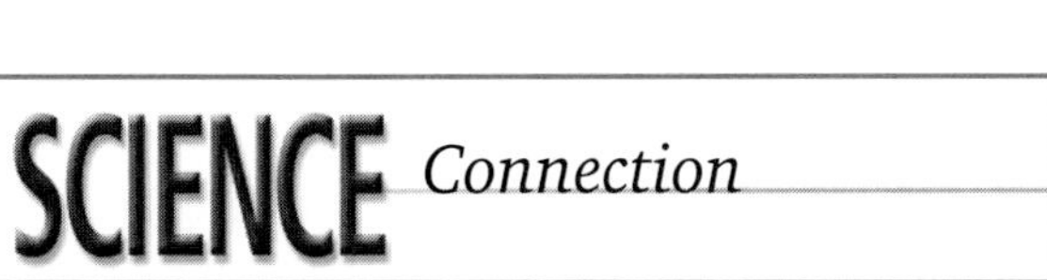

SCIENCE Connection

MATHEMATICS *Connection*

WOODWORKING TIPS

Woodworker's Handbook

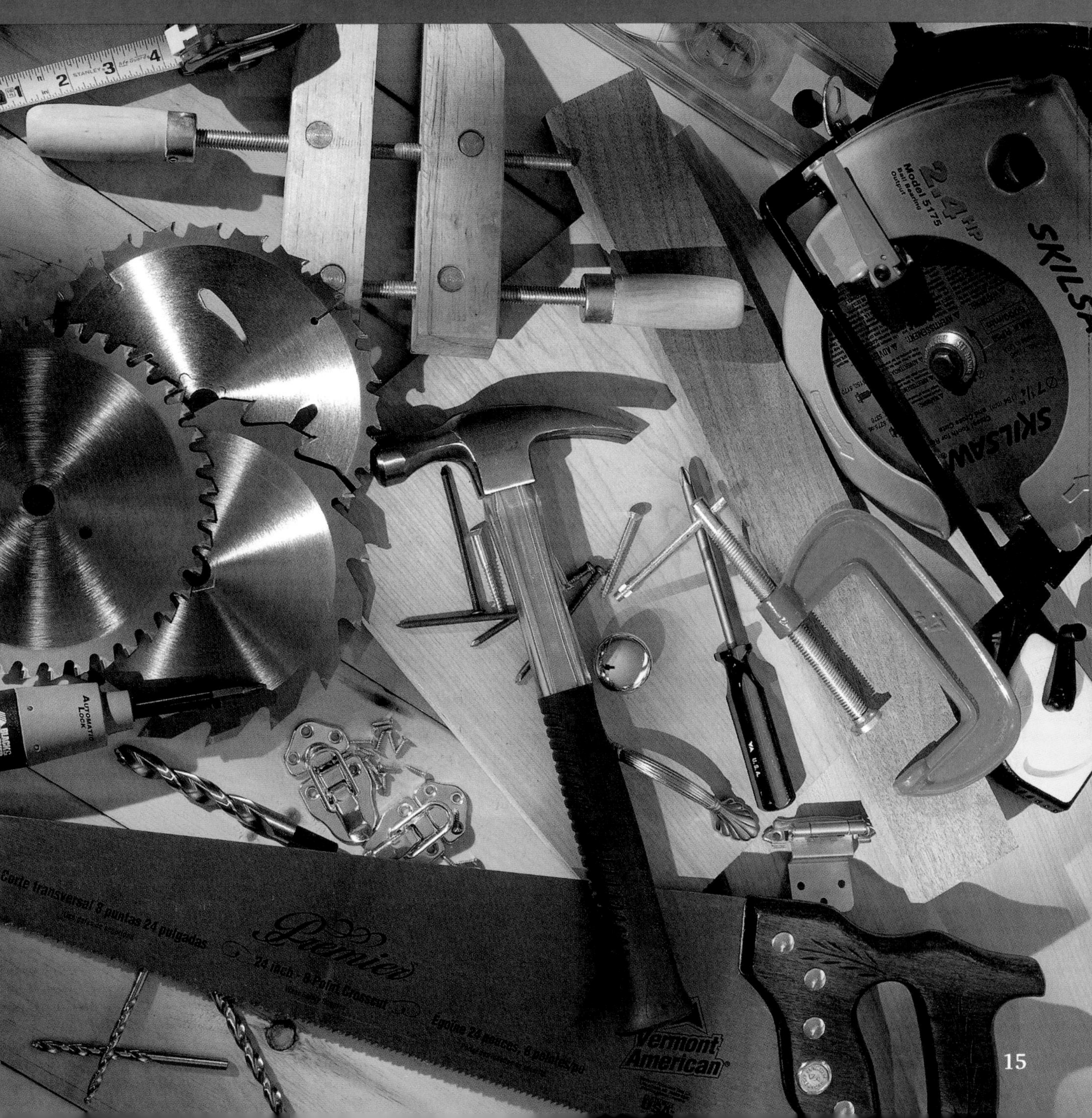

Measuring & Marking Tools

Accurate measurements are important on any woodworking project. A good selection of instruments for measuring lengths, widths, depths, and angles is therefore a must for a good workshop.

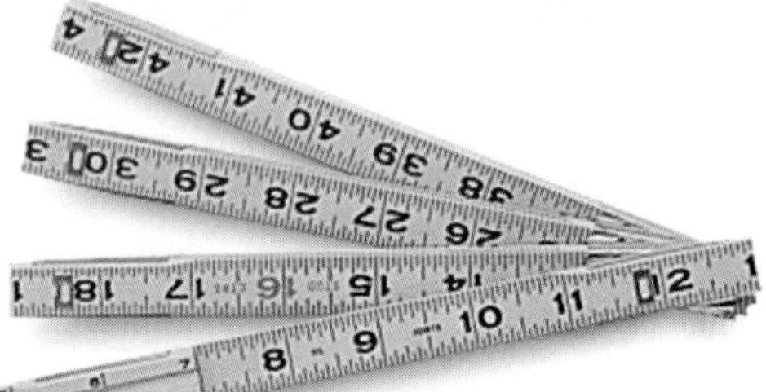

Folding wooden rule—usually 6 to 8 feet long, this rule is good for inside measurements. The readings on the brass extension can be added to the length of the rule itself.

Tape measure—a flexible tape that slides into a case. The tape has a hook on the end that adjusts to true zero.

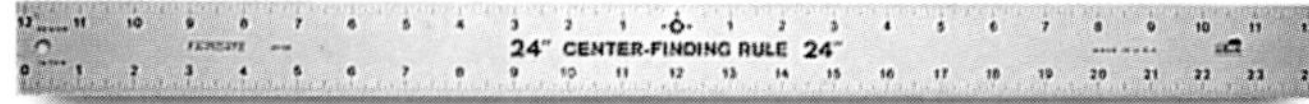

Bench rule—a steel or hardwood rule that is used for short measurements. One side is usually divided into eighths, the other into sixteenths.

Long tape measure—used for large measuring jobs such as measuring a building site or laying out a house.

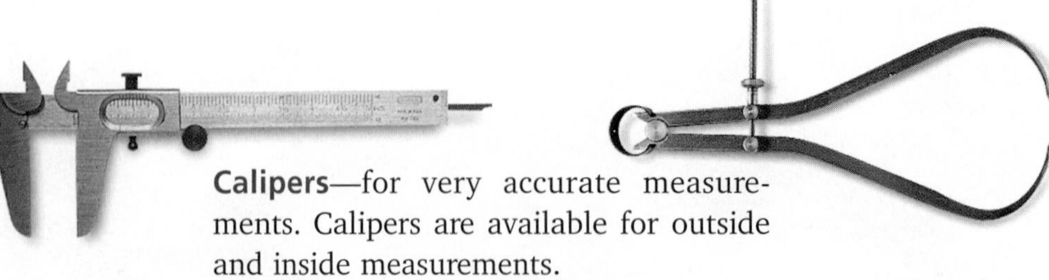

Calipers—for very accurate measurements. Calipers are available for outside and inside measurements.

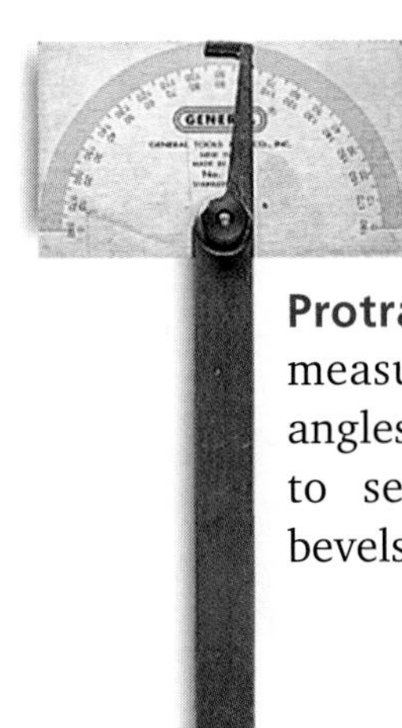

Protractor—used to measure and mark angles. It is often used to set and transfer bevels accurately.

Compass—for drawing circles and arcs. It can also be used to step off equal measurements.

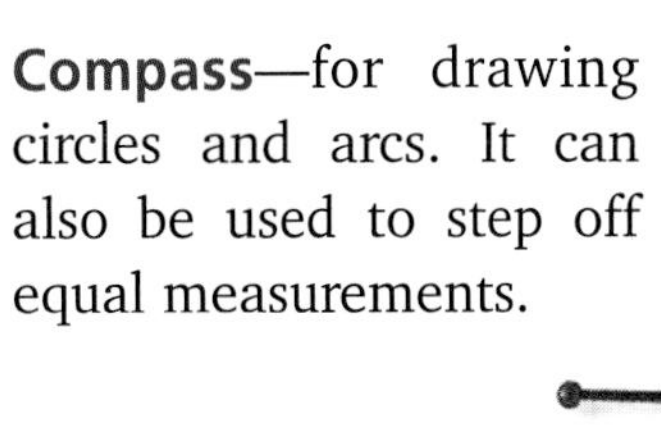

Flexible curve—bends to and holds any shape for drawing irregular curves.

Trammel points—used to draw circles and arcs when a compass is too small.

Sliding T-bevel—for checking and transferring angles. The blade pivots and can be locked to match any angle.

Profile gauge—used in finish carpentry to transfer an irregular shape or design. It is made of small, movable pins that, when pushed against an object, take on the contour of that object.

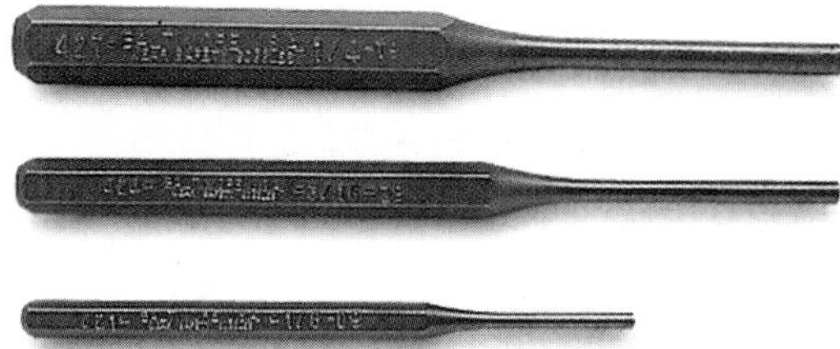

Punches—the center punch is used to make a starting point in wood or metal. The pin punch is flat on the end and is used to remove assembling pins.

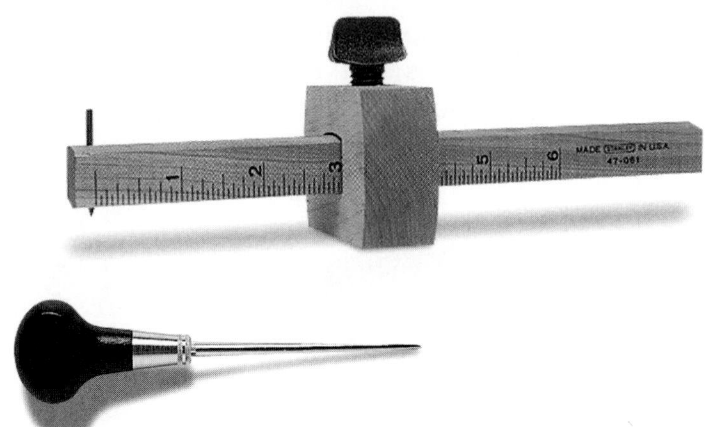

Marking gauge—accurately draws a line parallel to any straight edge at a distance you specify.

Scratch awl—used instead of a pencil to make fine layout lines, especially on metal. It can also be used to make small holes for starting drill bits, nails, or screws.

Chalk line—marks cutting or layout lines between two points. The case is filled with chalk and a long line of cord on a spool. Hold the line taut between the two points and snap it to leave a line of chalk.

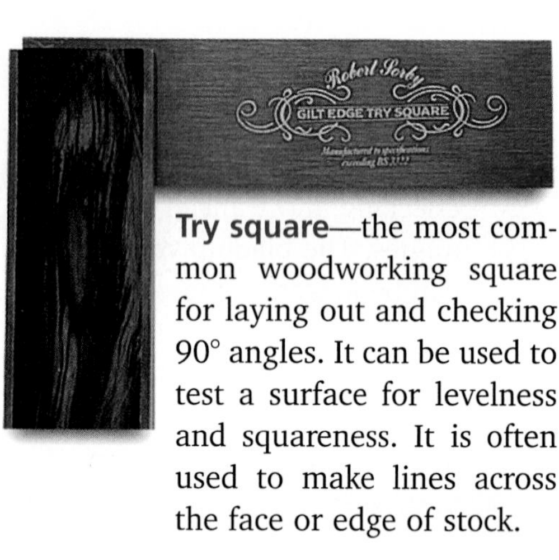

Try square—the most common woodworking square for laying out and checking 90° angles. It can be used to test a surface for levelness and squareness. It is often used to make lines across the face or edge of stock.

Carpenter's square—usually has a 2″ × 24″ blade and a $1\frac{1}{2}''$ × 16″ tongue. Made of metal, this square is used for laying out lines and squaring when a smaller square would not be as accurate. A framing square is a special version that has formulas printed on it for making quick calculations, usually for framing a roof.

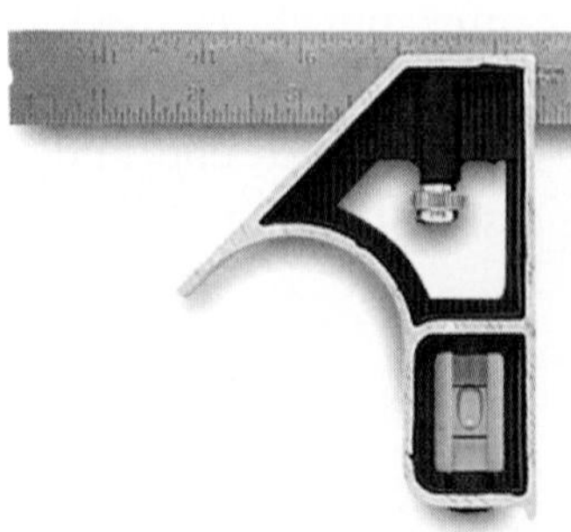

Combination square—used to check and lay out 90° and 45° angles. The handle slides on the blade so that it can be used as a depth gauge. A spirit level in the handle can be used to check level and plumb.

Rafter angle square—marked with degrees for fast layout. Its small size makes it handy for quick layouts.

Carpenter's level—used to check level and plumb. Digital models are also available. Standard lengths are 24″ and 48″.

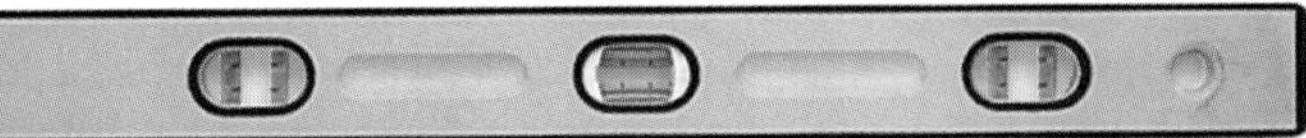

Torpedo level—preferred by plumbers. It is shorter than other levels and can be used in small areas.

Builder's level—often based on a laser for accuracy. Levels are used by surveyors to ensure plumb walls and in brick and block laying. Builders often use it to set cabinets in kitchens.

Plumb bob—establishes a plumb (vertical) line.

Stud finder—finds the studs in a wall by measuring the density of the wall.

Line level—a miniature level attached to a taut string line between two points. It is handy for landscaping, erecting fences, and doing masonry work.

Hammers & Mallets

Hammers come in a variety of shapes and sizes. The key is to find one that feels comfortable and is appropriate for the work you are doing. A good hammer has a forged steel head with a hardwood, graphite, or fiberglass handle. For most work other than finish work (it will mar the surface), a milled face on the head is preferable. It will help prevent glancing blows and flying nails.

Ripping hammer—has a wedge-shaped claw used for prying apart pieces that have been nailed together. It has a mesh-type face for rough framing work. It should not be used for finish work.

Tack hammer—a small, lightweight hammer that holds and sets tacks. It usually weighs only 5 to 8 ounces. It is used for picture framing, cabinetmaking, trim, and upholstery.

Claw hammer—the most commonly used hammer. The curved claw provides leverage for pulling nails. The face may be flat, bell-shaped, or checkered. For most work, a 13- or 16-ounce head weight is appropriate.

Dead-blow hammer—head is filled with lead shot and oil so that the energy of the impact is absorbed in the head. This eliminates any rebound in assembly work.

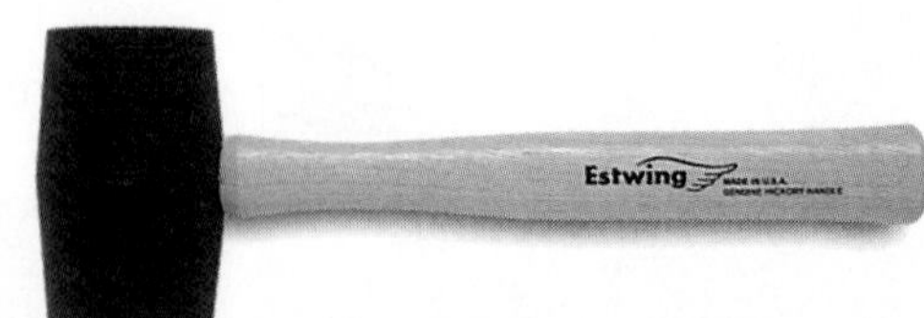

Rubber mallet—used mainly for assembling projects.

The Best Tool

Buy the best tool that you can afford. Think of it as an investment. Buying good-quality tools will save you money in the long run. Good tools can last a lifetime, and they hold their value much better than any other manufactured item you can buy. Don't just assume that more expensive tools are better, though. Ask for advice from experienced friends and professionals and check the trade magazines for reviews of the various kinds of tools.

Carpenter's mallet—used for assembling projects and striking chisels.

Ball-peen hammer—one face is used to strike cold chisels and punches; the other face is for shaping soft metal.

Power brad nailer—drives and countersinks brads without marring the surface of the wood.

Nail set—conceals the heads of finishing nails by driving them below the surface of the wood.

Pry bar—one end is a nail puller, and the other end is used for prying or separating materials that are nailed together.

Ripping bar—for pulling large nails and removing old materials during renovation or remodeling.

Nail puller—sometimes called a "cat's paw." It is driven under the head of a nail so that the nail can be pulled out more easily.

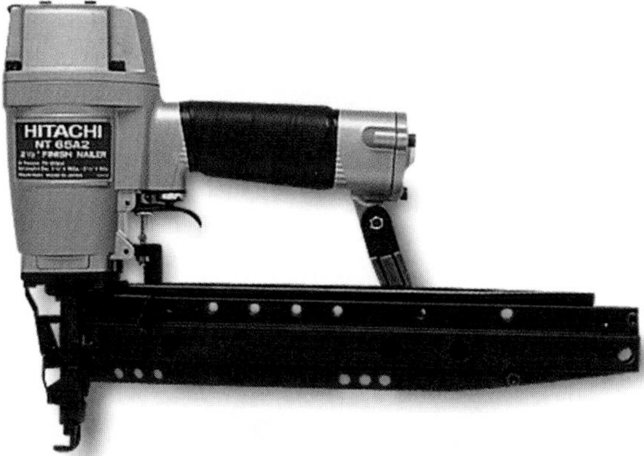

Finish nailer—drives and countersinks finish nails up to 2″ long. It allows controlled nailing because one hand is free to hold the work.

Drills & Drill Bits

In woodworking, holes are drilled as part of a project design, as pilots for screws and nails, or to hold bolts, dowels, or hardware. Hand drills are still used, but electric drills are used much more often because they are handier and more versatile. They can be operated very slowly, like a hand drill, for complete control. They can also be used at full force to "power" a big screw into a hard piece of wood.

Brace—bores larger holes in wood by hand. Special auger bits must be used with the brace.

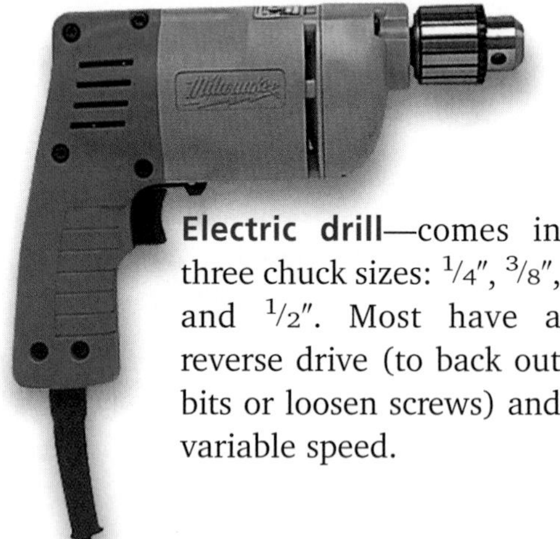

Electric drill—comes in three chuck sizes: $^{1}/_{4}$″, $^{3}/_{8}$″, and $^{1}/_{2}$″. Most have a reverse drive (to back out bits or loosen screws) and variable speed.

Hand drill—bores holes in wood, plastic, and soft metal. It uses twist drills with $^{1}/_{4}$″ shank bits.

Push drill—operates by pushing the handle up and down in a repetitive motion. It drills holes up to $^{11}/_{64}$″.

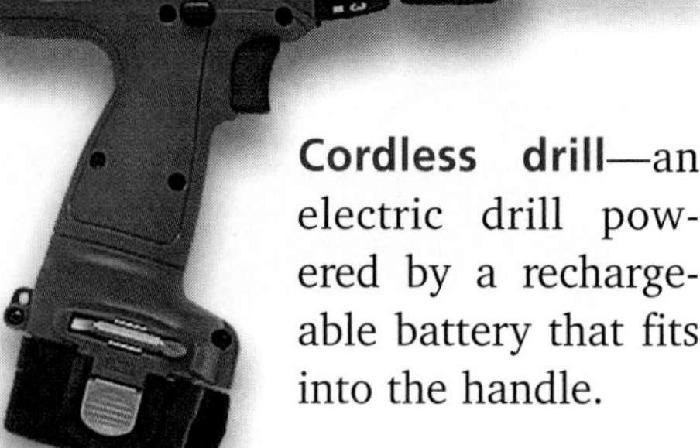

Cordless drill—an electric drill powered by a rechargeable battery that fits into the handle.

Variable speed rotary tool—can be used with many tool attachments, including small drill bits. It is used mostly in crafts.

Drywall driver—used specifically for driving drywall screws into drywall. It is designed so that it will not drive the screw in too far and damage the surface of the drywall.

Twist bit—designed for wood. If you use it with metal, lubricate it with machine oil.

Brad-point bit—has a center point to help guide the drill bit to the desired position. It drills a clean hole, like those needed for fine woodworking.

Forstner bit—drills a smooth, shallow hole. It has a small center spur, so it can drill a nearly flat-bottomed hole. Forstner bits should be used only in a drill press.

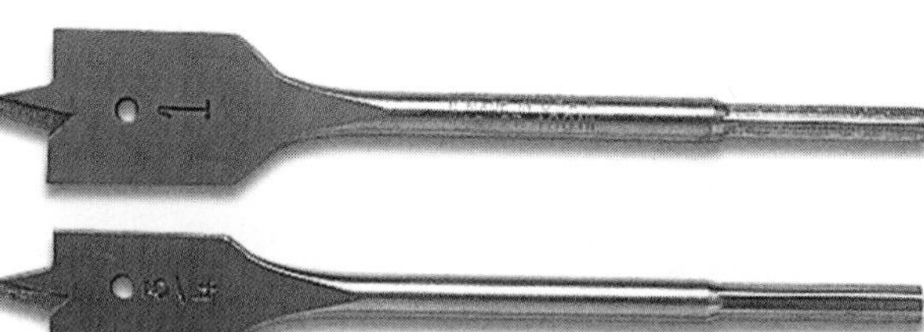

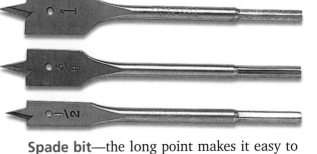

Spade bit—the long point makes it easy to locate the hole exactly where you want it. Start the drill at a slow speed as it enters the wood. If you are not careful, it will leave a splintered exit hole.

Fly cutter—also called a *circle cutter.* It should be used only in a drill press.

Expansion bit—can be adjusted to cut holes up to 3″. Should *not* be used with an electric drill.

Glass/tile bit—drills holes in glass and tile; must be used at a slow speed.

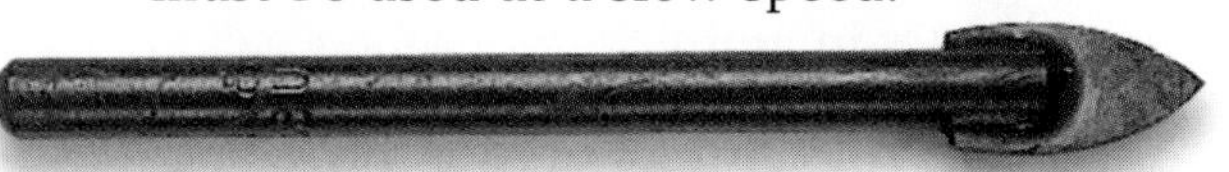

Step bit—has up to 13 diameters on it. The hole gets larger as the bit goes deeper into the wood.

Screw pilot bit—drills the shank and pilot hole for wood screws. The size of the bit needed is determined by the screw size.

Countersink bit—drills a neat taper for the head of a wood screw.

Hole saw—cuts large holes in wood, plastic, and thin metal. It is mounted onto a special arbor with a bit in the middle to guide the saw into the wood in the correct location.

Multispur bit—bores very clean holes at any angle. It should be used only in a drill press.

Keyless drill chuck—A chuck that does not require a key for tightening. Drill bits can be tightened by hand.

Auger bit—for an electric drill.

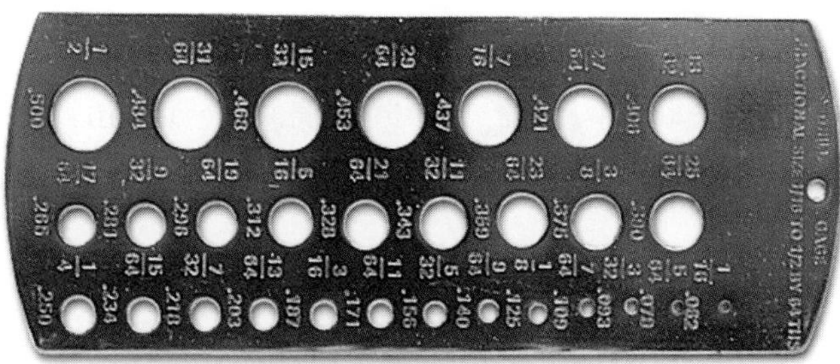

Drill gauge—measures the size of a drill bit. Insert the drill into the holes until you find a perfect fit.

Doweling jig—Centers dowel holes in the end or edge of wood. The jig clamps to the wood and centers itself on the edge.

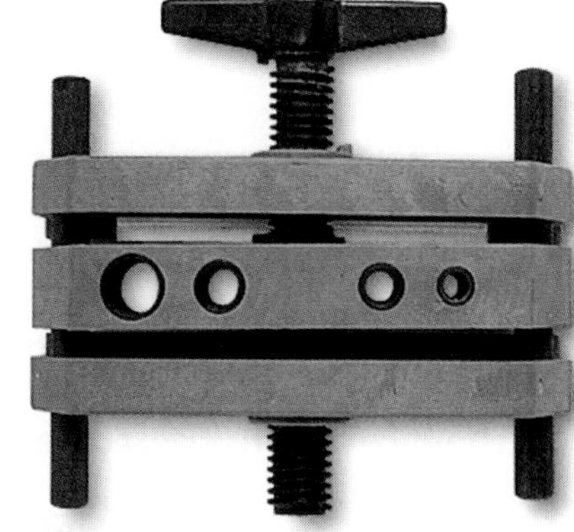

Flap wheel sander—converts an electric drill into a flexible sander; as the drill turns, the sandpaper strips conform to the contoured surface of the wood.

Depth stop—slides onto a bit to stop the bit at a certain depth in the wood.

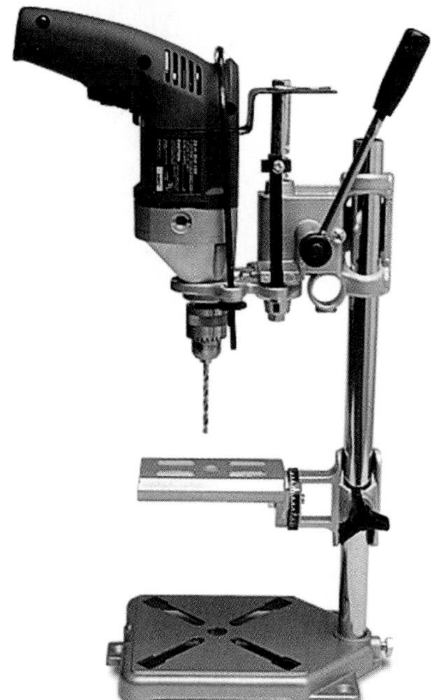

Vertical drill stand—converts an electric drill into a small benchtop drill press.

Precision drill stand—fits onto the drill in place of the chuck. It can drill perfectly perpendicular holes, or it can be set to drill holes at a specific angle.

Screwdrivers

Screwdrivers come in many different lengths, widths, and colors. They are available with different shanks and a variety of tips. They are used to insert or remove screws. Screws are driven into a wood piece by torque, not by pressure. The larger the handle on a screwdriver, the more torque can be generated. Screwdrivers should not be used as a pry bar or a chisel.

Cordless adjustable power screwdriver—does everything a hand screwdriver will do with little effort. It has interchangeable tips.

Spiral ratchet screwdriver—allows you to drive a screw by turning the handle of the screwdriver.

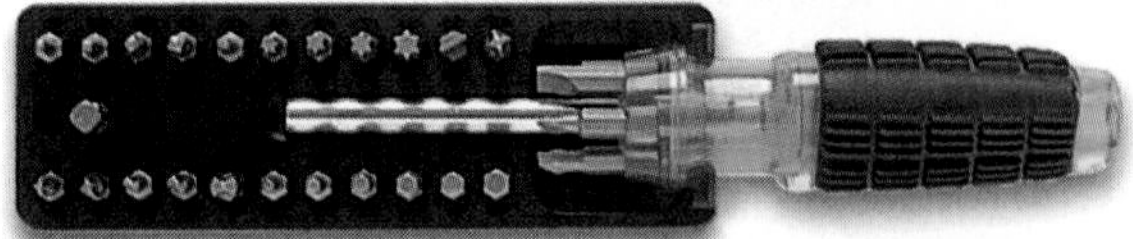

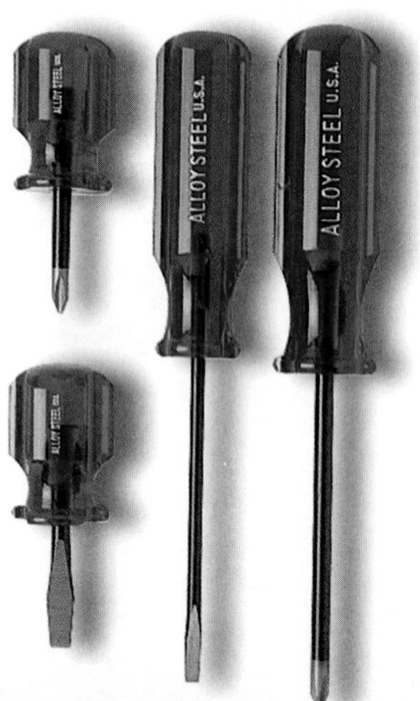

Screwdriver set—top and bottom left: Phillips-head and standard stubby screwdriver; middle: standard (slotted) screwdriver; right: Phillips-head screwdriver.

Offset screwdriver—used in hard-to-reach places where a normal screwdriver cannot be used; may be slotted or Phillips.

Nutdriver—acts like a screwdriver but turns hexagonal nuts and bolts instead of screws.

Knives & Snips

Sharp, accurate cutting tools are essential to good wood joints and project assemblies. Careful measurements and accurate cuts are the keys to good woodworking projects.

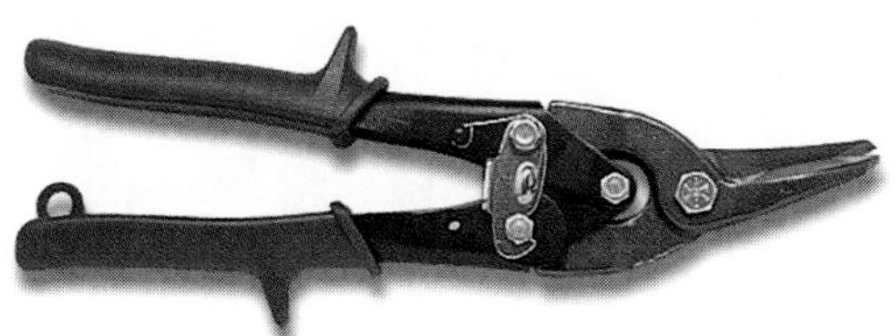

Aviation snips—easier to use on metal than tin snips; especially designed to make curved and straight cuts in metal.

Utility knife—used for many jobs around the woodworker's shop. For safety, the blade can be retracted into the handle when not in use.

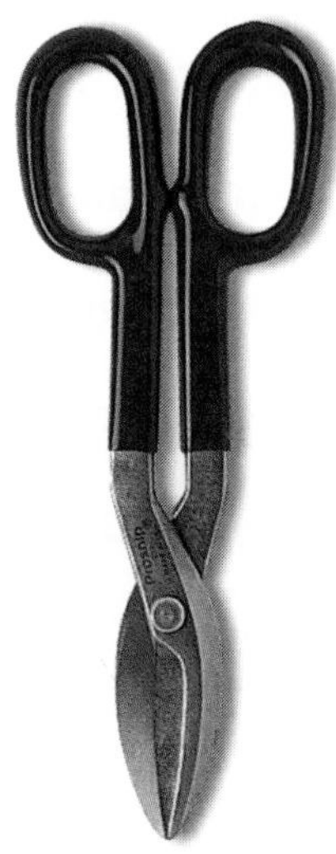

Tin snips—used to cut lightweight sheet metal; can make straight cuts and wide curves.

Putty knife—has a flexible blade to spread and smooth wood putty and filler; comes in several widths.

Razor-blade scraper—scrapes paint and stickers off windowpanes. It uses a single-edge retractable razor blade.

Tool Care

Sharp Tools

Working with sharp woodworking tools can be a great experience. Working with dull tools is not. Dull tools require more work, and the results are poor. Dull tools are actually unsafe and difficult to control. Learn how to sharpen your tools properly!

Pliers & Wrenches

All pliers have the same design: a pivot joint with a handle on each side like scissors. Use the various pliers for the job they are intended to do. Long-nose pliers are good for one thing only: to hold small objects. To use them for anything else is inefficient and might damage the work or the pliers. Use pliers to turn nuts only in an emergency. Chances are good that the pliers will strip the nut, making it very difficult to remove.

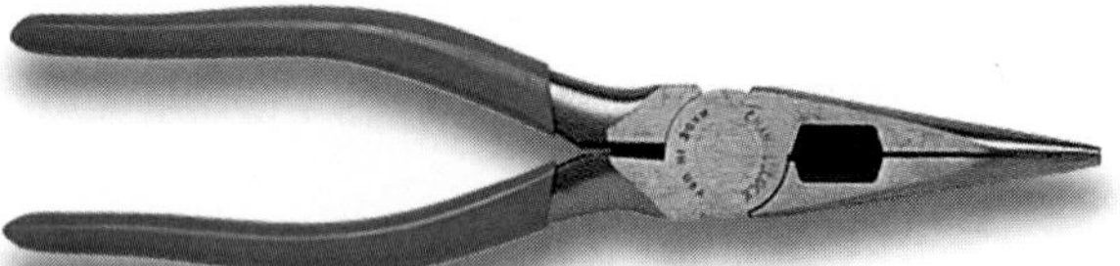

Long-nose pliers—used to hold a small object, especially in electrical work. Sometimes they have a wire cutter on the side for cutting small-gauge wire.

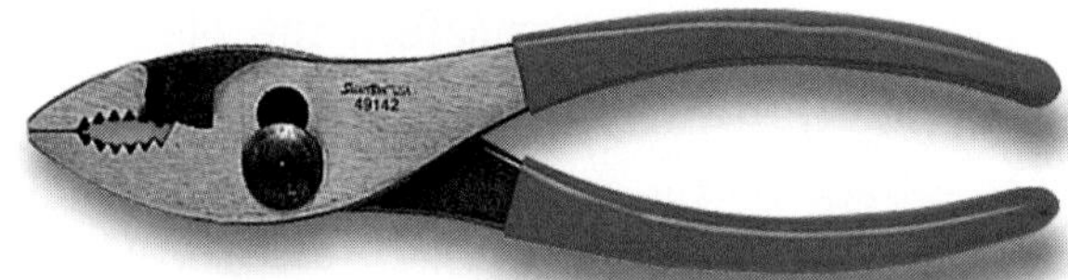

Slip-joint pliers—have small and large teeth to grip objects. The jaw size can be expanded by slipping the pivot into different positions.

Straight-jaw locking pliers—clamp firmly to an object. The jaws can be adjusted to clamp to objects of different sizes.

Lineman's pliers—used mainly for twisting and cutting wire.

Groove-joint pliers—grip objects that are round, square, flat, or hexagonal. By moving the jaws into the different grooves, you can expand them to five different sizes.

Diagonal-cutting pliers—sometimes called *side-cutting pliers.* They are designed for cutting wire and thin metal.

Wire stripper and cutter—designed to strip off the insulation on electrical wire; can also cut the wire.

Bolt cutters—heavy-duty pliers that are capable of cutting metal rods, wire, and bolts.

Adjustable wrench—can be used on a variety of bolts and nuts. It should be used only when a box or open-end wrench is not available.

Hex key—fits into setscrews. Using the short end gives you greater torque.

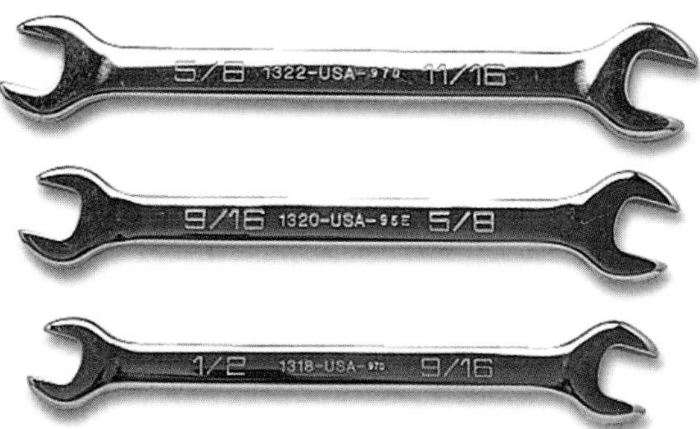

Open-end wrench—has a different size opening at each end.

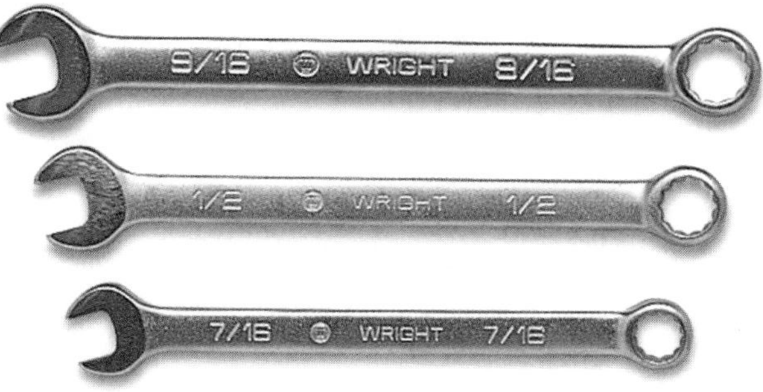

Combination wrench—has both box- and open-end heads. Both ends are for the same size bolt.

Clamps & Vises

Clamps are indispensable for holding workpieces firmly for fastening, drilling, shaping, and cutting. They keep pressure on joints while glue sets. They are like having an extra set of hands.

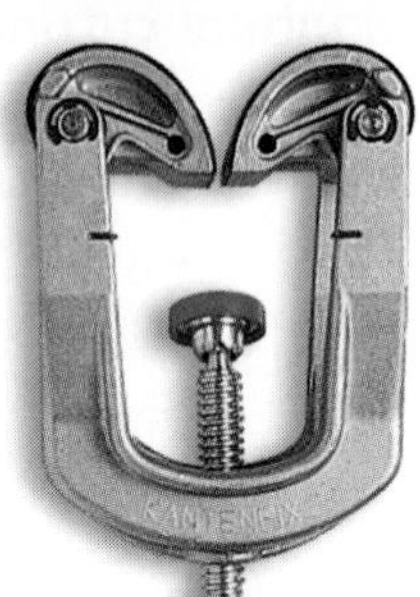

Edge clamp—applies right-angle pressure on the edge or side of a workpiece.

C-clamp—good for small jobs and when a clamp is needed in a localized area.

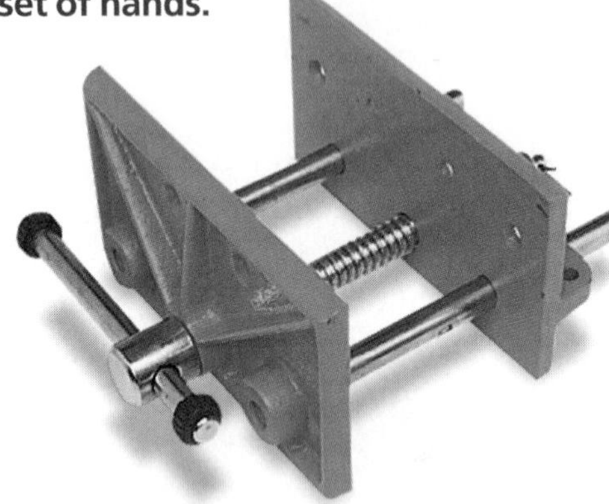

Woodworker's bench vise—mounts on the underside of a workbench. Wood inserts are usually used on the jaws to protect the wood held in the vise.

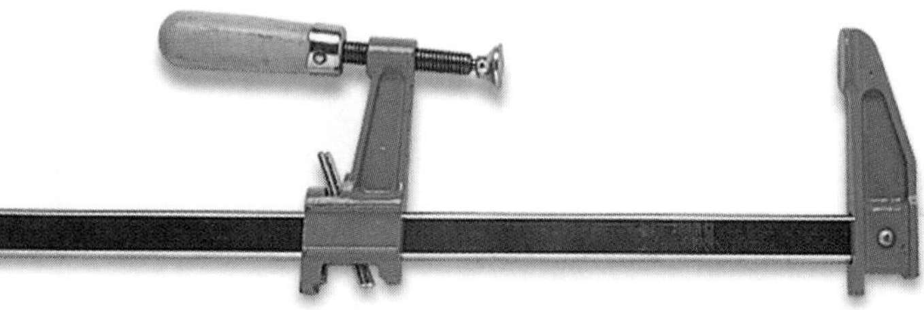

Fast-action clamp—small bar clamp that is quick and easy to operate; often used as a substitute for a C-clamp.

Machinist's vise—usually bolted to the top of the workbench. Unless the jaws have a wood protector, this vise is not used for woodworking.

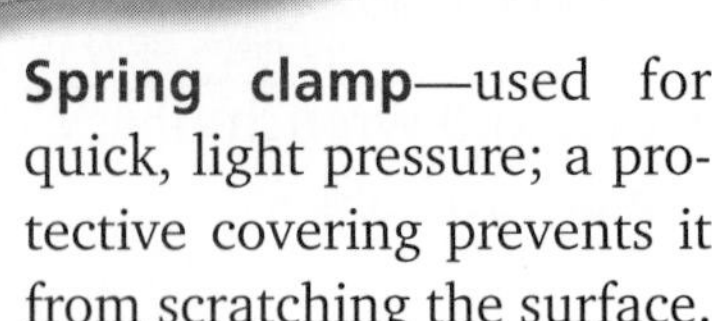

Spring clamp—used for quick, light pressure; a protective covering prevents it from scratching the surface.

Band clamp—clamps round, irregular, or box-shaped workpieces. Steel corners apply even pressure around irregular shapes.

Tool Care

Clamps

Quite often glue squeezed out from between pieces of wood will dry on the surface of your clamps. To prevent this buildup, apply a light coat of paste wax to the face of each of the jaws. The wax on the metal clamps will also prevent rust.

Pipe clamp—fittings will fit any length of pipe. The sliding jaw operates with a spring-lock device. Black, nongalvanized pipe is preferred.

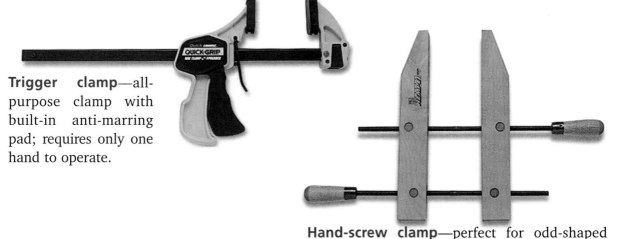

Trigger clamp—all-purpose clamp with built-in anti-marring pad; requires only one hand to operate.

Hand-screw clamp—perfect for odd-shaped assemblies; will even clamp round objects. Each jaw works independently, allowing the jaws to angle toward or away from each other.

Miter clamp—designed to hold the workpieces at exactly a 90° angle.

Saws

Saws range from handsaws and portable power saws to the more sophisticated—and more costly—stationary floor models. As witnessed by the furniture and houses built before electricity, all the cuts needed for woodworking and house construction can be done with handsaws. Look for the saw that meets your needs.

Wallboard saw—sometimes called a *drywall saw;* used to cut holes in drywall and plasterboard.

Crosscut saw—cuts across the grain of wood. Normal blade is 26″ long and has 10-12 teeth per inch.

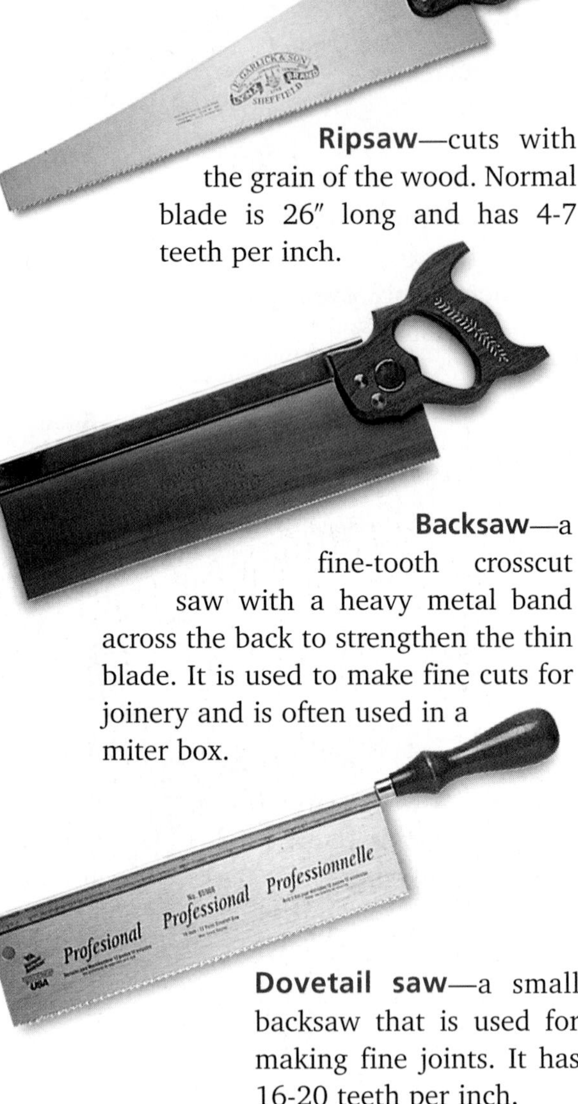

Ripsaw—cuts with the grain of the wood. Normal blade is 26″ long and has 4-7 teeth per inch.

Backsaw—a fine-tooth crosscut saw with a heavy metal band across the back to strengthen the thin blade. It is used to make fine cuts for joinery and is often used in a miter box.

Dovetail saw—a small backsaw that is used for making fine joints. It has 16-20 teeth per inch.

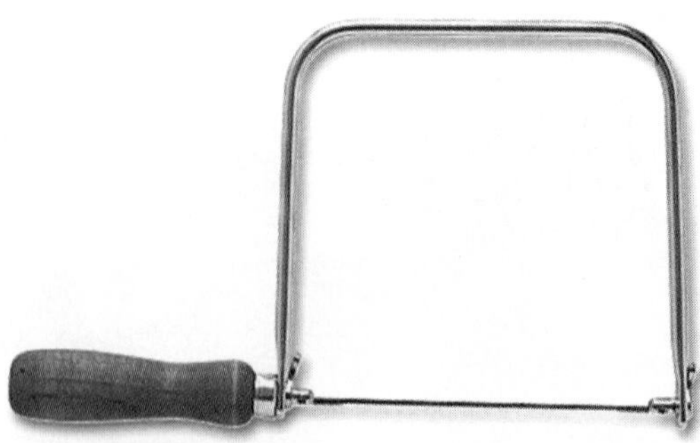

Coping saw—has a U-shaped frame; used to cut curves, scroll work, and molding as finishing trim.

Keyhole saw—sometimes called a *compass saw.* It is used to cut curves and holes in plywood and wallboard.

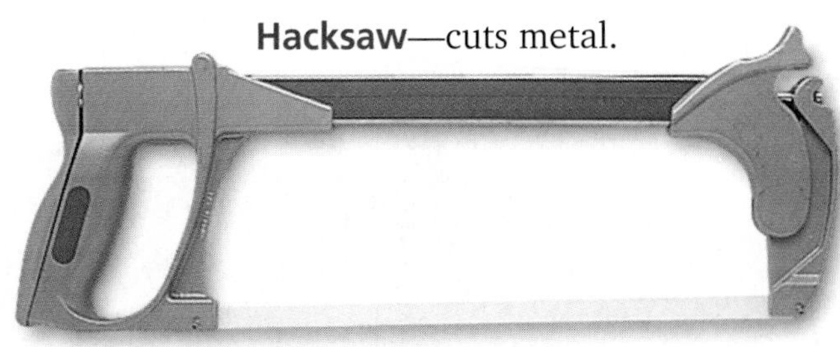

Hacksaw—cuts metal.

Jigsaw—makes straight and curved cuts.

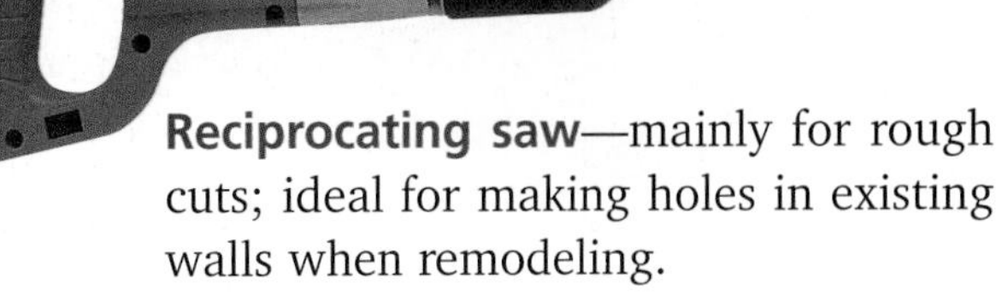

Reciprocating saw—mainly for rough cuts; ideal for making holes in existing walls when remodeling.

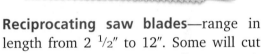

Band saw—cuts curves and resaws stock to thinner sizes. The size of a band saw is determined by the throat (distance from blade to vertical support).

Reciprocating saw blades—range in length from 2 ½″ to 12″. Some will cut very thick wood; others will cut metal. Use the shortest blade possible for the job.

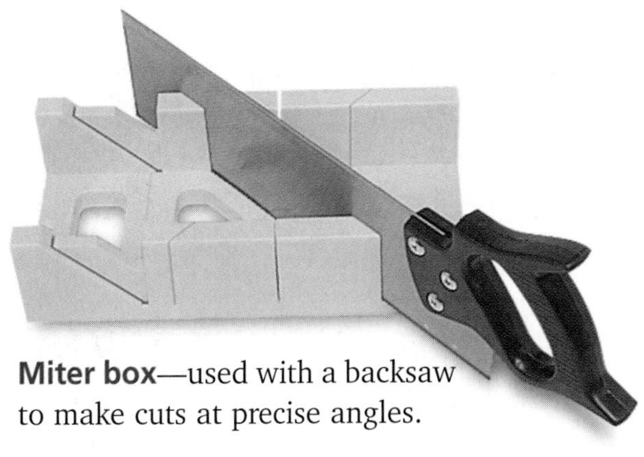

Miter box—used with a backsaw to make cuts at precise angles.

Table saw—the most commonly used saw. Its size is determined by the diameter of the largest blade it can use. The 10″ saw is the most common. The bench-top model (shown) is portable and useful for certain "on-the-job" applications, but a larger floor model usually makes smoother, more accurate cuts.

Pushstick—used to push wood through the saw while keeping fingers away from the blade.

Featherboard—holds the workpiece against the fence while you make rip cuts.

Buyer's Guide

Table Saw

When buying a table saw, consider these features:

- ***Blade size.*** In general, larger blades can cut thicker stock.
- ***Motor.*** The more powerful the motor, the less strain when cutting.
- ***Table surface.*** The table should have a smooth cast-metal surface.
- ***Accessories.*** What accessories can you add now or later?

Circular saw—a portable saw that cuts wood and sometimes other materials. Its size is determined by the largest blade it can accept.

Cordless circular saw—a circular saw that runs on battery power. It can make the same cuts as the bigger, more powerful corded models.

Sliding compound miter saw—makes compound miter cuts at angles up to 45° either side of center. Some models cut stock as thick as $4^{1}/_{2}''$ and as wide as $12''$.

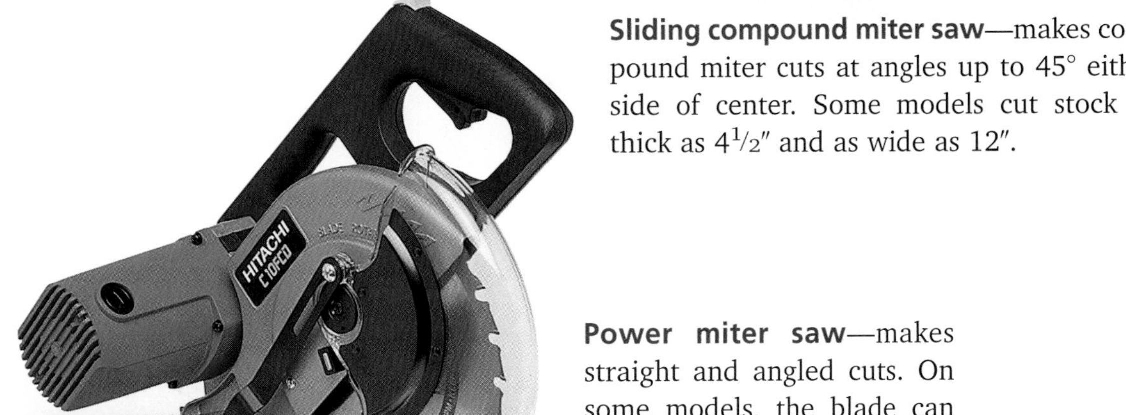

Power miter saw—makes straight and angled cuts. On some models, the blade can be tilted for compound miters.

Radial-arm saw—good for cross-cutting long boards. The blade can be angled, tilted, and rotated to make different cuts. It is also used for ripping and making dadoes.

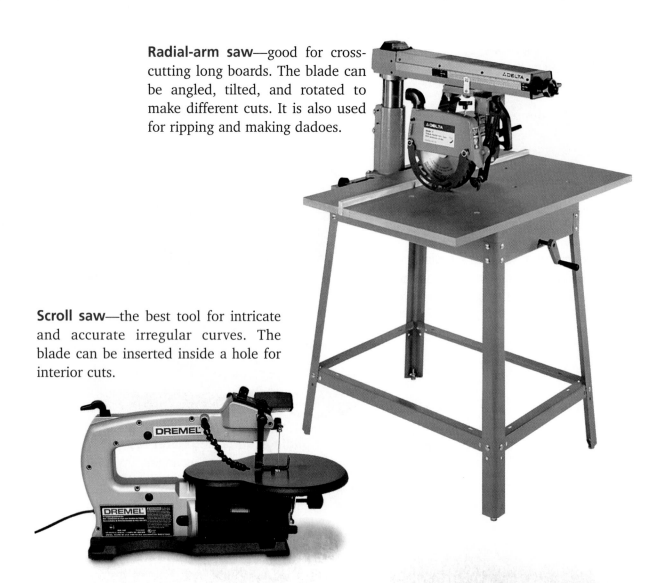

Scroll saw—the best tool for intricate and accurate irregular curves. The blade can be inserted inside a hole for interior cuts.

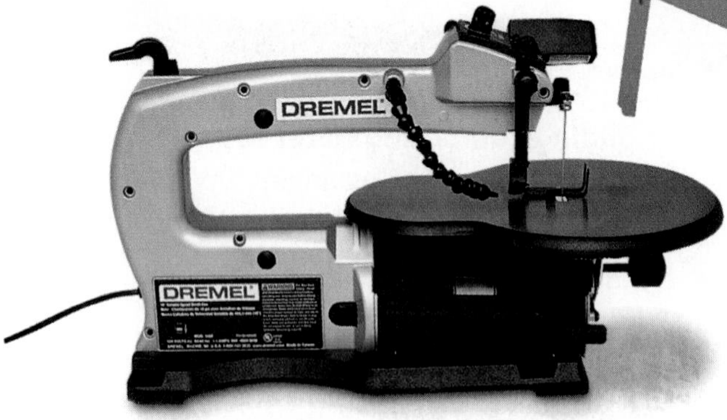

Tool Care

Saw Blades

Saw blades will last longer and cut better if you follow these tips:

- Follow the manufacturer's recommendations and clean the blades regularly.
- Store the blades properly.
- As soon as a blade becomes dull, have it sharpened.
- Use the proper blade for the job.

Saw blades—crosscut, rip, and plywood.

Routers

The router can be used to make decorative edges or surfaces, or it can be used to make very complex joints such as the dovetail joint. It can produce grooves, rabbets, and mortises with ease.

Standard router—a tool that can make decorative edges or surfaces, joinery, or freehand carvings. The size of the motor and the collet size determine how much work it can handle.

Plunge router—used for interior cuts. The router bit can be lowered into the wood and set to a specific depth. Some models have variable speeds.

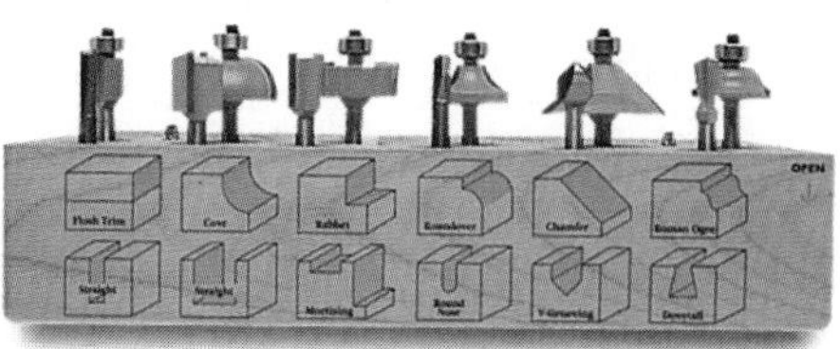

Router bits

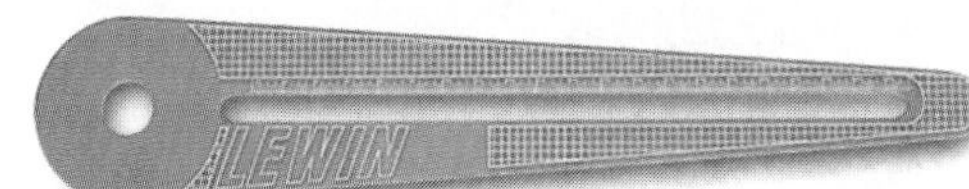

Router circle guide—attaches to the center of the desired circle and holds the router on course to make a perfect circle.

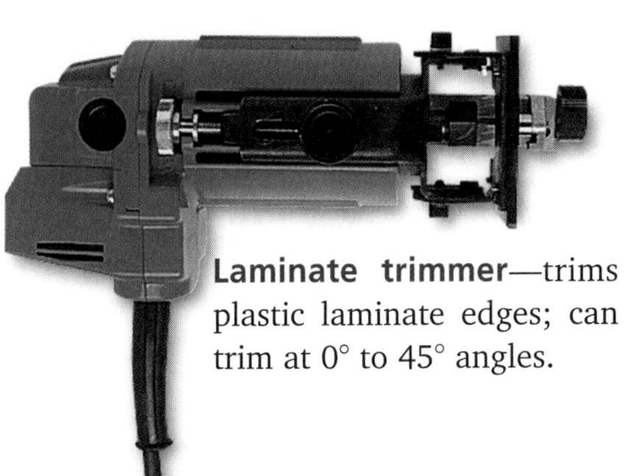

Laminate trimmer—trims plastic laminate edges; can trim at 0° to 45° angles.

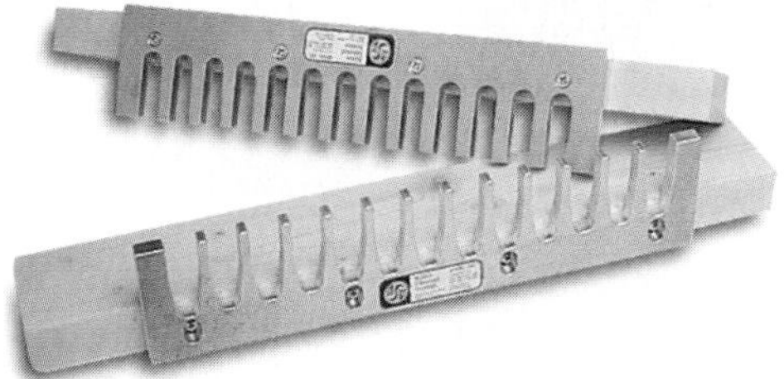

Dovetail template—accurately guides the router to make the tail and pin cuts for a dovetail joint. This model makes through dovetails.

Planes, Jointers, & Chisels

After the initial saw cuts, a workpiece usually needs additional cutting and shaping. Planes, jointers, and chisels are cutting tools that help prepare a project for completion.

Portable power planer—used to square and smooth the edges of boards. With a fence, it can plane specific angles.

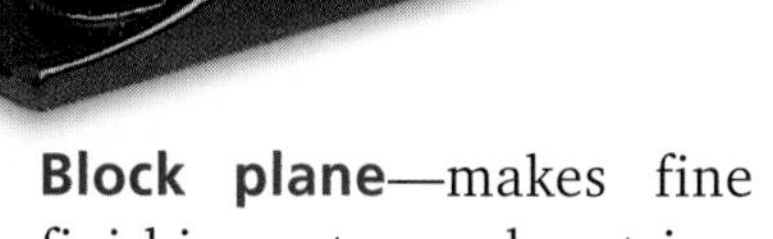

Block plane—makes fine finishing cuts; used on trim.

Thickness Planer—sizes the stock to the desired thickness after it has been straightened or flattened on a jointer.

Jack plane—has a wide iron for roughing cuts and for planing wide boards.

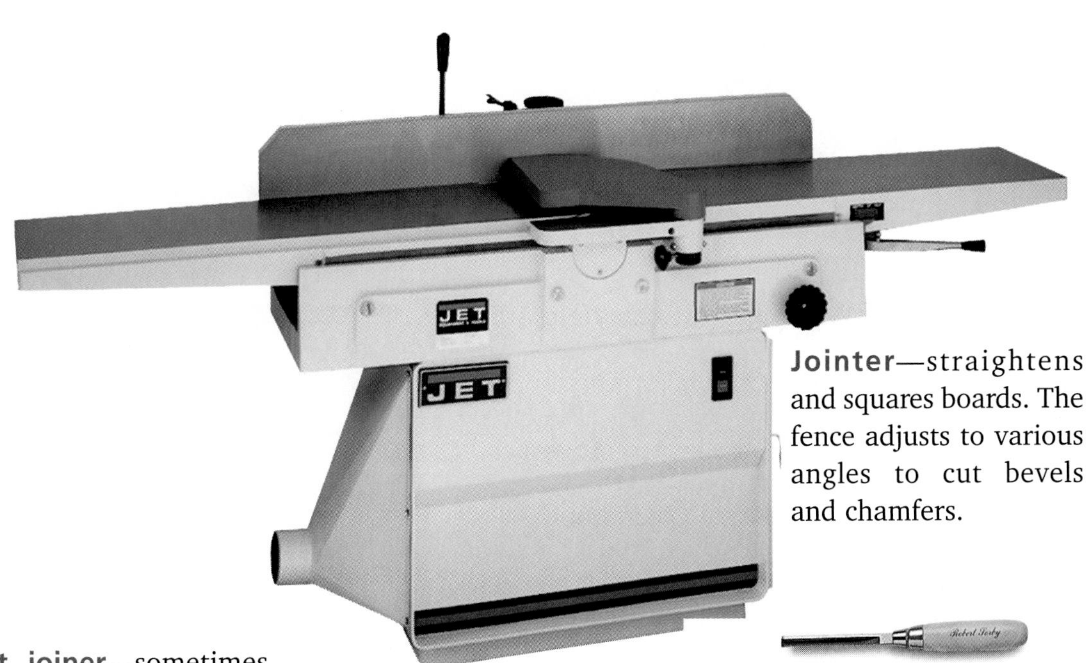

Jointer—straightens and squares boards. The fence adjusts to various angles to cut bevels and chamfers.

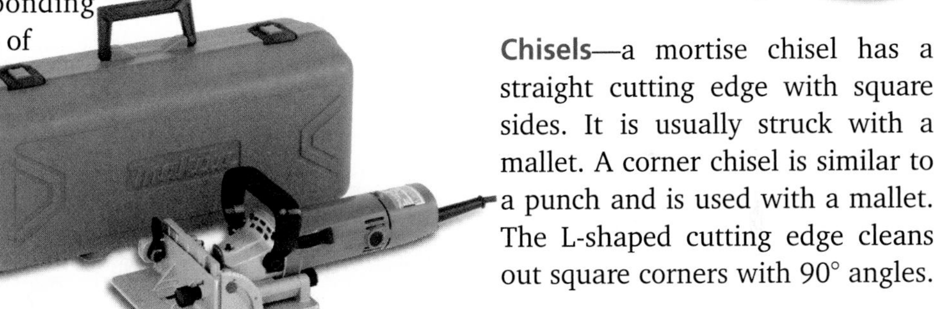

Biscuit joiner—sometimes called a *plate joiner.* The blade cuts corresponding slots into two pieces of wood so that biscuits can be inserted into them. The biscuits are used to strengthen joints.

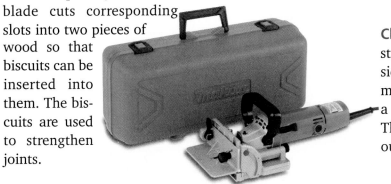

Chisels—a mortise chisel has a straight cutting edge with square sides. It is usually struck with a mallet. A corner chisel is similar to a punch and is used with a mallet. The L-shaped cutting edge cleans out square corners with 90° angles.

Butt chisel—easy to work with; often used to create dovetail joints. It can be driven by hand or with light strokes with a mallet.

Lathe—used for turning wood stock into cylindrical shapes. Special lathe chisels cut away the wood.

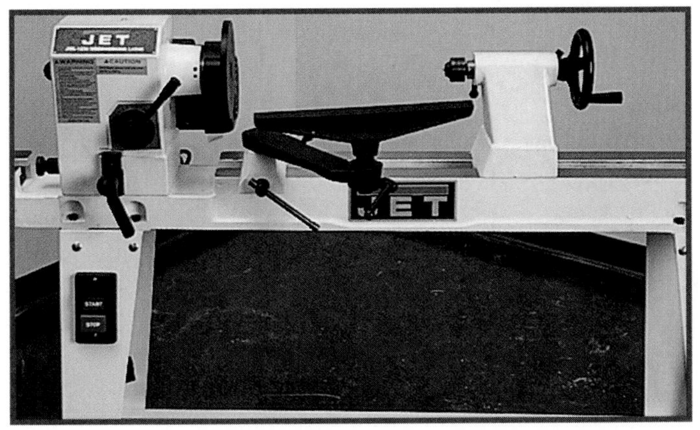

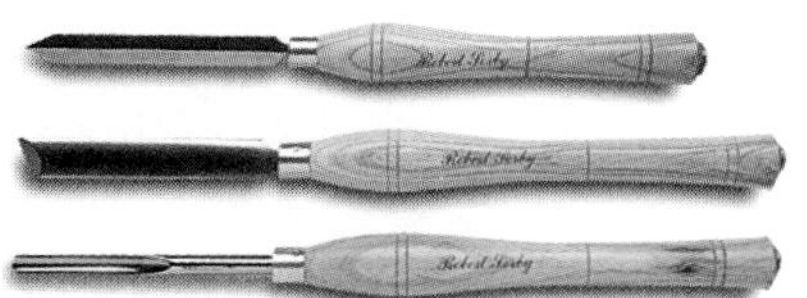

Lathe chisels

Sanders

Sanders are versatile tools that can help shape a workpiece and smooth it for finishing. For refinishing projects, they can remove old paint and other finishes from wood and other surfaces.

Surform plane—shapes and planes wood, plywood, and plasterboard. Blades are replaced, not sharpened, when they become dull.

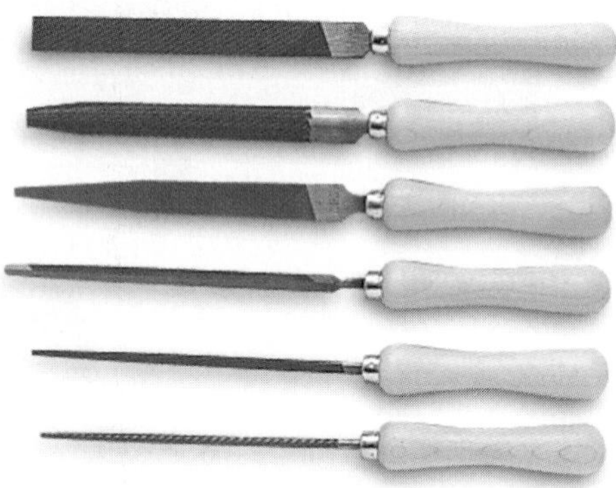

File set—a good set usually includes a flat file, cabinet file, triangle file, round file, and half-round file.

Random orbital sander—used for general sanding on flat surfaces; can be used to buff a finished surface.

Detail sander—used for areas larger sanders can't reach, such as inside corners.

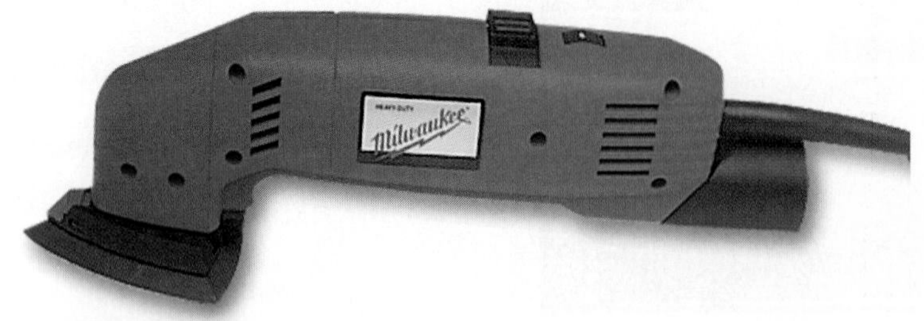

Portable belt sander—mainly for rough sanding; it "eats" the wood quickly. It is best for leveling over large areas.

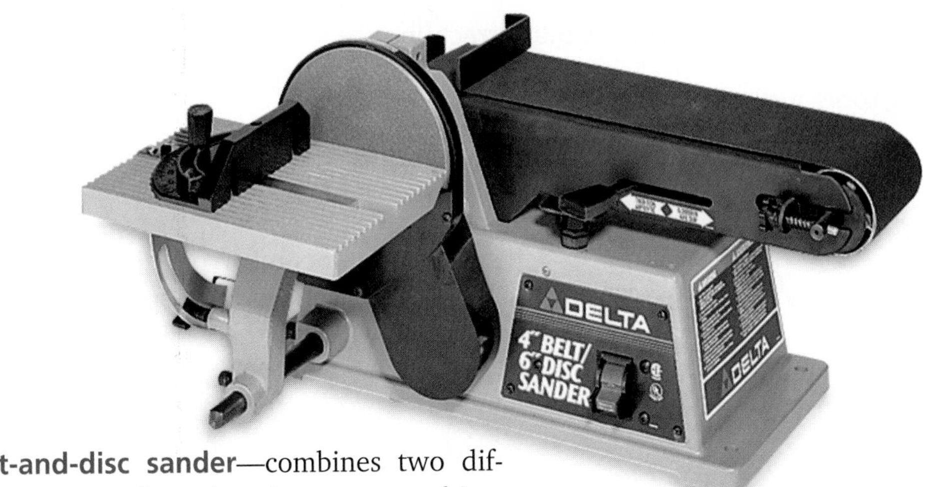

Belt-and-disc sander—combines two different types of sanders into one machine. More accurate sanding can be performed with the miter gauge and the fence.

Disc-shaped papers—used with disc sanders. Grits and backings are similar to those of abrasive sheets.

Abrasive sheets—have aluminum oxide, silicon particles, emery, garnet, or carbide particles or grit glued to a paper or cloth backing. The sheets can be folded or torn to size.

Steel wool—used to clean and strip wood and metal; number 4 grade is very coarse, 0 is fine, and 0000 is very fine.

Painting Tools

You can make even a so-so project look beautiful if you take the time to use the correct finishing tools and processes. Likewise, the most perfectly built project may be ruined if you do not finish it properly. Choose the appropriate tools to apply the finish to your project.

Wand and sponge applicators—mainly used for utility painting where a fine finish is not needed. The sponges are usually disposed of when the job is finished.

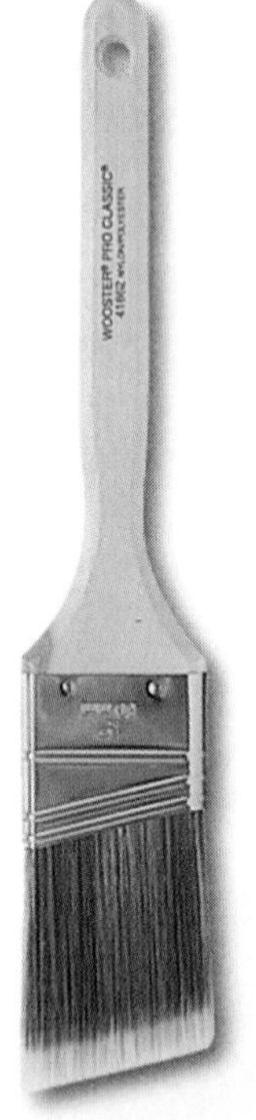

Chisel-edge brush—has angled bristles to paint a clean edge along trim.

Wall brush—used to cover a large area. Wall brushes are at least 4″ wide.

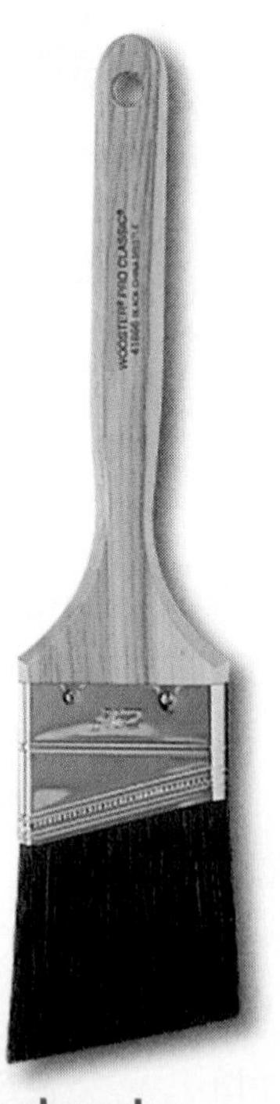

Trim brush—comes in widths up to 2″; useful in areas where rollers cannot reach.

Varnish brush—has natural "split end" bristles that hold more varnish and other finishes.

Nails & Fasteners

Nails, glue, or some kind of fastener is almost always needed to hold a project together. It is important to become familiar with the many types of fasteners because they are so often used in woodworking.

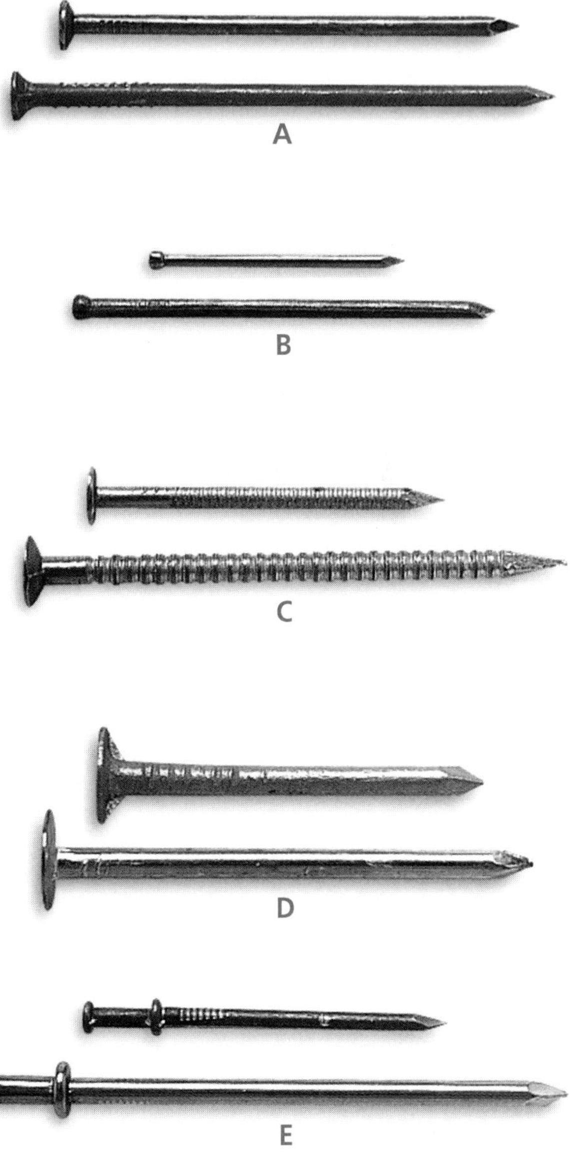

Common nail (A)—the most commonly used nail for building construction. It is a heavy-duty nail with a flat head that will not pull through the wood.

Finishing nail (B)—used when the nail head is to be concealed. It is set into the wood with a nail set.

Underlayment nail (C)—has deep, closely spaced rings; used for installing subfloors because of its good holding power.

Roofing nail (D)—usually galvanized and has a large head, so it will not pull through the roofing material.

Double-headed nail (E)—for nailing pieces together temporarily. The second head projects above the first so it can be pulled out easily.

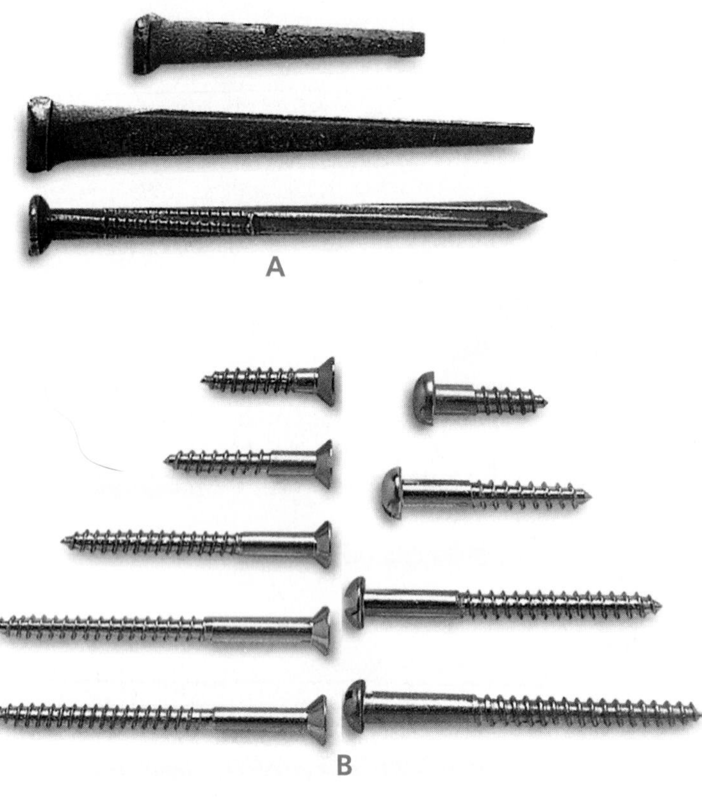

Masonry nail (A)—made of very hard steel that does not bend easily; used to nail into concrete.

Wood screws (B)—can have flat or round heads. They are usually zinc-plated or brass.

Sheet-metal screws (C)—have a very sharp, self-tapping tip; can pull two pieces of sheet metal together. (Round- and hex-head screws are shown here).

Drywall screws (D)—have a bugle-shaped head that cuts through the drywall and anchors in a wood stud.

Lag screws (E)—sometimes called *lag bolts;* a heavy-duty wood screw with a hex or square head so that it can be turned by a wrench or ratchet.

Thumb screws (F)—machine screws that can be turned by hand because of the wide, thin head on the screw.

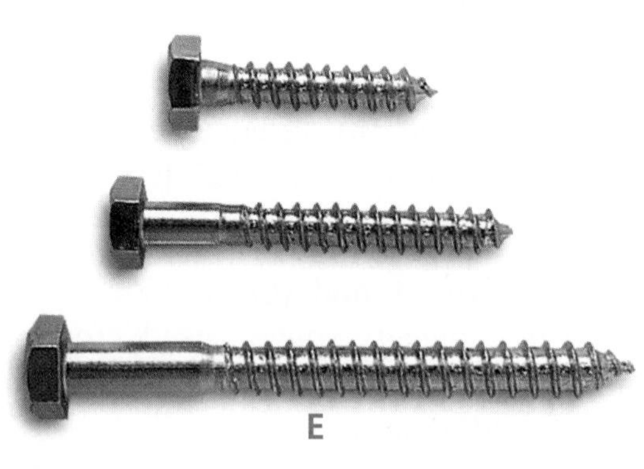

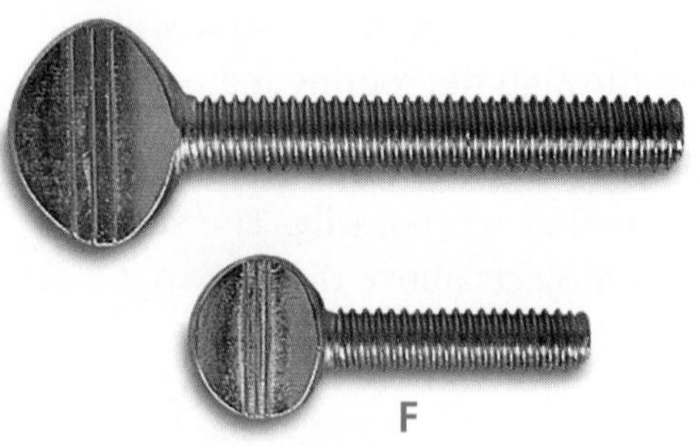

Hanger bolt (A)—has machine-screw threads on one end and wood-screw threads on the other.

Machine bolt (B)—a bolt with a square or hex head.

Stove bolt (C)—has a round or flat head that can be countersunk.

Machine screw (D)—similar to a wood screw, but threaded into metal.

Carriage bolt (E)—the square shoulder sinks into the wood and keeps the bolt from turning.

U-Bolt (F)—used on pipes and other round objects.

Eye-bolt (G)—holds wire or ropes in place.

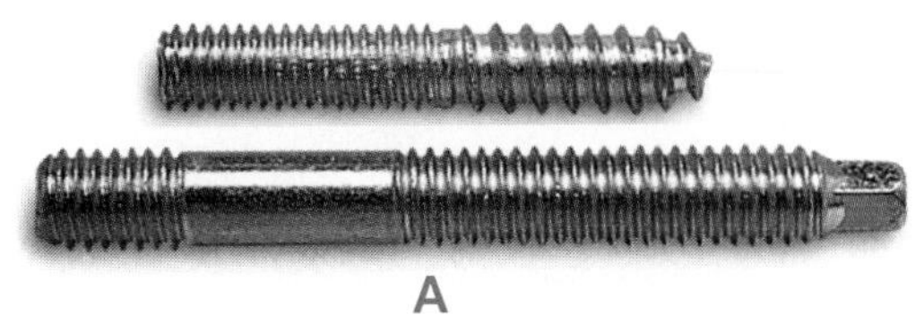

A

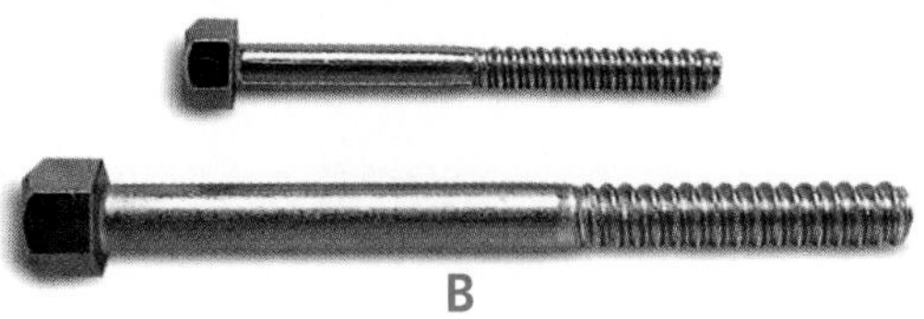

B

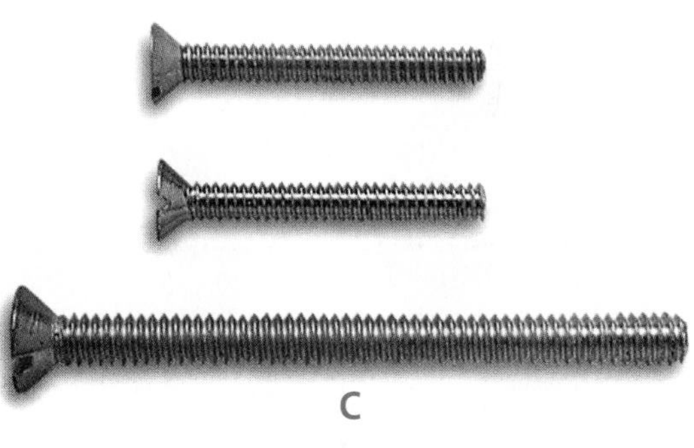

C

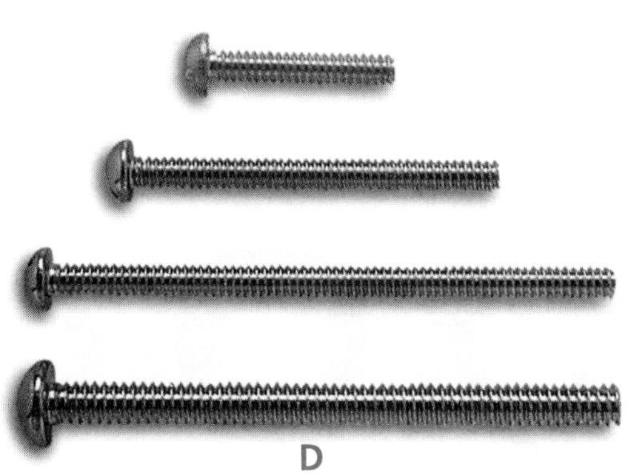

D

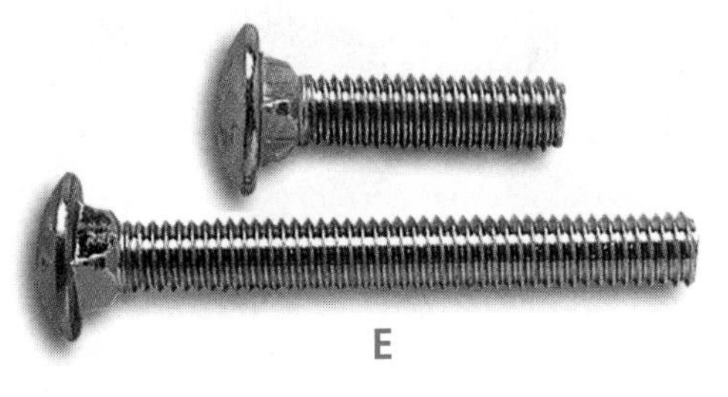

E

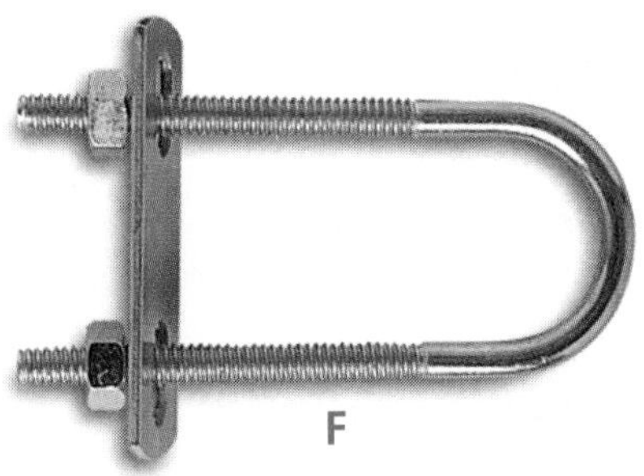

F

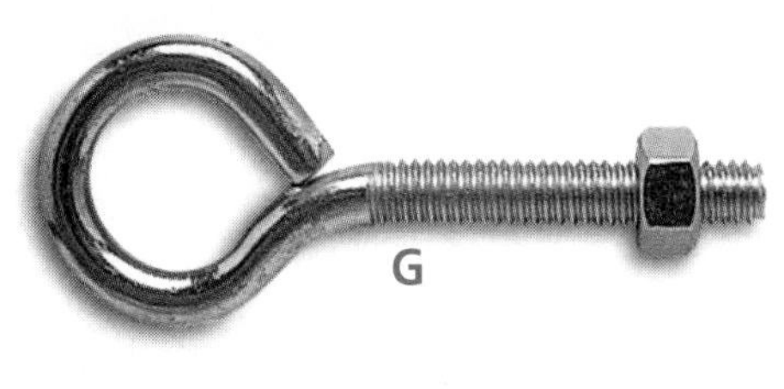

G

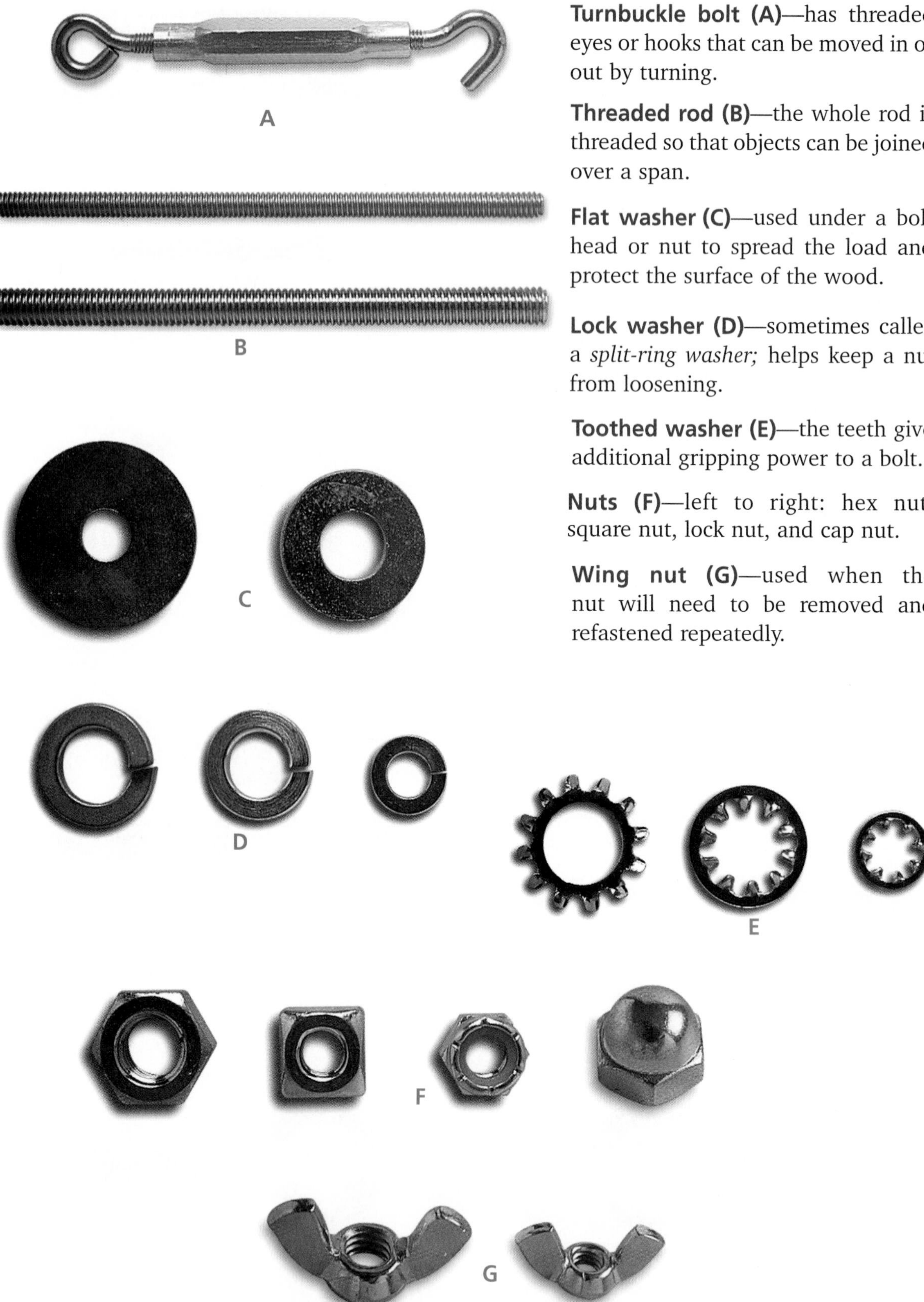

Turnbuckle bolt (A)—has threaded eyes or hooks that can be moved in or out by turning.

Threaded rod (B)—the whole rod is threaded so that objects can be joined over a span.

Flat washer (C)—used under a bolt head or nut to spread the load and protect the surface of the wood.

Lock washer (D)—sometimes called a *split-ring washer;* helps keep a nut from loosening.

Toothed washer (E)—the teeth give additional gripping power to a bolt.

Nuts (F)—left to right: hex nut, square nut, lock nut, and cap nut.

Wing nut (G)—used when the nut will need to be removed and refastened repeatedly.

Plastic or lead anchor (A)—inserted into a hole in a concrete or block wall; expands when a screw is driven into it to anchor the screw in place.

Hollow-wall anchor (B)—sometimes called a *molly bolt*. Push it into a pilot hole in the wall; it expands as a screw is driven into it to anchor the screw in place.

Toggle bolt (C)—a machine screw that has spring-loaded wings. The wings open behind the wallboard when pushed through a pre-drilled hole. It can then be tightened.

Corrugated fastener (D)—used on lightweight butt and miter joints.

Cotter pin (E)—holds a rod or shaft in place when put through a hole in the shaft.

Upholstery tack (F)—tacks that come with many different kinds of decorative heads; used to hold upholstery in place.

Staple (G)—different sizes and forms are used to hold fabric on furniture, for holding wire in place, and for insulation.

Screw hook (H)—screws into wood; open end can support various items.

screw eye (I)—similar to a screw hook, except the hook is closed.

Hook and eye (J)—often used with a screw eye to fasten a door or a gate.

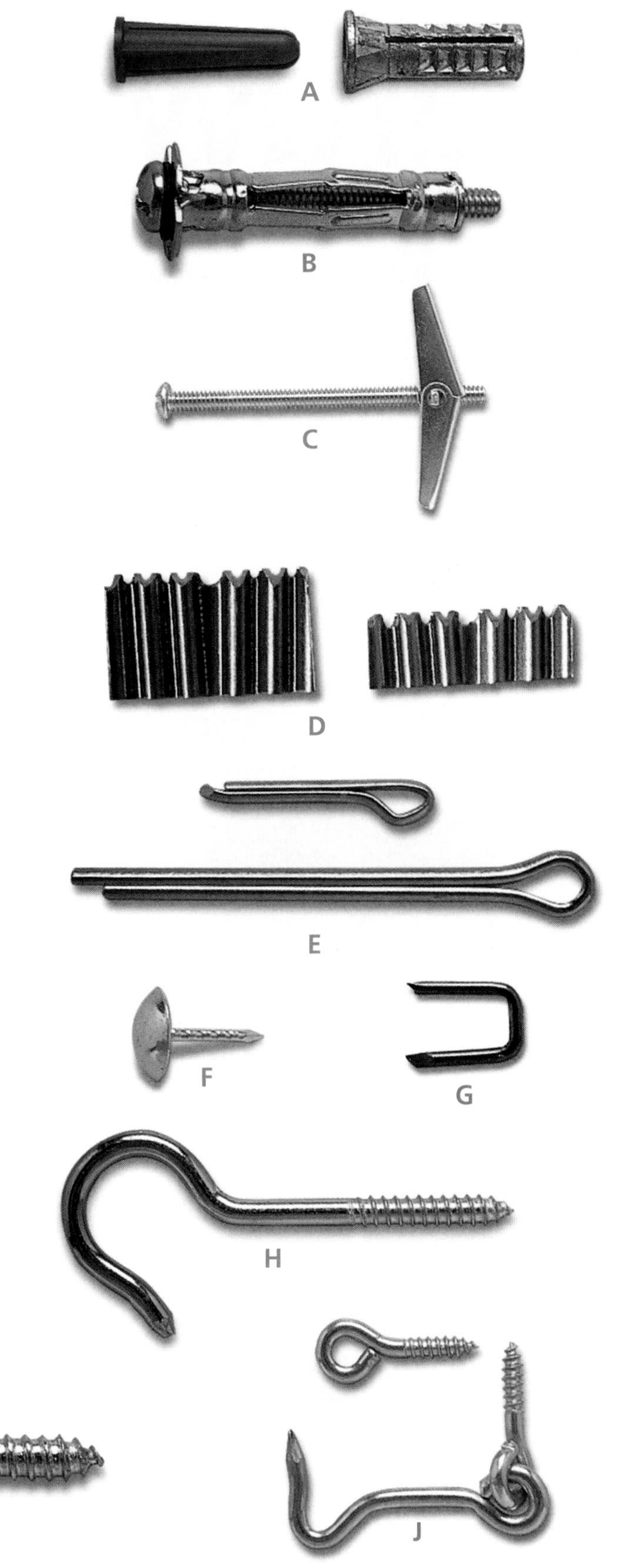

Hardware

Choosing the right piece of hardware for your project can make a big difference in how the finished product looks. Hardware is available in many different styles, from antique to ultra-modern, to suit the purpose and style of your project.

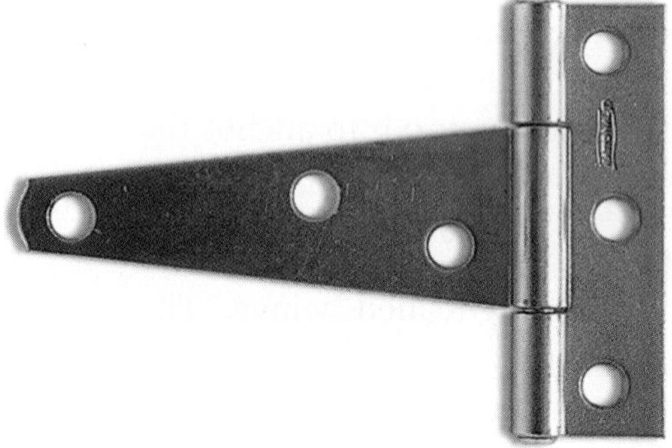

Single strap hinge—used on heavy-duty doors and gates; sometimes referred to as a *T-hinge.*

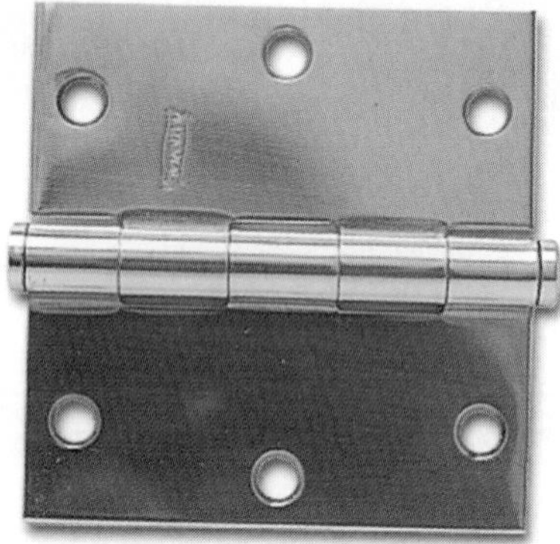

Loose-pin butt hinge—has a loose pin to allow for door removal.

Overlay hinge—mounted to the cabinet frame and the back of the door.

Cabinet butt hinge—for flush-mounted cabinet doors.

European-style concealed hinge—for inset and overlaying cabinet doors. A cup is recessed into the door, and the baseplate is attached to the cabinet side.

Hasp—perfect for holding lids closed on small boxes.

Flap stay—used in pairs to hold a desk lid or other flap in a 90° position.

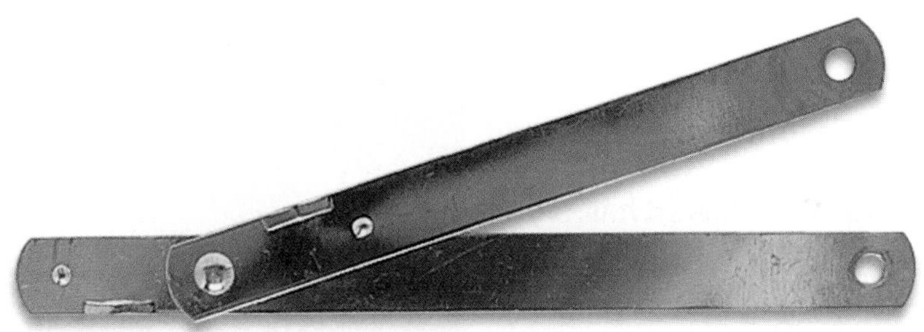

Lid stay—holds a box lid open.

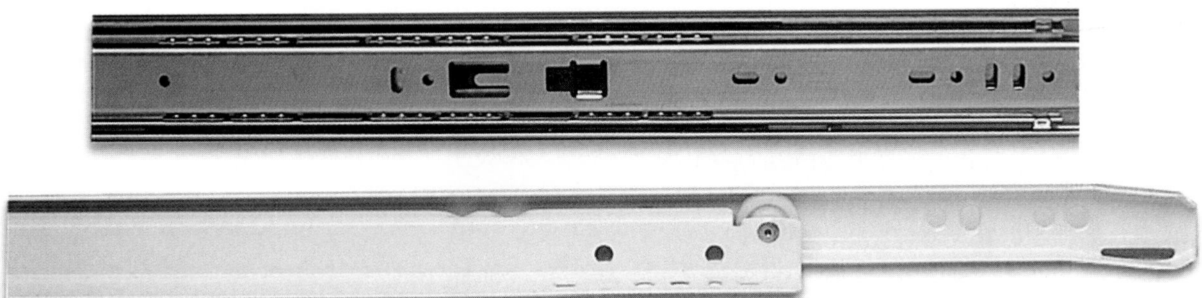

Drawer runner—used to slide drawers open and closed.

Grades of Wood

Buying lumber and veneers at your local woodworking store or at a lumber yard can be confusing. Why are some pieces more expensive than others? Lumber is graded by the amount of "clear" wood in the board and how many knots and other flaws it has. If you understand a little about how the wood is graded, it will help you buy the right wood for your project.

Check lumber before buying it.

Check lumber markings.

Softwoods

Most dimensional lumber is made of softwood. The best grade of softwood is Select. A Select board may be free of defects, or it may contain just a few knots. Finish grade lumber is often free of knots and suitable for a woodworking application. Common grade always has visible knots and other defects.

Pressure-treated lumber is treated under pressure with chemicals that resist decay. It is used outdoors for decks, railings, and fence posts. Redwood and cedar are also good choices for exterior projects. These woods are naturally resistant to decay and insects.

Dimensional lumber is sold according to nominal sizes, such as 2 × 4 and 2 × 8. The actual size is smaller than the nominal size.

Nominal vs. Actual Lumber Dimensions

Nominal	Actual
1 × 4	3/4″ × 3 1/2″
1 × 6	3/4″ × 5 1/2″
1 × 8	3/4″ × 7 1/2″
2 × 4	1 1/2″ × 3 1/2″
2 × 6	1 1/2″ × 5 1/2″
2 × 8	1 1/2″ × 7 1/2″
4 × 4	3 1/2″ × 3 1/2″

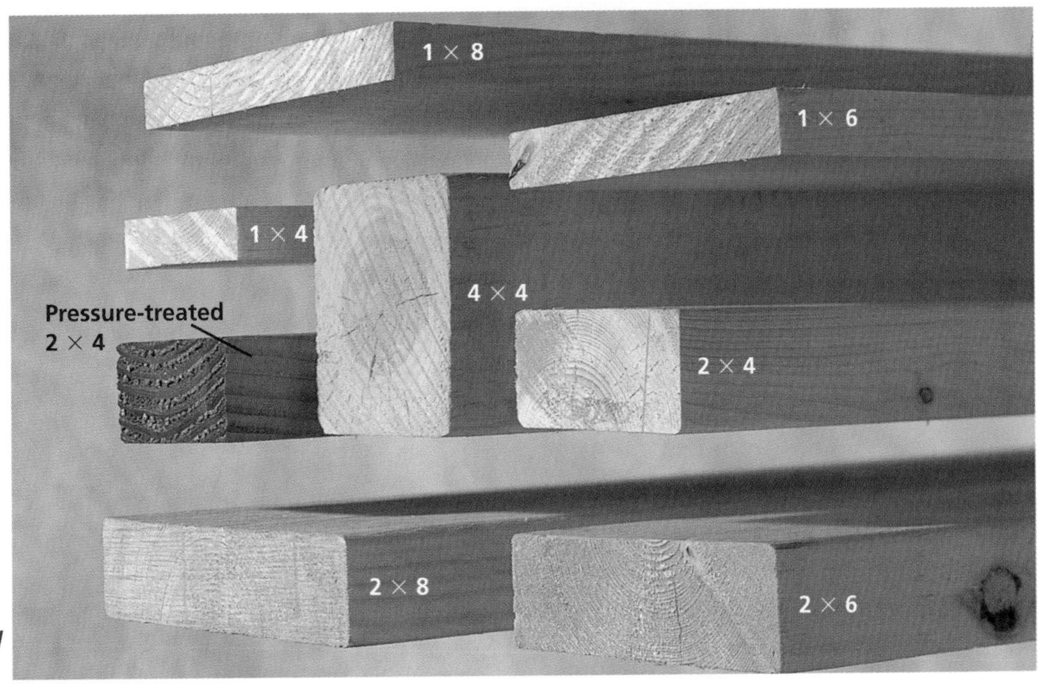

Dimensional lumber.

Fine Hardwoods

Hardwood grading guidelines are established by the National Hardwood Lumber Association (NHLA). Most hardwood used in woodworking falls into three categories: FAS (firsts and seconds), Select, and #1 Common.

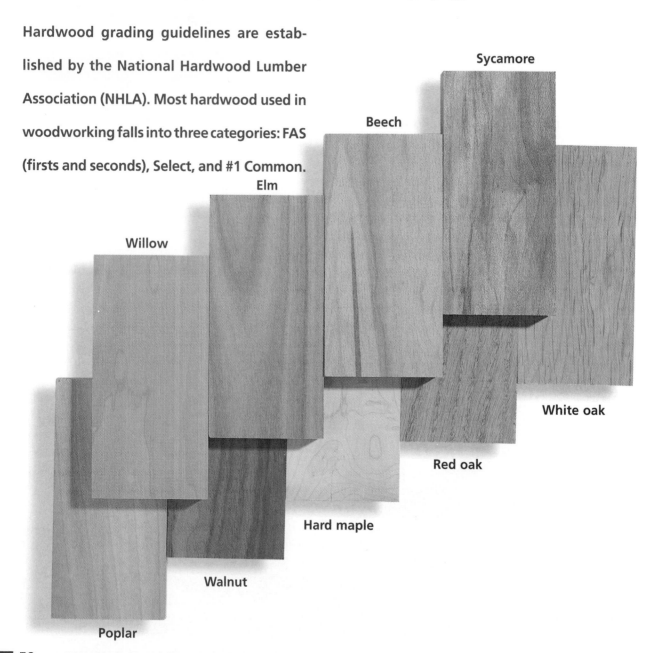

Sycamore

Beech

Elm

Willow

White oak

Red oak

Hard maple

Walnut

Poplar

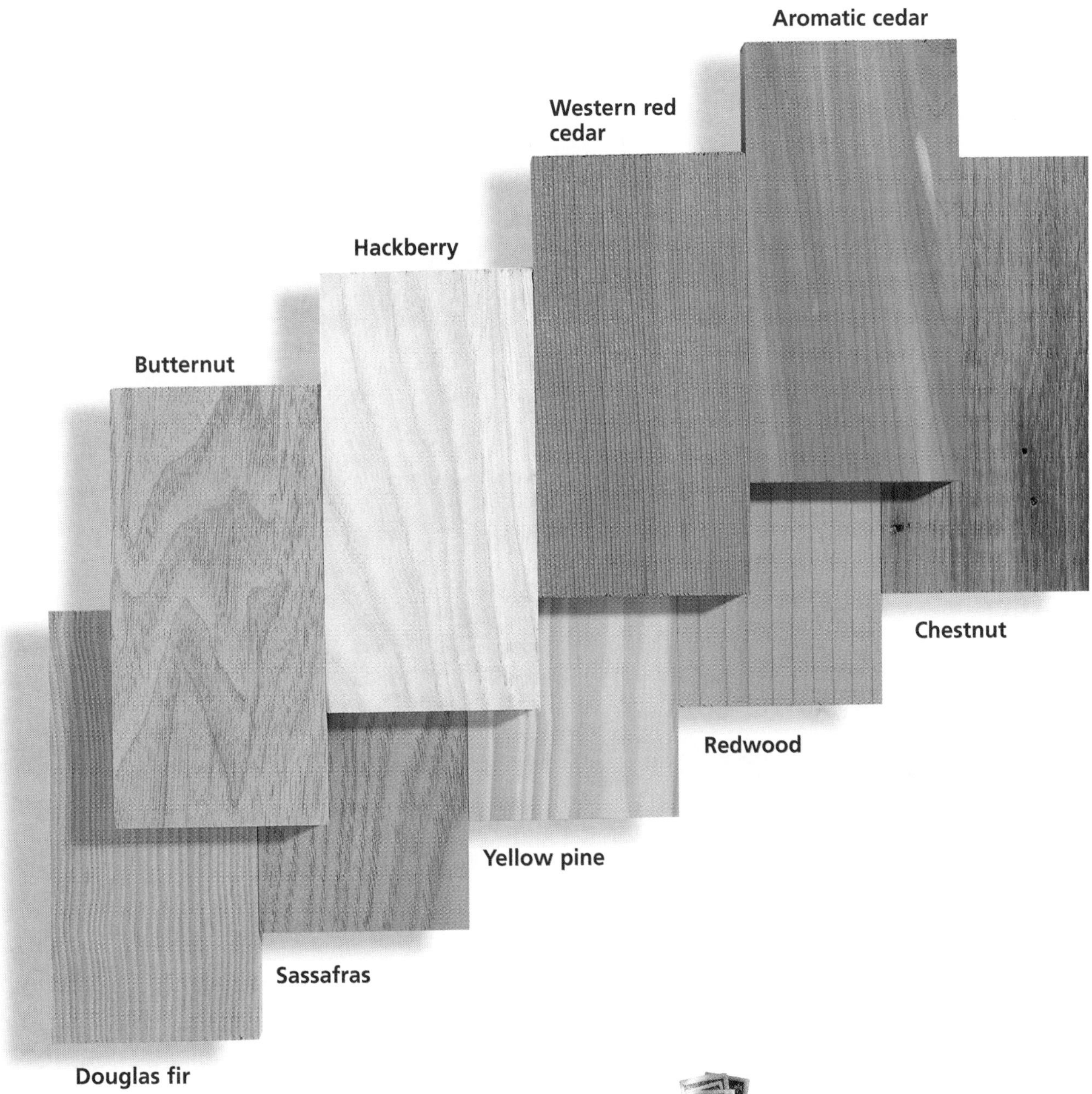

Aromatic cedar

Western red cedar

Hackberry

Butternut

Chestnut

Redwood

Yellow pine

Sassafras

Douglas fir

FAS has the biggest percentage of clear wood. It may have faults on up to 20 percent of the board. Select boards have the same amount of clear wood as the FAS boards on one face, but the back face may have small defects. #1 Common boards have more defects and up to 30 percent of the board may be waste.

Buying Lumber

When buying surfaced lumber, have it planed 1/16″ thicker than what you need. This will act as an extra protector of the wood until you are ready to use it. Then you can plane it down to its final dimension. It will be clean and free of stains and dents.

Exotic Hardwoods

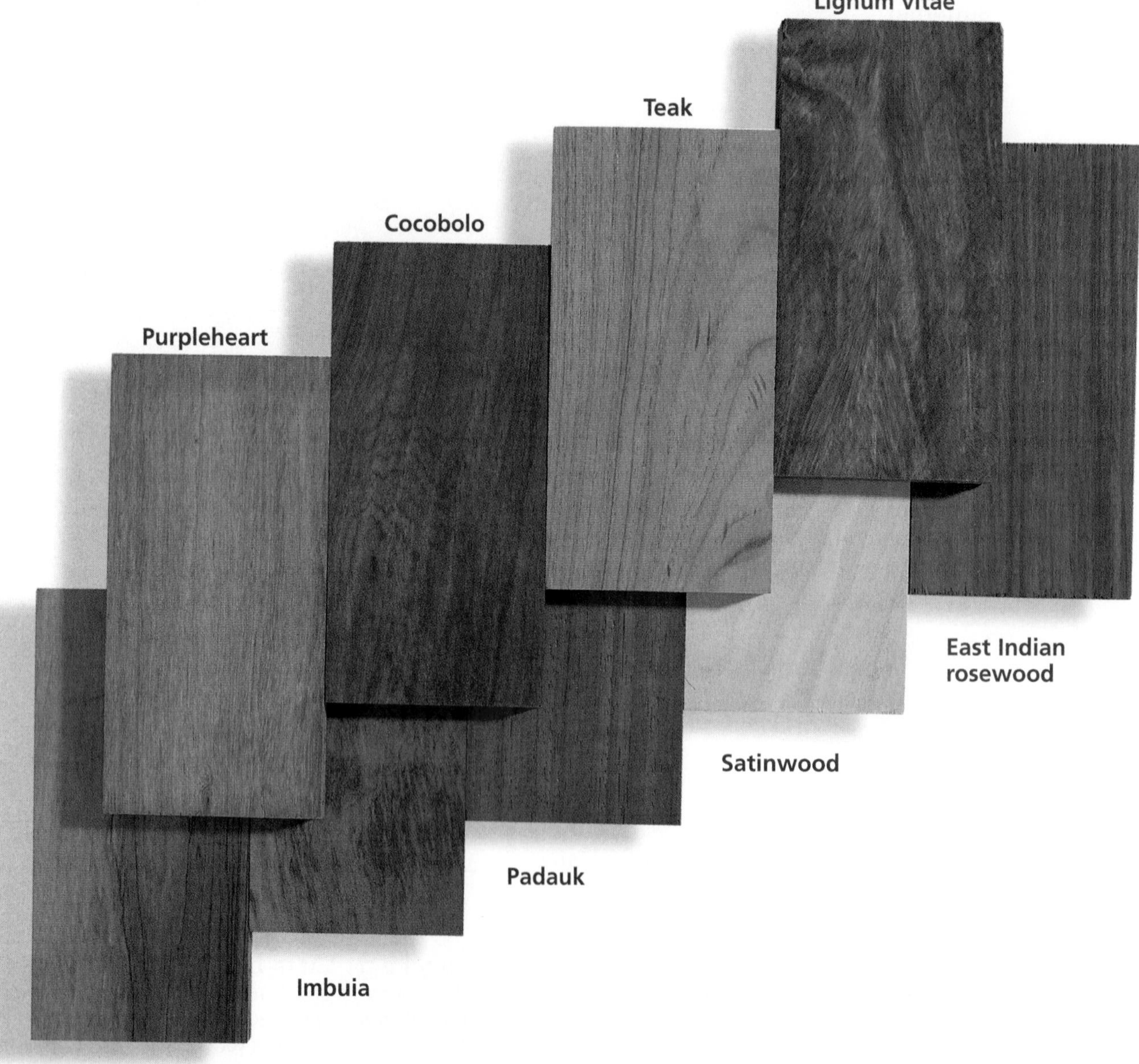

Exotic woods are usually more expensive and harder to find than native hardwoods and softwoods. Exotic woods are sometimes used for the entire project, if it is small, or to accent certain parts of the project, such as the front of a drawer. Some exotic woods require special considerations when finishing.

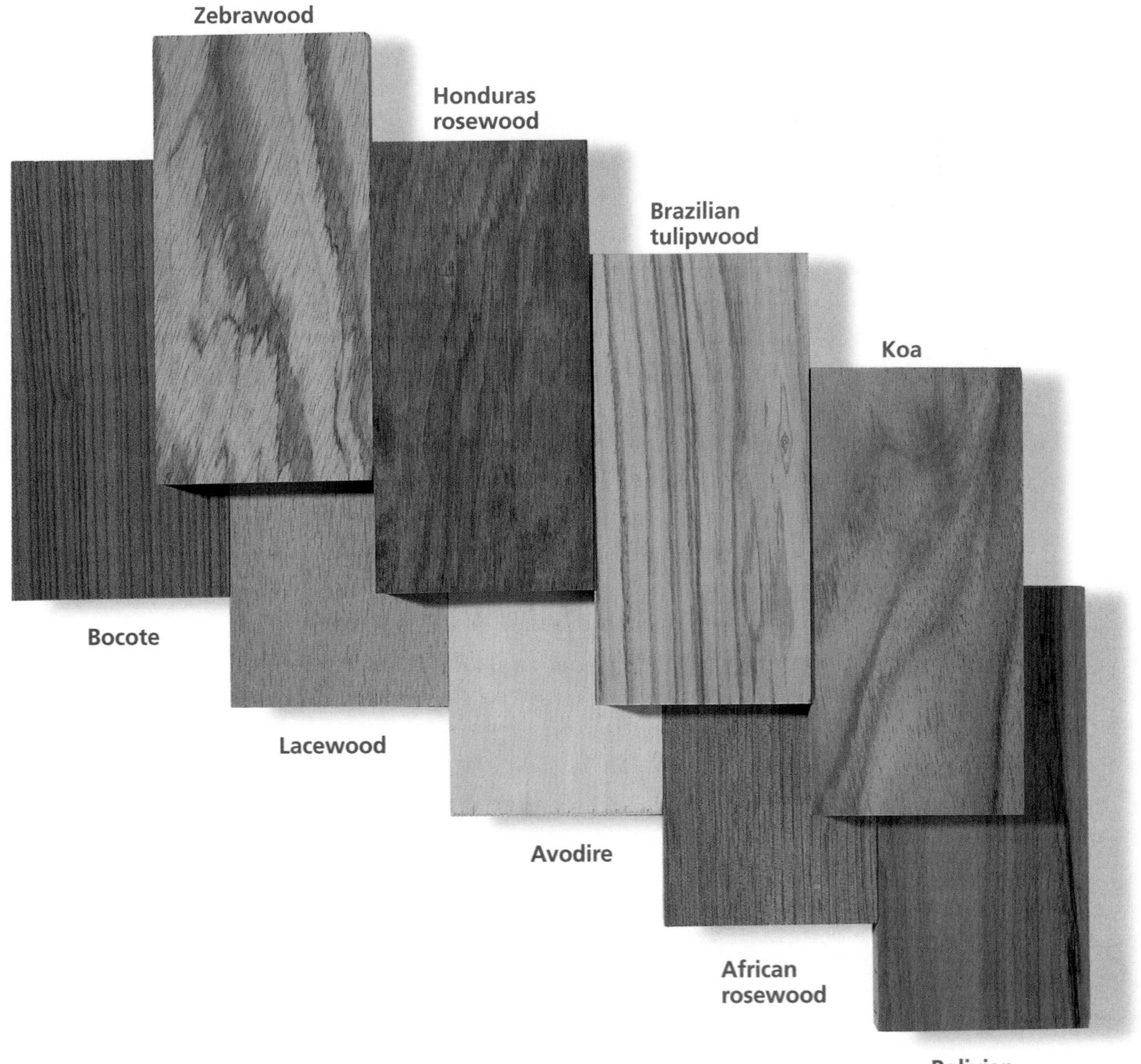

Veneers

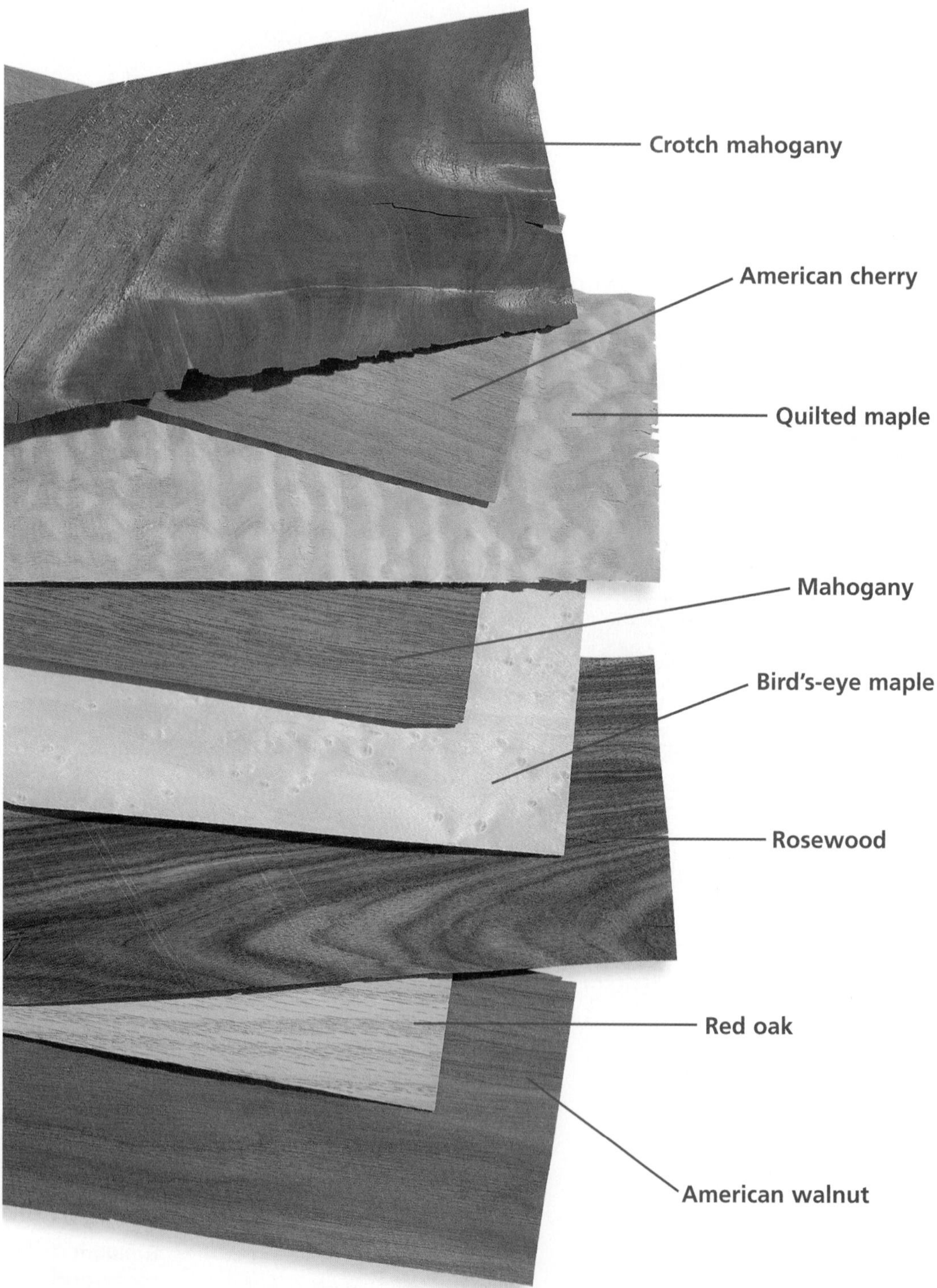
Crotch mahogany
American cherry
Quilted maple
Mahogany
Bird's-eye maple
Rosewood
Red oak
American walnut

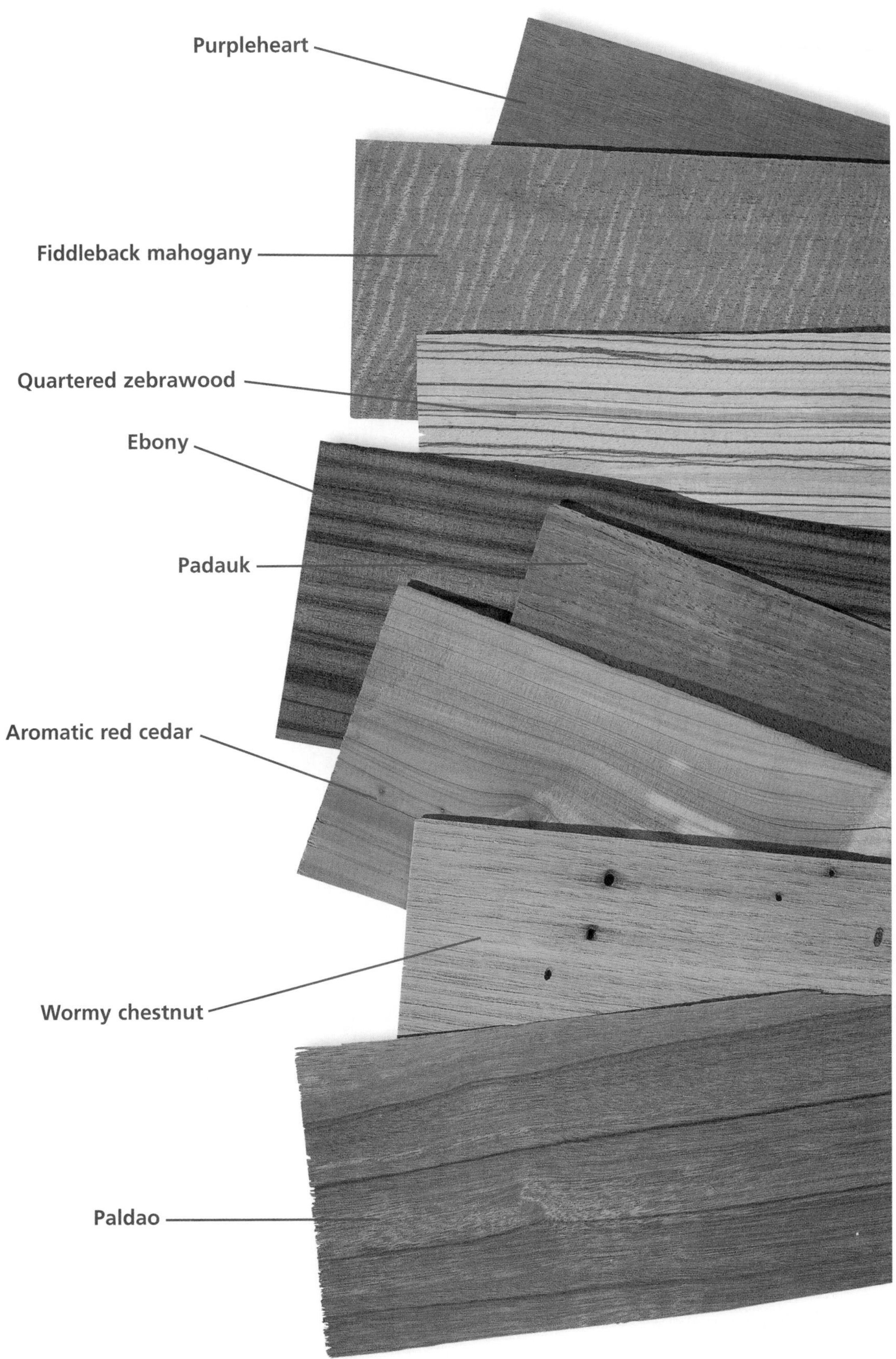
Purpleheart
Fiddleback mahogany
Quartered zebrawood
Ebony
Padauk
Aromatic red cedar
Wormy chestnut
Paldao

Plywood & Sheet Goods

Plywood **refers to sheet materials that are made by gluing thin layers of veneers (plies) to a core layer. Usually the core layer is an inexpensive solid wood, a composite material such as fiberboard or particleboard, or several layers of softwood glued together.**

Finish plywood can have one or both sides finished with a quality veneer. If only one side is a good veneer and the other side is a utility-grade ply, it is graded A-C. If it has a quality grade on both sides, it is graded A-A.

Sheathing plywood is mainly for structural use. It usually has a finish that is unsuitable for finished work. A common grade of sheathing plywood that is rough on both sides is C-D. It can also be designated *exterior grade,* which means it has a waterproof bond.

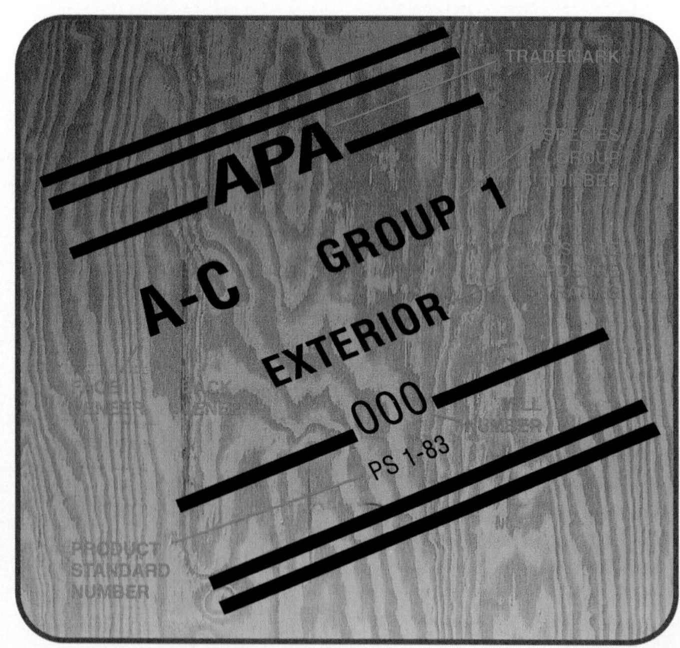

Finish plywood markings.

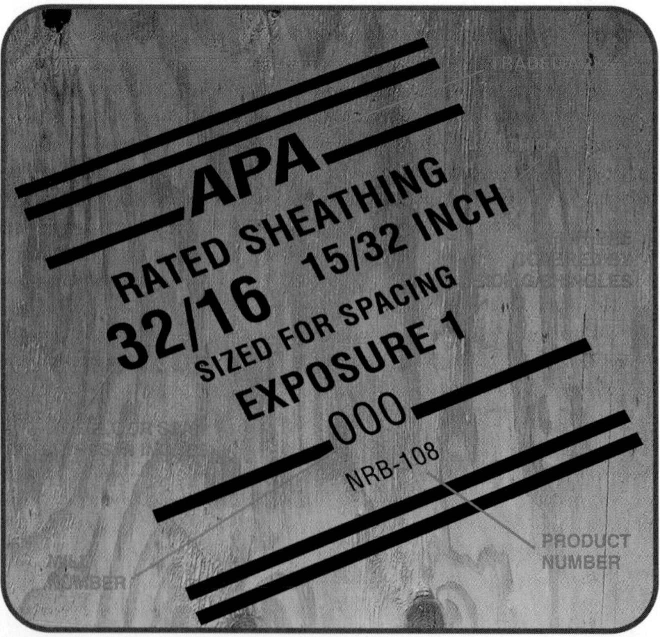

Sheathing plywood markings.

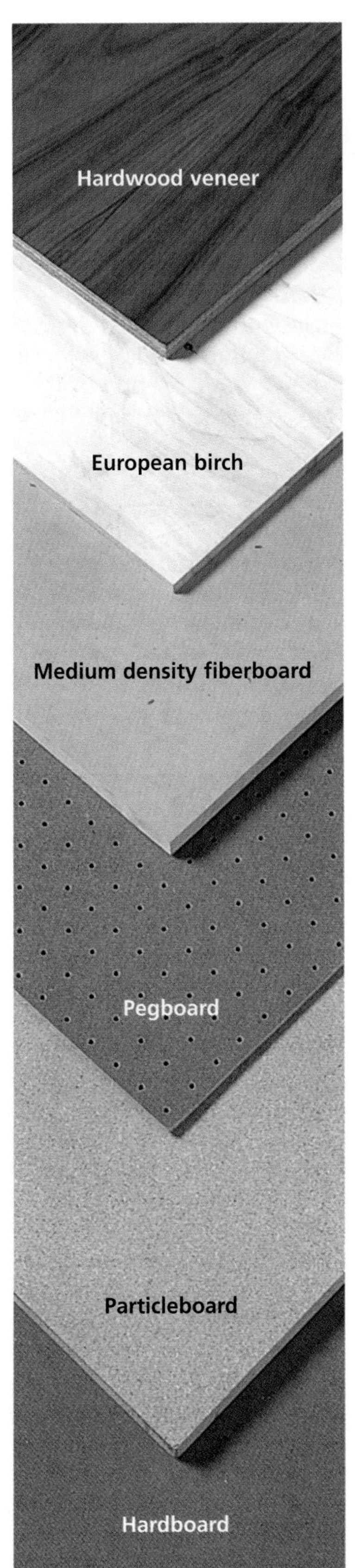

Hardwood veneer plywood has a thin hardwood surface veneer bonded to a softwood or composite core. It is used for fine cabinets and furniture. It is inexpensive compared to a solid hardwood the same size.

European birch plywood has layers of high-quality veneers with very few flaws. The plies are thinner, but there are more of them. It is often used for cabinets and drawers.

Medium Density Fiberboard (MDF) has resin-soaked particles that are glued together under very high pressure. It is very stable and remains stable and smooth. It takes paint well and can be shaped and handled like solid wood. It is good for furniture or cabinets that will be painted.

Pegboard is hardboard with many pre-drilled holes. The holes can be 1/8″ or 1/4″. It is often used for tool storage when accompanied by specially designed brackets for various tools.

Particleboard is made by bonding wood fibers and particles with resin under high pressure. This creates a substrate that is extremely smooth, tight, and grainless. It is often used as the substrate for countertops and for shelving.

Hardboard is heavier than plywood of the same thickness. There are two types: tempered and standard. Tempered hardboard is harder and more water-resistant. It is often used as an overlay for a workbench. Standard hardboard is softer and easier to bend.

Melamine is particleboard with a decorative paper resin coated to it under very high pressure. The result is a board similar to a plastic laminated board. It is very durable and is used for shelves and inexpensive furniture.

Medium Density Overlay (MDO) is an exterior grade plywood. It has a thin resin-soaked paper overlay that is perfect for painting. It is often used for outdoor signs.

Waferboard is made the same way as particleboard, only with bigger particles. It can be used for rough construction. It does not take paint well or give a smooth finish.

Trim Moldings

Molding can be both decorative and functional. It can be used to cover the edge of plywood, to cover gaps around the base or sides of cabinets, or just to add a little extra beauty to a project. Here are a few of the more popular types.

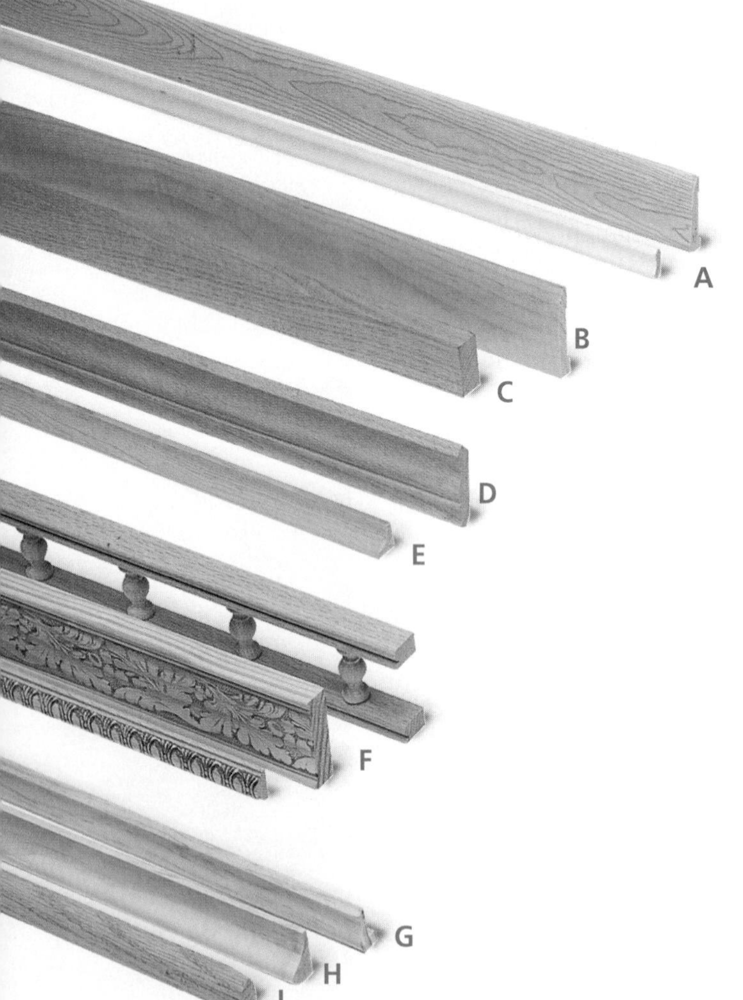

Synthetic trim (A) is made of wood composites and is covered with a layer of melamine. It is less expensive than wood, and it comes in many styles.

Baseboard molding (B) is used along the floor in a room or around cabinets.

Hardwood strips (C) are used on the face of projects. The most common are oak, pine, and poplar strips. They are available in different thicknesses and widths.

Crown moldings (D) are usually used to cover the gap between the top of a built-in project and the wall or ceiling. They are often used as decorative molding around the corner of the ceiling and wall in a room.

Cove molding (E) is a simple molding that fills a gap between a cabinet or other project and a wall or ceiling.

Ornamental moldings (F) are used mainly for decoration.

Door-edge molding (G) is used as a cap molding with finish grade plywood.

Shelf-edge molding (H) gives a decorative edge to plywood shelves. It also adds some strength to the shelf.

Base shoe molding (I) covers gaps around the sides, top, and bottom of cabinets and built-ins. It bends fairly easily, so it works well to cover the gap between uneven walls and floors.

Quick Reference

There are a few basic questions that must be answered before you can begin your woodworking project. The type of wood you select and the size and type of nails or screws you use are important to the success of the project. Use the charts on the following pages to answer some of the most frequently asked woodworking questions. For more detail, refer to the appropriate chapters in this book.

Nails are ordered by the penny number, or "d". Once based on the price per nail, this number now relates to length. Choose your nails according to the type of work you are doing. The information earlier in this handbook will help you. Use common nails or box nails if you need holding power. Nails with cement coatings and annular or spiral shanks provide extra holding power. If appearance is important, use casing or finish nails.

Estimate how many nails you will need for your project and then check the following chart for the number of pounds to buy.

PENNY NAIL GAUGE

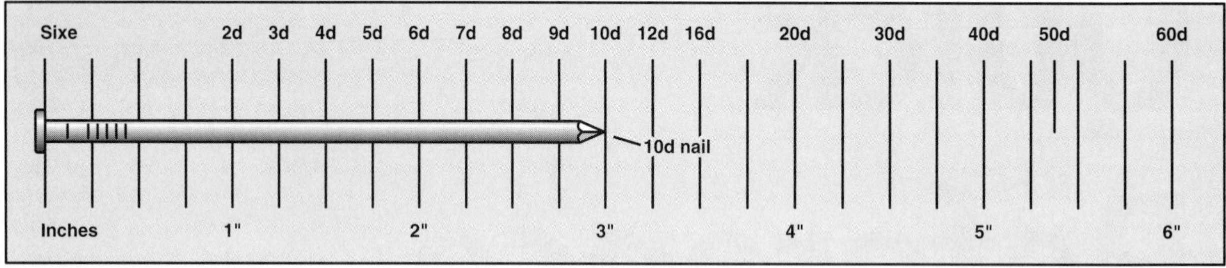

Actual size

NAIL SIZES AND WEIGHTS

PENNY NUMBER	LENGTH IN INCHES	COMMON NAILS (per lb.)	BOX NAILS (per lb.)	FINISHING NAILS (per lb.)
2	1	876	1010	1351
3	1¼	568	635	807
4	1½	316	437	548
6	2	181	236	309
8	2½	106	145	189
10	3	69	94	121
12	3¼	64	87	113
16	3½	49	71	90
20	4	31	52	62
30	4¾	20		
40	5			
50	5½			
60	6			

GUIDE FOR WOODSCREWS

Gauge		2	3	4	5	6	7	8	9	10	11	12
Shank-Hole Size		3/32"	7/64"	7/64"	1/8"	9/64"	5/32"	5/32"	11/64"	3/16"	7/32"	1/4"
Pilot-Hole Size	Hardwood	1/16"	1/16"	5/64"	5/64"	3/32"	7/64"	7/64"	1/8"	1/8"	9/64"	5/32"
	Softwood	1/16"	1/16"	1/16"	1/16"	3/64"	3/32"	3/32"	7/64"	7/64"	1/8"	9/64"
Available Lengths	1/4" 3/8" 1/2" 5/8" 3/4" 1" 1 1/8" 1 1/4" 1 3/8" 1 1/2" 1 5/8" 1 3/4" 2" 2 1/4" 2 1/2" 2 3/4"											

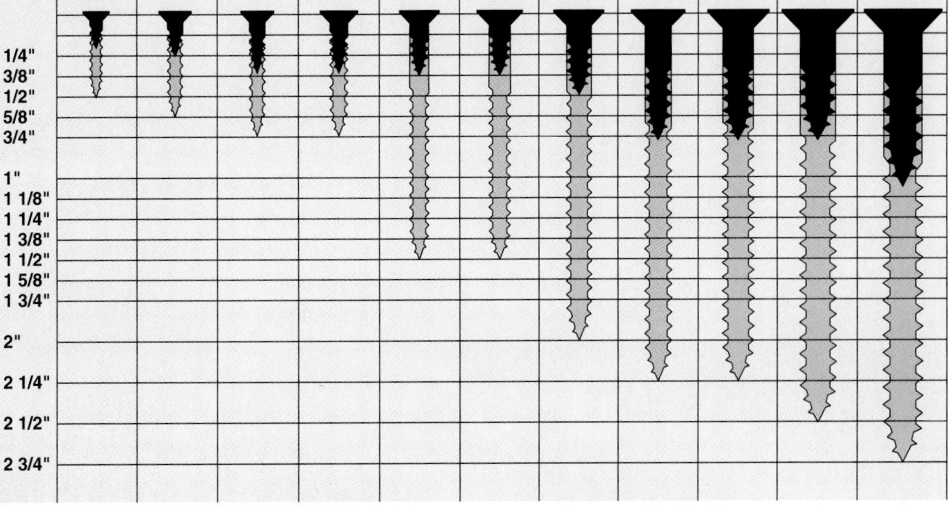

WORKING PROPERTIES OF SELECTED WOODS

TYPE	PLANING	SANDING	DRILLING	TURNING	GLUING
Ash, white	Average	Good	Good	Average	Average
Basswood	Good	Excellent	Excellent	Fair	Excellent
Beech, American	Excellent	Good	Excellent	Average	Excellent
Birch, yellow	Good	Poor	Good	Excellent	Good
Butternut	Excellent	Poor	Good	Average	Good
Cedar, aromatic	Fair	Average	Average	Good	Fair
Cherry, American	Excellent	Good	Good	Good	Good
Douglas-fir	Average	Average	Average	Average	Good
Hickory	Average	Good	Good	Good	Good
Maple, hard	Good	Excellent	Good	Excellent	Good
Oak, red	Average	Good	Fair	Average	Average
Oak, white	Average	Good	Fair	Good	Average
Pine, white	Good	Good	Good	Good	Good
Poplar	Good	Good	Good	Good	Good
Redwood	Excellent	Good	Good	Good	Excellent
Walnut	Good	Good	Good	Good	Average

CHAPTER 1

The Woodworking Industry

YOU'LL BE ABLE TO:

- Discuss the commercial importance of wood.
- Explain how wood is harvested and processed.
- Describe the different classifications for wood and wood materials.
- Understand and apply the problem-solving process.
- Describe several woodworking careers.
- Discuss ways in which to find and keep a job.

Look for These Terms

- clear cutting
- computer-aided drafting
- computer-aided manufacturing
- engineered wood
- entrepreneur
- hardwoods
- molding
- panel stock
- plywood
- problem-solving process
- selective cutting
- softwoods

Wood was one of the first resources early humans used, and it remains a vital resource today. Over thousands of years, people have developed ways to process this raw material for countless uses, from basic shelter to paper to plastic. (For more specific information about the kinds of wood materials described in this chapter, see the Woodworker's Handbook.)

Wood is not only one of the most versatile and economically important raw materials, it is also one of the most beautiful. It has inspired artists and craftspeople to produce works of enduring value. Fig. 1-1.

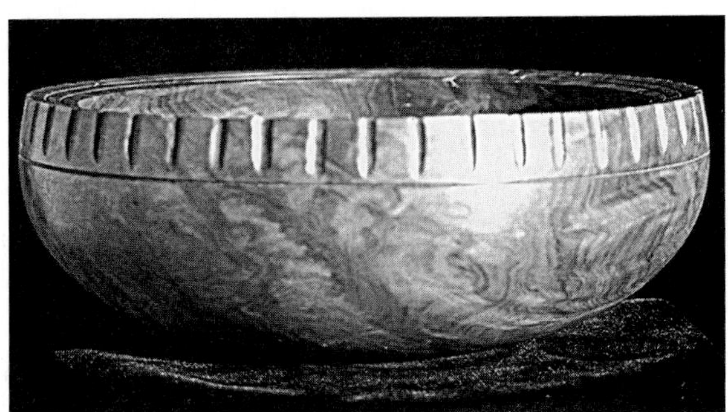

Fig. 1-1—*The natural beauty of the wood and the skill of the woodworker result in attractive products such as this bowl.*

THE COMMERCIAL IMPORTANCE OF WOOD

Wood's seemingly endless usefulness has helped determine its commercial importance. The United States exports about 800 million cubic feet of logs annually. More than 40 billion board feet of lumber are cut for domestic use each year, and that number is growing.

Of course, the most common uses of wood are in furniture and cabinetmaking and in building construction. Boards and other wood materials are used to frame and enclose many buildings. Inside the structures, wood may be used for wall paneling, finish flooring, cabinets, doors, and trim.

Trees unsuitable for lumber are ground up into pulp and used to make paper and cardboard. Wood chips and sawdust left over from cutting logs into boards can be processed to create new wood materials, such as pegboard for hanging tools. Even the bark removed when processing logs can be used as garden mulch.

Wood also has chemical uses. Rayon, a *synthetic* (man-made) fiber used in clothing, is created from the cellulose in wood pulp. Cellulose fibers can also be used in making photographic film and plastics.

Wood Detectives

The U.S. Forest Products Laboratory in Madison, Wisconsin, tests all kinds of wood products. This laboratory has identified woods taken from tombs of Egyptian pharaohs, sunken pirate ships, and even a beam that supported the Liberty Bell. Anyone having difficulty identifying a piece of wood can get help from this agency.

HARVESTING AND PROCESSING WOOD

The first step in getting the wood and wood products to consumers is harvesting the wood from the forests. Then the logs are sent to sawmills for processing.

Harvesting Trees

Two methods of harvesting trees are used: selective cutting and clear cutting. In **selective cutting,** only trees of a certain diameter and species are harvested. A number of smaller and larger trees are left behind. These remaining trees will drop seeds, which will regenerate the forest. (This is referred to as *natural regeneration.*)

Under certain conditions, **clear cutting** is chosen. In this method, loggers remove all trees, regardless of species, size, or age. Clearcut areas are then reseeded or allowed to naturally regenerate as the tree stumps develop sprouts. Clearcutting often destroys animals' natural habitats, so environmentalists disapprove of it.

Processing Lumber

The first step in lumber processing is called *barking.* In some mills, a large, knifelike jet of water peels the bark from the log. Barking can also be done by machine or by hand.

Newly barked logs are conveyed to a huge bandsaw, where they are sliced into large planks called *cants.* Cants are then passed through a series of saws that slice, edge, and trim them into various sizes. Fig. 1-2. Waste is taken to other plants to be made into lumber byproducts. (A *byproduct* is a secondary product made from things left over from making the main product.)

Once trimmed of all uneven edges, the lumber is placed on a moving belt called a *green chain*. There it is graded and sorted according to its characteristics and quality. Grades of lumber will be discussed later in this chapter.

Green lumber contains a lot of moisture, so it must be dried. In an outdoor storage yard, each piece is restacked so air can circulate between the pieces. This *air seasoning* may take from six months to two years. Wood that contains no more than 19 percent moisture is sold as *dry lumber.*

Following air drying, top grades of lumber are taken to huge ovens called *dry kilns* for final moisture removal. From the dry kiln, the lumber may go to cooling sheds and then to dry sheds for storage, or it may go to the planing mill for surfacing.

In the planing mill, lumber is turned into siding, paneling, and specialty products. Edges and ends are trimmed. *Faces* (top and bottom surfaces) are planed smooth or saw textured. The processed lumber is then bundled and stored in the planing mill, ready for shipment.

CLASSIFICATION OF WOOD MATERIALS

Lumber is wood that is sawed into smaller parts, such as beams and boards. *Rough lumber* comes just as it was cut at the sawmill. *Dressed (surfaced) lumber* has been put through a planer.

Panel stock (also called *sheet material*) is wood that has been processed further and formed into panels (or sheets). **Plywood,** for example, is made by gluing thin layers (plies) of wood together. Generally, each layer is placed at a right angle to the next. Hardboard, particleboard, waferboard, and oriented strand board (OSB) are panel stock made from wood chips, strands, or fibers that are mixed with adhesives and then bonded together under heat and pressure. Panel stock is most commonly found in 4′ × 8′ sheets, with thicknesses generally ranging from ¼ inch to 1 inch.

A **molding** is a narrow strip of wood shaped to a uniform curved profile throughout its length. Moldings have both practical and decorative uses. They are used to conceal joints as well as to ornament furniture and room interiors.

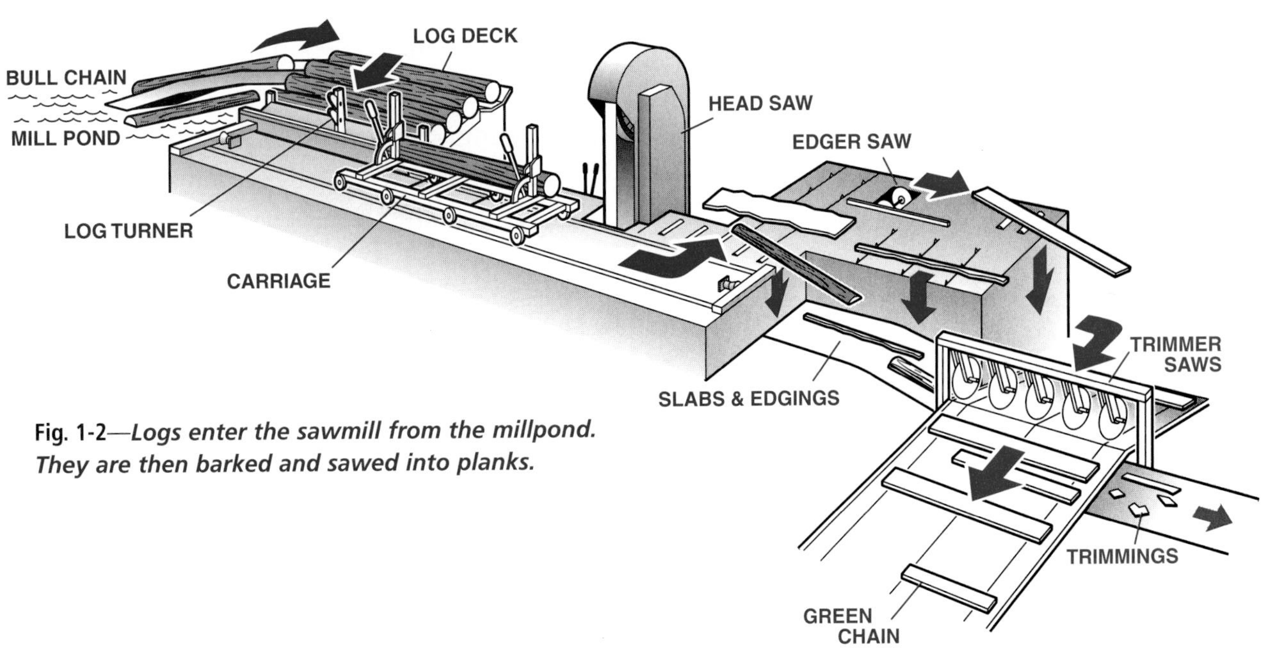

Fig. 1-2—*Logs enter the sawmill from the millpond. They are then barked and sawed into planks.*

Hardwoods and Softwoods

The terms *hardwood* and *softwood* are botanical terms that identify types of trees. The terms do not actually indicate the actual softness or hardness of the wood. Some softwoods are harder than some hardwoods, and vice versa.

Softwoods come from *coniferous,* or cone-bearing, trees. Common examples are pine, fir, cedar, and redwood. **Hardwoods** are cut from *deciduous* trees. A deciduous tree sheds its leaves annually. Some common hardwoods are walnut, maple, birch, cherry, and oak.

Cuts of Lumber

Most lumber is cut in such a way that the annual rings form an angle of less than 45 degrees with the top and bottom surfaces of the piece. Fig. 1-3. This lumber is called *plain-sawed* (when it is hardwood) or *flat-grained* (when it is softwood).

When wood is cut with the annual rings forming an angle of more than 45 degrees, it is called *quarter-sawed* or *edge-grained* or *vertical-grained.* Quarter-sawed lumber is usually considered more attractive than plain-sawed or flat-grained lumber, but it is more expensive.

Grades

Wood is graded according to its quality or purpose. Some defects in quality occur naturally as the tree grows. Others develop as lumber is cut, machined, and stored. Some common defects are described in Table 1-A. Both lumber and panel stock come in a variety of grades.

SCIENCE *Connection*

Humidity's Effects on Wood

Wood reacts to the relative humidity of the air. As relative humidity increases, the wood swells. As it decreases, the wood shrinks. If the wood contains too much moisture, cracks and splits will develop in the product as the wood dries, and any joints will become loose.

Ideally, when a product is built, the wood should be at the moisture content that it is expected to have, on average, during its use. Lumber used to make furniture and cabinets must contain no more than 6 to 7 percent moisture. Lumber for building construction must contain no more than 8 to 12 percent moisture.

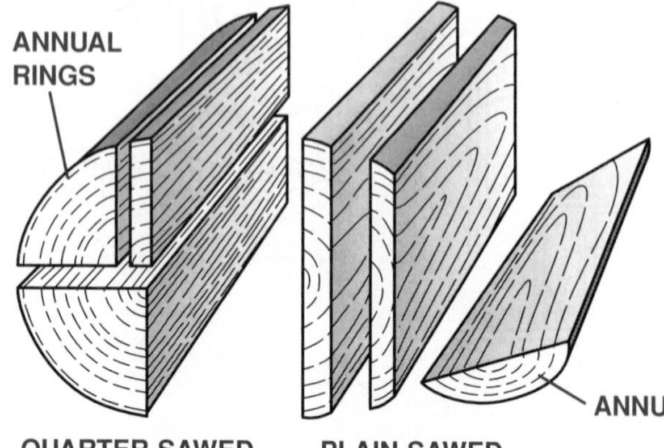

Fig. 1-3—*The two most common cuts of lumber are plain-sawed and quarter-sawed. Quarter sawing is a more expensive process and is usually reserved for wood for some types of furniture.*

Table 1-A. Common Defects in Wood

Defect	Description
Knot	Usually a dark, round or oval spot at the point where a branch had formed on the tree
Check	Separation of the wood, usually lengthwise across the growth rings; often appears at the end of a board. (*Honeycombing* is an area of checks not visible on the surface.)
Split	Lengthwise separation of the wood, like a large crack
Decay	Rotting of the wood
Stain	Discoloration on the surface
Wane	Slanting edge on a board created when the lumber is cut
Warp	Any variation from a straight, true (plane) surface. Types of warp include: *crook*, an edgewise bend; *bow*, a flatwise bend; *cup*, a curve across the grain so that the board is no longer flat across its width; *twist* or *wind*, a turning (winding) of the edges so that the four corners of any face are no longer in the same plane.

Lumber

Softwood lumber is graded first according to use. *Yard lumber,* cut for a wide variety of uses, may be common or select grade. *Common* yard lumber is suitable for rough carpentry. *Select* yard lumber has a good appearance and takes different finishes such as stain and paint. Select lumber is then further graded according to the quality of its appearance. Grades range from A to D. Grade A is free of almost all imperfections; lesser grades have increasingly more imperfections.

Factory (shop) lumber is usually cut up for manufacturing purposes. *Structural lumber* is used in construction; grades are based upon the strength of the pieces.

The top grade of hardwood lumber is labeled *FAS,* meaning "firsts and seconds." The remaining grades, Select and No. 1 common, contain more knots and defects.

Panel Stock

Softwood plywoods are divided into exterior (waterproof) and interior (nonwaterproof) types. For projects that are to be painted, interior plywood should be used.

For hardwood plywoods, the outside ply is a good hardwood such as birch or oak. The grade of the panel is determined by the quality of the front and back faces.

Hardboard, made by refining wood chips into fibers that are formed into panels under heat and pressure, is available in standard (untreated) and tempered (treated) grades. Tempered hardboard has been further hardened by being dipped in chemicals and then baked. On some hardboard, one face is smooth and the other is rough. Other types have two smooth surfaces.

MAKING WOOD PRODUCTS

Many wood products, from cutting boards to furniture, are made in factories using mass production methods. Large numbers of products are turned out at a relatively low cost.

In the factory, woodworkers cut and shape parts that are joined later to create a product. They may use traditional tools and equipment, or they may operate computerized machinery.

Technology in Manufacturing

Technology has introduced new methods to the woodworking industry, primarily by means of computers. Because computers can be very precise, they are especially useful for cutting operations. Fig. 1-4. Computers are used in manufacturing in the following ways:

- *Computer-numerical control (CNC).* CNC involves the use of computers and a numerical code to control the machines used to make parts for products.
- *Computer-aided drafting (CAD).* In **computer-aided drafting,** designers create drawings and product plans on a computer, using special software. With CAD, changes can be made to the design without extensive redrawing. Also, the product can be shown onscreen in various types of woods and finishes before a final design is chosen. Many CAD programs can also create three-dimensional computer models of an object. These models can be used to check sizes and other details before the item is actually manufactured.
- *Computer-aided manufacturing (CAM).* A CAD drawing can be sent directly to a computerized machine that then makes the part. This process, called **computer-aided manufacturing,** saves setup time and labor.
- *Computer-integrated manufacturing (CIM).* In computer-integrated manufacturing, an entire factory is linked together in one computer system. Designs are made, materials are ordered, and products are created and stored by means of computer control.
- *Robots.* Computer-controlled robots are usually used for jobs that are too dangerous or tedious for humans. In the woodworking industry, they are often used for finishing operations and for handling large sheets of material.
- *Lasers.* As shown in Fig. 1-4, computer-controlled lasers are often used for cutting, drilling, and etching. Lasers are also used to measure electronically.

Research and Development

The goal of research and development (R&D) is to improve existing products or create new ones. Large companies that produce wood products or finishing materials have their own research laboratories.

Although trees are a renewable resource, they take years to grow, and the current supply is being used up. One result of wood research and development has been to help stretch that supply by developing **engineered woods,** made from sawdust or small wood pieces and plastics. This process makes use of scrap materials and small, crooked trees that would not normally be harvested. Engineered wood is more expensive than natural wood, but it lasts longer and is much stronger.

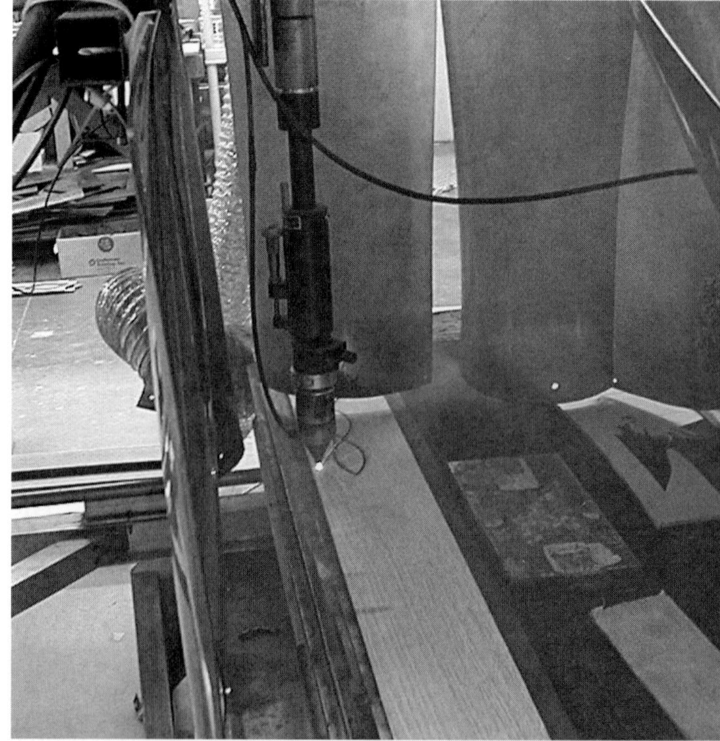

Fig. 1-4—*Because lasers are computer-controlled, the cuts made are very precise.*

THE BENEFITS OF A COURSE IN WOODWORKING

A course in working with wood offers many benefits. You learn about good design and how to judge the quality of wood products. You learn how to develop solutions to problems—a skill you will call upon throughout your life. You also learn to use many tools and machines, and at the same time, develop good safety habits that will help protect you and those around you, both now and in the future. In addition, as you work with fellow students on some of the activities and projects, you will develop teamwork skills, which will be helpful in your future workplace.

Using the Problem-Solving Process

During this course, you will often need to use the **problem-solving process** to develop solutions to the problems presented in the activities and projects. It consists of six basic steps.

1. *State the problem clearly.* Being able to identify the problem is often half the job of solving it.
2. *Collect information.* The more you know, the better your chances of finding a solution that works.
3. *Develop possible solutions.* Most problems have more than one possible solution. Propose several different ideas and consider them carefully.
4. *Select the best solution.* Look at the advantages and disadvantages of the solutions you've developed. Then pick the one that seems the most workable.
5. *Implement the solution.* Give it a try, but keep in mind that few solutions work perfectly the first time.
6. *Evaluate the solution.* Did it work as expected? How can it be improved?

Developing Teamwork and Leadership Skills

Certain activities and class projects will involve teamwork. As a member of a team, you will learn to cooperate as well as contribute. You will also learn the importance of *ethical behavior*—behaving according to a high standard.

Every team needs a leader. In the role of leader, you can develop communication skills, strength of purpose, and integrity.

SkillsUSA is a national organization serving high school and college students who are enrolled in technical, skilled, and service programs. This organization helps students develop their leadership, teamwork, and citizenship skills, as well as high standards of character. Ask your instructor about a SkillsUSA chapter at your school.

CAREERS IN WOODWORKING

More than two million people are employed in the broad areas of forestry, lumbering, millwork, furniture manufacturing, wood construction, and distribution of forest products. Over 5,000 industrial woodworking companies are located in the U.S. and Canada. Additional people are employed in business, sales, paper making, and wood science and technology.

Many people work in and around forests and sawmills where logs are converted into rough lumber. Others are employed in the grading and seasoning of lumber and in moving it from the forest to the mill and from the mill to the factory. In processing plants, workers convert wood materials into plywood, hardboard, and other manufactured products. Assembling and finishing operations supply jobs for many. Other workers create housing and other wood structures.

Career Categories

In general, careers in the wood industry can be divided into three main categories: craftspeople, technicians, and professionals. Table 1-B gives brief descriptions of jobs typical to each category. You can find even more information on the Internet at the Bureau of Labor Statistics' Web site.

Some workers associated with the wood industry are **entrepreneurs,** which means they own and are responsible for their own businesses. Architects are an example.

Finding a Job

Looking in the "Help Wanted" section of your local newspaper is not the only way to find a job or prepare for a career. The following suggestions can be especially helpful:

- Use library resources to find out more about fields in which you are interested. Find out which jobs are currently in demand or will be in demand in the future. The *Occupational Outlook Handbook* is a good source of this kind of information.
- Make the most of school resources. Talk to teachers and counselors about which courses and programs you should enroll in.
- Develop job contacts. Talk to neighbors and relatives about their jobs. They can often advise and help you. The offices of local unions can also offer information.
- Apply for an internship or other summer job in the field. You'll learn more about the field and your own aptitudes for it.
- Find out about industries in your community. Knowledge of the local job market can help you make your own plans.
- Use the Internet. Many companies list job opportunities on their Web sites.

Next, practice the skills you will need when applying for a job. For example, make a list of your education, skills, and experience. Then obtain a sample application form and practice filling it out. Ask a friend to help you practice your interview technique.

Keeping a Job

What makes a good employee? Following are the qualities most in demand by employers. How many could be used to describe you?

- *Positive attitude.* You try to view situations in a positive way.
- *Willingness to learn.* You're willing to try new skills and ways of doing things.
- *Ability to follow directions and obey rules.* You pay attention and understand that things are done a certain way for a reason.
- *Ability to solve problems.* You don't let problems stop you. You apply the problem-solving process.
- *Ability to accept criticism.* You understand that everyone makes mistakes, and you use others' criticism to improve your work.
- *Cooperation.* You work well with others.
- *Dependability.* You show up on time and you complete your assigned tasks.
- *Honesty.* You are trustworthy.
- *Initiative.* You look for things that need to be done and then do them.
- *Loyalty.* You recognize that no job is perfect and don't complain to others.

Table 1-B. Woodworking Careers

Career	Job Description	Training and/or Education	Employment Outlook Through 2006
CRAFTSPEOPLE			
Carpenter	Builds or repairs wooden structures, such as homes, offices, and boats	Apprenticeship training, completion of trade/ vocational school or community college program, or on-the-job experience	Many job opportunities
Precision Woodworker or Cabinetmaker	Highly skilled; builds cabinets, furniture. For some jobs, ability to operate computerized machines is required	Minimum of a high school education plus on-the-job training. Those with more education often become supervisors	High demand for skilled people
Millwright	Installs, maintains, repairs, and replaces tools and machinery	4 years of training on the job, as an apprentice, or in a college technology program	Slight decline overall, but good opportunities for skilled people
TECHNICIANS			
Drafter	Finishes plans begun by an architect or designer	Post-high school training in a drafting program; must meet company standards; CAD training highly desirable	Average growth
Forestry Technician	Works with foresters; compiles data on insect damage, conservation, etc.	Minimum of a high school education; 2-year degree in specialty desirable	Slow growth
Engineering technician	Works with engineers to solve technical problems in such areas as manufacturing, construction, and inspection	Minimum of a 2-year degree in engineering technology	Average growth
PROFESSIONALS			
Forester	Develops and cares for forested land belonging to individuals or governments	B.A. in forestry; advanced degree required for special research	Average growth Most opportunities are with state and local governments.
Architect	Designs buildings and other structures	Minimum of a 5-year B.A. degree in architecture, plus training period and licensing	Average growth 30% of architects are self-employed
Furniture Designer	Creates designs based on period furniture or current trends; makes detailed drawings	B.A. degree in design; CAD training essential	Average growth 40% of designers are self-employed
Teacher	Teaches high school courses in carpentry, construction, etc.	Minimum of a bachelor's degree, plus courses in education and a state license. Some states require an advanced degree	Faster than average growth

CHAPTER 1

Review & Applications

Chapter Summary

Major points in this chapter that you should remember include:

- Because of its usefulness, wood has great commercial importance.
- Trees are harvested using selective cutting or clear cutting.
- Wood may be processed into lumber, panel stock, or molding. It can also be classified by the type of tree it comes from, the way in which it is sawed, and its quality.
- Computers have become increasingly important in manufacturing wood products.
- The problem-solving process can aid you in finding workable solutions.
- There are a wide variety of careers in the woodworking industry.

Review Questions

1. Discuss three ways in which wood is commercially important.
2. Describe the two primary methods of harvesting trees.
3. What is the difference between hardwoods and softwoods? Give three examples of each.
4. Describe three defects that might affect the grade assigned to a piece of lumber.
5. Outline the six steps in the problem-solving process.
6. Describe the job of a drafter. What education and training is needed?
7. Name at least three characteristics employers look for in an employee.

Solving Real World Problems

You decide to remodel your kitchen into the gourmet's dream you've always wanted. As part of the project, you need to purchase a special cabinet for a built-in commercial stove.

Due to the heat generated by this stove, it must be insulated and ventilated according to very strict guidelines. There are no ready-built cabinets available in the size you need, and the ones that are available cannot be modified to fill your need. What group of woodworkers would you approach with the problem? Would you consider a furniture maker? Explain.

CHAPTER 2

Safety Practices

YOU'LL BE ABLE TO:

- Tell why safety is really an attitude.
- Discuss common woodshop hazards and how to prevent problems.
- Describe different types of personal safety gear and tell their purpose.
- Tell how to set up a safe workshop.
- Discuss the use of first aid for common workshop injuries.

Look for These Terms

- Environmental Protection Agency (EPA)
- ground-fault circuit interrupter (GFCI)
- hazards
- Material Safety Data Sheets (MSDS)
- National Institute for Occupational Safety and Health (NIOSH)
- Occupational Safety and Health Administration (OSHA)

Did you know that, each year, over two million people are injured in work accidents? More than 11,500 people die from those injuries. Even accidents in which no one gets hurt can damage equipment and waste materials. A project that took many long hours to create may be ruined.

Being concerned about safety is not the same thing as being "chicken." People who work with safety in mind want to get the most from their hours in the woodworking shop. They don't want to waste their time or turn the experience into something they or someone else will regret.

SAFETY IS AN ATTITUDE

How many times have you heard people say after an accident, "I just wasn't thinking," or "I had a feeling that wasn't going to work?" Either they didn't think about safety at all, or they ignored what their instincts and common sense were telling them. But dangerous machines and materials demand alertness and respect. Fig. 2-1.

Fig. 2-1—*TIP: Use your brain as well as your muscles. For example, when carrying large sheets of stock by yourself, use a length of rope (about 18 feet long) tied in a loop. Hook the loop over the lower corners of the stock. Then grab the rope to carry.*

Have you ever been bumped by someone who wasn't watching where he or she was going? Most accidents are the result of not paying attention. Other accidents may occur because someone made the wrong decision. Perhaps the person decided not to wear safety glasses this time and a wood chip flew into his or her eye.

The best way to avoid these and other unsafe situations is to develop a safe attitude. What is a safe attitude? It is a decision to take safety seriously. In general, you have a safe attitude if you:

- Think about the consequences of your actions, not only for yourself but also for the people around you.
- Decide to follow safe procedures, obey the rules, and act responsibly even when you don't feel like it.
- Keep in mind that no matter how lucky you may have been in the past, it won't last forever.

The general guidelines shown in Table 2-A are a good place to start. They will help you focus on safe behavior.

COMMON WOODSHOP HAZARDS

An important part of safety is being aware of the **hazards,** or dangers, that are present. Seven colors are used on labels and signs to indicate different kinds of hazards. The colors are listed in Table 2-B.

Common woodshop hazards include fire, electricity, and hazardous materials. The following discussions tell how to avoid injury from these hazards. In later chapters of this book, you will find safety rules for using specific tools and machines.

Table 2-A. General Safety Guidelines

Remember...

- Develop a safe attitude.
- Stay alert; be aware of your surroundings.
- Take the time to do the job right.
- Follow the manufacturer's instructions for proper use and care of woodworking tools and machines.
- Keep woodworking tools and machines in good working order.
- Keep your work area clean and neat.
- Anticipate problems before they occur and do something to prevent them.
- Dress for the job.
- Follow the rules.

Table 2-B. Colors Used for Safety

Color	Example of use	Meaning
Red	Panic buttons on machinery	Danger or emergency
Orange	Pulleys, belts, gears	Warning
Yellow	Machine controls	Caution
White	Traffic flow	Storage or boundary
Green	Safety equipment	First aid or safety
Purple on yellow	Signs in areas where radiation is present	Radiation caution
Blue	Bulletin boards	Information

Fire

Sawdust, wood chips, paint thinners and other solvents, chemical vapors, and tools that produce heat make fire a very real danger in the woodworking shop. Under the right conditions, even an explosion can occur.

Preventing Fires

You and your instructor or employer can take the following steps to prevent and anticipate fires in the workshop:

- Learn where the fire alarm is and how to operate it.
- Post the number of the fire department next to the nearest telephone.
- Install a smoke detector. Test it monthly to be sure the battery remains effective.
- ABC fire extinguishers are effective against the fires most likely to occur in the woodshop—wood, oil, solvent, chemical, and electrical fires. Install one in the shop, and be sure everyone knows where it is located and how to operate it. Fig. 2-2. Check the extinguisher periodically to be sure it is still active.
- Make a *fire safety plan* showing a map of the building and at least two escape routes. Be sure those routes are kept open at all times.
- Store all finishing products and solvents in airtight glass or metal containers away from heat.

USING FIRE EXTINGUISHERS

Class of Fire	A	B	C	D No icon designated for Class D fires
Material Burned	Combustibles: Wood, paper, cloth, etc.	Flammable liquids: Grease, gasoline, oil, etc.	Electrical equipment: Motors, switches, etc.	Combustible metals: Iron and magnesium
Type of Extinguisher				
Carbon Dioxide (Carbon Dioxide Gas Under Pressure)		●	●	
Pump Tank (Plain Water)	●			
Multi-Purpose Dry Chemical	●	●	●	●
Ordinary Dry Chemical		●	●	
Halon		●	●	
Dry Powder				●

Fig. 2-2—*This chart shows the different types of fires, the kind of extinguisher to use for each, and the appropriate symbols and colors.*

- Keep oily or paint-covered rags in approved containers.
- Be sure the workshop is well ventilated.
- Keep electrical cords in good condition and replace those that need repair.
- Do not overload electrical circuits.

Controlling a Fire

Most fires can be extinguished by one or more of the following methods.

- Reducing the heat, such as by throwing water on the fire. (Never use water on an electrical fire or on burning liquids.)
- Removing the source of fuel. For example, if the fire is electrical, turn off the power at the fuse box or circuit breaker.
- Preventing oxygen from reaching the fire. For example, use a fire extinguisher to spray the fire. If a person's clothing catches on fire, the person should roll around on the floor to smother the flames.

Electricity

Power machines and hand tools, heating systems, and light fixtures are all common in the woodworking shop. Because even a minor electrical shock can be dangerous, this electrical equipment is designed with safeguards. However, no safeguard can protect you against carelessness. Learn to take some basic precautions.

- Make sure there are enough outlets and circuits. Keep outlets in good condition. Fig. 2-3.
- Use only **ground-fault circuit interrupter (GFCI)** outlets and plugs. GFCIs protect you by detecting leaks to ground and breaking the circuit. Test these outlets monthly.
- Never remove a grounding prong from a three-prong plug. Instead, install a GFCI outlet to accommodate the plug.
- Check plugs and power cords regularly for cracks and fraying. Replace a damaged plug or cord.

Facts about Wood

Burning Question—Wood is a poor conductor of heat. Even in the hottest fire, the core of a thick beam will remain at a lower temperature. Wood also does not conduct electricity and can be used to push someone away from a source of electric shock.

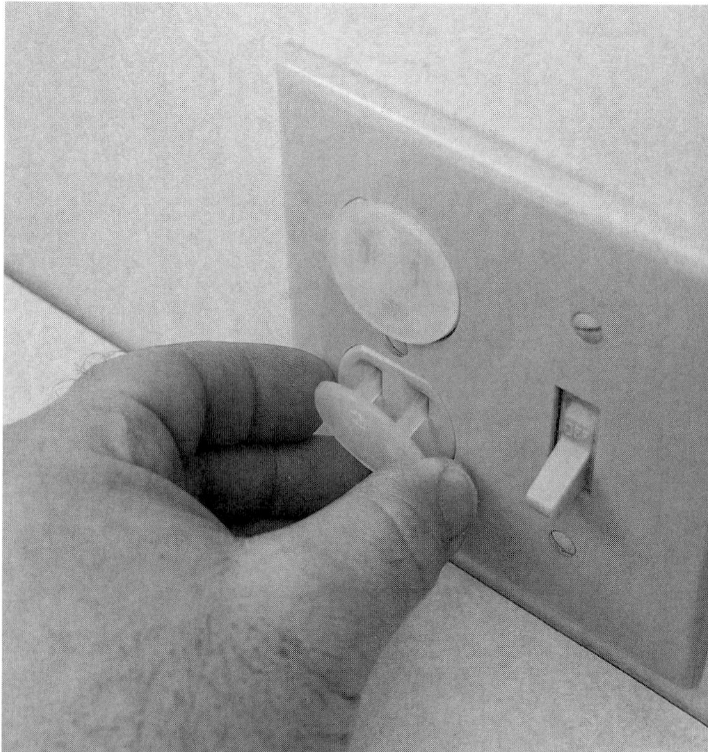

Fig. 2-3—*TIP: Plastic dust caps can be inserted into electrical outlets to keep out sawdust and dirt, which can otherwise cause short circuits and fires.*

- Avoid using extension cords whenever possible. If you must use them, be sure they are designed to carry the required amperage. Table 2-C.
- Always shut off the power switch on equipment and unplug it before making repairs.
- Never replace a fuse with one of higher amperage. Too much power will flow through and overload the circuit.
- If you see someone receiving a shock, turn off the power at the circuit breaker or fuse box. Do not touch the person or the appliance yourself, or you could become part of the circuit.

Hazardous Materials

Some liquids—and their vapors—can be hazardous. For example, solvents used in paint thinners, strippers, and finishes can cause headaches, nausea, dizziness, and skin and eye irritation. These dangerous chemicals can enter the body in several ways. They can be absorbed through the skin, inhaled as a vapor, and swallowed with food left around the shop.

Although most solvents do not cause harm when used occasionally, long-term use has been associated with cardiac arrest (stopping the heart), nervous system damage, brain damage, and birth defects. Read labels and be

Table 2-C. Extension Cord Gauges

Tool Amperage	Minimum Extension Cord Wire Gauge		
	25-ft. cord	50-ft. cord	100-ft. cord
0-6	18	16	16
6-10	18	16	14
10-12	16	16	14
12-16	14	12	Do not use

SCIENCE *Connection*

How Electricity Kills

Your brain sends instructions to the other parts of your body by means of electrical and chemical signals that travel through your nervous system. An electric shock overrides these signals. Particularly dangerous is a current that enters the heart and respiratory centers. When this happens, death occurs from either ventricular fibrillation or respiratory-center paralysis.

Ventricular fibrillation is very rapid or irregular beating of the heart. When the heart muscle is fibrillating, it can't pump effectively. The vital body organs are not supplied with fresh blood, and the person dies.

Respiratory-center paralysis occurs when the signal telling the lungs to work is interrupted. Breathing stops, and without oxygen, the person dies.

Electric shock also paralyzes other muscles, such as those of the hand. If the shock is coming from a power tool, the person may not be able to let go. This increases the person's exposure to the current.

Although many people think that the level of voltage is what determines danger from shock, this is not so. A shock of 10,000 volts may be no more deadly than a shock of only 100 volts. The real danger lies in the amount of current, which is measured in milliamperes. Any amount of current over 10 milliamperes can produce severe shock, and currents over 100 milliamperes are usually fatal.

aware of what's in the products you use. If possible, avoid using those that contain isobutyl ketone, methyl isobutyl ketone, toluene, xylene, methanol, and glycol ethers.

Every workshop should be well ventilated to keep fresh air circulating and prevent an accumulation of fumes. Open the windows and use an exhaust fan. Fig. 2-4. If you're using glues, strippers, and other toxic (poisonous) chemicals, wear a respirator. Make sure the respirator is designed to protect against toxic fumes.

It may surprise you to learn that some woods, such as South American mahogany and Western red cedar, can be toxic. They contain chemicals that can be inhaled in wood dust or that can get into cuts and scratches on your hands. These chemicals cause allergic reactions and symptoms such as headaches and irregular heartbeat. Use a respirator, gloves, and barrier creams on your hands when working with toxic woods.

Sawdust from any kind of wood can be hazardous to your lungs. When doing tasks such as sanding, wear a dust mask or respirator that filters out particles. (See page 82.)

Fig. 2-4—*TIP: If you don't have an exhaust fan, place an ordinary box fan in a shop window for ventilation. Be sure it faces outside so it will draw fumes from inside and expel them.*

Material Safety Data Sheets

Workers have a right to know about the dangers associated with their jobs. **Material Safety Data Sheets (MSDS)** are required by the **Occupational Safety and Health Administration (OSHA)** for all hazardous materials kept in the woodworking shop.

Before you use a hazardous material, review the information on the sheet. It will tell you whom to contact if you have questions about the material's use. It will also describe the ingredients in the material and its characteristics, including its ability to catch fire. The sheet will tell you about any health hazards, outline precautions to take, and indicate what conditions to avoid when using the material.

Disposing of Hazardous Wastes

Hazardous wastes are those that are dangerous to humans and/or the environment. Some examples are certain paints, stains, and clear finishes, brush cleaners, paint removers, paint thinners, cleaning solvents, and gasoline. The **Environmental Protection Agency (EPA)** has established limits as to the amount of hazardous waste permitted to accumulate at a work site and for how long.

Most hazardous wastes must be stored in special containers, such as 55-gallon drums, that are clearly labeled and that meet other EPA requirements. The EPA will provide information on all regulations that apply to a particular shop.

PERSONAL PROTECTIVE EQUIPMENT

Every woodworking shop should have adequate personal protective equipment (PPE) available, and every woodworker should learn how to use it. Fig. 2-5. Proper use of safety gear can prevent serious injury.

Feeding Stock into Tools

Pushsticks, pushblocks, and featherboards are helpers used to feed stock into stationary power tools. See "Make Your Own Shop Helpers" at the end of this chapter on page 85.

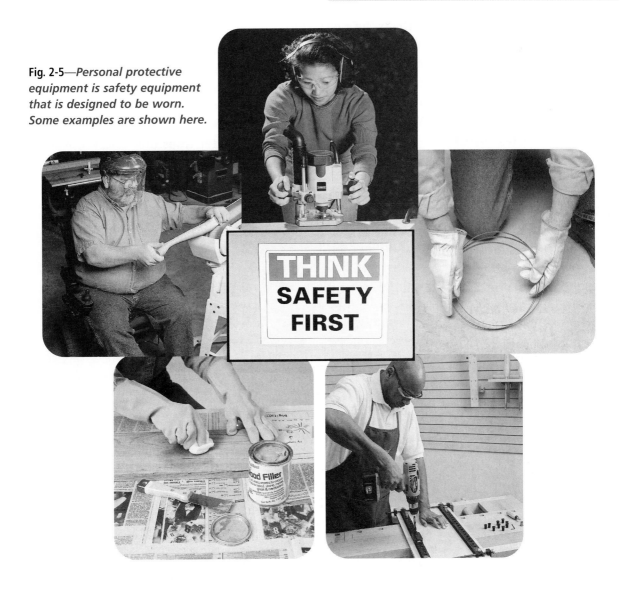

Fig. 2-5—***Personal protective equipment is safety equipment that is designed to be worn. Some examples are shown here.***

Clothing

Dress appropriately for working in the woodshop. Wear close-fitting clothing, and omit dangling jewelry, especially when you are using power tools. Anything loose can get tangled in the tool and cause injury. Roll long sleeves up above the elbow, and if your hair is long, tie it back.

Open-toed or canvas shoes do not offer enough protection to your feet. Wear durable boots with nonskid soles. Fig. 2-6.

Eye and Face Protection

Safety glasses and goggles are perhaps the easiest safety gear to use, and yet people often neglect to wear them. Although safety glasses give good general protection, goggles are preferred and can be worn over prescription eyeglasses. Use them when sawing, drilling, grinding, chiseling, and filing, and whenever dust, chips, or dangerous liquids might get into your eyes.

Goggles come in three basic types. Those with standard vents protect your eyes from flying objects and dust. Goggles with baffled vents or no vents protect against splashing liquids as well.

For complete protection of your eyes and face, use a face shield. Shields are adjustable and guard against both flying objects and splashes.

Hearing Protection

People don't often think of noise as a danger, but many power tools generate harmful noise levels. Some endanger hearing after only thirty minutes of exposure. See Table 2-D. Even short periods of loud noise, while not dangerous to hearing, can dull the senses and make you irritable and less alert. They can affect your heart rate and blood vessels, alter the natural rhythm of brain waves, and create stress.

To protect your hearing, wear earmuffs or ear plugs. Most earmuffs come with an adjustable headband and cover the entire outer ear. Ear plugs fit inside the ear. Some have a detachable neckband.

Protection Against Dust and Vapors

Wood dust is everywhere in the woodworking shop and can be hazardous when inhaled. The shop should be ventilated with an exhaust fan or by leaving doors and windows open.

Dust masks approved by the **National Institute for Occupational Safety and Health (NIOSH)** are recommended. The replaceable filter traps dust and keeps it from entering your lungs.

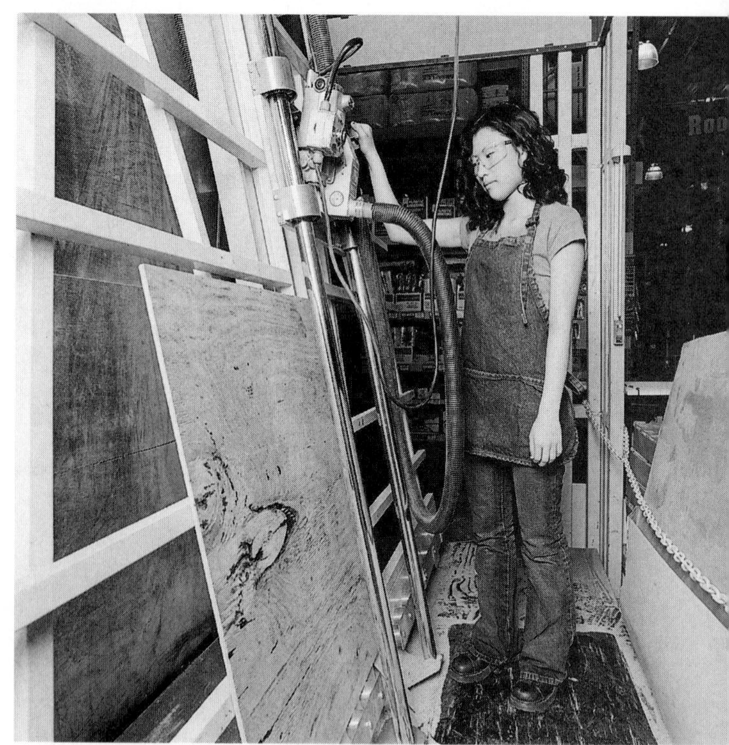

Fig. 2-6—***TIP: When you'll be standing on a hard shop floor for long periods, use an anti-fatigue mat. The mat protects leg and back muscles.***

Table 2-D. Noise Hazards

How Loud Is It?	
The intensity of sound is measured in decibels (dB). The softest sound audible by humans is 1 dB. Except for thunder and erupting volcanoes, nothing in nature is louder than 100 dB.	
quiet room	40 dB
½-hp drill press	87 dB
2-hp table saw	92 dB
½-hp jointer	101 dB
1-hp router	103 dB
2-hp circular saw	108 dB
jet airplane engine at 100 ft.	140 dB

A dual-cartridge, NIOSH-approved respirator should be worn for protection against chemical vapors and fine particles. The filter screens out dust, and the special cartridges designed for different applications purify the air. But a respirator must fit tightly against your skin or it can't do its job. Be sure to follow the manufacturer's instructions and test it for air leaks.

Hand Protection

Wear gloves to protect your hands when you're working with sharp edges and rough materials. The palms and fingertips should be of leather or thick fabric. The wrists should fit snugly so the gloves stay in place.

When handling solvents, adhesives, and finishes, wear plastic gloves designed to protect against these products. (Some plastics will dissolve when exposed to these chemicals.) The gloves prevent liquids from penetrating to your skin and causing burns or irritation.

However, don't wear gloves when you're using tools. Gloves can slip and cause you to lose your grip. Also, power tools can catch on them and pull your hand into the cutting area.

SETTING UP A SAFE WORKSHOP

Your safety depends not only on your own behavior and equipment but also on your surroundings. OSHA ensures that employers provide safe and healthful working conditions for all their employees. If you work in a woodshop, you should become familiar with OSHA regulations. It is also wise to heed OSHA's advice when setting up a home workshop.

Be sure there is enough light. In addition to overhead lights, you may also need individual, adjustable direct lighting in some work areas. Fig. 2-7.

Don't allow clutter to accumulate. Clutter not only contributes to dust, but it also gets in your way while you're working. On the floor it can cause tripping hazards. Be sure stock and equipment are placed where they cannot block safety exits.

Fig. 2-7—***TIP:*** ***Protect work lights from flying objects by covering them with window screening or wire mesh.***

A safe shop is also an efficient shop. Keep tools, safety gear, and other equipment organized so you know where to find it when you need it. Store hazardous materials in locked safety cabinets. Keep nails, screws, and other small items in labeled containers, such as empty coffee cans.

Be sure you use power tools near grounded outlets so you won't need extension cords. If extra cords are absolutely necessary, create a bridge over them where they cross traffic areas.

Be sure the shop has a smoke alarm and that the fire extinguisher is always within reach. Be sure the first aid kit is handy and well supplied.

WOODSHOP FIRST AID

Even the most careful woodworker can have an accident. Always have a well-supplied first aid kit in the shop and know what its various contents are used for.

Every woodworking shop must have a set procedure for emergencies. The teacher or manager will instruct you on these procedures.

Safety First

►*Too Much Strain, No Gain*

The human back is not designed to do the work of a forklift truck! Know its limits.

- Always lift with your legs and arms, not your back muscles.
- Get help when lifting heavy equipment and materials.
- Learn tricks for managing unwieldy items, like the method shown in Fig. 2-1.

If possible, take a first aid course that includes instruction in *cardiopulmonary resuscitation (CPR).* CPR is used to rescue victims who have stopped breathing and whose hearts may have stopped.

Some common woodworking shop accidents can be handled in the following ways. However, check with your teacher or supervisor about the procedures to use in your area.

- **Foreign object in the eye.** The victim should rotate the eye to find the object. Then the object can usually be gently wiped away using a clean tissue moistened with water. You can also wash out the particle using an eye irrigator. If the object is stuck to the eye or embedded in it, don't try to remove the object. Find medical help immediately.
- **Chemical in the eye.** Holding the eyelids open, rinse the eye under a gentle stream of cool water for at least 15 minutes. Cover the eye with a clean pad and find medical help at once.
- **Splinter under the skin.** Wash the area with soap and water. Sterilize a tweezers and needle with rubbing alcohol. Nudge the splinter out with the needle until it can be gripped with the tweezers and gently pulled out. Wash the area again with soap and water. If the splinter is metal or cannot be removed, the victim should see a doctor.
- **Cuts.** Wrap the cut with a clean cloth and press. Keep the wound elevated. After bleeding stops, wash the wound with soap and water and apply a bandage. If treating someone else's wound, wear gloves to avoid contact with blood. Severe bleeding requires immediate medical help.
- **Electrical shock.** If the victim cannot let go of the source of the shock, do not touch the person. Shut down the power at the source. If that is not possible, use a wood broom handle or a piece of dry wood to push the victim away. Call for medical help immediately.

Make Your Own Shop Helpers

A pushstick, pushblock, and featherboard are used with stationary power tools. Although they can be purchased, it's easy to make your own. Store them at the work station and have them within reach when using power equipment.

Pushsticks are used for feeding workpieces across stationary tool tables. Several designs are shown here. The stick should allow you to press down on the workpiece while keeping your hands away from a blade or cutter. The notch on the bottom should be deep enough to grip the workpiece without letting the stick contact the table. For use with a table saw, the pushstick should have a 45° angle between the handle and base. A pushstick with a smaller angle and a handle closer to the table works better with a radial-arm saw.

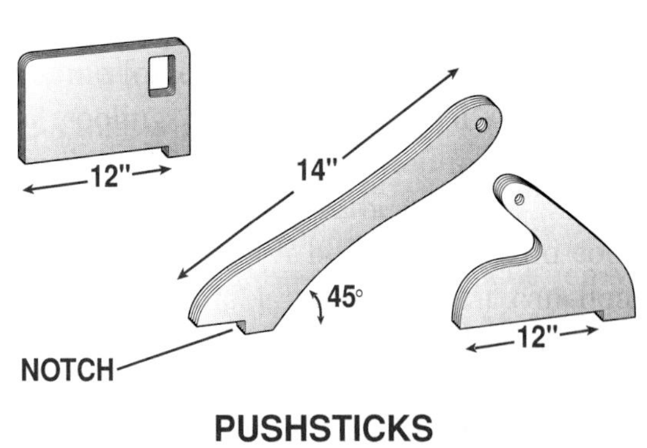

PUSHSTICKS

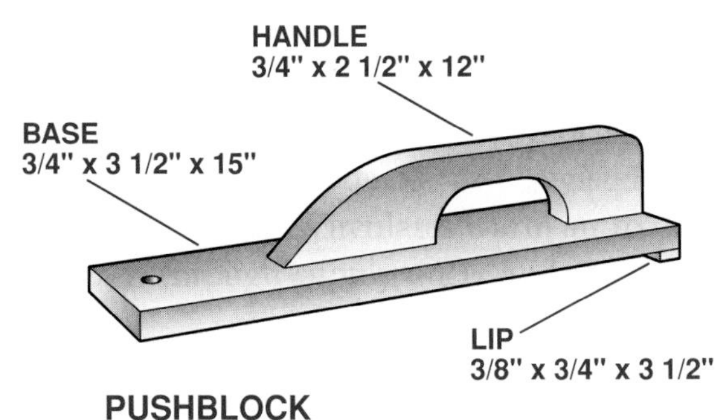

PUSHBLOCK

A pushblock helps you surface the face of a board on a jointer. An example is shown here, but you can make your own design. When attaching the handle, be sure to countersink any fasteners to avoid damaging the workpiece.

Featherboards, or fingerboards, help you press a workpiece against the fence or table of a stationary tool. Two possible designs are shown here. A standard featherboard is clamped to the fence or table. A notch in one edge fits a support board. The miter-slot design is clamped to the miter slot of a machine table.

FEATHERBOARDS

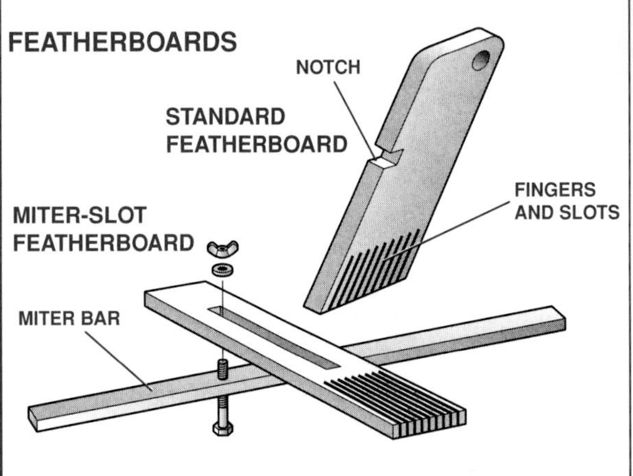

CHAPTER 2

Review & Applications

Chapter Summary

Major points from this chapter that you should remember include:

- Safety is an attitude. It is a conscious decision to take safety seriously.
- The most common woodshop hazards are fire, electricity, and hazardous materials.
- Personal protective equipment is available that protects eyes, ears, respiration, and hands. The right clothing should also be worn and long hair tied back.
- A safe shop has enough light, is free of clutter, and is well organized. Outlets are grounded, and a well-maintained smoke alarm is present.
- A well-stocked first-aid kit should be kept in the shop. Woodworkers should take a basic first-aid course and know how to handle minor emergencies.

Review Questions

1. What is a safe attitude?
2. List the seven safety colors and tell what they signify.
3. Which type of fire extinguisher should be kept in a woodworking shop?
4. What is the purpose of a ground-fault circuit interrupter (GFCI)?
5. What is the purpose of a Material Safety Data Sheet?
6. Name three types of personal protective equipment and tell their purpose.
7. How does clutter create unsafe conditions in the workshop?
8. What is the proper procedure for removing a splinter?

Solving Real World Problems

In Mr. Edwards' woodworking class, there is a need to change the blade on a table saw. The safety guard is moved out of the way. Everything not needed is cleared from the tabletop, and the new blade is inspected for damage. The necessary tools are gathered and checked for burrs that might cause a cut while they are used. Everyone is wearing his or her safety glasses. There is just one more safety step that must be performed before proceeding with the blade change. What is it?

CHAPTER 3

Designing and Planning

YOU'LL BE ABLE TO:

- List the three keys to good design.
- Describe at least three basic principles of design.
- Name the views shown in a three-view working drawing.
- Correctly read drawings to make a layout on materials.
- Make a bill of materials.
- Use a formula for calculating board feet to figure lumber needs.
- List the main steps in designing, planning, and completing a woodworking project.

Look for These Terms

- bill of materials
- board foot
- design
- dimension
- exploded view
- function
- layout
- pictorial drawing
- proportion
- scale
- standard stock
- stock-cutting list
- working drawings

Every manufactured object has been designed. Your home, furniture, cars, and even pencils have been developed from a design on paper. Information for building a designed product is provided in drawings.

Every step in the building process needs to be planned. Planning includes determining materials needed and figuring costs. All projects begin, however, with design. Fig. 3-1.

Fig. 3-1—*In your opinion, is this vase well-designed?*

DESIGN

A **design** is the outline, shape, or plan of something. No one will agree completely with everyone else as to what is good or bad design. We all look at things in slightly different ways and see in them things we like or dislike.

Keys to Good Product Design

The three keys to good design are function, appearance, and sound construction. **Function** refers to the purpose of the product. A lamp, for example, should provide the kind of light needed for a specific purpose. A lamp for reading would have a different function than a night-light. The function, or purpose, of a clock is to show the time. If the clock's face is poorly designed, it will be difficult to read.

Appearance, of course, refers to how the object looks. Are the materials and shapes combined to form an object that is pleasing to the eye?

Sound construction means that the product is well built and made of an appropriate material. The product should last a long time with a minimum of maintenance.

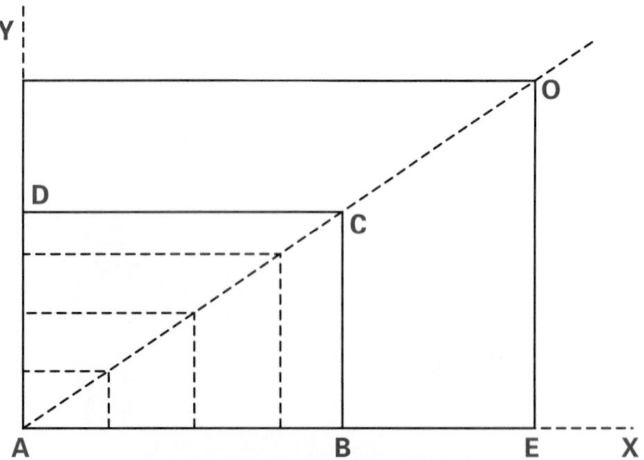

Fig. 3-2—*One way of reducing or enlarging a 5-by-8 proportion. This technique is useful in designing trays and picture frames. Make line AB eight units long and line BC five units long. Then lay off along the line AX any length you want, for example AE. The distance for width, then, would be EO.*

The design of woodworking projects can be divided into three groups:

- Trinkets or gadgets that are made to satisfy your own interests and desires.
- Practical objects that must be standard size if they are to be usable.
- Artistically designed objects, including furniture and accessories for the home. Design and style are particularly important considerations in making furniture.

Principles of Design

Basic principles of design should be considered in designing any item. These include the following:

- **Proportion** is the size relationship of parts or features. Size takes on more meaning when one thing is compared to another. Parts may differ in size, but the overall effect should be pleasing.
- Proportion in design is achieved when the height or length of an object appears appropriate for its width. In designing many wood products, the rectangular shape is used instead of a square one. This is simply because it *looks* better proportioned. Figure 3-2 shows one way of maintaining good proportions while enlarging a rectangular shape.
- *Balance* is achieved when parts or features are arranged to project a feeling of stability. There are two kinds of balance, formal and informal. *Formal balance* is present when both sides of an object are equal. One side appears to be the mirror image of the other. *Informal balance* is when design features are different but appear to have equal weight. Fig. 3-3.
- *Harmony* is the effect that is achieved when the parts, colors, shapes, and textures of an object work well together.

Fig. 3-3—*Examples of the two types of balance in design. (A) Formal. (B) Informal.*

A

B

- *Emphasis* is stress or accent on one feature. For example, the size or color of a feature may attract special attention.

Evaluating Product Design

In deciding whether something is well designed, ask yourself the following questions:

- Does the piece serve the purpose for which it is intended? Fig. 3-4.
- Does it perform its job efficiently?

Fig. 3-4—*Though designed as a game table, this table can be used for other purposes.*

Fig. 3-5—*Select a project suitable for your skills. (A) A good beginner's project. (B) A project suitable for someone with more woodworking experience.*

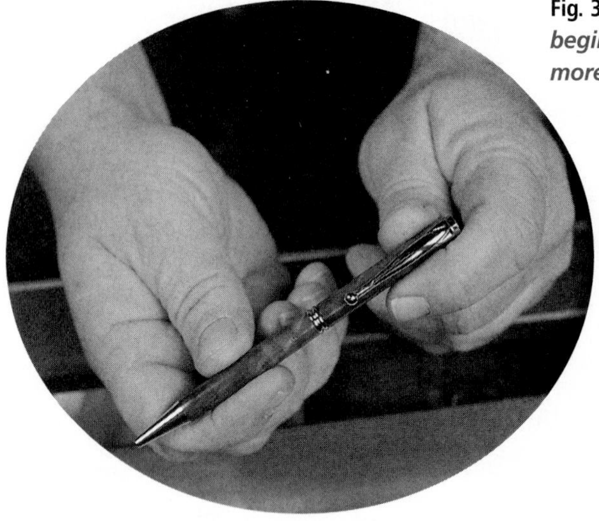

A

B

- Is it within your ability to construct and maintain? Fig. 3-5.
- Is it pleasing to the eye?
- Does it satisfy you and any other people you want it to satisfy?

If the design is a good one and suitable for the planned purpose, details of the design must be communicated in drawings.

DRAWINGS

A drawing or sketch is the map you follow in making a project. Fig. 3-6. It tells you the exact size of the article, the number and sizes of its pieces, the design of each part, and the way in which the pieces fit together. Understand the drawing *before* you begin to build.

In industry, everything to be produced is first drawn by a drafter either by hand or on a computer. The same general approach is also often used in school workshops. Figs. 3-7 and 3-8. *Pictorial* and *working* (view) *drawings* are the two types of drawings most commonly used in woodworking projects.

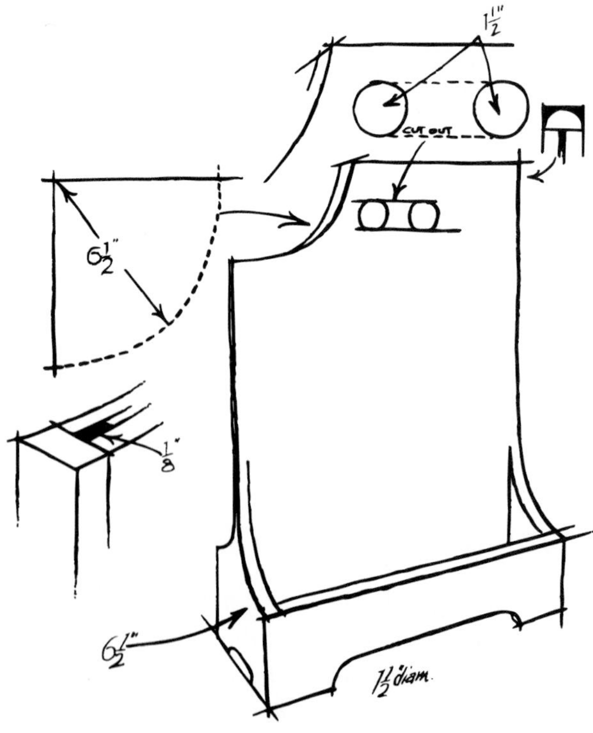

Fig. 3-6—*A sketch or drawing provides information for building the project.*

Fig. 3-7—*A few simple drafting instruments are all you need to make an accurate working drawing for a project.*

Fig. 3-8—*A computer-aided drafting (CAD) program makes it very easy to draw and dimension a very accurate working drawing.*

Fig. 3-9—*A pictorial drawing of several different plant containers.*

Pictorial Drawings

A **pictorial drawing** is also called a picture drawing or sketch. It shows the object the way it looks in use. Fig. 3-9. The most common kinds of pictorial drawings are isometric (equal angle), oblique, and perspective. All three are one-view drawings (the total object is shown in one view).

- *Isometric drawing:* Shows the object tilted so that its edges form three equal angles of 120 degrees each. One corner appears closest to you. Fig. 3-10.

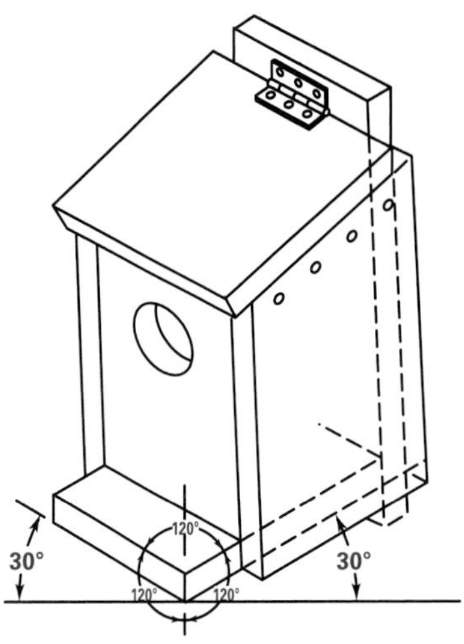

Fig. 3-10—*An isometric drawing of a birdhouse.*

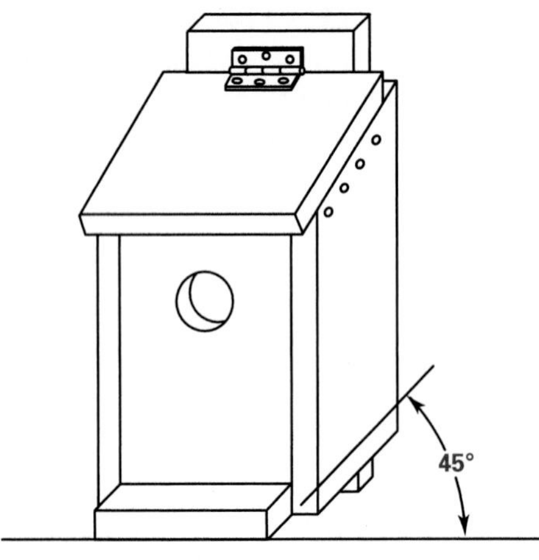

Fig. 3-11—*An oblique drawing of the same birdhouse shown in Fig. 3-10.*

- *Oblique drawing:* Shows the front view of the object as closest to you with the sides drawn at an angle, often 30 or 45 degrees. Fig. 3-11.
- *Perspective drawing:* Looks much like a photo of the object. However, lines that are parallel on the object are not drawn parallel. They are drawn so that, if extended, they would eventually come together at a *vanishing point.*

Perspective drawings may have either one or two vanishing points. The perspective with one vanishing point looks a great deal like an oblique drawing. The perspective with two vanishing points looks somewhat like an isometric drawing.

Working (View) Drawings

Working drawings are used for construction. Fig. 3-12. These drawings have one, two, three, or more views. These views show the article from different sides. Most projects require two or three views. In the three-view drawing, the lower left-hand view shows the way the article looks from the front. The top view, which is placed above the front view, shows how

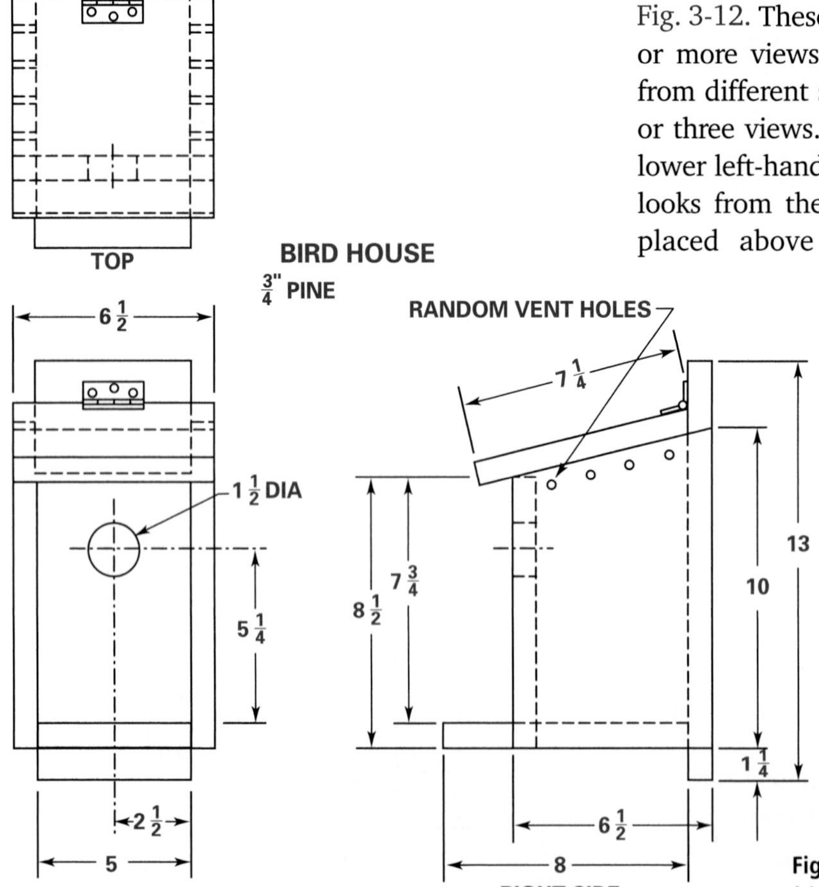

Fig. 3-12—*Working drawings of the birdhouse shown in Figs. 3-10 and 3-11.*

the article looks from the top. The view to the right is the right side or end view.

The views give the correct size measurements or **dimensions** of each piece. Fig. 3-13. The dimensions are placed on the views to be read from the bottom or right side. If only two views are included, either the front and top or front and side views are shown. Other views are provided as needed for construction. Figs. 3-14, 3-15, and 3-16.

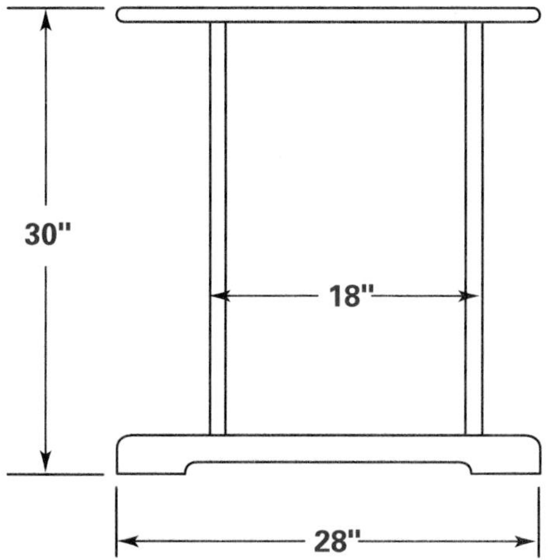

Fig. 3-13—*Use exact measurements for all dimensions.*

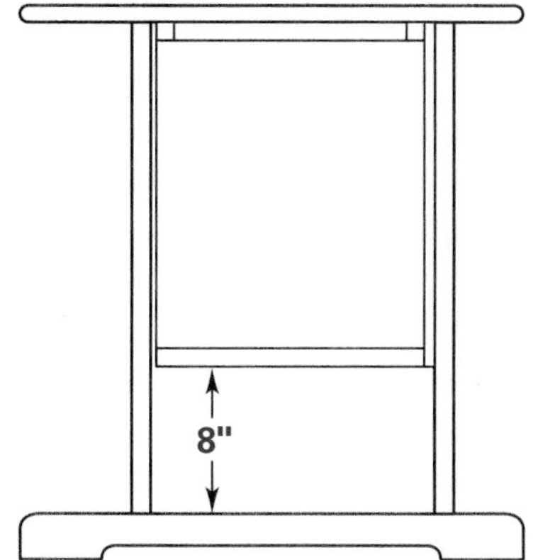

Fig. 3-14—*A section view will show parts not visible in conventional views.*

Drawings for Woodworking

Drawings for woodworking do not follow rigid rules. A drawing for a project is often partly a view drawing and partly a pictorial drawing.

Sometimes when view drawings are used, the views are not positioned as in working drawings. That is, the right side or end view is not always placed to the right of the front view.

Many isometric or perspective drawings are made as *exploded* (taken apart) *drawings.* Fig. 3-17 (page 94). An **exploded view** clearly shows the dimensions of each part and how the parts go together.

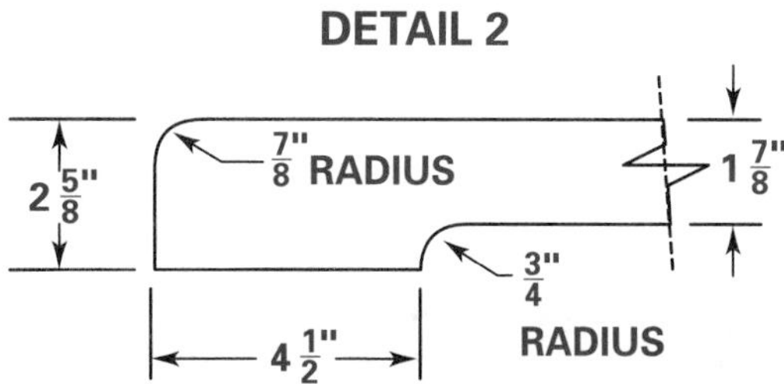

Fig. 3-15—*Close-up detail drawings are needed for complex cuts and operations.*

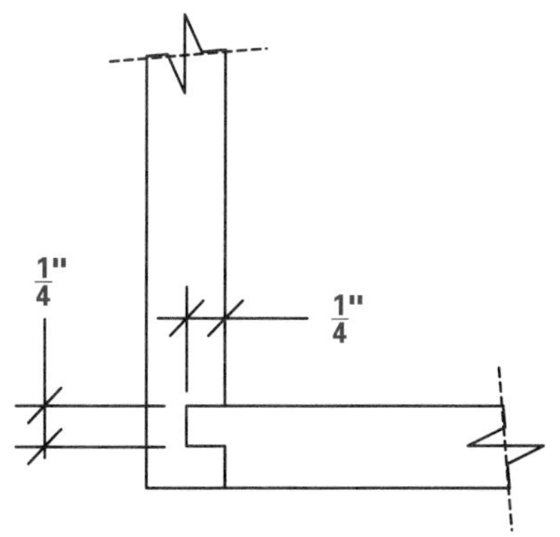

Fig. 3-16—*Detail drawings are good for showing joints.*

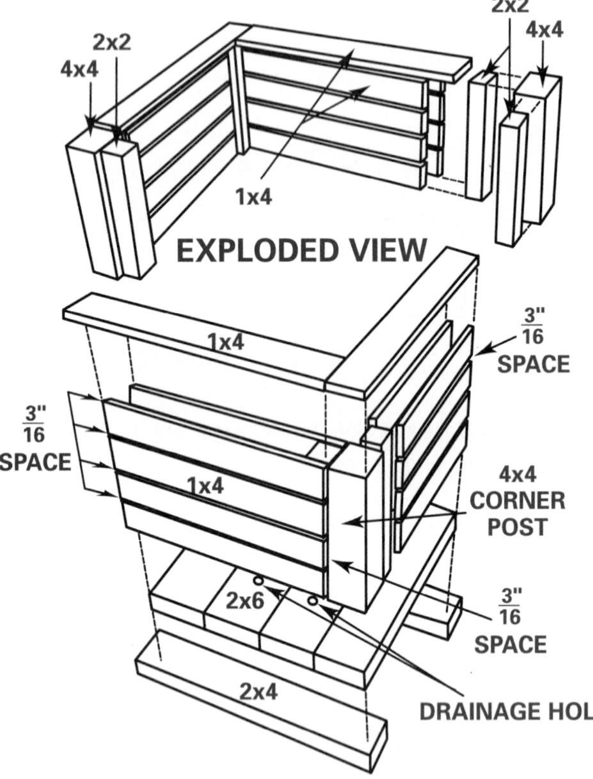

Fig. 3-17—*A pictorial drawing and an exploded view drawing.*

Reading Drawings

To build a project, you must be able to read and understand the drawings. They will give you the information you need to figure materials and make and assemble the parts. Fig. 3-18.

Meanings of Lines

In a drawing or sketch, different kinds of lines are used for different purposes. Fig. 3-19.

- *Visible line:* The major outline of the article.
- *Hidden line:* Indicates the outline that cannot be seen from the surface.
- *Centerline:* Shows the center or divides the drawing into equal, or symmetrical, parts.
- *Extension line:* Extends from the outline. It provides two lines between which measurements or dimensions can be shown.
- *Dimension line:* Runs between the extension lines. It usually has arrowheads at each end and is broken in the center. The size measurements or dimensions are given between the broken ends.

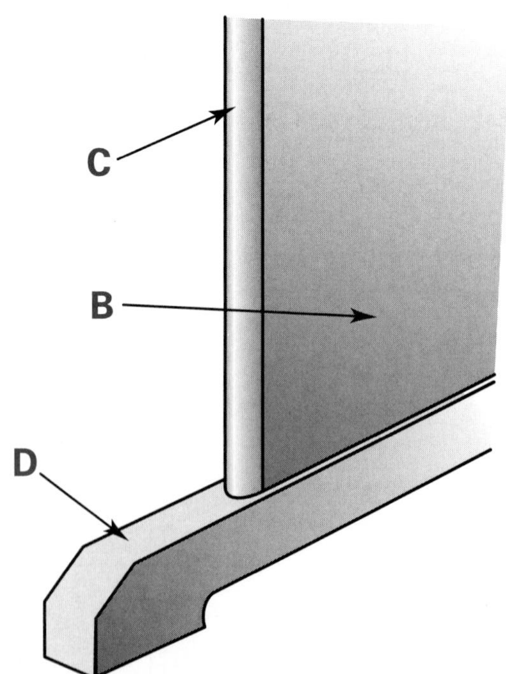

Fig. 3-18—*The parts of a project should be labeled with letters. These letters will be used to identify parts when determining materials needed.*

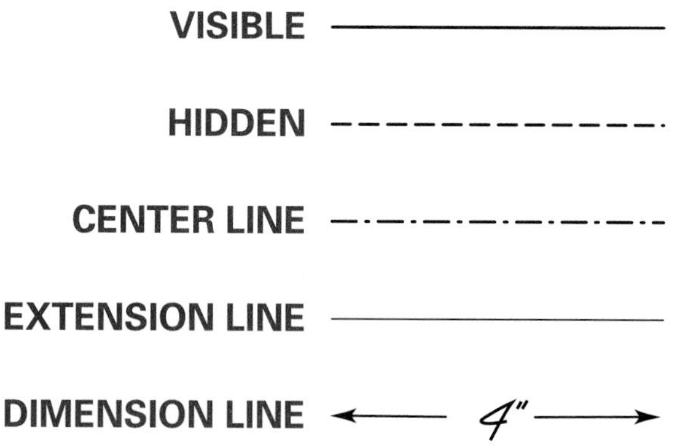

VISIBLE

HIDDEN

CENTER LINE

EXTENSION LINE

DIMENSION LINE 4"

Fig. 3-19—*Types of lines.*

Scale of the Drawing

Large objects must be drawn at a reduced size so the drawing will fit on one page. The drawing is made to **scale.** A small measurement is used to represent a larger measurement. For example, a typical woodworking drawing is made half size (6″ = 1′). If the object is one foot long, a 6-inch line will be drawn. For larger projects such as a house, a scale such as ¼ inch representing 1 foot (¼″ = 1′ 0″) may be used.

The scale (equation) is written on the drawing. The dimensions given are the actual measurements you will use when laying out and building the project.

Transferring Measurements for Layout

Layout is measuring and marking stock to size and shape. Measurements from the drawings are transferred to the stock from which you will make the part or object. For example, if the dimension given on the drawing is 6 inches, you will measure and mark 6 inches on the stock. Fig. 3-20 (page 96). See Chapter 4.

MATHEMATICS *Connection*

Sketching to Scale

Sketching your project to scale on graph paper is an easy way to get your ideas on paper without using drafting instruments. Simply let the length of one side of a square on the graph paper represent a measuring increment of the project. For example, one side of a square might equal 1 inch. To represent a dimension that is 6 inches long on the project, you would draw a line along six squares.

Graph paper comes in different sheet sizes and with squares of different sizes. Pick the paper best suited for representing the size of the project. Always use a pencil so that you can erase easily. Once you have finalized your plan on graph paper, you might want to make a formal drawing with drafting instruments or with a computer-aided drafting (CAD) system.

Take care in transferring the measurements to the wood. Fig. 3-21. Consider which way the grain should run in the part or object you are making and position the layout on the wood accordingly. If you are using a panel material, such as plywood, this is not a consideration.

Be sure that you read the correct dimension for the portion you are laying out. Always use the stated dimensions. Even if the drawing is made full size, never attempt to measure the drawing itself.

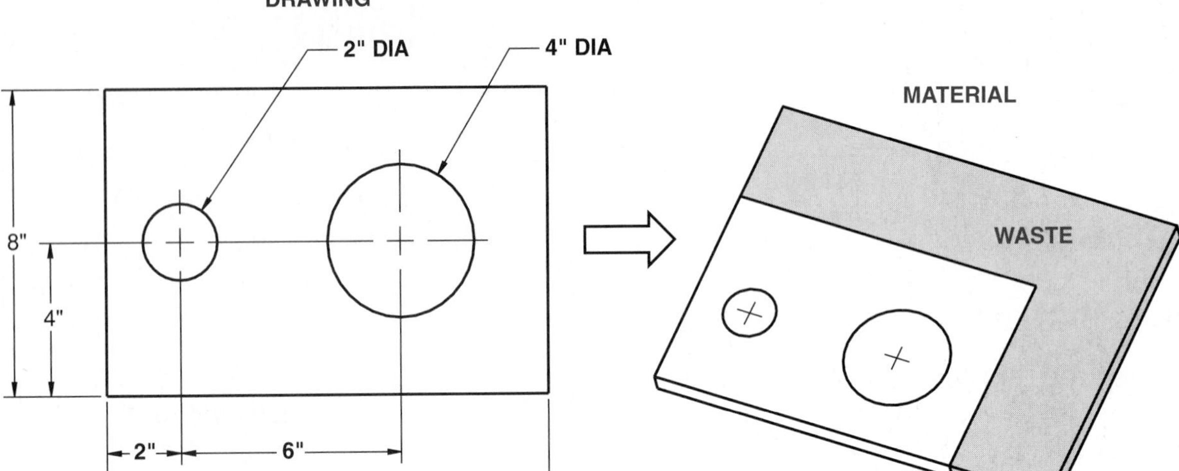

Fig. 3-20—*When making a layout, dimensions and shapes given on the drawing are measured and marked on the stock.*

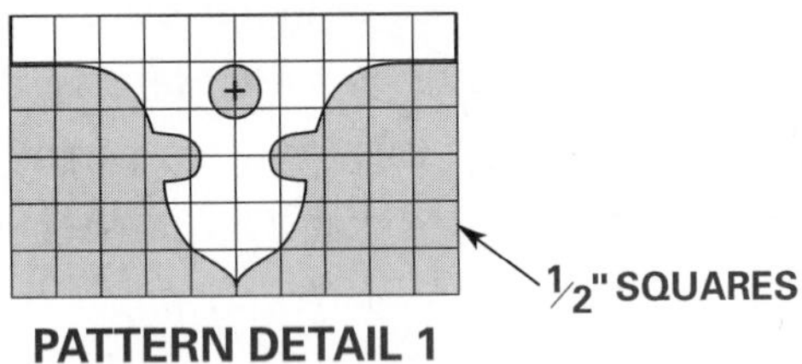

Fig. 3-21—*When transferring a pattern with curves, a scaled pattern grid is very useful.*

PLANNING YOUR WORK

Before you begin to work with tools and machines, you must plan carefully the steps you will follow in making the project. Planning is an important part of any successful business. Like good business, you should "plan your work; then work your plan."

Planning an Individual Project

In planning a project, you should follow three basic steps:

- Find out what materials you need.
- Determine what tools and equipment you need.
- List the steps you will follow in order.

If you fail to plan your work, you may, for example, waste lumber by forgetting to make the proper allowance for the joints needed. Many such mistakes can creep into your work.

Planning Procedures

In planning your work, you should first decide on the order in which the parts are to be made. Then list all of the things you will need to do to complete each particular part.

Decide what tools and equipment will be needed. Think through all of the steps in layout, cutting, shaping, making joints, fitting and assembling, and finishing the project. Then write the procedure on paper.

On some small projects, a written plan may not always be necessary. However, it is still an excellent idea. You should at least think through each step you will take before you begin to work. In this way, you avoid a great deal of trouble. Remember again: "Plan your work; then work your plan."

Making Materials Lists

A **bill of materials** is a complete list of material, fasteners, and accessories needed for the project. Fig. 3-22. It includes a description of the finished size of each wood part used in making the project. However, when these parts are cut from the stock, there is always some waste. You must allow for this when determining how much stock you will need to make the project.

Make a **stock-cutting list.** Fig. 3-23. In this list, add to the thickness, width, and length of the stock listed in the bill of materials to allow for cutting and other operations.

Determining Board Feet of Lumber

After you have made a stock-cutting list, figure the number of board feet in each piece or group of identical pieces.

A **board foot** of lumber is a piece 1 inch thick, 12 inches wide, and 12 inches long. Fig. 3-24. Stock less than 1 inch is figured as one inch. Stock more than one inch is figured by actual measurement.

KEY	NO. OF PIECES	PART NAME	THICKNESS	WIDTH	LENGTH	WOOD

Fig. 3-22—*Form for a bill of materials.*

STOCK-CUTTING LIST

KEY	QUANTITY	DESCRIPTION AND SIZE
A		
B		
C		
D		

Fig. 3-23—*Form for a stock-cutting list.*

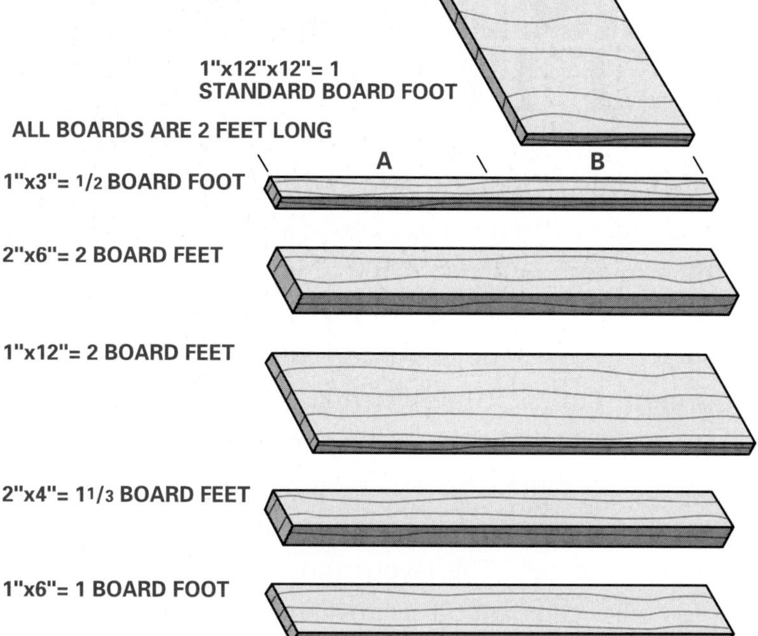

Fig. 3-24—*Examples of the number of board feet in various sizes of lumber.*

There are two formulas that can be used to find board feet. Using all inches, the formula is:

$$\text{Board Feet} = \frac{T \times W \times L(\text{all inches})}{144 \text{ square inches}}$$

That is, board feet equals the product of thickness (T) times width (W) times length (L) divided by the number of inches in one square foot (one square foot = 12 inches × 12 inches = 144 square inches).

Using inches and feet, the formula is:

$$\text{Board Feet} = \frac{T(\text{inches}) \times W(\text{inches}) \times L(\text{feet})}{12 \text{ inches}}$$

For example, a piece of stock 1 inch × 6 inches × 4 feet would contain 2 board feet of lumber (1 × 6 × 4 = 24 ÷ 12 = 2).

Figuring Lumber Needs

Using your stock-cutting list, figure the number of board feet you need. Group together all pieces made from the same thickness of lumber. Then determine the number of pieces of standard size lumber required.

Standard stock is lumber cut to a widely used size and shape. Softwood lumber comes in standard widths from 2 to 12 inches, increasing by 2-inch intervals. It comes in standard lengths from 6 to 20 feet, increasing at intervals of 2 feet.

Hardwood lumber comes in standard thicknesses by quarter inches beginning with ¾ inch. Because of its high cost, hardwood is cut in whatever widths and lengths are most economical and convenient. These random sizes are useful, though, for cutting parts of a wide variety of sizes and shapes.

Once you know the number of pieces of standard size lumber needed, you can make a lumber order to obtain the material from a lumber yard. Fig. 3-25.

Making a Cutting Diagram

Making a cutting diagram will help you determine how much material you will need for a project. A cutting diagram shows a layout of how the parts of a project can be cut efficiently from a piece of stock. In the cutting diagram, the parts should be arranged so that they can be produced with a minimum number of cuts. The layout should also waste as little material as possible.

Determining the Cost of Materials

The next step is to find the cost of items on the stock-cutting list. When you buy lumber, the price is usually quoted as so much per board foot, per hundred board feet, or per thousand board feet. For lumber sold by the board foot, multiply the number of board feet in each piece or group of identical pieces by the cost per board foot. Refer again to Fig. 3-25. When the cost of hardware items and finishing materials is listed, all items can be added to find the total cost of materials. Fig. 3-26.

Some lumber is not sold by the board foot. Dimensional softwood (straight and surfaced on four sides) is usually sold by the *lineal foot.* The word "lineal" refers to "line," in this case, length. For example, a 2 × 4 piece, 16 feet long has 16 lineal feet. Molding and special pieces are also sold by the lineal foot. Panel stock, including plywood, particleboard, and hardboard, is sold as full sheets, half sheets, or by the square foot. All prices vary according to quality.

LUMBER ORDER

NO. OF PIECES	T	W	L	KIND OF WOOD	NO. OF BD. FT.	COST PER BD. FT. OR LINEAL FT.	TOTAL COST	INSTR'S O.K.

Fig. 3-25—*A lumber order. To avoid having to make calculations for each individual item in the bill of materials, think of ways to group items. Several pieces can be cut out of one large board. Figure the board feet and cost of the large board instead of the pieces.*

OTHER SUPPLY COSTS

ITEM	QUAN.	SIZE	UNIT COST	TOTAL COST	INSTR'S O.K.	COST SUMMARY
						LUMBER COST ________
						SUPPLY COST ________
						TOTAL COST ________
						LESS ALLOW. ________
						AMOUNT DUE ________
						DATE PAID ________

Fig. 3-26—*A form for listing supplies such as fasteners and other hardware items, with a section for finding the total cost of all materials and supplies.*

DESIGNING, PLANNING, AND BUILDING A PROJECT

After you have reached a certain skill level, you may want to design, plan, and build a woodworking project of your own. Follow these steps:

1. Get the idea for the project you would like to build. Perhaps you have a problem that a constructed item could solve. For example, if you need storage space, perhaps you could design and build shelves.

   Maybe you have something in mind that you have always wanted to make. If not, magazines and books with projects in them are good sources of ideas. Another way to get ideas is to visit stores selling furniture, hobby supplies, or sporting goods. You might like to build:

   - Toys, models, games, puzzles, and other hobby equipment.
   - Pet shelters; feeders or houses for birds.
   - Things for your room or home, such as lamps, bookends, or shoe racks.
   - Kitchen items, such as cutting boards and shelves.
   - Sports equipment.
   - Furniture, including chairs, tables, chests, and desks.

2. After you've decided what to make, consider how its purpose will affect its design. For example, a book rack is supposed to hold books conveniently. How large are the books? What must be the depth of the shelves? What should be the distance between shelves?

   If you want to build a piece of furniture, decide on the style or design. It should blend with other furnishings in your room or home. Also, certain standards, especially of height, must be observed.

3. Make a sketch of what you would like to build to see how it will look. Suppose you decide to make a wall rack. Two sketches of possible designs are shown in Fig. 3-27. Let's suppose you decide on sketch A.
4. Make a working drawing of the project to determine the exact size of each part and how it is to be made. Fig. 3-28.
5. Plan the building of the project. Decide what jobs must be done and in what order jobs will be performed. Fig. 3-29.

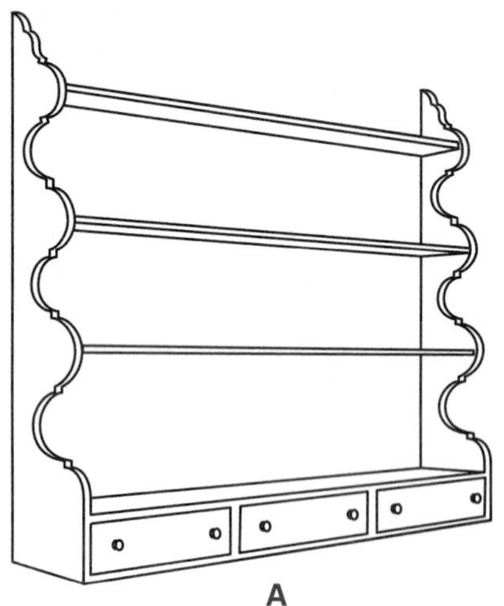

Fig. 3-27—*Two sketches of possible wall rack designs. Either of these would make a well-designed project.*

Fig. 3-28—*Working drawings of wall rack A in Fig. 3-27.*

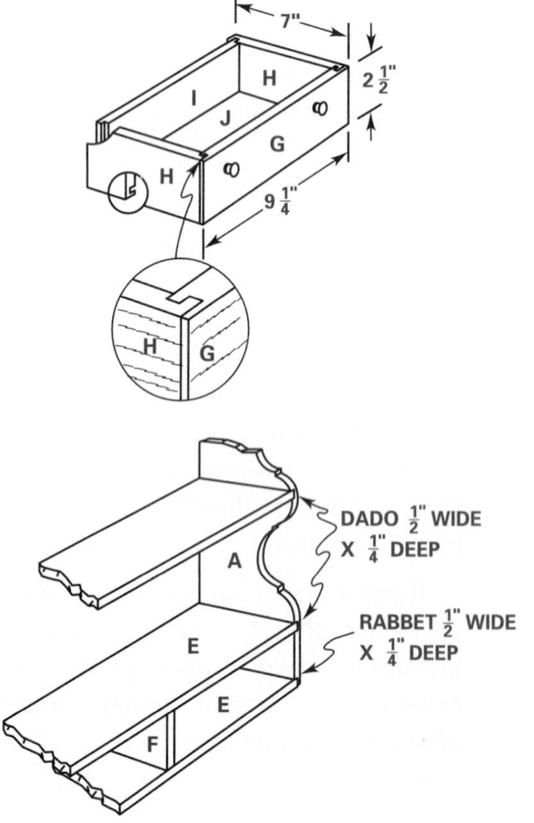

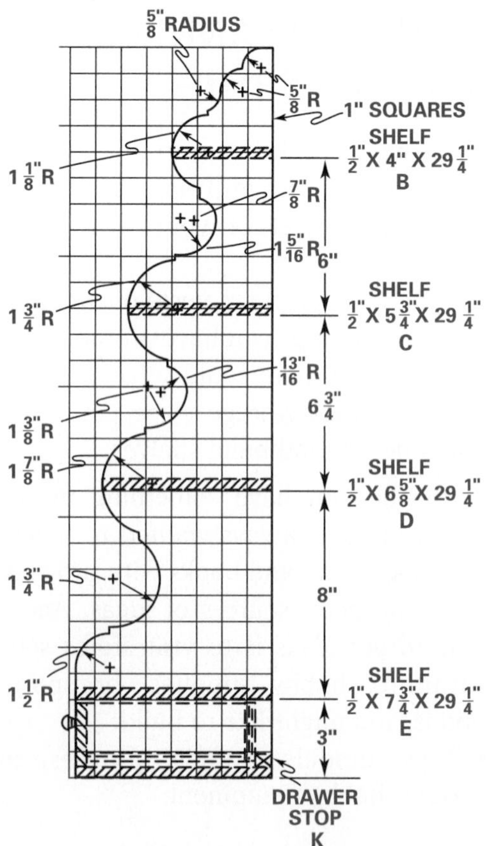

6. Decide on what tools and machines you will need. Your future experiences in woodworking will help you decide on how to use the equipment.
7. Make a bill of materials and a stock-cutting list. Figs. 3-30 and 3-31.
8. Figure lumber needs and costs. Obtain the necessary materials.
9. Build the project. This is the time to display your skill by doing a fine job on each part.
10. Judge your work. Is the finished item useful, attractive, and economical? Was it completed in a reasonable length of time? Did you learn by making it? Does it accomplish its purpose? If you can answer yes to these questions, the project was successful.

PROCEDURE:
1. Lay out pattern of the ends on paper and trace on wood. Cut out on a jigsaw.
2. Cut the dadoes for the shelves and the rabbet for the bottom board.
3. Saw the shelves and bottom board.
4. Dowel and glue the shelves and bottom in place. Before the glue sets be sure the entire structure is square.
5. Make drawer separators; install dowels and glue in place.
6. Cut the drawer fronts and fit into each opening. Then complete the drawers, using the joints suggested in the detailed drawing—or make a simple rabbet joint to fasten the sides and front, and a dado joint to fasten the sides and back. Glue drawer stops in place so the drawer fronts will be flush.
7. Sand edges to give a worn appearance.
8. Apply an antique pine finish, and add knobs.

Fig. 3-29—*Plan the procedure for building the wall rack.*

CUTTING LIST

KEY	QTY.	DESCRIPTION & SIZE
A	2	ENDS, 1/2"X 7 3/4"X 28"
B	1	SHELF, 1/2"X 4" X 29 1/4"
C	1	SHELF, 1/2"X5 3/4"X29 1/4"
D	1	SHELF, 1/2"X6 5/8"X29 1/4"
E	2	SHELVES, 1/2"X7 3/4"X29 1/4"
F	2	DRAWER SEPARATORS, 1/2"X2 1/2"X7 3/4"
G	3	DRAWER FRONTS, 1/2"X2 1/2"X9 1/4"
H	6	DRAWER SIDES, 3/8"X2 1/2"X7"
I	3	DRAWER BACKS, 3/8"X2 1/2"X8 7/8"
J	3	DRAWER BOTTOMS, 1/4"X6 1/4"X8 3/4"
K	6	DRAWER STOPS, 1/2"X1/2"X1"

NOTE: ALL WOOD IS PINE.
MISC.: GLUE, DOWELS, #8X1"
ROUNDHEAD AND FLATHEAD WOOD SCREWS

Fig. 3-31—*Stock-cutting list for the wall rack.*

BILL OF MATERIALS

IMPORTANT: ALL DIMENSIONS LISTED ON DRAWINGS, EXCEPT FOR LENGTH OF DOWEL, ETC., **FINISHED** SIZE

KEY	NO. OF PIECES	PART NAME	THICKNESS	WIDTH	LENGTH	WOOD
A	2	ENDS	1/2"	7 3/4"	28"	KNOTTY PINE
B	1	SHELF	1/2"	4"	29 1/4"	KNOTTY PINE
C	1	SHELF	1/2"	5 3/4"	29 1/4"	KNOTTY PINE
D	1	SHELF	1/2"	6 5/8"	29 1/4"	KNOTTY PINE
E	2	SHELVES	1/2"	7 3/4"	29 1/4"	KNOTTY PINE
F	2	DRAWER SEPARATORS	1/2"	2 1/2"	7 3/4"	KNOTTY PINE
G	3	DRAWER FRONTS	1/2"	2 1/2"	9 1/4"	KNOTTY PINE
H	6	DRAWER SIDES	3/8"	2 1/2"	7"	KNOTTY PINE
I	3	DRAWER BACKS	3/8"	2 1/2"	8 7/8"	KNOTTY PINE
J	3	DRAWER BOTTOMS	1/4"	6 1/4"	8 3/4"	PLYWOOD
K	6	DRAWER STOPS	1/2"	1/2"	1"	CLEAR PINE

Fig. 3-30—*A bill of materials for the wall rack.*

CHAPTER 3

Review & Applications

Chapter Summary

Major points from this chapter that you should remember include:

- A design is the outline, shape, or plan of something.
- Drawings provide information needed to build a project.
- In scale drawings, small measurements are used to represent larger measurements.
- "Plan your work; work your plan."
- In planning a project, you need to determine what materials, tools, and equipment are needed, and what steps you will follow in what order.
- A bill of materials is a complete list of everything you will need to build a project.

Review Questions

1. What are the three keys to good design?
2. Name and describe at least three basic principles of design.
3. What two types of drawings are commonly used in woodworking?
4. List the views shown in a three-view working drawing.
5. Briefly explain how a layout is made.
6. What information is included on a bill of materials?
7. Determine the board feet in a piece of wood that is 1½ inches thick, 8 inches wide, and 10 feet long.
8. Briefly outline the main steps in designing, planning, and completing a project.

Solving Real World Problems

James wants to build a bookshelf unit for the wall in his bedroom. It will have two shelves 48″ long and 12″ wide and will be made from ¾″ stock. There will be two vertical dividers that support the upper shelf above the lower. The vertical dividers will be spaced 12″ in from each end and must be jointed into the shelf above and below. The total height without the brackets will be 16″. The unit will be supported by three brackets beneath the lower shelf. The three brackets will be made 10″ tall by 10″ deep from 2″ stock.

Make a three-view sketch of the shelving unit and a bill of materials for James to take to a lumberyard. This must include all lumber needed in standard dimensions as well as any nails or screws. What type of joints should James use for his shelves?

CHAPTER 4

Measuring and Cutting

YOU'LL BE ABLE TO:

- Accurately read measurements on a customary rule and a metric rule.
- Select and use the correct measuring tool for a specific measuring task.
- Correctly measure and mark stock for cutting.
- Name the basic types of cuts made with saws.
- Properly use an appropriate handsaw or portable power saw for a specific cutting task, observing all safety rules.

Look for These Terms

- backsaw
- bevel cut
- circular saw
- crosscut
- customary system
- jigsaw
- kerf
- miter cut
- reciprocating saw
- rip cut
- rule
- SI Metric System
- sliding T-bevel
- square

Accurate measurement (along with layout) and cutting are key processes in successful woodworking. When making measurements, you must answer these questions: How thick? How wide? How long? What are the angles? What are the diameters and depths?

Avoid making mistakes by making sure everything is properly measured and laid out. Fig. 4-1. Remember this old saying: "Measure

Fig. 4-1—*To avoid mistakes, always measure at least twice before cutting.*

twice, cut once." Also, use the proper cutting tool and techniques for the type of cut that you are making.

A number of tools are used for measuring, marking, and cutting stock. The main ones are discussed and shown in this chapter. Refer also to the Woodworker's Handbook beginning on page 15 of this book.

CUSTOMARY AND METRIC MEASUREMENT

The system of measure used most in the United States is the **customary system.** Most other countries of the world use the **SI Metric System.** (SI stands for System International.) Many products used in the United States are made using metric sizes and dimensions. It is becoming increasingly important to have a working knowledge of both the metric and the customary systems of measurement.

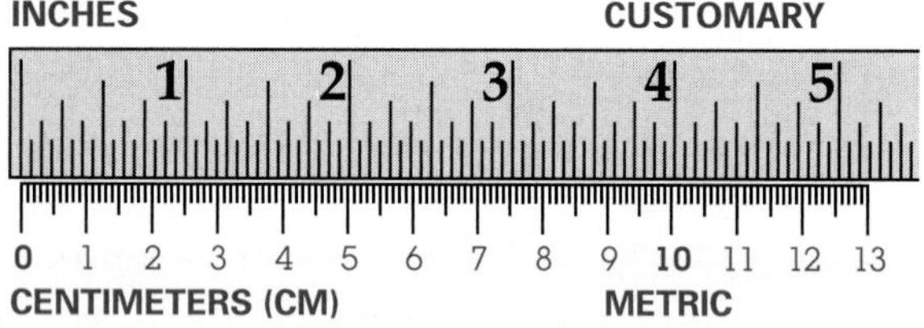

Fig. 4-2—*A comparison of the metric rule with the customary rule. NOTE: Rules are not shown to scale; however, their relative sizes are correct.*

In woodworking, you need to be able to make accurate measurements. In the customary system, the units of measure most used are the *inch* and *foot* and sometimes the *yard.* You also need to be able to measure *fractions* of an inch (half, quarters or fourths, eighths, and sixteenths).

In the metric system, the units most used are the *millimeter, centimeter,* and *meter.* Fractions are not required in metrics. It is a *decimal system* like the U.S. money system.

When comparing units in the two systems, the meter (m) is a little longer than the yard (39.37 inches). There are 2.54 centimeters (cm) in one inch. One millimeter is slightly smaller than 1/16 inch. Fig. 4-2.

Measures in one system can be converted to measures in the other. Common conversions are given in Table 4-A. In woodworking, projects are made using one system or the other.

MEASURING AND MARKING TOOLS

Rules

In woodworking, most measuring is done using **rules.** Rules are available with three kinds of graduation: *customary, metric,* or a *customary-metric* combination.

Reading a Customary Rule

The most common customary rules are divided into major inch divisions. Each inch is

Table 4-A. Common Conversions

Customary to Metric	Metric to Customary
1 INCH = 25.40 MILLIMETERS-MM	1 MILLIMETER = 0.03937 INCH
1 INCH = 2.540 CENTIMETERS-CM	1 CENTIMETER = 0.3937 INCH
1 FOOT = 30.480 CENTIMETERS-CM	1 METER = 39.37 INCHES
1 FOOT = 0.3048 METER-M	1 METER = 3.2808 FEET
1 YARD = 91.440 CENTIMETERS-CM	1 METER = 1.0936 YARDS
1 YARD = 0.9144 METER-M	

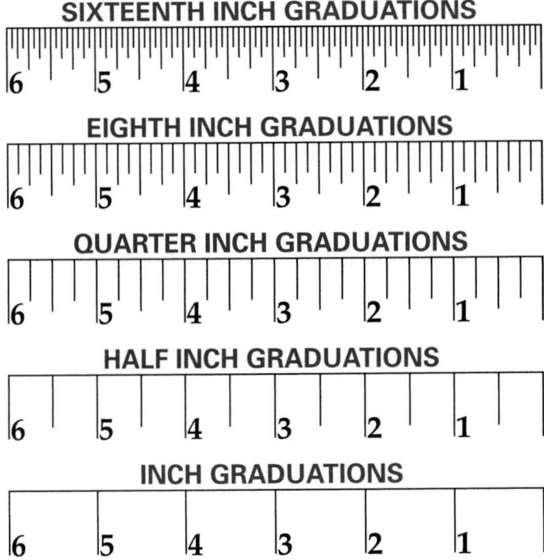

Fig. 4-3—*The customary bench rule is graduated in sixteenths of an inch on one side and eighths of an inch on the other. Other graduations are shown for comparison.*

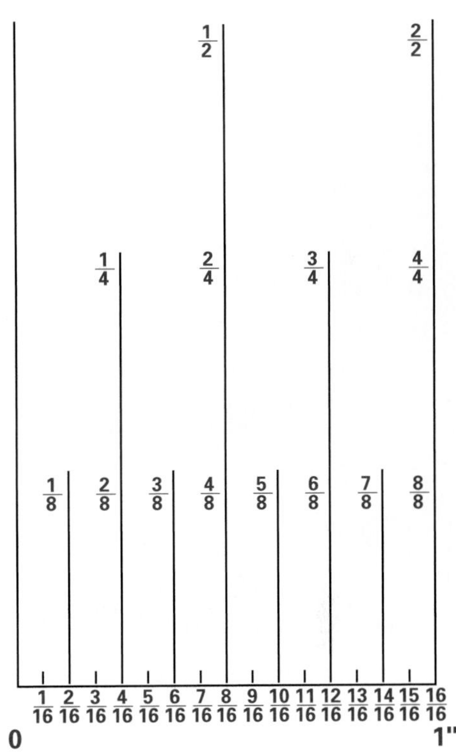

Fig. 4-4—*Fractional measurements represented in one inch on a customary rule.*

divided into smaller measurements (fractions of an inch). Either eighths or sixteenths are the smallest. Fig. 4-3.

Examine a rule. Notice that the longest line between 0 and 1 inch divides the inch into two equal parts. It is the $\frac{1}{2}$-inch mark ($\frac{2}{2}$ = 1). Each half inch is divided into two equal parts by the next longest line, the $\frac{1}{4}$-inch mark ($\frac{2}{4} = \frac{1}{2}$). The quarter inch is divided by the next longest line, the $\frac{1}{8}$-inch mark ($\frac{2}{8} = \frac{1}{4}$). The eighth inch is divided by the shortest line, the $\frac{1}{16}$-inch mark ($\frac{2}{16} = \frac{1}{8}$). Therefore, four-sixteenths ($\frac{4}{16}$) inch equals two-eighths ($\frac{2}{8}$) inch or one-quarter ($\frac{1}{4}$) inch. Fig. 4-4.

Reading a Metric Rule

A metric rule shows centimeters and millimeters. The smallest division on a metric rule is 1 millimeter. Refer again to Fig. 4-2. There are 10 millimeters in 1 centimeter. The fifth mark between two centimeter marks is longer than the others, but they all mark millimeters. The longer line makes it easier to count.

When you read a metric rule in centimeters, give the number of whole centimeters and express the number of millimeters over that in tenths. For example, a measurement of 3 millimeters past 5 centimeters would be expressed as 5.3 centimeters.

In woodworking, projects may be measured entirely in millimeters. This avoids the use of decimals. With ten millimeters in each centimeter, the 5.3 centimeters can be expressed as 53 millimeters (5.3 × 10 = 53).

Rules for Measuring Short Lengths

Two types of tools are used most often for measuring small pieces and marking short distances: the bench rule and the folding rule. A customary *bench rule* is 1, 2, or 3 feet long. Fig. 4-5 (page 106). The *folding rule* is 8 inches long when folded, but can be unfolded to various lengths up to 6 feet. Fig. 4-6.

Fig. 4-5—*Measuring with a one-foot bench rule.*

Rules for Measuring Long Lengths

Long lengths are best measured using a *tape measure.* It is a metal tape rolled up inside a case. Fig. 4-7. A small catch at its end slips over the end of a board. This makes it easier to pull out the tape. The tape is retracted (pulled back) by pressing a button on the case. The tape measure typically comes in lengths of 6 to 30 feet.

Squares

Squares are used for checking the squareness of stock, cuts, or joined pieces. Pieces are square when they form right (90-degree) angles. The *try square* has a metal blade and a handle of metal or wood. The blade is marked in eighths of an inch on one side. Some blades are marked in millimeters on the other. Fig. 4-8.

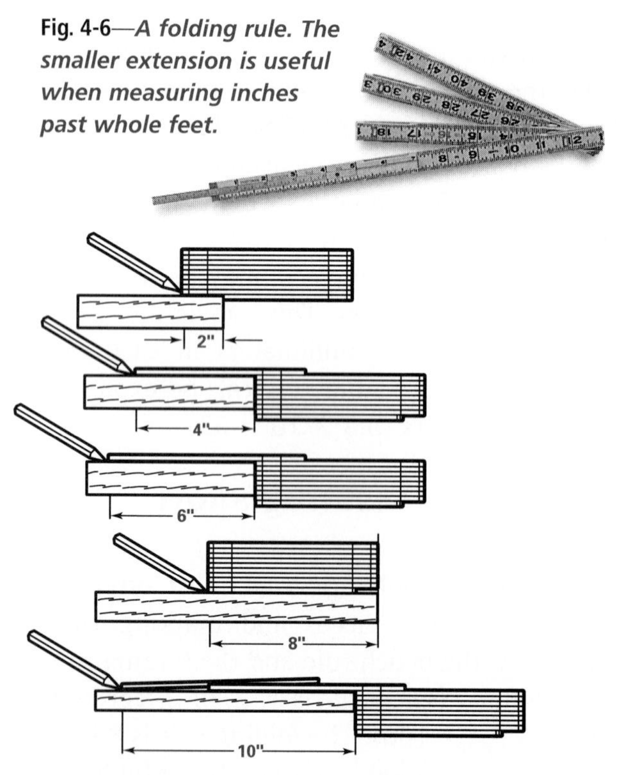

Fig. 4-6—*A folding rule. The smaller extension is useful when measuring inches past whole feet.*

Making Accurate Inside Measurements

Making an accurate inside measurement between two pieces is sometimes difficult with just a tape measure. One easy way to make an accurate measurement is to use your combination square *and* a tape measure. Butt the square against one inside surface and use the tape measure to measure from the opposite edge to the end of the blade of the square. Then just add 12 inches (the length of the square) to the tape measure reading.

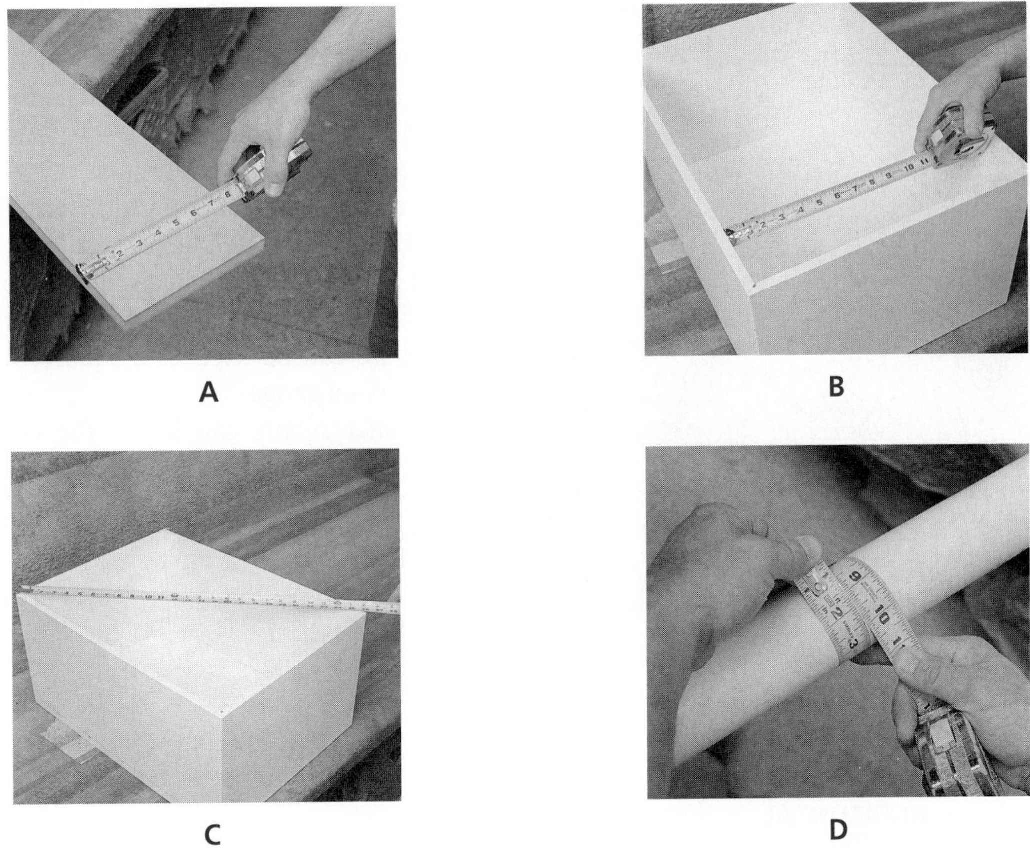

Fig. 4-7—*The retractable tape measure can be used to measure accurately in a number of ways. (A) External dimension. (B) Internal dimension. Add the length of the case to the inches shown on the tape. (C) Squareness. If the diagonals are equal, the box is square. (D) Circumference. Align the tape with the 2-inch mark and subtract 2 inches for the correct measurement.*

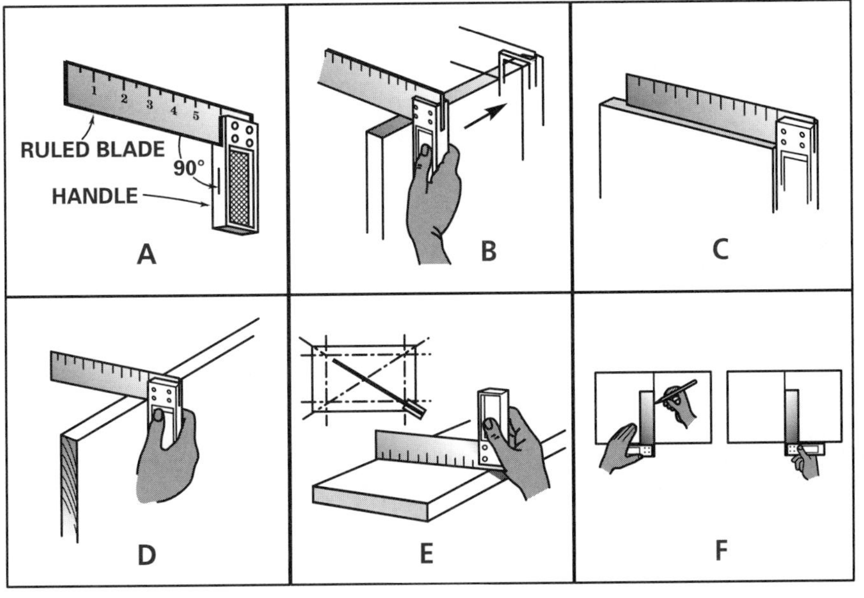

Fig. 4-8—*Parts and uses of the try square. (A) Parts of a square. (B and C) Checking end of board for squareness. (D) Checking the edge for squareness. (E) Checking for flatness. (F) Technique for marking a wide board.*

Fig. 4-9—*Using a framing square and a pencil to mark wide stock.*

The *framing square* (carpenter's square) is used mostly in general building work. Tables on the square help carpenters make building calculations. In woodworking, it is used when marking wide stock. Fig. 4-9.

The *combination square* has many uses. Fig. 4-10. Most models feature a level and a scratch awl. The sliding head adjusts for both 45- and 90-degree angle edges. The combination square can be used as a square, marker, level, rule, or gauge.

The *center square* can locate the center of any size circle. It can also be used for checking 90-degree measurements and as a protractor. Fig. 4-11.

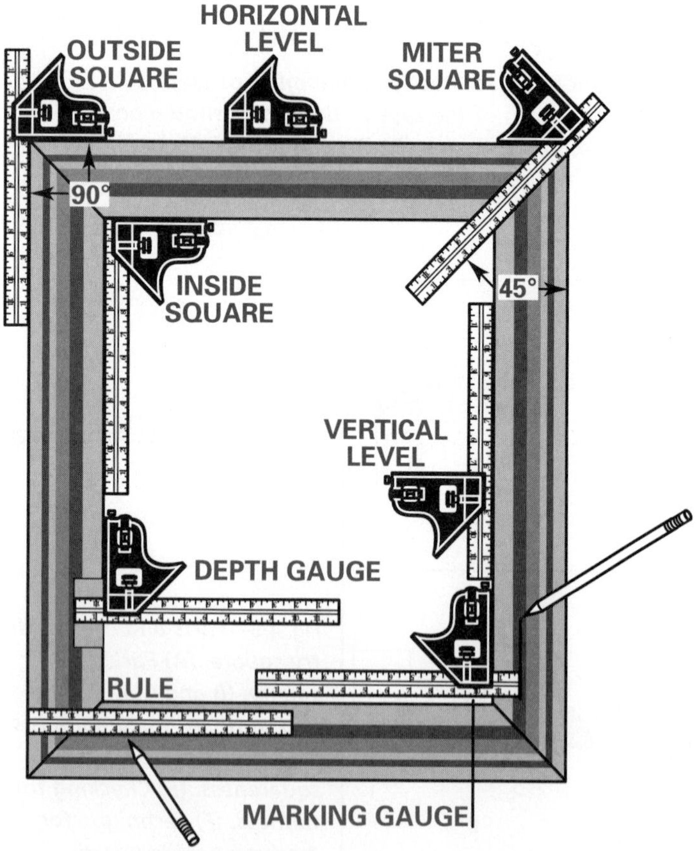

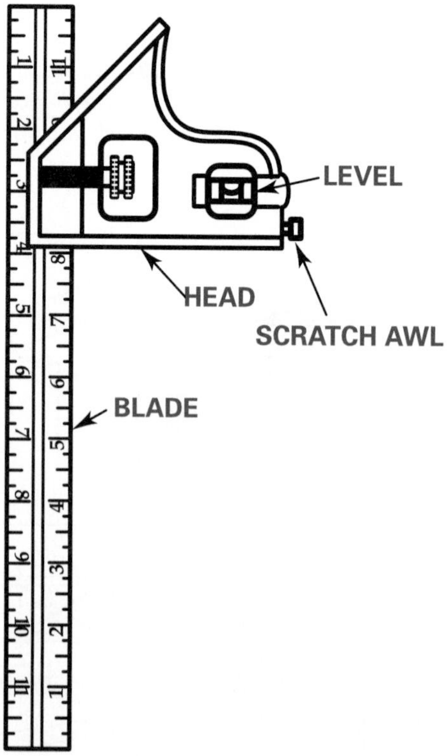

HORIZONTAL LEVEL
OUTSIDE SQUARE
MITER SQUARE
90°
INSIDE SQUARE
45°
VERTICAL LEVEL
DEPTH GAUGE
RULE
MARKING GAUGE
LEVEL
HEAD
SCRATCH AWL
BLADE

Fig. 4-10—*A combination square can be used for measuring and marking in many ways.*

Other Measuring Tools

A *level* is used to check a horizontal surface to see if it is level or a vertical surface to see if it is *plumb* (straight up and down). Fig. 4-12. Levels have hardwood or aluminum frames. Inside the frame are glass tubes or vials that contain a liquid with an air bubble in it. Marks on the glass show where the bubble should be located when the surface is level or plumb.

Fig. 4-11—*A center square.*

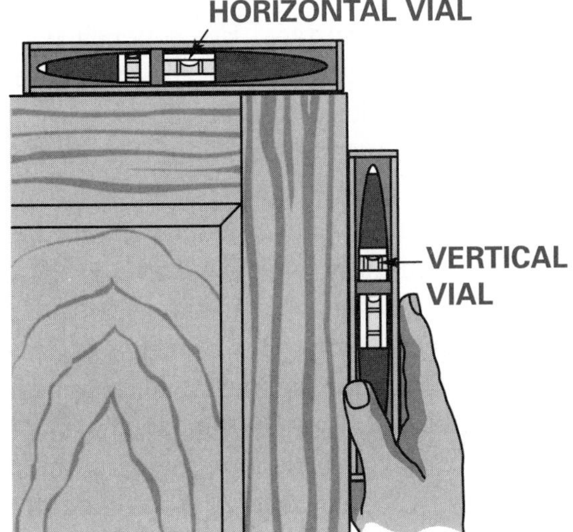

Fig. 4-12—*Using a level.*

A small wood *caliper* is used to measure cylindrical (round) stock, mostly diameters. Fig. 4-13.

The **sliding T-bevel** has an adjustable blade in a handle. It is used for laying out all angles other than right angles. Fig. 4-14 (page 110). To lay out a 45-degree angle, for instance, the adjustable T-bevel can be set with the framing square. To set the tool to such angles as 30 and 60 degrees, the bevel can be checked with the 30/60 degree triangles used in drawing. For other angles, the tool can best be set with a protractor. The sliding T-bevel can be used to lay out wood joints.

MATHEMATICS *Connection*

Laying Out Angles

Angles of 30 and 60 degrees can be set on a sliding T-bevel by first laying out a right triangle on paper. Make one side one unit long and the *hypotenuse* (side opposite the 90-degree angle) two units long. For example, one unit might be 2 inches. The hypotenuse, then would be twice that or 4 inches long. The resulting angles are 30 degrees and 60 degrees.

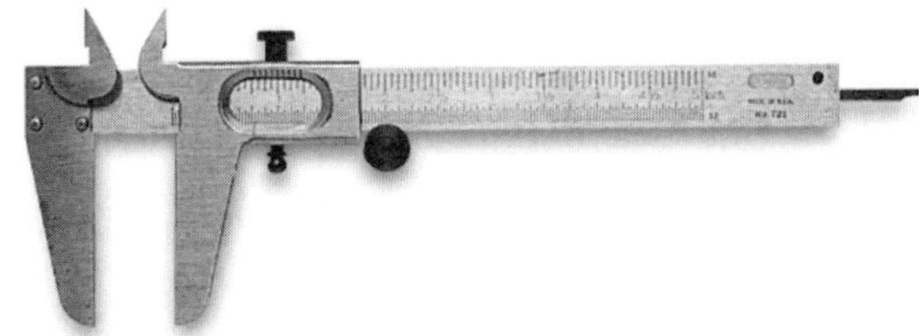

Fig. 4-13—*Use a caliper to measure round stock.*

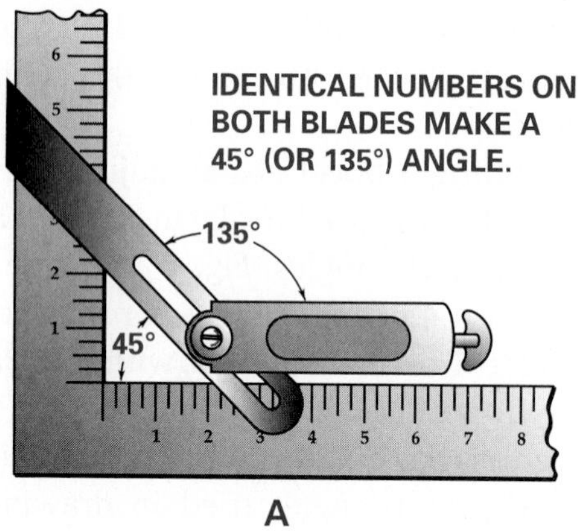

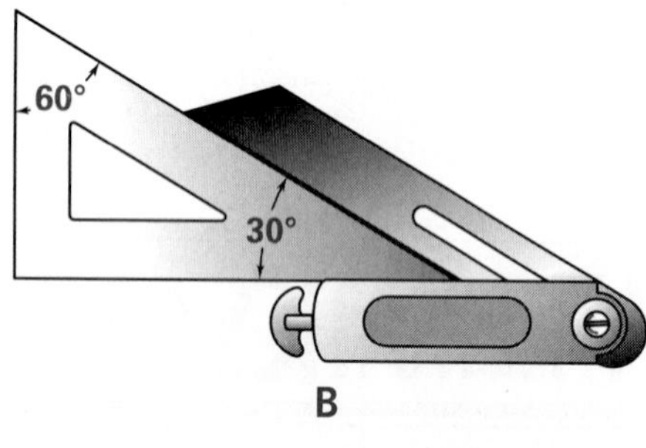

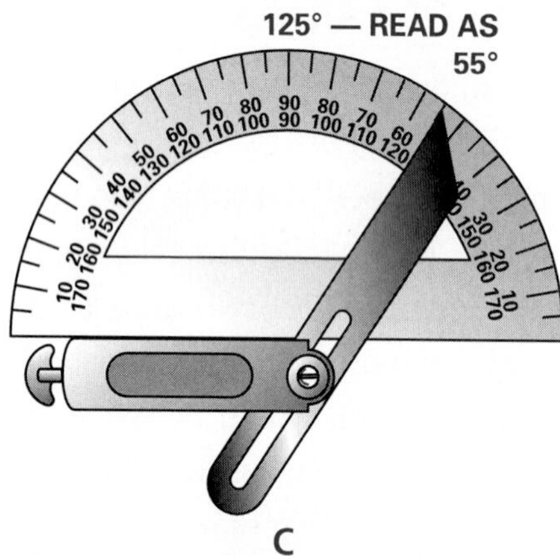

Fig. 4-14—*Methods of setting a sliding T-bevel. (A) Using a framing square. This shows the bevel set at a 45-degree angle. (B) Using a triangle. Angles of 30 or 60 degrees can be set. (C) Using a protractor. Any angle can be accurately set.*

Safety First

► *Using Marking Tools*

- Use all tools properly.
- Take special care with tools that are sharp.

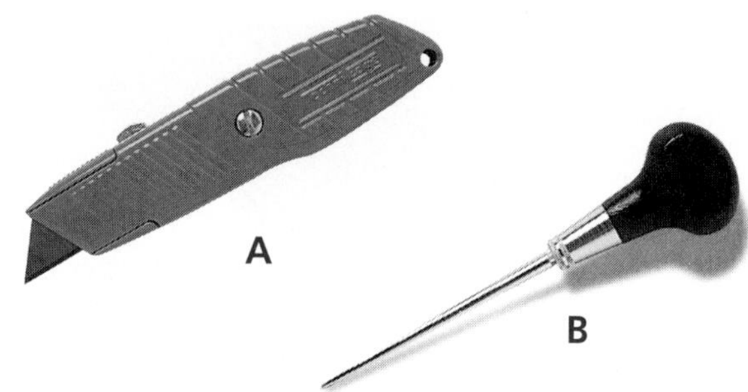

Fig. 4-15—*Besides the pencil, marking can be done with a utility knife (A) or a scratch awl (B).*

Tools for Marking

An ordinary lead pencil is the most common marking tool. Its mark can be seen easily on both rough and finished lumber. The mark is easy to remove, and the pencil does not scratch or mar the wood surface. Use a pencil with hard lead for laying out fine, accurate lines. Keep the pencil sharpened in the shape of a chisel so the point can be held directly against the edge of the rule or square.

Tools used occasionally for marking are the *utility knife* and the *scratch awl.* Fig. 4-15. The utility knife is a sharp tool that is good for very accurate marking. However, use it only when you know that the mark will disappear as the wood is cut, formed, or shaped.

The scratch awl is a slender, metal, pointed tool with a wooden handle. It is good for marking and punching the location of holes to be drilled.

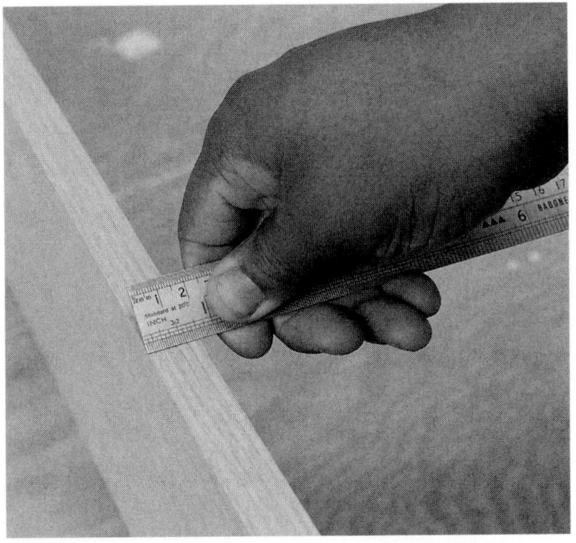

Fig. 4-16—*Measuring the thickness of stock with a bench rule.*

Fig. 4-17—*Measuring the width of stock.*

MEASURING STOCK

The thickness, width, and length of stock must be measured carefully. Use the following procedures to measure accurately.

Measuring Thickness and Width

Measure the *thickness* of stock by holding the rule across the edge. The thickness is the distance between the two lines on the rule that just enclose the stock. Fig. 4-16.

Measure the *width* by holding the left end of the rule (or the inch mark) on one edge of the stock. Turn the rule on edge. Hold the end of the rule even with the side of the stock. Slide your thumb along the rule until the width is shown. Fig. 4-17.

Measuring Length

Select the end of the stock from which the length measurement is to be taken. Check its squareness by holding a try square against the truest edge. Make sure that the end is not split or checked. If it is, square off and cut the end of the wood. Take the measurement from the sawed end. If a *short* length of stock is needed, hold the rule on edge and mark the length. Fig. 4-18.

Fig. 4-18—*For measuring short lengths, place the end of the rule directly over the squared end of the stock.*

For measuring *long* stock, use a tape measure. Using a long measuring tool will eliminate errors that result from moving a short rule several times. Make a small mark at the point to be squared.

MARKING STOCK FOR CUTTING

Like measuring, marking stock for cutting requires precision. Careful measurement means nothing without accurate marking and layout.

Marking for Length

Use a tape when marking long stock. This will eliminate measuring errors that come from moving a short ruler several times. Fig. 4-19.

Use a framing square on *wide* stock. Tip the blade slightly. Then hold it firmly against the truest edge while you mark across the stock. Refer again to Fig. 4-9. To be sure of trueness, square a line across the edges from the face line.

Marking Pieces of Equal Lengths

If several pieces must be measured and marked to equal lengths, do them all at the same time. Place the pieces side by side. Hold a try square across the ends to align them. Then measure and mark the correct length across all the pieces. Fig. 4-20.

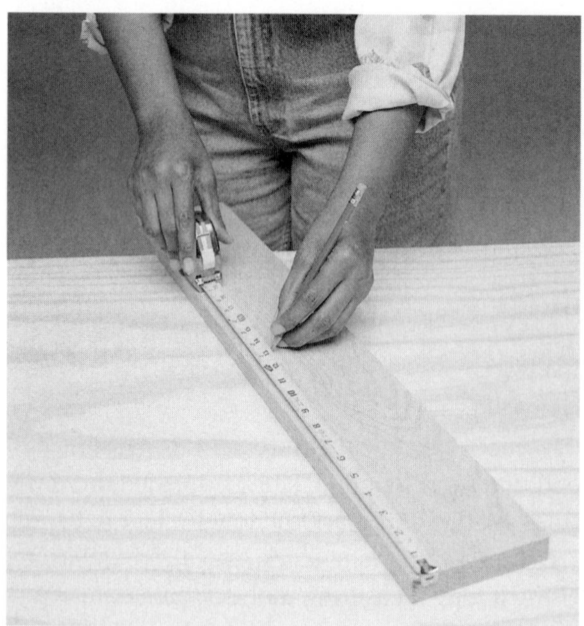

Fig. 4-19—*Marking narrow stock for cutting.*

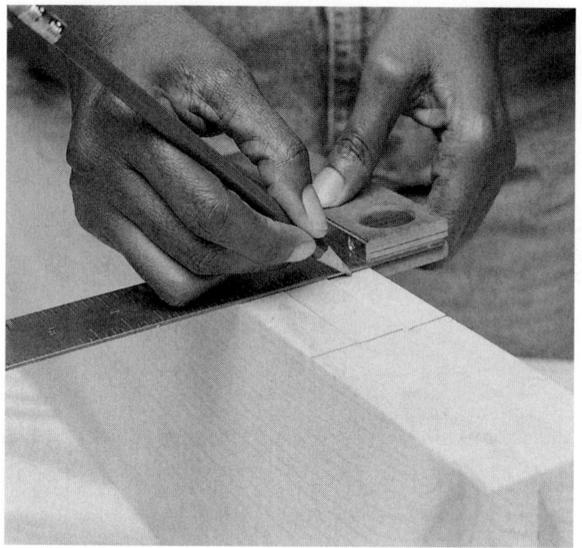

Fig. 4-20—*Marking multiple pieces for cutting to equal lengths.*

Marking Stock for Width

Decide on the width of stock you need. Hold a rule at right angles to the truest edge of the stock. (This can also be done with a try square, combination square, or framing square.) Measure the correct width and mark it with a pencil. Do this at several points along the stock. Then hold a *straightedge* over these points. (A straightedge is any tool or object, such as a board, along which a straight line can be drawn.) Draw a line connecting the points. The cut will be made along this line.

Another method can be used when working with sheets of plywood. Make a mark at the desired width at each end of the plywood. Then, using a *chalk line,* snap a line connecting the marks to guide cutting. Fig. 4-21.

Dividing a Board into Equal Parts by Width

To divide a board into two or more equal parts by width, hold a rule at an angle across the face of the stock. Turn the rule until the inch

Fig. 4-21—*Using a chalk line to mark a long cutting line.*

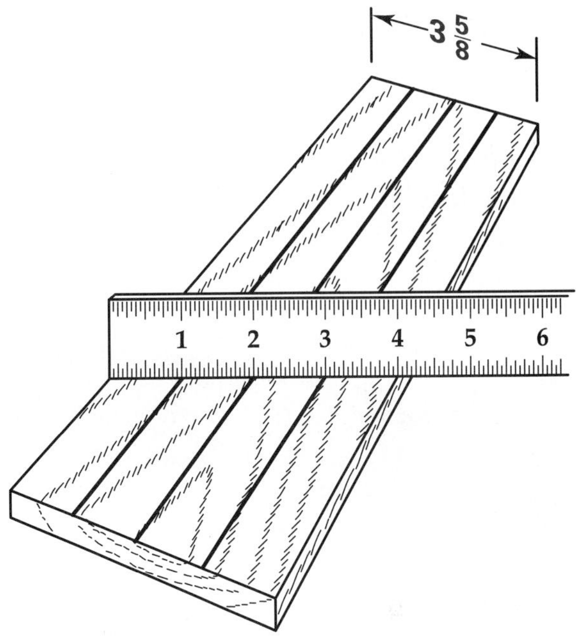

Fig. 4-22—*The proper method of dividing a board into several equal parts by width. This piece is divided into four equal parts.*

marks evenly divide the space for the number of parts desired. Fig. 4-22. Mark the stock as described in the previous section. For this work, the board must be true along both sides.

CUTTING WITH HANDSAWS

Four basic types of cuts are made with saws: crosscuts, rip cuts, miter cuts, and bevels. **Crosscuts** are made across the wood grain to cut stock to length. **Rip cuts** are made with the grain to cut stock to width. **Miter cuts** are angled cuts across the face of the stock. **Bevel cuts** are angled cuts made along the edge or end of the stock. Either handsaws or various power saws can be used to make these cuts.

Handsaws used most often in woodworking are the crosscut saw, ripsaw, and backsaw. Fig. 4-23 (page 114). The dovetail saw, coping saw, and compass saw are good for more specialized purposes. When using any handsaw, cut with the good side of the workpiece up.

Safety First

► *Using Handsaws*

- Use all saws properly, as instructed by your teacher.
- Keep your hands and fingers away from the sharp edge of the blade.
- Wear eye protection.
- Concentrate on your work. Do not become careless. Do not allow yourself to be distracted.

Crosscut Saw

Probably the most used handsaw in woodworking is the *crosscut saw.* It is used to make cuts across the grain, mainly when cutting stock to length.

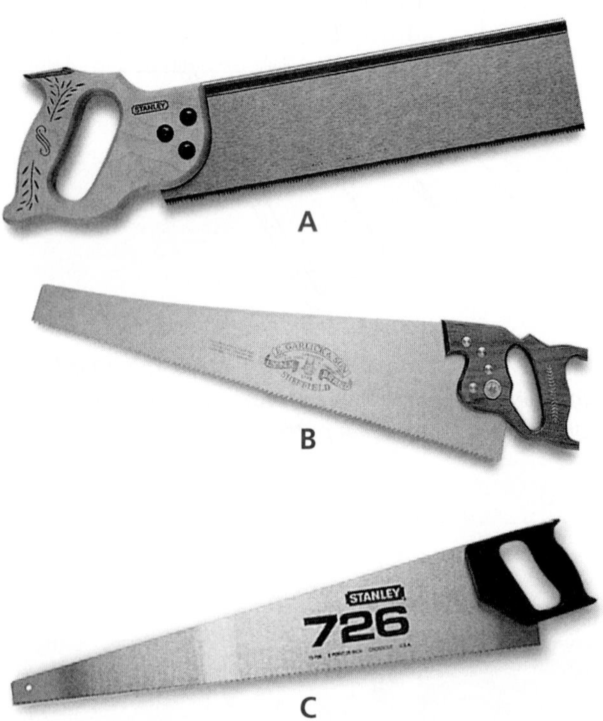

Fig. 4-23—*Saws used most often for general types of cutting: (A) Backsaw. (B) Ripsaw. (C) Crosscut saw.*

The teeth of the crosscut saw are shaped like small knife blades. They are bent alternately to the right and to the left. This is called the set of the saw. The set makes the **kerf** (saw cut) wider than the saw itself. Fig. 4-24. This keeps the saw from binding.

Saw blades are described by the number of points to the inch. *Points* are the tips of saw teeth. The more points, the finer the cut. For hard, dry wood, a fine saw with perhaps seven, eight, or nine points to the inch is best. *Note:* There is always one more point to the inch than there are teeth. Fig. 4-25.

Crosscut saws are available in many different lengths. The easiest sizes to handle are about 20 to 26 inches long.

Using the Crosscut Saw

When cutting with a crosscut saw, follow these steps:

1. Measure the desired length, and mark a line across the stock to guide cutting.
2. Place the stock on a workbench or across sawhorses. Fig. 4-26.
3. If you are right-handed, use your left knee to hold the stock. (Use your right knee, if left-handed.)
4. Position the blade just inside the cutting line on the waste side. Fig. 4-27.

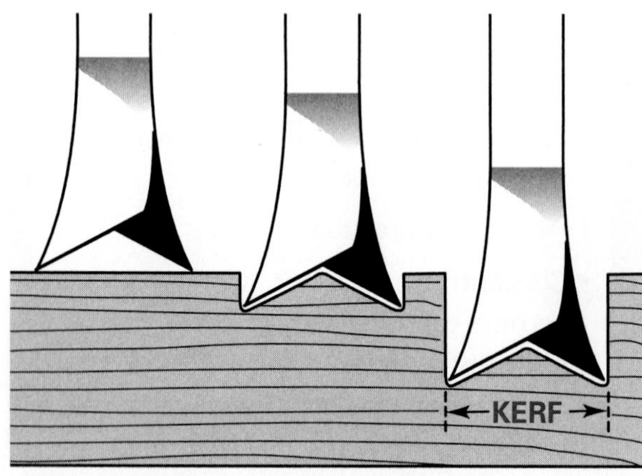

Fig. 4-24—*Crosscut saw teeth cutting a kerf.*

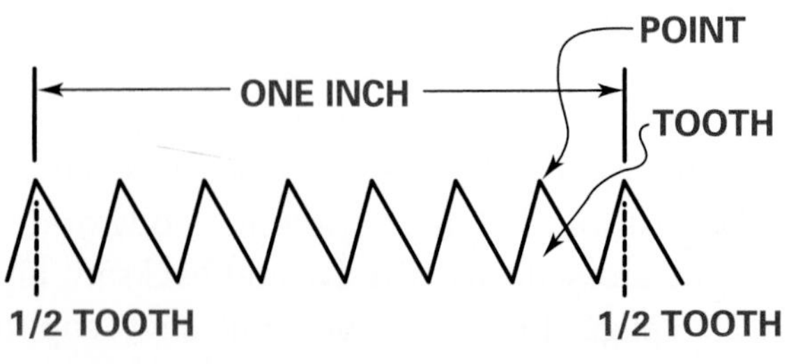

Fig. 4-25—*Points and teeth on a saw. The inch is measured from the point of one tooth to the point of another tooth. Both points are counted as being within the inch, but not both teeth.*

Safety First

► *Using a Crosscut Saw*

- Keep your thumb and fingers away from the teeth of the blade.

Fig. 4-26—*When cutting long stock, place it over two sawhorses with the cutting line extending just beyond one of the horses. Never try to make a cut between the supports.*

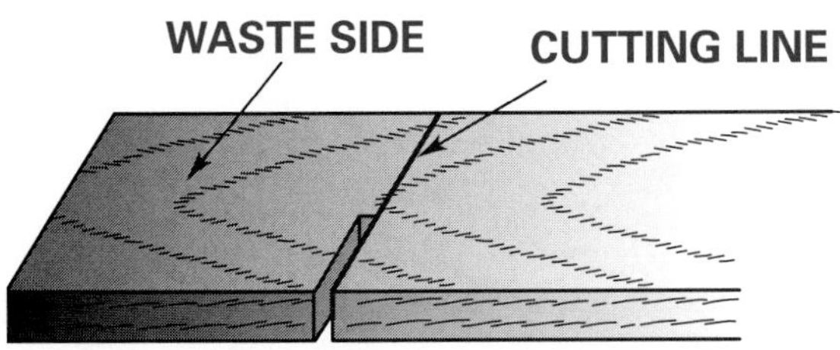

Fig. 4-27—*Cut on the waste side of the cutting line. If you cut directly on the line, part of the good side will be taken by the kerf. The board will be too short.*

5. With your free hand, grasp the stock beside the saw blade.
6. Hold the saw at a 45-degree angle. Using your thumb to guide the blade, begin the cut by drawing back on the saw. Fig. 4-28. Repeat this two or more times until a kerf is started.
7. Begin cutting, applying light pressure on the push cuts. Make sure the cut is square (at a right angle) to the face surface. Check this with a try square. Fig. 4-29 (page 116).
8. Continue cutting, using long, even strokes to produce a smooth cut.
9. Follow the cutting line as you saw. If the blade begins to move away from the line, use shorter strokes and twist the saw handle slightly. When the cut is back on line, resume taking long strokes.
10. As you complete the cut, take shorter strokes. Also, provide support for the part of the stock being cut off or ask someone to hold it. Otherwise, the wood may splinter as the piece falls off.

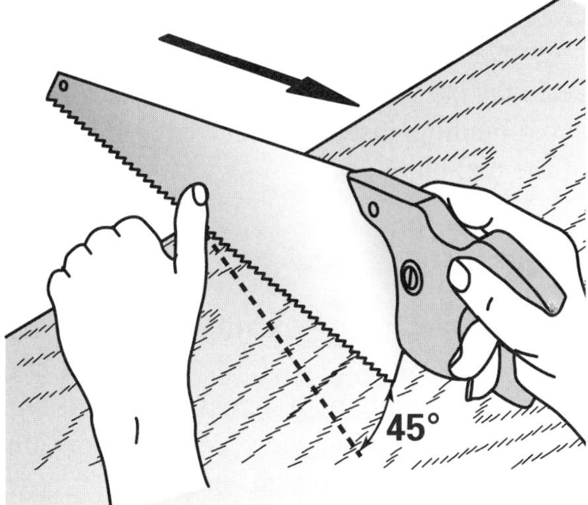

Fig. 4-28—*When crosscutting, start a cut by pulling the blade towards you. Be sure to keep your thumb on the smooth surface of the blade, away from the sharp saw teeth.*

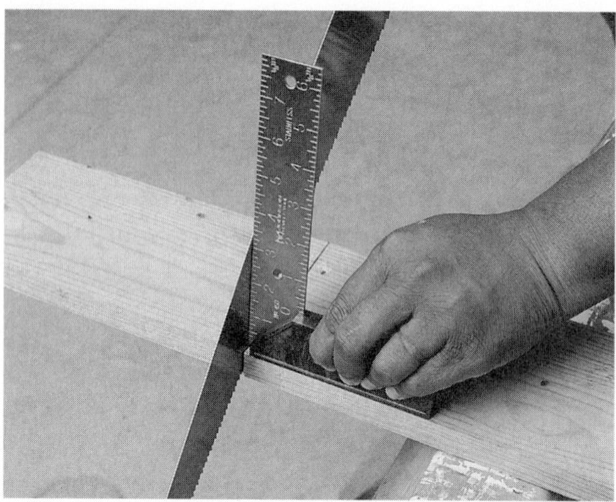

Fig. 4-29—*To make sure the cut is square, place the handle of the try square firmly on the face of the stock and against the blade of the saw.*

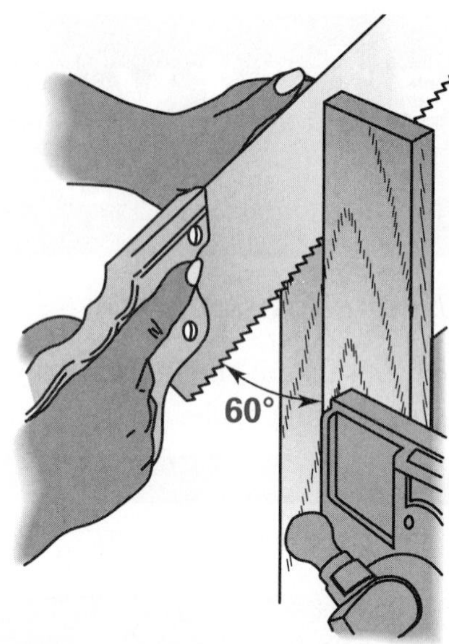

Fig. 4-30—*Ripping short stock with the work held in a vise. Saw close to the vise jaws, but make sure the saw does not strike any tools on the workbench or cut into the bench.*

Ripsaw

Ripsaws are used to cut stock to width. Refer back to Fig. 4-23. They have chisel-like teeth. Ripsaws used for ordinary woodworking should be 24 to 26 inches long with 5½ points per inch.

Rip cuts are made with the grain. To cut with a ripsaw, hold the saw at a 60-degree angle. Begin the rip cut by making several short forward thrusts. Then follow the same steps as for crosscutting. Fig. 4-30.

After a rip cut has extended a few feet, the kerf may close. This causes the saw to bind. To avoid binding, insert a small wedge at the start of the cut. Fig. 4-31.

Backsaw

The **backsaw** has a very thin blade with fine teeth. Refer back to Fig. 4-23. This saw is used to make fine cuts both across grain and with the grain. It is limited, however, as to how long or deep a cut it can make. Fig. 4-32.

The backsaw gets its name from an extra band of metal across its back. This band stiffens the saw. The backsaw is often used in a miter box. See Chapter 12.

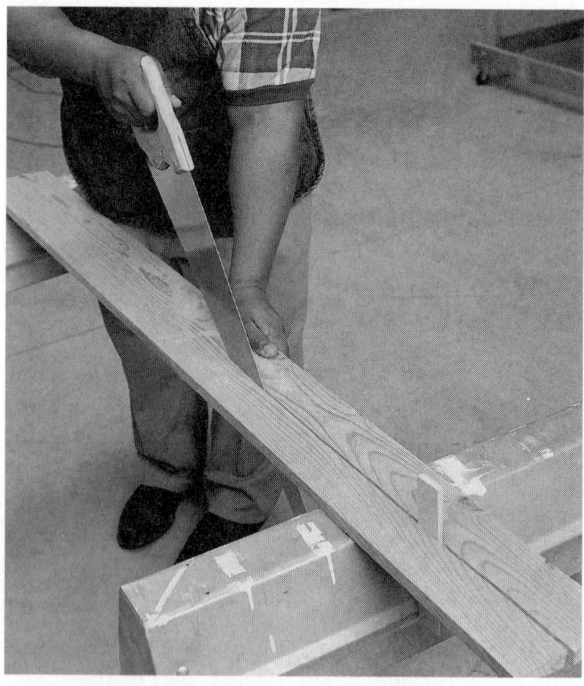

Fig. 4-31—*Insert a small wedge at the start of a rip cut to hold the sides of the kerf apart.*

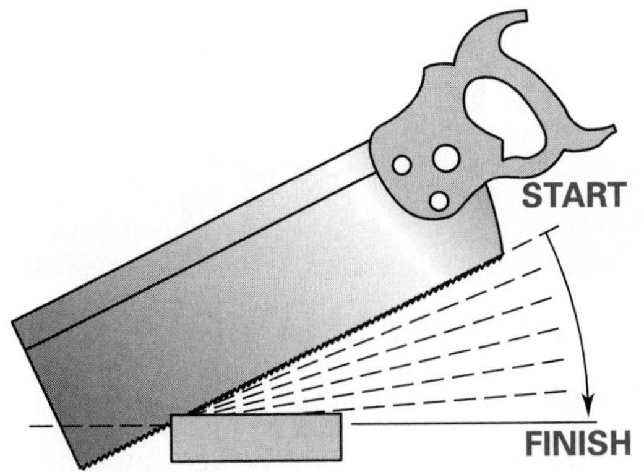

Fig. 4-32—*Hold the backsaw at an angle to start the kerf. Then lower the handle gradually until it is parallel with the top of the workbench.*

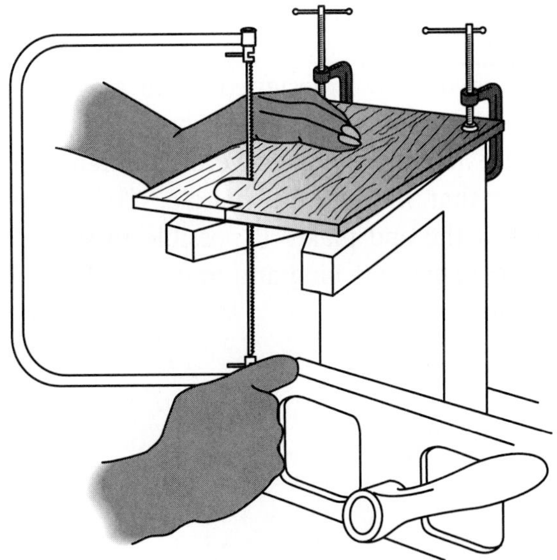

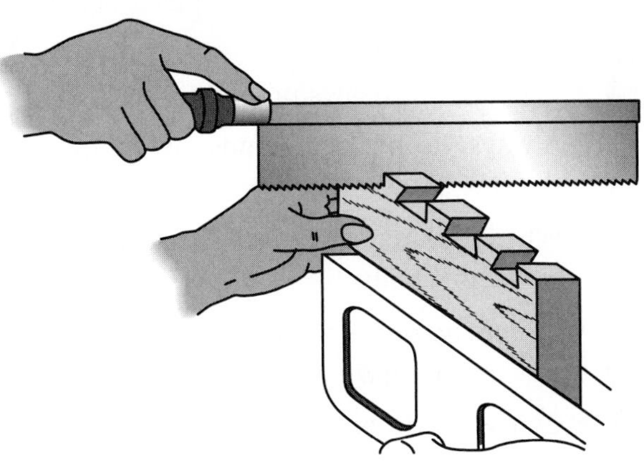

Fig. 4-33—*Dovetail saws are designed for cutting dovetails and tenons. They are also used for other types of precision work, such as building models and patternmaking.*

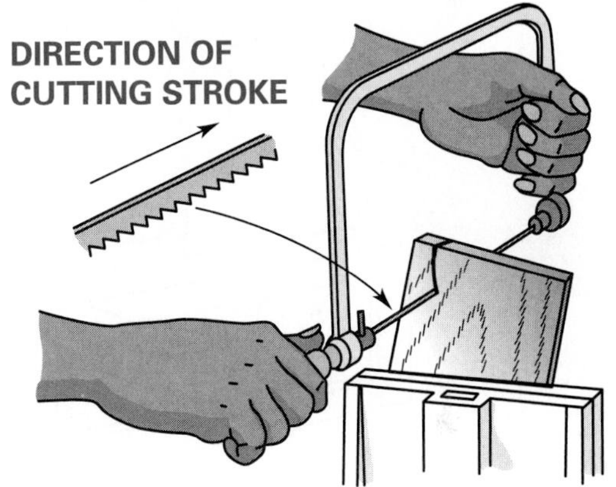

Fig. 4-34—*Using a coping saw. (A) Cutting stock held over a saw bracket. Shift the position of the work to accommodate the saw most easily. Teeth should be pointed toward the handle. (B) Cutting heavy stock held in a vise. Hold the saw with both hands to provide more support.*

Dovetail Saw

The *dovetail saw* is very similar to the backsaw except that it has a narrower blade and finer teeth. It cuts a true, smooth, and narrow kerf. The dovetail saw is used for extremely accurate work. Fig. 4-33.

Coping Saw

The *coping saw* has a U-shaped frame, a wood handle, and a replaceable blade with ripsaw-like teeth. The blade should be inserted with the teeth pointing away from the handle when cutting thick stock and toward the handle for thin stock. Fig. 4-34.

When cutting with a coping saw, take full, easy strokes. Guide the blade carefully. This will help prevent blade breakage. In cutting some types of wood (plywood in particular), draw the blade through beeswax from time to time. This will make cutting easier.

Compass Saw

The *compass saw* is designed to cut a wide kerf for sawing curves or irregular shapes. Fig. 4-35. Usually it comes with several different-sized blades that fit into the handle. The narrow point at the end makes it possible to start the tool in a small opening and to cut small curves and circles.

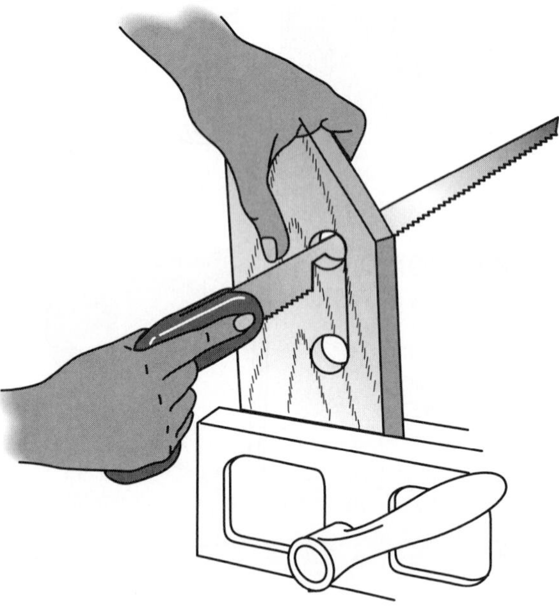

Fig. 4-35—*Using a compass saw to cut an internal opening in which two drilled holes are a part of the pattern.*

PORTABLE POWER SAWS

Power saws cut wood much faster than handsaws. Care must be taken, however, to set up the work properly.

Circular Saw

The portable **circular saw** is used for both crosscutting and ripping. It can be moved freehand or used with guides. Accurate finish cuts can be made quickly using a circular saw. Fig. 4-36.

Safety First

► *Using Portable Power Saws*

- Wear eye protection.
- Do not operate any power saw before receiving and understanding instructions from your teacher.
- Always get permission from your teacher before you use a power saw.
- Be sure all guards are in place before you plug in the saw.
- Use all power saws properly.
- Always keep your hands away from the saw blade.
- When making cuts, move the saw away from your body.
- Be sure no one is standing in front of the saw as you make a cut.
- Concentrate on your work. Do not become careless. Do not allow yourself to be distracted.
- Unplug power saws before making any adjustments.
- Do not set a saw down until the blade has stopped turning.
- Be careful not to cut the electric cord.

Using the Circular Saw

Follow these steps to use a circular saw:

Setting up:

1. Attach the correct blade for the cut being made. Position the blade so the teeth on the bottom point forward and up.

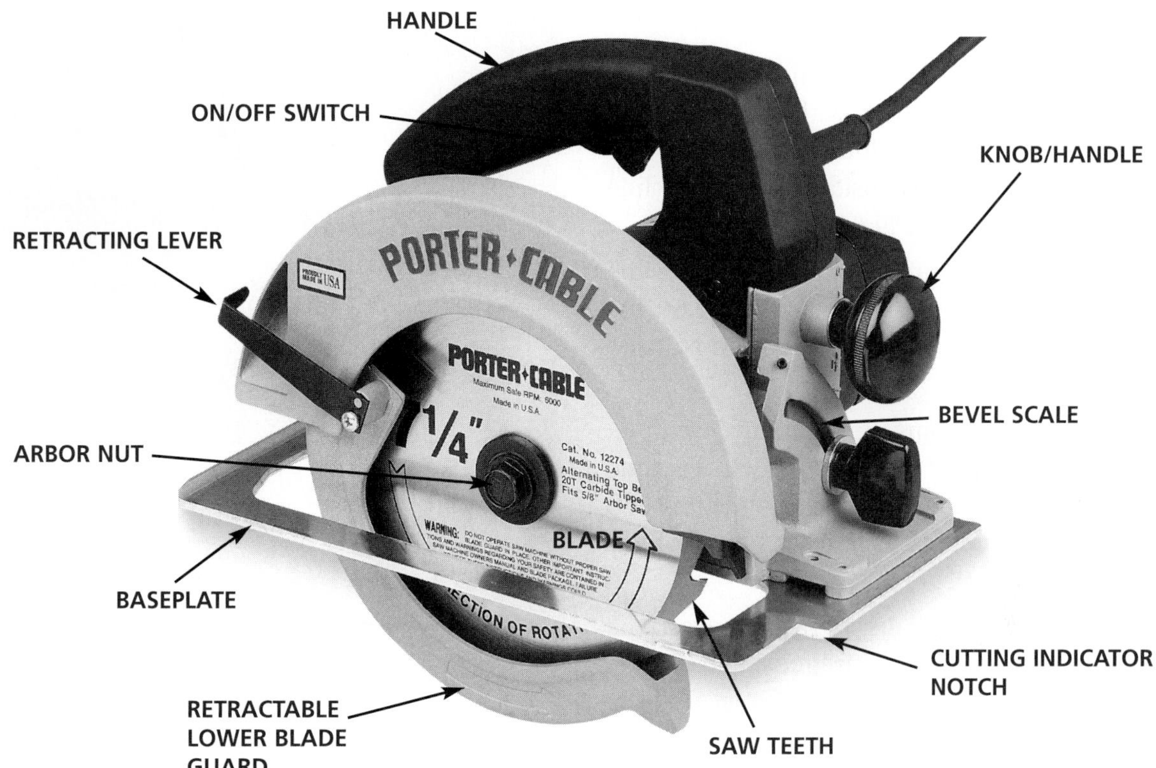

Fig. 4-36—*A circular saw. Note how the saw teeth point forward and up. In operation, the blade turns in a counterclockwise direction, cutting upward as the saw is moved forward.*

2. Adjust the angle of the blade.

   Most cuts require a 90-degree angle. Set this angle by loosening the angle adjustment lever, moving the base plate into the horizontal position, and then retightening the lever.

   For bevel cuts, set the blade according to the bevel degree scale on the body of the saw. Check the setting by making a cut on a piece of scrap wood.

3. Set the blade depth by first loosening the depth adjustment lever. Set the blade so it extends about ⅛ inch below the bottom side of the stock being cut. Fig. 4-37. A tape measure can be used to determine this setting.

4. Position stock for cutting on sawhorses or clamp it securely to a workbench. Place the good face surface down. Since the blade cuts upward, the top side of the cut is likely to splinter somewhat.

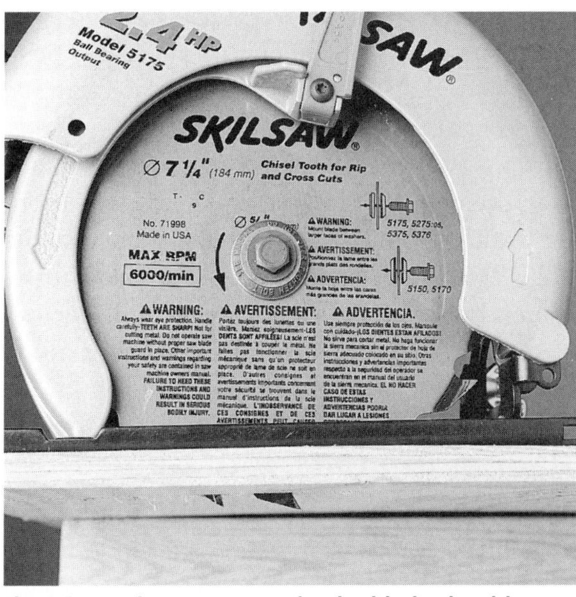

Fig. 4-37—*When set correctly, the blade should extend ⅛ inch below the stock.*

Making the cut:

5. To line up the cut, place the base plate of the saw on the stock without letting the blade touch it. Position the saw so the blade will cut just inside the cutting line on the waste side. Fig. 4-38.
6. After making sure the cord is not in the cutting path, press the trigger. Let the motor reach cutting speed before starting the cut.
7. Begin to cut using firm, steady pressure. Stay on line by keeping the cutting line beside the blade or by tracking it through the cutting indicator notch on the front of the saw.
8. When nearing the end of the cut, grasp the front handle with your free hand. When crosscutting, move the saw faster near the end of the cut to avoid splintering.
9. For safety, let the blade come to a complete stop before you move the saw up and away from the work or set it down.

Fig. 4-38—***Line up the blade so it cuts on the waste side of the cutting line.***

Making Crosscuts with Guides

To make sure your crosscuts are straight, use a cutting guide. A guide can be purchased or a straight piece of scrap wood can be used. Fig. 4-39. The guide is clamped to the stock, positioned so that the saw's base plate can be moved against it to make the cut.

To position the guide, measure the distance from the edge of the base plate to a saw tooth set towards that edge. Place the guide on the stock at that distance away from the cutting line and parallel to it. Clamp the guide securely to the stock. Make sure the clamps don't interfere with saw movement.

Fig. 4-39—***Crosscutting with a straightedge clamped to the workpiece as a guide.***

Making Rip Cuts with Guides

When making relatively wide cuts, use a long, straight piece of scrap wood as a guide as for crosscutting. Another method is to construct a ripping jig.

When making a rip cut near the edge of the stock, use a *ripping fence.* Fig. 4-40. To prepare for the cut, attach the fence loosely and set the saw blade at its least depth. At the end of the stock, line up the blade with the width layout line and set the blade depth as for crosscutting. Tighten the fence against the edge of the stock.

When ripping, move the saw slowly forward, away from your body. When you need to move, pull the saw back about an inch and let the blade stop.

During long rip cuts, the kerf tends to close in behind the saw. This can cause binding. To avoid this, use a *kerf splitter.* This device holds the sides of the kerf apart. Stop the saw and move the splitter forward from time to time as you make the cut.

Fig. 4-40—***Making a long rip cut using a ripping fence as a guide. Note the kerf splitter that keeps the saw from binding.***

Jigsaw

The **jigsaw,** or saber saw, is lightweight and portable. Fig. 4-41. It is perfect for making both curved cuts and straight cuts in hard-to-reach places.

The Largest and Oldest Living Things

The world's tallest tree—named Libby—grows in Redwood National Park in northern California. Libby is 367 feet tall. Its weight is estimated at more than one million pounds.

The oldest tree is believed to be a bristlecone pine. These trees grow in the White Mountains of California. Some of these trees are more than 5,000 years old.

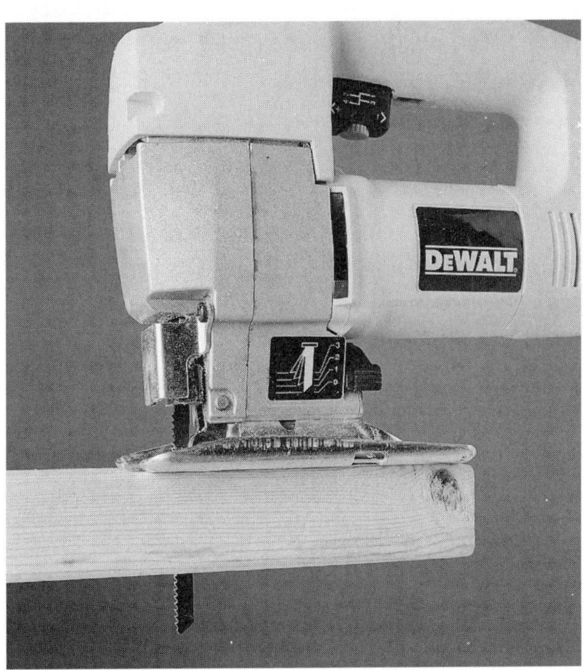

Fig. 4-41—***A good jigsaw will cut wood up to 2 inches thick.***

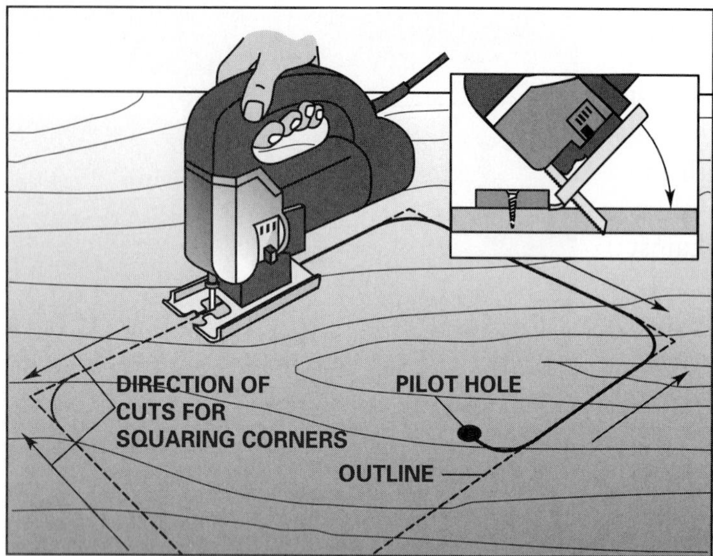

Fig. 4-42—*To make a cutout, first drill a small pilot hole in the waste area. Then place the blade of the jigsaw in the hole and begin the cut. The inset shows how to start a cutout without using a pilot hole.*

When cutting curves, use the most narrow blade you can. Line up the blade beside the cutting line on the waste side. For inside cuts, drill a *pilot hole* in the waste area. Fig. 4-42. Begin cutting. Move slowly to cut curves. In general, the tighter the curve, the slower the cut should be made. For a really tight curve, part of the waste area may need to be removed to create space to work.

Reciprocating Saw

The **reciprocating saw** is used mainly for rough cutting. Fig. 4-43. It is used in carpentry and remodeling more than in fine woodworking. The back-and-forth (or up-and-down) motion of the blade will cut wood, plaster, metal, and other materials. It is perfect for jobs like cutting an opening in a wall to install a new window or door.

Fig. 4-43—*The reciprocating saw is good for making inside cuts because of the back-and-forth motion of the blade.*

CHAPTER 4

Review & Applications

Chapter Summary

Major points from this chapter that you should remember include:

- Customary rules show inches and fractions of an inch; metric rules show centimeters and millimeters.
- Squares are used for measuring and marking.
- The sliding T-bevel is used for laying out all angles other than right angles.
- The thickness, width, and length of stock must be measured accurately.
- Types of cuts made with saws are crosscuts, rip cuts, miter cuts, and bevel cuts.
- Handsaws used most often are the crosscut saw, the ripsaw, and the backsaw.
- Portable power saws commonly used are the circular saw, jigsaw, and reciprocating saw.

Review Questions

1. Look at Fig. 4-2. Find 56 millimeters on the metric rule. To what measurement does it correspond on the customary rule?
2. What type of tool would you use to check the right angles of stock?
3. Which tool would you use to locate the center of a circle? Which tool would you use to mark it for drilling?
4. Name the types of cuts made with saws.
5. How are crosscut saws and ripsaws different?
6. What tasks should be done while power saws are unplugged?
7. Briefly outline the steps in using a portable circular saw.
8. Which portable power saw would you use to cut curves?

Solving Real World Problems

Tim and Shelley are cutting a large number of parts on a miter saw. The parts will be used to make a number of magazine racks for a school fund-raising event. All the parts for the sides must be the same length. Shelley, using a tape measure, marks each cut with a pencil and gives the part to Tim for cutting. As they start to work, they discover that the parts are shorter than they intended them to be. Why? What can be done to fix the problem?

CHAPTER 5

Nailing

Look for These Terms

- box nail
- casing nail
- claw hammer
- common nail
- finishing nail
- nailer
- nail set
- penny number
- ripping bar
- toenailing

YOU'LL BE ABLE TO:

- Identify different types of nails.
- Demonstrate the correct technique for driving nails into wood.
- Describe the technique of toenailing.

You have learned how to saw the boards to length and width. Now let's nail them together. Nails are the easiest and least expensive way of joining wood.

TOOLS FOR NAILING

The **claw hammer** has a head of drop-forged steel and a handle of hickory, steel, or fiberglass. Fig. 5-1. The metal handle is covered with hand grips of plastic, leather, or rubber. The face of the hammer should be slightly domed to prevent hammer marks. Fig. 5-2.

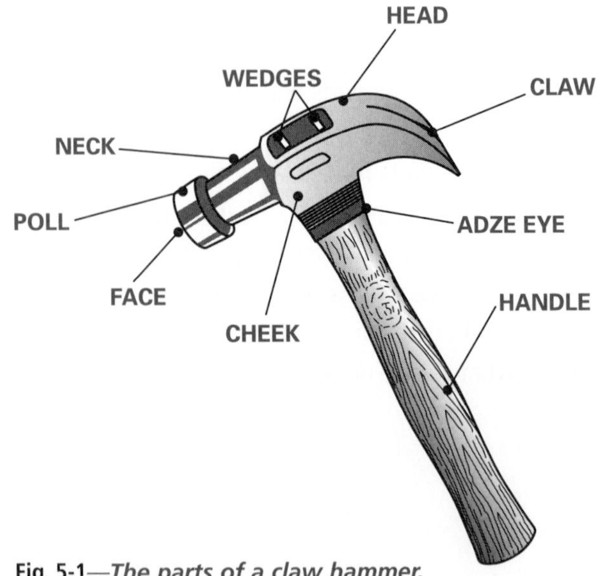

Fig. 5-1—*The parts of a claw hammer.*

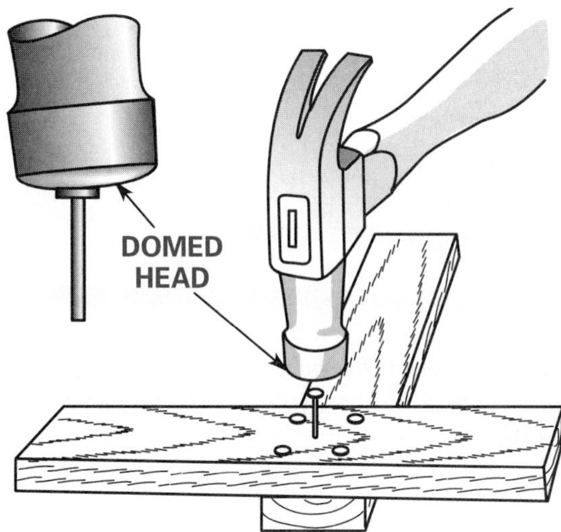

Fig. 5-2—*The head of a hammer is domed slightly. This helps keep the nail from bending if you don't strike it exactly square. It also concentrates force at the contact point and helps prevent hammer marks.*

Fig. 5-3—*The nail set is used to drive the nail head below the surface of the wood. This hole is then filled with putty or similar material before the project is finished.*

Hammers come in weights ranging from 5 to 20 ounces. The 16-ounce hammer is best for everyday use.

A **nail set** is a short metal punch with a cup-shaped head. It is used to drive the head of the nail below the surface of wood. Fig. 5-3.

A **ripping bar** has a gooseneck with a nail slot on one end and a chisel-shape on the other. It is used for ripping down buildings, opening crates, and similar jobs. A *pry bar* and *nail puller* are similar to a ripping bar. Fig. 5-4 (page 126).

SCIENCE *Connection*

Types of Levers

When you think of a hammer, chances are you don't associate it with a lever. Yet a hammer is a type of lever.

A *lever* is a bar that turns or pivots on a point called a *fulcrum*. Levers make it possible to exert a great deal of force for cutting, holding, or lifting. There are three kinds of levers: first-class, second-class, and third-class. In the first-class lever, the fulcrum is between the weight and the effort. Examples of first-class levers are pliers and pry-bars. In the second-class lever, the weight is between the fulcrum and the effort. Examples are the nutcracker and the wheelbarrow. The third-class lever has the effort between the fulcrum and the weight. Hammering a nail is an example of the third-class lever.

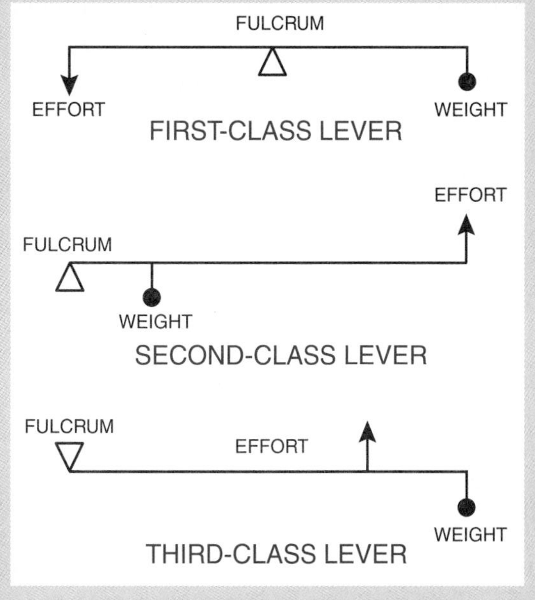

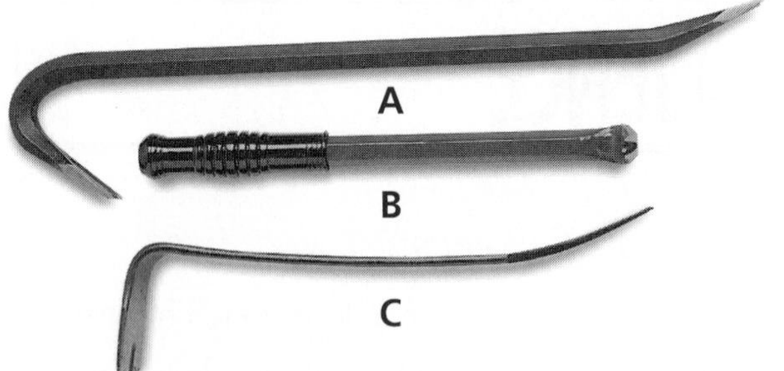

Fig. 5-4—*(A) Ripping bar. (B) Nail puller (sometimes called a "cats paw"). (C) Pry bar.*

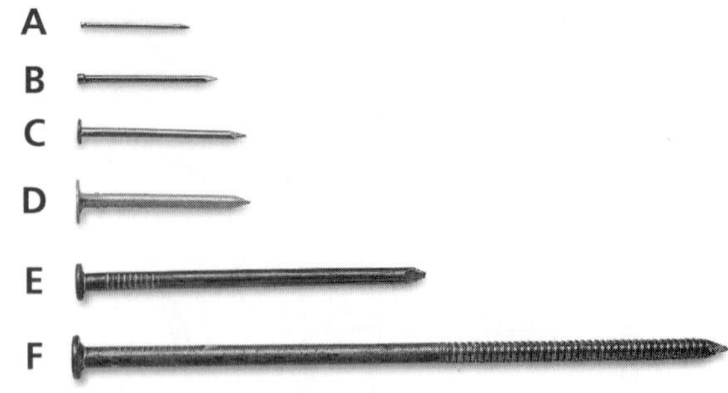

Fig. 5-5—*A few of the more popular nails: (A) wire brad; (B) finish; (C) box; (D) roof; (E) common; and (F) spike.*

KINDS OF NAILS

There are nails to meet every specific need. Fig. 5-5. Nails are distinguished by their heads, shanks, points, surface finish, and the material from which they are made. See "Nails" in the *Woodworker's Handbook* at the front of the book.

Common nails have large flat heads. These nails are used in building construction for rough carpentry. The larger sizes are called *spikes.* **Box nails** are similar to common nails except they have a smaller diameter. They are used mostly for light carpentry and for construction of packing cases. The **casing nail** has a small head. It is a rather heavy nail for more-finished carpentry. The **finishing nail,** the finest of all nails, is used for all fine cabinet and construction work.

Nails are ordered by the **penny number,** or *d.* Once based on the price per nail, this number now relates to length. Nail sizes start at 2d, which is 1 inch long, and range up to 60d, which is 6 inches long. Table 5-A and 5-B.

When selecting nails, choose nails with small diameters for thin stock. For heavy stock, choose large diameter nails.

Besides the nails mentioned, other kinds of small metal fasteners include *wire brads, wire nails, tacks, staples,* and *upholstery nails.* Fig. 5-6. *Escutcheon nails* are small brass nails with round heads used to decorate small projects. *Corrugated fasteners* are used in rough construction. Fig. 5-7.

Table 5-A. Nail Sizes

		Number Per Pound		
Penny Number	Length in Inches	Common Nails	Box Nails	Finishing Nails
2	1	876	1010	1351
3	1¼	568	635	807
4	1½	316	437	548
6	2	181	236	309
8	2½	106	145	189
10	3	69	94	121
12	3¼	64	87	113
16	3½	49	71	90
20	4	31	52	62
30	4¾	20		
40	5			
50	5½			
60	6			

DRIVING NAILS

Nails can be driven straight into wood. For a tighter joint, they may be driven at a slight angle. If two pieces are to be nailed together, as on the corner of a box, drive one or two nails through the first piece. Then hold this piece over the second piece and drive the nails in place.

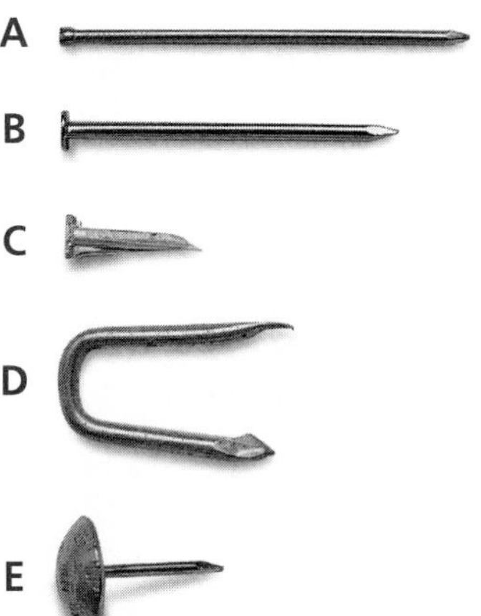

Fig. 5-6—*Small metal fasteners: (A) wire brad; (B) wire nail; (C) tack; (D) staple; and (E) upholstery nail.*

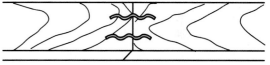

Fig. 5-7—*Corrugated fasteners, or "wiggle" nails, are used in place of standard nails for certain purposes. These include repair work and box-and-frame construction. They are also used to hold miter and butt joints together.*

Table 5-B. Nail Characteristics

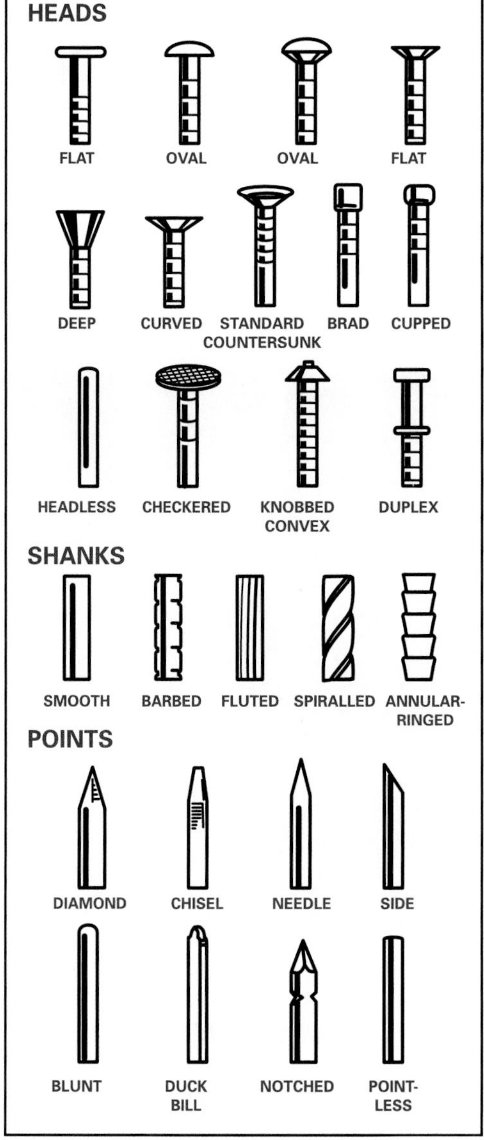

Safety First

► *Nailing Hazards*

Other than hitting your finger, nailing may not appear to be a real safety issue. That simply is not true. A nail, if not struck squarely on the head with a hammer, can go flying off like a bullet. You, as well as the people around you, should all be wearing safety glasses. A nail can easily fly up into the nailer's face or can hit someone several feet away. Special caution should be taken when nailing with a hammer or with a nailer.

To start a nail, hold it in one hand between thumb and forefinger and close to the point. Grasp the hammer near the head and tap the head of the nail with the hammer. Fig. 5-8.

Remove your fingers from the nail as you continue to strike it with firm, even blows. Hold the hammer near its end. Use wrist as well as elbow and arm movement, depending on the size of the nail being driven. Fig. 5-9.

Watch the head of the nail, not the hammer. Drive the nail with a few well-placed blows, rather than with many quick taps. If a nail begins to bend, remove it. Start over with a new nail.

Choose the location for nails wisely. Do not put several nails along the same grain. This will split the wood. A few well-placed nails in a staggered pattern will hold more strongly than a larger number placed carelessly.

When using casing or finishing nails, do not drive the heads completely down to the surface. Complete the driving with a nail set.

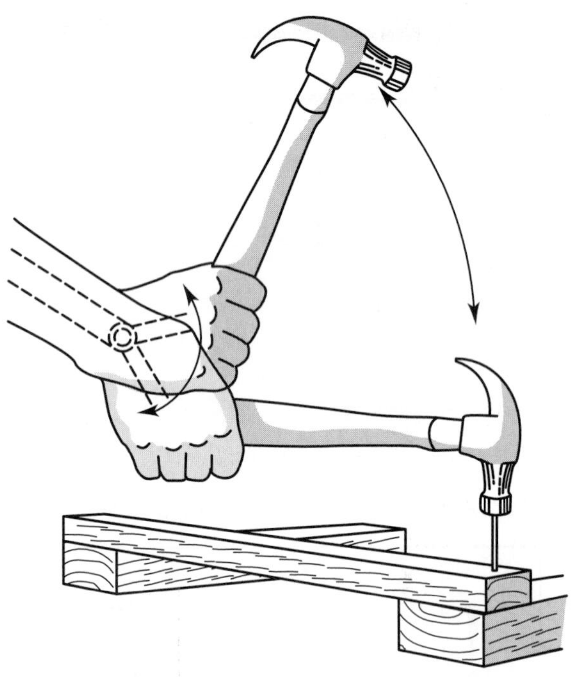

Fig. 5-9—*Use wrist movement for driving small nails. For driving large nails, use elbow movement, as well.*

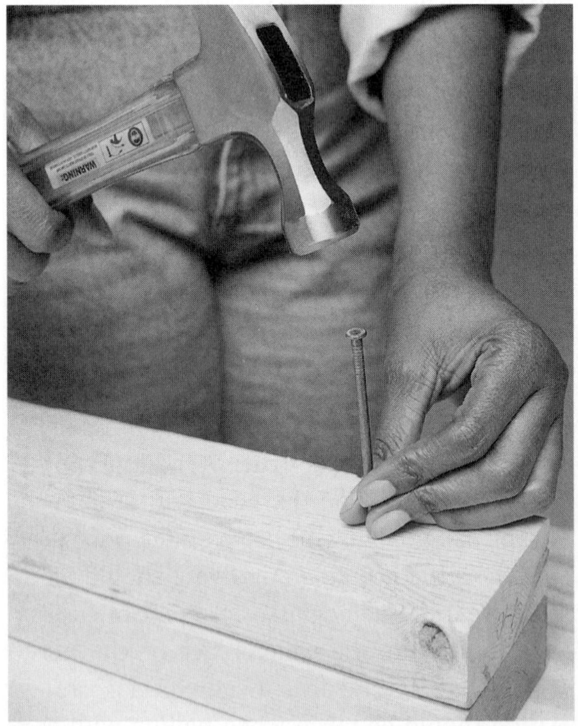

Fig. 5-8—*Starting a nail. Hold the nail between the thumb and forefinger. Grasp the hammer close to the head.*

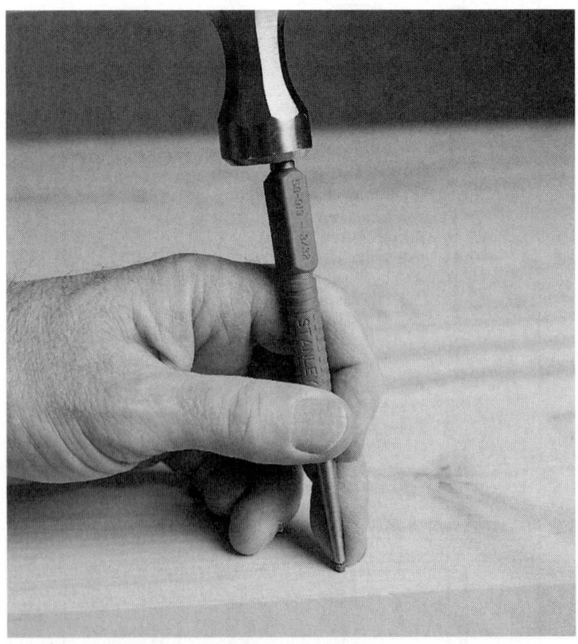

Fig. 5-10—*Using a nail set. Hold the nail set between the thumb and forefinger. Guide it with the other fingers. This helps keep it from slipping off the nail head and marring the wood surface.*

Hold the nail set in one hand with the middle finger against the surface of the work and the side of the nail. Then drive the nail in until it is about 1⁄16 inch below the surface. Fig. 5-10. Cover the small hole with a wood filler.

When nailing hardwood, drill holes that are slightly smaller than the diameter of the nail. Apply a little wax to the nail and drive it in.

To nail the end of one piece of wood to the side of another piece, drive the nails into both sides of the piece at an angle. This is called **toenailing.** Fig. 5-11.

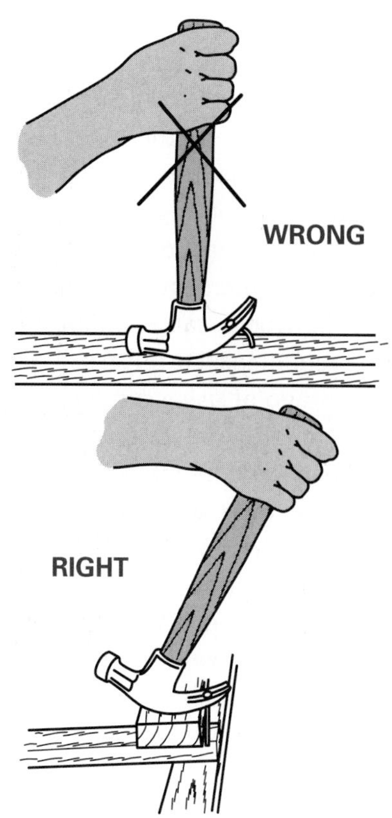

Fig. 5-12—*Removing a nail. After the nail is partway out, do not continue to pull it out. It will bend the nail and ruin the surface of the wood. Place a scrap of wood under the hammer head. This will provide leverage for drawing the nail out straight. It will also protect the surface of the wood.*

REMOVING NAILS

To remove nails, force the claw of the hammer under the head of the nail and pull on the handle. When the nail is drawn partway out, slip a piece of scrap wood under the hammer head (to protect the surface) before continuing to draw out the nail. Fig. 5-12.

NAILERS

A **nailer** is a nice addition to the woodworker's toolbox. Fig. 5-13. Nailers have been around for many years but only recently have become common in the woodworker's shop.

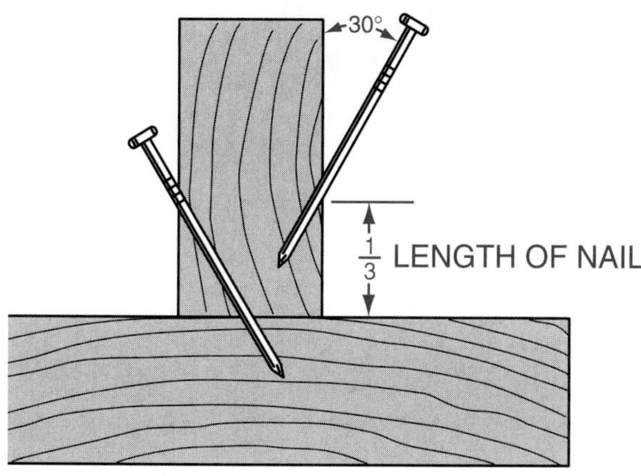

Fig. 5-11—*Toenailing.*

Fig. 5-13—*A nailer makes nailing easy, and there are no hammer marks on the wood.*

There are basically four types of nailers: framing, roofing, finish, and brad. Nailers can handle nails as long as 3½ inches and as short as ⅜ inch, depending on the type. Fig. 5-14.

Finish and brad nailers are the type most often used in woodworking. The serious woodworker will have at least one of each of these. Finish and brad nailers can eliminate a lot of dent marks in wood; more importantly, they can also prevent smashed fingers.

A big advantage of using a nailer is that it is easier and faster to use than a hammer and nail. It eliminates the hammer and the nail set. Nailers will not only drive the nail into the workpiece, but they will also set the nail. Better accuracy when assembling a workpiece is another advantage of a nailer. For example, you can position the workpiece with one hand while using the other hand to squeeze the trigger to drive the nail into the wood. Fig. 5-15.

Avoiding Split Ends

Driving a nail into wood is like driving a wedge between the wood fibers. If you do not take some precautions, this will cause the wood to split. One thing that helps is to blunt the point of the nail before you drive it in. Just stand the nail on its head and tap the point with your hammer. Blunting the point will cause the nail to tear through the wood fibers instead of wedging between them.

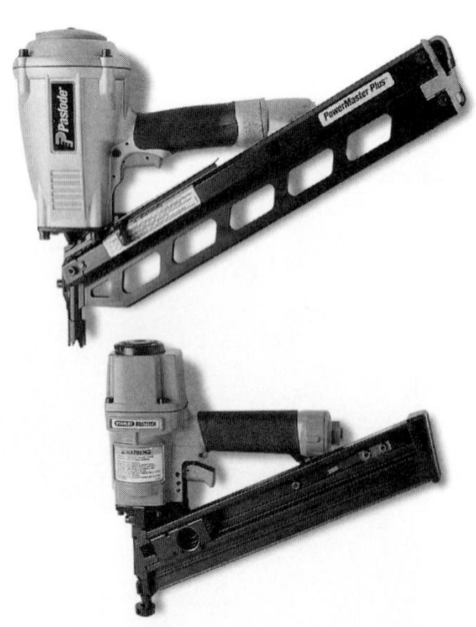

Fig. 5-14—***A framing nailer will drive 3½-inch nails. The brad nailer will handle brads as short as ⅜ inches.***

Fig. 5-15—***One of the advantages of a nailer is that one hand is free to hold the workpiece in place as you are nailing.***

CHAPTER 5

Review & Applications

Chapter Summary

Major points from this chapter that you should remember include:

- Claw hammers, nail sets, and ripping bars are tools used for nailing.
- There are nails to meet every specific need, including common, box, casing, and finishing nails.
- Proper nail driving technique includes: holding the hammer near its end; using wrist, elbow, and arm movement; watching the head of the nail; striking the nail with firm, even blows.
- When removing a nail, slip a piece of scrap wood under the hammer head to protect the workpiece.
- An advantage of using a nailer is that it's easier and faster than a hammer and nail.

Review Questions

1. Name the parts of a claw hammer.
2. Explain how to use a nail set.
3. Name and describe four types of nails.
4. What are the larger sizes of common nails called?
5. Based on illustrations in this chapter, describe how a brad is different from a finishing nail.
6. How is a nail started?
7. In hammering a nail, should you watch the head of the hammer or the nail?
8. How are finishing nails set?
9. Hardwood is very difficult to nail. What would you do to overcome this difficulty?
10. Explain the technique of toenailing.

Solving Real World Problems

Ambra is building an oak TV stand. She has gathered all the materials and hardware and has cut everything to size. She has used a drill bit that is slightly smaller than the nails to pre-drill the hole for each nail. Ambra discovers that when she "sets" a nail or misses a nail with the hammer, there is a dimple left in the surface of the wood. What can she do to protect the wood from the hammer? How should Ambra be setting the nails?

CHAPTER 6

Drilling

YOU'LL BE ABLE TO:

- Drill holes with a variety of hand tools as well as with a power drill.
- Describe the qualities needed in a power drill for woodworking.
- Identify several accessories available for power drills and explain how they are used in woodworking.

Look for These Terms

- auger bit
- automatic drill
- brace
- brad point
- chuck
- depth stops
- doweling jigs
- hand drill
- pocket hole jigs
- tang
- twist drill
- variable-speed (drill)

Many woodworking projects involve cutting round holes. Technically, holes that are ¼ inch or smaller in diameter are said to be *drilled,* while larger holes are said to be *bored.* However, the term *drilling* is commonly accepted as meaning cutting holes of any diameter.

TWIST DRILLS AND BITS

Twist drills come in fractional-sized sets, ranging from $\frac{1}{64}$ to ½ inch, in steps of $\frac{1}{16}$ inch. The size of the twist drill is indicated on its *shank,* the round, noncutting end of the twist drill. Fig. 6-1. The shank of the twist drill is fastened into the drill's **chuck**, which has jaws that are adjusted to tightly grip the shank. Fig. 6-2.

Fig. 6-1—*A small set of twist drills. The size of each drill is stamped on its shank.*

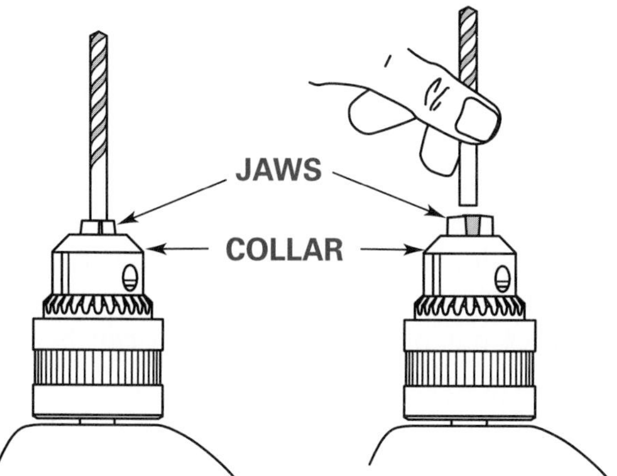

Fig. 6-2—*The jaws of the chuck are adjusted to tightly grip the shank of the twist drill.*

Fig. 6-3—*A single-twist auger bit. The tang of the bit is gripped by the chuck of a bit brace. The brad point helps draw the bit into the wood.*

Auger bits are used to cut holes ¼ inch to 1 inch in diameter. Auger bits come in sizes ranging from No. 4 to No. 16. This number indicates the size in sixteenths of an inch—a No. 4 is 4⁄16 inch, a No. 5 is 5⁄16 inch, etc. The **brad point** (or feed screw) on the cutting end of the auger bit helps draw the bit into the wood. Fig. 6-3. The noncutting end of the auger bit is a square **tang,** which is fastened into a bit brace. (See next page.)

HAND TOOLS FOR DRILLING

There are a number of hand tools available for drilling. Three of the more common examples are the automatic drill, hand drill, and brace.

Automatic Drill

The **automatic drill** (or *push drill*) with a set of drill points provides a quick and easy way to drill small holes. The drill points, which are stored in the handle of the drill, are numbered from one to eight. Number 1 is 1⁄16 inch; 2 is 5⁄64 inch; 3 is 3⁄32 inch; 4 is 7⁄64 inch; 5 is ⅛ inch; 6 is 9⁄64 inch; 7 is 5⁄32 inch; and 8 is 11⁄64 inch.

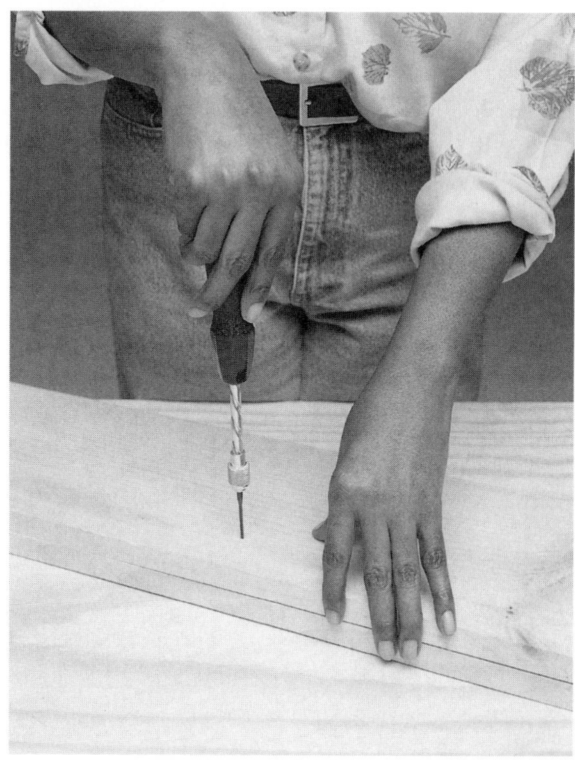

Fig. 6-4—*Secure the desired size drill point in the chuck. Locate the position for the hole. Then use one hand to push the drill handle down a few times, allowing it to spring back after each stroke.*

After tightening the desired size drill point in the drill's chuck, place the point where the hole is needed. Then, using one hand, simply push down a few times, allowing the handle to spring back after each stroke. Fig. 6-4. Make sure the drill is held square with the work.

Hand Drill

The **hand drill** is used with twist drills to drill small holes, usually ¼ inch or less in diameter. Fig. 6-5. The handle of the drill is held with one hand, while the other hand turns the crank evenly to drive the twist drill into the wood. Because the small twist drills can break easily, take special care to keep the drill square with the stock throughout the drilling operation.

Brace and Bits

A **brace,** such as the one shown in Fig. 6-6, is used with an auger bit. The auger bit is driven into the wood as you turn the handle of the brace. Fig. 6-7.

The size of a brace is determined by the size of its *sweep,* which is the amount of space it takes to sweep (or swing) the handle of the brace around one full revolution. A 10-inch brace is a good choice for general use.

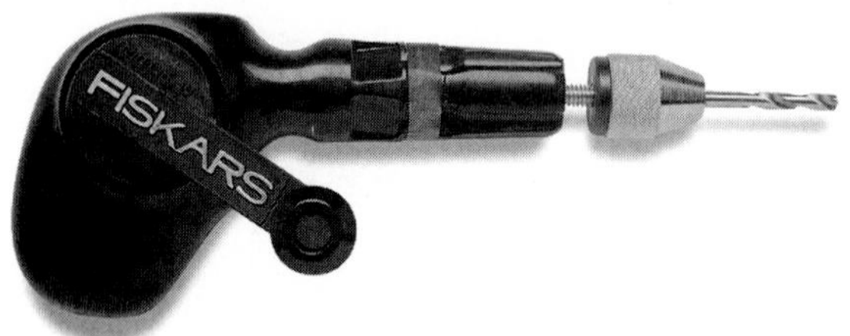

Fig. 6-5—*A hand drill.*

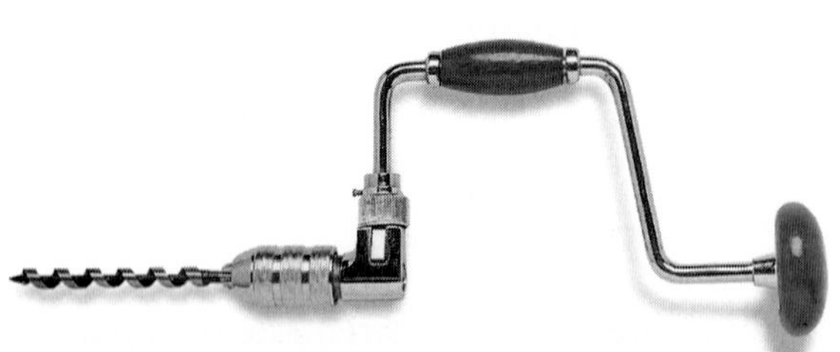

Fig. 6-6—*The bit brace is used for holding auger bits, Forstner bits, and other tools with rectangular shaped shanks. Most braces have a ratchet arrangement, making it possible to bore in corners and otherwise inaccessible places.*

Most braces have a ratchet mechanism that allows you to drill in corners or in close quarters where you cannot make a full sweep of the handle. When the ratchet is turned to the right (clockwise), the brace can turn the auger bit to the right as far as space allows. The ratchet slips when the brace handle is turned back to the left, so the auger bit is not reversed out of the hole and you can begin another partial sweep. In this way, you can drill the hole a half sweep or less at a time.

POWER DRILLS

Power drills may be electric plug-in models or cordless, with rechargeable batteries. Fig. 6-8. Most cordless drills can drill hundreds of holes or insert hundreds of screws on one charge. A fully-charged cordless drill offers about the same amount of power as its plug-in counterpart.

SCIENCE *Connection*

Friction

Friction is resistance to motion that is caused by the molecules of the two objects being slightly attracted to each other. When you drill, the friction between the bit and the wood generates heat. You can create much the same effect by rubbing your hands together briskly.

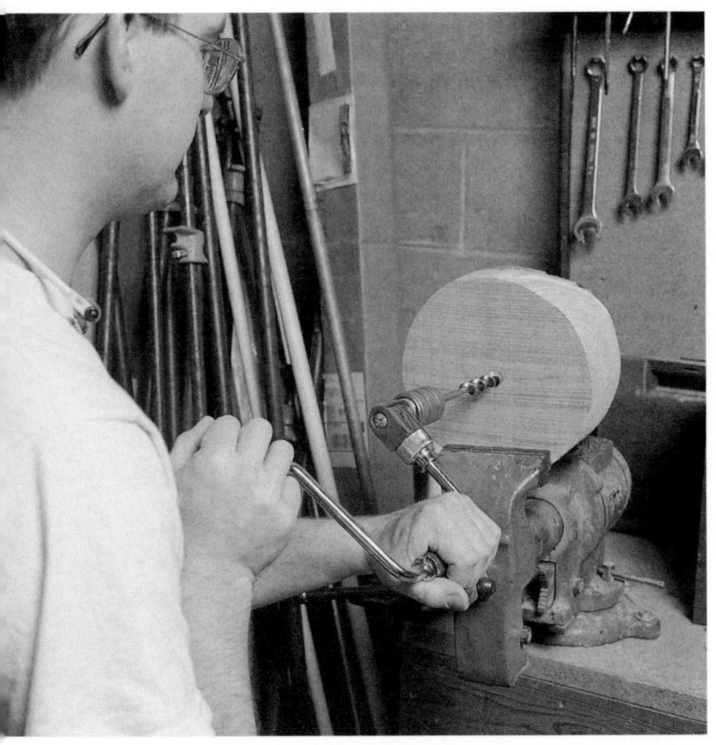

Fig. 6-7—*Horizontal drilling with a brace and bit. While using one hand to turn the handle clockwise, cup the other hand over the head of the brace and apply pressure with your body to keep the bit at a steady right angle to the stock.*

The size of the power drill is determined by the size of its chuck. The larger the chuck, the more torque the drill can provide. (*Torque* is the force that produces a rotating motion.) However, as the torque increases, the drill speed decreases. A $\frac{3}{8}$-inch power drill provides a good balance between torque and speed for most woodworking needs.

A power drill for woodworking should be variable-speed and reversible. The **variable-speed** feature allows you to adjust the speed by the amount of pressure you apply to the trigger. The *reversible* feature allows you to remove screws or back out of a hole.

Most bits for power drills have rounded shanks. The size of the shank varies according to the size of the bit. On some models, a chuck key is used to tighten the jaws of the chuck around the bit. Fig. 6-9. Increasingly, however, the chuck is keyless. With a keyless chuck, you rotate the body of the chuck by hand to tighten or release the drill bits. Fig. 6-10 (page 136).

Fig. 6-8—*A plug-in electric drill and a cordless drill, which is powered by a rechargeable battery pack.*

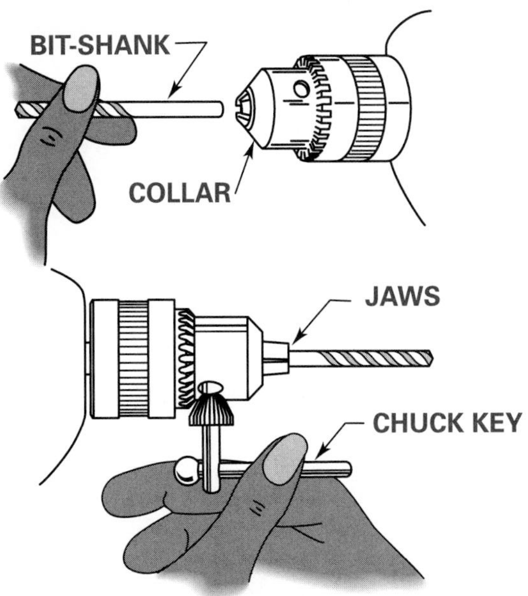

Fig. 6-9—*On some power drills, a chuck key is used to tighten the jaws of the chuck. NOTE: Be sure to remove the chuck key before starting the drill.*

Accessories for the Power Drill

A number of accessories are available for the power drill. Some of these accessories increase the tool's versatility, while others help ensure precise cutting.

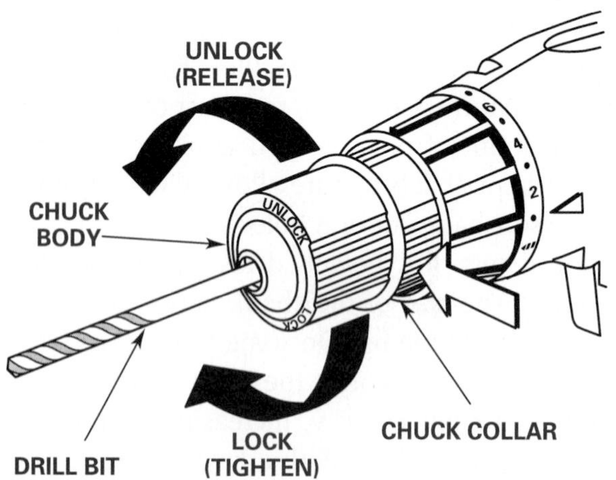

Fig. 6-10—*The keyless chuck makes it faster and easier to change bits.*

All portable power drills can be turned into power screwdrivers or they can drill special holes by inserting various bits into the chuck. Fig. 6-11. Accessories are also available to turn your power drill into a sander when the workpiece is too large or too awkward to sand with a regular sander. Fig. 6-12. A drill stand can be used to convert a portable drill into a small drill press. Fig. 6-13.

A number of guides and jigs are available to help you drill holes with precision. Some of these accessories are described below.

- *Drill guides* help keep the bit centered on the workpiece. Fig. 6-14.
- **Depth stops** are used to control the depth of the hole being drilled. There are a number of different types available for purchase. Fig. 6-15.

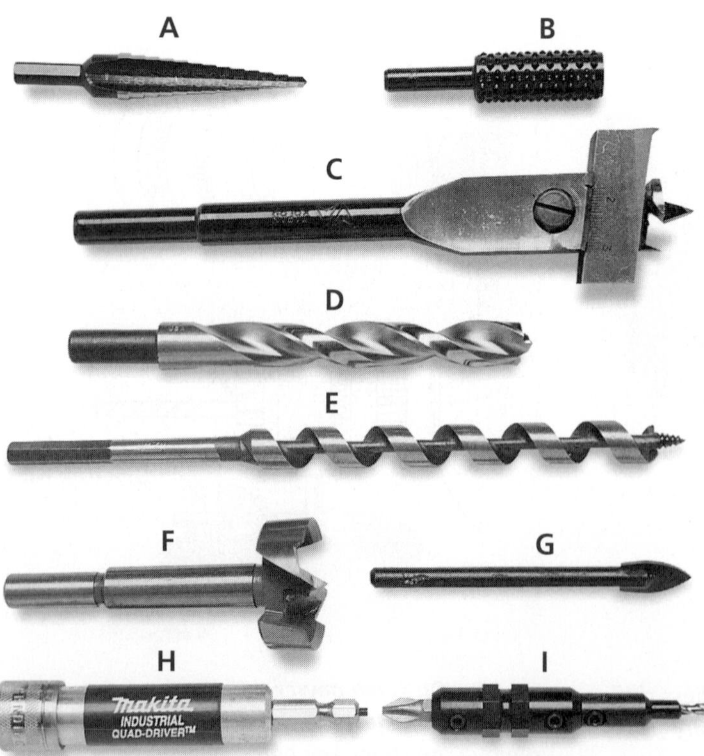

Fig. 6-11—*(A) Step bit. (B) Rotary rasp. (C) Expansion bit (not used in portable drills). (D) Brad-point bit. (E) Power auger bit. (F) Forstner bit. (G) Glass/tile bit. (H) Drill and driver shaft for a quick connector. (I) Countersink/driver insert for drill and driver shaft of quick connector.*

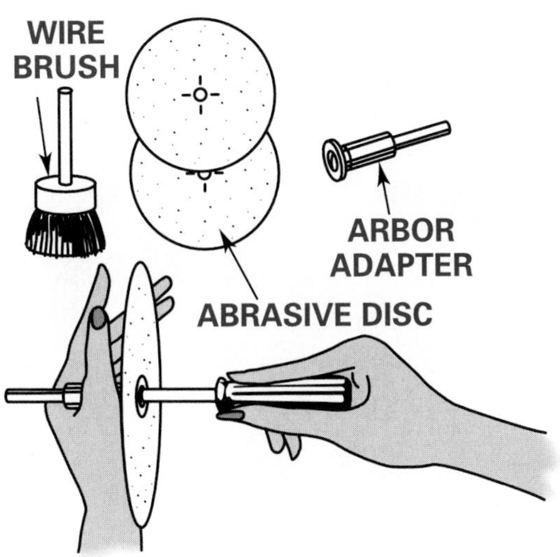

Fig. 6-12—*A wire brush usually has a shank that is installed directly into the chuck. Sanding discs and other abrasive and polishing devices must be attached to an arbor adapter before being installed in the chuck.*

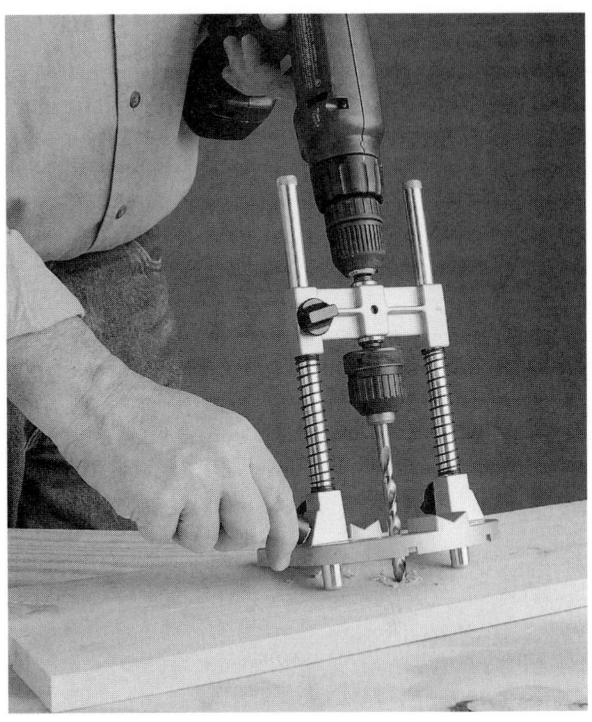

Fig. 6-14—*Drill guides help drill perfectly aligned holes. They can also be used for drilling holes at an angle.*

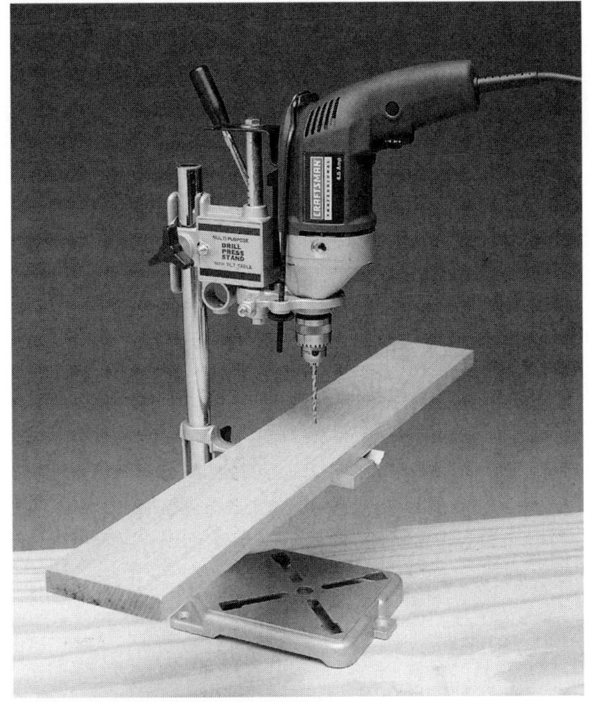

Fig. 6-13—*With a drill stand, you can convert your power drill into a small drill press.*

Fig. 6-15—*A commercial depth stop.*

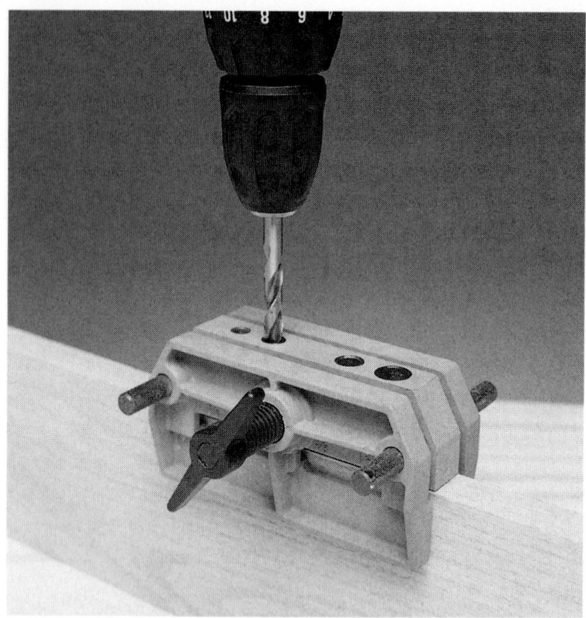

Fig. 6-16—*A doweling jig guides the drill to cut straight perpendicular holes.*

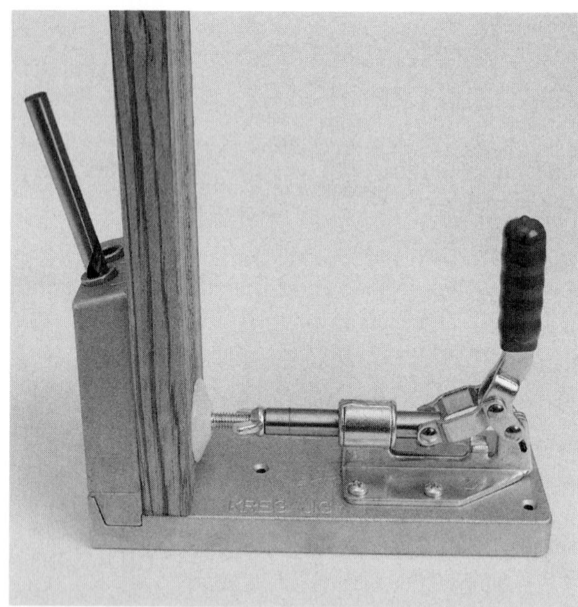

Fig. 6-17—*A pocket-hole jig.*

- **Doweling jigs** attached to the edge of the stock guide the drill to cut dowel holes that are perfectly straight and perpendicular. Fig. 6-16.

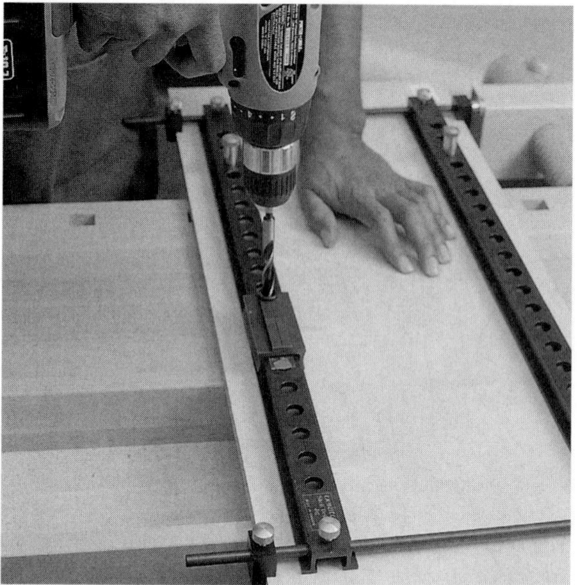

Fig. 6-18—*The shelf-drilling jig makes it very easy to accurately drill evenly spaced holes.*

- **Pocket hole jigs** can be used as an aid for drilling holes that hold two pieces of wood together, such as the rails to the underside of the table. Fig. 6-17.
- *Shelf-drilling jigs* assist in drilling a series of evenly spaced holes, such as for shelf pins for adjustable bookshelves. Fig. 6-18.

DRILLING A HOLE

Whether drilling with hand tools or with power drills, the basic steps are the same. The steps for straight drilling are described below.

1. Mark the location of the center of the hole. First, use a measuring tape, straight edge, and a pencil to locate and mark the center of the hole with two crossed lines (+). The center of the crossed lines indicates the center of the hole. Then use a scratch awl to punch a small hole in the center of the crossed lines. Fig. 6-19. This starter hole makes it easier to start the drill bit on track.

WOODWORKING TIP

Controlling the Depth of Holes

While there are a number of depth stops that are available commercially, you can also make a simple depth stop in the shop from a piece of dowel rod. Cut the dowel rod to cover part of the drill bit like a sleeve, so that the length of the exposed portion of the bit equals the desired depth of the hole to be cut. Fig. A.

A simpler, but less accurate, quick-fix depth stop can be made by placing a piece of masking tape on the drill bit to indicate the correct depth. Fig. B.

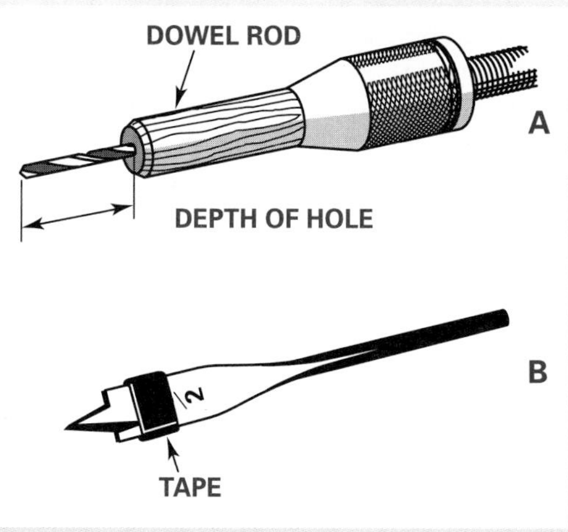

Safety First

► *Drilling Holes*

- If using an electric drill, be sure the drill is disconnected from the power source before installing drills, bits, or other tools. If using a cordless power drill, remove the battery pack.
- Make sure that the twist drill or bit is properly sharpened before installing it in the chuck.
- Be sure the drill or bit is tightly fastened into the chuck. Rotate the chuck by hand to make sure the tool runs straight.
- If a chuck key is used, make sure it has been removed before starting the drill.
- Clamp the workpiece in a vise or to the top of a workbench before drilling.
- Use a scratch awl or a punch to mark a starter hole.
- When drilling a through hole, be sure to attach a piece of scrap wood to the exit side of the workpiece.
- Wear eye protection.

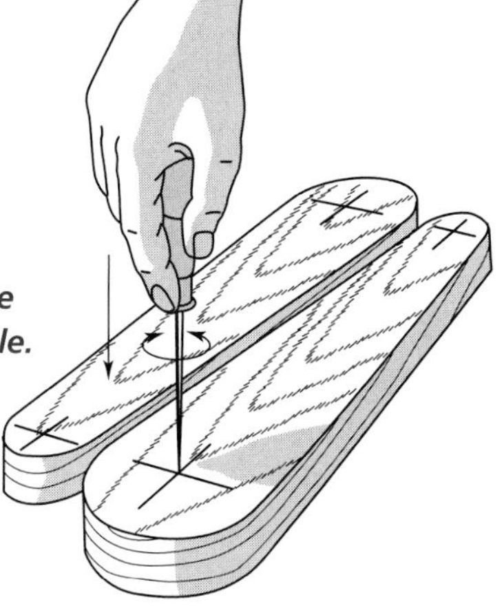

Fig. 6-19—*After marking off the center of the hole, use a scratch awl to make a starter hole.*

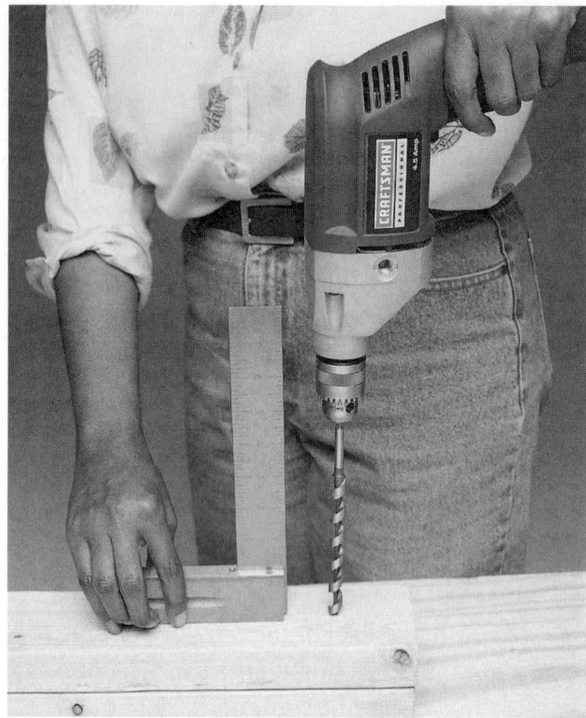

Fig. 6-20—*Checking to make sure the drill bit is square with the work.*

2. Choose the correct size bit and fasten it in the drill. If drilling to a desired depth, attach a depth stop.
3. Clamp the workpiece in a vise or to the top of the workbench.
4. Place the point of the bit in the starter hole. Pressing straight down (or straight ahead, if drilling horizontally), begin to drill. Make sure you keep the drill at a right angle to the workpiece throughout the drilling operation. Use a try square to check the angle between the drill bit and the workpiece. Fig. 6-20.
5. When drilling a through hole, you will need to prevent tearout as the drill nears the bottom of the workpiece. Clamp a piece of scrap wood onto the exit side of the workpiece. If using a twist drill, drill through the workpiece into the scrap wood. Fig. 6-21. If using an auger bit, stop drilling when the brad point begins to enter the scrap wood. Then remove the scrap wood; turn the workpiece over; and begin to drill from the exit side, using the brad point exit hole as your starter hole. (Refer again to Fig. 6-21.)
6. If drilling a large, deep hole, periodically remove the bit from the hole to remove the waste wood and to allow the bit and the wood to cool.

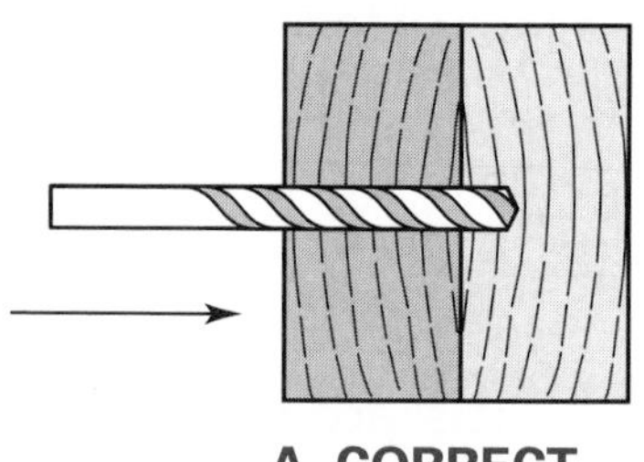

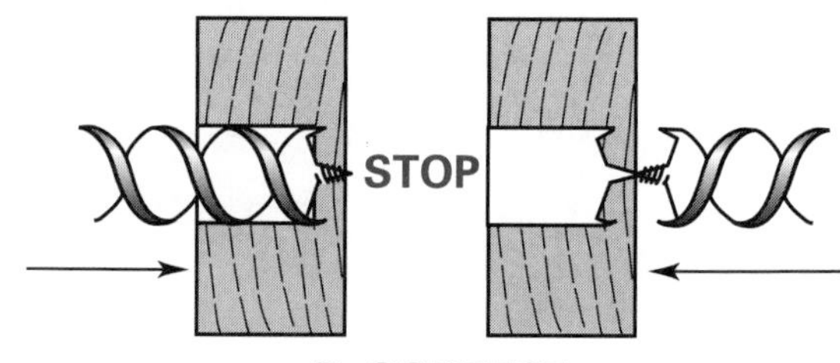

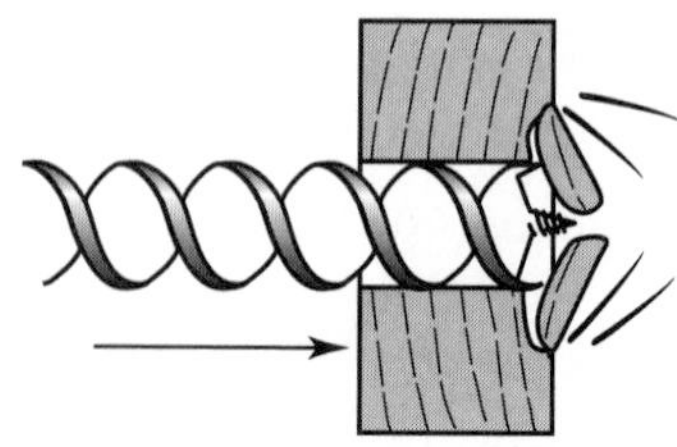

Fig. 6-21—*When drilling a through hole, always clamp a piece of scrap wood to the exit side of the workpiece. (A) If using a twist drill, drill through the stock into the scrap wood. (B) If using an auger bit, stop when the brad point begins to enter the scrap wood. (C) Don't drill completely through the stock when using an auger bit.*

CHAPTER 6

Review & Applications

Chapter Summary

Major points from this chapter that you should remember include:

- Hand tools for drilling include the automatic (or push) drill, the hand drill, and the brace. These tools drive a drill point, twist drill, or auger bit into the wood to cut the desired size hole.
- Power drills may be electric plug-in models or cordless models with a rechargeable battery pack.
- A number of accessories are available to make a power drill more versatile or to help ensure precision drilling.
- The basic steps for drilling are the same, whether using hand tools or power drills.
- There are a number of safety precautions to keep in mind when drilling.

Review Questions

1. Describe an automatic drill and explain how it is used.
2. What is a good size power drill for woodworking? What are the benefits of the variable-speed and reversible features on a power drill?
3. What is the purpose of a depth stop? Describe one commercial and one handmade depth stop.
4. What power drill accessory can you use to help drill a series of evenly spaced holes?
5. How can you check to make sure you are drilling straight?
6. Explain how you drill a through hole when using an auger bit.

Solving Real World Problems

Eric is building a bookcase with five movable shelves. Each shelf will be supported on pins that are placed in holes drilled inside the bookcase sideboards. Eric has to drill a long line of ¼″ holes spaced one inch apart. The holes must not go through the board that makes up the side, and the drill must cut a very clean hole on the surface because each hole will be visible. What can Eric use to give him the proper spacing and location for drilling the holes? How can he control the depth of the holes? What type of drill bit should he use to get the cleanest possible cut?

CHAPTER 7

Planing, Chiseling, and Sanding

YOU'LL BE ABLE TO:

- Plane the surface of a piece of stock using proper planing techniques.
- Use a chisel correctly, observing all safety rules.
- Sand the surface of a piece of stock, using proper sanding procedures.
- Operate a portable belt sander, using proper sanding techniques and observing all safety rules.

Look for These Terms

- belt sander
- chisel
- detail sander
- grit
- hand plane
- hand scraper
- orbital sander
- random-orbit sander
- sanding
- warp

Planing and chiseling are processes used to shape wood by removing excess material. Planing also helps to smooth surfaces. Sanding is a process used to smooth surfaces and prepare them for finishing.

PLANING

Most wood has been surfaced on two sides (S2S) at the mill. However, if you look closely at the surfaces, you will see small mill or knife marks (waves) made by the rotating cutter of the planer, or surfacer. One way to remove these marks is to use a plane. If you do not, these marks will show up after finishing. Fig. 7-1.

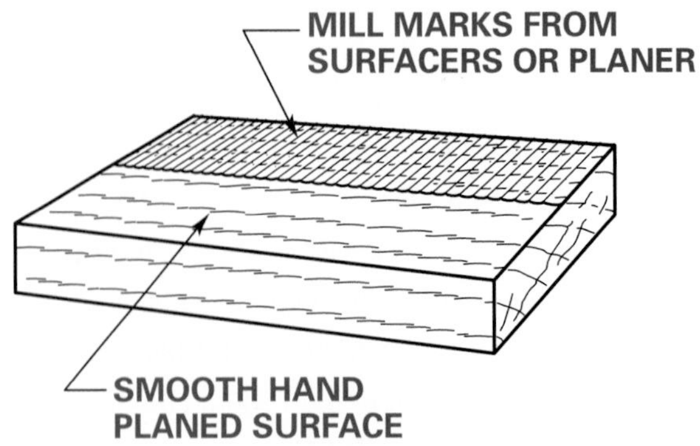

Fig. 7-1—*Mill or knife marks made by the surfacer at the mill must be removed by planing.*

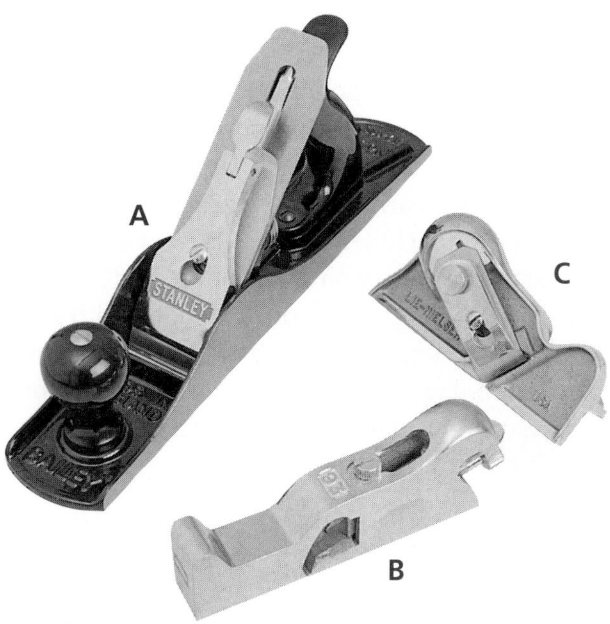

Fig. 7-2—*Kinds of planes. (A) Jack plane. (B) Rabbet plane. (C) Edge plane.*

Fig. 7-3—*A portable electric plane makes planing the edge of a piece of wood much easier and more accurate.*

Kinds of Planes

Hand planes are cutting tools used to shape and smooth stock. Fig. 7-2. Four kinds of hand planes are commonly used for planing surfaces and edges. *Jack planes,* from 11½ to 15 inches long, are used for general planing. *Smooth planes,* from 7 to 9 inches long, are used for general planing and finishing work. *Fore planes,* from 18 to 24 inches long, are good for planing large surfaces and edges. *Jointer planes* also range from 18 to 24 inches in length. They are especially useful for planing edges straight when jointing (making joints on) long pieces of stock.

Portable power planes are also available. These can be used to perform most planing operations quickly and accurately. Fig. 7-3. They can make cuts up to 3 ¼ inches wide and ⅛ inch deep.

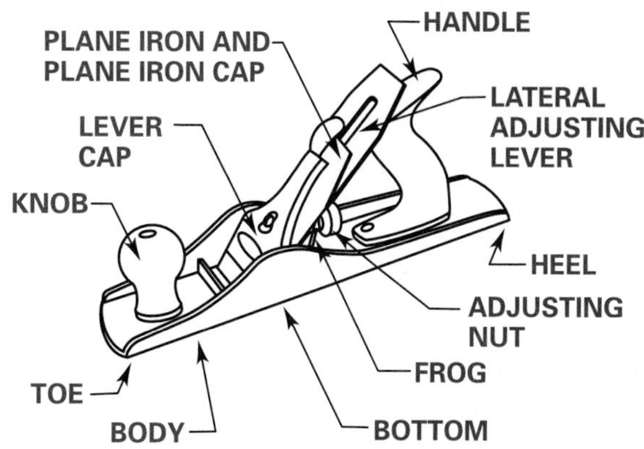

Fig. 7-4—*Parts of a jack plane.*

Assembling and Adjusting a Plane

Assemble the plane as shown in Fig. 7-4. To make adjustments, turn the plane upside down with the bottom about eye level. Fig. 7-5 (page 144). Turn the brass knurled nut until the *plane iron* (cutter) appears just beyond the bottom of the plane. Then, using the lateral adjustment lever, move the blade to one side or the other until it is parallel to the bottom.

Test the plane on a piece of scrap stock to see how it cuts. When truing up a surface and making it smooth, a light cut that forms a feathery shaving is best. For rough planing and when much stock is to be removed, set the plane to take a deeper cut.

Safety First

► *Using a Plane*

- Wear eye protection when planing.
- Secure stock in a vise.
- Make adjustments carefully. The plane iron is sharp.

Fig. 7-5—***Adjusting a plane.***

Planing a Surface

The face surface should be free of flaws and have the most interesting grain. Before beginning to plane, inspect the surface for *warp.* **Warp** is any variation from a true, or plane, surface. It includes crook, bow, cut, wind (twist), or any combination of these. Fig. 7-6. By using a straightedge or the blade of a try or framing square, you can check to see if a board is cupped. Fig. 7-7. Consider how to move the plane on the board to correct its faults. Then begin planing.

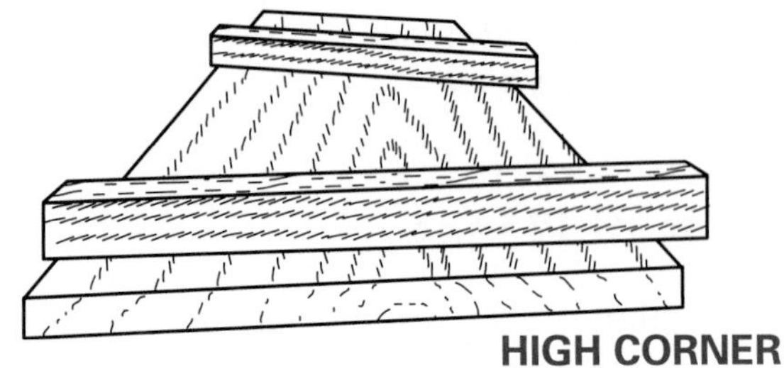

Fig. 7-6—***Stock placed across the ends of the board will show wind or twist.***

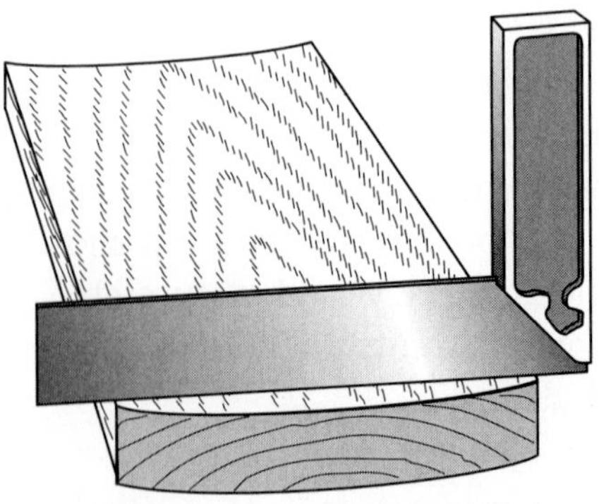

Fig. 7-7—***Using a try square will show you if the board is cupped.***

If you are right-handed, grasp the knob of the plane in your left hand and the handle in your right hand. Stand just behind the work with your left foot forward. If you are left-handed, do just the opposite.

Always plane with the grain. If you plane against it, you will roughen the surface. Fig. 7-8.

Place the toe, or front, of the plane on the board. Apply pressure to the knob at the start of the stroke. As the whole base contacts the wood, apply even pressure to the knob and handle. Then, as the plane begins to leave the surface, apply more pressure to the handle. Fig. 7-9.

Lift the plane off the board on the return stroke. Don't drag it back. This will dull the blade. Sometimes the plane will cut more easily if you take a shearing cut at an angle.

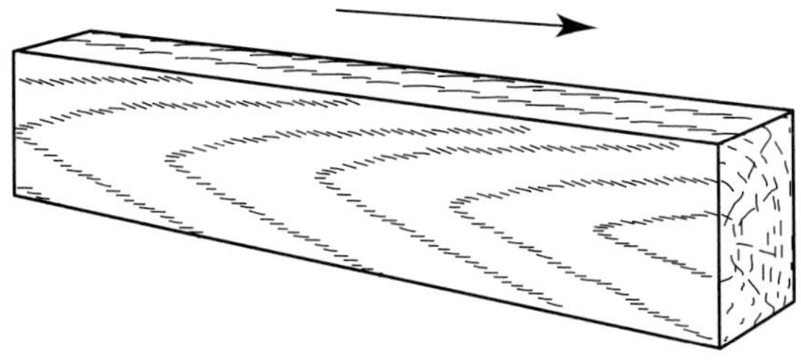

Fig. 7-8—*Plane with the grain.*

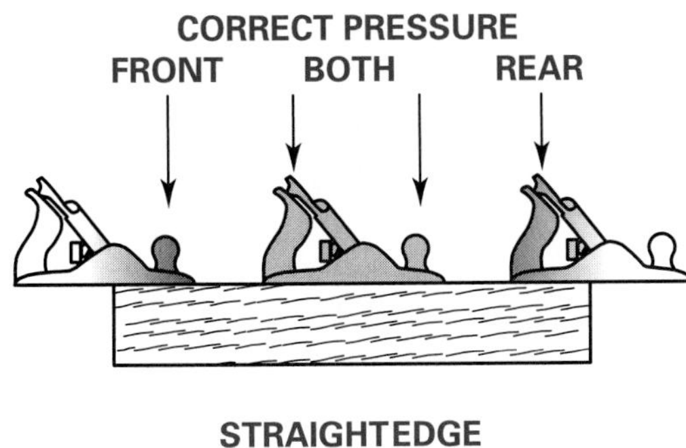

Fig. 7-9—*The proper method for applying pressure when planing.*

Swing your body back and forth as you plane. At the same time, use a forward motion with your arms. Work across the board gradually. High points will take more planing than other areas.

Once the planed area gets smooth, check it with a straightedge. The straightedge will touch any high points. Figs. 7-10 and 7-11.

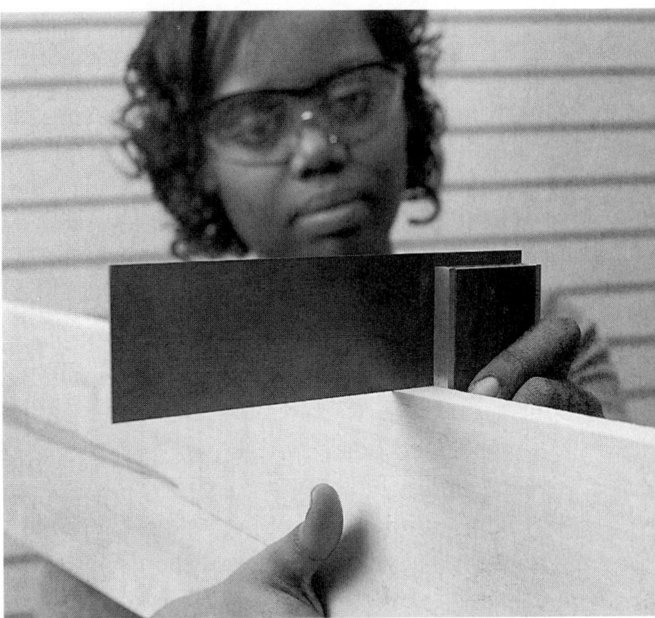

Fig. 7-10—*Using a try square to check the joint edge for squareness. Move the try square to make certain that the edge is square along its total length.*

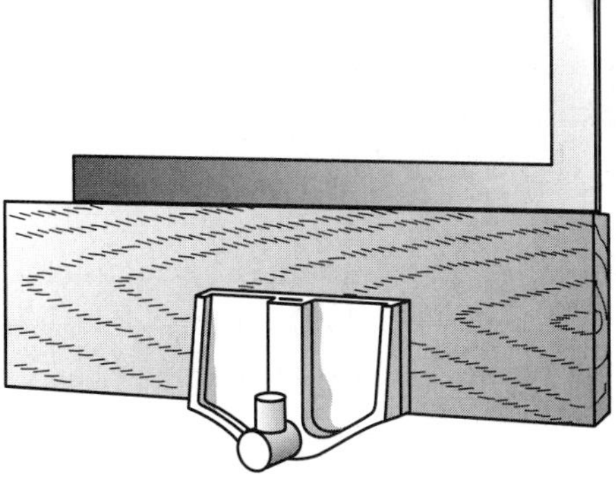

Fig. 7-11—*Using a square to check an edge for straightness. This is especially important on long stock. Check to see whether any light shows between the square (or any straightedge) and the planed edge.*

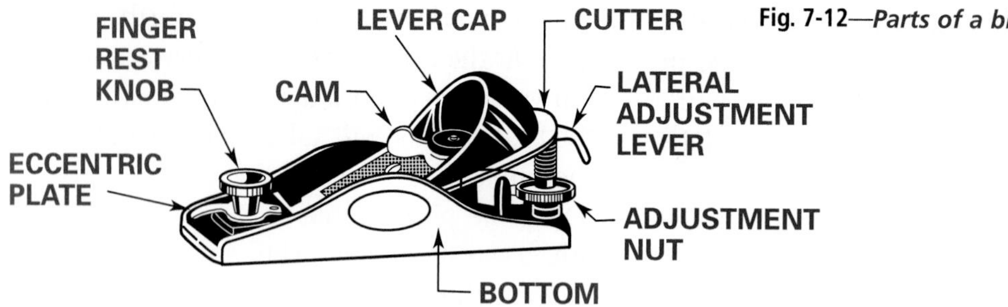

Fig. 7-12—*Parts of a block plane.*

Planing End Grain

End grain is produced when stock is cut across the grain. In planing end grain, you actually cut off the tips of the wood fibers. This takes a very sharp plane iron. One of the best planes for this is the *block plane.* Fig. 7-12.

Skill in cutting end grain properly takes effort and experience. You should take shallow cuts. This keeps the plane from jumping. Planing completely across the end will usually split the wood. To avoid this, use one of the following methods:

- Plane about halfway across the stock. Then lift the handle of the plane slowly. Begin at the other edge and do the same thing in the opposite direction. Fig. 7-13. Check for squareness with the working face and working edge. Fig. 7-14.
- Plane a short bevel on the waste edge of the stock. Then begin from the other edge to plane all the way across.
- Get a piece of scrap stock exactly the same thickness as the piece you are working. Lock it in the vise just ahead of the piece you are planing. By doing this, you have actually extended the end grain. Then you can plane all of the way across the end grain of the first board without splitting the piece.

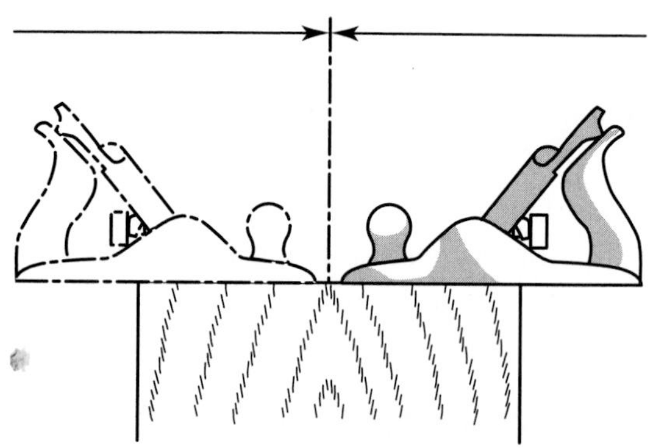

Fig. 7-13—*Plane end grain halfway across the stock first in one direction and then in the other.*

Fig. 7-14—*Checking an end from the face surface.*

Scraping a Surface

For a really fine surface on open-grain wood such as oak, mahogany, and walnut, plane the wood first. Then scrape it with a **hand scraper.** Fig. 7-15. Scraping removes the small imperfections left by the plane. A scraper will take a wide shaving that is so thin you can see through it. Fig. 7-16. It will smooth irregularities prior to sanding.

Safety First

► *Using a Scraper*

- Wear eye protection.
- Use special care when working with tools that are sharp.

Fig. 7-15—*A hand scraper.*

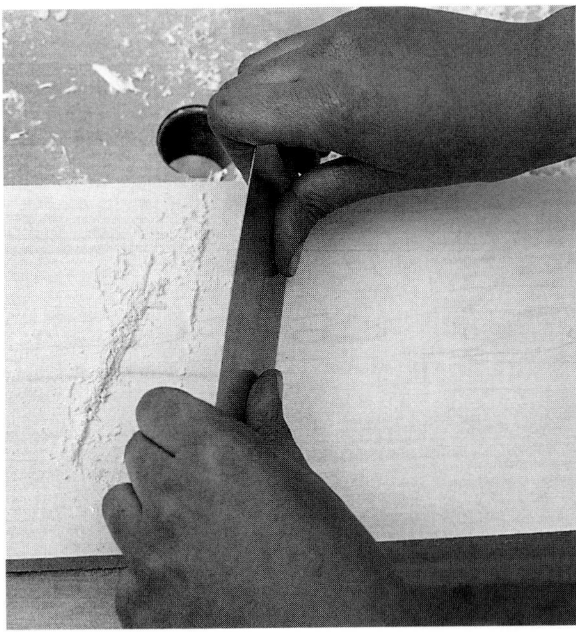

Fig. 7-16—*Scraping a surface with a hand scraper. Note that the scraper is held in both hands and tipped at an angle to the wood surface.*

Some woods, such as curly maple and cedar, can't be planed very well. However, they can be scraped very smooth. It is always necessary to sharpen the scraper before each use, and frequently during its use.

CHISELING

A **chisel** is a straight-edged cutting tool used to shape and trim wood. The three most common chisels are the mortise, butt, and paring chisels.

- *Mortise chisels* are very sturdy tools with thick blades ground flat on the sides. They come in widths from ⅛ to ⅝ inch. They are used for clearing out the mortise to fit a tenon. See page 205.
- *Butt chisels* have short blades about 3 to 4 inches long with beveled sides. They come in widths from ½ to 1½ inches. These chisels are used to shape joints and to cut recesses for hinges. Fig. 7-17.
- *Paring chisels* are lightweight chisels for final trimming and fitting. Fig. 7-17.

Having several chisels of different sizes will help you complete chiseling tasks efficiently. Different chiseling techniques are used to cut with the grain and across the grain. Figs. 7-18, 7-19, 7-20, and 7-21 (pages 148-149).

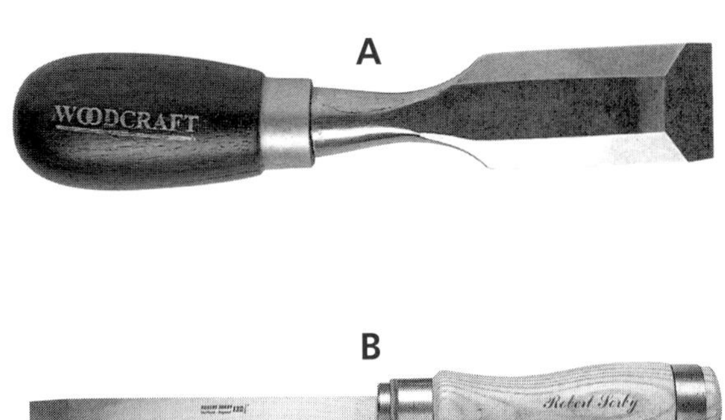

Fig. 7-17—*Kinds of chisels. (A) Butt chisel. (B) Paring chisel.*

Safety First

► *Using a Chisel*

- Secure the workpiece with clamps so that it cannot move.
- Keep both hands behind the cutting edge and the chisel away from the body.
- Hold the tool correctly. If you are right-handed, use your right hand to push the handle while the left hand guides the blade. When using a *mallet,* the left hand should hold the chisel, while the right hand taps it with a mallet. If you are left-handed, do just the opposite.
- Protect the handle by using a wood, rubber, or plastic mallet (Fig. 7-22) or a *soft-faced hammer* (Fig. 7-23). Never use a metal hammer.
- Always hit the tool squarely on top of the handle.
- Never allow the edge to touch other tools. Avoid dropping the chisel on the floor or any other hard surface.
- When finished, store the chisel safely in a tool rack.

SCIENCE Connection

Tool Steel

A chisel is only as good as the steel used in the blade. Chisels and other edge tools are usually constructed of steel that has a high carbon content. A carbon content of .55 percent to .95 percent is preferred. Higher carbon content results in harder tool steel.

Tool manufacturers often include specific metals such as manganese and chromium in the steel. These metals make the tool hard enough to hold an edge, but not so hard that it becomes brittle and breaks easily.

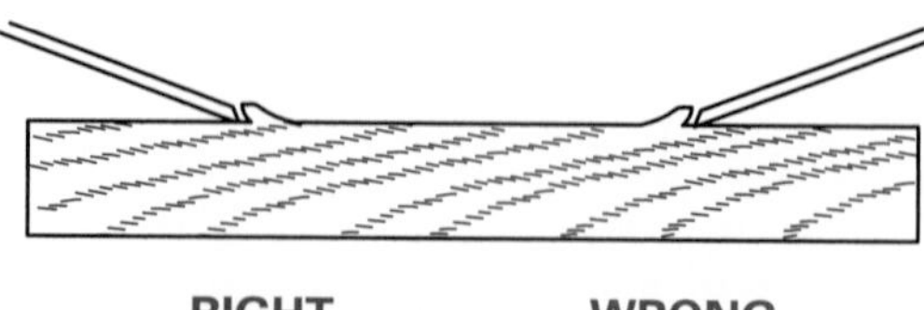

Fig. 7-18—*Cutting action of the chisel when cutting horizontally with the grain. Always cut with the grain, never against it, for this procedure.*

Fig. 7-19—*Cutting vertically across the grain. One hand guides the chisel while the other hand applies pressure to the handle. The chisel should be tilted slightly to give a sliding action.*

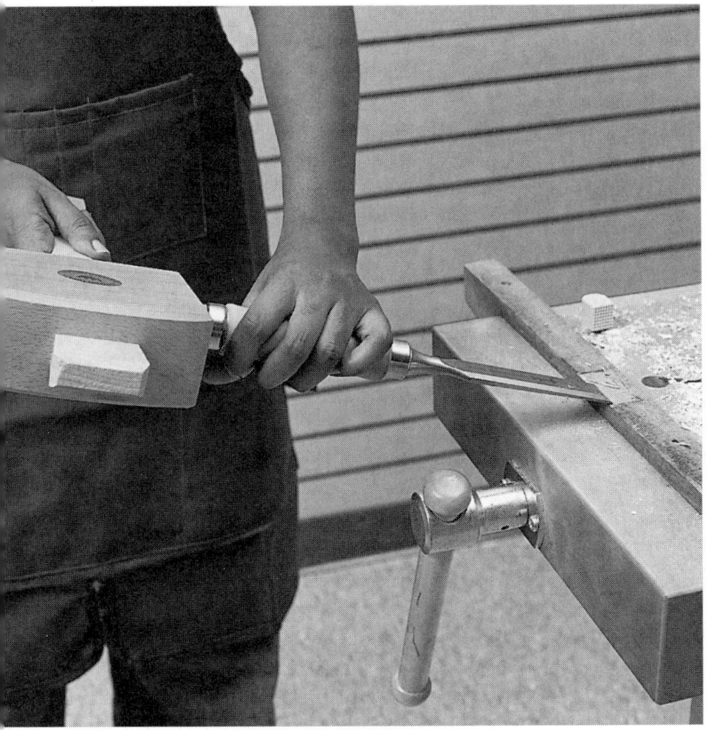

Fig. 7-20—*Roughing out a cut across grain. Hold the chisel with the bevel side down. Use a mallet for driving.*

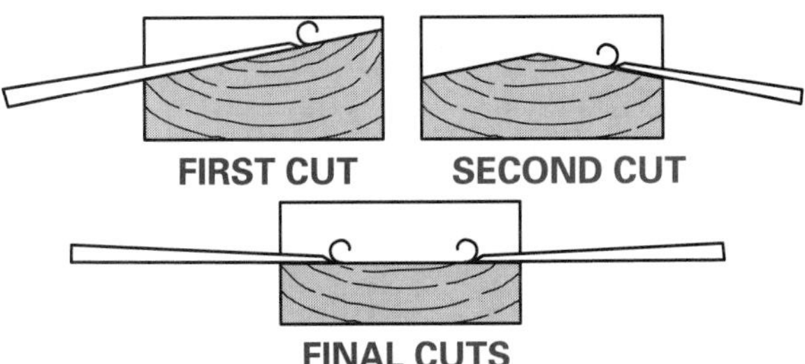

Fig. 7-21—*Proper method of cutting a lap joint from both sides, leaving the center high and then trimming the center down. See Chapter 11.*

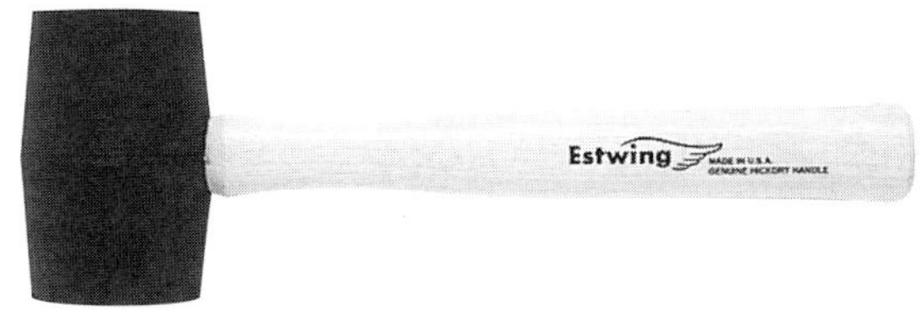

Fig. 7-22—*A rubber mallet.*

Fig. 7-23—*A soft-faced hammer.*

SANDING

Sanding is the process of smoothing wood by rubbing it with an abrasive. It is done to prepare the wood surface for finishing, not to form or shape it. All planing, cutting, and forming should be completed before sanding. Sanding should never be done in place of using cutting tools.

Abrasives

Abrasive sheets are produced by bonding (gluing) abrasive material to paper or cloth backing. There is really no such thing as "sand" paper, though the name is given to several types of abrasive sheets. Fig. 7-24 (page 150). Different types of abrasives are used. *Garnet* is quite durable and fairly inexpensive. *Aluminum oxide* is generally more expensive but it is also durable and cuts quickly. A still sharper abrasive is *silicon carbide.* However, it tends to be brittle.

Fig. 7-24—*Abrasives are produced in many forms to serve different purposes and to fit various machines.*

The usefulness of the abrasives depends in part on the weight of the backing to which they are mounted. The backing papers of sandpaper are produced in four weights: A, C, D, and E. A is the lightest, and E is the heaviest. Abrasives with heavier backing generally last longer when attached to soft pads.

Grades of Abrasives

Abrasives are made in a wide range of grades from very coarse to very fine. The grades are based on the size of the grains of the abrasive. The size of the grains is referred to as **grit.**

Grades range from 40-grit (coarse) to 600-grit (very fine). Most woodworking jobs require abrasives in the 40- to 180-grit ranges. Begin with fairly coarse grit and use increasingly finer grit as you work.

When you begin sanding, use abrasives in the 40- to 60-grit range to level surfaces and remove deep scratches or other imperfections. Then use abrasives in the 80- to 120-grit range to begin smoothing and to remove the lighter scratches left by the coarser grits. Finally, use abrasives in the 150- to 180-grit range to prepare the surfaces for finishing processes.

Types of Sandpapers

Abrasives mounted on paper are used most often. There are two types of sandpapers. These are *open coat* and *closed coat.* On open-coat sandpapers, abrasives are spaced far apart. These sandpapers are good for working with softwoods. They do not clog easily.

The abrasives on closed-coat sandpapers cover the surface. These are good for working with hardwoods.

Information regarding weights, classifications, and other information about the various sandpapers are printed on the back of the sandpaper. Fig. 7-25.

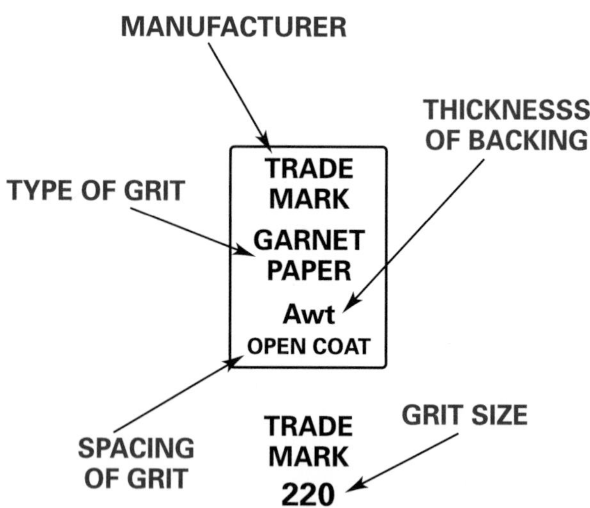

Fig. 7-25—*The label printed on the back of sandpaper provides information about the sandpaper.*

Safety First

► ***Sanding Stock***

- Wear eye protection.
- Protect your lungs by wearing a dust mask.

Sanding Tips

Careful sanding makes a big difference in the final appearance of a project. The following tips will help:

- Always sand with the grain, except when working on end grain.
- Sand end grain in only one direction. Lift the sandpaper on the return stroke.
- If hand sanding, exert even pressure over the entire surface being sanded. Fig. 7-26. This is especially necessary on thin veneers.
- Take special care to prevent a round corner or wavy appearance on edges and corners.
- Never use a grit coarser than necessary. Work from a coarser paper to a finer one.
- Always dust, wipe, or blow off the stock being sanded when changing grits. Projects that are to be finished must be absolutely dust free.
- Be careful when sanding an edge. It is important to sand an edge square. After the edge has been sanded, round the sharp edge slightly by drawing the sandpaper over it.
- A convex (rounded outward) surface such as a rounded end can be sanded more satisfactorily by holding the paper in your fingers or the palm of your hand. Fig. 7-27. Special blocks can be used for curved surfaces. Fig. 7-28 (page 152).

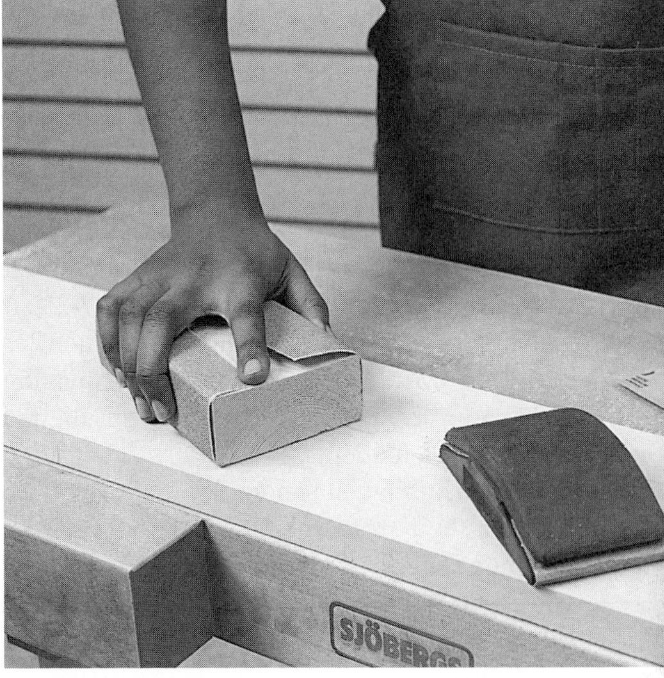

Fig. 7-26—***Attaching sandpaper to a sanding block makes it easier to hold while sanding.***

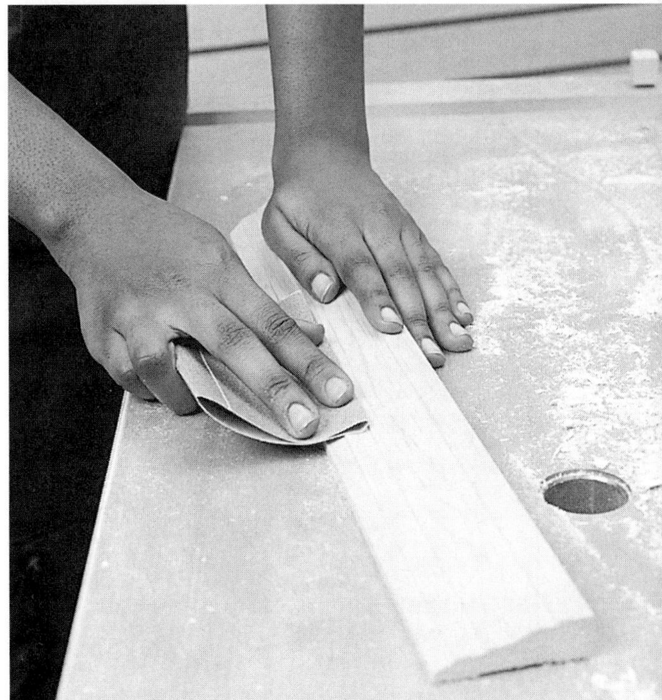

Fig. 7-27—***Sanding an uneven surface.***

Portable Power Sanders

There are three main types of portable power sanders. These are the belt, orbital, and random-orbit sanders.

Belt Sander

A **belt sander** has a replaceable abrasive belt that is turned around two rollers. Fig. 7-29. It can perform rough sanding tasks such as removing waste wood as well as fine finishing tasks. You can use it to even out misaligned surfaces and fix dents.

When using a belt sander, let the rear of the moving belt touch first. Then level the machine. Keep it flat on the work surface. If you don't, depressions could result. To remove stock evenly, keep the sander moving steadily. Fig. 7-30.

The rotating action of the belt causes the abrasives to cut in a straight line. Because of this, you should align and move the sander with the wood grain.

Safety First

► *Using Portable Power Sanders*

- Make sure the abrasive sheet or belt is in good condition and properly installed.
- Use a dust mask to filter out the harmful, fine particles produced by sanding.
- Wear goggles or a face shield.
- Tie back long hair.
- Be sure the switch is off before connecting the sander to a power source.
- Keep your hands and body clear of the moving pad or belt.
- Make sure the cord is out of the way.
- Start the tool above the work surface and lower it slowly. With a belt sander, let the rear of the moving belt touch first. Then level the machine.
- Hold the sander firmly, but let the weight of the sander do the work.

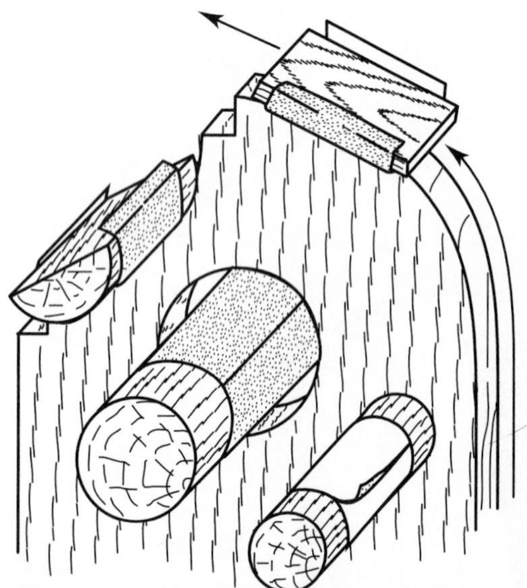

Fig. 7-28—*Sanding a piece with inside and outside curved surfaces.*

Fig. 7-29—*Belt sander.*

Fig. 7-30—*The correct method to begin sanding with a portable belt sander.*

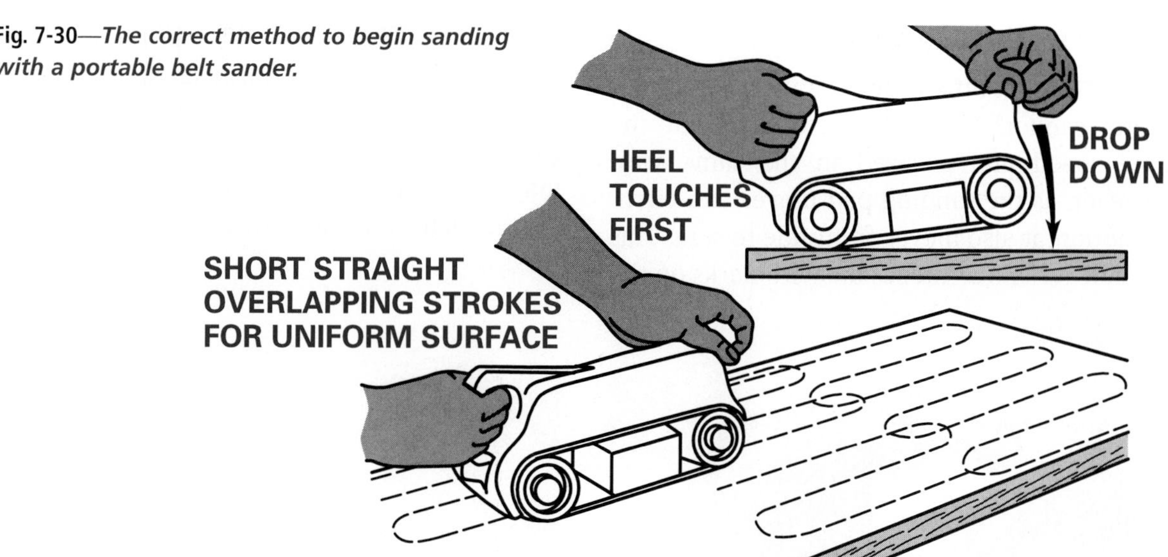

Sanding Square Corners

For pieces that are joined at square corners, it is not entirely possible to move a belt sander with the grain. You will need to first sand one board with the grain, smoothing the other against the grain on the joint end. Then sand the other board with the grain without touching the first board.

The belt sander tends to round edges. To avoid this, allow the belt to extend just beyond the edge. Be careful that the sander does not tilt.

Orbital Sander

The **orbital sander** is so-named because its sanding pad has a circular motion. It can make up to 12,000 orbits per minute. Fig. 7-31.

The two basic sizes of orbital sanders are referred to as ½ and ¼ sanders. The ½ uses one-half of a standard sheet of sandpaper, and the ¼ uses one-quarter of one standard sheet. The ½ sander is controlled with both hands. It is good for sanding large surfaces. The ¼ sander can be controlled with one hand. It is good for sanding in tight spaces such as the insides of drawers.

Fig. 7-31—*Orbital sander.*

Random-Orbit Sander

The **random-orbit sander** is a variation of the orbital sander. Fig. 7-32. It operates at the same speed and is sized in the same way. However, as the sanding pad moves in a circular motion, it also moves from side to side. This action leaves no scratches or swirl marks on the work surface. Because of this, the random-orbit sander is excellent for finish sanding. You may still, however, wish to lightly hand-sand the surface with the grain. Also, a **detail sander** may be needed for spots larger sanders can't reach, such as inside corners. Fig. 7-33.

Fig. 7-32—*Random-orbit sander.*

Fig. 7-33—*A detail sander is perfect for sanding in hard-to-reach places.*

CHAPTER 7

Review & Applications

Chapter Summary

Major points from this chapter that you should remember include:

- Hand planes shape and smooth stock.
- Planing should be done with the grain.
- Planing completely across end grain will usually cause the wood to split.
- Scraping is done to remove imperfections left by the plane.
- Three common chisels are the mortise, butt, and paring chisels.
- Sanding is done to prepare the wood surface for finishing.
- Abrasives are made in a wide range of grades from very coarse to very fine.
- The main types of portable power sanders are the belt, orbital, and random-orbit sanders.

Review Questions

1. Describe how you would plane the surface of a board.
2. Describe the three ways of planing end grain to avoid spitting the stock.
3. In what order would you perform these tasks: sanding, planing, scraping?
4. Name three types of chisels and tell how each is used.
5. Describe briefly how to use a mallet with a chisel.
6. In what sequence would you use the grades of sandpaper to smooth a surface?
7. Describe the technique for operating the portable belt sander.
8. Describe the sanding actions of the three types of portable power sanders.

Solving Real World Problems

Todd's mother has always wanted a mission style sofa table, so Todd decides to build her one for her birthday. To make the large number of mortise-and-tenon joints needed in such a piece, Todd knows he will need some special chisels and a hand plane to fit the tenons to the mortises. When Todd goes to the woodworking tool store, he finds that a wide selection of chisels is available. The selection of hand planes is no smaller. What kind of chisel does he need for the work he wants to do? How is it different from the rest? Looking at block planes, jack planes, smoothing planes, and rabbet planes, Todd is really confused—which does he need?

CHAPTER 8

Butt, Biscuit, and Dowel Joints

Look for These Terms

- biscuit joint
- butt joint
- dowel
- dowel center
- dowel joint
- dowel rod
- edge butt joint
- end butt joint

YOU'LL BE ABLE TO:

- Identify the types of butt joints and tell how a butt joint can be strengthened.
- Make an edge biscuit joint.
- Make an edge dowel joint.
- List the steps in making a dowel joint on a frame.

The simplest joints to make are butt joints. They can be strengthened by adding biscuits or dowels.

BUTT JOINTS

Butt joints are made by joining together the edge, end, or face surface of one piece of wood with the edge, end, or face surface of another piece of wood. Adjoining surfaces must fit flush against each other.

Butt joints are easy to make, but they are the weakest of all wood joints. However, they can be strengthened in various ways.

End Butt Joints

In an **end butt joint,** the end of one piece is connected to the face surface, edge, or end of the second piece. Figs. 8-1 and 8-2. The end piece must be square to the face and edge.

Before assembly, clean the surface. Apply a coating of glue to the mating parts. Position the pieces so that they will line up perfectly and clamp until the glue is dry.

Reinforcing the Joint

An end butt joint is weak but can be strengthened in several ways. Always use fasteners that will give the greatest holding strength. Corner blocks or irons can be used. Fig. 8-3.

Fig. 8-1—*This quilt rack was assembled using end and edge butt joints.*

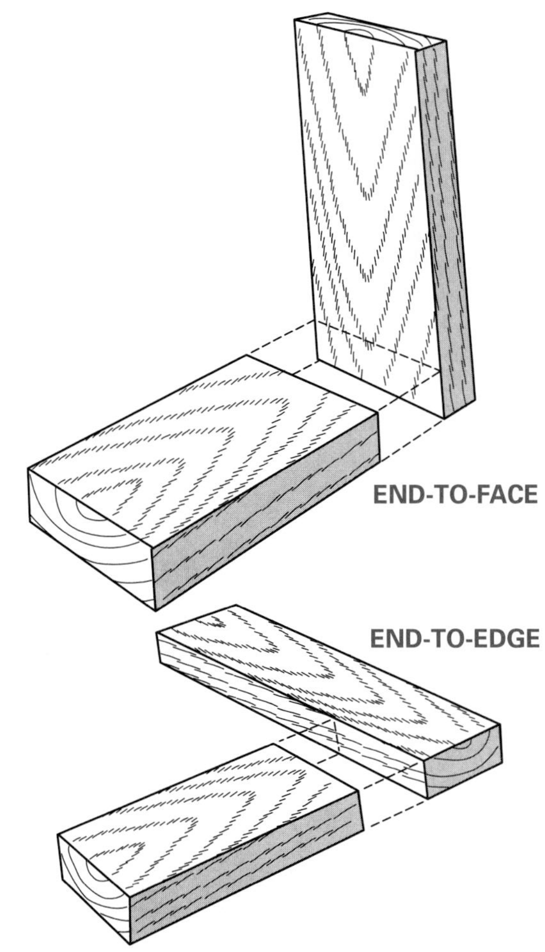

Fig. 8-2—*Butt joints like these are used to join the ends and rails of the quilt rack.*

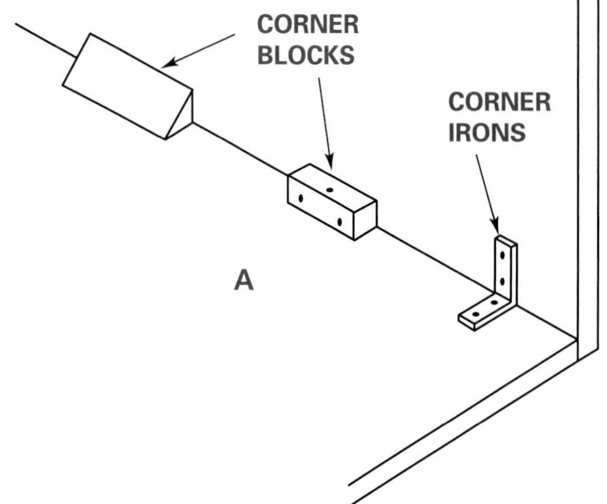

Fig. 8-3— *(A) Corner blocks and irons. (B) It is a good idea to use full length corner blocks when thin material is assembled with butt joints.*

Metal fasteners like nails and screws provide good reinforcement if installed properly. Nails hold best when driven in at an angle. In wood that splits easily, drill a *pilot hole* before driving in the nail. Screws grip better in cross grain than they do in end grain. Select screws that are long enough to provide ample gripping surface within the end grain. Fig. 8-4 (page 158).

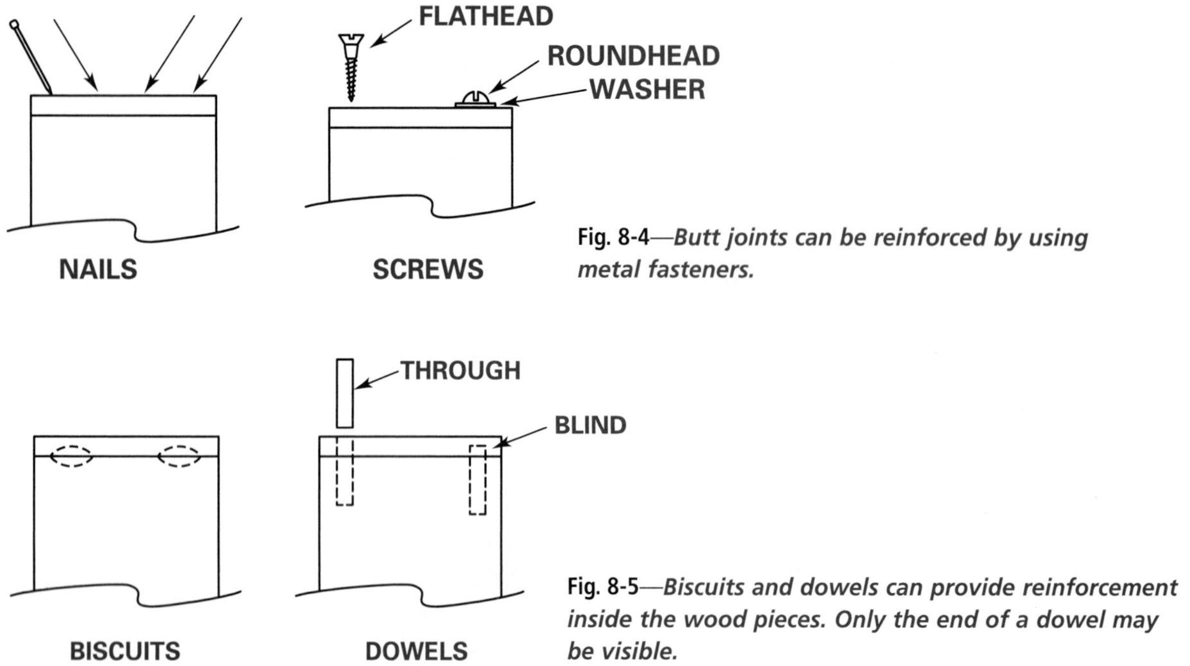

Fig. 8-4—*Butt joints can be reinforced by using metal fasteners.*

Fig. 8-5—*Biscuits and dowels can provide reinforcement inside the wood pieces. Only the end of a dowel may be visible.*

Joints can also be reinforced with biscuits or dowels. These are particularly useful if appearance is important because they can be hidden. Fig. 8-5. Biscuits and dowels are discussed later in this chapter.

Edge Butt Joints

In an **edge butt joint** (edge-to-edge), the edges of two pieces are fastened together. Fig. 8-6. The edge butt joint is used for joining narrow boards to make wider widths for tabletops and other large parts. Strengthening this joint with biscuits or dowels is a good idea.

BISCUIT JOINTS

Biscuit joinery is the ideal way to strengthen butt joints. A **biscuit joint** is a wood joint strengthened by a football-shaped piece of wood called a biscuit. Fig. 8-7.

Biscuits are made of compressed beech. The most commonly used sizes are numbers 0, 10, and 20. Fig. 8-8. Other sizes are available for special needs or to fit special equipment.

Increasing the Holding Power of Screws

The hold of screw threads in the end grain of wood can be strengthened. One way is to insert a dowel into the piece of wood perpendicular (at right angles) to the end grain. The screws go through the end grain and into the cross grain of the dowel. This greatly increases the holding power of the screws in the joint.

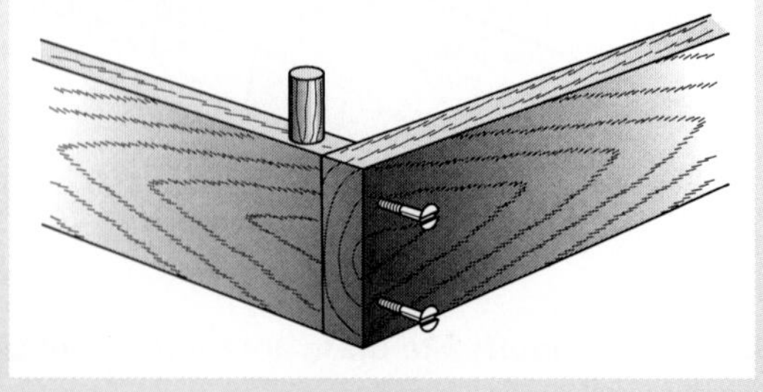

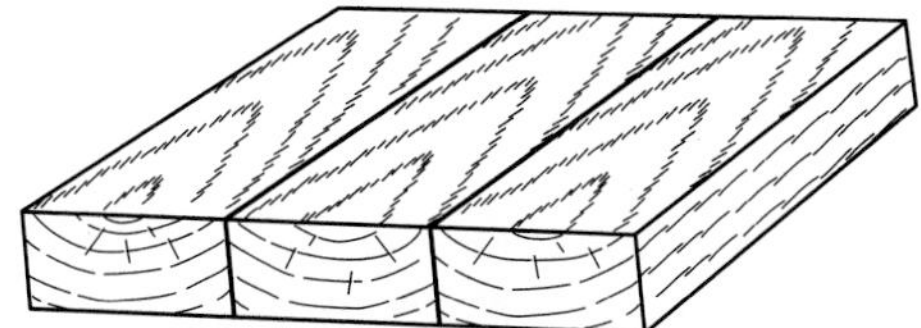

Fig. 8-6—*Edge butt joints.*

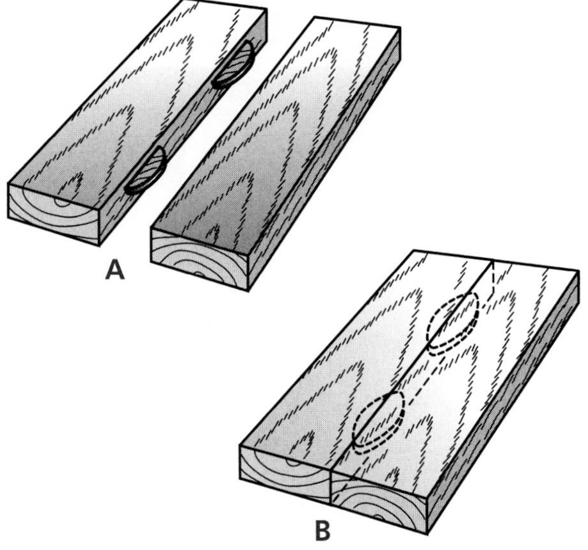

Fig. 8-7—*Edge butt joints can be strengthened using biscuits.*

Fig. 8-8—*Commonly used sizes of biscuits.*

A biscuit machine is used to cut the slots for the biscuits that fit into those slots. Fig. 8-9 (page 160). The head is spring-loaded and has adjustable stops for different depths of cut. The fence can be adjusted to angles other than 90 degrees.

SCIENCE Connection

Storing Biscuits

When beech shavings are compressed to form biscuits, most of the moisture is pressed out. When you add glue to a biscuit, the moisture in the glue causes the biscuit to expand. Unfortunately, glue isn't the only form of moisture that affects the biscuits. Exposure to any type of moisture, including high humidity, can cause them to expand. Therefore, you should keep biscuits in a dry, air-tight container to prevent them from swelling.

Safety First

► *Using a Biscuit Joiner*

- Before using the biscuit joiner, read and understand its operation manual.
- Keep the cutters sharp. Dull cutters require extra pressure when cutting.
- Use clamps so that you can use both hands to guide the tool.
- When operating the joiner, hold it by the D handle and motor housing.

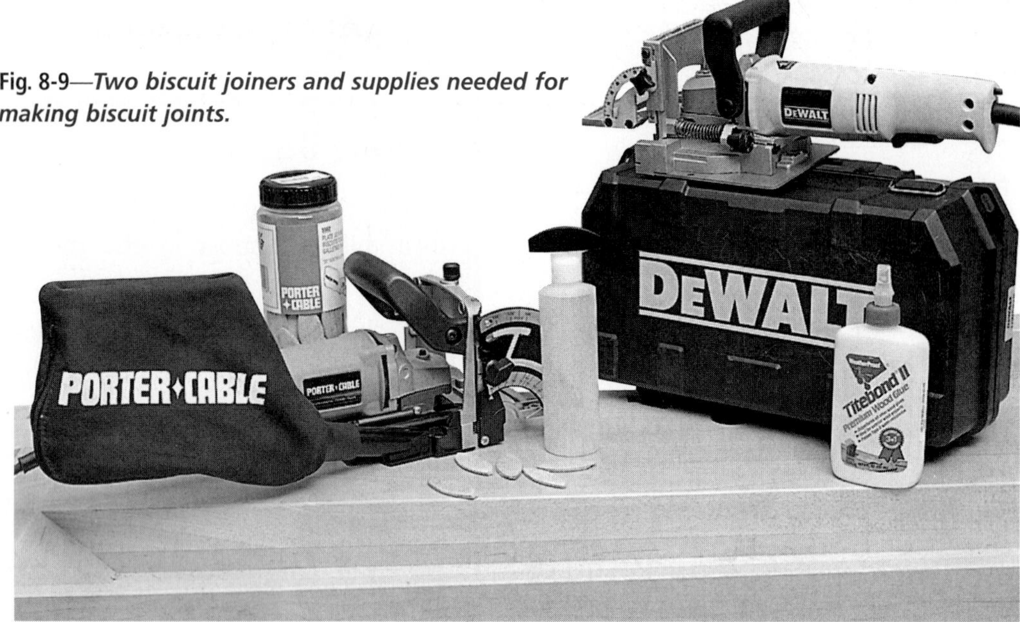

Fig. 8-9—*Two biscuit joiners and supplies needed for making biscuit joints.*

Making a Biscuit Joint

When making a biscuit joint, first lay out the location of the joint on the two adjoining surfaces. Fig. 8-10. Adjust the fences of the machine for vertical location of the slots. For most joints the fence should be at 90 degrees. Adjust for depth of cut, depending on the size of biscuits that are selected. Half of the biscuit fits into each of the matching slots on the pieces being joined.

Next, clamp the two pieces of wood securely to the top of the bench using a vise or clamp. Align the mark on the front of the fence or faceplate with the layout lines on the wood. Turn on the machine and, using a steady forward motion, push the blade into the wood. Fig. 8-11. Slots should be slightly longer than the biscuit that will be used. Repeat for each adjoining slot.

Finally, assemble the joint. Be sure to use the correct size of biscuits for the slots. Apply water-based glue to the slots and the surfaces to be joined. Fig. 8-12. Insert the biscuits into the slots of the first piece and then fit the second piece to it. Because the slots are cut slightly longer than the biscuits, the two parts can be shifted slightly for alignment. (This isn't possible when using dowels.) Clamp the pieces together. Fig. 8-13. The compressed biscuits expand as moisture permeates (soaks into) the wood, making a tight, perfectly aligned joint. Leave clamped until the glue is dry.

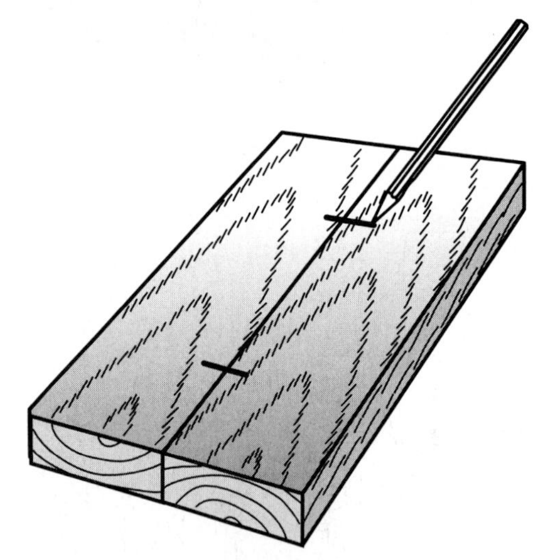

Fig. 8-10—*Marking the edges to show where biscuits will be located.*

Fig. 8-11—*Align the mark on the machine with the layout lines on the wood and make the cut. Be sure the wood is clamped.*

Fig. 8-13—*Assemble the pieces and clamp them together until the glue is dry.*

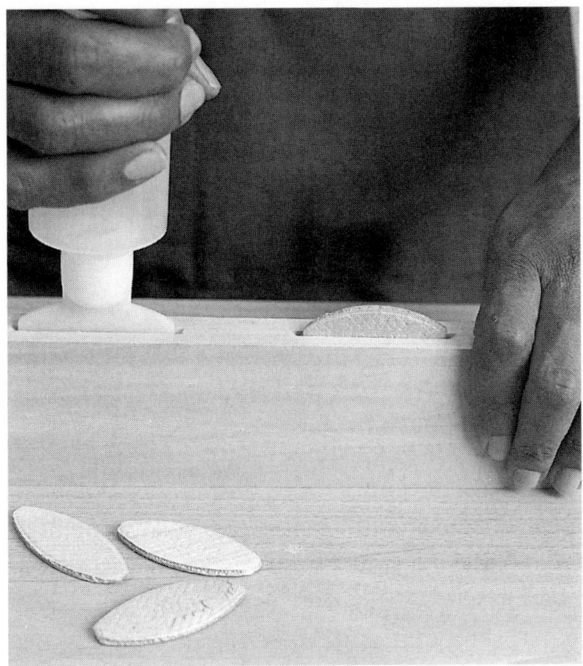

Fig. 8-12—*Apply glue to the biscuits and to the edge surfaces.*

DOWEL JOINTS

Dowels are wood or plastic pins placed in matching holes where the two pieces of a joint are joined. Fig. 8-14 (page 162). They add strength to the joint. Dowels may be used on an end or edge butt joint or a *miter joint* (See Chapter 12.) When dowels are added, the joint may be called a **dowel joint.**

Dowel Materials and Tools

Dowel rod is usually made of birch in diameters from ⅛ to 1 inch in 3-foot lengths. Small dowel pins can be purchased. These are made with spiral or straight grooves and beveled (angled) ends. Fig. 8-15. The grooves allow the glue to flow more freely. The beveled ends make the dowels slip into the holes easily.

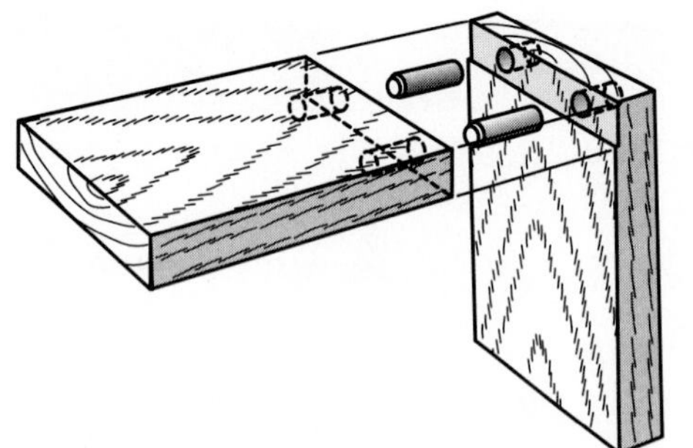
The corner butt with dowels is sometimes found in case construction.

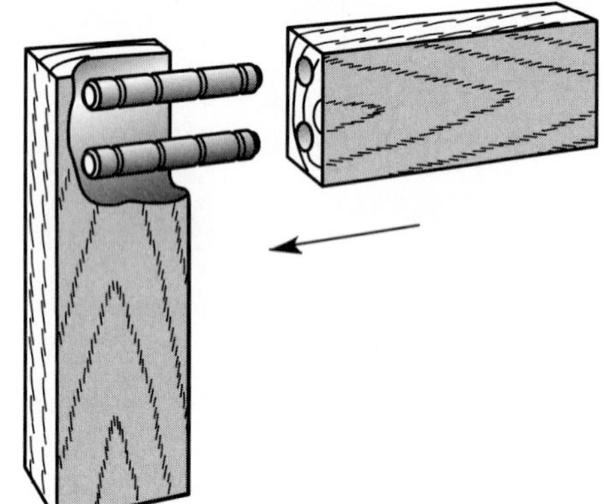
Usually two dowels are used in each corner of a butt-frame joint.

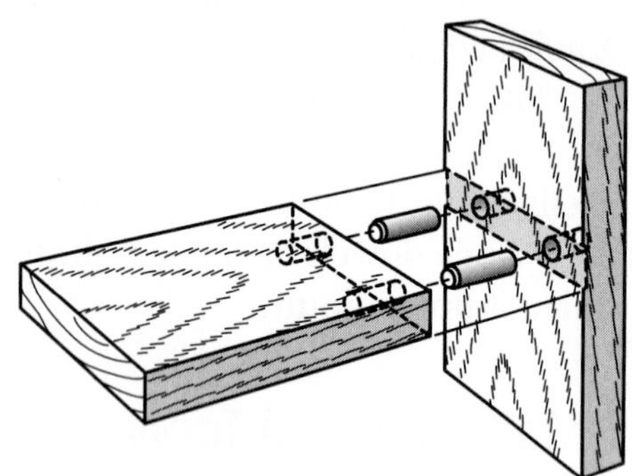
The middle-rail butt with dowels.

Fig. 8-14—*Typical dowel joints.*

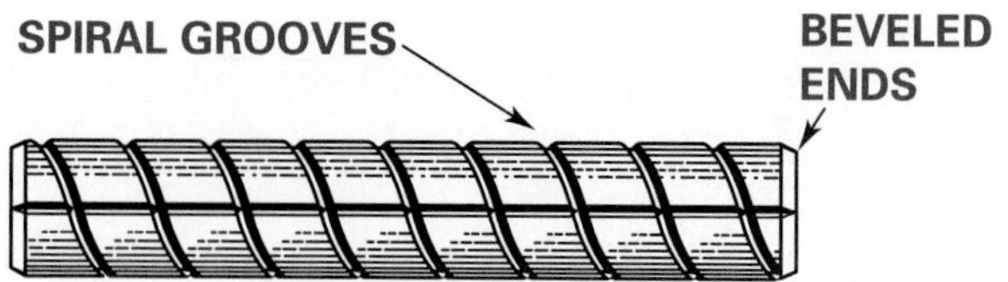

Fig. 8-15—*Dowel pins usually have spiral grooves that help the glue to flow.*

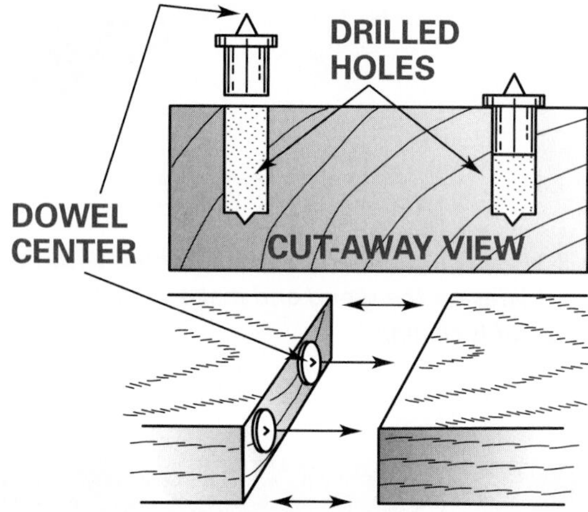

Fig. 8-16—*The locations for the dowels are marked on the first piece and drilled. The dowel centers are put in place. When the two pieces are held together, the dowel centers make indentions on the second piece to show where holes should be located.*

Dowel centers are small metal pins used for matching and marking the location of holes on the two parts of a joint. Figs. 8-16 and 8-17. A *doweling jig* helps identify hole locations and can be used as a guide when drilling the holes. Fig. 8-18.

Making an Edge Dowel Joint

To make an edge dowel joint, clamp the two pieces to be joined with the edges flush and face surfaces out. Using a try square, mark across the edges of both pieces at the several points where the dowels will be. Fig. 8-19. Next, set a marking gauge to half the thickness of the wood. Mark the center locations of the dowel joints.

Decide on the size of the dowel you want to use. The diameter should never be more than half the thickness of the wood. Usually, the dowel should be no longer than 3 inches.

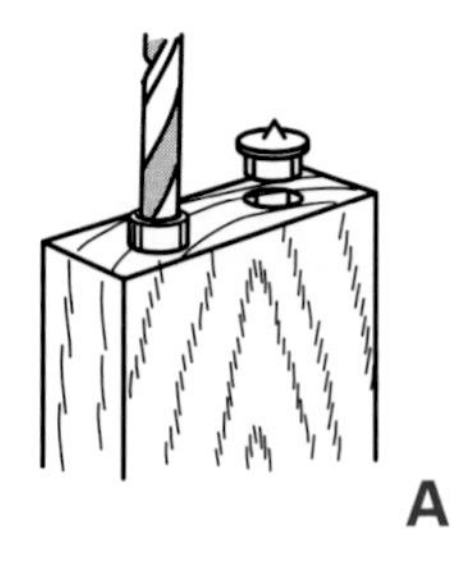

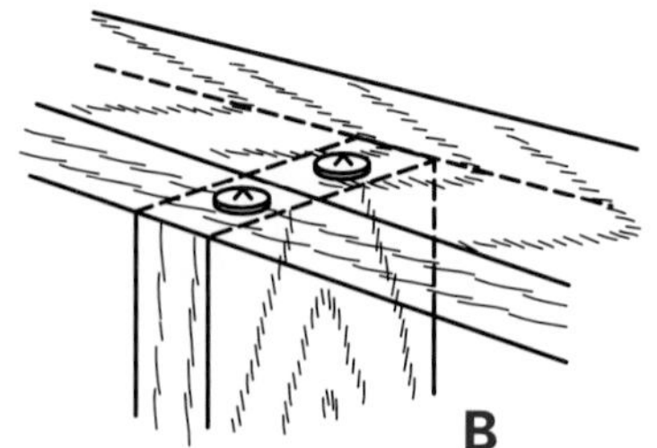

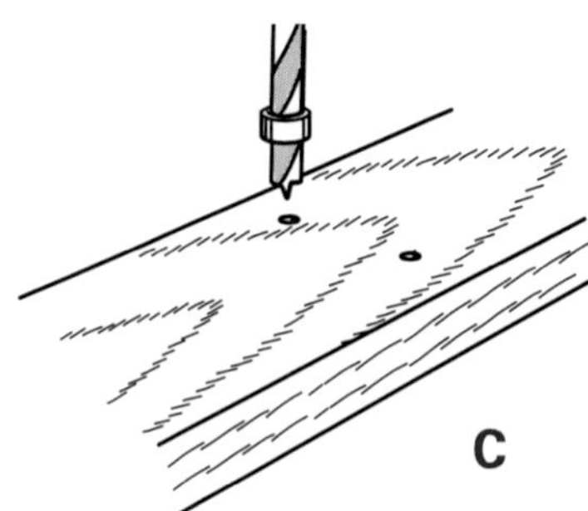

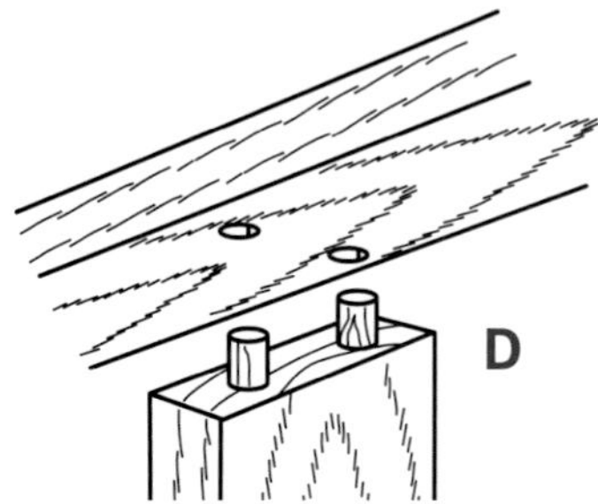

Fig. 8-17—*Installing dowels for an end-to-face butt joint. (A) Note that the holes are drilled in the end pieces first. (B) Then place dowel centers in the holes to locate the holes in the face pieces. (C) Drill these holes to the proper depth. (D) Insert dowels and test for proper fit and positioning.*

Fig. 8-18—*A doweling jig will help locate the position of holes and guide the drill bit.*

Fig. 8-19—*Marking the edges to show where the dowels will be located.*

Safety First

► *Drilling for Dowels*

See Chapter 6 for information about using a drill safely.

Determine the depth to which the dowel will go. Half the dowel will extend into each piece. Add about $\frac{1}{8}$ inch for clearance at the bottom of each hole. (The additional space holds glue and helps prevent a dry dowel joint.) For example, if the dowel is 3 inches long, a hole will be drilled about $1\frac{5}{8}$ inches deep in each piece.

Using a scratch awl, make a small dent at dowel locations. Select a drill the same size as the dowel rod. Carefully drill the hole to the proper depth, making sure that you are working square with the edge of the stock. Use a *depth gauge* as a guide. If available, also use a doweling jig. Fig. 8-20. This tool enables you to drill the holes square and in the right places.

After drilling the holes, saw the dowel to the chosen length. Put a slight bevel on each end.

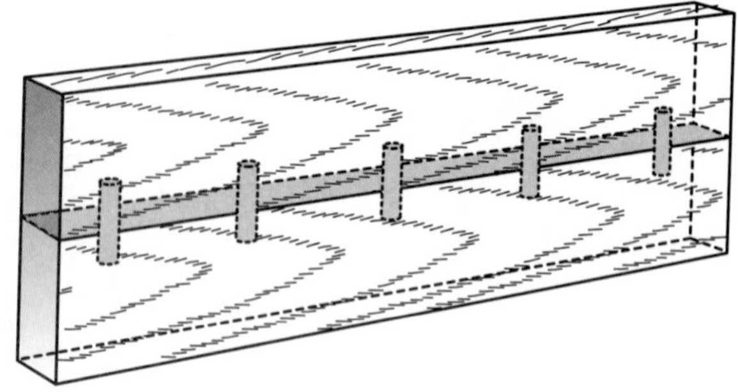

Fig. 8-21—***Edge dowel joints with the dowels installed.***

When gluing, dip the dowels about half their length into the glue. Drive the glue-covered ends into the holes on one edge. Then coat the other ends and the edges of the wood pieces and assemble. Fig. 8-21.

Fig. 8-20—***Drill the holes to the proper depth.***

Making a Dowel Joint on a Frame

One way to strengthen a butt joint on a simple frame is to install two or more dowels at each corner. Fig. 8-22. Do this as follows:

1. Square up pieces of wood that are to go into the frame. Carefully saw the ends.
2. Lay out the frame. Mark adjoining corner pieces with corresponding numbers for identification.
3. Indicate on the face surfaces the location and number of dowels. Fig. 8-23.
4. Place the two pieces that are to form one of the corners in a vise. The butting end and the butting edge should be exposed, with the face surfaces of the two pieces out and the end and edge flush.

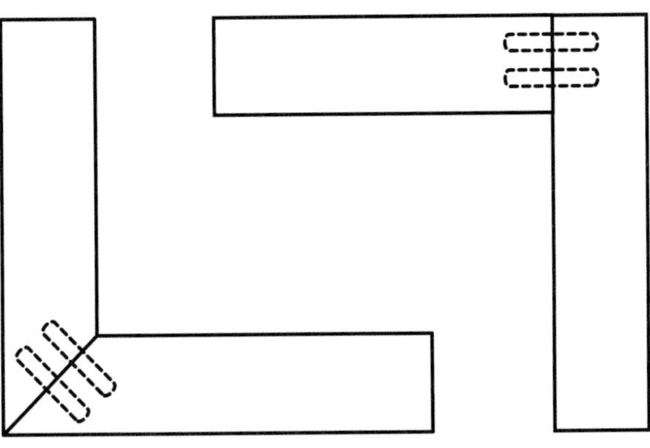

Fig. 8-22—*Dowel joints for a frame.*

5. Hold a try square against one of the face surfaces. Mark lines across the edges and ends to show the location of the dowel rods. Fig. 8-24.
6. After setting a marking gauge to half the thickness of the wood, hold it against the surface of each piece. Using a scratch awl, mark the exact location.
7. Drill the holes and complete the joint as for an edge dowel joint described previously.

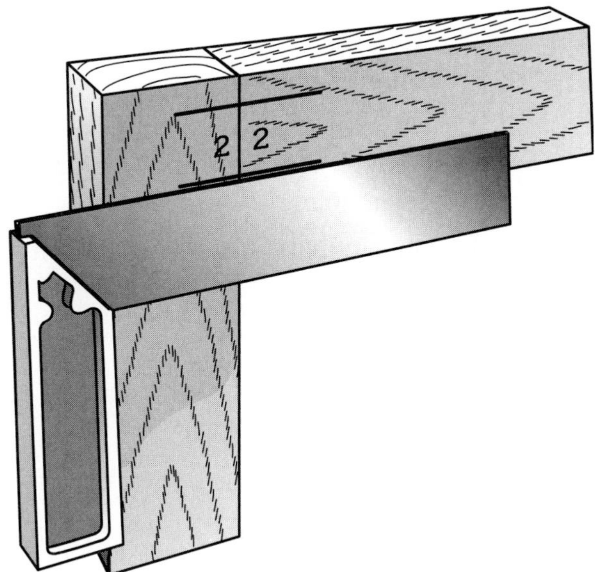

Fig. 8-23—*Using a try square to mark the locations of the dowels on the face surfaces. Both pieces have been marked with the same number (2) to indicate that these two pieces meet at this joint.*

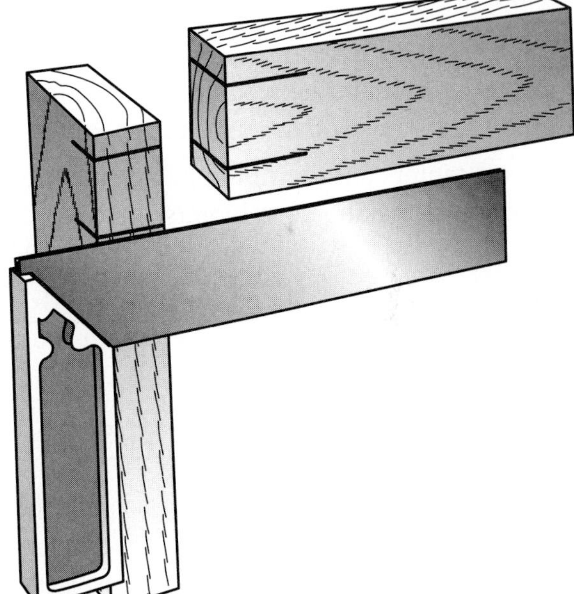

Fig. 8-24—*Squaring a line across the edge and end of both pieces.*

CHAPTER 8

Review & Applications

Chapter Summary

Major points from this chapter that you should remember include:

- The butt joint is the simplest but weakest type of joint. A variety of techniques can be used to strengthen butt joints.
- Biscuit joinery is one way to strengthen end and edge butt joints.
- Biscuits are pieces of compressed wood that fit into slots cut into adjoining pieces of wood using a biscuit machine.
- Dowels are cylindrical pins glued into matching holes in the wood pieces making up a joint. They add strength to the joint.
- Dowel centers and doweling jigs are devices used when making dowel joints.

Review Questions

1. Name two types of butt joints.
2. How can a butt joint be strengthened?
3. Describe how to make a biscuit joint.
4. What would be the result if the biscuit slots of two joint pieces were slightly misaligned when joined?
5. From what kinds of wood are biscuits and dowels usually made?
6. Why are dowels often used in making an edge joint?
7. State the rule for choosing the correct dowel diameter.
8. Briefly describe the procedure for making an edge dowel joint.
9. List the steps in making a dowel joint on a frame.

Solving Real World Problems

Jerry's mother wants to grow flowers in planter boxes on the patio. Jerry built the box shown here, and he wants to build two more. The problem with the first box is that when dirt is put into it, the sides pull apart a little at the corners. When his mother waters the plants, the water seeps out and makes a mess. Jerry wants to avoid this problem with the next two boxes. The drawings show two techniques Jerry thought might work. Can you think of other solutions?

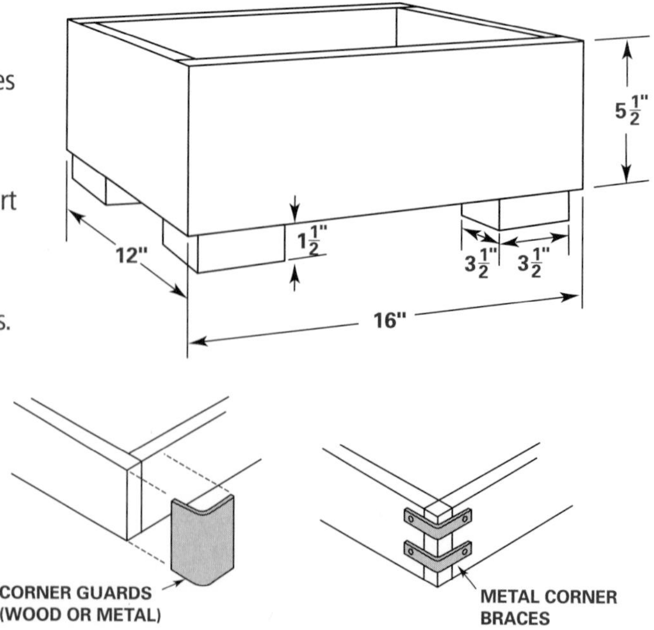

CHAPTER 9

Rabbet Joint

Look for These Terms

- rabbet
- rabbet joint

YOU'LL BE ABLE TO:

- Lay out a rabbet joint.
- Make a rabbet joint using hand tools.
- List power tools that can be used to cut rabbets.
- Make a rabbet joint using power tools.
- Assemble a rabbet joint.

A **rabbet** is an L-shaped cut along the end or edge of a board. **Fig. 9-1.** Hand or power tools can be used to cut a rabbet. A **rabbet joint** is formed by fitting the end or edge of one piece into a rabbet cut at the end or edge of another piece. **Fig. 9-2 (page 168).** This joint is commonly used in the construction of drawers, cases, and cabinet frames.

LAYING OUT A RABBET JOINT

The width and depth of the cut made for a rabbet joint is determined by the thickness of the stock. Check the ends of the pieces to be joined to be sure that they are square.

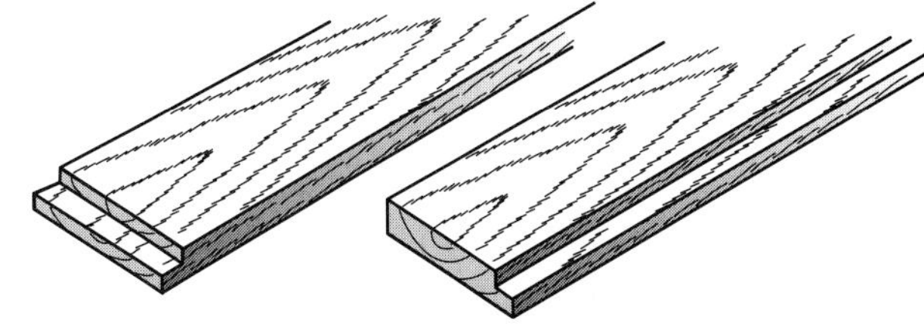

Fig. 9-1—*Rabbet cuts. A rabbet joint can be on end grain or edge grain.*

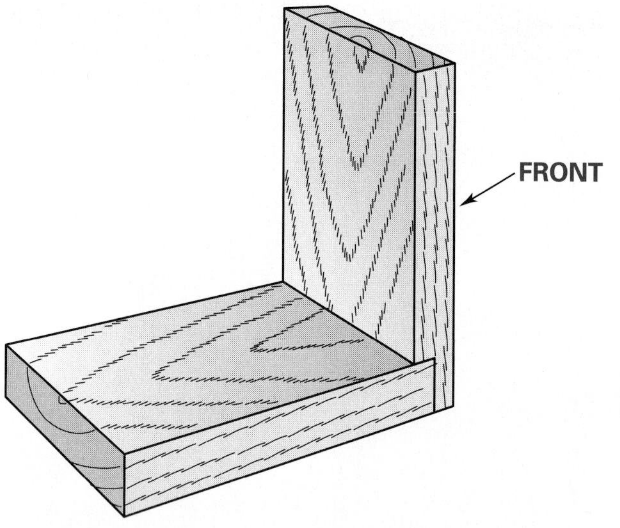

Fig. 9-2—*A rabbet joint. The cut should be made so the joint is hidden from the front.*

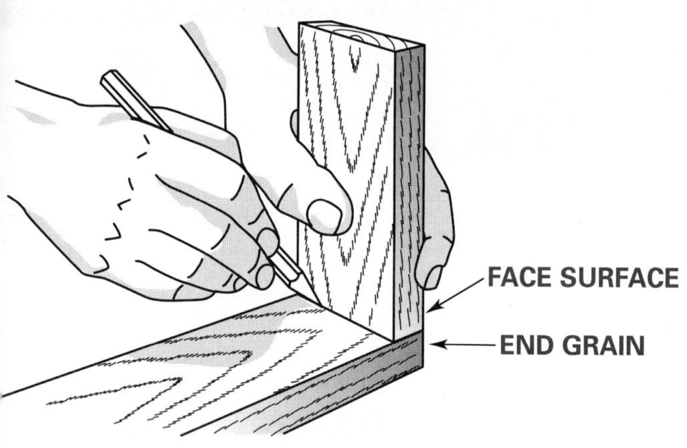

Fig. 9-3—*Marking the width of the rabbet.*

Place the board on which the rabbet will be cut on the workbench with the face surface down. Place the end of the second piece directly over the end of the first piece. The face surface of the second piece should be flush with the end grain of the first. Fig. 9-3. With a sharp pencil or knife, mark the thickness of the second piece. This will be the *width* of the rabbet. Set the second piece aside.

Safety First

►*Using Hand Tools*

- Follow all rules of good safety when using hand tools. Take special care with tools that are sharp.

For more safety information on using hand tools, see Chapters 5, 6, and 7.

With a try square held on the joint edge of the first piece, square the line just made. Continue this line down each edge. Measuring from the surface to be cut, mark the depth of the rabbet on the sides and end. The *depth* is usually one-half to two-thirds the thickness of the stock. Fig. 9-4.

CUTTING THE RABBET

Rabbets can be cut using either hand tools or power tools.

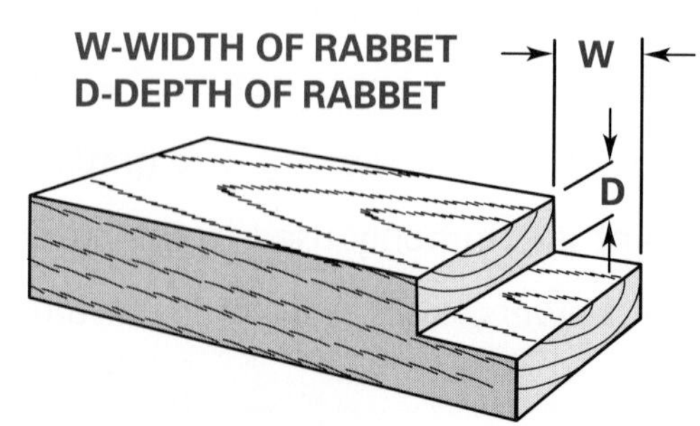

Fig. 9-4—*The width of the rabbet must be equal to the thickness of the stock. The depth of the rabbet is usually one-half to two-thirds the thickness.*

Using Hand Tools

To cut the rabbet by hand, clamp the workpiece securely to the top of the workbench. Clamp a squared piece of scrap stock on the workpiece with the edge directly over the layout line. Then hold a backsaw against this edge to make the shoulder cut in the waste stock just inside the layout line. Fig. 9-5. Cut the joint to the proper depth, as indicated by the layout line.

The excess stock from the joint can either be sawed out or pared out with a chisel. If you saw it out, lock the workpiece in a vise with the joint showing. With a *backsaw,* carefully make several saw cuts to remove the excess stock. If you use a *chisel,* leave the workpiece clamped to the top of the bench and pare out the excess stock. Fig. 9-6.

Mark both pieces of the joint with corresponding numbers for proper assembly. Make a final check with a try square to be sure the joint is square.

Fig. 9-6—*Using a chisel to trim excess stock. Note that the blade of the chisel is held between the thumb and the forefinger.*

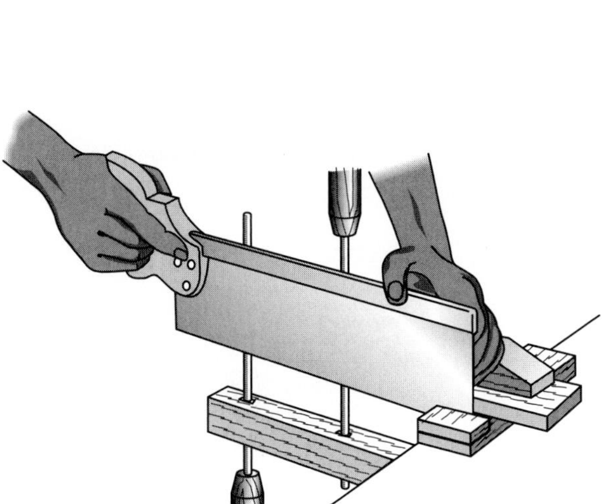

Fig. 9-5—*Making a shoulder cut on a rabbet joint. Notice that a piece of scrap stock is clamped over the layout line with hand screws. The backsaw is held against the edge of the scrap stock. This helps ensure a square cut and prevents the saw from jumping out of the kerf and damaging the wood.*

Using Power Tools

A number of power tools may be used to cut rabbets. These include the table saw, radial arm saw, router, and jointer. See Chapters 21, 22, 23, and 28.

Using Power Saws

To cut a rabbet with a table saw, use two passes. On the first pass, make the shoulder cut for the inset. Since there is no cutoff involved, use the rip fence as a guide. Position the fence so that the distance from the outside of the blade to the fence is the same as the *width* of the rabbet. Set the saw blade so that the amount that it projects above the surface is the same as the *depth* of the rabbet. Place the workpiece face up on the table with the edge against the fence. Make the cut.

The second pass is made to remove the excess stock. Set the saw-blade projection to the *width* of the rabbet. Set the fence to its depth. Place the workpiece on its end with the edge toward the saw blade and the face toward the fence. The material to be removed must be facing out so it can be safely separated from the workpiece. Fig. 9-7.

To cut a rabbet in one pass, a dadoing tool can be used. The workpiece is moved along the fence. A *radial-arm saw* can also be used with a dadoing tool to cut a rabbet in one pass. Fig. 9-8.

Using a Router and a Jointer

A *router* is a good power tool to use for cutting rabbets. Fig. 9-9. A rabbet bit is used most often in the router to make the cut. The width of the cut is controlled by the pilot end of the bit. The pilot end moves against the edge of the workpiece as the cut is made. If the bit used has a straight shank, a straightedge guide is needed.

Safety First

►*Using the Table Saw and Radial-Arm Saw*

- Always use the saw guard.
- Do not allow your fingers to come closer than 5 inches to the saw blade on a table saw. Maintain at least a 6-inch margin of safety from the blade on a radial-arm saw.

For further safety information about using the table saw and radial arm saw, see Chapters 22 and 23.

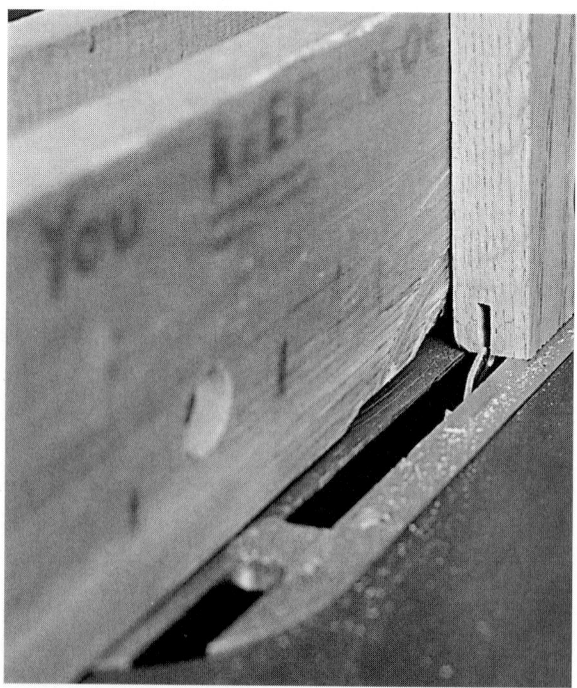

Fig. 9-7—*Using a table saw to cut a rabbet. A featherboard should be used to hold the workpiece against the fence.*

Fig. 9-8—*Using a dado cutter on a radial-arm saw to cut a rabbet.*

Safety First

►*Using Power Tools*

- Always wear ear and eye protection when using a router.
- Keep your hands clear of the rotating cutters on a router.
- Never allow your hand to pass directly over the cutterhead of the jointer.

See Chapters 21 and 28 for safety rules for using the machines discussed here.

Fig. 9-9—***Using a router to cut a rabbet.***

Rabbets can be cleanly and accurately cut on a *jointer.* Most jointers have a special ledge that is used for this purpose. Fig. 9-10.

Fig. 9-10—***Note the rabbet ledge, If you need to remove the guard from the jointer, clamp a scrap piece to the fence to cover the cutterheard.***

Safety First

►*Avoiding Kickback*

When using a table saw to cut a rabbet, be sure the material to be removed is facing out and is not against the fence. Waste trapped between the saw blade and the fence will cause *kickback.*

ASSEMBLING THE JOINT

A rabbet joint is usually assembled with glue or with both glue and nails, screws, or dowels. Simply glued, this joint is not very strong. Nails, screws, and dowels add strength. These are driven through the side into the end grain of the front. If the joint is nailed, drive the nails in at a slight angle to help them fasten the joint more tightly. Fig. 9-11. Screws should be long and thin so they will have more holding power in the end grain.

APPLICATIONS

Rabbet joints are used in the construction of many wood products. Fig. 9-12. They are particularly useful in making drawers. The joints between the front and the sides of a drawer are usually rabbet joints. These can either be flush or the sides can be set in from the edges of the front. Setting them in gives the front a *lip* which will overlap the frame around the drawer opening. Fig. 9-13.

Back panels of cases and cabinet frames may have rabbets cut along the back edges of the frame components. This allows the back panel to be inset. Fig. 9-14. The rabbets should be cut somewhat deeper than the thickness of the panel. Later, the back edges of the case or cabinet can be trimmed flush.

Fig. 9-11—***Nailing a rabbet joint. Driving the nails in at a slight angle gives them more holding power.***

SCIENCE *Connection*

Laser-Equipped Routers

Although most routers for home use are electric, some commercial models employ a carbon dioxide (CO_2) laser to make a clean, sharp cut. Laser light is *coherent* (all the light waves travel in the same direction). Because of this, the beam of light can be controlled precisely to perform very accurate work.

Laser-equipped routers do not work like standard routers. Instead, the laser beam *vaporizes* the wood in its path, or changes it from a solid to a gas. The gas disperses, leaving a clean "cut" in the remaining stock. The cut is made without any contact with the wood, so wear on the machine is minimal.

Fig. 9-12—*Rabbet joints were used to put together this silverware chest.*

Fig. 9-13—*A rabbet joint used to join the front of a drawer to the side.*

WOODWORKING TIP

Smoothing a Rabbet Joint

A smooth surface across a rabbet joint can be achieved by first cutting the rabbet slightly wider than necessary. Glue and assemble the pieces. After the glue is dry, sand off excess material so the joined surfaces are flush and smooth. See the figure below.

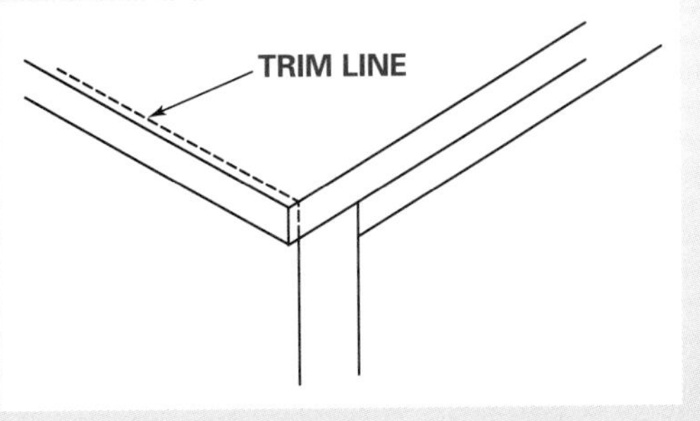

Fig. 9-14—*The back of a cabinet fits into a rabbet cut in the frame.*

CHAPTER 9

Review & Applications

Chapter Summary

Major points from this chapter that you should remember include:

- A rabbet joint is formed by fitting the end or edge of one piece into a rabbet cut in another piece.
- The width and depth of cut made for the rabbet joint is determined by the thickness of the stock.
- A backsaw is used to make the shoulder cut for the rabbet joint by hand.
- Power tools that can be used to cut rabbets are the table saw, radial arm saw, router, and jointer.
- A rabbet joint is assembled with glue or glue and nails, screws, or dowels.
- Rabbet joints are particularly useful in drawer construction.

Review Questions

1. Describe a rabbet joint.
2. How is a try square used when laying out a rabbet joint?
3. Explain how you would use hand tools when cutting a rabbet for a rabbet joint.
4. List the power tools that can be used to cut a rabbet.
5. When cutting a rabbet with a table saw, what cut is made on the first pass? The second pass?
6. When assembling a rabbet joint, what can you do to increase its strength?
7. In what wood products do you typically find rabbet joints?

Solving Real World Problems

Janna has a drawing of a modular home entertainment center that she would like to build. Each of the five modules is constructed with 3/4″ rabbet joints in the top and bottom boards. Each module also has a 1/4″ rabbet on the back for the back panel. Janna has a table saw and dado set as well as a router, but she is unsure which is the better choice. She is also concerned about using a dado set right against the fence. Which would you use and why? How should Janna solve the fence problem if she uses the table saw?

CHAPTER 10

Dado Joint

Look for These Terms

- blind dado joint
- dado
- dado joint
- rabbet-and-dado joint

YOU'LL BE ABLE TO:

- Lay out and cut a dado.
- Make a blind dado joint.
- Make a rabbet-and-dado joint.
- Explain how to cut dadoes with power tools.

A **dado** is a channel or a groove cut across the grain of wood. (Often, the terms *groove* and *dado* are used interchangeably. Technically, however, a groove is cut *with* the grain, while a dado is cut *across* the grain—at right angles to the edge of the stock.) In a **dado joint,** a dado is cut across one board to receive the end of another board. **Fig. 10-1.**

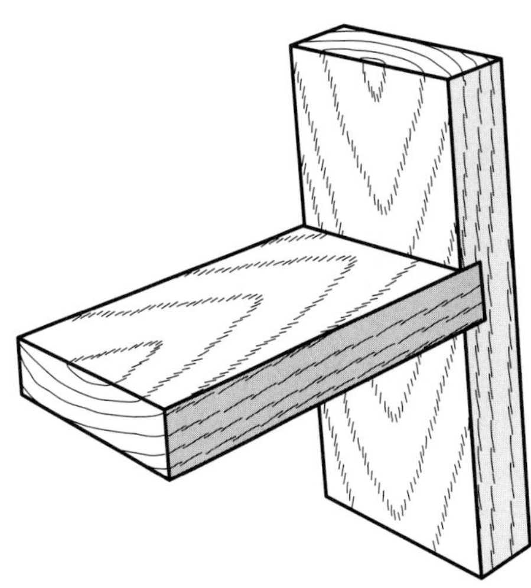

Fig. 10-1—*The width of the dado cut depends on the thickness of the board being inserted into the cut. The joint must be snug.*

When snugly fit, the dado joint is quite strong. The strength of a joint is largely determined by how much the surface area of one piece is in contact with the adjoining piece.

As you can see in Fig. 10-1, the dado joint not only provides a larger shared surface area to be glued, it also forms a supporting lip or ledge. Because of its strength, this joint is often found in pieces in which the crossmembers must support considerable weight, such as in bookshelves, drawers, cabinet shelves, ladders, and steps. Fig. 10-2.

LAYING OUT A DADO JOINT

To lay out a dado joint, follow these steps:

1. From the end of the board, measure up the correct distance to one side of where the dado is to be cut. Then square off a line across the face surface of the workpiece at this point.
2. *Superimpose* the end of the second board across the face of the first board. In other words, hold the second board in the position it will have in the finished joint—it should be at a right angle with the face of the first board, with one edge of its end lined up directly with the line you made in Step 1. With a sharp knife or pencil, mark the correct width of the dado. Fig. 10-3.
3. Remove the second board and square off a line across the face of the first board to show the proper width of the cut to be made. Continue both lines down both of the edges. Fig. 10-4.
4. Lay out the correct depth of the joint. The depth is usually one-half the thickness of the workpiece.

Fig. 10-2—***This CD rack has a number of dado cuts in it.***

Fig. 10-3—***Superimpose the end of the second board across the face of the first board. Then use a knife or pencil to mark off the width of the dado.***

Fig. 10-4—*Laying out a dado joint. One line is laid out on the face surface. Then, by superimposing, the width of the dado is marked. Finally the lines are drawn across the surface and down the edge of the stock.*

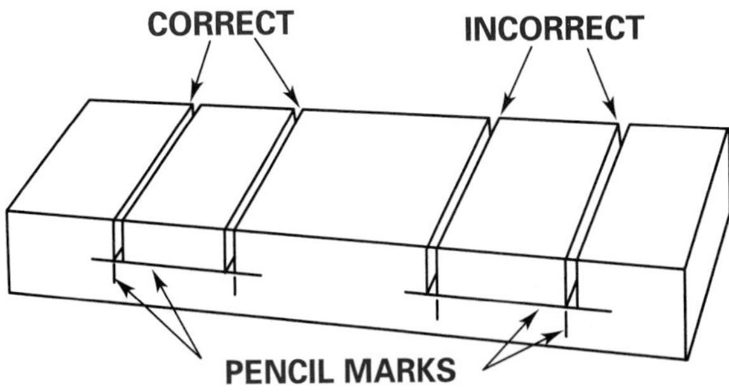

Fig. 10-5—*Make sure that the saw kerfs of the dado are made inside the layout lines.*

CUTTING A DADO

To cut a dado, follow the steps below.

1. Using a backsaw, cut the dado to the proper depth at both layout lines. (As stated above, the depth of the cut is usually one-half the thickness of the workpiece.) Make sure that the kerfs are in the waste stock and not outside the layout lines. Fig. 10-5.
2. Use a chisel to remove the waste stock. Fig. 10-6. If the dado is very wide, you may need to make several saw cuts to depth so that the waste stock can be easily trimmed out.
3. With a combination square, check the dado to make sure that it has the same depth throughout. Fig. 10-7 (page 178).

Fig. 10-6—*Trim the bottom of the dado with a chisel.*

4. Check the dado joint by inserting the second piece in the dado. Fig. 10-8. The second piece should fit snugly but smoothly into the dado. *Do not force the second piece into the dado.* If the fit is just slightly too snug, wrap a piece of

fine sandpaper around a narrow block of wood to smooth out the channel. Fig. 10-9. If more material must be removed, you may have to plane the sides of the second piece slightly to make it fit into the dado. If the dado has been cut too wide, about all you can do is to use a shim to fill the gap.

Fig. 10-7—*The depth of a dado can be checked with a combination square set to the correct depth.*

Depth Guide

When cutting the shoulder cuts with a backsaw here is a tip to make the cuts more accurate and at a uniform depth. Clamp a strip of wood to the side of the saw at the proper depth for the cut. The strip acts as a stop and will not let you cut any deeper.

The dado joint is usually assembled either with glue alone or with glue and nails or screws. Fig. 10-10.

Fig. 10-8—*Checking the fit of a dado joint. The second piece is inserted in the dado. If necessary for a proper fit, it is simpler to remove a little stock from the second piece than to cut the dado wider on the first piece.*

Fig. 10-9—*Fine adjustments can be made by sanding the dado with fine sandpaper.*

Fig. 10-10—*Assembling a shelf with dado joints.*

Safety First

► *Using Hand Tools*

Follow all rules of good safety when using hand tools. Take special care with tools that are sharp, such as handsaws, chisels, and planes. See Chapters 4 and 7.

HIDING THE JOINT

One disadvantage of the dado joint is that the joint lines are clearly visible. One simple method of hiding the joint is for the design to include face trim, which will effectively cover the joint.

If face trim is not used, another method of hiding the joint lines in, for example, a bookcase, is to have the front ends of the shelves extend beyond the sides of the case. Fig. 10-11. This doesn't eliminate the joint lines, but it does make the joints less visible. In addition, as shown in Fig. 10-11, decorative cuts can be made on the projecting corners or along the front edge of the shelves.

SCIENCE *Connection*

Allowing for Humidity

Humidity is a factor that must be considered in woodworking. For example, wooden drawer bottoms should never be glued to the drawer frames. The bottoms are slid into grooves cut along the front and sides (and sometimes the back) of the drawer frame. The bottoms are cut slightly narrower or shorter (usually about $\frac{1}{16}$ inch) than the total distance between the grooves. This way, the bottoms remain free-floating to allow for the swelling and shrinkage caused by changes in humidity levels.

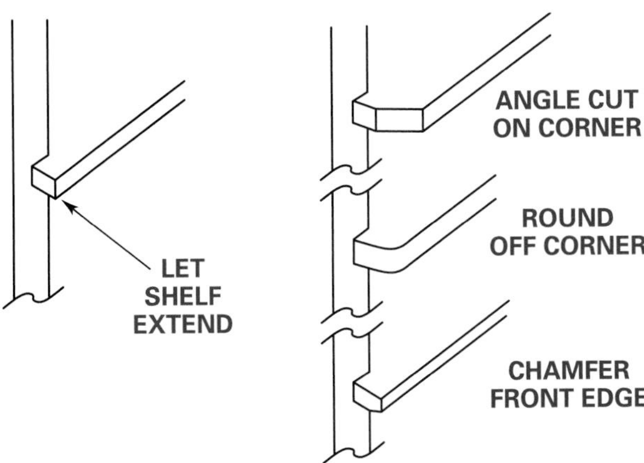

Fig. 10-11—*The ends of bookshelves can be cut wider at the front to extend beyond the sides of the case. Various decorative cuts can be made on the ends or edges of the shelves to add visual appeal.*

Making a Blind Dado Joint

Another method of concealing the joint is to make a variation of the simple dado joint. In a **blind dado joint,** the dado is cut only partway across the board and the piece that fits into the dado is notched so that the joint does not show from the front. Fig. 10-12.

To lay out a blind dado joint, lay out the width of the dado as described before, but mark the depth of the dado on the back edge only. Also lay out the length of the dado from the back edge to within ½ to ¾ inch of the front edge. The dado can be cut by drilling a series of holes in the waste stock and then trimming out with a chisel.

MAKING A RABBET-AND-DADO JOINT

A **rabbet-and-dado joint** is used when additional strength and stiffness are needed. This joint consists of a rabbet with a tongue that fits into the dado. Fig. 10-13.

To make a rabbet-and-dado joint, first lay out and cut the rabbet. (See Chapter 9.) Then lay out the position of the dado joint by superimposing the tongue of the rabbet. Mark the width of the dado. Make the dado as described earlier in this chapter. Fit the tongue of the rabbet into the dado.

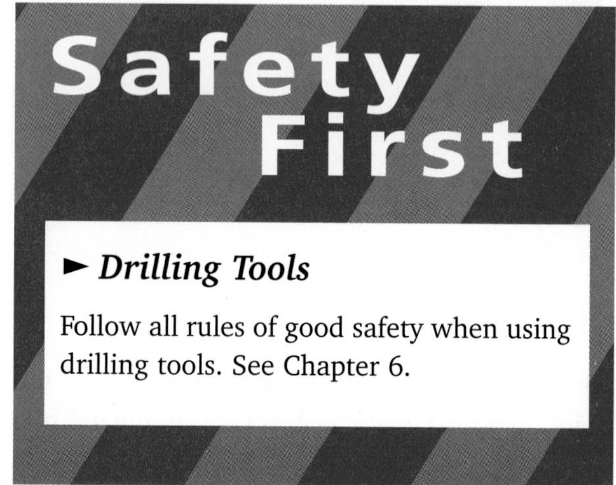

Safety
First
► *Drilling Tools*
Follow all rules of good safety when using drilling tools. See Chapter 6.

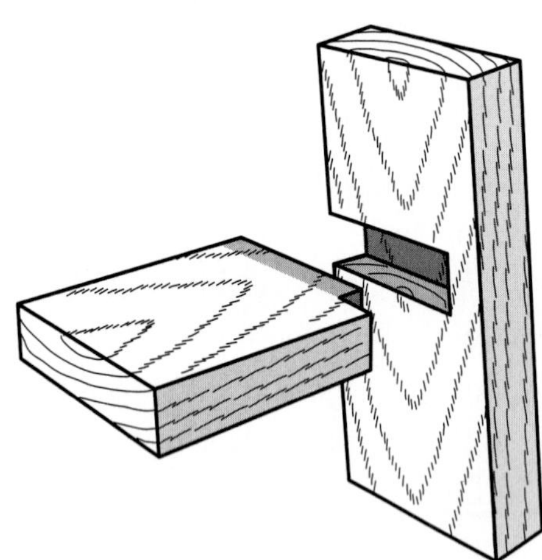

Fig. 10-12—*A blind dado joint has the same strength advantage as the dado joint as well as the added advantage of concealing the joint. It is built into bookcases and other projects of this type when a neat appearance at the front of the shelves is desirable.*

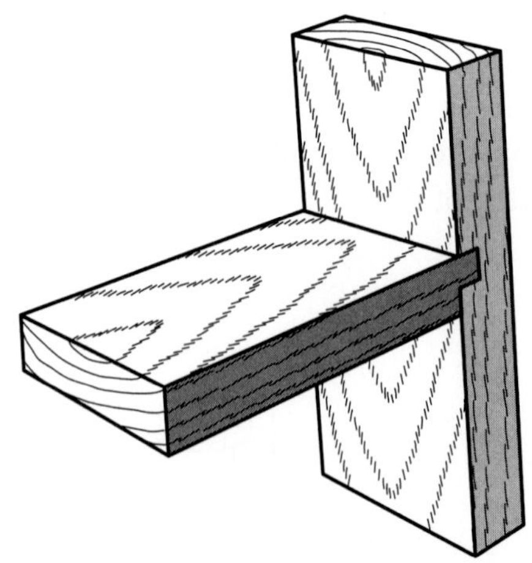

Fig. 10-13—*A rabbet-and-dado joint gives added strength and rigidity. It is commonly used for the back corners of drawers.*

CUTTING DADOES WITH POWER TOOLS

Dadoes can also be made very easily with power tools. A dado cutter mounted on a table saw or a radial-arm saw makes quick work of cutting a dado. Figs. 10-14 and 10-15.

If only one or two dadoes need to be made, you can make several repeated passes with a regular saw blade instead of using a dado cutter. Make the shoulder cuts first. Then make overlapping passes between the shoulder cuts to clear away the waste stock. Fig. 10-16.

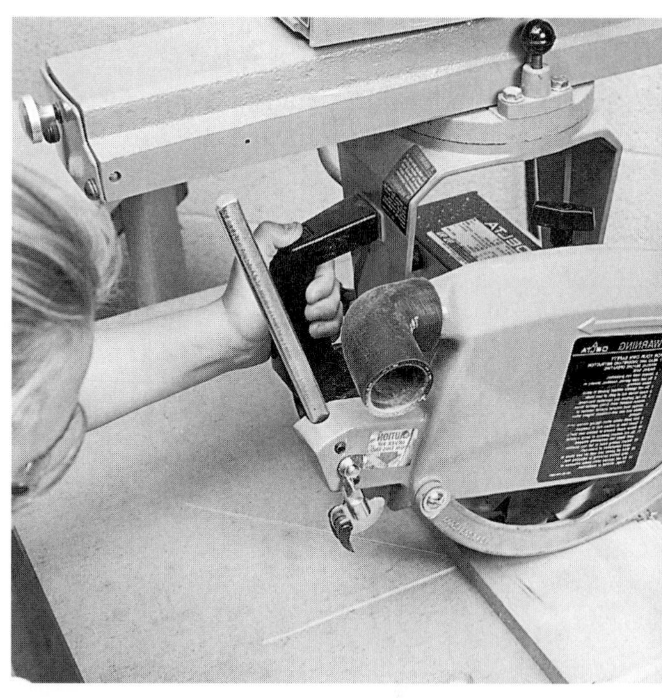

Fig. 10-15—*Using a dado cutter on the radial-arm saw.*

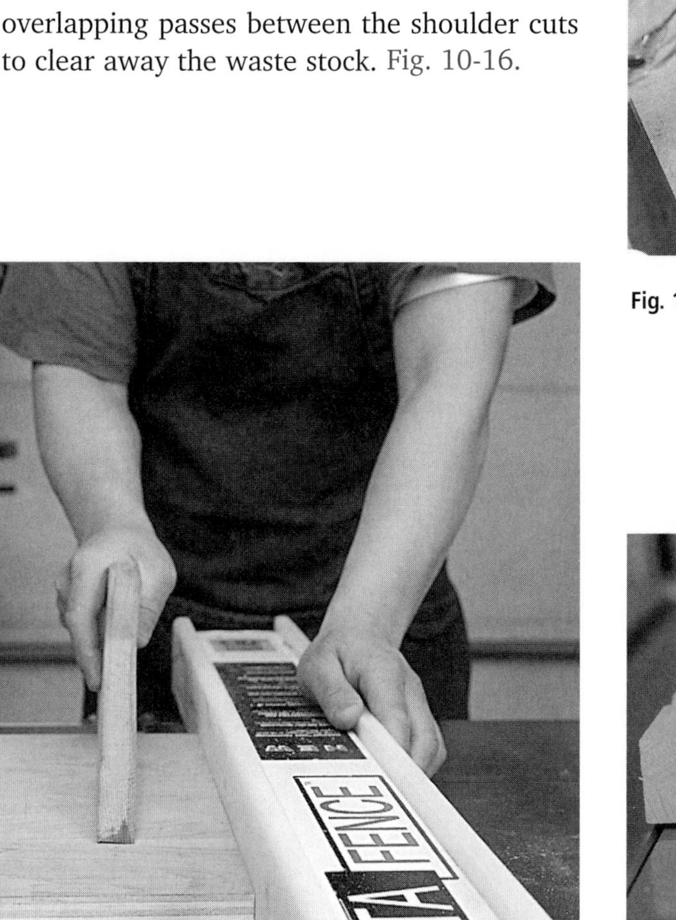

Fig. 10-14—*Using a dado cutter on the table saw.*

Fig. 10-16—*To cut a dado or a groove with a regular saw blade, make the shoulder cuts first. Then make overlapping passes between the shoulder cuts to clear away the waste stock.*

When a straightedge is provided, a portable router can make very accurate dadoes. Fig. 10-17. The router bit should be a straight shank. The size of the bit can usually be the exact size of the desired dado.

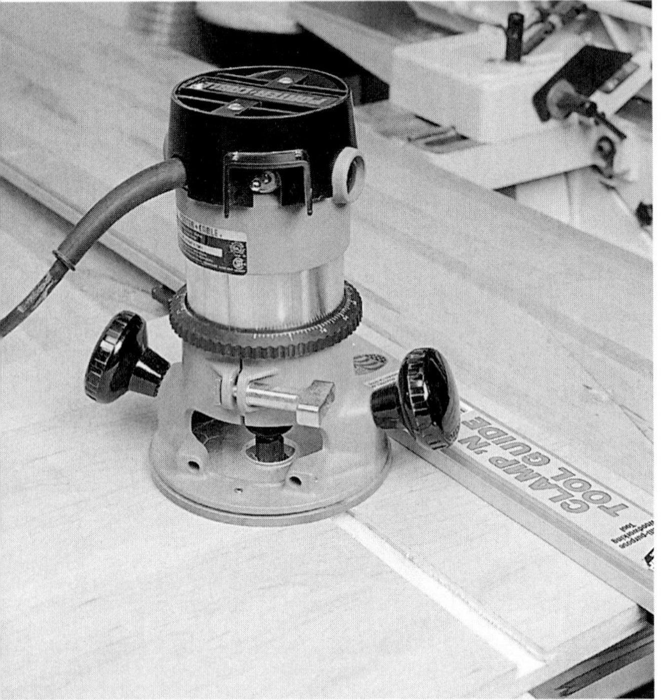

Fig. 10-17—*Using a router to cut a dado.*

Safety First

► *Using Power Tools*

Follow all rules of good safety when using power tools. See Chapters 22, 23, and 28 for safety rules for using the machines discussed here.

Easy Drawer Slides

One simple way to mount a drawer is to fasten wooden slides to the side of the cabinet, as shown in the illustration below. The slides fit into grooves in the sides of the drawer. The drawer will slide in and out very accurately in the groove. This is inexpensive and easy to do.

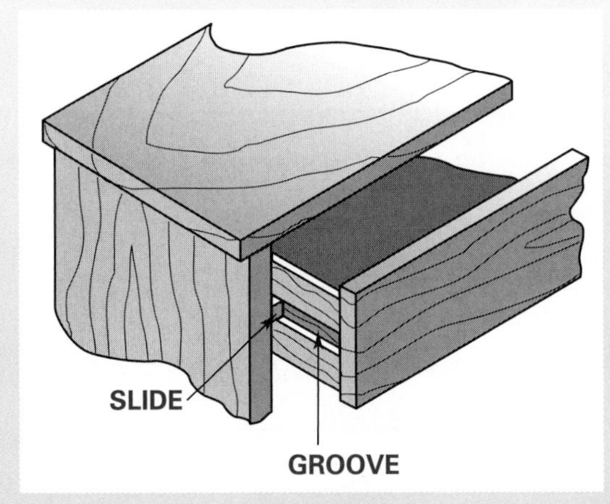

CHAPTER 10

Review & Applications

Chapter Summary

Major points from this chapter that you should remember include:

- Because of its strength, a dado joint is often used when crossmembers of a piece must support a lot of weight.
- After it is laid out, the dado can be cut out with a backsaw or with power tools such as a table saw, a radial-arm saw, or a router.
- After it has been cut, the dado should be checked to make sure it has the same depth throughout.
- The second piece should fit snugly but smoothly into the dado.
- There are several methods of concealing dado joints.
- A rabbet-and-dado joint is used when additional strength and stiffness are needed.

Review Questions

1. What is a dado?
2. Explain what makes the dado joint a strong joint.
3. Describe how to lay out a dado joint.
4. What is the general guideline for the depth of the dado?
5. When cutting a dado, where should the saw kerf be formed?
6. How can you check a dado to make sure it has the same depth throughout the cut?
7. What are two things you can do if the fit between the second board and the dado is too tight?
8. Sketch a blind dado joint.
9. Sketch a rabbet-and-dado joint.
10. Explain how to cut a dado using a regular saw blade on a table saw.

Solving Real World Problems

In two modules of an entertainment center John is building, there are two 3⁄4″ shelves. John knows these must be dadoes jointed into the sideboards. While considering his options, he dismisses the use of sliding dovetail joints because he lacks the equipment and jigs needed for such a complex joint. He finally decides that the 3⁄4″ × 1⁄4″ dado is the joint he wants to use. He will use his router to make the joint. John gathers his tools, a router, a 3⁄4″ router bit, a guide clamp, and a tape measure. How should he proceed to lay out and cut the dadoes?

CHAPTER 11

Lap Joint

YOU'LL BE ABLE TO:

- List major types of lap joints.
- Lay out and make a cross-lap joint.
- Make a half-lap joint.
- Make a full-lap joint.
- Make a finger-lap joint.

Look for These Terms

- cross-lap joint
- edge half-lap joint
- end-to-end half-lap joint
- finger-lap joint
- full-lap joint
- half-lap joint
- lap joint
- surface-lap joint

A **lap joint** is basically one piece laid (lapped) over another with the two pieces fastened together in the contact area. There are several different types of lap joints. These look much alike but vary in strength and in how they are made. **Fig. 11-1.**

When pieces are simply joined without additional processing, the joint is called a **surface-lap joint.** If only glued, this type of joint is not strong. It needs to be reinforced with nails or other fasteners. Surface-lap joints are used most often in unrefined wood products such as crates and pallets.

Other types of lap joints are stronger and more attractive. They are used in furniture and cabinet construction. These include cross-lap, half-lap, full-lap, and finger-lap joints.

CROSS-LAP JOINT

The **cross-lap joint** is by far the most common type of lap joint. An equal amount of material is removed from the area of contact on each piece. When the pieces are fit together, the surfaces are flush. A cross-lap joint is often made in the exact center of two pieces that cross at a 90-degree angle. However, this type of joint can be made at any angle.

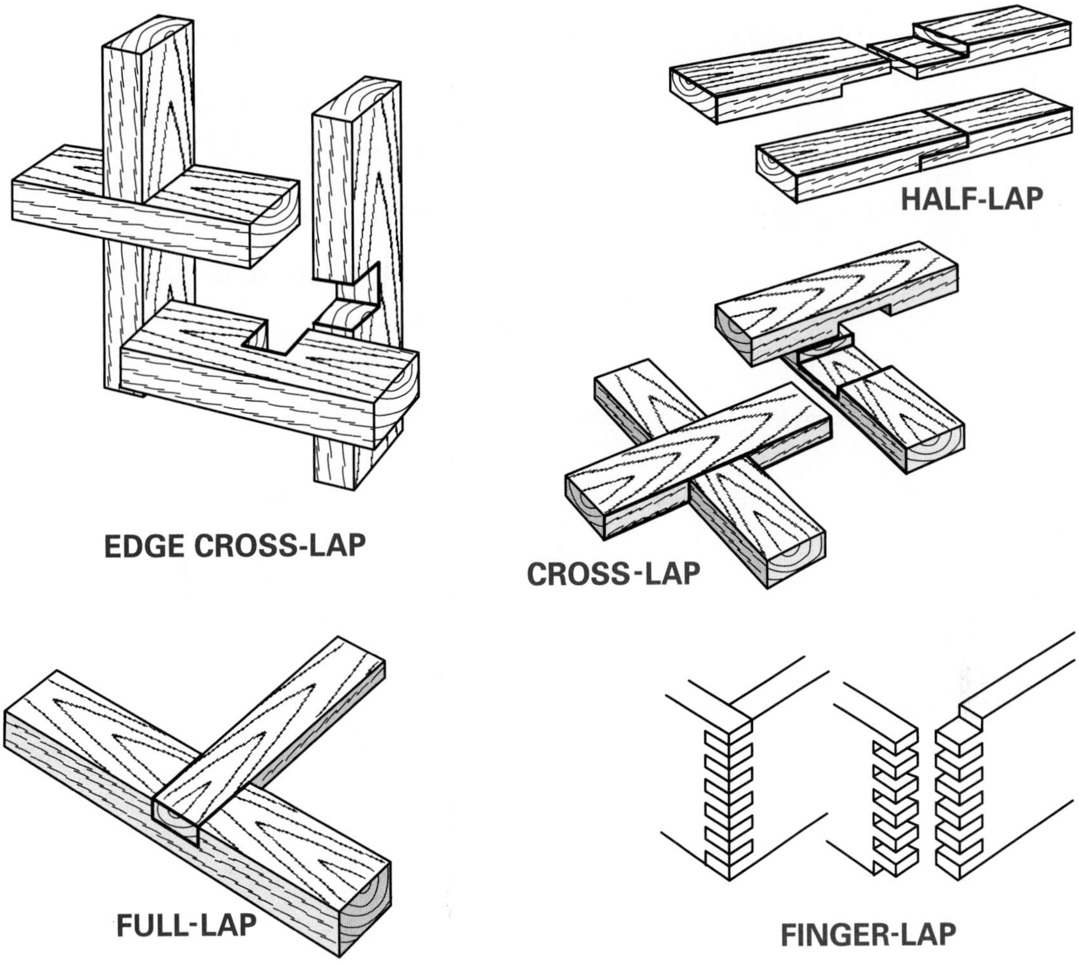

EDGE CROSS-LAP

HALF-LAP

CROSS-LAP

FULL-LAP

FINGER-LAP

Fig. 11-1—*Several types of lap joints.*

Laying Out the Joint

For a cross-lap joint in the center of both pieces, the two pieces must be exactly the same thickness and width. Lay the two pieces on the workbench side by side. The face surface of one (piece A) and the opposite surface of the other (piece B) should be upward. Fig. 11-2. Divide the length of each into equal parts. Lay out a centerline across the two pieces. Measure the width of the stock and divide this measurement in half. Lay out a line this distance on each side of the centerline. The space between the lines will be the location of the joint.

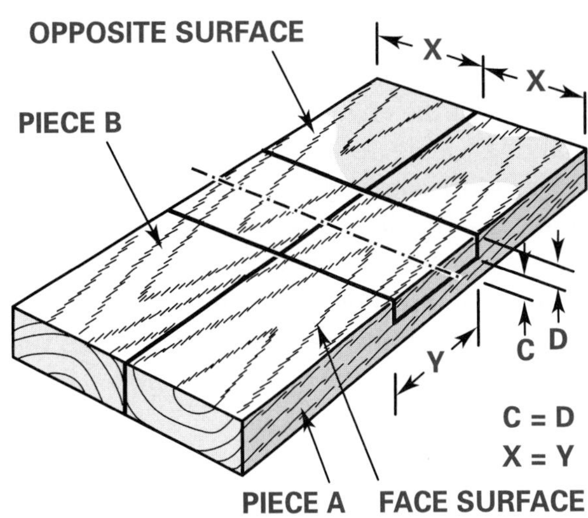

Fig. 11-2—*Correct layout for a center cross-lap joint. Both pieces should be marked at the same time.*

Fig. 11-3—*Checking the layout by superimposing Part B over Part A. Note that a try square is being used to keep the pieces at right angles.*

Check this measurement by superimposing piece B over piece A at right angles. Place it between the lines, exactly in the position that the joint will be when assembled. Fig. 11-3. The layout line should just barely show beyond each edge of piece B.

Continue the lines showing the width of the joint down the edge of each piece. Next, set a marking gauge to half the thickness of the stock. Mark along the edge on each side to show the depth of the joint. Mark the two pieces so that they will be flush when the joint is made. You will be cutting the joint from the face surface of one piece and the opposite surface of the other.

Cutting the Joint

The cuts needed for the cross-lap joint can be made using either hand tools or a power saw. Use the following procedures.

Safety First

► ***Using Hand Tools***

- Follow all rules of good safety when using hand tools.
- Always take special care with tools that are sharp.

For more safety information on using hand tools, see Chapters 5, 6, and 7.

Using Hand Tools

Clamp one workpiece to the top of the workbench. Cut with a backsaw to the depth of the joint just *inside* each of the layout lines. If the joint is wide, make several cuts in the waste stock. This helps to remove the waste. It also acts as a guide when you chisel it out. Cut the other piece in the same manner.

Next, use a chisel to remove the waste stock. Work from both sides of each piece, tapering up toward the center. If you chisel across the stock from only one side, you will chip the opposite side. After you have brought the joint down to the layout line on each edge, continue to pare (cut) the high point in the center of the joint. See Chapter 7.

Complete the chiseling on both pieces. Then fit the pieces together. Fig. 11-4. They should fit snugly. They should not be so loose that they fall apart or so tight that they must be forced together. If the fit is too tight, plane a little from the edge of one piece. This technique works better than trying to enlarge the cut by trimming the shoulder.

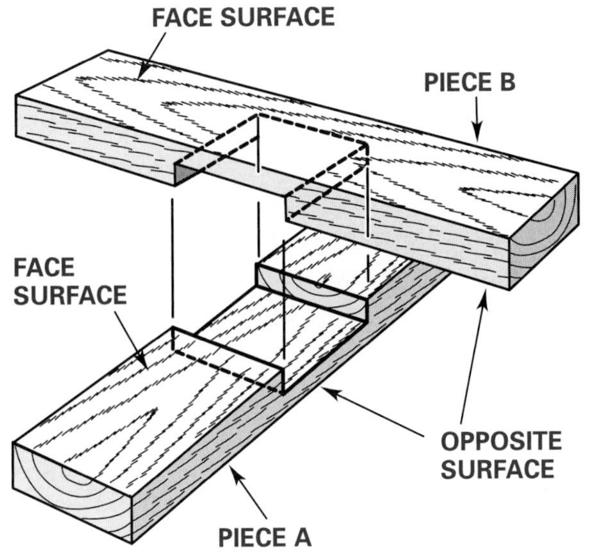

Fig. 11-4—*Fitting a cross-lap joint.*

Safety First

► ***Using the Table Saw***

- Always use the saw guard.
- Do not allow your fingers to come closer than 5 inches to the saw blade.

For further safety information about setting up and operating the table saw, see Chapter 22.

Using a Table Saw

A table saw can be used to make the cuts for the cross-lap joint. Either a dadoing tool or regular saw blade can be used. Set the saw blade to the depth of the cut. Clamp the pieces together edge to edge to cut both at the same time. Be sure to cut one on the face side and the other on the opposite side. Cut each shoulder. Then make several cuts in between to remove the waste stock.

Assembling the Joint

When assembled, the surfaces of both pieces should fit flush with one another. Fig. 11-5. Often, both glue and nails or screws are used. Nails or screws should be installed from the underside so they will not show. Fig. 11-6 (page 188).

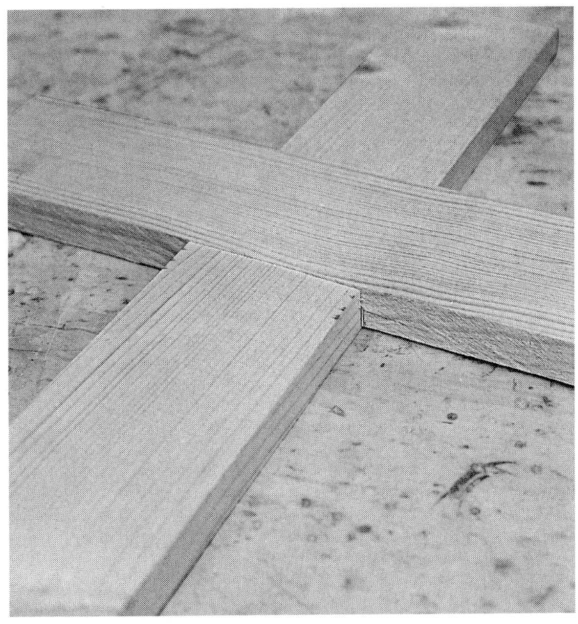
Fig. 11-5—*A finished cross-lap joint.*

HALF-LAP JOINT

In a **half-lap joint,** the pieces joined are the same thickness. Half the thickness of each piece is removed from the contact area. When assembled, the pieces interlock and the surfaces are flush. To make a half-lap joint, follow the same general procedure as for the cross-lap joint.

In an **edge half-lap joint,** dadoes are cut in the edges of the stock and not into the surfaces. Most of these dadoes are relatively deep and narrow. For this reason, they are typically referred to as *notches*. The width of a notch, or dado, should be the same as the thickness of the piece. Its depth should be one-half the width of the piece. Fig. 11-7.

FULL-LAP JOINT

A **full-lap joint** is used when pieces with different thicknesses are joined. The procedure for making this type of joint is much like that used to make a cross-lap joint. However, when making a full-lap joint, a seat cut (rabbet or dado) is made in the thicker piece. The dimensions of the cut must be the same as the width and thickness of the thinner piece. In assembly, the thinner piece is laid in the cut to form a flush surface. Fig. 11-8.

Fig. 11-6—***The decorative latticework on this jewelry box is a good example of how lap joints are used in fine furniture making.***

FINGER-LAP JOINT

The **finger-lap joint** consists of series of fingers and notches that mesh together. Because of the extensive glue area, it is a very strong joint. It is also visually attractive. These qualities make the finger-lap joint useful in both hidden and exposed areas of construction. It's commonly used for building drawers and boxes and is often referred to as a *box joint*.

Extending the Usefulness of Wood Pieces

Short pieces of wood are often thrown away as waste. However, these can be made into a longer, more usable piece by joining them with an **end-to-end half-lap joint.** The greater the overlap, the stronger the joint will be.

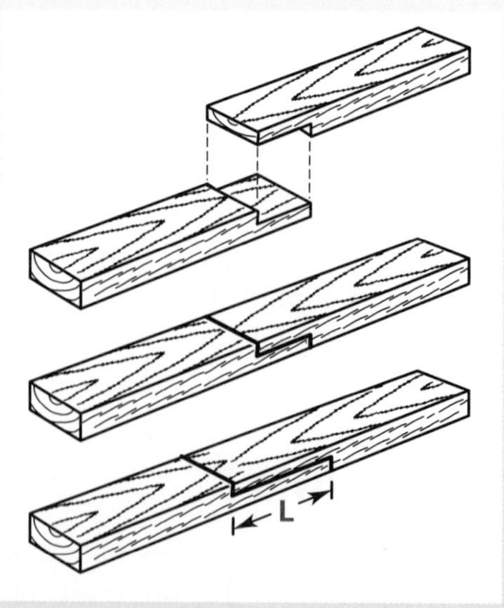

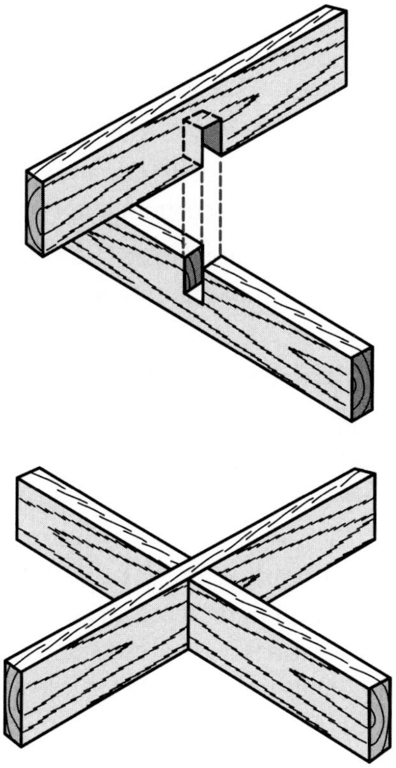

Fig. 11-7—*An edge half-lap joint. The dadoes cut for this type of joint are deep and narrow and are usually referred to as notches.*

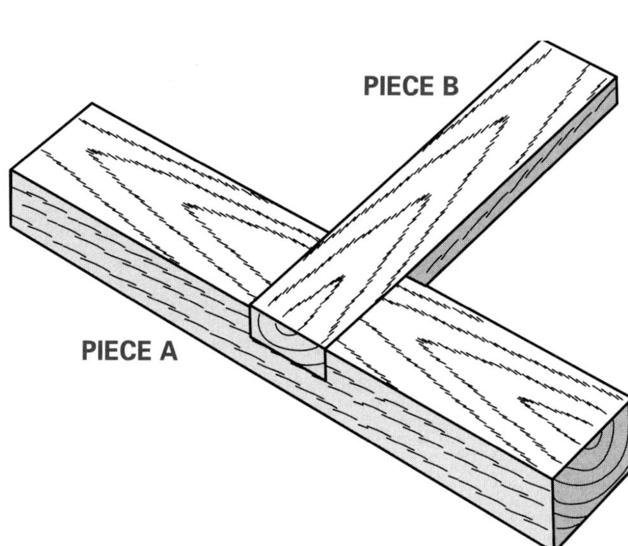

Fig. 11-8—*A full-lap joint. Note that the only cut made is in the thicker piece (A). Its dimensions are the full thickness and width of the thinner piece (B).*

Designing the Joint

One popular design technique is to make the width of the fingers and notches the same as the thickness of the stock. However, making them half the thickness or less provides more glue area resulting in a stronger joint. It also gives a better appearance. Fig. 11-9.

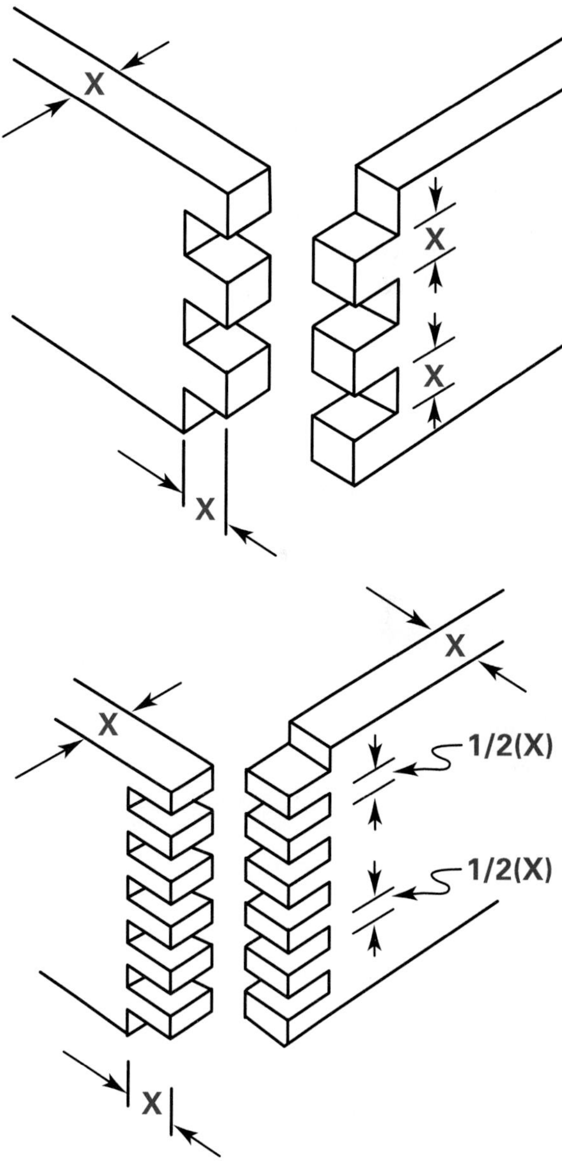

Fig. 11-9—*The width of the fingers and notches of a finger-lap joint are often made equal to the thickness of the pieces or to half the thickness.*

Cutting the Joint

Either hand tools or a power saw may be used to make the cuts for the finger-lap joint. Use the following procedures.

Using Hand Tools

Follow these steps to cut the joint.

1. Figure the width and depth of the cuts.
2. Lay out the cuts on the workpieces.
3. Place the pieces together, offsetting them by the width of one finger.
4. Fasten the workpieces securely in a vise.
5. To prepare for drilling out the bottom of the notches, clamp a block of scrap wood to the pieces behind the joint location. The block will prevent splintering when the drill bit goes through the workpieces.
6. Select a drill bit that is the same width of the notches or slightly smaller.
7. Hold the drill at right angles to the surface. Working slowly, drill holes at the bottom of the notches. Remove the wood block.
8. Staying slightly inside the layout lines, make the shoulder cuts and remove the waste stock. Use either a backsaw or a dovetail saw.
9. Square the bottom of the notches with a chisel. Clean the corners carefully.
10. Smooth the cuts using fine sandpaper wrapped around a small wood block.

Using a Table Saw

A layout can be followed to cut a finger-lap joint using a dado blade on a table saw. Fig. 11-10. However, to cut more accurately, make and use a jig with a notch. The height of the notch guide should be the same as the depth of the notches. It should be positioned one finger-width away from the nearest side of the saw blade. The length of the notch guide should be at least double the thickness of the stock. You will also need to make a space holder. It should be the same width as one finger and long enough so you can hold it without getting your hands near the saw blade.

Safety First

► ***Using Hand Tools***

- Follow all rules of good safety when using hand tools.
- Take special care with tools that are sharp.

For more safety information on using hand tools, see Chapters 5, 6, and 7.

MATHEMATICS *Connection*

Finger-Lap Joints

Finger-lap joints were used for a small box made out of ¾-inch stock that is 4 inches wide. The finger-lap joints are spaced at half the thickness of the boards. There are 11 fingers in each corner. What is the total surface area glued in all four corners? Give your answer in square inches. (Hint: Be sure to include both horizontal and vertical glue surfaces in your calculations.)

To make the cuts:

1. Place the space holder between the notch guide and the edge of one piece. Hold the workpiece against the space holder.
2. Hold the space holder and the workpiece flat against the jig. The workpiece must be perpendicular to the work surface.
3. Make the first notch on the edge of the workpiece by moving the jig forward to feed the piece into the saw blade.
4. Set the space holder aside.
5. Place the notch that you just cut over the notch guide.
6. Place the second piece in front of the first. The second will be offset from the first by one finger-width. Press both firmly against the guide and the jig.
7. Move the jig forward to cut a notch in both pieces.
8. Place the notches just cut over the guide and cut the next pair of notches.
9. Continue working in this manner until all notches are cut.

Complete the finger-lap joint by mating the two pieces together with glue.

Fig. 11-10—***Cutting a finger-lap joint on a table saw using a dado blade and miter gauge.***

Safety First

► *Using the Table Saw*

- Always use the saw guard.
- Do not allow your fingers to come closer than 5 inches to the saw blade.

For further safety information about setting up and operating the table saw, see Chapter 22.

CHAPTER 11

Review & Applications

Chapter Summary

Major points from this chapter that you should remember include:

- A lap joint is basically one piece laid over another with the two pieces fastened together in the contact area.
- Cross-lap joints can be made at any angle.
- In a cross-lap joint, the cut on one piece is made on the face surface and the cut on the other piece is made on the opposite surface.
- In a half-lap joint, the pieces joined are the same thickness and half the thickness of each piece is removed from the contact area.
- A full-lap joint is used when pieces of different thicknesses are joined.
- The finger-lap joint consists of series of fingers and notches that mesh together.

Review Questions

1. List four types of lap joints used in furniture and cabinet construction.
2. Which type of lap joint is most common?
3. Describe how to lay out the contact area for a cross-lap joint in the center of two pieces of wood.
4. Explain how to use a table saw to make the cuts for a cross-lap joint.
5. Describe a half-lap joint.
6. How is a full-lap joint different from other lap joints?
7. Which lap joint is also called a box joint?
8. How are hand tools used to cut the notches of a finger-lap joint?
9. What device is helpful when making cuts for a finger-lap joint using a table saw?

Solving Real World Problems

Art photographers like Sam and Rochelle always need picture frames. Neat frames that are unusual are very hard to find. They keep their eyes open for new ideas. While visiting a gallery, they see some great-looking frames that are simply made by joining four pieces of wood together with lap joints. After some discussion, they decide to use walnut for the vertical parts of the frame and maple for the horizontal parts. They have access to a good woodworking shop. How should they proceed to make a lap joint for their frames?

CHAPTER 12

Miter Joint

Look for These Terms

- miter box
- miter joint

YOU'LL BE ABLE TO:

- Explain the importance of accuracy when cutting miter joints.
- Lay out, cut, and assemble miter joints to create a picture frame.
- Describe how miters are cut with various power tools.

A **miter joint** is an angle joint that hides the end grain of both pieces. **Fig. 12-1.** The end of each piece is usually cut at 45 degrees to form a right angle when the two pieces are joined. The joint can be either *flat* or *on edge.*

Fig. 12-1—*Miter joints are used in the crown molding in this wardrobe.*

Miter joints on simple projects that get little wear and tear, such as lightweight picture frames, are often fastened with only glue and finishing nails. However, because it is a relatively weak joint, the miter joint is often strengthened with a spline, dowel, key, or biscuit. Fig. 12-2.

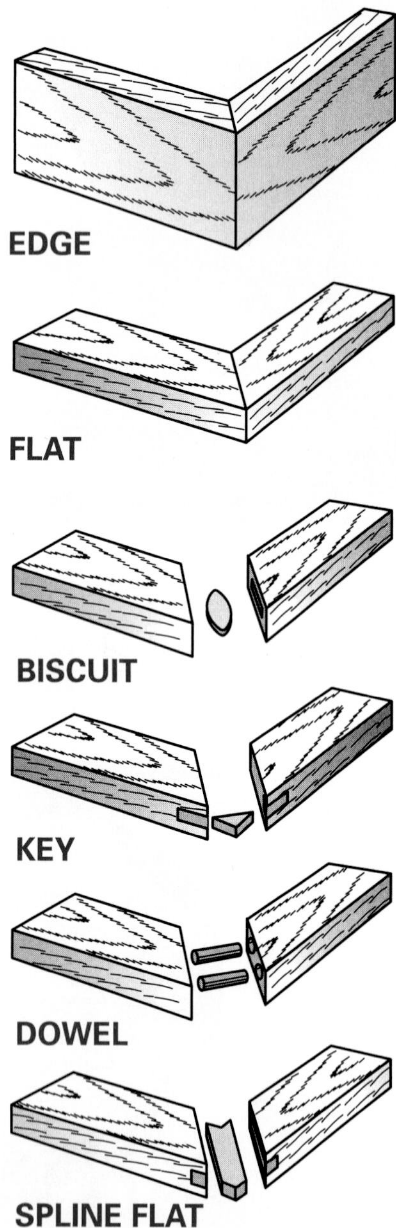

Fig. 12-2—*The miter joint is weak by itself. It can be strengthened by adding dowels, a spline, a key, or a biscuit.*

IMPORTANCE OF ACCURACY

A **miter box,** like the one shown in Fig. 12-3, can be adjusted to cut any angle from 30 to 90 degrees. Miter joints cut at different angles can form a variety of different shapes. Fig. 12-4.

It is extremely important to make very precise angle cuts when making any miter joint, no matter what the angle. Even being one or two degrees off on each piece can result in a gap in the joint line and cause difficulty in keeping even the simplest project, such as a picture frame, square. Imagine what would happen if you were a few degrees off on each piece of an octagon-shaped clock. Small mistakes in measuring and/or cutting can add up to big problems when the components of a project are assembled.

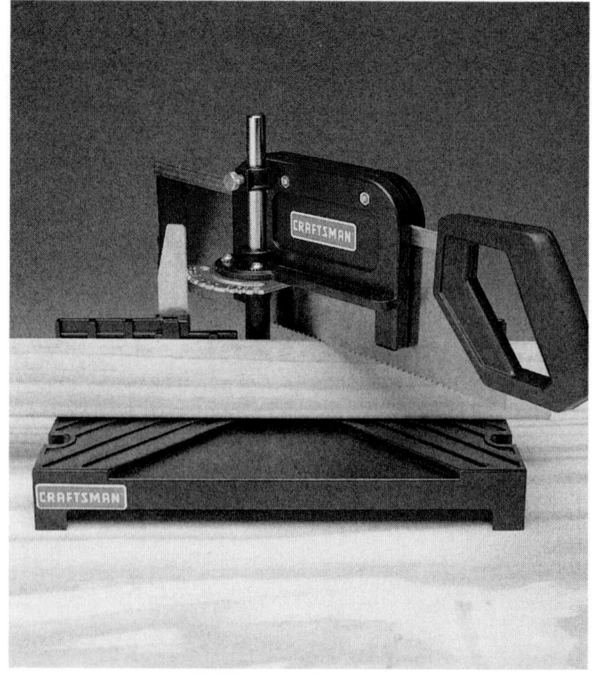

Fig. 12-3—*A commercial miter box. The guide that holds the saw can be adjusted to cut a variety of angles.*

MAKING A PICTURE FRAME

The most common use of a miter joint is in making a picture frame. You can purchase framing material with a pre-cut rabbet edge to hold the glass in place. You can also choose from among the hundreds of patterns and sizes of moldings that are available to create your own framing material. Fig. 12-5.

If using molding to make your frame, you will need to glue up two pieces of molding in a manner that will form a rabbet edge about ⅛ inch wide. Fig. 12-6. After applying the glue, clamp the strips of molding together until the glue is thoroughly dry.

Fig. 12-5—*These moldings are typical of the more than 250 patterns and 400 sizes available.*

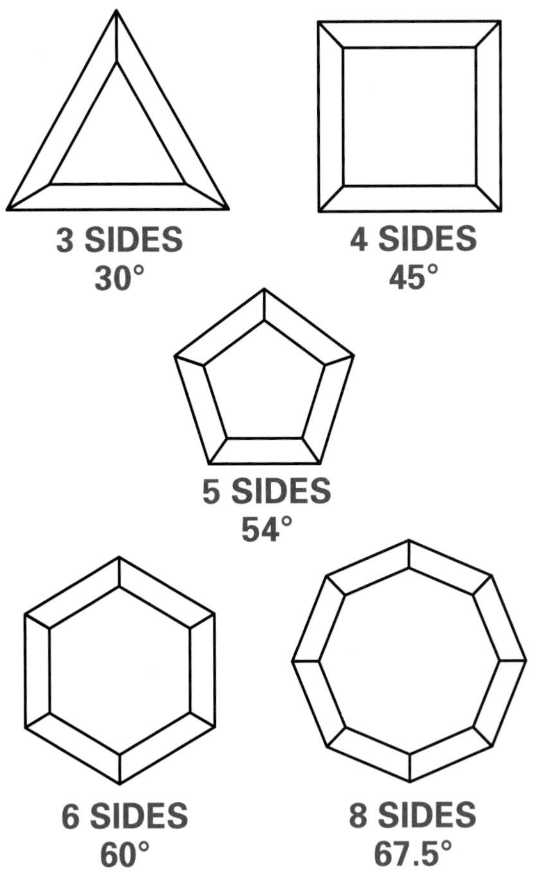

Fig. 12-4—*Not all miter joints are formed by pieces cut at 45°. A miter joint can be cut at any angle. These are a few of the more common ones.*

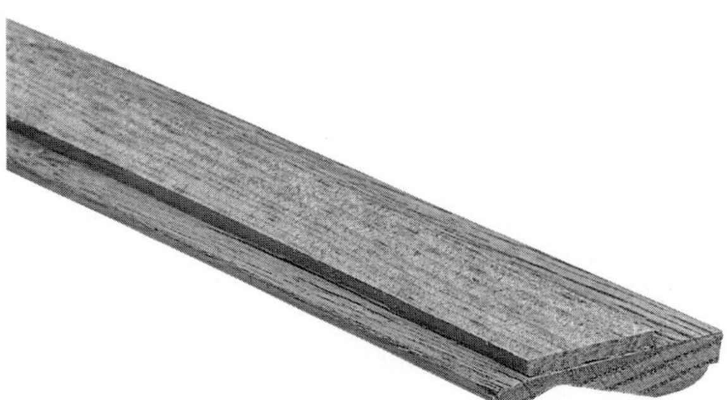

Fig. 12-6—*Glue two strips of molding together so that they form a rabbet edge. Generally, the rabbet for a picture frame is about ⅛ inch wide.*

Laying Out the Frame

Once you have your framing material, the steps for laying out, cutting, and assembling the frame are basically the same. Follow the steps below to lay out a picture frame.

1. Measure the length and width of the glass or picture to be framed.
2. Measure the width of the framing material from the shoulder of the rabbet to the outside edge of the stock. Fig. 12-7 (page 196).

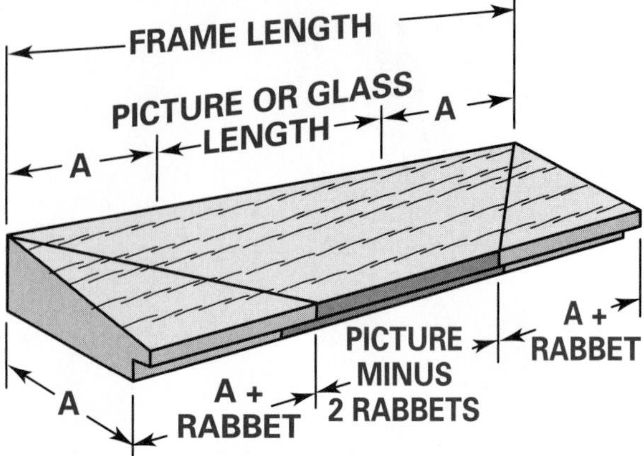

Fig. 12-7—***Laying out the proper measurement for cutting miter joints for a picture frame. The length of the frame is equal to the length of the glass plus twice the width of the stock measured from the rabbet to the outside edge.***

3. Add twice the distance you calculated in Step 2 to the length of the glass or picture. The sum equals the length to lay out for the sides of the frame. Refer again to Fig. 12-7.
4. Add twice the distance you calculated in Step 2 to the width of the glass or picture. This sum equals the width to lay out for the ends of the frame.

Cutting the Frame

Follow these steps to cut the picture frame.

1. If a commercial miter box is available, swing the saw to the left and set it at 45 degrees. If you are using a homemade wood miter box, place the saw in the slot that is to the left as you face it. Fig. 12-8.
2. Place one side of the frame in the box, with the rabbet edge down and toward you.
3. Hold the stock firmly, with the thumb of one hand against the side of the box. Fig. 12-9.
4. Carefully bring the saw down on the stock and cut the angle, using uniform strokes. Do not let the stock slip when starting the cut. This could ruin the frame.
5. Swing the saw to the right and set it to cut at 45 degrees.
6. Clamp the stock to the miter box with a hand screw. Then operate the saw with your favored hand.
7. Repeat Steps 1-6 for the second side and both ends.

Assembling the Frame

Before assembling the frame, check to make sure the corners fit properly. Lay the sides and ends in place on a flat table or bench. Use a try square to check the fit at each corner. Using a rule, measure the diagonals—top left to bottom right and bottom left to top right. Both measurements must be the same. Make sure there are no high spots.

Once you are sure all the joints fit properly, you are ready for the final assembly. Basically, this is just a matter of applying glue to each mitered edge, nailing each corner, and holding the assembled frame in a clamping device until the glue is thoroughly dry. The procedure varies somewhat, depending on the type of clamping device being used.

Safety First

► *Using Hand and Power Tools*

Follow all rules of good safety when using hand tools such as handsaws and nailing equipment. See Chapters 4 and 5.

If you are using a homemade clamping device like the one shown in Fig. 12-10 (page 198), you will need to glue and nail each corner together and fully assemble the frame before you can put it in the clamp. Follow these steps:

1. Drive one or two finishing nails or small brads partway into one piece. Apply glue to the mitered edge of this piece.
2. Apply glue to the edge of an adjoining piece. Then lock this piece in a vise in a vertical position.

MATHEMATICS Connection

Angles for Polygon Miters

To calculate the correct angle to cut each piece of a three- to 12-sided project, use the method below.

Step 1: 180 ÷ number of sides = x

Step 2: 90° − x = angle to cut

Example: A twelve-sided polygon

Step 1: 180 ÷ 12 = 15

Step 2: 90° − 15 = 75°

For a twelve-sided polygon, each piece would be cut at an angle of 75 degrees.

Fig. 12-8—*If using a homemade miter box, make your first cut with the saw in the 45° slot that is on your left when you are facing the saw.*

Fig. 12-9—*Make sure the stock is held firmly against the back of the box. This will keep it from slipping when the saw kerf is started. Start the cut with a careful backstroke.*

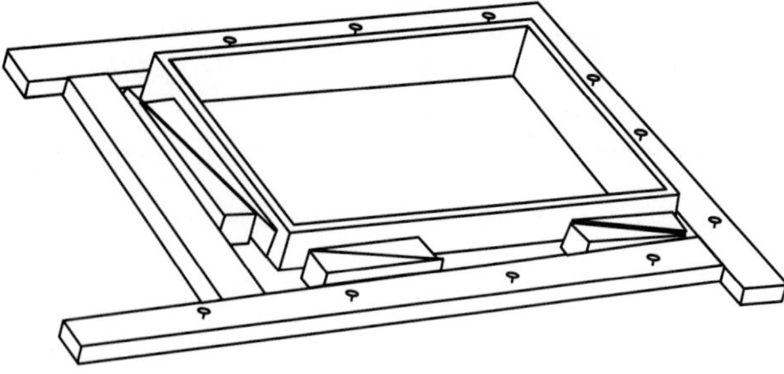

Fig. 12-10—*You can make a simple frame clamp by fastening pieces of wood to a piece of scrap and then using wedges to hold the frame securely until the glue dries.*

Miter Joints of Pieces with Different Widths

Miter joints can even be used to join two pieces of different widths. Place the narrower piece over the wider piece, as shown in the illustration. Mark the width of the superimposed narrower piece. Then draw a line from this mark on the wider piece up to that piece's corner to mark the cutting line.

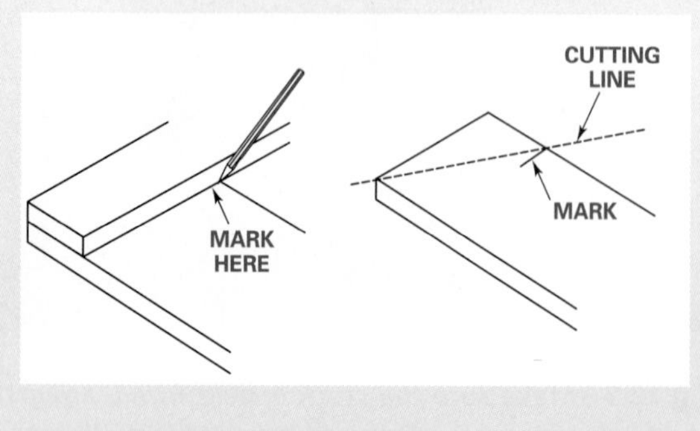

3. Hold the first piece over the vertical piece with its corner extending somewhat outside the edge of the vertical piece. Fig. 12-11. As you nail the corners together, the top piece will slip down until it fits squarely. Allow the nail heads to protrude $\frac{1}{16}$ inch; then use a nail set to drive the nail heads $\frac{1}{16}$ inch below the surface. Fig. 12-12. Wipe off any excess glue.
4. Repeat Steps 1-3 for each corner.
5. Secure the assembled frame in the clamp, making sure that all corners remain square. Allow the glue to dry thoroughly before removing the clamp. Fill the nail holes and sand lightly when dry.

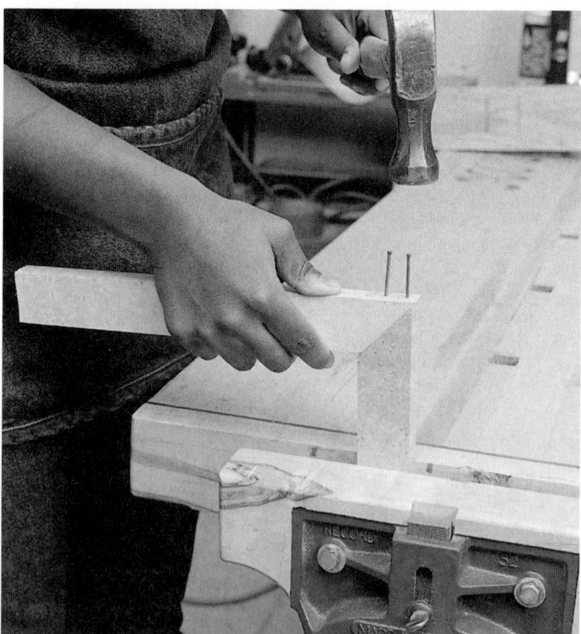

Fig. 12-11—*Nailing a miter joint. One piece is locked vertically in a vise and the second piece is held over it, with its corner extending slightly beyond the outside edge of the vertical piece.*

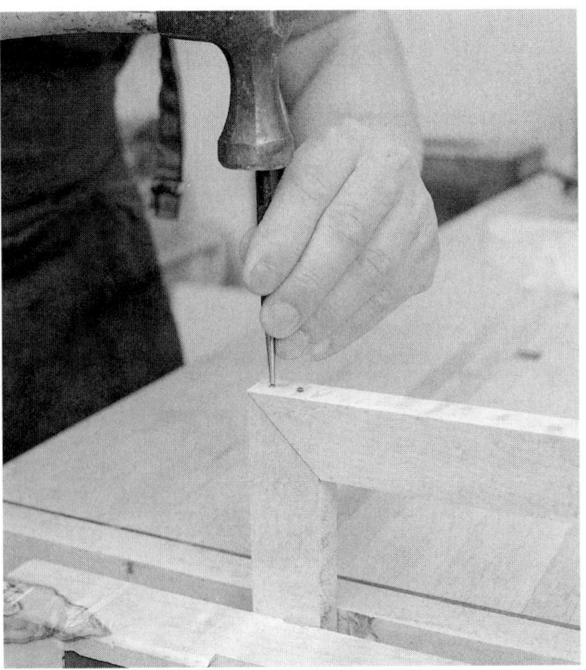

Fig. 12-12—*Use a nail set to drive the nail heads 1/16 inch below the surface.*

Fig. 12-13—*A miter vise with a tilting base. This vise can clamp moldings up to 4 inches wide. It will join any frame larger than 3½ × 3½ inches.*

If you are using a commercial clamp like the one shown in Fig. 12-13, follow these steps:

1. Apply glue to the mitered edges of one side piece and one end piece. Insert them in a corner clamp. Align the joint carefully and tighten the clamp. Wipe away any excess glue with a warm, damp rag.
2. Drive two or more brads or finishing nails into each side of the corner. Allow the heads of the nails to protrude slightly. Use a nail set to drive the nail heads below the surface. Fill each hole. After the glue has dried, sand this filled surface lightly.
3. Repeat Steps 1 and 2 for each corner.

Using a Power Miter Box

The power miter box has an attached, pivoting motor that drives a table saw blade. Fig. 12-14. Built-in *index plates* (locked stops) allow you to lock the saw quickly into position at a variety of left and right angles.

Fig. 12-14—*A power miter box makes quick work of cutting miters at a variety of angles. It features index plates and an adjustable clamp.*

Using Power Saws

To cut miters for frames on the table saw, set the miter gauge to the correct angle. Use a miter gauge clamp to keep the work in place as you cut through the stock. Fig. 12-15.

The setup for cutting cross miters on the table saw is the same as for crosscutting (see Chapter 22), except the blade is tilted to the desired angle. Place the stock with its good side up and secure it with a gauge clamp. Make the cut more slowly than in regular crosscutting.

Radial-arm saws can also be used to cut miters quickly and accurately. Swing the overarm of the saw to the desired angle, and pull the blade through the workpiece the same as you would for simple crosscutting. Fig. 12-16.

The sliding compound miter saw operates much like a radial-arm saw. Once the saw unit is pivoted to and locked in at the correct angle, it is pushed through the workpiece to make the miter cut. Fig. 12-17.

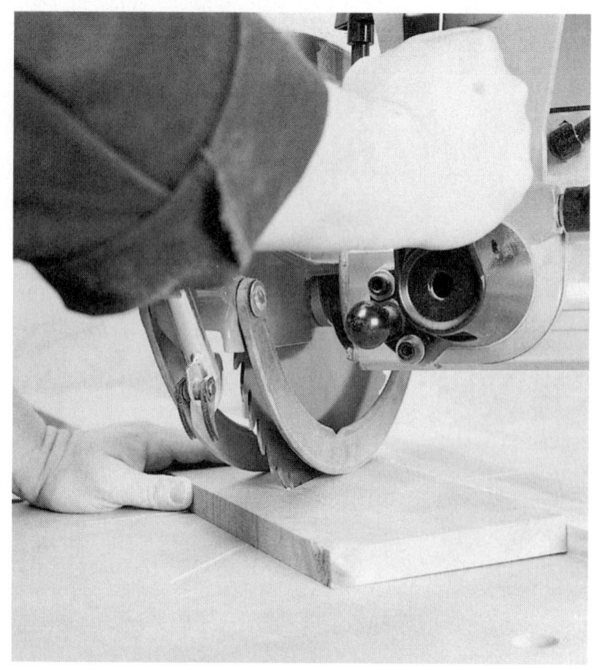

Fig. 12-16—*Making a miter cut on a radial-arm saw.*

Fig. 12-15—*Use a miter gauge with a clamp to cut a miter joint on the table saw. NOTE: The guard was omitted in order to show details. Always use a guard when doing this operation.*

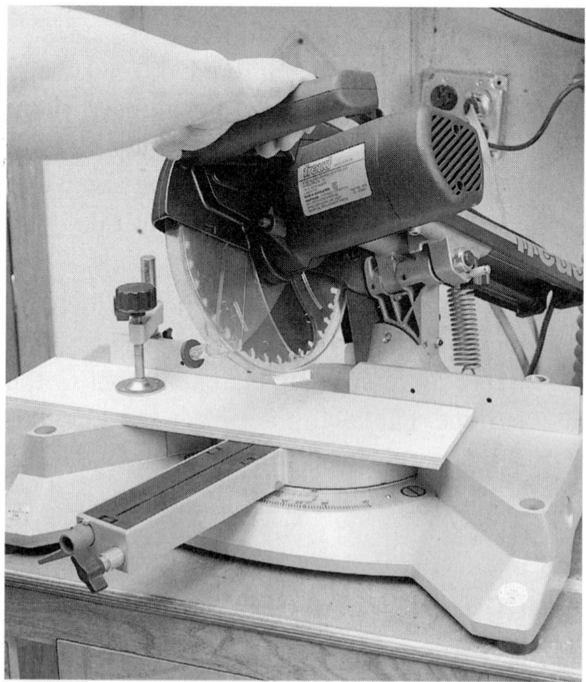

Fig. 12-17—*A sliding compound miter saw.*

CHAPTER 12

Review & Applications

Chapter Summary

Major points from this chapter that you should remember include:

- The miter joint is a fairly weak joint that hides the end grain of the joined pieces.
- Miter joints can be cut at many different angles. The most common miter joint is cut at 45 degrees.
- Miter joints can be cut by hand using either a commercial or homemade miter box.
- The most common use of the miter joint is in making picture frames. There are special procedures for accurately laying out, cutting, and assembling the frames.
- Power miter boxes and sliding compound miter saws cut miters quickly and accurately. Table saws and radial-arm saws can also be set up to cut accurate miters.

Review Questions

1. How can miter joints be strengthened?
2. Explain why accuracy is so important when laying out and cutting miter joints.
3. Describe how to lay out a picture frame.
4. Outline the steps for cutting a frame on a commercial miter box.
5. Before assembling the pieces of the frame, how can you make sure that the corners all fit properly?
6. Describe a power miter box.
7. Explain how to cut cross miters on a table saw.
8. Explain how a sliding compound miter saw operates.

Solving Real World Problems

Joanne has an old painting of the original homestead that her grandfather built, where she now lives. It is a wonderful painting that is now in need of a new frame. Because of the age, condition, and subject matter, and because the room where it will be hung is very rustic, she does not want an expensive commercial frame. Instead, she wants to make a frame from old barn boards she has salvaged from the barn. These boards were hand-cut and planed by her grandfather when he was her age. Joanne has some basic carpentry skills her grandfather taught her, but she has only his old hand tools. How can she make the frame?

CHAPTER 13

Mortise-and-Tenon Joint

YOU'LL BE ABLE TO:

- Describe a mortise-and-tenon joint.
- Explain how the length, width, and thickness of the tenon are determined.
- Explain how hand tools are used to make a mortise-and-tenon joint.
- Name the power tools used to make a mortise-and-tenon joint.
- Assemble a mortise-and-tenon joint.

Look for These Terms

- blind mortise
- mortise
- mortise-and-tenon joint
- tenon
- through mortise

Mortise-and-tenon joints are very strong joints found in fine furniture. **Fig. 13-1.** They can be used in leg assemblies on tables and for assembling doors, cabinets, and a number of other products.

A **mortise** is a rectangular hole cut in wood. The **tenon** is a projecting piece of wood shaped to fit into the mortise. Fig. 13-2.

The main advantage of the mortise-and-tenon joint is that it resists pressures that would force other joints apart. It not only provides a strong interlock but also has considerable area for gluing. The mortise-and-tenon joint can be formed and put together in a number of ways. Fig. 13-3.

Fig. 13-1—*This butcher-block cabinet was built using several mortise-and-tenon joints.*

DESIGNING A MORTISE-AND-TENON JOINT

When designing the mortise-and-tenon joint, you will need to determine the length, width, and thickness of the tenon. Fig. 13-4 (page 204). It must be sized and shaped to fit snugly within the mortise.

The *length* of the tenon depends on whether the mortise is blind or through. A **blind mortise** is not cut all the way through the piece. Therefore, the end of the tenon cannot be seen when the parts are assembled. Make the tenon slightly shorter than the depth of the blind mortise. Also round the edges slightly. Doing these things allows space for extra glue.

A **through mortise** is cut all the way through the piece. Cut the tenon slightly longer than needed. After assembly, the excess material can be sanded away, and the surfaces will be flush.

When planning the *width* and *thickness* of the tenon, consider the sizes of the cutting tools used to cut the mortise. A mortise sized according to the size of the cutting tool can be cut quickly and efficiently. The tenon can be sized to fit the mortise.

The *width* of a tenon should be 5 inches or less. If too wide, the piece in which the mortise is cut could be weakened. Instead of having a wide tenon, cut two or more tenons. Leave enough space between them to maintain strength. Fig. 13-5.

The *thickness* of the tenon should be between one-third and one-half the thickness of the piece in which the mortise will be cut. The width of the mortise is the same as the thickness of the tenon.

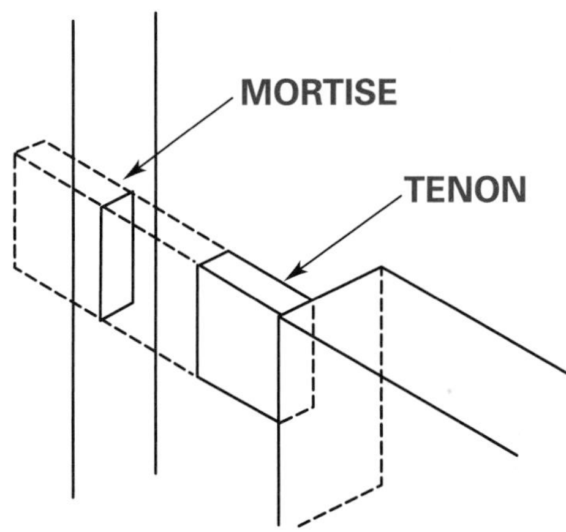

Fig. 13-2—*Components of a mortise-and-tenon joint.*

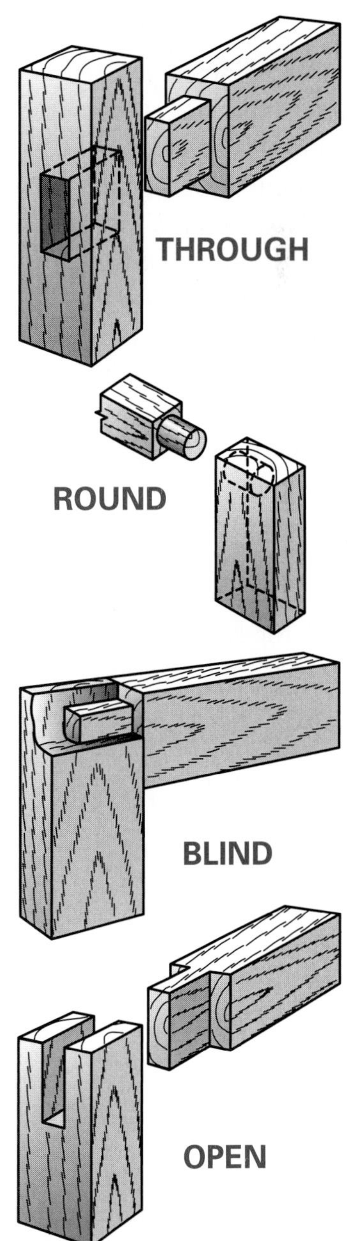

Fig. 13-3—*Common types of mortise-and-tenon joints.*

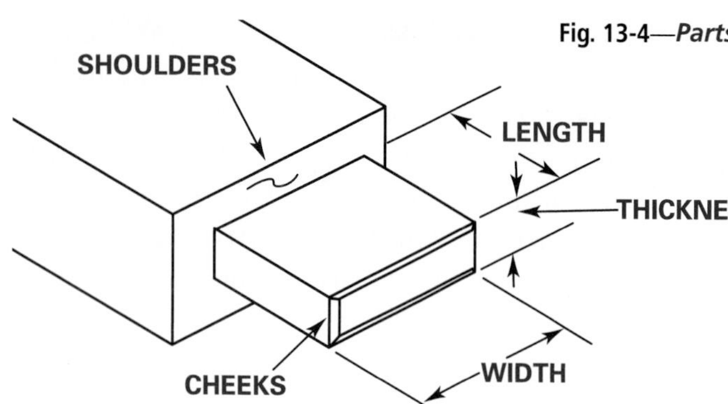

Fig. 13-4—*Parts of the tenon.*

An assembled mortise-and-tenon joint looks like a simple butt joint. When pieces are the same size, all surfaces must be flush. When the pieces to be joined are not the same size, the part less likely to show is inset and a shoulder is left. Fig. 13-6. Insetting has another advantage. Under the stress of use, the joint may separate slightly. For example, a slight crack may form between a leg and rail on a footstool. Such cracks are not as noticeable when one piece is inset.

Use a square and pencil to lay out the mortise and the tenon on the stock. Make layout marks to outline the thickness and width of the tenon. Fig. 13-7. *Be especially careful.* These marks for sawing must be accurate in order to make a snugly fitting tenon.

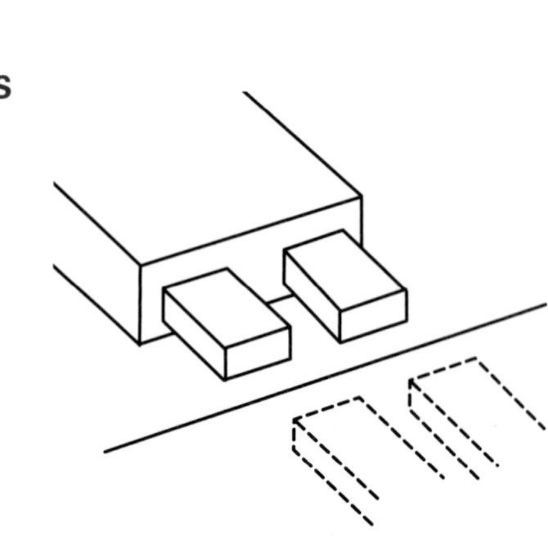

Fig. 13-5—*Use multiple tenons when one tenon would be too wide.*

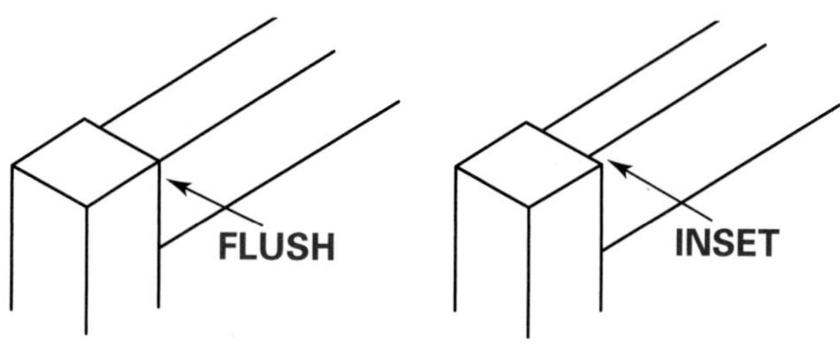

Fig. 13-6—*Separation of the joint is less likely to show when one part is inset.*

FORMING THE PARTS OF THE JOINT

Once the dimensions are decided, preparations can be made for cutting. Plan to cut the mortise first. The reason for this is that it is easier to make size adjustments on the tenon.

Using Hand Tools

Making a mortise-and-tenon joint by hand is difficult and time-consuming. However, its strength and appearance can make the effort worthwhile.

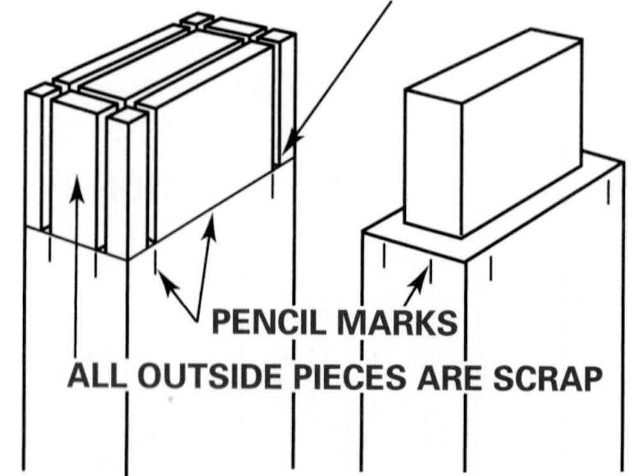

Fig. 13-7—*The four cuts that shape the thickness and width of the tenon. NOTE: Make the cuts in waste stock.*

Cutting the Mortise

With a *drill,* make a series of holes to remove most of the waste stock from the opening. Fig. 13-8. Use a *depth gauge* set to make the holes slightly deeper than the length of the tenon. With a chisel, trim the sides and ends of the opening to the layout line. Finish the ends with a narrow chisel.

Another method is to remove all of the waste stock with a *mortising chisel.* The width of the blade should be exactly the width of the opening. Fig. 13-9. Clamp the workpiece firmly to the bench. Begin to cut at the center of the mortise. Hold the chisel in a vertical position with the bevel side turned away from you. Cut out a V-shaped notch to the depth required.

Safety First

►*Using Hand Tools*

- Follow all rules of good safety when using hand tools.
- Always take special care with tools that are sharp.

For more safety information on using hand tools, see Chapters 4, 6, and 7.

Fig. 13-8—***Drilling out a mortise. Note that the bit selected has the same diameter as the width of the mortise (or thickness of the tenon).***

Fig. 13-9—***The blade of the mortising chisel should be exactly the same width as the mortise.***

Then continue to remove the waste by driving the chisel down with a *mallet.* Draw down on the handle to remove the chips. Stop when you are within about ⅛ inch of the end of the opening. Turn the chisel around with the flat side toward the end of the mortise and cut out the remainder of the waste. Fig. 13-10.

Cutting the Tenon

Lock the stock in a *vise* with the marked tenon showing. Using a *backsaw* or fine *crosscut saw,* make four saw cuts in the waste stock.

Next, remove the stock from the vise and clamp it to the top of the workbench. Make the shoulder cuts to remove the waste stock and form a tenon of the correct thickness and width. Fig. 13-11.

The tenon must fit snugly within the mortise without being either too tight or too loose. Cut a small chamfer around the end of the tenon to help it slip easily into the mortise opening.

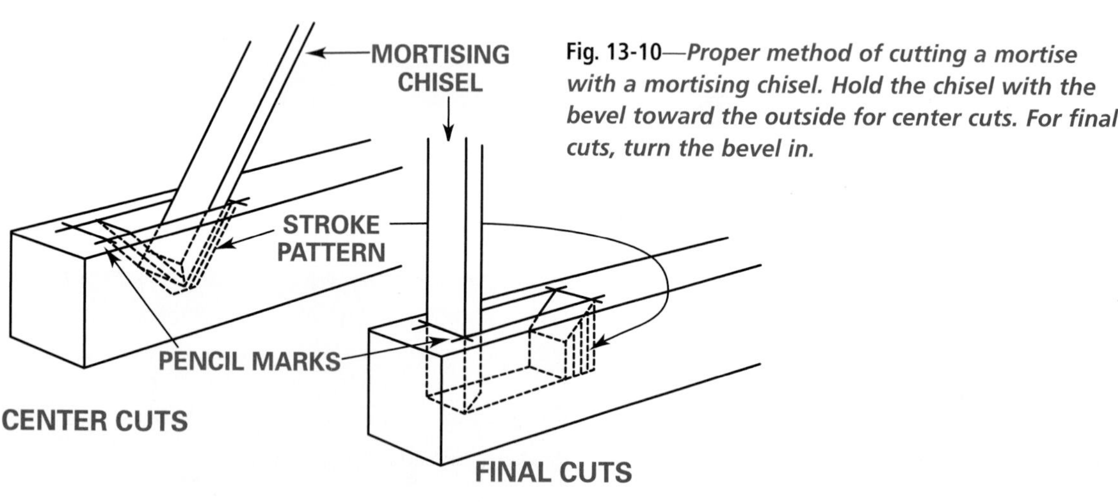

Fig. 13-10—*Proper method of cutting a mortise with a mortising chisel. Hold the chisel with the bevel toward the outside for center cuts. For final cuts, turn the bevel in.*

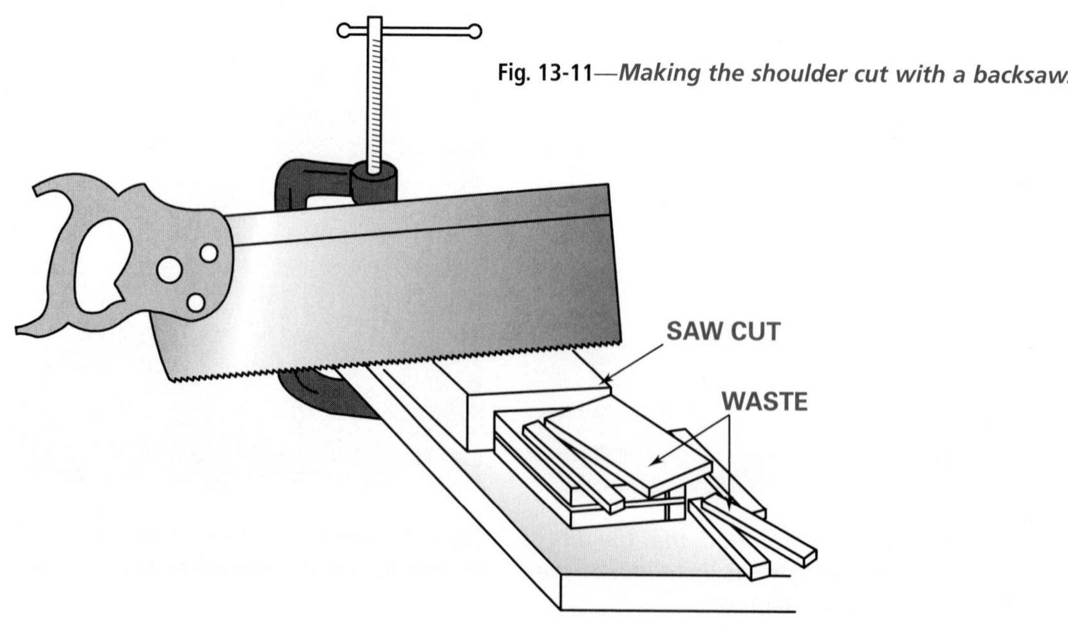

Fig. 13-11—*Making the shoulder cut with a backsaw.*

Safety First

►*Using the Drill Press or Mortising Machine*

- Keep your fingers at least 4 inches away from the rotating tool.
- Remove chips and shavings with a brush or stick of wood—never your fingers.

For further safety information about operating the drill press, see Chapter 27.

Using Power Tools

A *drill press* can be used to make the cuts for a mortise-and-tenon joint. A special mortising accessory is used. It includes a hold-down clamp, a fence, and a device that is mounted on the spindle of the drill press. A mortising machine may also be used. Fig. 13-12. For cutting, a mortising chisel and bit is attached to this device. The square mortising chisel is hollow. Inside it is a bit that cuts and removes waste stock as the four-sided chisel cuts out the corners to make a square hole. When setting up, make certain the width of the chisel is square with the piece.

Mortises should be sized according to the standard sizes of mortising chisels. These sizes are ½, ⅜, 5⁄16, and ¼ inch.

The speed at which the machine is run depends on the size of the chisel and the wood's density. As you work, determine which speed works best for the wood you are using. When feeding the tools into the wood, use a firm, steady pressure.

The tenon can be easily cut on the *table saw* with a tenon jig. Fig. 13-13 (page 208).

Facts about Wood

Wooden Steamboats—The paddle-wheel steamboat was developed in the nineteenth century. It was used to haul cargo and passengers on America's rivers—the Ohio, the Mississippi, and the Missouri. Steamboats were built of woods such as oak and walnut. A steamboat could float in such a small amount of water that some joked that it could run on a wet handkerchief. Their design ensured that they would not scrape bottom in shallow rivers. The boilers on a steamboat were fueled by wood. Each boat had along its route a series of stations at which fresh supplies of wood could be taken on. From the early 1830s until the early 1880s, the steamboat was a common sight on America's major rivers.

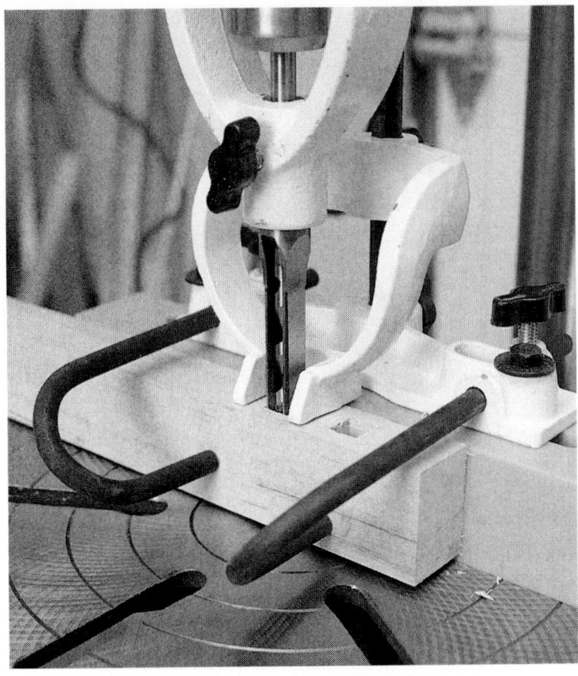

Fig. 13-12—***When using a mortising machine, make the end cuts first. Finish by cleaning out the waste between the ends.***

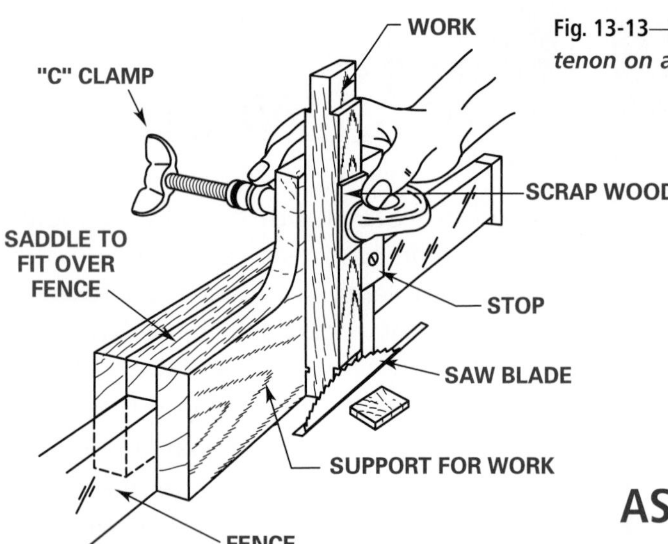

Fig. 13-13—*A tenon jig is helpful when cutting the tenon on a table saw.*

ASSEMBLING THE JOINT

Before assembling the joint, check the fit of the pieces and adjust them if necessary. To adjust the size of the tenon, use a chisel to pare off stock from its thickness and width as needed. You should be able to insert the tenon into the mortise with a moderate amount of pressure. Make sure that the shoulder of the tenon fits squarely against the face of the mortise piece.

When you are confident that the pieces fit correctly, fasten them together. Mortise-and-tenon joints are usually fastened permanently with adhesives.

SCIENCE *Connection*

Importance of Clean Gluing Surfaces

Before you apply glue (adhesive) to a mortise and tenon (or any other joint), be sure the area to be glued is clean. Be especially sure to brush away any wood shavings or loose wood fibers that result from cutting the mortise and tenon. These small fibers tend to soak up the glue before it can penetrate the solid wood.

One reason for this is that the individual fibers have more surface area than the solid wood. The greater surface area allows greater adhesion between the glue and the wood. *Adhesion* is a molecular attraction between two different objects—in this case, the wood or wood fibers and the glue.

Safety First

►*Using the Table Saw*

- Always use the saw guard.
- Do not allow your fingers to come closer than 5 inches to the saw blade.

For further safety information about setting up and operating the table saw, see Chapter 22.

CHAPTER 13

Review & Applications

Chapter Summary

Major points from this chapter that you should remember include:

- Mortise-and-tenon joints are very strong.
- The length of the tenon depends on whether the mortise is blind or through.
- A mortise should be sized according to the size of the cutting tool used to cut it.
- When joined pieces are the same size, all surfaces should be flush. When unlike sized pieces are joined, one part is inset and a shoulder is left.
- The mortise should be cut slightly deeper than the length of the tenon.
- The tenon must fit the mortise exactly.
- Adhesives are used to fasten the mortise-and-tenon joint permanently.

Review Questions

1. Why is a mortise-and-tenon joint used in making better-quality furniture?
2. Describe a mortise-and-tenon joint.
3. Name two types of mortises.
4. How does the depth of the mortise compare to the length of the tenon?
5. What is the rule for determining the thickness of the tenon?
6. How do the width of the mortise and the thickness of the tenon compare?
7. Is the mortise or the tenon cut first? Why?
8. How are hand tools used to cut a mortise?
9. Which types of hand and power saws can be used to cut the tenon?
10. Describe the mortising accessory used when cutting a mortise with a drill press.

Solving Real World Problems

Li is a big fan of arts-and-crafts style furniture designed by Gustov Stickly in the early 1900s. In fact, he likes the style so much that he bought a home designed by Stickly in 1901. He now wants to fill it with arts-and-crafts style furniture. Li is an accomplished woodworker and is capable of building all the furniture himself. He has most of the tools he needs. While researching design parameters for the dining room set, he decides that the number of mortise-and-tenon joints necessary will require him to use some sort of machine to make the mortises. There are just too many to do it manually. There are two machines he might use. What are they, and which would you choose? Why?

CHAPTER 14

Dovetail Joints and Casework

YOU'LL BE ABLE TO:

- Make a dovetail joint using a dovetail jig and a router with a dovetail bit.
- Build a project using simple casework construction.
- List five methods of installing shelves within a bookcase.
- Construct a drawer.
- Make a paneled door.

Look for These Terms

- casework
- dovetail joint
- drawer guide
- flush door
- flush drawer
- lip door
- lip drawer
- rail
- shelf standard
- sliding door
- stile

When you handcraft a piece of furniture, you will have something uniquely yours. No two handcrafted items are exactly alike. Yet they are all made using principles and techniques developed and passed down by previous generations. The dovetail joint, for example, is an interlocking joint that was developed during a time when good glues and mechanical fasteners were not available. **Fig. 14-1.** You can use these tried-and-true methods to build solid and attractive furniture of your own.

Fig. 14-1—*The dovetail joint is a strong, interlocking joint.*

DOVETAIL JOINTS

The **dovetail joint** is a sign of quality in woodworking. It is the preferred joint in drawer construction primarily because of its strength. Fig. 14-2. The "fingers" of the joint are wedge-shaped and resemble the spread of a dove's tail. These fit into matching sockets to form the interlocking joint. Figs. 14-3 and 14-4.

Making a dovetail joint with hand tools requires a great deal of time and skill. Today the easiest and most accurate way to make a dovetail joint is to use a dovetail jig with a router and dovetail bit. Fig. 14-5 (page 212). The time and skill now required is devoted to preparing the setup accurately. Once you have the proper setup, however, you can cut as many dovetails as you need, quickly and accurately.

Using a Dovetail Jig in Drawer Construction

Typically, a half-blind dovetail joint is used in drawer construction. It is a very professional-looking joint. The jig for cutting these joints is probably the cheapest and easiest to set up of the dovetail jigs. Regardless of the width or thickness of the wood, you can clamp both pieces of the joint in the jig and cut them with a router at the same time. This ensures that tails fit the sockets.

Fig. 14-2—Dovetail joints are often used in drawer construction. Shown here is a half-blind dovetail joint. It can be seen from the side but not from the front.

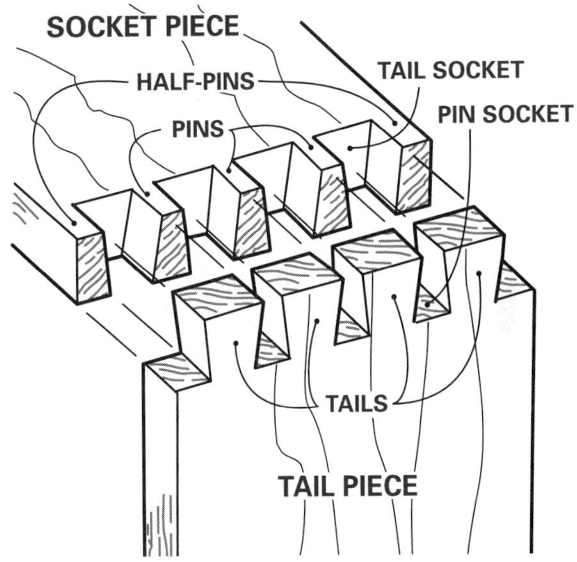

Fig. 14-3—Parts of a dovetail joint. This is a through dovetail joint. Both parts of the joint can be seen.

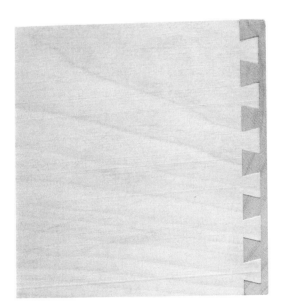

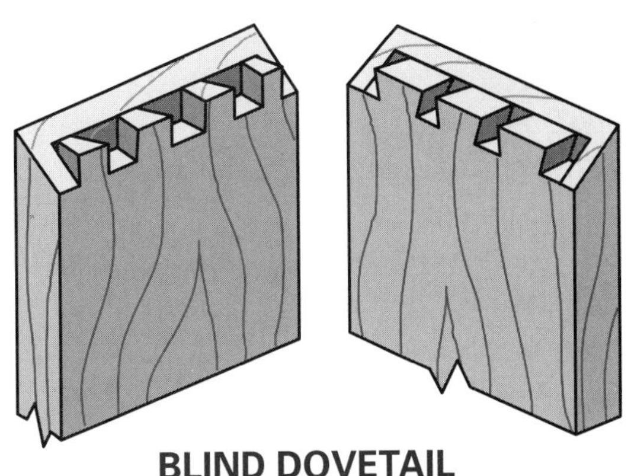

Fig. 14-4—A blind dovetail joint. All parts of the joint are completely hidden when assembled.

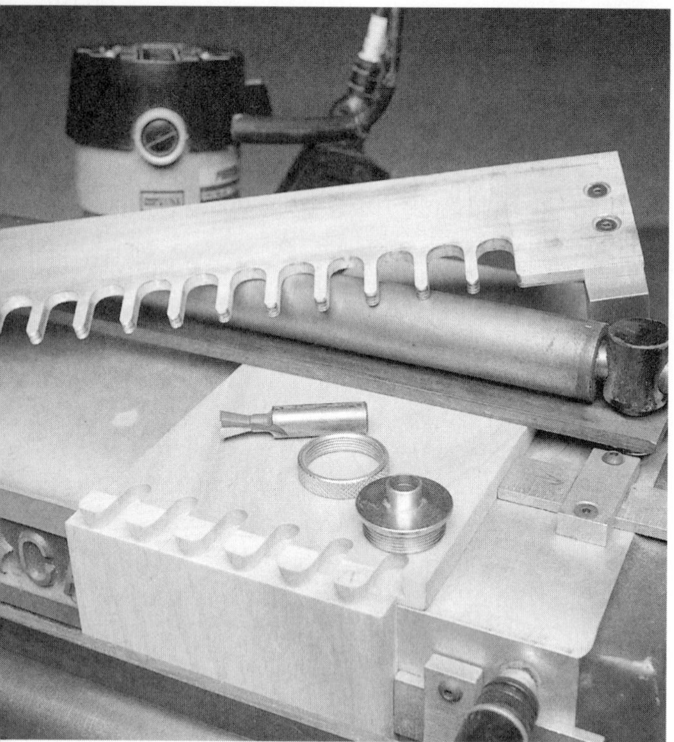

Fig. 14-5—*A router, dovetail bit, and a dovetail jig. The shape of the dovetail bit, not the shape of the jig, determines the shape of the tails and sockets.*

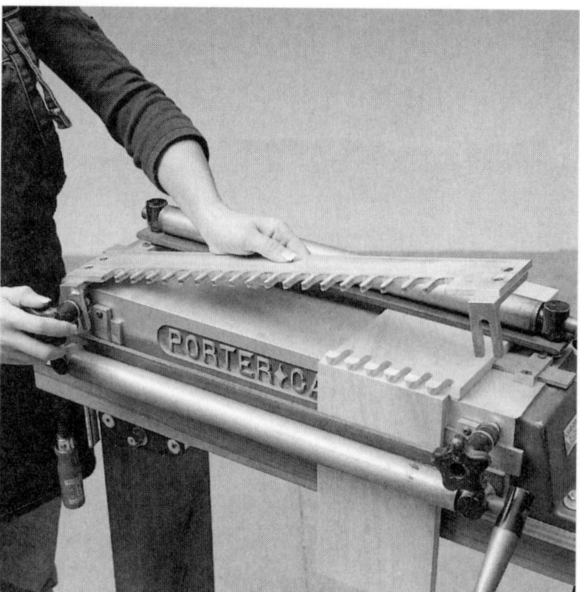

Fig. 14-6—*Be sure the pieces are clamped in the jig correctly before positioning the template.*

Setting Up the Jig

Before setting up most jigs, you will need to attach the jig to a base. Plywood ¾ inch thick works well. Clamp the base to the edge of a workbench.

Prepare your setup using test pieces of wood. Doing this will allow you to make corrections in the setup without ruining your final workpieces.

Select one piece to be the socket piece and the other to be the tail piece. Clamp the socket piece to the *top* of the jig. Then clamp the tail piece to the *front* of the jig. Fig. 14-6. Butt the socket piece against the face surface of the tail piece. Be sure the end of the tail piece is flush with the top surface of the socket piece. Both should fit snugly against the alignment stops of the jig.

Place the template over the test pieces. Slip the brackets over the rods on the jig base and tighten the stop nuts. Tighten the lock-down knobs.

Safety First

►*Using a Router*

- Always wear eye protection when using a router.
- Keep your hands and clothing away from the rotating cutters at all times.
- After turning off the power, wait until the machine comes to a complete stop before setting it on its side.

For further safety information about operating a router, see Chapter 28.

Make the test cuts. Using a router with a dovetail bit, begin cutting from left to right, going in and out of each slot. Fig. 14-7. When you complete the cut, go through the process again more quickly.

Without unclamping any parts, examine your work. If any more cutting needs to be done, do it now, before you move the pieces. Remember, the tails and pins must fit the sockets exactly. Accuracy is important.

When you are sure all cutting is complete, remove the template and unclamp the test pieces. Make a test assembly of the joint.

Refining the Setup

When you examine your test assembly, you will often find that some fine adjustments need to be made. Fig. 14-8. If the parts fit too loosely, you will need to increase your depth of cut. If the fit is too tight, you will need to decrease your depth of cut. Whatever problems you find, refer to the instruction sheet for using the jig. This sheet will tell you what to do to correct the problems.

Making Final Cuts

When all final adjustments have been made, you are ready to cut your actual workpieces. One final concern is to make sure the workpieces are clamped correctly in the jig. The sides that you should see are the sides that will be *inside* the final assembly. Therefore, you can mark those sides because they will eventually be hidden. Figure 14-9 (page 214) shows how you can label the parts and position them on the jig. It's helpful to remember that the front and back pieces of a drawer will always be clamped to the top of the jig. Side pieces will always be clamped to the front of the jig.

Note also that certain joints must be cut on certain sides of the jig. Joints on the left side of the drawer are cut on the left side of the jig. Joints on the right side of the drawer are cut on the right side of the jig. Label the jig as well as the workpieces so that all parts are matched correctly.

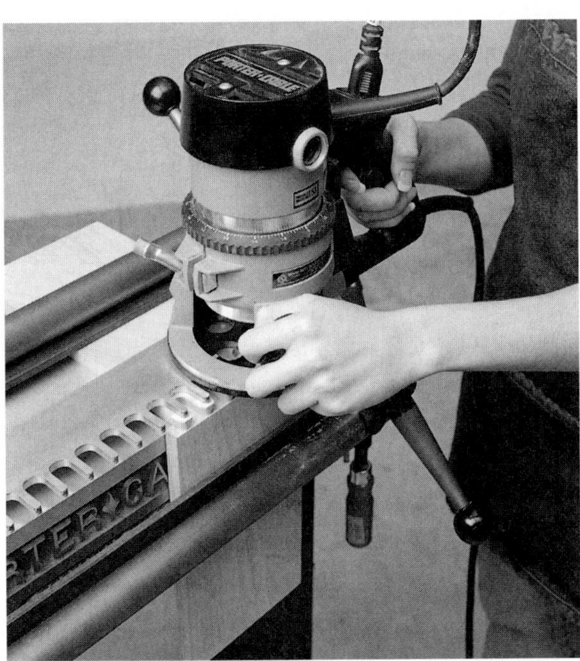

Fig. 14-7—*Be sure to get clean cuts in each slot so the parts will fit together correctly.*

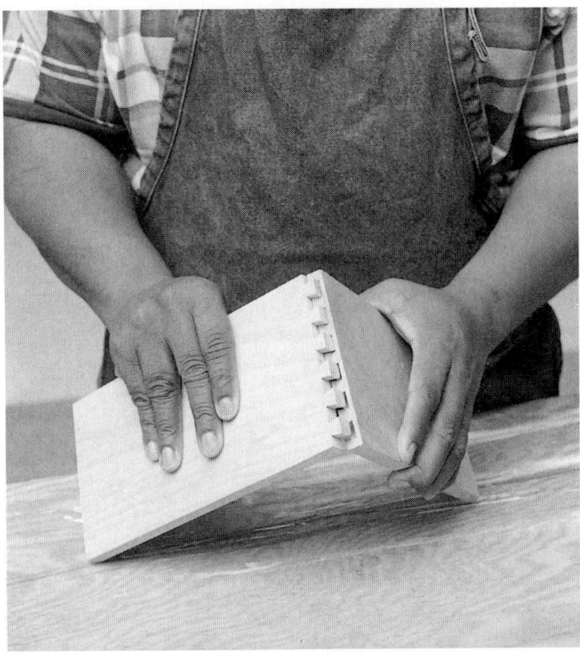

Fig. 14-8—*Check the test assembly to determine whether the parts of the joint fit snugly together.*

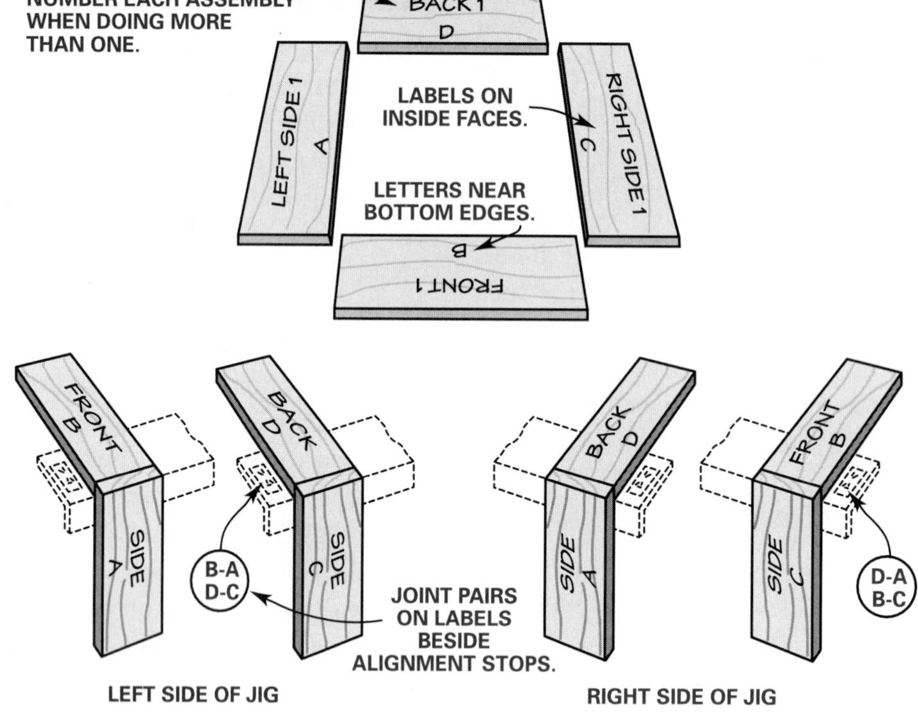

Fig. 14-9—*Label the parts and the jig so that the joints are cut correctly.*

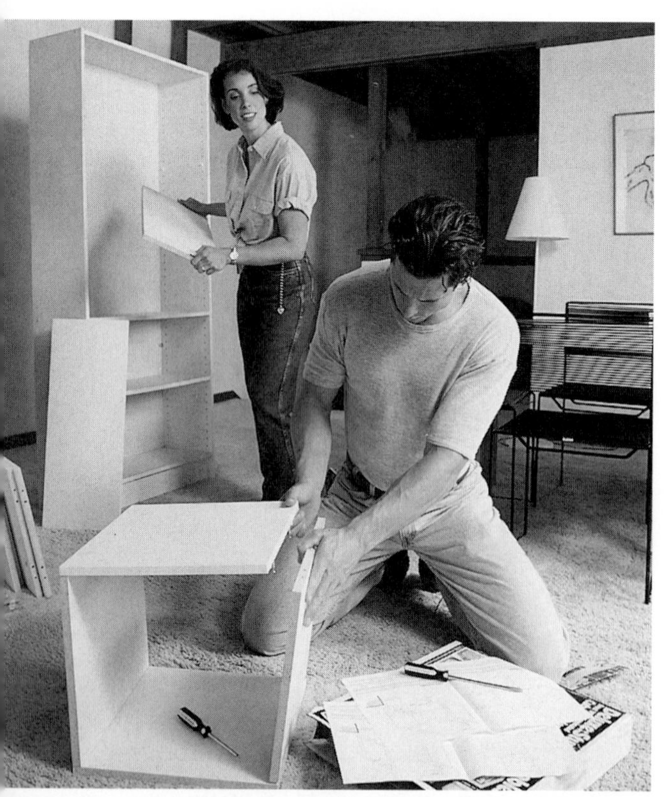

Fig. 14-10—*An example of simple casework.*

CASEWORK

Simple **casework** is a box turned on its end or edge. Fig. 14-10. The construction of many types of furniture is based on solid panel casework. Bookcases, desks, and chests of drawers are examples.

The Case

In simple casework, four panels are joined to form the basic case and a back is fit to the case. The grain of the panels should all run in the same direction. Then, if panels swell or shrink, they will do so together. Separate spaces can be formed inside the case by installing horizontal shelves or vertical dividers.

There are many ways of fitting a back to the case. The most common is to cut a rabbet around the back edge. As you may recall from Chapter 9, a *rabbet* is an L-shaped cut. You can then fit a back panel of thin plywood or hardboard into the recess and fasten it in place with nails. Fig. 14-11.

Most simple casework is constructed of softwood, plywood, medium-density fiberboard (MDF), or particleboard. The case can be painted or covered with plastic laminate.

Advanced casework may have web frames, face frames (sometimes called face plates), drawers, and doors in addition to shelves and dividers. Such casework is often used for cabinets in kitchens and bathrooms. Fig. 14-12.

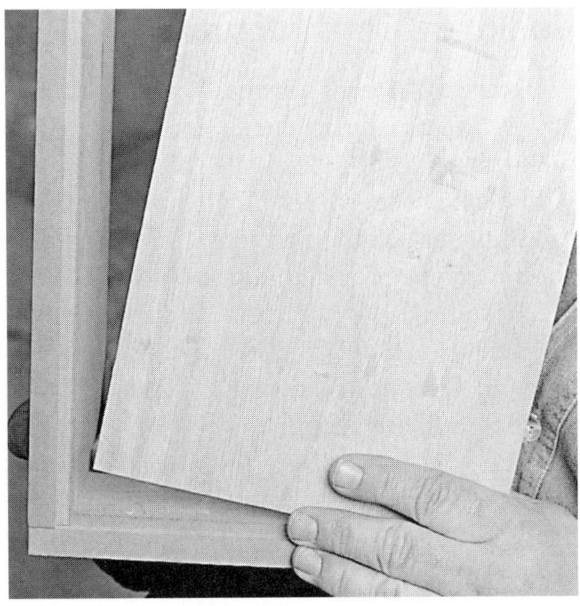

Fig. 14-11—***Fitting the back of a case into a rabbet cut.***

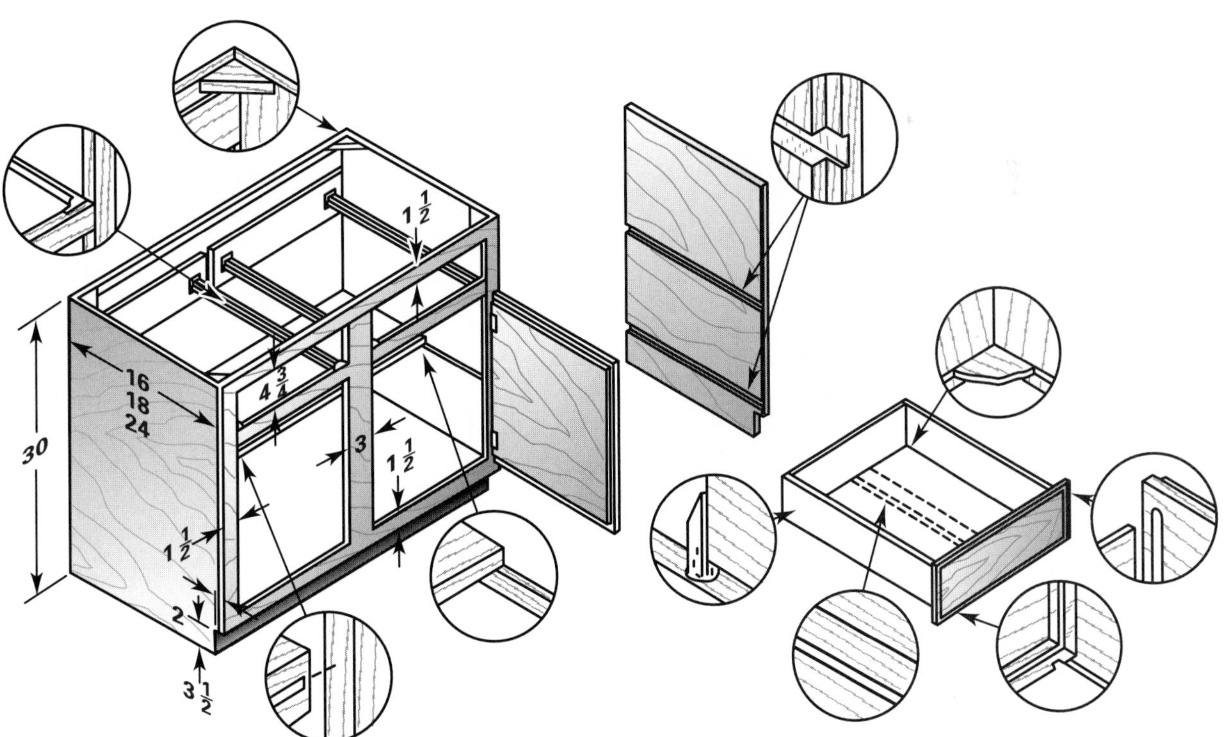

Fig. 14-12—***Typical construction for quality kitchen cabinets.***

Bookcase Construction

Bookcase construction is simple casework. Fig. 14-13. It usually includes either fixed or adjustable shelves. Fixed shelves are installed with dado or other types of joints, wood cleats, or shelf brackets. Support for adjustable shelves can be provided in many ways. Various metal and wood commercial **shelf standards** (slotted strips) and brackets or clips are available for all kinds of shelving. Standards can be purchased in any length. Fig. 14-14.

Metal shelf standards with snap-on clips work well in bookcases. You will need two on each side. The best way to install them is to cut grooves into the sides so standards placed in the grooves fit flush with the inside surface. Then cut the shelves to the correct lengths. If you mount standards without grooves, the ends of each shelf must be notched to fit around the standards.

Another way to provide shelf support is to drill a series of holes in the sides of the case. Fig. 14-15. Then insert metal or plastic shelf pins or dowels into the holes to hold the shelves. Fig. 14-16.

BUILDING FURNITURE

A good first project is to build a small piece of furniture such as a table, stool, or cabinet. When you have more experience, you can build larger furniture for a living room, dining room, or bedroom.

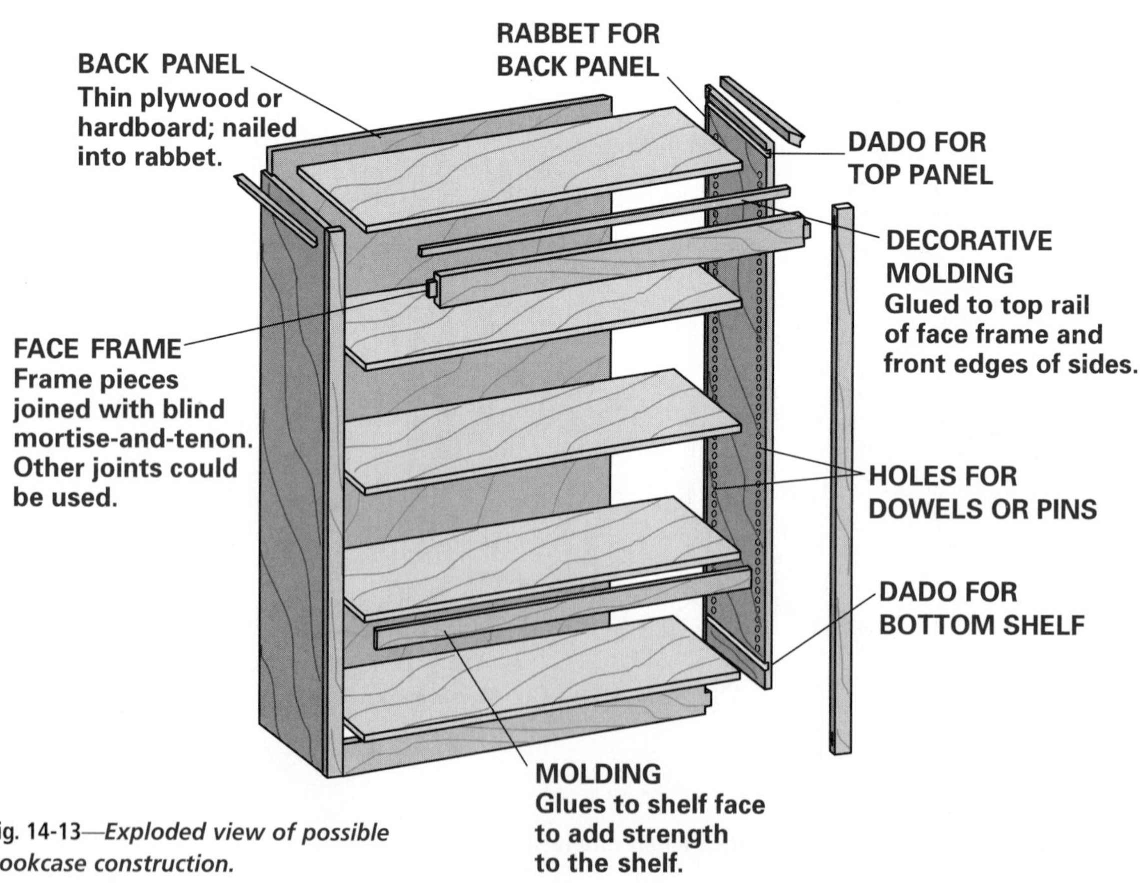

Fig. 14-13—*Exploded view of possible bookcase construction.*

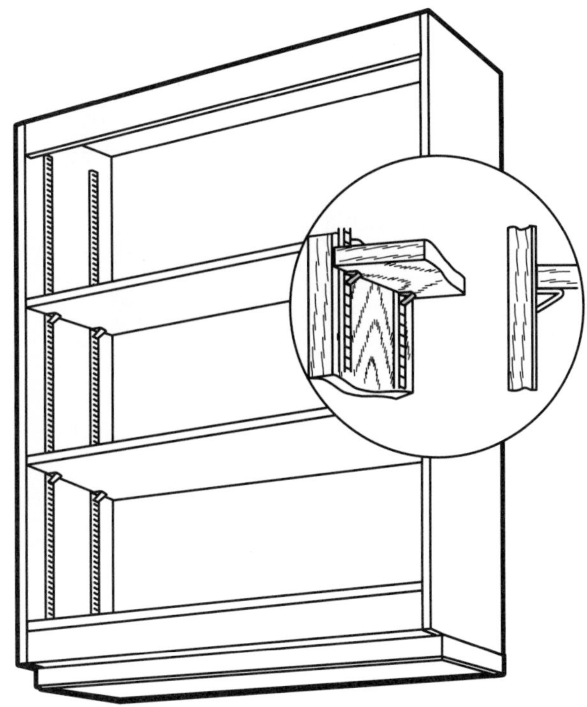

Fig. 14-14—*A bookcase with adjustable shelves. The metal standards are recessed into grooves so that the ends of the shelves can be cut square. Clips are snapped into the standards to hold the shelves.*

WOODWORKING TIP

Building Stronger Shelves

You can do several things to make your shelves stronger:

- Change the dimensions. Make the shelves shorter, wider, or thicker to add strength.
- Use a more rigid wood. The more rigid a wood is, the more strength the shelf will have. From most to least rigid, you could choose hickory, birch, hard maple, oak, ash, walnut, soft maple and poplar, cherry and basswood, butternut, plywood, or composites.
- Use fixed shelves instead of adjustable shelves. Mounting the ends of a shelf into the sides makes it more rigid.
- Install a vertical brace below the center of the shelf.

Fig. 14-15—*A special jig makes it easier to drill holes that are equally spaced.*

Fig. 14-16—*Shelf pins or dowels can be placed in holes to hold shelves in place.*

Table, Chair, and Stool Construction

Most tables, chairs, and stools are made with four legs joined by *rails*. Fig. 14-17. **Rails** are horizontal parts of a frame. The legs and rails are usually fastened together with either dowel or biscuit construction or mortise-and-tenon joints. Fig. 14-18. To strengthen the joining parts, a wood or metal corner block is inserted. Fig. 14-19. The corner blocks help to hold the furniture square. They also add support at its weakest points. Fig. 14-20.

Legs for tables, chairs, and stools can be made of wood in various designs. Matching wood, metal, or plastic legs can be purchased. Usually, these are sold in pairs.

Fig. 14-17—*A small footstool is a simple project made by joining four legs with rails.*

Fig. 14-18—*A double dowel joint used to fasten the leg and rail together.*

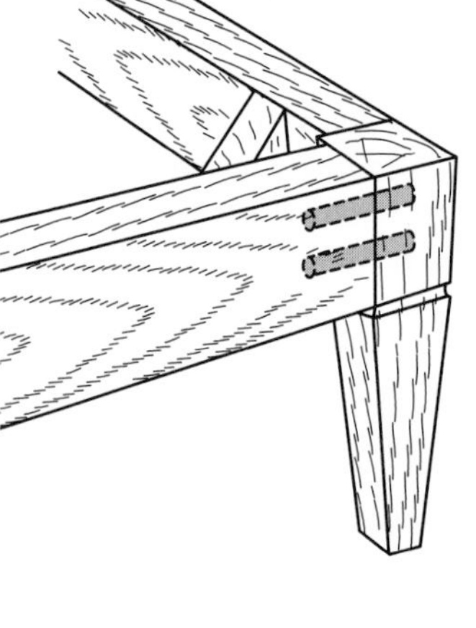

SCIENCE *Connection*

Moisture Content and Seasoning

When a tree is cut down, it contains moisture. Before the wood can be used, a large part of this moisture must be removed. This is done by air and/or kiln drying. In drying, free water inside the cells is removed first. When all of this is gone, the wood is said to be at the *fiber-saturation point.* The wood still contains about 23 to 30 percent moisture.

Lumber for furniture construction should have only about 6 to 10 percent moisture content. If wood has too much moisture when the project is built, the wood continues to dry out, causing warpage and cracked joints.

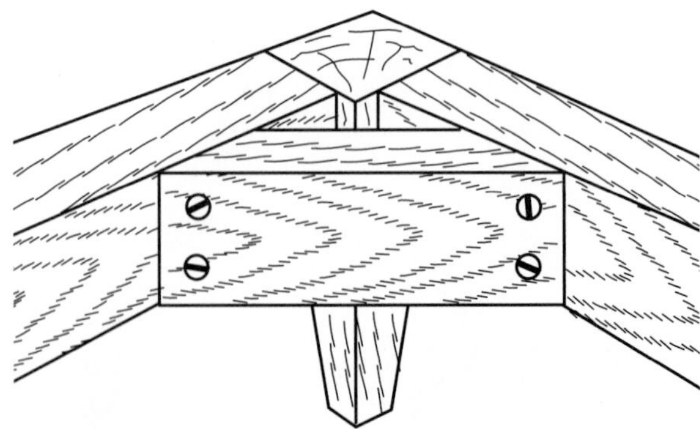

Fig. 14-19—*Wood corner blocks should be installed using glue and screws.*

Safety First

►*Using Tools*

- Always wear eye protection when working in the shop.
- Use all tools properly.
- Always take special care with tools that are sharp.

Fig. 14-20—*These chairs have four legs and rails joined with dowels. The corners are strengthened with corner blocks.*

A tabletop is usually made of plywood, particleboard, or MDF. It is often covered with either wood veneer or plastic laminate. The edge may be trimmed with solid wood or with plastic laminate or veneer. Figure 14-21 shows three common methods of attaching a tabletop to the rails.

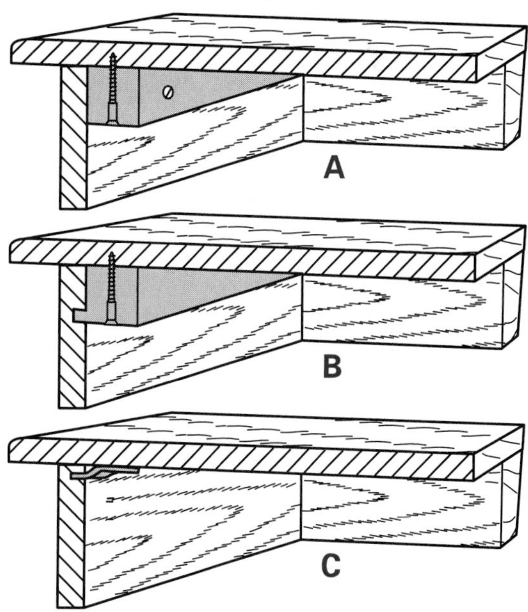

Fig. 14-21—*Ways of attaching a tabletop to the rails. (A) Square cleat. (B) Square cleat with rabbet and groove. (C) Metal tabletop fastener.*

Drawer Construction

Most cabinets, chests, and desks include drawers. There are three basic steps in installing a drawer in a table: cutting the rail to receive the drawer, making the drawer, and installing drawer guides. In desk and chest construction, only the last two steps are needed.

Cutting the Rail

Before cutting the table rail to receive the drawer, first decide on the exact size of the drawer. Then cut an opening that is $\frac{1}{16}$ inch wider and $\frac{1}{8}$ inch longer than the drawer front so the drawer will slide easily.

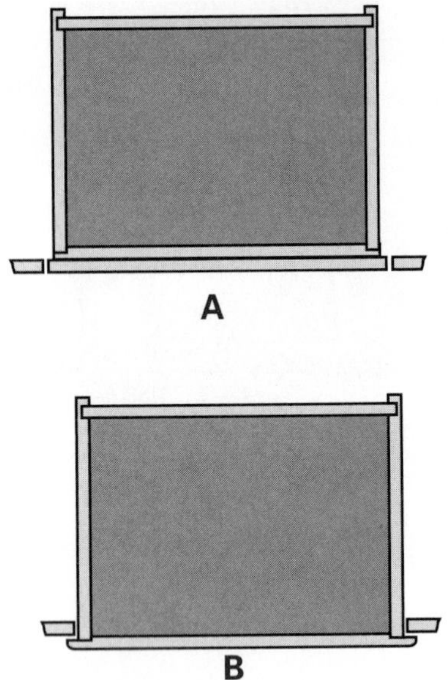

Fig. 14-22—*Two basic types of drawer fronts. (A) Flush drawer. (B) Lip drawer.*

Making the Drawer

The drawer may be made to fit flush with the opening (a **flush drawer**) or it may be a **lip drawer.** Fig. 14-22. To make a lip drawer, a 3/8-inch rabbet is cut around the inside edge of the drawer front. The rabbet edge, or lip, overlaps the frame.

The front is usually made of 3/4-inch material. The grain and color usually match the material used in the project. The sides and back of the drawer are made of 1/2-inch stock such as pine, birch, or maple. The bottom is usually made of 1/4-inch fir plywood or hardboard. Cut all stock to size.

A common way of joining the front to the sides is to use a rabbet joint. Fig. 14-23. The rabbet is cut to a width of two-thirds the thickness of the sides. This allows some clearance for the drawer. Other kinds of drawer joints can also be used. Although a little more challenging, the strongest way is to use a dovetail joint. Fig. 14-24. The back is joined to the sides with a butt or dado joint.

For the drawer bottom, cut a groove 1/4 inch by 1/4 inch about 1/2 inch above the lower edge and on the inside of the front and sides. Sometimes a groove is also cut across the back. Cut the bottom about 1/16 inch smaller than the width between the grooves. This allows for shrinkage and swelling.

Assemble the drawer. Glue and nail the sides to the front. Never put glue into the grooves for the bottom. Slip the bottom in place. Then glue or nail the sides to the back.

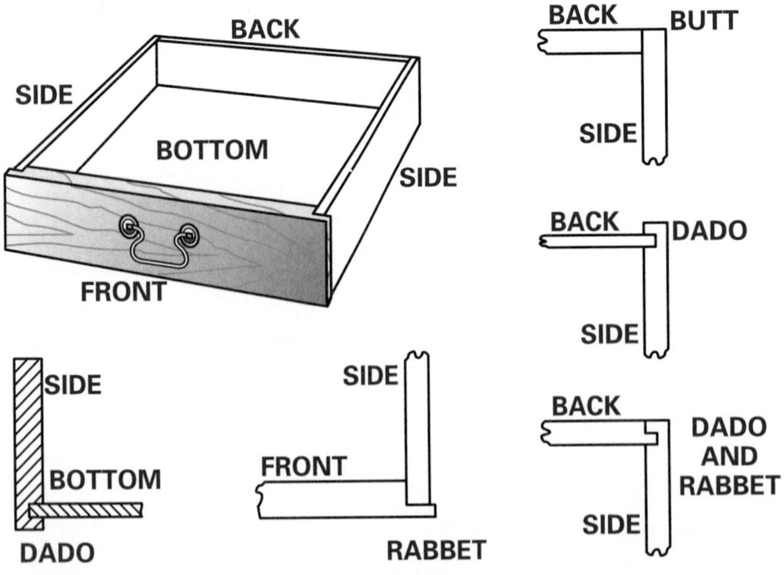

Fig. 14-23—*Simple ways of fastening the sides to the front and back in drawer construction.*

Fig. 14-24—*A dovetail joint used to join the front and side of a drawer.*

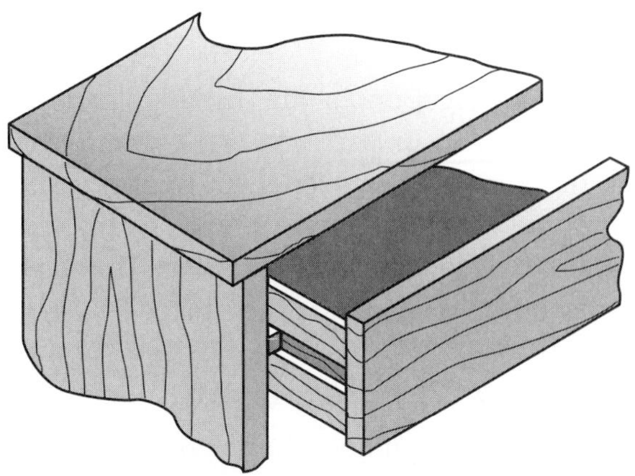
Fig. 14-25—*A simple drawer guide.*

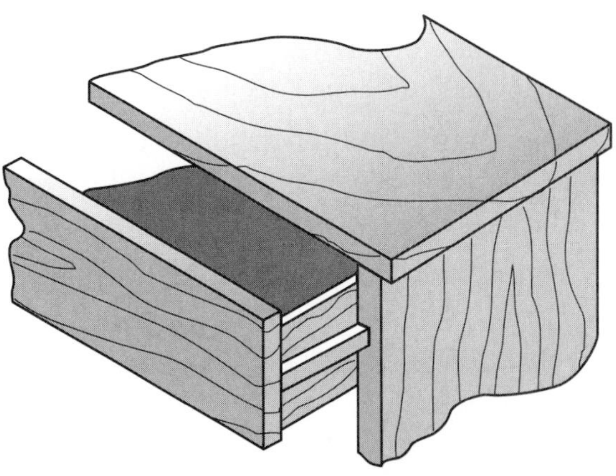
Fig. 14-26—*Side guide and runner with the dado cut in the case and a strip attached to the drawer.*

Installing Drawer Guides

Drawer guides are devices, usually tracks, on which the drawer slides. Fig. 14-25. The three most common types are the slide-block guide and runner (simple drawer guide), the side guide, and the center guide. Drawer guides support the drawer. They keep it from slipping from one side to the other. They also allow the drawer to open and close more easily. Another method is to attach a runner to each side of the drawer and then cut matching grooves in the project. Fig. 14-26.

On better-quality furniture, a center rail is attached to the project. Then a guide with a groove cut into it is attached to the bottom of the drawer. Fig. 14-27. If a drawer sticks, rub a little paraffin at the tight spots.

One of the easiest ways to install drawers is with commercial metal drawer guides. See Chapter 17. Commercial guides come in many sizes and styles.

Fig. 14-27—*A center guide is fastened to the bottom of this drawer. Small glue blocks are placed along the two adjoining surfaces to help keep the guide from becoming loose when used repeatedly.*

Door Construction

Common hinge doors are usually made of either plywood or particleboard covered with plastic laminate or veneers. Better-quality doors are of panels because they warp less. Only the frame can change in size. The panel inside is free to expand or contract.

Making a Paneled Door

To make a paneled door, follow these steps:

1. Lay out and cut the stock for the frame. Allow extra length for making the joints.
2. Square up the stock.
3. Cut a groove along the joint edges of each piece into which the panel will fit. This groove should be as deep as it is wide.
4. Use mortise-and-tenon joints to connect the vertical side pieces, called **stiles,** with the rails. Fig. 14-28.
5. Fit the panel temporarily into the frame to check it.
6. Take the frame apart.
7. Cover the edge of the panel with the soap or wax to help keep glue from getting into the groove or the edge of the panel.
8. Apply glue to the joints and fit the pieces together around the panel.
9. Clamp the frame together until the glue sets. See Chapter 16.

Mounting Doors

Doors are mounted into the frame of a piece of furniture. They are made to fit the frame in two ways. A **flush door** fits inside the frame, flush with the surface. Like the lip drawer, a **lip door** is made to cover part of the frame. It has a rabbet cut around the inside edge of the door on three or four sides. **Sliding doors** are used in places in which it is difficult to have doors that swing open and shut. Sliding doors are usually made of plywood, hardboard, or glass. Fig. 14-29.

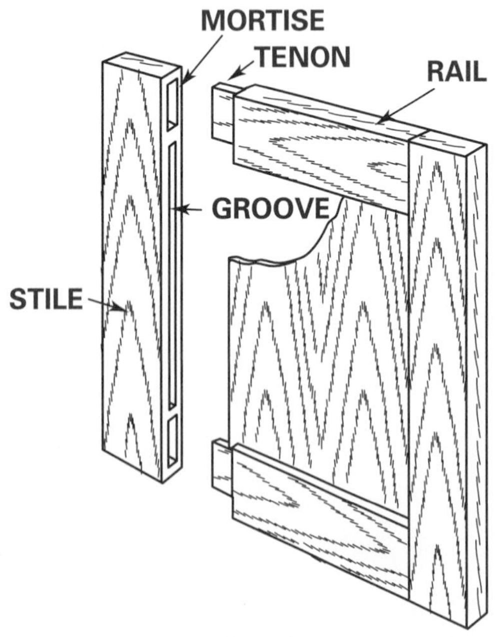

Fig. 14-28—***A mortise-and-tenon joint for a panel door.***

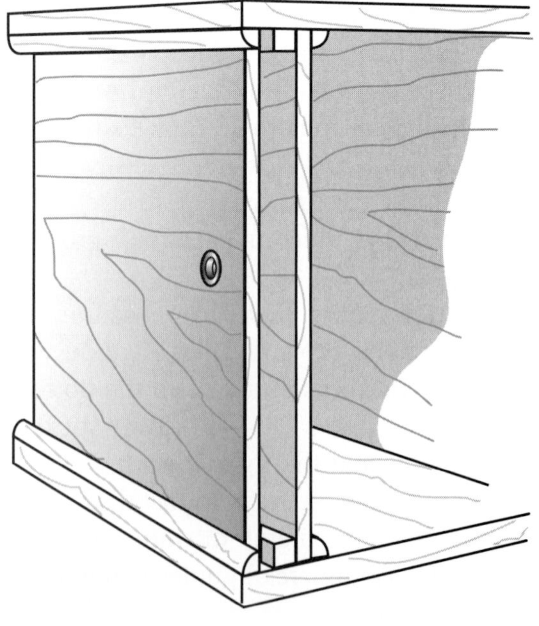

Fig. 14-29—***One method of installing sliding doors.***

CHAPTER 14

Review & Applications

Chapter Summary

Major points from this chapter that you should remember include:

- The dovetail joint is a strong, interlocking joint often used in drawer construction.
- Prepare and try out your dovetail jig setup using test pieces of wood.
- Casework consists of a box turned on its end or edge.
- Shelf standards with clips or brackets are devices used to support adjustable shelves.
- Most tables and stools are made with four legs joined by rails.
- A drawer may be made to fit flush with the drawer opening or to overlap the frame.
- The frame of a paneled door consists of vertical stiles and horizontal rails.
- Swinging doors may be flush or lip doors.

Review Questions

1. Describe the dovetail jig setup and process for cutting a dovetail joint with a router.
2. In simple casework, how do you build a basic case with a back?
3. What materials are typically used in simple casework?
4. List three ways of installing fixed shelves in a case.
5. Name four ways in which support can be provided for adjustable shelves in a case.
6. Describe briefly how you would construct a flush drawer.
7. What are the three common types of drawer guide?
8. How do you prepare stock and assemble the parts when making a paneled door?

Solving Real World Problems

Shelving is very limited in the Wilson family home. The Wilsons would like to make a wall rack to hold their favorite magazines, paperback books, and audio CDs. They want the shelves to be deep and strong. The wall space where the rack would fit is 30″ wide and 36″ high.

Design a wall rack for the Wilsons. What are some construction techniques that would make the shelves strong? How should the shelves be spaced?

CHAPTER 15

Using Fasteners

YOU'LL BE ABLE TO:

- Discuss tips and guidelines to be followed when working with screwdrivers and screws.
- Explain how a clearance hole should be drilled.
- Describe the process of countersinking for flathead screws.
- Demonstrate how to drive a wood screw.

Look for These Terms

- American Screw Wire Gauge
- countersink
- countersinking
- offset screwdriver
- Phillips-head screwdriver
- plain screwdriver
- screwdriver bit
- screw-mate counterbore
- screw-mate drill and countersink
- wood screw

You read about nails in Chapter 5; now let's discuss the screw, a much more secure fastener than the nail. A **wood screw** is a fastener with a groove twisting around part of its length. The wood screw cuts its own threads into the piece it is fastening. It is a strong fastener that can be removed to take the product apart. **Fig. 15-1.**

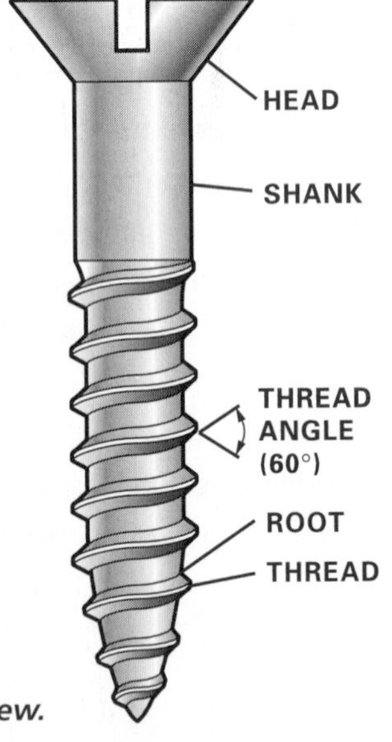

Fig. 15-1—*A typical wood screw.*

SCREWDRIVERS

The **plain,** or regular, **screwdriver** is used to install *slotted-head screws.* The size depends on the length of the blade and the width of the tip. Fig. 15-2. Make sure that the tip of the screwdriver is the same width as the diameter of the screw head. If it is wider than the screw it may mar the surface of the wood as the screw is set in place.

Other screwdrivers are made to install screws with *recessed heads.* The **Phillips-head screwdriver** is made for driving *crossed-head screws.* Fig. 15-3. The square-head (Robertson) screwdriver is also used on wood screws. It works better than the slotted but not as well as the Phillips. The spiral ratchet screwdriver is usually sold in a set with several plain bits and a Phillips bit.

Offset screwdrivers are used to install or remove screws located in tight places where a standard screwdriver cannot be used. They are available for slotted or Phillips-head screws. Fig. 15-4.

Figure 15-5 shows proper and improper screwdriver tips. If not ground to a square edge, the tip tends to slip out of the slot. This can mar the surface of the wood or damage the head of the screw.

The easiest screwdriver to use is the power screwdriver. Fig. 15-6 (page 266). For large screws, a power drill with a **screwdriver bit** will make the job much easier.

Here are some hints for using screwdrivers:

- Select a screwdriver with a length and tip fitted to the work. The tip should fit the screw slot or recess snugly. A plain screwdriver should not be wider than the screw head.

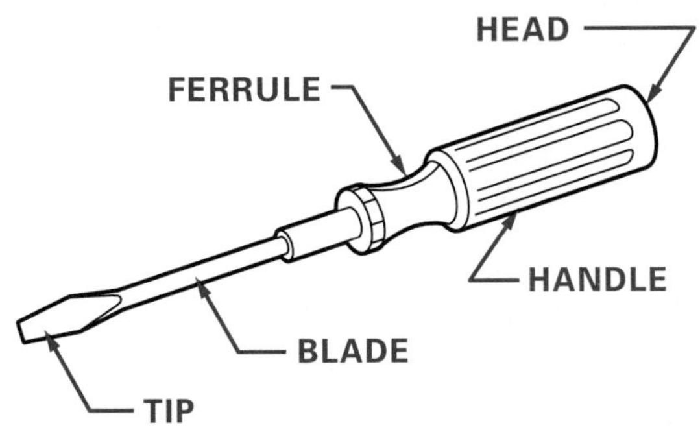

Fig. 15-2—*Parts of a plain screwdriver.*

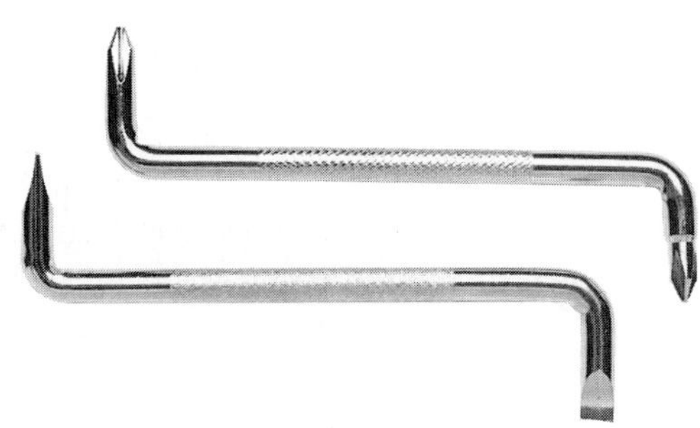
Fig. 15-4—*Offset screwdrivers.*

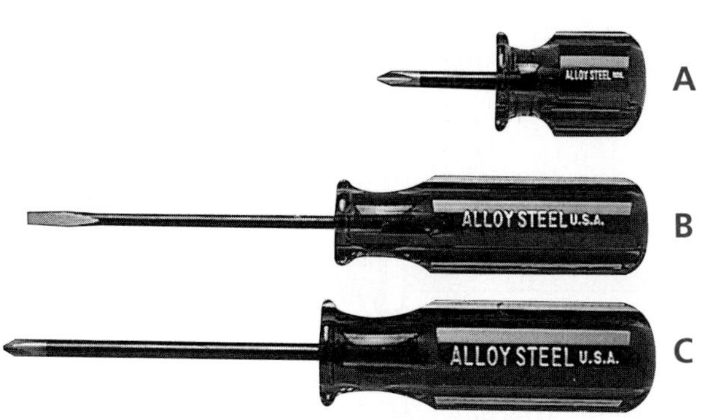

Fig. 15-3—*Screwdrivers used for wood screws. (A) Stubby Phillips. (B) Standard (slotted) screwdriver. (C) Phillips.*

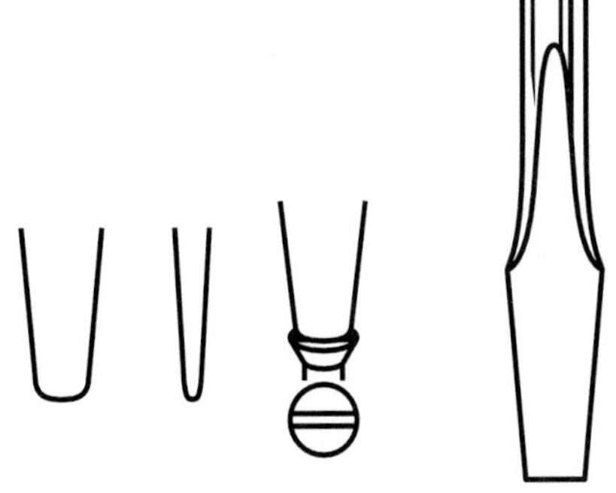
Fig. 15-5—*If the tip is rounded or beveled, the driver will rise out of the slot, spoiling the screw head. Regrind or file the screwdriver tip to make it as shown above on right.*

- Use the widest-tipped screwdriver convenient for the work. Wider tips provide more power. Also, there is less danger of wider tips slipping from the screw.
- Hold the handle firmly in the palm of one hand with thumb and forefinger grasping the handle near the ferrule. Use the other hand to steady the tip and keep it pressed into the screw. Fig. 15-7.

In working with screws, always try to follow these guidelines:

- When screws must be driven into hard-to-reach places, use a screwdriver with steel jaws that hold the screw firmly in place for easy starting. Fig. 15-8.
- When driving brass screws into hardwood, first use a steel screw of the same size to complete pilot holes. This reduces the risk of shearing the heads off the brass screws.
- If the screw tends to bind, back off and enlarge the pilot hole or rub paraffin or wax on the screw.
- Heating the head of a screw with a soldering iron makes the screw easier to remove after the screw cools.
- Use wood screws instead of nails:
  - To avoid splitting the wood
  - For greater holding power
  - For ease in taking the product apart later
  - For better appearance
- Never try to install screws partway with a hammer.
- Screw length should be selected so that the screw comes no closer than ⅛ inch from the bottom of the bottom (second) piece.

Fig. 15-6—*The power screwdrivers and cordless drills with screwdriver bits make your work a lot easier.*

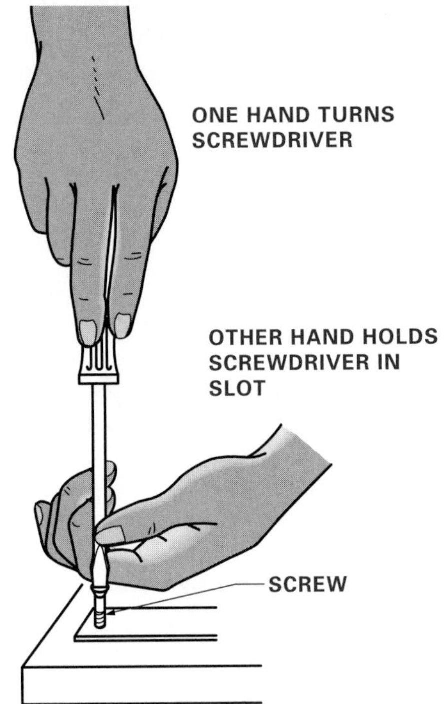

Fig. 15-7—*Hold the screwdriver properly to start the screw.*

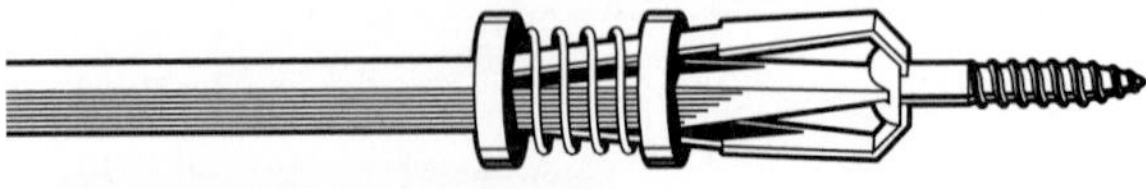

Fig. 15-8—*Steel jaws hold the screw firmly for easy starting. After the screw is started, pull back the screwdriver and the jaws will automatically withdraw from the screw. This screwdriver is ideal for driving and seating screws in hard-to-reach places.*

Rust-Free Screws

When building anything that will get wet, use stainless steel nails and screws; they are least likely to corrode. The next best rust-resistant fastener is a hot-dipped galvanized screw or nail. These fasteners are dipped repeatedly in a vat of molten zinc. Zinc-plated or "hot-galvanized" fasteners provide the least protection against rust. In the plating process, the fasteners are tumbled in a barrel of hot zinc powder. Always look at the box the screws or nails come in to see what kind of surface corrosion treatment was applied.

SCREWS FOR WOODWORKING

Screws used for woodworking vary in length, driver type, head style, material, finish, and gauge size. Fig. 15-9. Most wood screws are made of mild steel. Screws made of brass, aluminum, stainless steel, and silicon bronze are used primarily for boat construction or wherever moisture would rust other screws. Most screws have a bright finish.

Sheet-metal screws are ideal for fastening thin metal to wood products. They are used, for example, to attach metal legs to a plywood top. The sheet-metal screw has excellent holding power. Unlike wood screws, the threaded shank of a sheet-metal screw is the same diameter throughout its length. Fig. 15-10 (page 228).

WOOD SCREW SIZES

Wood screws come in different sizes. Both length and diameter vary. Fig. 15-11. The **American Screw Wire Gauge** indicates the screw shank diameter with numbers from 0 to 24. The smallest gauge number is 0. A number 0 screw has a shank diameter of 0.060 inch. The diameter of each succeeding gauge number is 0.013 inch larger. For example, a number 5 screw is 0.125 inch, or ⅛ inch, in diameter.

Two screws can be the same length but have a different gauge size. For example, a 3-inch screw comes in three gauge sizes; a 1½-inch screw comes in ten gauge sizes. Fig. 15-12.

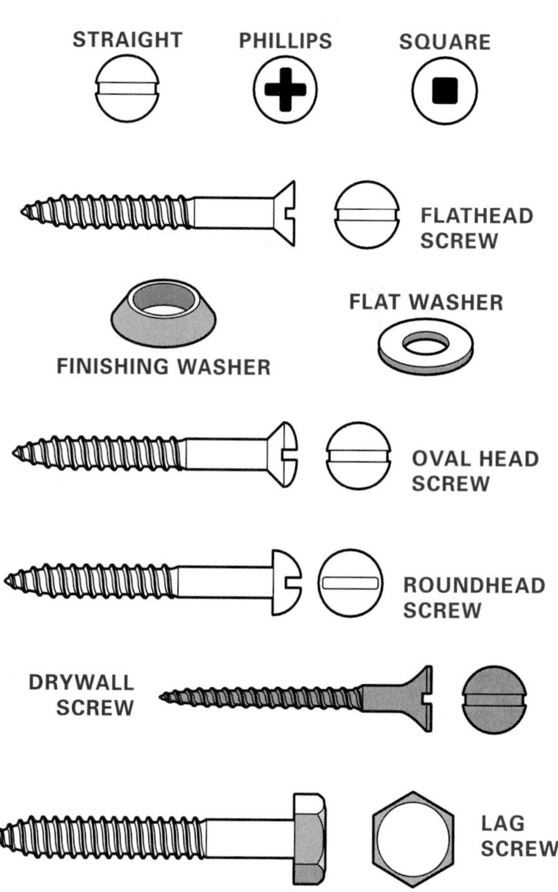

Fig. 15-9—*Screws vary in length, driver types, head styles, materials, finishes, and sizes.*

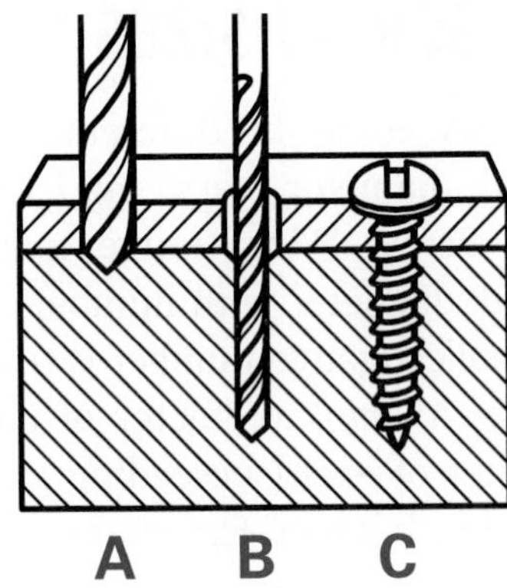

Fig. 15-10—*Installing a roundheaded sheet-metal screw: (A) Drilling the clearance hole. (B) Drilling the pilot or anchor hole. (C) Screw installed.*

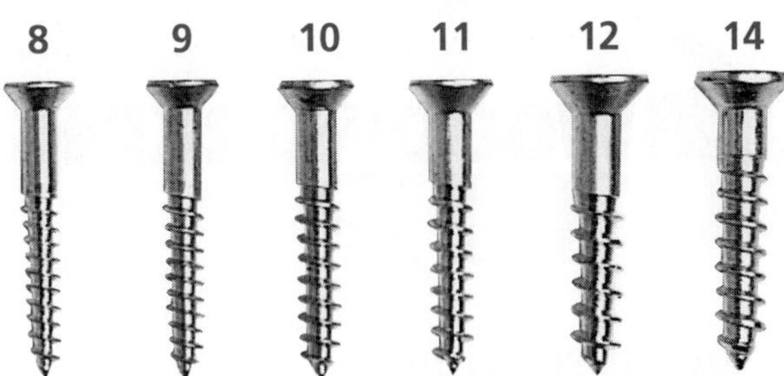

Fig. 15-12—*Different gauge sizes of 1¼-inch screws. Wood screws range in length from ¼ to 6 inches and in gauge sizes from 0 to 24. Of course, each length is not made in all gauges.*

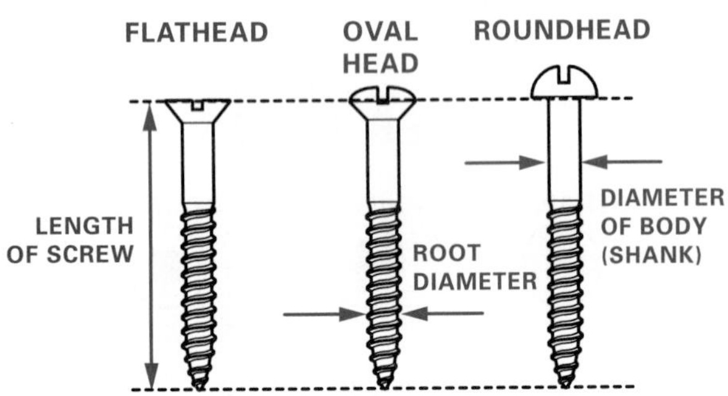

Fig. 15-11—*How to measure screw length and shank diameter.*

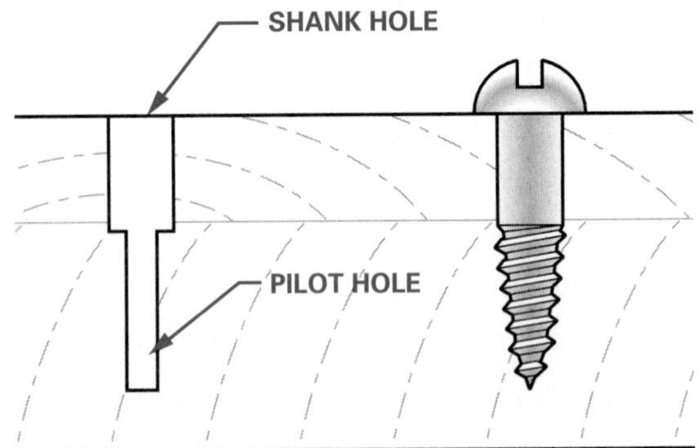

Fig. 15-13—*Here the shank hole and pilot hole are properly drilled, and the screw is installed.*

The size of the screw is usually shown in the woodworking drawing. For example, No. 8 R.H. 1½ means that the screw is number 8 gauge size, roundhead, and 1½ inches long. If the size isn't shown, choose a screw that will go at least two-thirds of its length into the *second* piece. If the second piece is end grain, the screw should be even longer, since end grain does not hold well.

PREDRILLING HOLES

Prior to installation, wood must be predrilled for screws. This involves making clearance holes and pilot holes. Fig. 15-13. Note in the illustration the two drill sizes required. The first one is for the *shank clearance hole,* drilled in the first piece. The second one is for the *pilot hole,* drilled in the second piece. The shank clearance hole should be the same size or slightly larger than the shank of the screw. In this way, the screw can be inserted in the first piece without forcing. Table 15-A.

Drill the shank clearance hole in the first piece of stock. Then hold this piece over the second piece. Mark the location for the pilot hole with a scratch awl. When assembling softwood pieces, drill the pilot hole only about half the depth to which the screw will go. When drilling hardwood, make sure that it is drilled to the total depth of the screw.

Table 15-A. ***Common Wood Screw Sizes***

The bottom part of the table shows the correct size of drills and/or auger bits needed to install the screws.

Length	Number of Screw Size																	
Gauge	**0**	**1**	**2**	**3**	**4**	**5**	**6**	**7**	**8**	**9**	**10**	**11**	**12**	**14**	**16**	**18**	**20**	**24**
¼ inch	0	1	2	3														
⅜ inch			2	3	4	5	6	7										
½ inch			2	3	4	5	6	7	8									
⅝ inch				3	4	5	6	7	8	9	10							
¾ inch					4	5	6	7	8	9	10	11						
⅞ inch							6	7	8	9	10	11	12					
1 inch							6	7	8	9	10	11	12	14				
1¼ inch								7	8	9	10	11	12	14	16			
1½ inch							6	7	8	9	10	11	12	14	16	18		
1¾ inch									8	9	10	11	12	14	16	18	20	
2 inch									8	9	10	11	12	14	16	18	20	
2¼ inch										9	10	11	12	14	16	18	20	
2½ inch													12	14	16	18	20	
2¾ inch														14	16	18	20	
3 inch															16	18	20	
3½ inch																18	20	24
4 inch																18	20	24
Diameter In Inches At Body	.060	.073	.086	.099	.112	.125	.138	.151	.164	.177	.190	.203	.216	.242	.268	.294	.320	.372
Shank Hole Hard & Soft Wood	1/16	5/64	3/32	7/64	7/64	1/8	9/64	5/32	11/64	3/16	3/16	13/64	7/32	1/4	17/64	19/64	21/64	3/8
Pilot Hole Soft Wood	1/64	1/32	1/32	3/64	3/64	1/16	1/16	1/16	5/64	5/64	3/32	3/32	7/64	7/64	9/64	9/64	11/64	3/16
Pilot Hole Hard Wood	1/32	1/32	3/64	1/16	1/16	5/64	5/64	3/32	3/32	7/64	7/64	1/8	1/8	9/64	5/32	3/16	13/64	7/32

COUNTERSINKING FOR FLATHEAD SCREWS

Countersinking is a way of enlarging the top portion of a hole to a cone shape so that the head of a flathead screw will be flush with the surface of the wood. Figure 15-14 shows a **countersink** that can be used to cut the recess in the upper surface of the first piece of stock. Check the depth of the countersunk hole by turning the screw upside down and fitting it in the hole. Fig. 15-15.

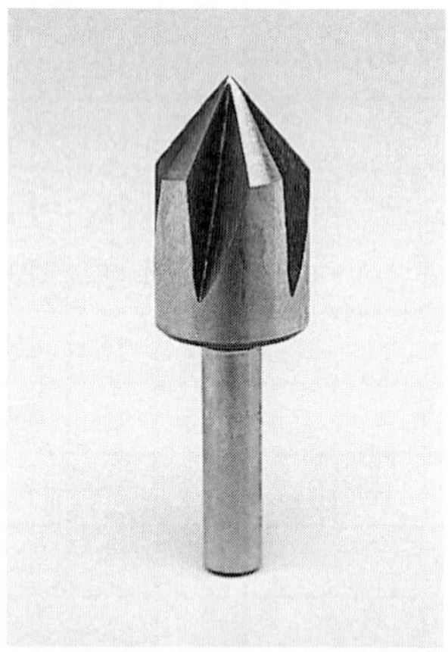

Fig. 15-14—*Quick and accurate chamfers can be made with an 82-degree countersink.*

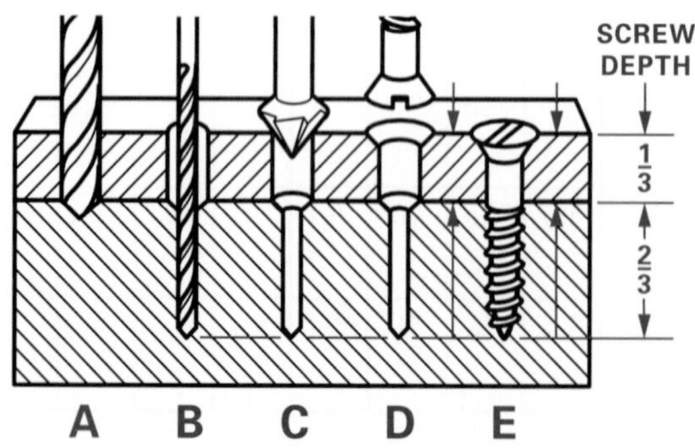

Fig. 15-15—*Steps in installing a flathead screw: (A) Drill the shank hole. (B) Drill the pilot or anchor hole. (C) Countersink. (D) Check the amount of countersink with the screw head. (E) Install the flathead screw.*

MATHEMATICS *Connection*

Working with Wood Screws

Refer to Table 15A. Answer the following questions.

1. What is the diameter in decimal inches of a No. 14 wood screw?
2. What size twist drill is needed for the shank hole for a No. 7 screw?
3. How much larger in diameter is a No. 10 screw than a No. 5 screw?
4. What size twist drill is needed for the pilot hole for hardwood when using a No. 7 screw?

A **screw-mate drill and countersink** can be used with flathead screws. Fig. 15-16. A **screw-mate counterbore** can do all the operations performed by the screw-mate drill and countersink. In addition, it can drill plug holes for wooden plugs. Fig. 15-17.

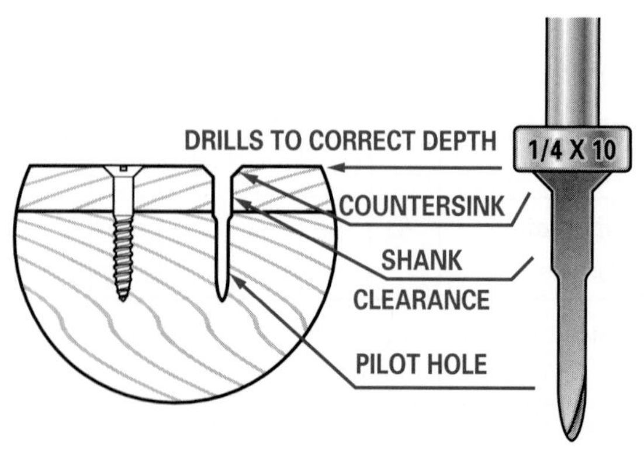

Fig. 15-16—*The screw-mate drill and countersink does four things.*

PLUGGING SCREW HOLES

In most furniture construction, screws are not supposed to show. Choose a drill bit the same size as the head of the screw. Counterbore a hole about 3/8 inch deep in the first surface. The screw will then be below the surface of the wood. After the parts are assembled, this hole can be filled with plastic wood or surface putty, or you can make a little screw plug with the tool shown in Fig. 15-18. For a more decorative look, you can buy pre-made rounded-top plugs.

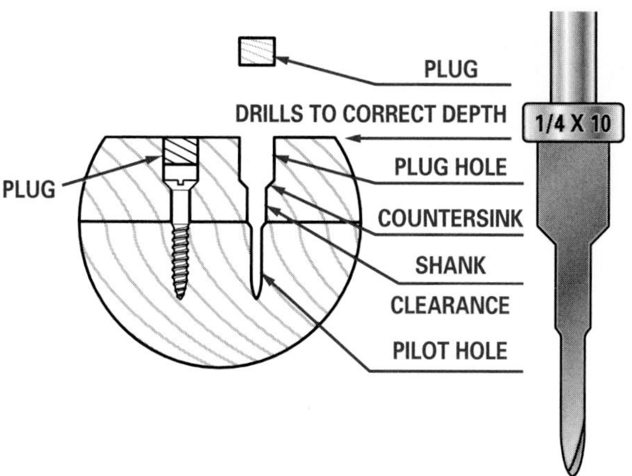

Fig. 15-17—***A screw-mate counterbore does five things at once. A wood plug can be used to cover the screw head.***

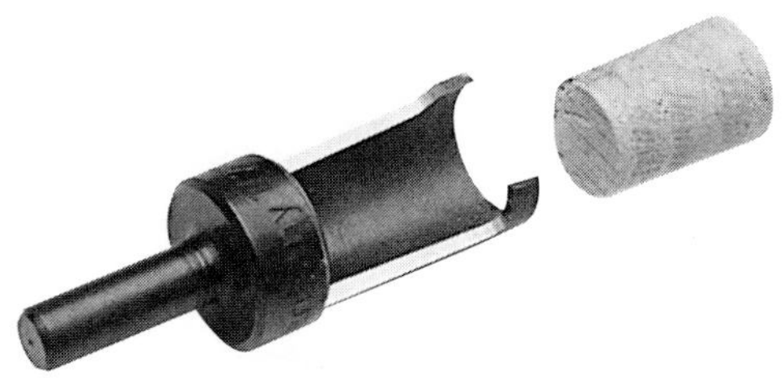

Fig. 15-18—***If you do much furniture work in which screws are countersunk or counterbored, it pays to use a plug cutter. This tool permits cutting perfect plugs from the same stock of which the item is built. The plug cutters are made in sizes 6, 8, 10, and 12 to match the commonly used screw sizes. The plugs are a snug fit in the counterbored holes.***

POCKET SCREWS

When using pocket screws, you really only need to know how to make a simple butt joint. No grooves or dadoes are needed. Fig. 15-19. Using pocket screws, you assemble panels with tight fitting joints without clamps. The alignment clamp used when drilling the holes is all that is needed for perfect joints. Fig. 15-20.

Fig. 15-19—***Pocket screws work well in butt joints.***

Fig. 15-20—***A special pocket screw jig is needed for pocket screws.***

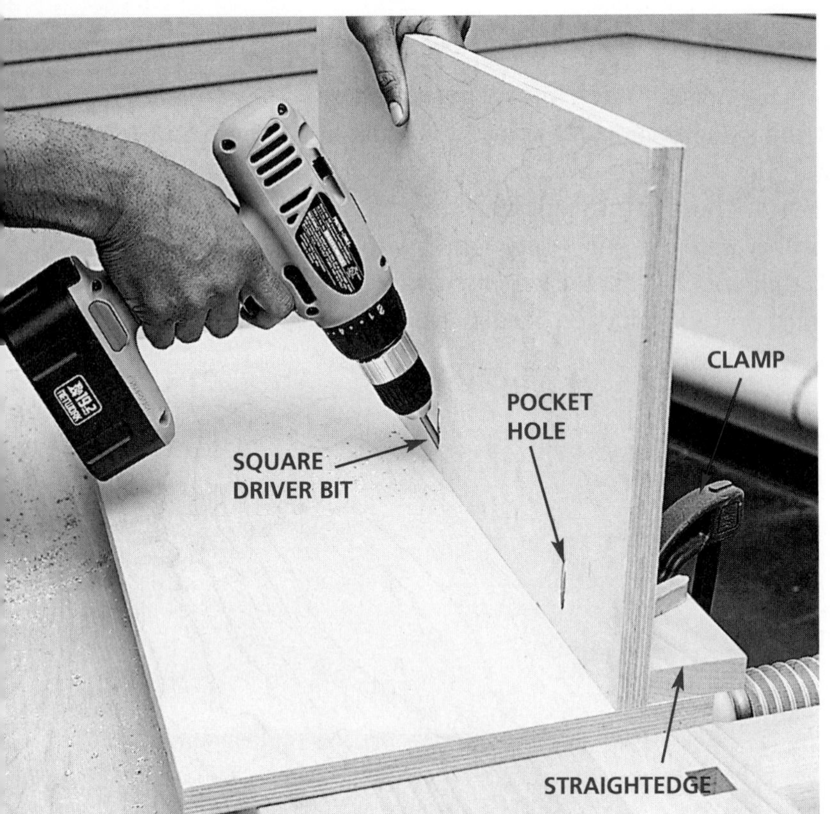

Fig. 15-21—*A square drive screw bit is generally used for pocket screws.*

The screws actually provide the clamping action while the glue dries. Fig. 15-21. One disadvantage of pocket screws is that the pocket, an oblong hole, does not look very good. However, special wood plugs can be used to fill the unsightly holes.

SPECIAL FASTENERS

Standard wood screws do not work well in composition panels, such as particleboard, waferboard, and oriented-strand board. They tend to chew up the particles rather than groove them for holding power. Special wood screws that have widely spaced threads are available for these materials. Also, twin-threaded utility screws hold well in plywood and particleboard. When driven into the material, the thread allows more of the particles to remain between the turns. Fig. 15-22. These screws may have a square drive, a Phillips-head drive, or a combination of both. Another good screw for this purpose is the *drywall screw* that is threaded throughout its length.

Several other kinds of special fasteners are used in assembling wood projects. Fig. 15-23. The *chevron wood fastener* has permanent spring tension. When it is driven into a joint, it holds the joint together without warping and twisting.

Fig. 15-22—*This heat-treated steel wood screw for composite materials has a combination square and Phillips head. The thread surfaces are nearly at right angles to the screw, the body diameter is small, and the point is sharp.*

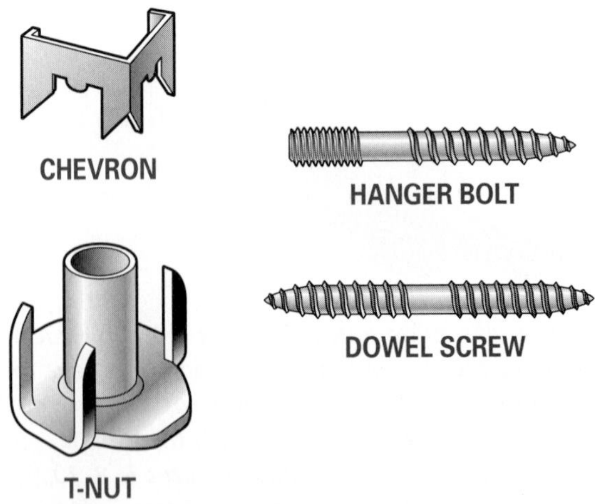

Fig. 15-23—*Fasteners for special uses.*

Another type of special fastening device is a T-nut with a hanger bolt. The *T-nut* has a round base with prongs and a threaded hole in its center. A *hanger* bolt has a wood-screw thread on one end and a metal-screw thread on the other. A hole is bored into the underside of a table top, for example. The hole is just large enough for the center portion of the T-nut. Then the nut is driven into the wood. The wood-screw end of the hanger bolt is then installed in the end of the table leg, and the metal-screw end is fitted into the T-nut. This method of fastening makes it quick and easy to disassemble the table for shipping or storage.

A *dowel screw* has a wood-screw thread on each end. Matching holes are drilled into two pieces to be assembled. The screw is driven an equal distance into both pieces.

In addition to the fasteners mentioned, others are available for special applications in woodworking. Included among these are machine screws, bolts, and corrugated fasteners. Figs. 15-24 and 15-25.

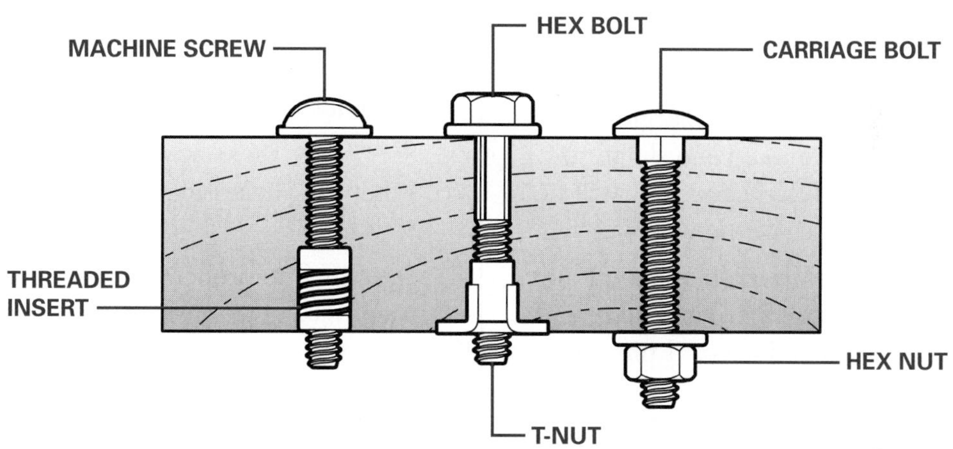

Fig. 15-24—*Machine screws and bolts are often used as fasteners in woodworking.*

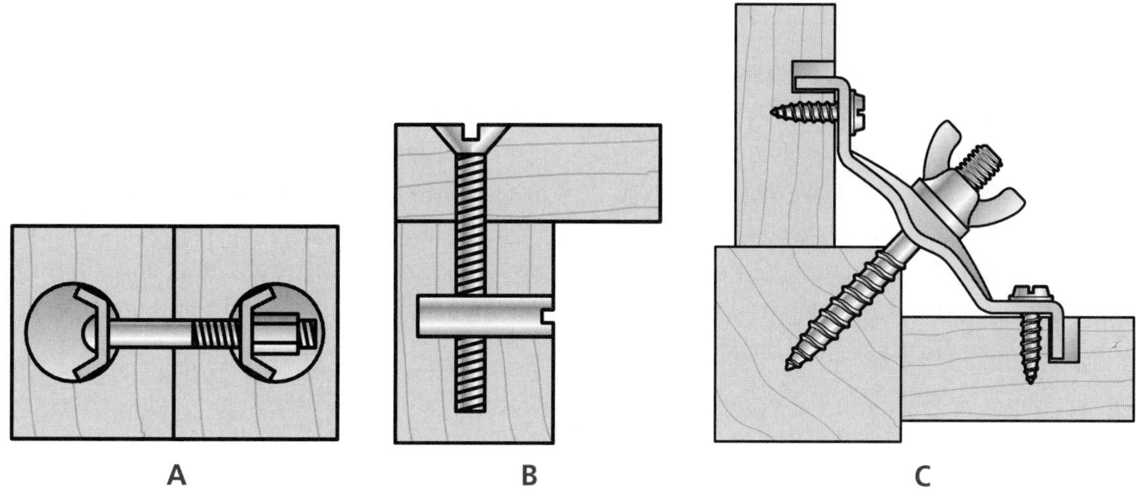

Fig. 15-25—*Other special fasteners. (A) Panel connectors are used to join panels edge to edge. (B) The cross dowel and bolt are used to fasten right-angle joints. (C) Corner brackets make it easy to fasten legs to the apron on a table.*

CHAPTER 15

Review & Applications

Chapter Summary

Major points from this chapter that you should remember include:

- Several different types of screwdrivers are available, including plain, Phillips-head, and offset screwdrivers.
- Wood screws vary in length, driver type, head style, material, finish, and gauge size.
- Prior to installing screws, you should drill clearance holes and pilot holes.
- Countersinking is a way of enlarging the top portion of a hole to a cone shape so that the thread of a flathead screw will be flush with the surface of the wood.
- When using a screwdriver, one hand holds the handle while the other hand steadies the screwdriver tip and keeps it pressed into the screw.

Review Questions

1. How do you know what size screwdriver to choose?
2. When should you use a screw instead of a nail?
3. As the gauge number increases, how is the diameter of the screw affected?
4. What is a general rule for selecting screw lengths?
5. What is a shank clearance hole, what purpose does it serve, and how should one be drilled?
6. When is it necessary to make a countersink hole?
7. Describe the proper method of installing a screw.

Solving Real World Problems

Joshua stands before the fastener display at his local hardware store in complete dismay. The instructions for building the computer desk called for "2" screws." Here before him are 2" drywall screws, cap screws, pan head screws, button head screws, Allen head screws, flathead screws, machine screws—arggg! "What will work?" he finally asks Geoff, a worker in the store. Geoff asks him what he is building, what material it is constructed from, and what joints Joshua is using. Joshua tells Geoff that the desk he is building will be made of plywood and that most of the joints will be dadoes or rabbet joints, but some are butt joints. Geoff then reaches over to the wall and takes a package of screws down for Joshua. What are they? Why did Geoff choose them?

CHAPTER 16

Gluing and Clamping

Look for These Terms

- adhesion
- adhesive
- clamp
- cleat
- double spread
- lamination
- single spread
- trial assembly

YOU'LL BE ABLE TO:

- Select the correct adhesive for specific gluing jobs.
- Select appropriate clamps for holding glued parts.
- Correctly glue up and clamp an edge joint.
- Prepare a laminate for a wood project.
- List the advantages of making a trial assembly.

The term "gluing" means to assemble parts with *adhesive* (glue). **Fig. 16-1.** An **adhesive** is a substance used to hold other materials together. **Clamps** are holding tools often used to hold glued materials together while the adhesive *cures* (sets).

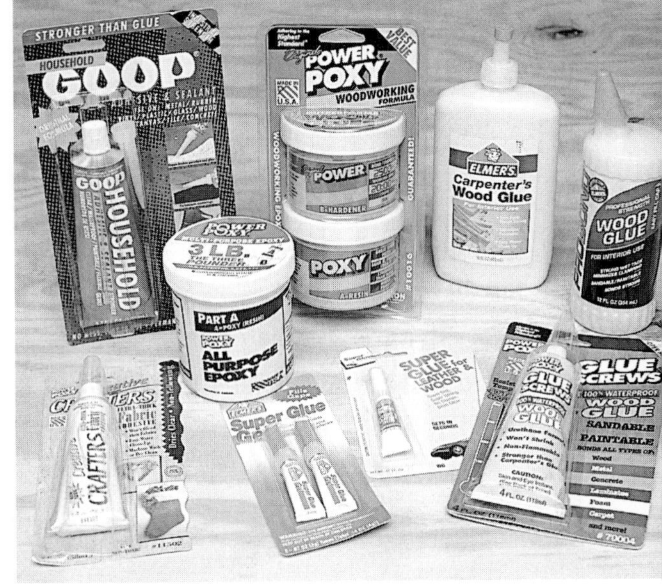

Fig. 16-1—*A variety of adhesives are used in woodworking. Using the proper adhesive and gluing techniques can make joints that are stronger than the wood itself.*

GLUING

Two pieces of wood are held together with an adhesive because of *adhesion*. **Adhesion,** or bonding, is a combining or uniting force that develops between the adhesive and the wood. During the gluing process, an adhesive is applied to adjoining wood surfaces. The adhesive soaks into the wood. The molecules of the adhesive surround the natural wood fibers. When the surfaces are held together, the adhesive's molecules lock into each other. This links the surrounded fibers together as the adhesive cures and hardens.

Kinds of Adhesives

Eight kinds of adhesives are commonly used in woodworking. Table 16-A. They are available under many different trade names.

Safety First

►*Handling Adhesives Safely*

- Always wear a charcoal-filtered respirator to protect your lungs from fumes when using adhesives.
- Be sure the filters in the respirator are new.
- Work in a well-ventilated area.
- Read and follow all safety instructions on the package. Some solvent-based adhesives can become explosive under certain conditions.

White Liquid-Resin Glue

White liquid-resin (polyvinyl) *glue* is a general all-purpose glue. It is considered by some to be the hobbyist's glue. This glue works well for small woodworking projects. It is always ready for use and is nonstaining, economical, and odorless. However, it is not moisture resistant and will not hold in products exposed to weather.

SCIENCE *Connection*

Chemical Structure of Resorcinol

From a chemical perspective, resorcinol is a relatively simple substance. Its molecular formula is $C_6H_6O_2$. This means that a molecule of resorcinol contains 6 carbon (C) atoms, 6 hydrogen (H) atoms, and 2 oxygen (O) atoms.

The carbon and hydrogen atoms form a six-sided polygon (hexagon) known as a *benzene ring*. A diagram of the molecular structure is shown below. Each point on the hexagon represents a carbon atom. One hydrogen atom is attached to each carbon. Two of the carbons in resorcinol "share" their hydrogens with the oxygen atoms to form the OH, or *hydroxy,* portions of the molecule.

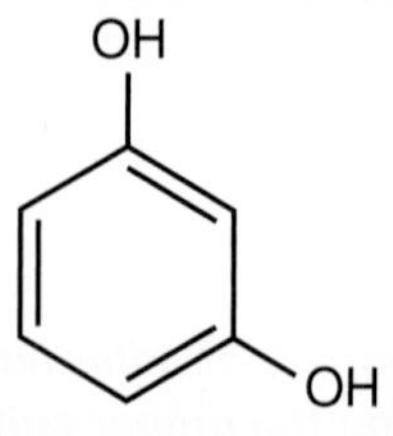

Table 16-A. Wood Adhesives

Type	Description	Recommended Use	Care in Using	Correct Use
Liquid resin (white) polyvinyl glue	Comes ready to use. Clean working, quick setting. Strong enough for most work.	First choice for small jobs where clamping or good fit may be difficult.	Not sufficiently resistant to moisture for outdoor furniture or outdoor storage units.	Use at any temperature. Spread on both surfaces, clamp at once. Sets in 1½ hours.
Liquid resin (yellow) polyvinyl glue	Same as white above.	Best all-purpose glue. Good moisture resistance. Good sandability.	Gets stringy with age and should not be used. Also cannot be used if it has been frozen.	Same as for white above.
Resorcinol (waterproof)	Comes in powder plus liquid; must be mixed each time used. Dark colored, very strong, completely waterproof.	This is the glue to use with exterior type plywood for work to be exposed to extreme dampness.	Expense, trouble to mix, and dark color make it unsuitable for jobs where waterproof glue is not required.	Use within 8 hours of mixing. Work at temperature above 70°F. Apply thin coat to both surfaces. Allow 16 hours to dry.
Urea-resin adhesive	Comes as powder to be mixed with water and used within 4 hours. Light colored.	Almost waterproof. First choice for work that must stand some exposure to dampness.	Needs well-fitted joints, tight clamping, and room temperature 70°F or warmer.	Make sure joint fits tightly. Mix glue and apply thin coat. Allow 16 hours drying time.
Hide glue	Comes in flakes to be heated in water or in prepared form as liquid hide glue. Very strong, tough; light color.	Excellent for furniture and cabinetwork. Gives strength even to joints that do not fit very well.	Not waterproof; do not use for outdoor furniture or anything exposed to weather or dampness.	Apply glue to both surfaces in warm room and let glue become tacky before joining. Clamp 3 hours.
Contact cement	Comes in a can as a light tan liquid.	Excellent for bonding veneer, plastic laminates, leather, plastics, metal foil, or canvas to wood.	Adheres immediately on contact. Parts can't be shifted once contact is made. Position accurately. Temperature for working must be 70°F or above.	Apply to both surfaces. Let dry for 30 minutes. Apply second coat. Allow to dry for 30 minutes. Test for dryness by pressing wrapping paper to surface. If paper doesn't stick, the surfaces are ready for bonding.
Epoxy cement	Comes in two tubes, or cans, that must be mixed in exact proportions.	Excellent for attaching hardware to wood. Good for extremely difficult gluing jobs. Will fill large holes.	Epoxies harden quickly. Use at temperatures above 60°F. Keep epoxy compounds separate. Don't reverse caps.	Mix small amounts. Clean and roughen the surfaces. Apply to surfaces with putty knife. Press parts together.
Hot-melt glues	Cream-colored polyethylene-based adhesive in stick form.	Quick bonding adhesive; best for small areas.	Glue hardens quickly. Work fast before glue cools.	Apply small amounts. Don't spread. Press surfaces together for 20 seconds.

Yellow Liquid-Resin Glue

Yellow liquid-resin (polyvinyl) *glue* is an excellent all-purpose glue. It is much like the white glue described previously. A yellow dye is added to it to show that the two are different.

Like the white glue, yellow liquid-resin glue has high bonding strength, sets rapidly, and is easy to clean up. However, it has greater moisture resistance and tack (stickiness) and better sandability. This glue becomes unusable in a relatively short time when stored.

Resorcinol

Resorcinol is made by mixing liquid resin with a powder catalyst. It comes in a can divided into two compartments. This glue should be mixed only as needed. Follow the manufacturer's directions. This glue does not require much pressure when curing. It will fill gaps and can be used for gluing poorly fitted joints. It provides complete protection from both fresh and salt water.

Urea-Resin Adhesive

Urea-resin adhesive is made from urea resin and formaldehyde. It comes in powder form and is mixed with water to the thickness of cream. It does not stain woods, is waterproof, and dries to a light color. The manufacturer's directions must always be followed for mixing and curing. Urea-resin adhesive is very good for cabinetwork and for bonding plywood.

Hide Glue

Hide (animal) *glue* is made from hoofs, hides, bones, and other animal parts. It is refined and purified, and then formed into flakes or prepared in liquid form.

Hide glue makes a stronger-than-wood joint, but it is not waterproof. It has long been used as an all-purpose furniture glue. If used dry, a glue pot or double boiler is needed to prepare it. The liquid form does not need to be heated or mixed.

Contact Cement

Contact cement is a ready-mixed, rubber-type bonding agent. It bonds practically all like materials and combinations of materials. There is no need for clamps, nails, or pressure. See Chapter 18 for more information about contact cement.

Epoxy Cement

Epoxy cement is a two-part adhesive that sticks to most materials. It can be used on wood, plastics, leather, metal, ceramics, and other materials. Epoxy cement comes in two containers. One holds a special resin and the other a chemical hardener. These are mixed together at the time of use. Epoxy cement produces strong, waterproof joints without clamping.

Hot-Melt Glue

Hot-melt glues are good for use on small areas. They are supplied in stick or chunk form for use with an electric glue gun. A stick of hot-melt glue fits into the gun itself. Fig. 16-2.

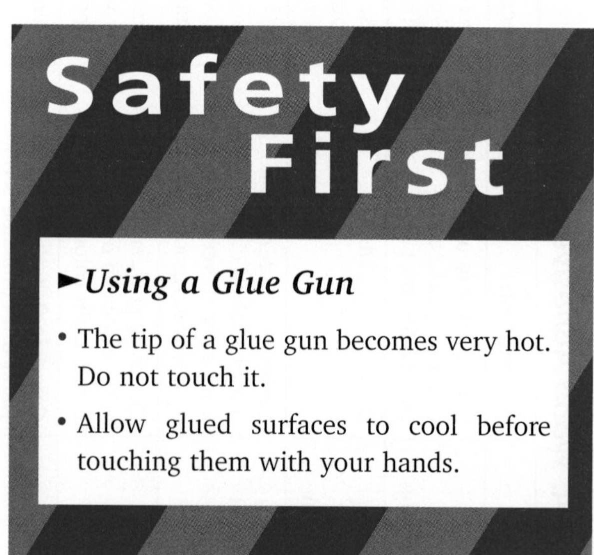

►*Using a Glue Gun*

- The tip of a glue gun becomes very hot. Do not touch it.
- Allow glued surfaces to cool before touching them with your hands.

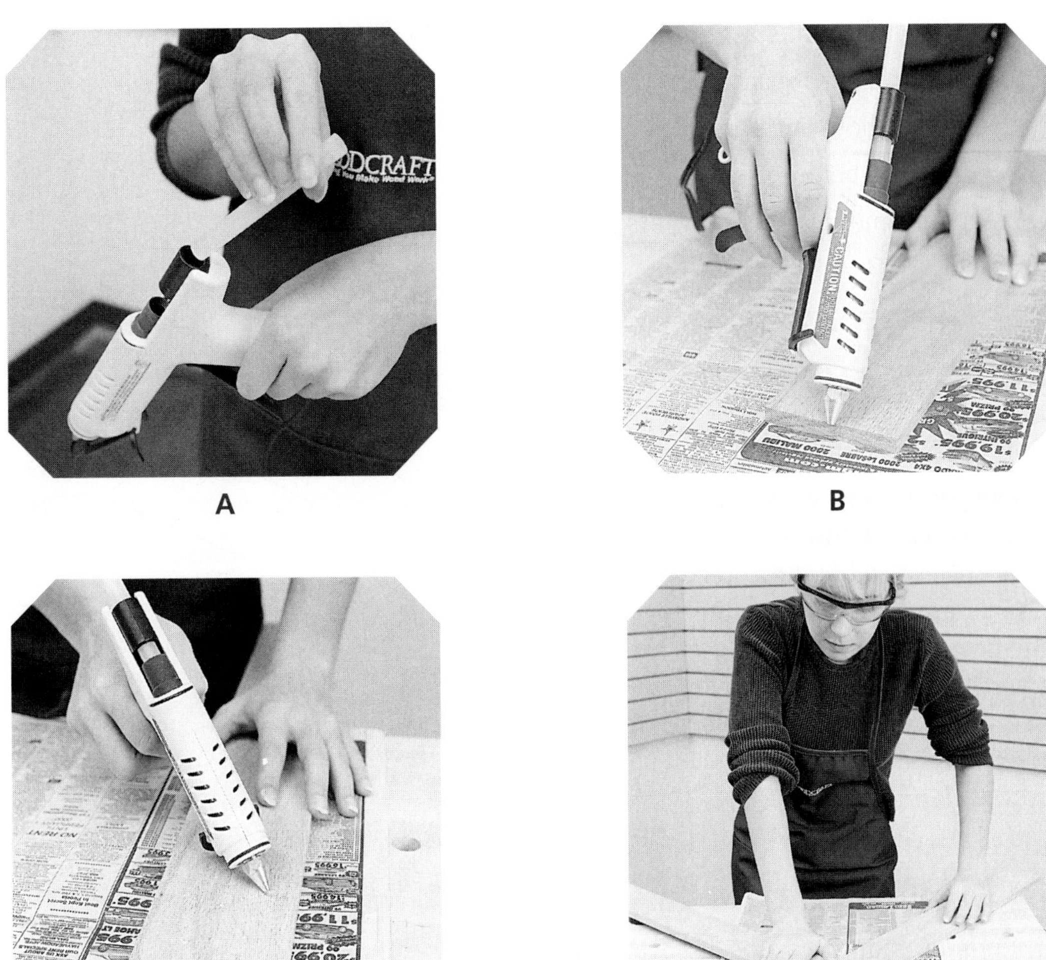

Fig. 16-2—*Using a hot-melt glue gun. (A) Insert glue stick in gun. (B) Open flow control valve by tapping nozzle. (C) Apply the glue. (D) Press glued surfaces together for 20 seconds.*

To use a glue gun, follow these steps:

1. Make sure that the flow-control valve in the gun tip is closed.
2. Insert a hot-melt glue stick into the gun.
3. Plug in the cord and allow about three minutes for the gun to heat.
4. Clean the surface to be glued.
5. Soften the end of another glue stick on the hot tip of the gun. Insert this stick into the gun. The softened end will fuse it with the stick already in the gun.
6. Open the flow-control valve by tapping the valve pin against any hard surface.
7. Feed the glue onto the work by pressing the glue stick with your thumb.
8. Press the glued pieces together for approximately 20 seconds.

Gluing Tips

- Work in a well-ventilated area.
- Make well-fitting joints.
- Make sure the surfaces are clean and dry.
- Choose the correct glue for the job.
- Mix the glue to proper thickness.

- Mark the pieces for correct assembly.
- Have the proper clamps ready.
- Apply the glue to both surfaces of the joint.
- Remove extra glue before it sets.
- Allow plenty of time for glue in the assembly to set.

CLAMPING GLUED WORKPIECES

When choosing clamps, consider the size and shape of the pieces and the amount of pressure needed for successful bonding. Remember that most clamps apply direct pressure to a small area. The jaws of the clamp could mar the surface to which it's attached. To avoid this, you may need to place scrap wood pieces or other protective material or pads between the surface and the clamp's jaws. Fig. 16-3.

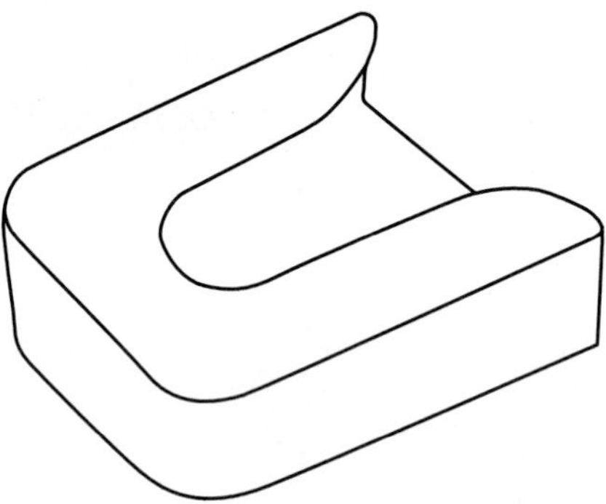

Fig. 16-3—***Plastic bar clamp pads slip onto the faces of bar and pipe clamps. They provide a protective cushion between the finished wood and the metal of the jaw clamps.***

Kinds of Clamps

Several kinds of clamps can be used to hold glued pieces together while the adhesive cures. The most common ones are described below.

Bar Clamps

The *bar clamp* (cabinet clamp) is used when gluing up large surfaces edge to edge and for clamping parts together during assembly. Fig. 16-4. It is made in lengths from 2 to 10 feet and in several styles. One end is adjusted to length by friction or by a *pawl* (movable jaw). The other end is moved in and out by a screw.

When using a bar clamp, turn the screw out completely. Then move in the pawl or friction end until the clamp is slightly wider than the total width of the stock to be clamped.

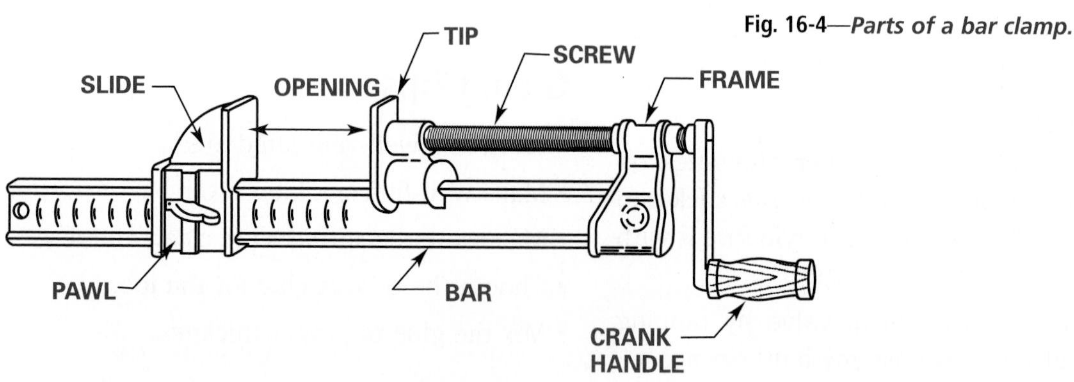

Fig. 16-4—***Parts of a bar clamp.***

When using bar clamps on finished stock, protect the surface of the wood. Place small pieces of scrap stock between the clamp jaws and the wood or put plastic pads on the clamp itself.

Speed bar clamps are very convenient to use. They can be adjusted instantly for quick assembly. Fig. 16-5.

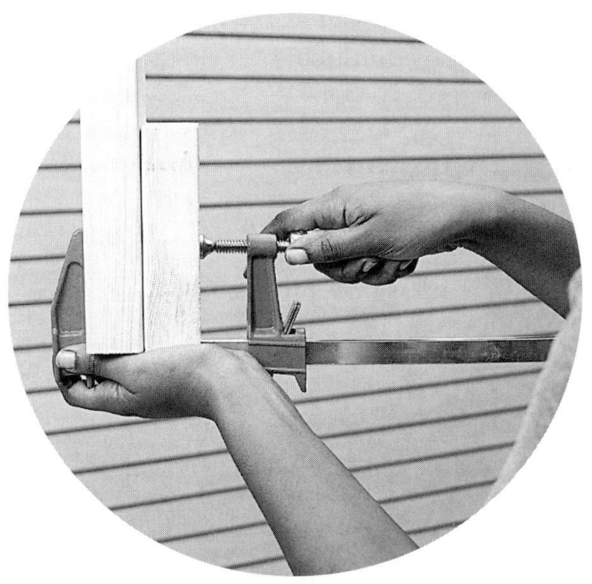

Fig. 16-5—***Speed bar clamp.***

Fig. 16-6—***Pipe clamps are made with single or double pipes. Shown here are single-pipe clamps.***

Pipe Clamps

Pipe clamps are made with either a single pipe or double pipes. Fig. 16-6. The clamp units that fit the steel pipe are purchased so that the clamps can be made to any length. Both parts of double pipe clamps apply equal pressure to stock being glued up.

Hand Screws

Hand screws are wooden parallel clamps about 6 to 20 inches long. They open from 4 to 20 inches. Fig. 16-7.

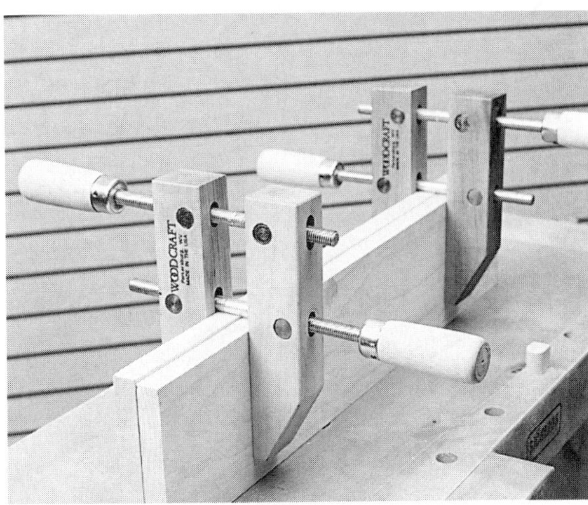

Fig. 16-7—***Incorrect and correct ways of clamping with hand screws. The clamps in the top photograph are not parallel. They will not apply pressure correctly.***

When using hand screws, hold the center screw in one hand and the outside screw in the other hand. Open and close the clamp by twisting the handles in opposite directions.

The hand screw is used for gluing stock face to face. It is also used for clamping together any work that is within the size range of the clamp jaws.

C-Clamps

A *C-clamp* resembles the letter "C." Fig. 16-8. C-clamps come in many sizes. The size is the maximum width of the opening between the jaws. One jaw is moved by turning the screw handle. This jaw swivels to fit flat against irregularly shaped pieces or slanted surfaces.

Spring Clamps

Spring clamps are easy to use. Some types have pivoting jaws made of stainless steel with double rows of serrated teeth along the pressure edge. Those toothed jaws hold the surface of parts so that miter joints and other oddly shaped pieces can be held together. Fig. 16-9.

Edging Clamps

An *edging clamp* is designed to hold moldings, veneer, and laminates to the edge of a workpiece. It has three screws. There is a screw at the top and at the bottom of the jaw opening. An additional screw tightens in from the side. Fig. 16-10.

Band Clamps

A *band clamp* (web clamp) is a nylon strap that tightens around projects. It is used to glue up multisided projects such as a chair frame. Fig. 16-11.

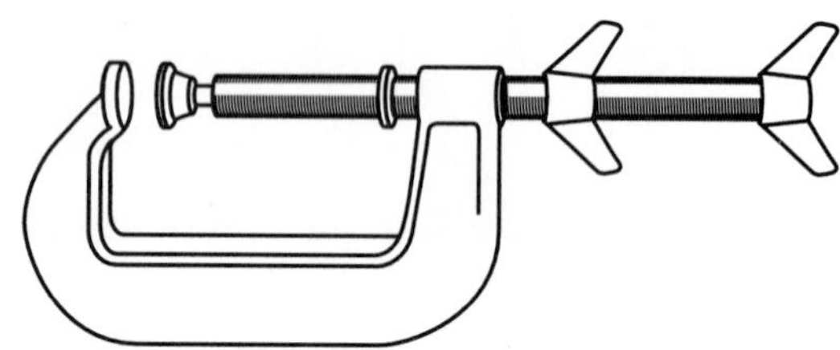

Fig. 16-8—*C-clamp.*

Fig. 16-9—*Spring clamp.*

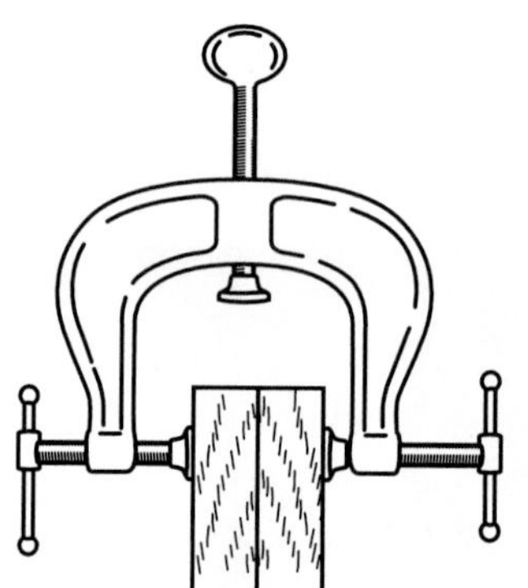

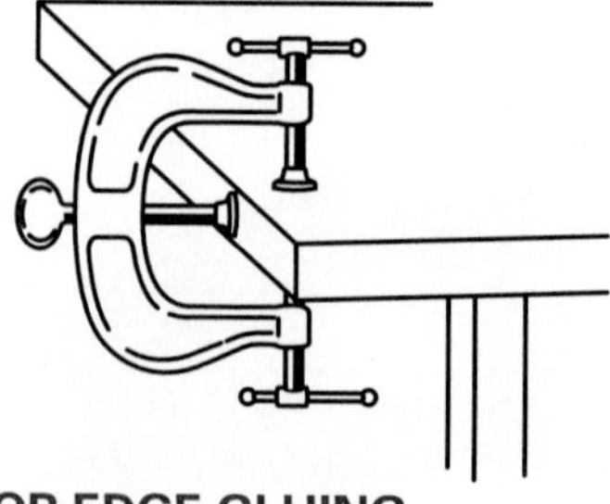

Fig. 16-10—*Edging clamp.*

Fig. 16-11—*Band clamp.*

Dust It Off

It is normal for some glue to squeeze out when you clamp a wood joint. The glue is difficult to remove once it has dried. One way to blot up the freshly squeezed glue is to use sawdust. Put coarse sawdust on the joint and then wipe it or scrape it up immediately. Then sprinkle some fine sawdust on the joint. Rub the fine sawdust into the area to pick up any remaining glue. Again, wipe or scrape it immediately.

Clamping Tips

- Dry-clamp all your workpieces before gluing. This will help you make sure the joints fit properly and that you have enough clamps for the job.
- Before final gluing up and clamping, mark the adjoining pieces with matching numbers or lines to show how they should go together for assembly.
- When gluing edge to edge, alternate bar clamps—one above, one below, every 10 to 15 inches—to prevent buckling.
- Use *cauls* (small pieces of scrap wood) between clamp jaws and the wood surface to prevent metal jaw faces from marring the surface.
- Don't apply too much pressure. This will force the glue out of the joint, causing a weak joint. Clamp down until the pieces are snug, but not too tight.

MAKING EDGE JOINTS

The edge joint is one of most commonly used joints in woodworking. Edges of stock are glued together to form a single piece with a larger face surface.

Preparing the Stock

Select pieces of stock that will form the larger surface. Pieces wider than 8 to 10 inches should be ripped into narrower strips and the strips glued together. Doing this prevents warping and results in a stronger piece. After cutting the pieces, arrange them in their correct order. Remember the following:

- Make sure that the grain of all pieces runs in the same direction. Planing will be easier after the parts are glued.

- Alternate the pieces so the annular rings on the ends face in opposite directions. Fig. 16-12. This will help prevent the surface from warping.
- Try to match the pieces to form the most interesting grain arrangement.

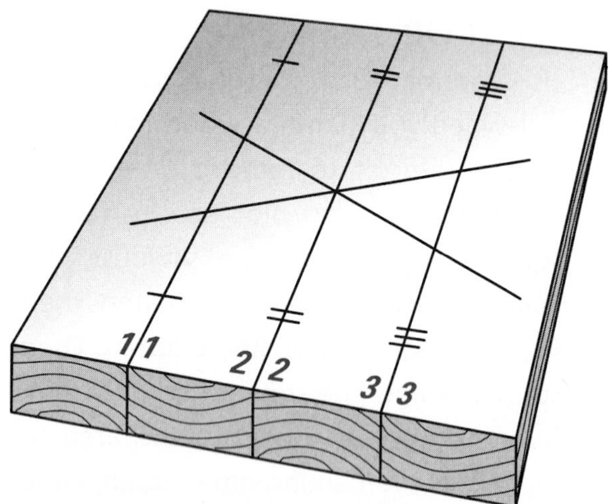

Fig. 16-12—*Arrange stock before gluing pieces together. Note how the direction of the annular rings on the ends of the stock is alternated. Also, mark the adjoining pieces so that they will be easy to assemble.*

Fig. 16-13—*Using a straightedge to check stock for bowing.*

After deciding how parts should be arranged, place matching marks on adjoining faces of each joint. Make the marks in a place where they can easily be seen. Refer again to Fig. 16-12.

Plane one surface of each piece to remove wind or warpage and to help you see grain direction. If any pieces are running in opposite directions, reverse them and re-mark the ends.

Make sure both edges are square with the face surface. Hold a straightedge against the pieces to make sure they do not bow. Fig. 16-13. Tap the top piece with your finger to see that it does not rock. If needed, plane the face surface and the two edges of each piece until adjoining edges fit against each other exactly and the face surfaces are even.

When joints with additional strength are needed, dowels or biscuits can be added. See Chapter 8.

Preparing for Gluing and Clamping

After all joint edges have been formed, prepare the work area for gluing and clamping. Short stock can be assembled on a workbench. Clamps for holding long stock should be placed over two sawhorses. Cover the workbench or floor below the sawhorses with wrapping paper to protect the surface from glue.

Place the pieces in position. Use three or more bar clamps, depending on the length of the stock. Carefully set all the clamps to the proper openings. They should be ready to clamp the stock as soon as the edges are glued. Have a rubber mallet nearby.

Gluing Up and Clamping the Stock

Hold two matching edges to be glued side by side so they are flush. Spread the glue with a brush, stick, or roller. Fig. 16-14. Make sure that both edges are completely covered. This is

A

B

Fig. 16-14—*Two methods of spreading glue: (A) Spreading cold glue with a roller after it is applied with a brush. (B) Applying hot glue with a brush.*

called **double spread.** (For some jobs, glue is applied to only one surface and the surfaces are then rubbed together. This is called **single spread.**) Do not apply too much glue. It will squeeze out of the joint when it is put together. You will need to remove this excess glue later.

WOODWORKING TIP

How Many Clamps?

Uniform pressure must be applied all along glued edges to press them snugly together with no gaps. The number of clamps needed to accomplish this depends on the width of the pieces being glued. The wider the wood, the fewer clamps you need. Pressure applied by clamp jaws fans out through the wood. As shown in the illustration, the farther the pressure travels the wider the force becomes. When in doubt, use more clamps than you need. Don't use too few.

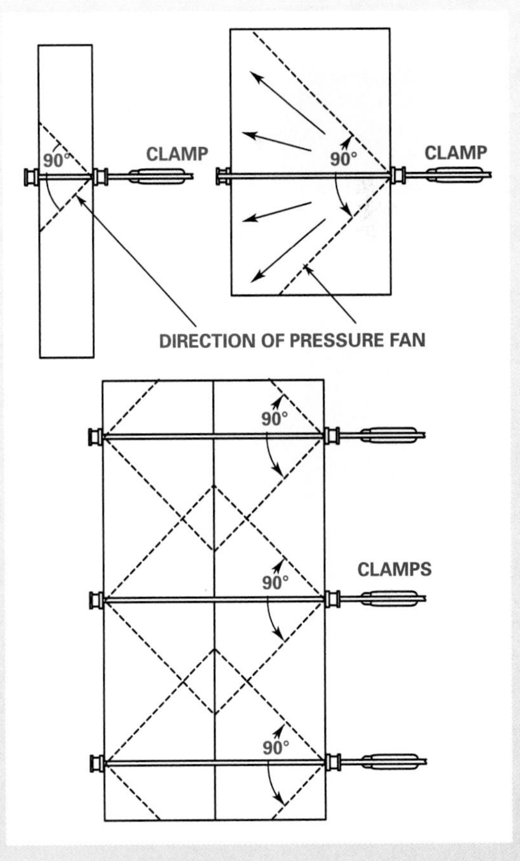

When you have glued all the edges, rapidly lay the pieces on the bar clamps until all are in place. If possible, rub the pieces of the joints together to work the glue into the wood. Tighten the outside clamps lightly. Then place another clamp upside down on the stock at the midpoint. Tighten this clamp lightly. Fig. 16-15.

Use the rubber mallet to tap the ends and face surfaces to line them up. Check all the joints to make sure that the face surfaces are flush and the ends are in line. Then tighten each clamp until it applies firm pressure.

Place a wood **cleat** (piece to hold the work in place) above and below the surface at each end. (Waxed paper under the cleats will keep them from sticking.) Clamp these in place with hand screws. Cleats will help keep the surface true and free from warpage. Before the glue begins to set, wipe off the excess with a damp sponge or cloth.

Fig. 16-15—*Tighten the clamps lightly. Note that scrap pieces of wood are used to protect the surface from being marred while the glue is curing.*

LAMINATING WOOD

Lamination is the process of building up the thickness or width of material by gluing and clamping together several layers. *Veneers* are used in lamination. A veneer is a thin layer of wood of good quality. (See Chapter 19.) Layers used in laminating are usually $\frac{1}{16}$ to $\frac{1}{8}$ inch thick. Fig. 16-16.

When laminating narrow strips, the grain of all the layers must run in approximately the same direction. When laminating larger surfaces, such as for chair seats or backs, the grain direction can be alternated to add strength. An odd number of layers is required for this. The grain direction should be the same in the top and bottom layers.

When making a lamination, glue is applied to both sides of the inside layers and to the inside surface of the outside layers. The finished sides must be free of glue. Flat laminations are placed between flat pieces of wood or other material and clamped for bonding. Position the clamps and tighten them to the degree needed for the lamination to have even thickness throughout. For curved surfaces, forms (or dies) are made.

Fig. 16-16—*A lamination is made by gluing together several layers of thin wood veneers like these.*

Wood laminations resist warpage. They are stronger than regular wood and develop fewer splits and checks. Laminating is done to produce a variety of products. Arches, beams, and furniture parts are a few examples. Small formed projects such as salad servers and bent bookends can also be made using lamination. Fig. 16-17.

Making a Laminated Wood Project

The processes for making a laminated wood project include planning and designing the project, making a form for laminating, laminating the project, and making the project.

Planning and designing:

1. Decide on the project and design you want.
2. Decide on the number of thicknesses needed. Usually the number is odd, such as three, five, or seven.
3. For a curved project such as a salad server, draw a full-size pattern of the curve on a piece of heavy paper. To make a form, or die, for laminating the pieces in the curved shape, cut the curve through a piece of wood. Fig. 16-18.

Making a form for laminating:

4. Select a piece of hard maple or birch that is wide enough and long enough to enclose the veneer sandwich. It should be thick enough to have at least 1 inch of material above and below the lamination.
5. To make the form, transfer the full-size curve to the side of the wood, following the grain direction.
6. Cut the curve on the band saw. (See Chapter 24.) Keep the cut smooth. Avoid as many irregularities as possible. The two halves of the form, or die, should fit perfectly. Refer again to Fig. 16-18.

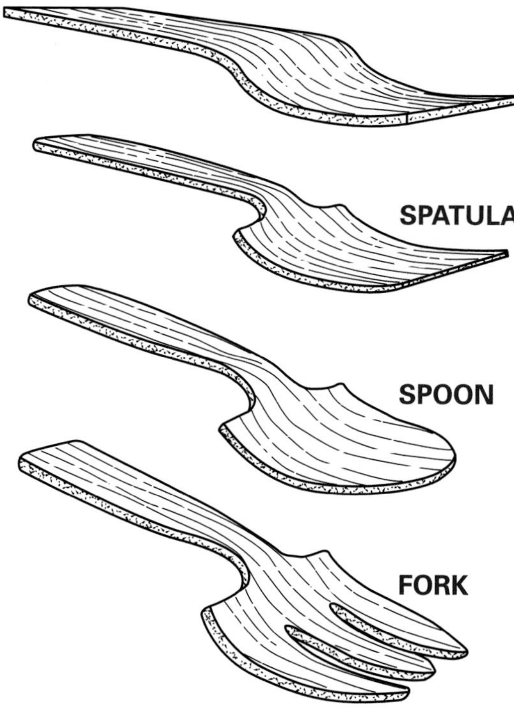

Fig. 16-17—***These projects are typical of the laminated products you can make. The rough laminated panel can be made into any of the objects shown beneath it.***

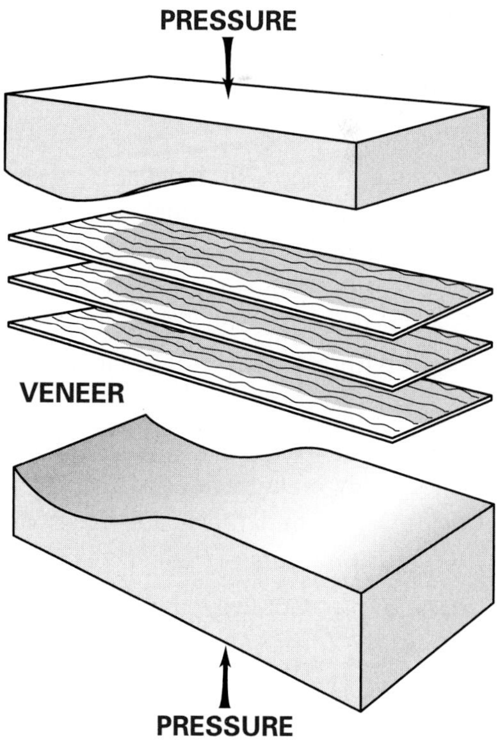

Fig. 16-18—***Form for making the rough laminated panel shown in Fig. 16-17.***

Safety First

►*Using a Band Saw*

- Keep your fingers away from the blade.
- All needed guards should be in place.
- Wear eye protection.

For further safety information about using the band saw, see Chapter 24.

7. Sand very lightly with fine paper to remove saw marks, if necessary.
8. It is recommended that you apply a thin coat of flexible material such as thin rubber or plastic to both sides of the form. Doing this provides a better rough lamination. If rubber is used, it should be tacked onto the form. The surface should then be covered with wax.

Laminating:

9. Cut several pieces of veneer. They should be large enough and thick enough combined to make the project. One type of dark wood, such as mahogany or walnut, may be used. The layers might be alternated with a lighter wood, such as birch or maple, for contrast.
10. Select a good-quality glue. Spread the glue evenly on both sides of each piece of inside layers and the inside faces of the two outside layers of veneer.
11. Place a piece of waxed paper over one side of the form.
12. Stack the layers of veneer on the form.
13. Place another piece of waxed paper over the last layer.
14. Place the other half of the form over the pieces. Clamp the halves together with standard wood clamps. Fig. 16-19.
15. Allow the sandwich to remain under pressure for at least twenty-four hours.

Making the project:

16. Remove the lamination from the form and pull off the waxed paper.
17. Trace the outline of the project on the rough lamination.
18. Cut it out using a coping, band, or scroll saw.
19. Sand and smooth the edges.
20. Apply penetrating finish to the object as soon as possible.

Safety First

►*Finishing*

If the laminated project is to be used around food, use mineral oil as a sealer.

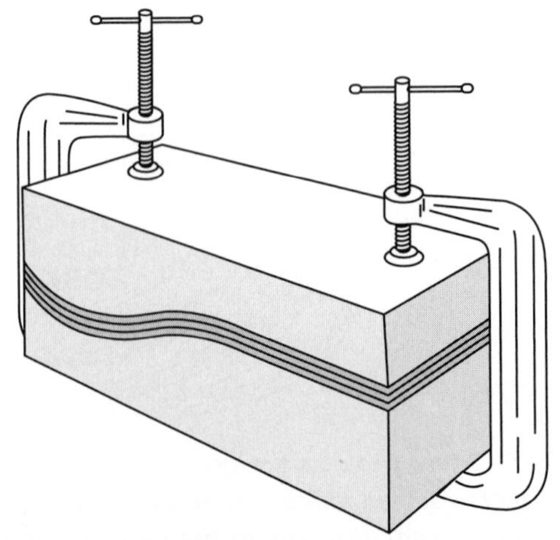

Fig. 16-19—***Clamp the halves together.***

Safety First

►*Using a Jigsaw*

- Keep your fingers away from the blade.
- All needed guards should be in place.
- Wear eye protection.

ASSEMBLING PROJECTS

Assembling a project can be difficult or easy depending on the size and complexity of the project. However, certain things must be done during the assembly process for all projects.

Making a Trial Assembly

To make a **trial assembly,** put all the parts together temporarily before gluing. Use bar clamps and hand screws. Doing this will give you an idea of how the pieces fit together and whether they fit properly. You can see if any corrections need to be made.

This step is simple if you are assembling a project such as a bookcase or a hanging wall shelf. These are made with parallel sides and crosspieces usually joined with dado or rabbet joints. With these types of projects, all you need are flat pieces of scrap stock, bar clamps that fit across the project on the front and back, and a pair of clamps for each shelf. Projects such as end tables, stools, and small desks usually have legs and rails with corners made with mortise-and-tenon joints or dowel joints. In either case, you will need clamps to go across the ends and bar clamps to go across the sides.

After assembling the project with clamps, check with a square to make sure that the project is squared up. Fig. 16-20. Using a straightedge, measure across the corners and up and down to check that the sides and ends are parallel and that the project has the same height throughout. Fig. 16-21. By shifting a clamp or tapping a side or leg with a mallet, you can bring it into place.

This trial assembly allows you to adjust all clamps to the correct width. It prepares the project for final assembly.

Fig. 16-20—*Checking a trial assembly for squareness.*

Fig. 16-21—*Checking the project with a straightedge.*

Gluing the Parts

Assemble the project on a glue bench or in a gluing room. If a gluing room is not available, place wrapping paper over the bench or floor where the gluing is to be done. Carefully remove the clamps from the trial assembly. Work from top to bottom. Lay the clamps on the bench in order so that you can pick them up easily. Place protective pieces of scrap wood next to the clamps so that everything will be on hand as needed.

If you can assemble the entire project at one time, do so. If you cannot, you will need to glue and assemble sections and then do the final assembly later.

In addition to the project parts, clamps, scrap pieces, and glue, have ready a rubber mallet, a rule or tape measure, and a square.

With a brush, bottle, stick, or glue gun, carefully apply glue to both parts of the joint. Do not use too much. Excess glue can be a problem to remove later. As quickly as possible, assemble the joints.

Clamping Assembled Parts

Place scrap pieces at points where clamps will hold the workpiece. Adjust the clamps to apply light pressure. Do this until all the clamps are in place. Then gradually tighten all the clamps, adjusting each clamp a little at a time. Check often to see that the project is squared. Fig. 16-22. Measure the distance between parallel surfaces to be sure they are the same distance apart at all points. Using a straightedge, check for flatness. Fig. 16-23. Check also that the diagonals are equal. Fig. 16-24. You may need to change the position of a clamp or use the rubber mallet to tap a joint into place.

As soon as the project is clamped together, clean off the excess glue with a damp sponge. Do this *before* the glue dries. Be careful not to mar the surface of the project.

Place the project in a safe place where no one will bump it. For most projects, you will need to allow twelve to twenty-four hours for the glue to cure. Once curing is complete, carefully remove each of the clamps and scrap pieces.

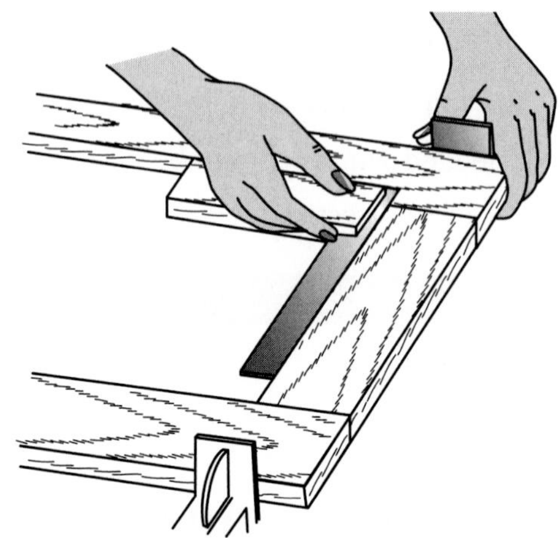

Fig. 16-22—*Checking the assembly with a try square.*

Fig. 16-23—*Use a straightedge to check across several places to make sure the surface is flat.*

Fig. 16-24—*Checking across corners to make sure the distance is equal in both directions.*

CHAPTER 16

Review & Applications

Chapter Summary

Major points from this chapter that you should remember include:

- Adhesives bond materials together. Clamps hold materials together while the adhesive cures.
- Eight types of adhesives are commonly used.
- White and yellow liquid-resin glues are excellent all-purpose adhesives.
- The type of clamp used for a particular job depends on the size and shape of the glued pieces and the amount of pressure needed for successful bonding.
- In an edge joint, the two adjoining edges must fit against each other exactly and the face surfaces must be even.
- When making a trial assembly, all parts are clamped together before gluing.

Review Questions

1. Name at least five of the eight kinds of adhesives and give one use for each.
2. Which kinds of adhesive might you use in building an outdoor bird feeder? Why?
3. List the steps in using a glue gun.
4. For what types of gluing tasks are spring clamps used?
5. Which type of clamp can apply pressure from three different directions at once?
6. Explain how you would glue up and clamp an edge joint.
7. Explain briefly how to prepare a lamination for a curved wood project.
8. Why is making a trial assembly useful?
9. What things must you check when tightening the clamps during final assembly?

Solving Real World Problems

Stan and Cherie are building a big, old-style farmer's table for their kitchen. The top is 48″ wide and 8′ long. It is made of 1″ × 6″ boards. Each board will be joined to the next with dowels and glue along the entire seam. Stan places each board in such a way that the grain of each board is running in the same direction as the board next to it. After the boards are all laid out in order, Cherie applies a layer of yellow wood glue to each dowel hole and along the edge of the board. After the glue is applied, Stan inserts the dowels into one board and squeezes the next board to the first, closing the joint. After all the boards have been assembled, Stan and Cherie must clamp the table top to ensure flatness. What type of clamps could they use? How should they be applied?

CHAPTER 17

Installing Hardware

Look for These Terms

- cabinet hardware
- catch
- gain
- hardware
- hinge
- repair plate
- structural hardware

YOU'LL BE ABLE TO:

- Name and give examples of the two basic types of hardware needed to build a project.
- Select an appropriate type of hinge to serve a specific purpose.
- Install drawer knobs or pulls.
- Select and install the appropriate type of repair plate for a specific purpose.

Most woodworking projects require some kind of hardware. The simplest and most frequently used are fasteners such as nails and screws. Quite often, though, hardware such as hinges, drawer or door pulls or knobs, drawer guides (or slides), and joint support pieces are needed. Exterior hardware can also be used to decorate and add interest to a project. **Fig. 17-1.**

Fig. 17-1—*Hardware can make a woodworking project more interesting and attractive.*

At first, the installation of hardware may seem difficult. However, once you learn basic installation techniques for each piece, you will see that it really is not hard at all. For example, just a couple simple steps are needed to attach drawer guides for a smoothly operating cabinet drawer.

KINDS OF HARDWARE

Hardware is a term used to refer to parts (usually metal) needed to:

- Complete a project.
- Make a project usable.
- Provide structural support within the project.

Two basic kinds of hardware are *cabinet hardware* and *structural hardware.* **Cabinet hardware** includes items such as hinges, pulls, and catches. These items make a project with drawers and doors more usable. Cabinet hardware should be selected to match the style of furniture. Hardware catalogs are available that show many different styles of hardware.

Structural hardware includes items such as repair plates and corner blocks. These are used to strengthen joints and hold unseen parts together.

Hinges

A **hinge** is a piece of hardware that is used as a joint. It allows one of two joined parts to move. Hinges are used mainly on doors, gates, and chest lids. Many types of hinges are available. Fig. 17-2.

Most hinges are made of steel. These are available in a number of different finishes. Black, brass, bronze, chrome, and copper are some examples. You can even buy steel hinges that are coated with primer for painting.

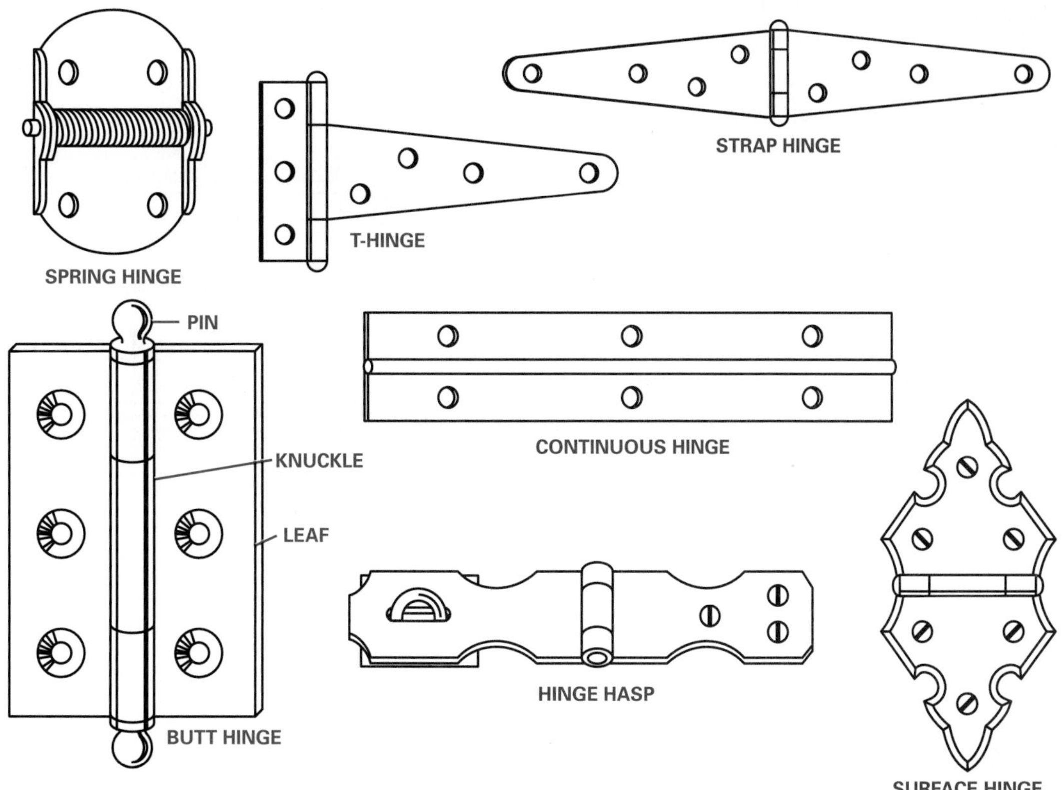

Fig. 17-2—*Common types of hinges. The main parts of a hinge are identified on the butt hinge.*

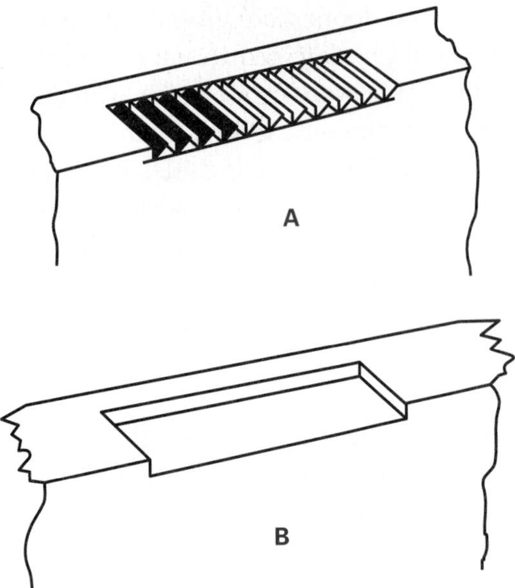

Fig. 17-3—*A chisel is used to cut a gain. (A) Chisel cuts in the stock to be removed to form the gain. (B) The gain is cut and ready to have the hinge installed.*

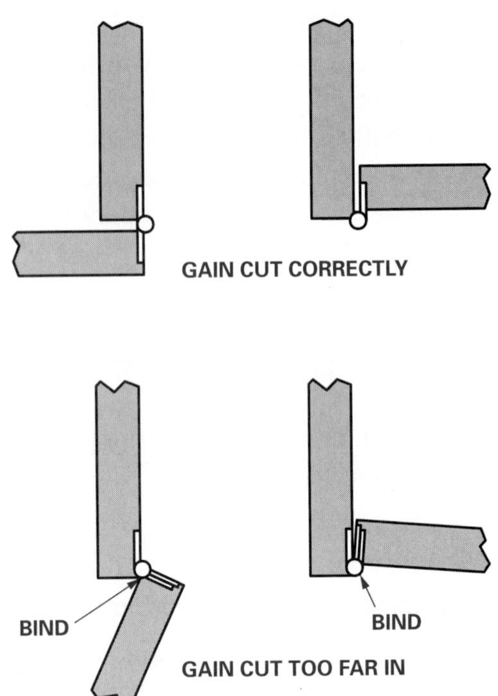

Fig. 17-4—*The gain must be cut correctly or the door will not close properly.*

Butt Hinges

The *butt hinge* is used for flush doors and some overlay doors. It requires that a recess, or **gain,** be cut. In most cases, a gain is cut both in the door and in the frame on which the door is fastened. Fig. 17-3. Sometimes one deeper gain is cut either in the frame or in the door. Thus, the hinge is recessed in only one part of the two surfaces. In both cases, when the door is closed, the surfaces fit closely together. Only the knuckle of the hinge can be seen. Fig. 17-4.

A butt hinge may have either a fixed or a loose pin. The leaves may be either straight or *swaged.* Swaged leaves are shaped at the knuckle to allow for a closer fit. One or both leaves may be swaged. The tightest fit is achieved when both leaves are swaged. Fig. 17-5.

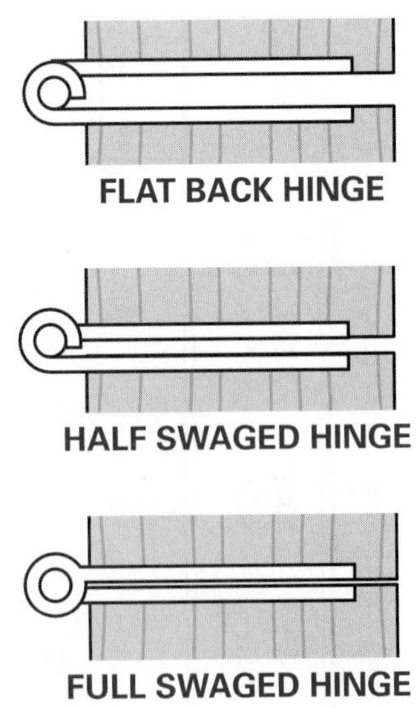

Fig. 17-5—*Straight and swaged leaves.*

Surface Hinges

Surface hinges are different from butt hinges in that no gain is required. Leaves are installed on the surfaces. A *flat surface hinge* is used on flush doors. An *offset surface hinge* fits over the outside of lip doors.

Fig. 17-6—*A semi-concealed hinge for a lip door has an offset that fits the rabbet cut in the door.*

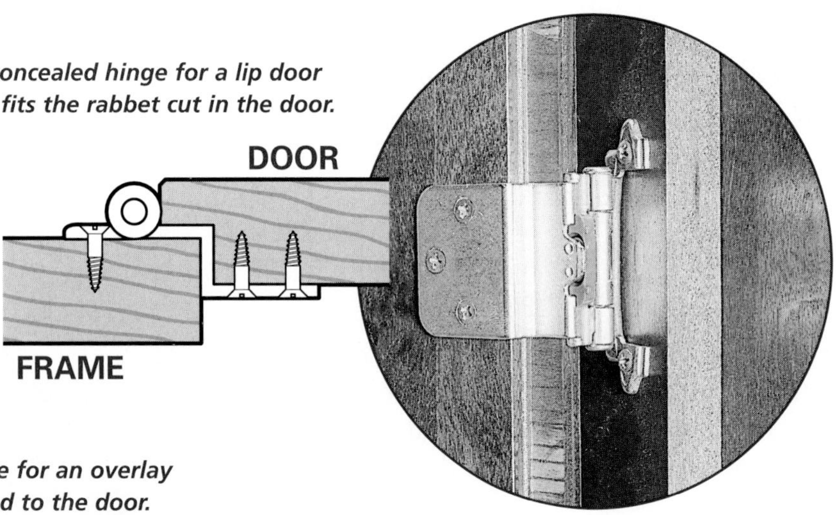

DOOR

FRAME

Fig. 17-7—*A semi-concealed hinge for an overlay door with a straight leaf attached to the door.*

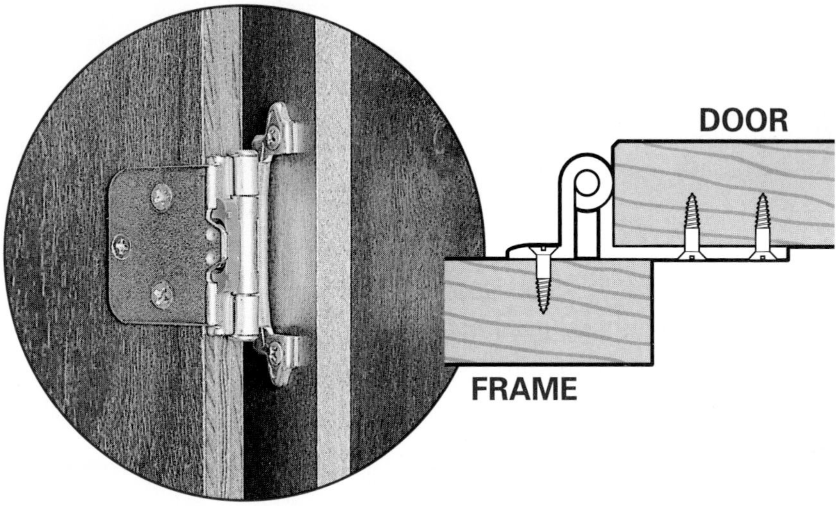

DOOR

FRAME

Semi-Concealed Hinges

Semi-concealed hinges are used on lip or overlay doors. Those used on lip doors have a sharp bend in one of their leaves that fits through the rabbet edge. Fig. 17-6. On an overlay door, a semi-concealed hinge with a straight leaf is used. Fig. 17-7. These hinges are available with spring actions that cause the door to close automatically and stay closed. No catch is needed.

Specialized Hinges

Many specialized kinds of hinges are available. These include:

- *Pivot hinges.* These are usually installed on the top and bottom of an overlay door. They are designed for installation on either the horizontal or vertical surface of a door and frame. Fig. 17-8.

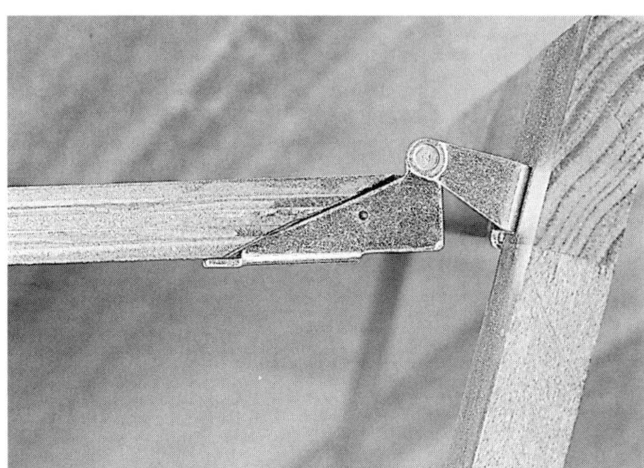

Fig. 17-8—*A pivot hinge.*

- *Knife hinges.* These are used much like pivot hinges but are not as strong. Fig. 17-9.
- *European hinges.* One side of this hinge is installed inside a cabinet and the other into a hole drilled in the door. These hinges are easy to install on any kind of door. Use a Forstner bit the width of the hinge body (typically 35 mm) to drill the hole. Fig. 17-10.
- *Invisible hinges.* These are used on both flush and overlay doors. They are installed in recesses cut in the frame and door. They cannot be seen when the door is closed. Fig. 17-11.
- *Concealed hinges.* These are installed in holes drilled in the edges of the frame and flush door. When the door is closed, only the pivot point can be seen. Fig. 17-12.

Catches

A **catch** is a device for holding a door closed. Fig. 17-13. Magnetic catches are used most often. These consist of two main parts: the magnet assembly and a metal plate called a *strike.* The magnet assembly is installed inside the cabinet. The strike is attached to the door. Fig. 17-14.

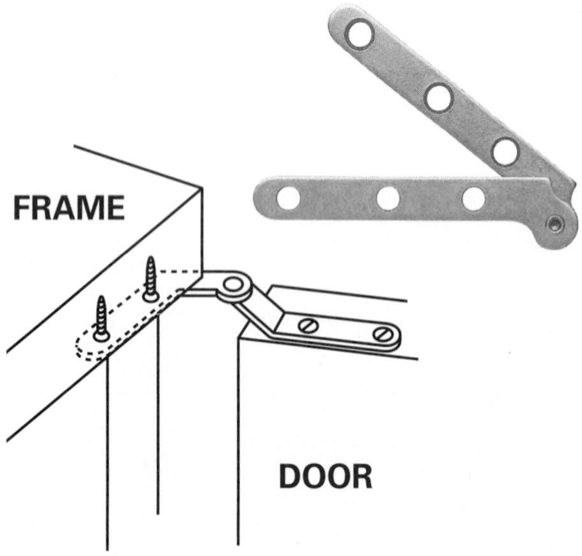

Fig. 17-9—*Knife hinges are installed on the top and bottom edges of a door.*

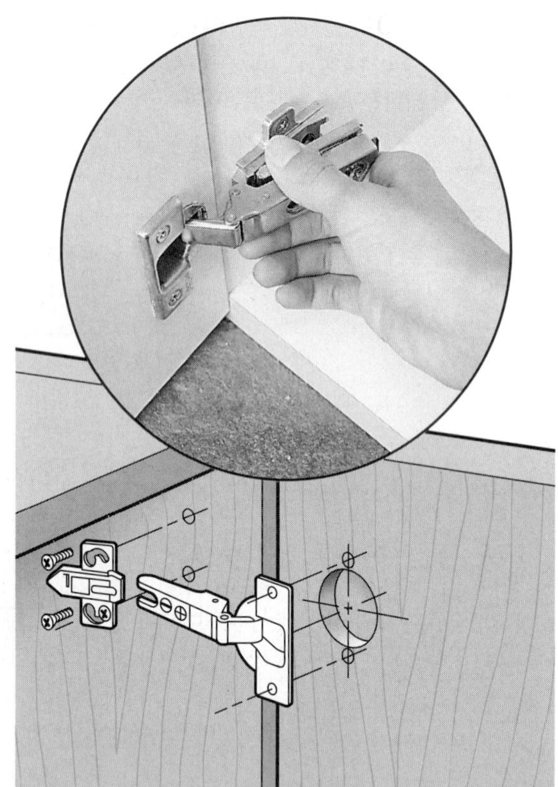

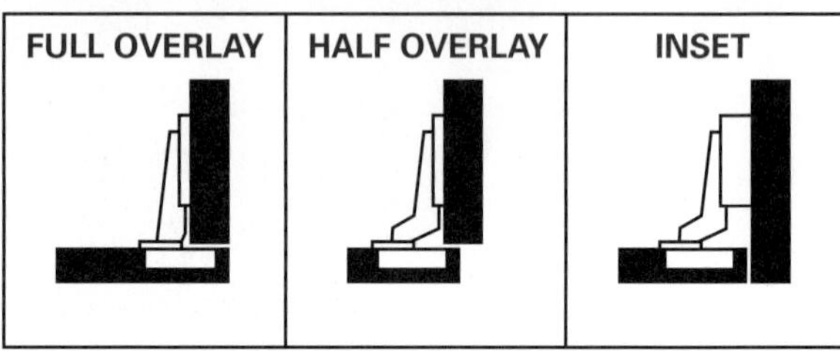

Fig. 17-10—*Installing a European hinge.*

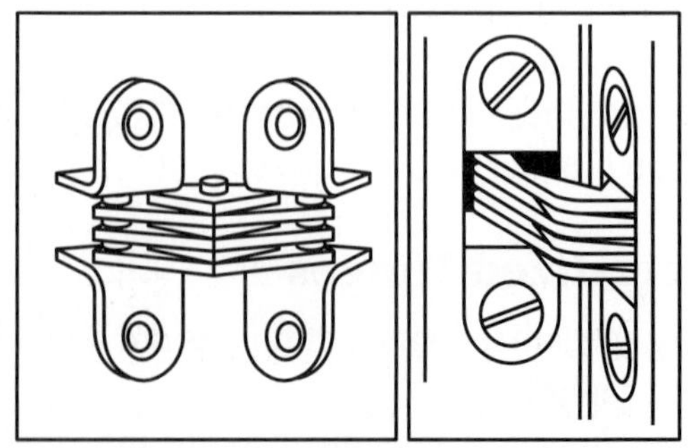

Fig. 17-11—*An invisible hinge.*

Fig. 17-12—*Installing a concealed hinge.*

Magnetic catches can be used on most furniture and kitchen cabinets. They do not work well for large or heavy doors.

Drop-Door Supports

Some furniture or cabinets may have drop doors. These open outward and down. A variety of supports are available for these. Some have a telescopic setup. Fig. 17-15 (page 258). Folding supports allow an opening of exactly 90 degrees. Other types allow the amount of opening to be controlled. An automatic support uses a braking system to keep the door from simply dropping open.

Door and Drawer Pulls and Knobs

Door pulls and knobs are usually installed near the opening edge. They should be positioned at a height that is easy to reach by most people. Drawer pulls are usually placed in the center of the drawer face. Long drawers may have two pulls or knobs located toward the ends of the drawer face.

Pulls and knobs come in a variety of sizes and styles. Matching sets are available that also include hinges. Select a design that is appropriate for your furniture style. For example, hardware appropriate for contemporary furniture would look out of place on furniture with a more traditional design.

Fig. 17-13—*In the roller catch, the rollers are spring loaded.*

Fig. 17-14—*A magnetic catch.*

INSTALLING DRAWER HARDWARE

Pulls and Knobs

Installation procedures may vary slightly depending on the type of pull or knob. In general, measure and mark the location for the pull or knob on the drawer front. Drill a hole through the drawer front from the outside to the inside. The hole should be the same size as the machine screw used to fasten the pull or knob. Hold the pull or knob in position and insert the screw into the hole from the inside. Tighten the screw with a screwdriver. Fig. 17-16.

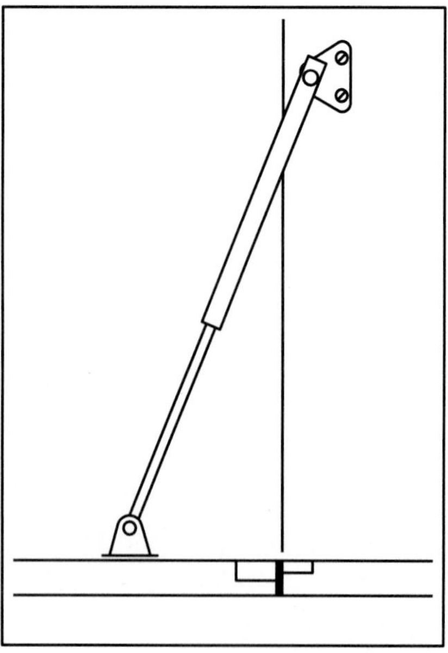

Fig. 17-15—*A drop-door support that has a telescoping action.*

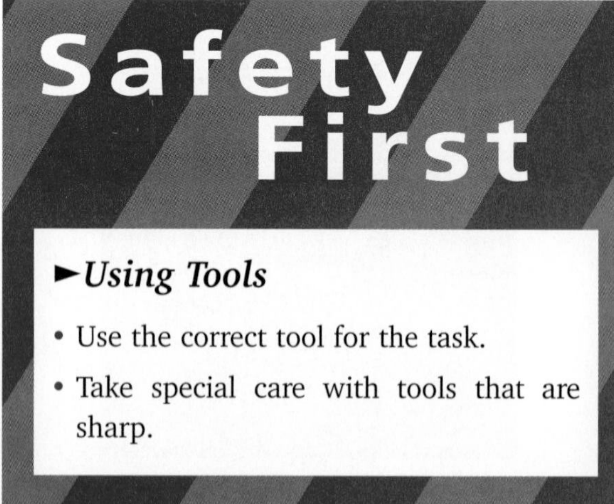

►*Using Tools*

- Use the correct tool for the task.
- Take special care with tools that are sharp.

SCIENCE *Connection*

Magnetic Properties

Some materials, such as iron, cobalt, and nickel, have natural magnetic properties. They are commonly known as *permanent magnets* because they retain magnetic properties for a long time. In a magnet, each atom exerts force on the atoms around it. This force causes the atoms to align their magnetic poles so that all the north poles are facing the same direction. All the south poles face the opposite way. The result is a magnetic field around the object that can affect other objects, such as the metal strike plate in a magnetic catch.

When the strike enters the magnetic field around the magnet, the combined force of the magnet's aligned atoms causes the atoms in the strike plate to become temporarily aligned. They are attracted to the atoms in the magnet, and the catch closes. When you open the catch by moving the strike away from the magnet, the atoms in the strike soon revert to their original pattern.

Fig. 17-16—*Installing a drawer pull.*

Guides

The best way to support the average drawer is to use a metal ball-bearing guide set. Guide sets are made to be attached to either the sides or bottom of the drawer. Fig. 17-17. Side guides are usually the best choice. They can carry more weight and generally operate more smoothly than bottom guides.

Extracting a Broken Screw

Sometimes the head of a screw twists off when it is being driven into the wood. This is especially true of brass screws. Here is a way to remove the screw and repair the damage.

- With a small bit, drill holes all around the screw. Make the holes a little deeper than the length of the screw.
- Using needlenose pliers, twist the headless screw out.
- With a larger bit, drill out the damaged area and glue in a hardwood dowel of the appropriate size.
- Drill the correct-size pilot hole into the dowel for the new screw.

Once the screw is in, you will hardly notice the repaired area.

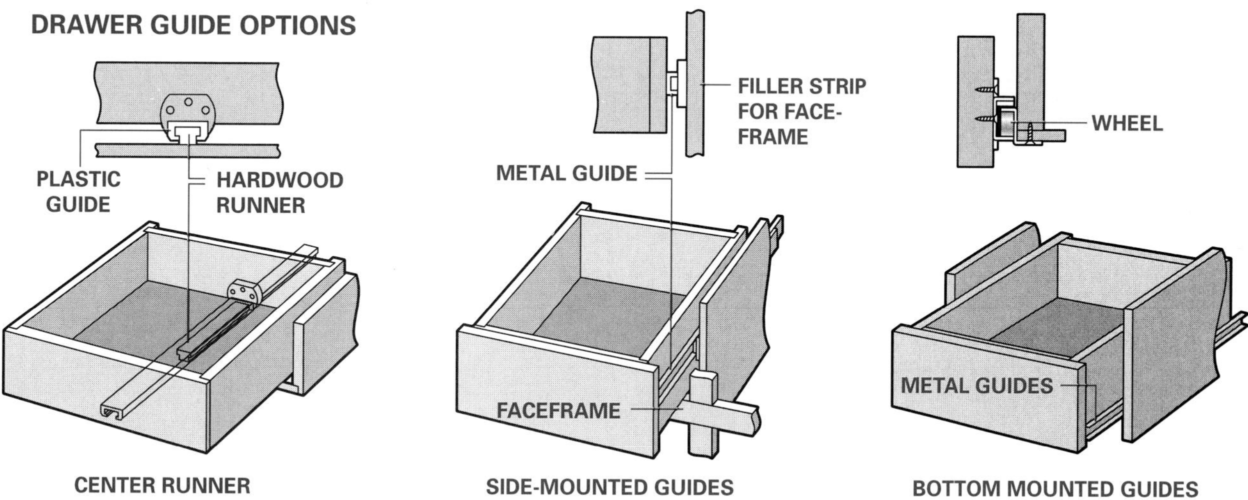

Fig. 17-17—*Installation setup of various types of drawer guide sets.*

The clearance for bottom guides is 3⁄16 inch on both the top and the bottom and 1⁄8 inch on each side. Side guides require 1⁄2 inch clearance on each side. Filler strips should be installed to make the mounting surface flush with the face frame. Be sure that the case is not twisted. Check to be sure it is level and plumb.

USING REPAIR PLATES

Repair plates come in many sizes and shapes. Fig. 17-18. *Mending plates* are used to strengthen a butt or lap joint. The *flat corner iron* is used to strengthen corners of frames such as a screen door or window. The *bent corner iron* can be applied to shelves and the inside corners of tables, chairs, and cabinets. It can also be used to hang cabinets and shelves. *T-plates* are used to strengthen the center rail of a frame.

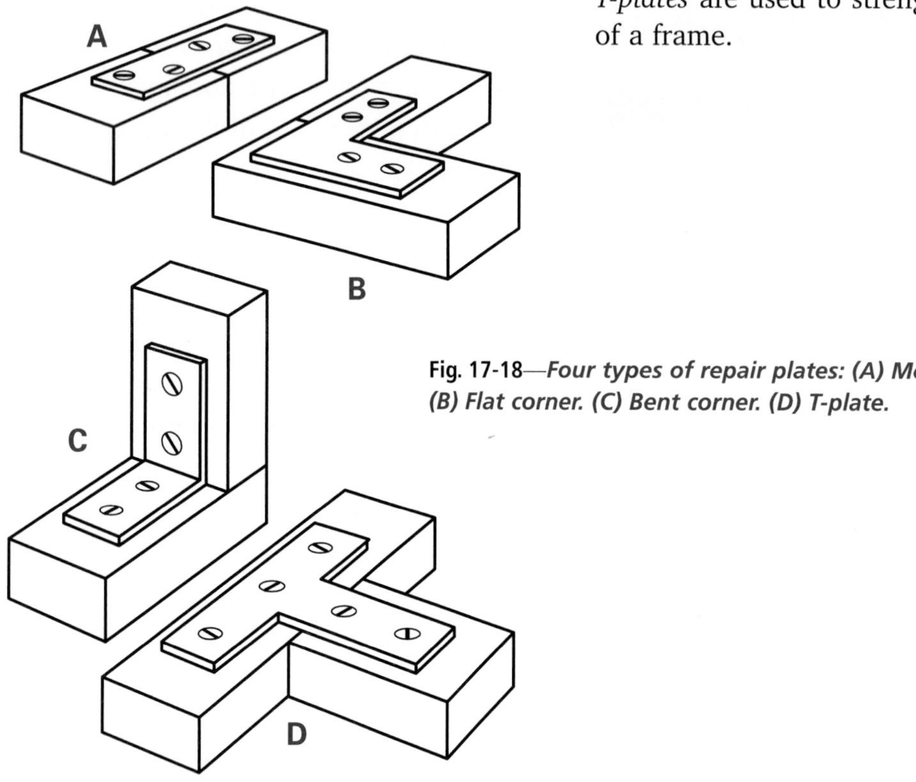

Fig. 17-18—***Four types of repair plates: (A) Mending. (B) Flat corner. (C) Bent corner. (D) T-plate.***

CHAPTER 17

Review & Applications

Chapter Summary

Major points from this chapter that you should remember include:

- Hardware is used to complete a project, make a project usable, and provide structural support within the project.
- A hinge is a piece of hardware that is used as a joint.
- A butt hinge requires that a gain be cut. A surface hinge does not.
- Most hinges are made of steel.
- Magnetic catches are used most often on furniture and kitchen cabinets.
- A metal, ball-bearing guide set provides the best support for most types of drawers.
- Repair plates are installed to provide structural support.

Review Questions

1. What three main purposes does hardware serve?
2. Name the two basic types of hardware and give at least two examples of each.
3. What is the main difference between a butt hinge and a surface hinge?
4. What types of hinges might you use on a lip door?
5. List the two main parts of a magnetic catch.
6. In what ways are side guides better than bottom guides for drawers?
7. Describe how you might install a knob on the center of a drawer face.
8. What type of repair plate would you use to strengthen a butt or lap joint? the inside corner of a chair?

Solving Real World Problems

Angela is adding the final touches to her new kitchen cabinets by installing knobs and drawer pulls. She has 28 knobs to install. Worried that she might not get each knob's mounting hole drilled the same, she devises a small jig with a hole in it to place on the corner of each door. After lightly clamping the jig to the door, Angela can use the hole in the jig as a guide to drill the mounting hole in the same place each time. The drawer pulls require two holes, but since each drawer is the same size, a jig can be used here also. Each pull requires two holes spaced 4″ apart, and the total width of each drawer is 20″. Draw a sketch of the jig to do this job.

CHAPTER 18

Plastic Laminates

Look for These Terms

- contact cement
- laminate
- plastic laminate
- substrate

YOU'LL BE ABLE TO:

- Cut plastic laminate to appropriate size.
- Adhere plastic laminate to a substrate using contact cement.

A **laminate** is a material made up of several different layers that are firmly united together. The tops of tables, cabinets, desks, and chests are often covered with an extremely hard **plastic laminate. Fig. 18-1.** You may know this material by one of its trade names. Panelyte™, Formica™, and Wilsonart™ are examples.

Fig. 18-1—*Plastic laminate was used in the construction of this kitchen counter.*

Facts about Wood

What's in a Name—In 1913 a Westinghouse chemical engineer invented plastic laminate. He developed it to be used as an electrical insulator. Westinghouse was not very interested in his invention, so he and another engineer started their own company to manufacture plastic laminates. One of their first orders was from an electric motor company for commutator V-rings. Until then, the V-rings were always made of mica, another kind of insulator. Because plastic laminate was a substitute for the mica, the new company was named "Formica." Even though other companies make plastic laminates, for many years the general public called all plastic laminates Formica.

Plastic laminate is made primarily from layers of kraft paper. This paper is impregnated (filled) with resin. A rayon surface paper is covered with another kind of resin. These layers are placed under high heat and pressure to produce a 1/16-inch sheet of material that is very hard. It wears well and is scratch-resistant. Coffee, ink, iodine, crayon wax, and most other substances have no effect on this surface. Soap and water can be used to clean it.

LAMINATE MATERIALS

On some furniture, plastic laminates that give the appearance of wood grain are used. However, a wide selection of laminates is available. You can choose among solid colors, various textures, and a wide variety of patterns. If you will need to use a number of sheets, keep in mind that there will be butt joints where the pieces meet. Dull finishes and patterns with straight lines hide joints better than glossy finishes and solid colors.

You can buy sheets of plastic laminate in sizes up to 5 feet by 12 feet. Using large sheets will often enable you to cover large areas without having joints. Plan how you will lay out your laminate so that you can cover the most area with the fewest joints and with the least wasted materials.

WORKING WITH PLASTIC LAMINATE

Plastic laminates are usually applied to a **substrate** (underlayment) of plywood, particleboard, or medium-density fiberboard (MDF). Fig. 18-2. When using a laminate, begin by cutting the substrate from sheet stock.

Fig. 18-2—*Types of substrate. Top to bottom: particleboard, MDF, and plywood.*

Building Up the Substrate

Install build-up pieces on the substrate. To do this, turn the piece(s) upside down. Cut and install stretcher strips of ¾-inch stock 4 inches wide around the edges. Place short crosspieces across the ends and at intervals of about 2 feet along the length. Also place them across any substrate joints and around any cutouts. Use glue and 1¼-inch self-tapping screws to install the strips. Fig. 18-3.

Molding will need to be installed on the facing edge of your project. If the laminate will extend over any part of it, the molding should be installed now, before the laminate is applied. Fig. 18-4. Later in this chapter you will see how the edge is finished.

Fig. 18-3—***Build-up pieces are glued to the underside of the substrate before screws are installed.***

Cutting Laminate

Plastic laminates are hard but very brittle. They must be well supported before being cut. Since a sheet is only 1⁄16 inch thick, it cracks easily. Any of the standard woodworking cutting tools can be used to cut plastic laminate. However, most woodworkers prefer carbide-tipped tools. These are very hard and remain sharp longer than ordinary tools. A router with carbide-tipped bits works well for cutting plastic laminate. See Chapter 28.

Safety First

►*Using Tools*

- Always wear eye protection when working in the shop.
- Use all tools properly.
- Always take special care with tools that are sharp.

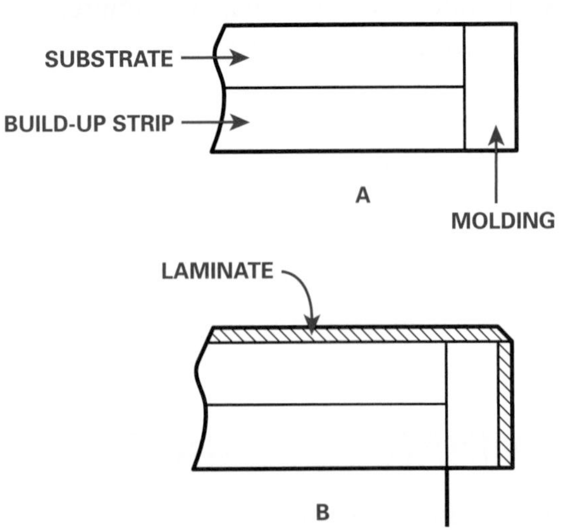

Fig. 18-4—***(A) If laminate will extend over the molding, install the molding first. (B) One example of how laminate may be applied over molding.***

SCIENCE Connection

Warpage in Plastic Laminates

Like wood, plastic laminates are subject to changes in humidity. They contract when humidity decreases and expand when it increases. The amount of change is different in one dimension (width) than it is in the other (length). This change is known as *dimensional movement.* In most plastic laminates, the width of the laminate changes twice as much as the length.

There are several ways to minimize the effect of humidity on plastic laminates:

- Use backer sheets to help protect the substrate from changes in humidity.
- Use a substrate that has dimensional movement that is similar to that of the laminate.
- Allow the laminate, substrate, and adhesive to acclimate for at least 48 hours in the same conditions (heat and humidity) under which they will be used.
- Use the same laminate on both sides of a panel.
- Use the same adhesive and application techniques on both sides of a panel.

Safety First

►*Using a Router to Cut Laminate*

- *Always* wear eye protection. Sharp chips fly from laminate when it is being cut.
- Keep your hands and clothing away from the rotating cutters at all times.
- After turning off the power, wait until the machine comes to a complete stop before setting it on its side.

For further safety information about operating a router, see Chapter 28.

Use a router with a top-bearing bit to cut the laminate to rough length and width. The piece should be at least ½ inch longer and wider than the substrate.

Position the sheet on several ¾-inch strips to provide clearance for the router bit. Clamp a straightedge across the laminate and use it as a guide when cutting. Fig. 18-5. After making the first cut, go over it again to smooth and even out the cut.

Fig. 18-5—*Using a router to cut laminate.*

Edges that will butt against each other must match. Cut these at the same time. Overlap the ends about 1 inch. Place a cutting guide over these and clamp them all together. Fig. 18-6. Make the cut using a router with a top-bearing bit. The cut edges of both pieces should match very closely.

Contact Cement

When adhering laminate to substrate, use **contact cement.** Contact cement is an adhesive that is coated on the two surfaces to be joined and then allowed to dry. Once it has dried, it will not adhere to any surface not coated with contact cement. When placed in contact with another coated surface, however, it bonds immediately and permanently.

Applying Contact Cement

Before applying contact cement, be sure the surfaces to be joined are clean and free of sawdust. Apply the cement to the underside of the laminate and the top of the substrate. A good-quality brush or roller could be used. The cement may also be poured on the surface and spread with a trowel. Spread the cement evenly over each surface. Allow this first coat to dry to the point that kraft paper will not stick to it. Then apply a second coat.

When contact cement is ready for bonding, the dry coats will have a glossy appearance. If any part appears dull, apply more cement. Let the second coat dry, again to the point at which kraft paper will not stick. When this point is reached, the parts are ready to be joined.

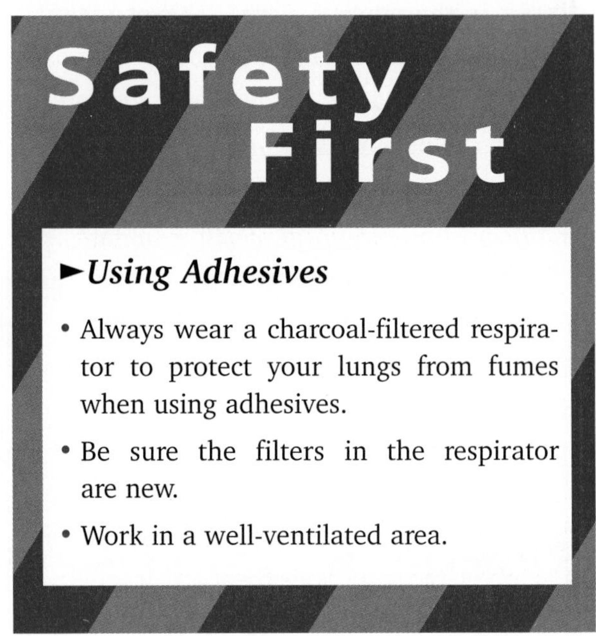

►*Using Adhesives*

- Always wear a charcoal-filtered respirator to protect your lungs from fumes when using adhesives.
- Be sure the filters in the respirator are new.
- Work in a well-ventilated area.

Fig. 18-6—*Setup for cutting parts of a laminate butt joint.*

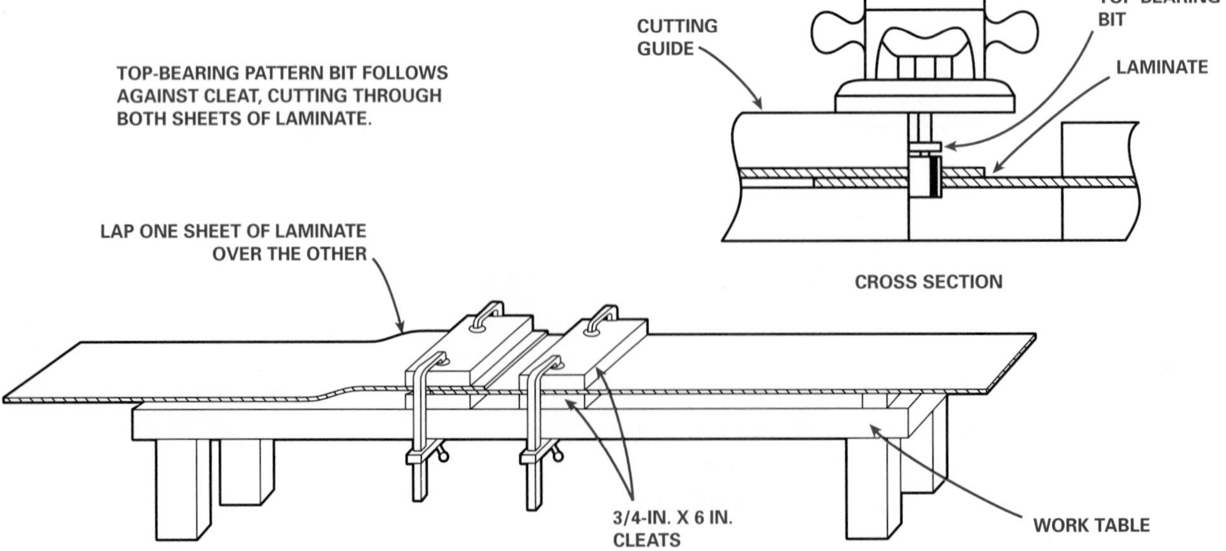

Adhering Laminate to Substrate

Remember that contact cement adheres *immediately* when contact is made between the prepared surfaces. Because of this, you will need separators between the surfaces as you position the laminate over the substrate. Lay ⅜-inch or ½-inch dowels every 6 to 8 inches across the substrate. The dowels should be long enough to extend past the edges of the pieces to be joined.

Lay the laminate on the dowels. Position it so that it extends past all edges of the substrate. If the laminate has a pattern with straight lines, these should align with the edges or ends for good appearance.

Begin the gluing process by removing the dowels. Fig. 18-7. As each dowel is removed, press down on the laminate so that contact is made with the substrate. Work your way toward the ends, removing a dowel first from one side and then from the other. Smooth the laminate on the substrate to keep air pockets from forming. Fig. 18-8.

Strengthening the Bond

To make sure the cement achieves its full strength, you will need to apply pressure over the entire surface. It's best to use the small rubber roller specifically designed for this purpose. Fig. 18-9 (page 268). Larger rollers do not work well. Less pressure can be applied because it is spread over a larger area.

If a small roller is not available, use a mallet and a 3-inch square block with chamfered edges. Slide the block over the surface, tapping it with the mallet. If glue is forced out along the edges, use your fingers to remove it while it is still in a rubbery state.

Fig. 18-7—*When installing the laminate, remove the dowel in the middle first. Press down on the laminate and smooth it in place.*

Fig. 18-8—*Smoothing the laminate toward the ends prevents the formation of air pockets.*

Trimming the Laminate

Trim the laminate to the same size as the substrate. Use a router with a lower-bearing flush-trim bit. Fig. 18-10. Remember to follow all safety rules.

As you cut, the bit tends to become clogged with cement and dust. To prevent this, make a few cuts and stop. Allow the cutters to stop and spray both the cutters and the bearings with aerosol oil such as WD-40 or CRC. Repeat the process as you work.

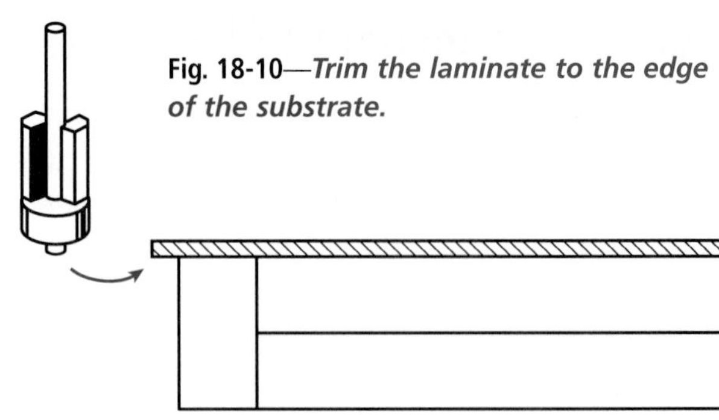

Fig. 18-10—*Trim the laminate to the edge of the substrate.*

Installing Molding and Making Finish Cuts

To provide a finished appearance, molding is installed on the facing edges of objects topped with plastic laminates. If the molding was installed before the laminate, cut a decorative shape, or *profile,* along the top edge. Fig. 18-11. Pre-shaped molding is installed after the laminate has been applied.

Fig. 18-9—*Using a rubber roller to apply pressure over the entire surface.*

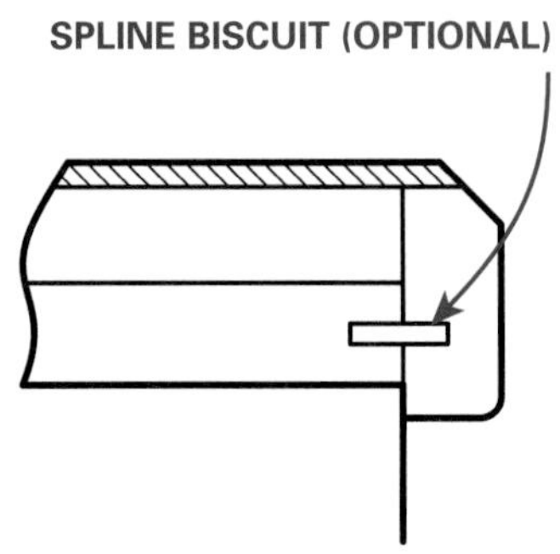

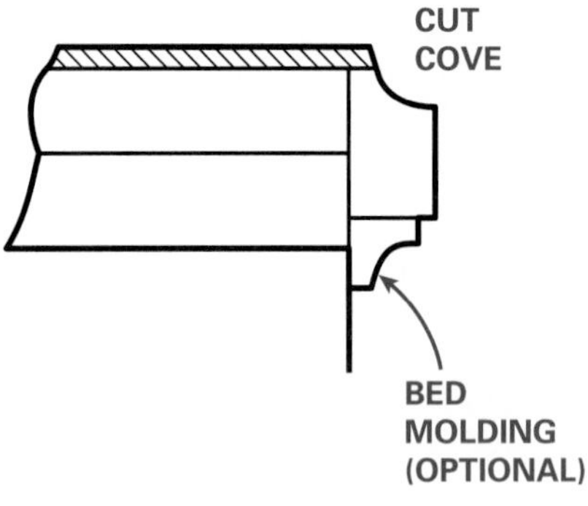

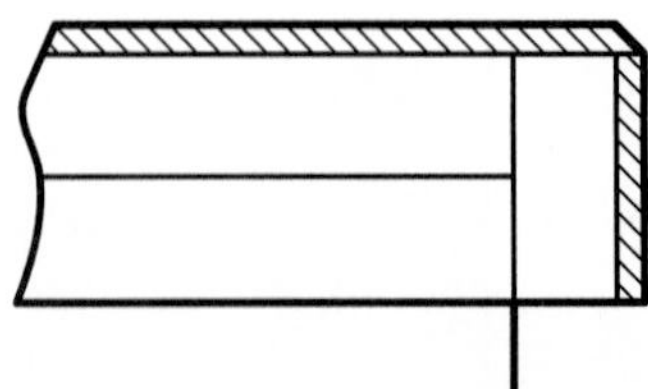

Fig. 18-11—*A decorative profile can be cut along the top edge of the molding. Note also the optional structural features.*

CHAPTER 18

Review & Applications

Chapter Summary

Major points from this chapter that you should remember include:

- Plastic laminate is a very hard, brittle material used to top counters and furniture.
- Plastic laminates are usually applied to a substrate.
- A router with carbide-tipped bits is good for cutting plastic laminate.
- Contact cement is an excellent adhesive to use when adhering laminate to substrate.
- To install laminate, begin in the middle. Remove the dowel separators one at a time, pressing the laminate against the substrate.
- After adhering laminate to substrate, apply pressure over the entire surface to achieve the strongest bond possible.

Review Questions

1. What materials make up plastic laminate? How are they joined?
2. Why are carbide-tipped tools good for cutting plastic laminates?
3. What type of breathing protection device should you wear when working with contact cement? Why?
4. What are the adhesive tendencies of contact cement that has dried on a surface?
5. How can you tell when contact cement is ready for bonding?
6. Describe the procedure for installing laminate on substrate.
7. What can be done to give the facing edge of a project topped with plastic laminate a finished appearance?

Solving Real World Problems

David is remodeling the master bath in his 20-year-old home. The old orange plastic laminate simply has to go! David manages to peel, scrape, and sand the old laminate and glue off the surface of the vanity. Following the directions of a friend, David prepares the surface for the new laminate. However, when he purchases the new material, he finds that it must be spliced. The vanity is 16′ long, and plastic laminate is only available in 12′ lengths. How can he make a splice that will not show badly?

CHAPTER 19

Veneering

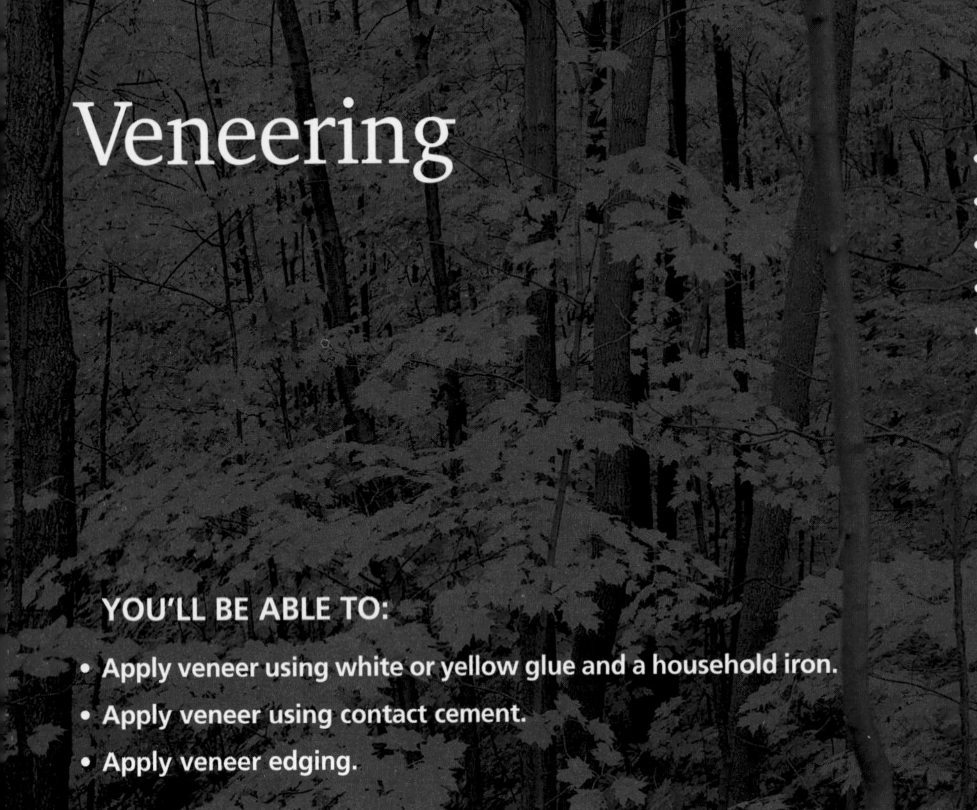

Look for These Terms

- flat cutting method
- flitch
- rotary cutting method
- veneer
- veneering

YOU'LL BE ABLE TO:

- Apply veneer using white or yellow glue and a household iron.
- Apply veneer using contact cement.
- Apply veneer edging.

Veneering is the process of applying a thin layer of fine wood, called **veneer,** to the surface of a wood of lesser quality. Veneer adds beauty and character to furniture and other wood projects. **Fig. 19-1.**

Veneering techniques have been used in woodworking for at least 3,000 years. In the last century, veneer has added a touch of elegance to items such as the dashboard in luxurious automobiles and the surface of magnificent dining tables.

Fig. 19-1—*A well-designed table is made even more attractive by veneer.*

Woodworkers today can still use the design freedom veneer allows to add beauty as well as quality to tables, chests, and other projects. Figured woods such as burled walnut or bird's-eye maple, exotic woods such as rosewood or ebony, or fine cabinet woods such as walnut or cherry can be used as veneer to create interesting, attractive designs. Fig. 19-2.

Fig. 19-2—*Veneers were used to add an interesting and unique beauty to a project.*

VENEER

A sheet of veneer is very thin, usually $\frac{1}{28}$ inch thick. The wood material to which it is applied is called the *substrate.* Besides covering solid woods, veneer can be applied to plywood, medium-density fiberboard (MDF), or particleboard. Veneer not only makes the surface more attractive, it also enhances strength.

Veneer has some advantages over solid wood. It doesn't swell or shrink as humidity increases or decreases. It doesn't cup or warp like boards. You don't have to worry about joints coming apart. Veneer will remain flat even when covering a wide area.

Probably the most important thing about veneer, though, is that good wood is used efficiently. Veneer sliced from an average size log can cover up to 40 times more area than boards cut from a log the same size.

Cutting Veneer

Only the best trees are chosen to be made into veneer. The trunks should be long, straight, and free of knots. Of course, these are generally more expensive than an average log. Veneer is cut using either a rotary method or a flat method. Fig. 19-3.

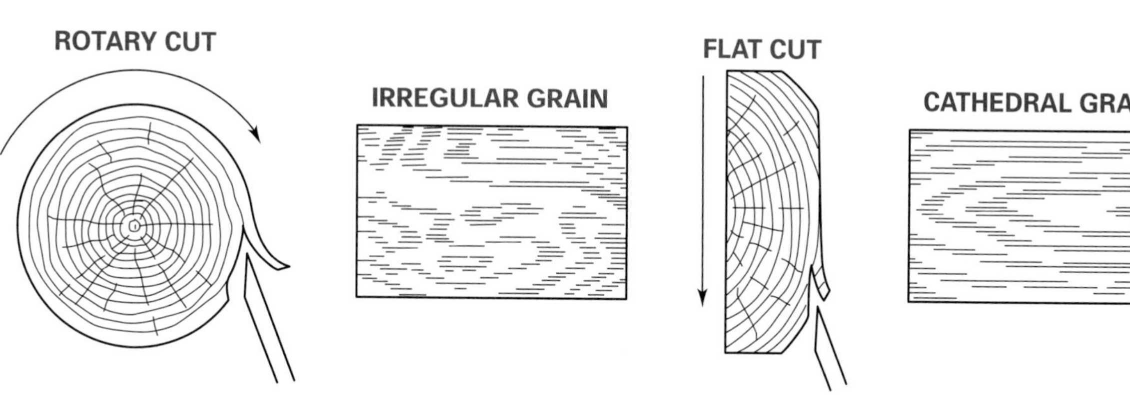

Fig. 19-3—*Methods of cutting veneers from a log.*

Rotary Cutting Method

The **rotary cutting method** is the faster of the two. A log is mounted in a huge lathe and turned against a blade. A continuous sheet of veneer as wide as the log is long is cut as the log is rotated.

Rotary-cut veneers are generally less attractive and expensive than veneers cut by the flat method. The grain pattern is spread out and irregular. These veneers are most useful when covering large areas.

Flat Cutting Method

In the **flat cutting method,** a log is clamped into a carrier. The log is moved back and forth across a blade and cut into slices. Most veneers cut by this method are narrow. The widest slice is the diameter of the log.

All slices from one log are kept in order in bundles called **flitches,** or books. The color and grain pattern of neighboring slices match each other closely. Keeping them together is helpful to the woodworker when creating designs.

Fig. 19-4—*These 18 × 6-inch pieces of veneer are typical of those available for small projects.*

The flat cutting method is preferred for furniture veneers. Grain patterns of flat-cut veneers are more clearly defined and attractive than those of rotary-cut veneers, and they match up better in design patterns.

Veneer Sizes

Most veneers are 1⁄28 inch thick. The normal flitch is about 12 to 14 inches wide and 16 feet long. However, veneers are also available in standard sizes such as 18 × 24 inches, 18 × 6 inches, and 24 × 48 inches. These sizes are designed specifically for veneering small projects. Fig. 19-4.

Facts about Wood

History of Veneering—Over the centuries the popularity of veneers has come and gone. They were popular as far back as the ancient Egyptians. In the 18th and 19th centuries, high-style furniture used fine veneers almost exclusively. The craftspeople veneered only over a solid wood base. They used hot glue made from animal hides, blood, and bones. They developed special hammers to smooth the veneer. Some woodworkers today still use this method, but most choose a more modern veneer press.

With the advent of production machinery in the 19th century, veneering declined. It came back into favor in the 20th century when manufactured board technology and improved adhesives became available. Furniture that was very expensive with solid exotic woods can now be veneered at a much more affordable cost.

Veneer edging is available in matching woods. It comes in widths of ½ inch, ⅝ inch, ¾ inch, 1 inch, and 2 inches. Fig. 19-5.

APPLYING VENEER

Veneers are often applied to furniture in design patterns. These patterns are made by some type of *matching*. Fig. 19-6. The pieces of veneer must be cut and glued edge to edge before being applied to the substrate.

For small projects such as jewelry boxes and chess boards, veneering is simpler. The project is built of some relatively inexpensive material

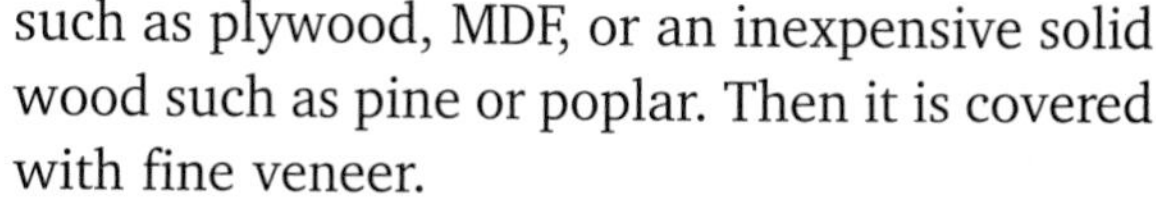

such as plywood, MDF, or an inexpensive solid wood such as pine or poplar. Then it is covered with fine veneer.

Tools

A veneer saw, craft knife, or a very sharp utility knife can be used to cut veneer. Fig. 19-7. A paper cutter or heavy-duty shears can also be used. A veneer roller is best for applying pressure to veneer surfaces. A household iron is useful for certain applications.

Fig. 19-5—***Veneer for edging.***

Fig. 19-7—***A veneer saw is made specifically for cutting veneer.***

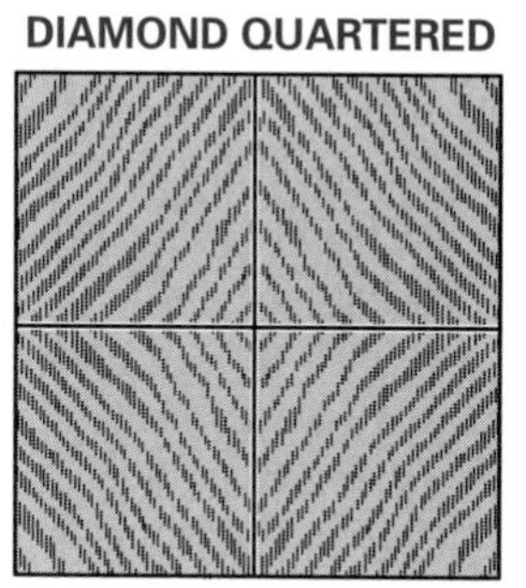

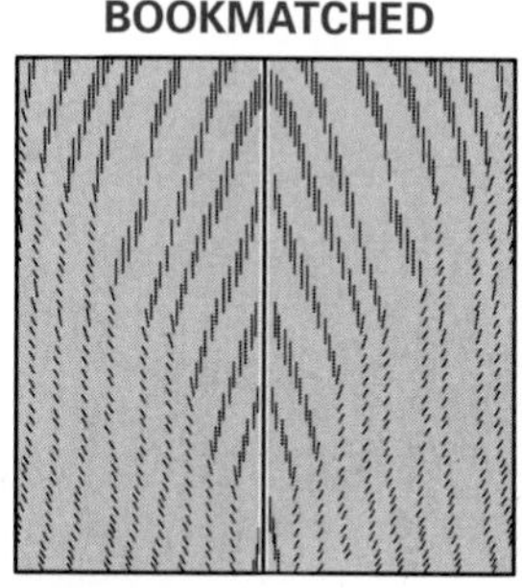

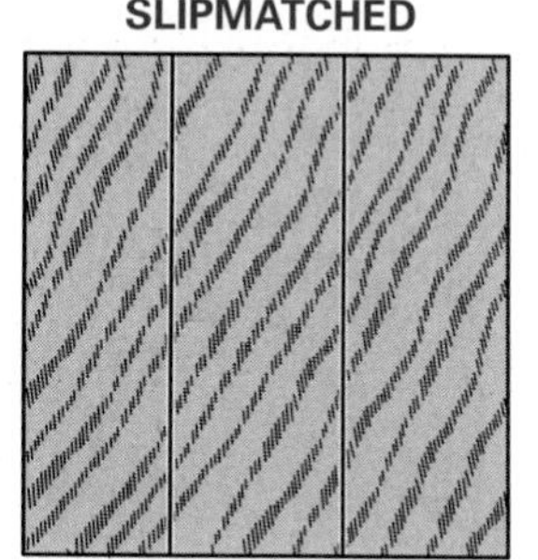

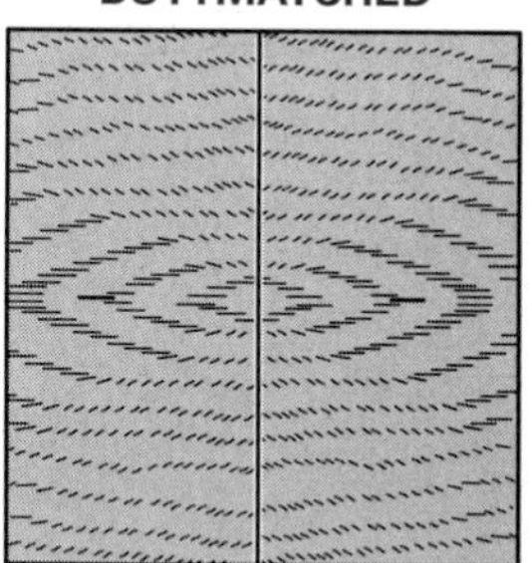

Fig. 19-6—***Basic veneer matching effects.***

Veneer must often be clamped to apply pressure during the gluing process. For the serious woodworker, the most popular method of applying pressure is with a vacuum press. Fig. 19-8. This system will put as much as 1,500 pounds per square foot of pressure on the veneer when adhering it to the substrate. When the vacuum is turned on, a plastic bag draws up around the material and holds it tightly in place until the glue sets. Fig. 19-9. It applies even pressure all over the veneer.

Safety First

►Using Tools

- Always wear eye protection when working in the shop.
- Use all tools properly.
- Always take special care with tools that are sharp.
- Take special care with tools or devices that become hot.

SCIENCE Connection

Vacuum

In science, a true vacuum is a space without matter, including gases. Air is a mixture of gases, so to create a true vacuum, all of the air must be removed from the space.

The vacuum used in a woodworker's vacuum press is actually an efficient partial vacuum. To create the partial vacuum, a pump is used to remove most of the air from inside a heavy plastic bag. As the volume of air inside the bag decreases, the pressure increases proportionately. This relationship between the volume and pressure of gases (at a constant temperature) was first stated by Robert Boyle in 1662. It is now referred to as Boyle's law.

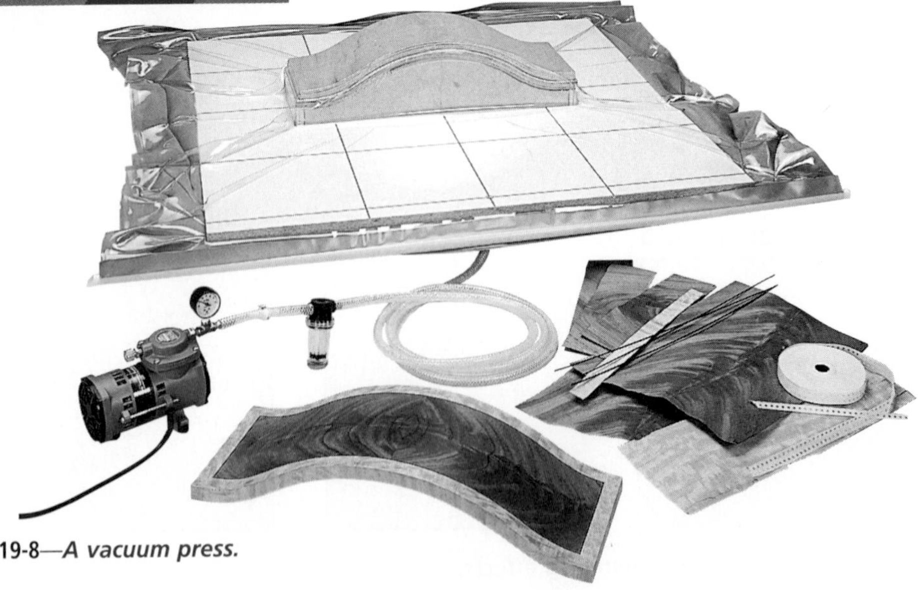

Fig. 19-8—*A vacuum press.*

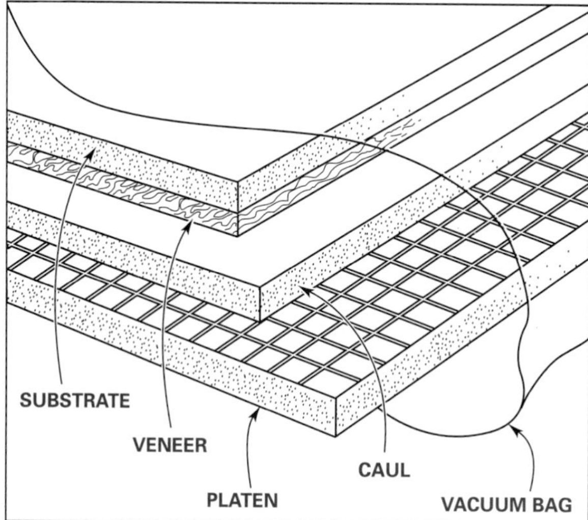

Fig. 19-9—*Order of layers inside a vacuum bag.*

Safety First

►*Using a Hot Iron*

- Use care and common sense when handling a hot iron. Don't touch the heated portion with your hands, and let the ironed surface cool before touching it.
- When you have finished using the iron, turn it off and set it aside in an upright position on a flat surface.

Heat-Adhesive Method

One of the easiest ways to apply veneer is with white or yellow glue. The glue is applied to the veneer and substrate and allowed to dry. Then the veneer is put in place and pressed with a hot household iron.

In making small projects, the main advantage of using the heat-adhesive method is that the pieces can be cut and fitted for a design after adhesive is applied. Also, a design can be put together piece by piece, because the veneer will not adhere until heat is applied.

Heat-Adhesive Procedure

Apply the glue by brushing it on both surfaces. To make it spread more easily, you may want to thin it somewhat with water. The glue should have the consistency of heavy cream. Usually two coats are enough. Let it dry.

Place the veneer on the surface and align it. Use a regular household iron set at a high heat. Starting at one corner, move the iron slowly over the veneer. Fig. 19-10. The heat must penetrate the veneer to liquefy the adhesive so it will bond. Cover the surface with a very smooth block of wood to hold the veneer in place while the glue sets. If more pressure is needed, clamp the wood in place. Be sure the pressure is applied evenly over the entire glued area.

Fig. 19-10—*Using a household iron set at a high heat to apply veneer.*

After the glue has set, trim any excess veneer at sharp right angles to the edge. Make sure that the edge is square, smooth, and free of sawdust. Then apply the edging. Methods of edging are discussed later in this chapter.

Safety First

►*Using Cutting Tools*

- Always take special care with tools that are sharp.
- Cut down and away from your body.

Using Contact Cement

Contact cement works well for bonding veneer to a substrate. Cut sheets of veneer for each major surface of the project. Each sheet should be about ¼ inch larger than the surface on all sides. Coat the surface of the project and one surface of the veneer with contact cement. Brush thoroughly. Allow it to dry about 30 minutes or until the gloss is gone.

Hold the veneer above the surface and align it. Then lower it in place. Remember that contact cement bonds immediately and permanently once the coated surfaces make contact. The veneer cannot be moved after it touches the surface.

A surface coated with contact cement will not adhere to any surface not coated with contact cement. If a relatively large area is to be covered, use separators to keep the coated surfaces apart while you align the veneer. Use dowels or sticks as separators. Lay them lightly across the coated surface of the project and place the veneer on top. When you have the alignment you want, remove one separator at a time and press the veneer to the surface.

Apply pressure to the surface with a small roller. Place a block of softwood over the veneer and strike with a hammer until the veneer is in complete contact with the surface.

Finish the edges by first trimming excess material. Sand edges lightly so they are square. Remove all sawdust before applying the edging.

APPLYING VENEER EDGING

Veneer edging is usually applied last. Three types of veneer edging are available: plain, heat-adhesive, and pressure-adhesive.

Plain edging is simply attached with adhesive. White glue, yellow glue, or contact cement can be used. If you use contact cement, remember to let the coated surfaces dry before applying the edging.

If *heat-adhesive edging* (veneer tape) is used, remove the paper liner and apply the veneer tape to the edge with hand pressure. Press a hot iron on the tape until the heat penetrates the veneer, activating the adhesive. Follow directly behind the iron with a block of wood to hold the edging in place until the adhesive cools and sets.

Pressure-adhesive edging has a backing of pressure-sensitive adhesive covered with paper. To glue this edging, strip the cover paper away and fasten the edging in place.

CHAPTER 19

Review & Applications

Chapter Summary

Major points from this chapter that you should remember include:

- Veneering is the process of applying a thin layer of fine wood, or veneer, to a substrate.
- Veneer can be cut using either a rotary or flat method.
- Veneers come in standard sizes as well as in flitches.
- When using the heat-adhesive method, veneer can be cut and fitted for a design after adhesive has been applied.
- White glue, yellow glue, or contact cement can be used to bond veneer to a substrate.
- Veneer edgings are usually applied last.

Review Questions

1. To what types of substrate can veneer be applied?
2. What advantages does veneer have over solid woods?
3. If you were building a coffee table, would you choose rotary-cut or flat-cut veneer?
4. In what standard sizes are veneers commonly available?
5. What tools can be used to cut veneer?
6. Describe how you would apply veneer using the heat-adhesive procedure.
7. Describe the process of using contact cement to apply veneer.
8. List the three types of veneer edging.
9. Describe how you would attach edging using contact cement.

Solving Real World Problems

Elena has purchased a few pieces of different veneers that she plans to apply to the top of a jewelry box. She has created a design for a pattern and has already rough-cut the pieces to size. She plans to use a heat-setting glue as an adhesive, but she does not know which one would be best. What would you suggest? Why?

CHAPTER 20

Planer

Look for These Terms

- backing board
- chip breaker
- clip
- cutterhead
- infeed roll
- outfeed roll
- planer
- pressure bar

YOU'LL BE ABLE TO:

- Identify the major parts of the planer.
- Surface a board to thickness.
- Explain the special procedure for planing thin stock.
- Plane several short boards to the same thickness.

The thickness **planer,** or *surfacer,* is designed to surface boards to thickness and to smooth rough-cut lumber. **Fig. 20-1.** It will not straighten a warped board. The cutting head is mounted above the table, so only the top of a board is surfaced.

The size of the planer indicates the maximum width and thickness that can be surfaced. For example, if the capacity is 12½ by 6 inches, like the portable planer shown in Fig. 20-2, the largest piece that can be surfaced is 12½ inches wide and 6 inches thick.

PARTS OF THE PLANER

The major parts of the planer include a motor, table (or bed), cutterhead, infeed and outfeed rolls, chip breaker, and pressure bar. Fig. 20-3.

The planer is self-feeding. After stock has been fed into it, it will continue through the machine by itself. As the stock is fed in, the upper corrugated **infeed roll** grips the stock and moves it toward the cutterhead. The **chip breaker** presses firmly on the top of the wood to prevent the grain from tearing out. The rotating **cutterhead** surfaces the board to the desired thickness. The **pressure bar** holds the stock firmly against the table. The **outfeed roll** helps move the stock out of the back of the machine.

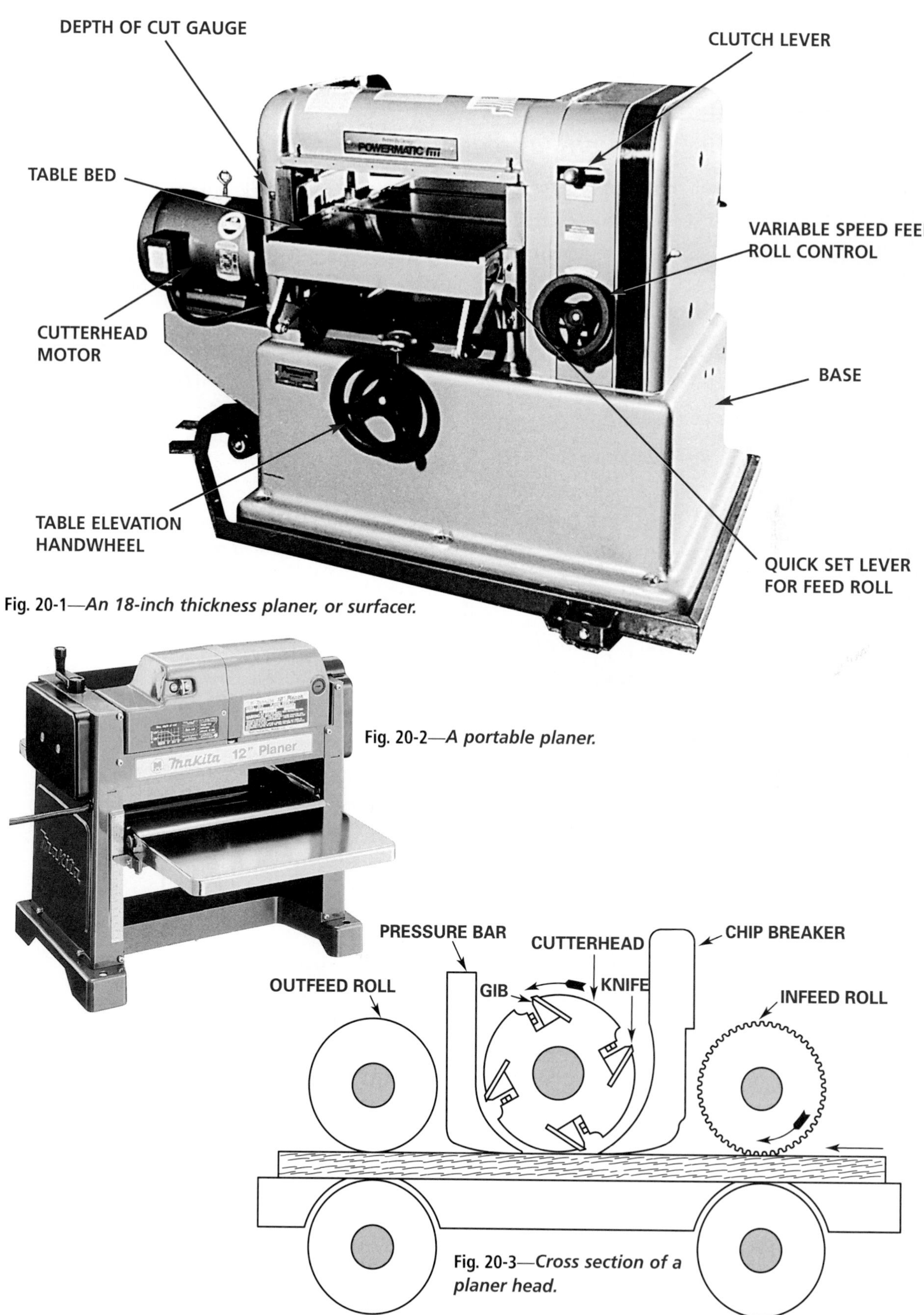

Fig. 20-1—*An 18-inch thickness planer, or surfacer.*

Fig. 20-2—*A portable planer.*

Fig. 20-3—*Cross section of a planer head.*

MATHEMATICS *Connection*

Effect of Cutting Speed on Finish

The cutting speed of a planer is the rate at which the wood moves through the machine. Cutting speeds are specified in feet per minute (fpm). At a given cutting speed, higher rates of cutterhead revolution result in more cuts per foot. This produces mill marks that are closer together, which results in a smoother finish.

You can figure out how many cuts are made per foot of stock using the following formula:

$$\frac{\text{Number of blades} \times \text{rate of revolution (rpm)}}{\text{fpm}}$$

Example: A planer operates at 8,000 rpm, has 3 blades, and has a cutting speed of 26 fpm. How many cuts does it make per foot?

$$\frac{3 \times 8{,}000 \text{ rpm}}{26 \text{ fpm}} = 923 \text{ cuts per foot}$$

Determine the cuts per foot for each of the following planer specifications.

1. 4,200 rpm, 2 blades, 10 fpm
2. 16,000 rpm, 4 blades, 26 fpm
3. 5,200 rpm, 3 blades, 33 fpm
4. 4,500 rpm, 3 blades, 16 rpm

On small planers, the controls are relatively simple. They consist of the following:

- A *switch* to turn on the machine.
- A *handwheel* that elevates or lowers the table.
- The *pointer* on the table that indicates thickness of the stock after it has been fed through the machine.
- A *feed control* that regulates the rate of feed from slow to fast. Fig. 20-4. (Some machines do not have this feature.)

The planer should be connected to a dust collection system to carry away the chips. Fig. 20-5.

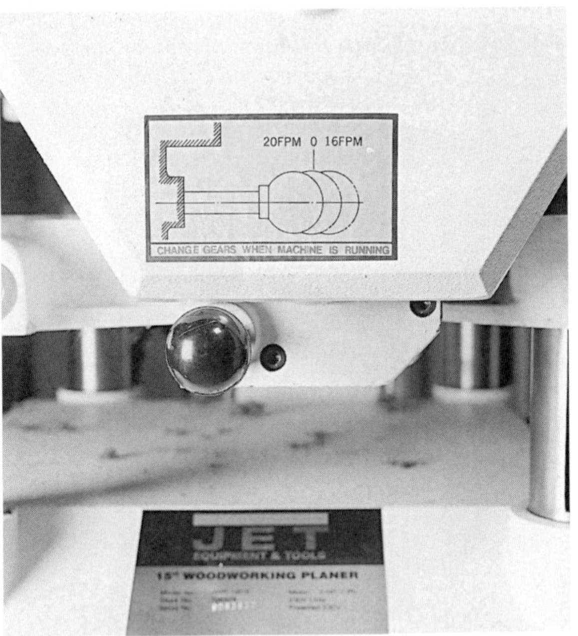

Fig. 20-4—*The variable-speed cutting feed makes it possible to select a feed range in feet per minute.*

Fig. 20-5—*The dust collection system helps keep the area around the machine free of sawdust.*

Safety First

►*Using a Planer*

- Check the board to be sure it is free of nails, loose knots, and other imperfections.
- Wear eye protection.
- Make sure that the board is at least 2 inches longer than the distance between the feed rolls. For a small planer, this usually means a board should be at least 14 inches long.
- Plane a warped board only after one surface has been trued on a jointer.
- Never attempt to plane more than one thickness at a time. If several boards of different thicknesses are to be surfaced, always plane the thickest one first, until it is about the same thickness as the others.
- Always stand to one side when planing. Never stand directly behind the board.
- Keep your hands away from the board after it starts through the planer.
- Never look into the planer as the board is passing through. Loose chips may be thrown back with great force.
- If a board sticks, turn off the switch. Wait for the cutterhead to stop. Then lower the table and remove the board.
- *Never reach over the planer.* If the board is long, have a helper support the surfaced stock as it is being fed out of the planer.
- Whenever helping to "tail off," simply support the board and keep it level. *Never pull on a board being planed.* Allow the outfeed rollers to feed the stock out of the planer.
- Stop the planer to make any adjustments or to clean or oil it.

USING THE PLANER

Before surfacing a board, make sure that it has one flat face. If necessary, use a jointer to plane the surface flat. (See Chapter 21.) Fig. 20-6.

1. Measure the thickness of the board at its thickest point.

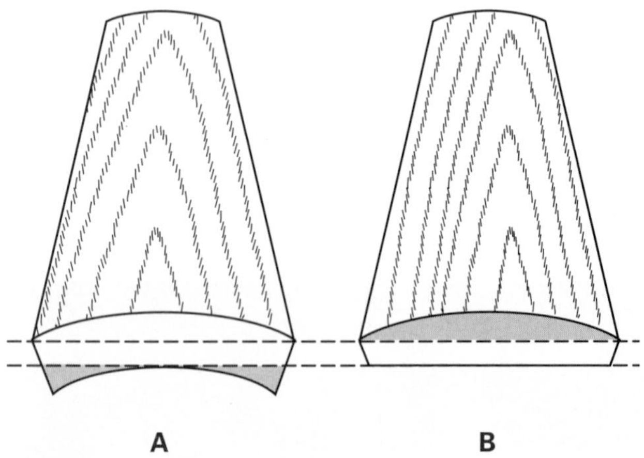

Fig. 20-6—***(A) First, surface plane the cup out on the jointer. (B) Then, thickness plane to dimension.***

2. Adjust the machine for the correct thickness—about $\frac{1}{16}$ to $\frac{1}{8}$ inch less than the measurement obtained in Step 1. (The maximum adjustment for a tabletop planer may be slightly less than $\frac{1}{8}$ inch, depending on the width of the board.) Never try to remove more than $\frac{1}{8}$ inch in thickness for rough work or $\frac{1}{16}$ inch for finished work.
3. Turn on the power and engage the feed control. Place the working face of the stock on the table so that the stock will feed with the grain. Fig. 20-7.
4. Start the stock into the planer. As soon as it takes hold, remove your hands.
5. After at least half the board has passed through the machine, walk around to the back. Hold the surfaced end of the board up as the remainder of the board passes through the machine. On long stock, it is a good idea to have a helper support and guide the board as it comes off the table.
6. Check the stock and then reset for thickness if another cut is to be made. The last cut should remove not more than $\frac{1}{32}$ to $\frac{1}{16}$ inch of material.

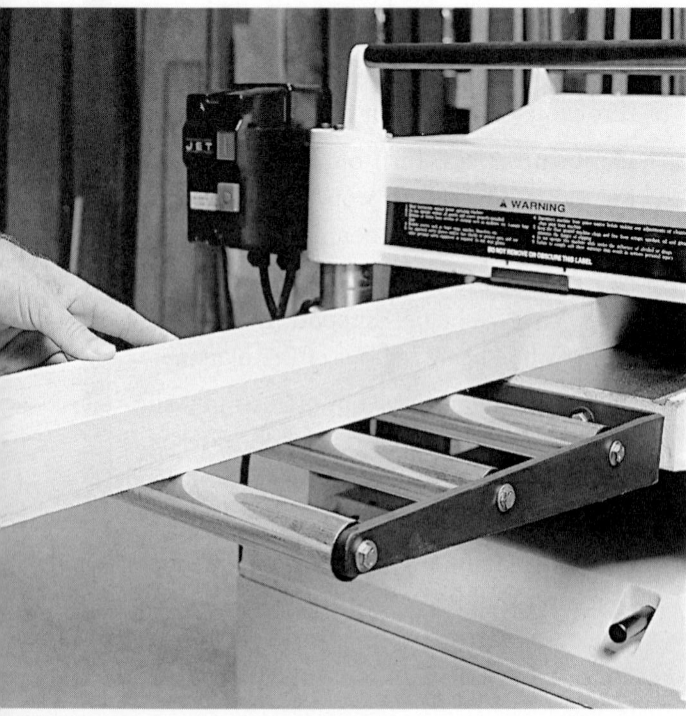

Fig. 20-7—***Be sure to feed the stock into the planer so that the cutter will be cutting with the grain.***

SQUARING LEGS

The planer is often used to square stock to be used for legs such as on the table shown in Fig. 20-8.

1. Begin the squaring operation by cutting the stock to rough size.
2. For each leg, joint one face and then one edge 90 degrees to the jointed face. (This face and edge are shown as sides A and B in Fig. 20-9.) Mark the jointed face and edge of each leg for identification.
3. Place the stock with the jointed surface (Side A) on the planer table and the jointed edge (Side B) to the right. Fig. 20-9. Set the planer for the necessary cut on Side C. Make the cut. When taking the stock from the outfeed table, take care not to alter the position of the piece.
4. Place the stock on the infeed table in the same position as for the first cut. Then turn the piece one-quarter turn clockwise to make the cut on Side D. Refer again to Fig. 20-9. Do not change the thickness setting. Feed the stock through the planer for this second cut.
5. Without changing the thickness setting, plane each side of the other three legs the same way as you did the first leg.

Remember that each side of all four of the legs has to be of identical width. If more cuts need to be made to get the desired dimensions, adjust the planer for the new cuts and repeat Steps 3 through 5.

PLANING THIN STOCK

If very thin stock must be planed, it is a good idea to use a **backing board.** This is true for all stock ⅜ inch or less in thickness. Make sure that the backing board is true, smooth, and at least ¾ inch thick.

Fig. 20-8—*A planer was used to square up the legs for this table.*

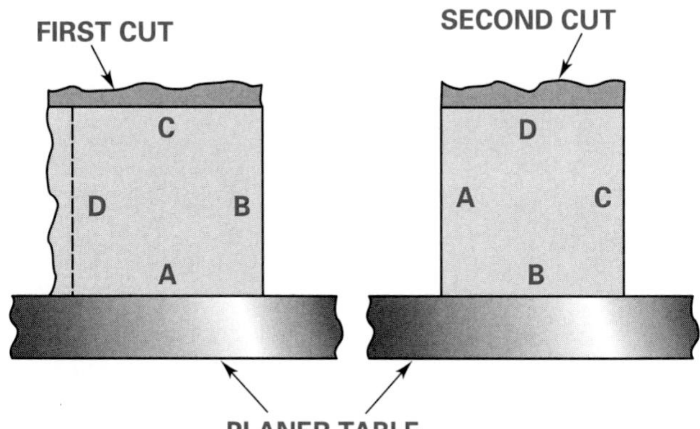

Fig. 20-9—*Squaring stock on the planer.*

Place the backing board on the infeed table. Then put the thin stock on top of the backing board. Adjust for the correct depth of cut, taking into consideration the thickness of the backing board. Then run the two boards together through the planer. Fig. 20-10.

SURFACING SEVERAL SHORT PIECES

If several short pieces of the same thickness are to be surfaced at the same time (such as four rails for a table), butt the ends together as they are fed through the planer. This helps keep the wood moving. It also eliminates the possibility of a clip or snipe dip. A **clip,** or *snipe,* is a small, concave cut at the end of the stock.

The last board should always be a longer board. It should be at least 2 inches longer than the distance between the feed rolls. Feed the pieces at a slight angle to improve the quality of the surface. Fig. 20-11.

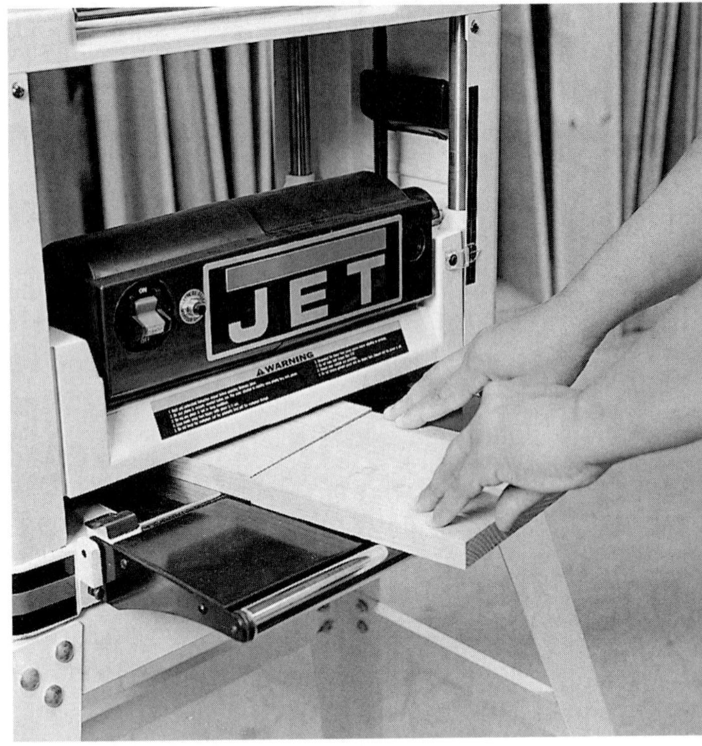

Fig. 20-11—*Surface short stock by feeding one board behind the other at a slight angle.*

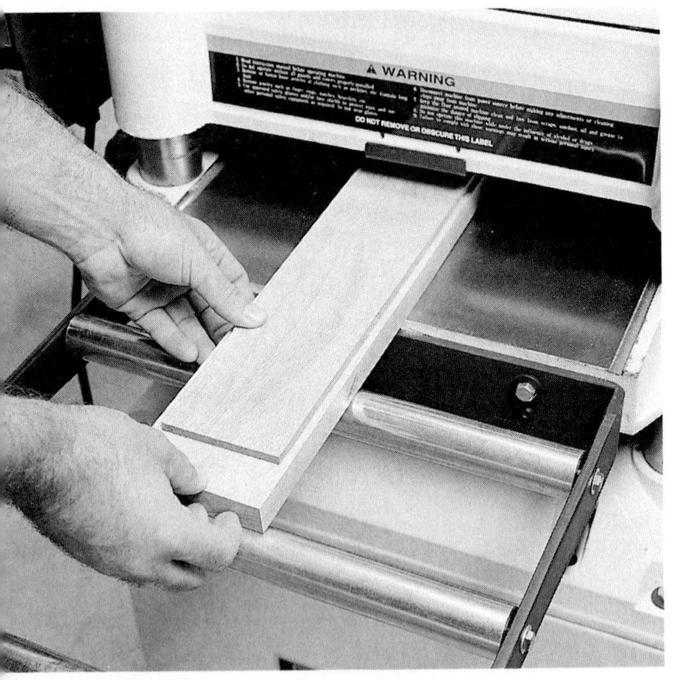

Fig. 20-10—*Surfacing thin stock. Use a backing board so that the piece will not split.*

What to Do with a Snipe

The snipe on the end of a board is usually only .0005 to .0015 inch deep. It can be trimmed off if the finished dimensions will allow. When trimming is not an option, the snipe can be sanded out with a palm sander if it is only .0001 inch deep or with a random-orbit sander if it is .0015 inch deep.

CHAPTER 20

Review & Applications

Chapter Summary

Major points in this chapter that you should remember include:

- The planer is designed to surface boards to a desired thickness and to smooth rough-cut lumber.
- Any board being surfaced must have one flat face.
- Only thin amounts of stock can be removed in each pass through the planer, so several cuts may be needed to reach the desired final thickness.
- The planer can be used to make square legs for tables.
- Special procedures are needed when planing thin stock and when surfacing several short pieces to the same thickness.

Review Questions

1. Identify the basic parts of a planer and describe the function of each.
2. List eight safety rules to follow when using the planer.
3. Describe the procedure for planing a board with one flat face.
4. What is the maximum amount of stock that should be removed at one cut for rough work? For finished work?
5. Explain how to plane thin stock.
6. Explain how to plane several short boards to the same thickness.

Solving Real World Problems

Howard wants to build a shadow box for his sister. It will measure 18″ × 24″ × 2″ thick. The outer frame will be made from 1/2″ stock. The shelves will be 3/8″ thick and 1 1/2″ wide. All assembly will be made using butt joints, and a thin sheet of plywood will be applied to the back. Howard has a board that measures 9/16″ × 6″ and is 8′ long. If Howard planes the board to 1/2″, how many 2″ wide strips will he need to make the shadow box frame? How much more stock does he need to plane off the remainder to make the shelves for the shadow box? Will he have enough for six 24″ shelves?

CHAPTER 21

Jointer

Look for These Terms

- face planing
- infeed table
- jointer
- kickback
- outfeed table
- pushblock
- ways

YOU'LL BE ABLE TO:

- Identify the main parts of the jointer.
- Describe face planing with a jointer.
- Joint an edge.
- Cut a bevel on a jointer.
- Adjust the jointer to cut a rabbet.

One of the most useful tools in the workshop is the *jointer.* **Figs. 21-1** and **21-2.** A **jointer** is a machine used to true up stock. That is, you can straighten, smooth, square up, and size boards. **Fig. 21-3.** Stock should be processed in this way before beginning any project in order to ensure good results.

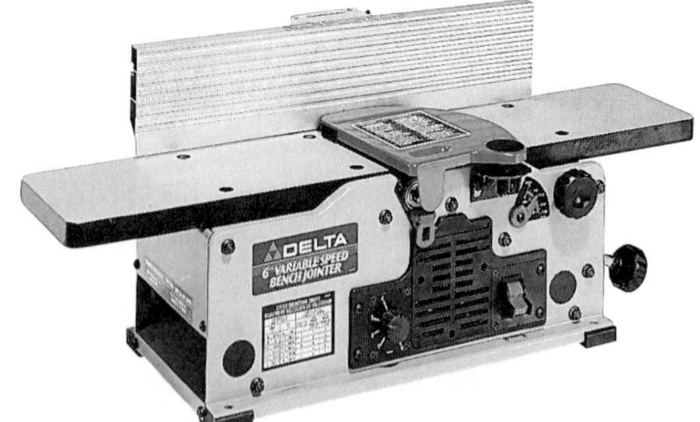

Fig. 21-1—*A bench jointer.*

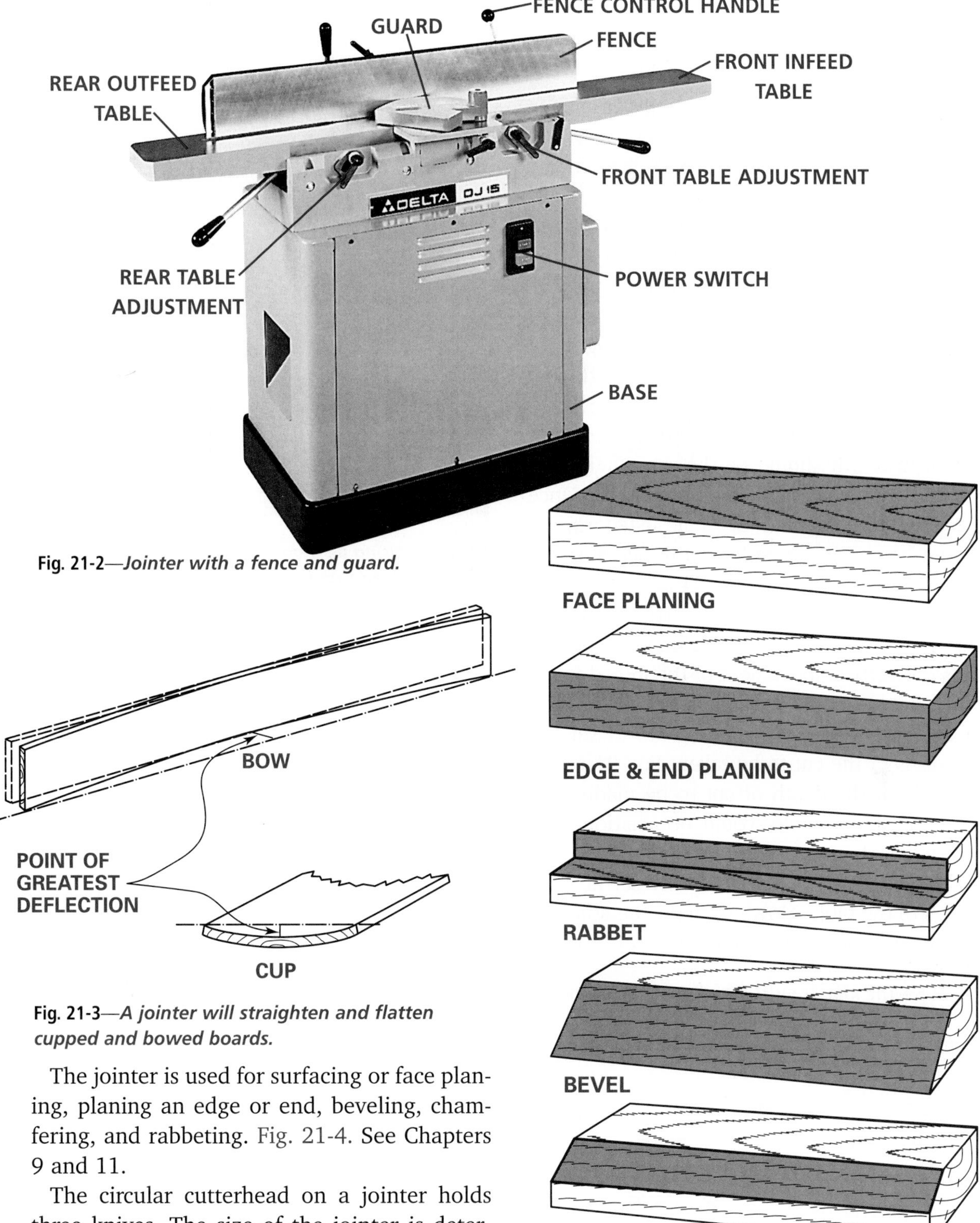

Fig. 21-2—*Jointer with a fence and guard.*

Fig. 21-3—*A jointer will straighten and flatten cupped and bowed boards.*

The jointer is used for surfacing or face planing, planing an edge or end, beveling, chamfering, and rabbeting. Fig. 21-4. See Chapters 9 and 11.

The circular cutterhead on a jointer holds three knives. The size of the jointer is determined by the length of these knives. The most common sizes are the 6- and 8-inch machines.

Fig. 21-4—*Common processes that can be done on the jointer.*

PROCESSING ON THE JOINTER

During processing on the jointer, the *cutterhead* rotates in a clockwise direction. When stock passes over the cutterhead, it is cut by the moving knives.

Stock is moved on **infeed** and **outfeed tables.** These tables are mounted on sliding **ways** (guiding surfaces) so that they may be raised and lowered to make height adjustments. The difference between the heights of the two tables determines the depth of the cut on the stock.

The outfeed table supports the work *after* it has been cut. It should be the same height as the cutting knives at their highest point. Once the outfeed table has been adjusted to the proper height, it can be locked in position and should not be changed. Fig. 21-5.

The infeed table supports the work *before* it is cut. The infeed table adjusting handwheel controls the vertical movement of the infeed table. The height of this table is adjusted in relation to the height of the outfeed table to set the thickness of the cut to be taken. Fig. 21-6. For example, if the depth of cut to be made is 1⁄8 inch, the infeed table would be positioned 1⁄8 inch lower than the outfeed table.

The *fence* of a jointer is usually set at right angles to the table. However, it can be adjusted to any angle depending on the type of process.

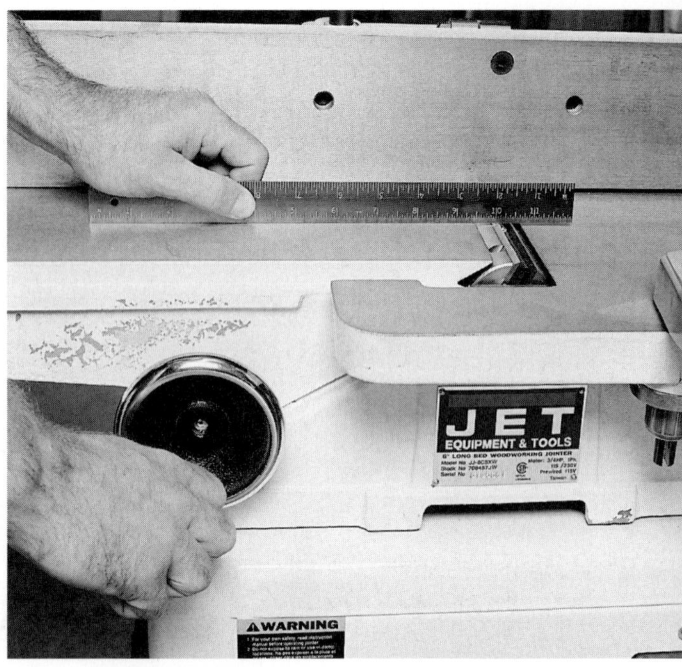

Fig. 21-5—***Using a straightedge to check the position of the outfeed table so that it is level with the knives. Be sure the power is off (not just the operating switch).***

Fig. 21-6—***The jointer must be adjusted so the outfeed table is at exactly the same height as the cutterhead knife at its highest point. Note that the cutterhead rotates in a clockwise direction.***

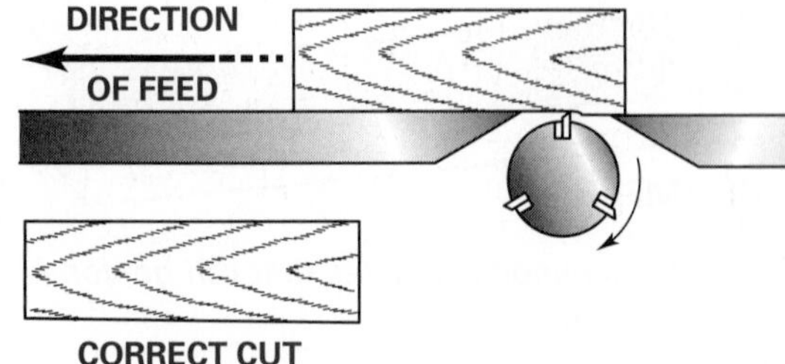

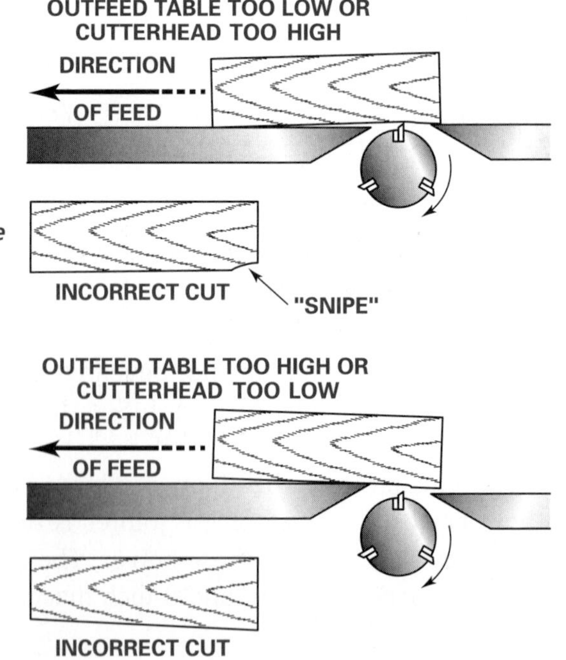

Safety First

►*Using a Jointer*

- Wear eye protection.
- Adjust the tables for the depth of the cut before turning on the power. Turn off the jointer before making any adjustments.
- Never allow your hand to pass directly over the cutterhead.
- Do not use a jointer to process stock less than 2 inches wide or 12 inches long or ½-inch thick.
- Always have the guard in place over the knives while the jointer is being operated. (The only exception to this rule is when rabbeting stock.)
- Stand to the left of the jointer, never directly behind it.
- Feed the stock into the machine in the direction opposite the direction of the grain. Fig. 21-7.
- Do not allow your fingers to come any closer to the rotating knives than 5 inches when jointing stock. Always keep your hands on top of the work, away from the danger zone. Fig. 21-8 (page 290).
- Do not use the jointer for cuts heavier than ⅛ inch. A cut that is too heavy may cause **kickback.** That is, the stock may suddenly and violently be thrust upward and back. It is safer to take several lighter cuts.
- Feed (push) work into the rotating knives firmly and at a rate that will not overload the machine.
- Use a pushblock (wood block or stick) to push stock through the machine when planing the face of a board. Fig. 21-9.
- Keep your hands on top of the stock and not on the ends. A sudden jar, caused by a knot, might dislodge your hands or jab the corners of the stock into your palms.

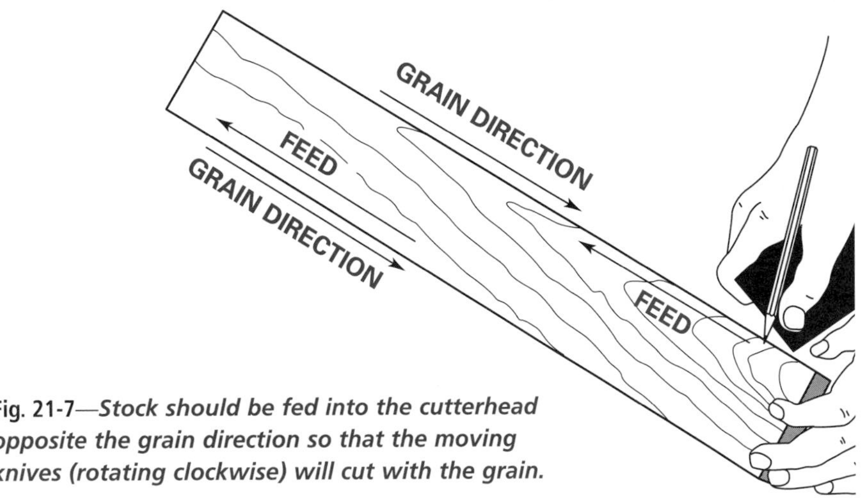

Fig. 21-7—*Stock should be fed into the cutterhead opposite the grain direction so that the moving knives (rotating clockwise) will cut with the grain.*

Fig. 21-8—***The danger zone of a jointer is the area directly over the rotating cutterhead. Keep your hands outside this area throughout the cutting process.***

Fig. 21-9—***Using a pushblock. Never attempt to surface thin stock or joint the edge of narrow pieces without using a pushblock to keep your fingers away from the rotating cutterhead.***

SCIENCE *Connection*

Why Wood Warps

You have probably seen "growth rings" in tree stumps. As a tree grows, two new rings (one light and one dark) appear each year. In wet years, when the tree receives more moisture, the rings are wider than those made in dry years.

When a tree is cut into lumber, the rings appear in cross-section to form the *grain* of the wood. Wider growth rings become "open" grain, and narrower ones are "tight." If a board has a tighter grain on one side than on the other, the tighter side "pulls" the open side, causing the wood to warp (cup or bow). This effect becomes more obvious as the wood dries out.

The Way of the Grain

Sometimes it's hard to determine the direction of the grain in a piece of wood. Try this: Run your hand along the face of the board. The board will feel relatively smooth when your hand is moving *with* the grain. It will feel relatively rough when moving *against* the grain. *Caution:* Take care that you don't run a splinter into your hand. Be sure to stroke lightly in the middle of the board, away from the edges.

Face Planing or Surfacing

Face planing, or surfacing, means planing the surfaces to true up stock. Most often, only stock that is rough on all four sides is face planed on the jointer. Only stock that is less in width than the width of the knives of the jointer should be face planed.

If face planing is used, set the infeed table to take a very thin cut. Hold the stock firmly against the front of the table. Use a **push-block** to move the stock with one hand as the other hand holds the front of the stock down. Slowly push the board through the cutterhead. Fig. 21-10. Then apply equal pressure to both front and rear. When the major portion has passed the outfeed table, apply more pressure to the rear of the board to keep the board level.

Fig. 21-10—*The correct method of face planing. 1. Start the cut by applying pressure on the stock with both hands against the infeed table. 2. As the stock passes over the cutterhead, move one hand to the outfeed table and apply pressure to finish the cut.*

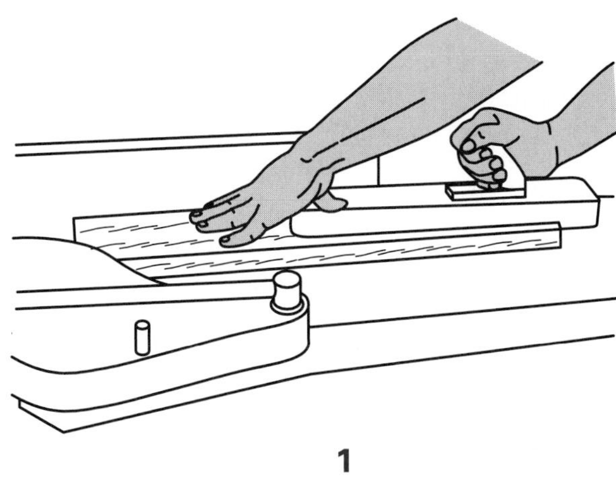

1

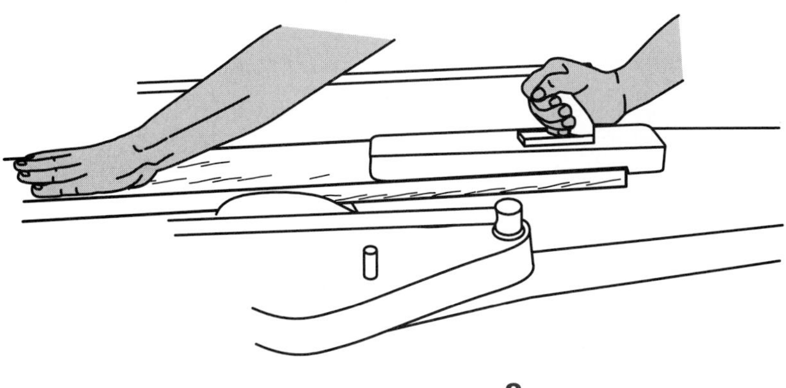

2

Jointing an Edge

Jointing an edge is the most common use of the jointer. Jointing is smoothing and straightening an edge to make it square with the face surface. To joint an edge, follow this procedure:

1. Make sure that the fence is at right angles to the table. Check this with a try square.
2. Hold the stock on the infeed table with the face surface against the fence.
3. Use one hand to guide the stock and the other hand to apply forward pressure. Fig. 21-11. Do not push the stock through the jointer too fast. Doing so will cause little ripples to be formed by the rotating cutterhead.
4. Hold the stock firmly against the infeed table at the beginning of the cut.
5. Apply even pressure against both tables during the middle portion of the cut.
6. Apply pressure against the outfeed table while completing the cut.

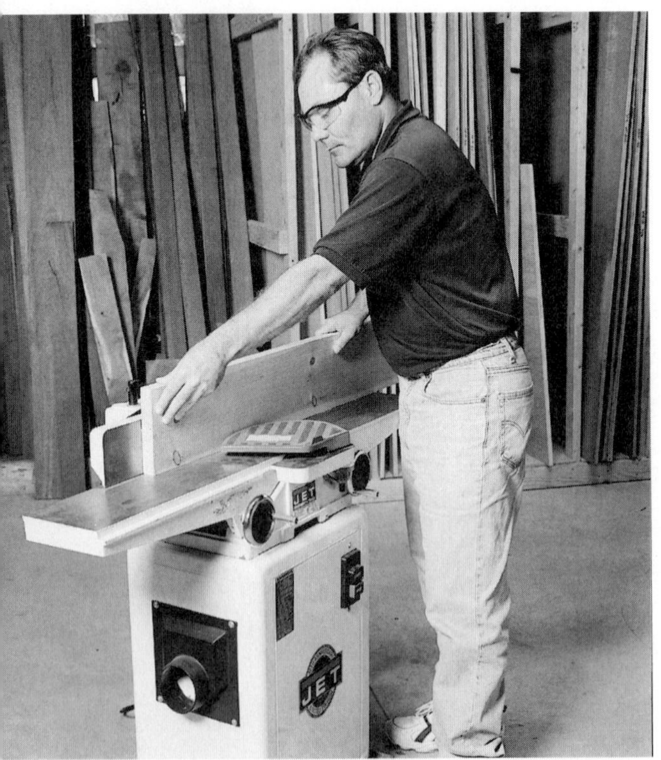

Fig. 21-11—*Jointing an edge.*

Safety First

►*Edge Jointing*

Hold the work face of the stock flat against the fence throughout the edge jointing operation.

End Jointing

Most wood parts do not require that the end grain be planed. Usually an end that has been sawed is accurate enough. End planing is quite difficult because you are actually cutting off the ends of the wood fibers. Therefore, very light cuts should be made.

In end jointing there is always a tendency for the wood to split out at the end of the cut. For this reason, it is a good idea to make a short, light cut along the end grain for about 1 inch. Then reverse the board to complete the cut. Fig. 21-12.

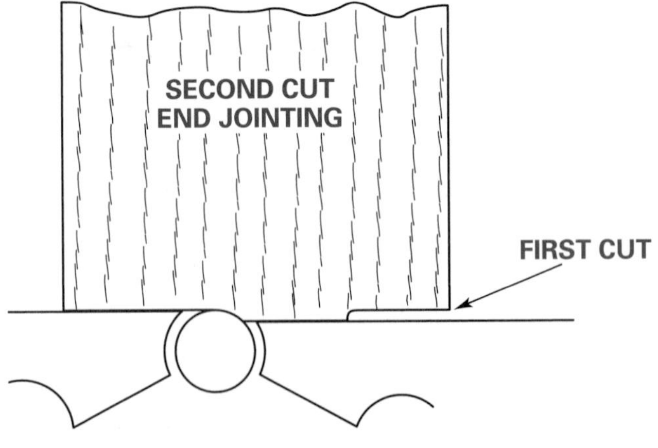

Fig. 21-12—*Cutting technique for preventing splintering when planing end grain.*

Beveling and Chamfering

Basically, a *bevel* is any angle other than a right angle. It can also refer to an angled surface. A *chamfer* is the angled surface that results when a sharp edge on a piece of stock is cut off at a slight bevel. To cut a bevel or chamfer on an edge with a jointer, set the fence at the proper angle to the table. The fence may be tilted in or out. However, it is safer to cut with the fence tilted in. Fig. 21-13. Check the angle with a sliding T-bevel. Then proceed as in jointing an edge.

Rabbeting

A *rabbet* is an L-shaped cut along the end or edge of a board made when making a rabbet joint. See Chapter 9. It can be cut with the grain. First, adjust the fence to an amount equal to the width of the rabbet. Then adjust the infeed table to an amount equal to the depth of the rabbet. If the rabbet is quite deep, it may be necessary to cut it in more than one pass.

Safety First

►*Cutting a Rabbet*

- The guard must be removed to cut a rabbet. Work very carefully.
- Clamp an auxiliary piece to the fence to cover the cutterhead. Fig. 21-14.

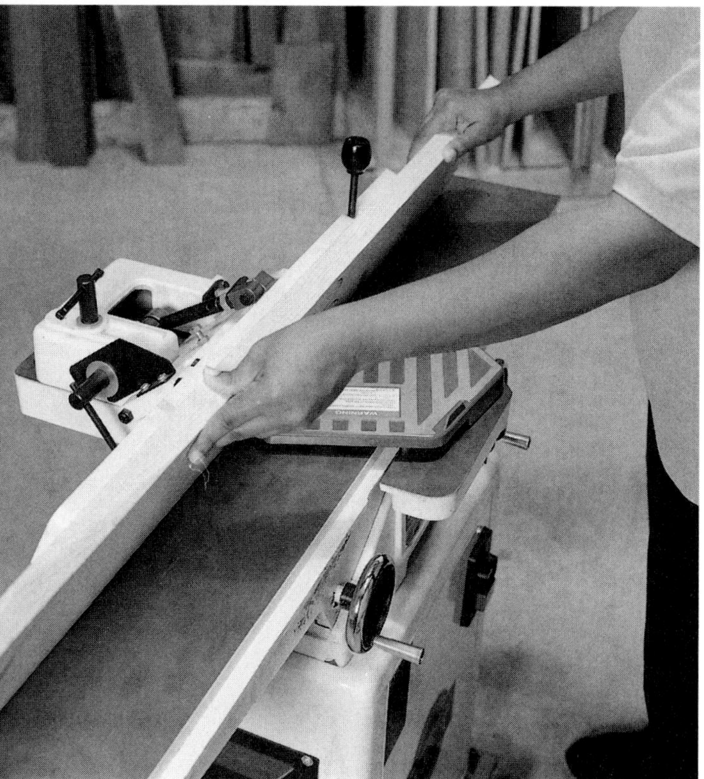

Fig. 21-13—***Cutting a bevel or chamfer. The fence of the jointer is tilted for these operations.***

Fig. 21-14—***Cutting a rabbet on the jointer. Note the auxiliary piece attached to the fence.***

CHAPTER 21

Review & Applications

Chapter Summary

Major points in this chapter that you should remember include:

- A jointer is used to straighten, smooth, square up, and size stock.
- The difference between the heights of the infeed table and the outfeed table determines the depth of cut on the stock.
- Keep the guard over the cutterhead or, when cutting a rabbet, use an auxiliary piece to cover it.
- Face planing is the process of truing up stock by planing the surfaces.
- Make light cuts when squaring an end.
- Beveling and chamfering are the cutting of angled surfaces on a piece of stock.

Review Questions

1. Name five main parts of a jointer.
2. Explain how to adjust the tables to set the depth of cut.
3. Where is the danger zone when working on a jointer? Why should it be avoided?
4. Describe face planing with a jointer.
5. List the steps in jointing an edge.
6. What causes ripples to be formed on a board when it is run through the jointer?
7. Describe the cutting technique that will keep the wood from splitting while you are end planing.
8. How would you cut a bevel on a jointer?
9. How would you adjust a jointer in order to cut a rabbet?

Solving Real World Problems

After carefully cutting six boards for a small tabletop, Jerome lays them out on a workbench in the order they will be assembled. Noting that the seams are less than perfect, Jerome sets up his jointer to remove 1/64″ per pass. He uses a square to make sure that the fence is correctly set, and he adjusts the table featherboards for a tight fit. Jerome then makes two passes on each edge of each board. He lays the boards back out in order and checks the seams again. This time they are perfect. However, when he measures the assembled top, he discovers that it is now too narrow for the base. Removing too much stock with the jointer must have caused this. What was the total amount removed by Jerome with the jointer? How much wider should he have made each board?

CHAPTER 22

Table Saw

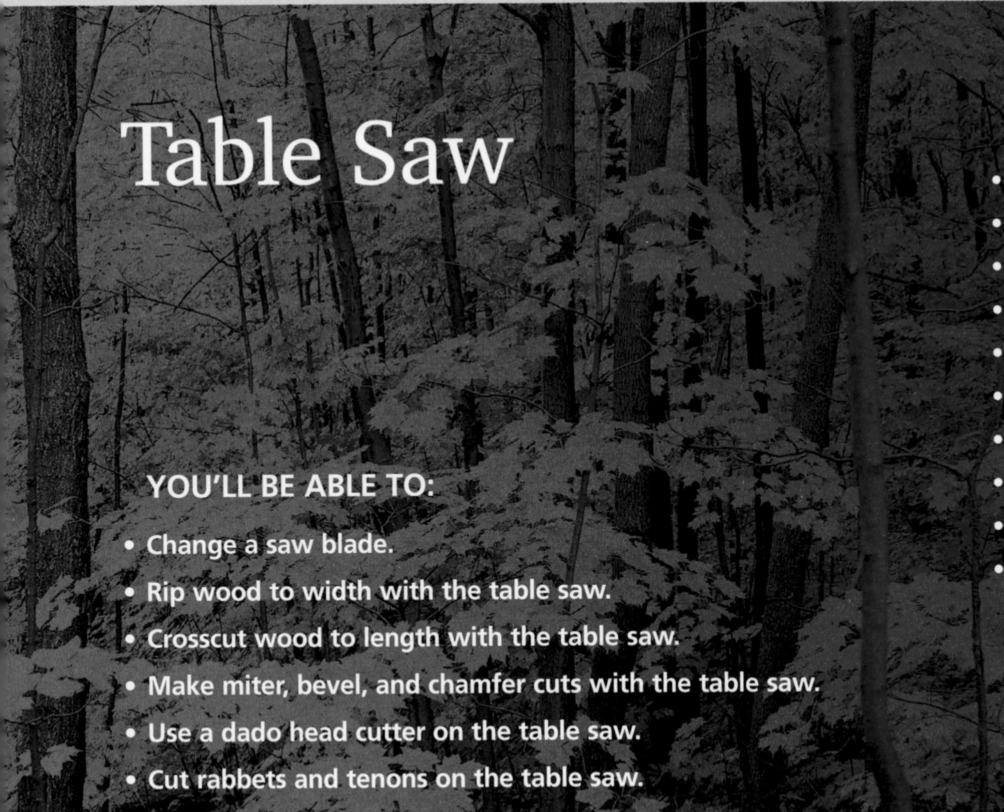

YOU'LL BE ABLE TO:

- Change a saw blade.
- Rip wood to width with the table saw.
- Crosscut wood to length with the table saw.
- Make miter, bevel, and chamfer cuts with the table saw.
- Use a dado head cutter on the table saw.
- Cut rabbets and tenons on the table saw.

Look for These Terms

- antikickback pawls
- arbor
- carbide tip
- dado head
- featherboard
- miter gauge
- pushstick
- rip fence
- splitter
- table saw

The **table saw** is the machine most often used in a woodworker's shop. This saw is used for ripping boards to width and crosscutting them to length. Most table saws have a fixed, horizontal table and a blade that can be raised, lowered, and set at an angle.

PARTS OF A TABLE SAW

The parts of a table saw are shown in Fig. 22-1 (page 296). This machine should always have a *guard,* a *splitter,* and *antikickback pawls* (fingers) to protect the operator from injury. The *saw-raising handwheel* is usually under the front of the table. It is used to raise and lower the saw blade. The *saw-tilt handwheel,* usually on the left side, is used to tilt the arbor for cutting at an angle. The **arbor** is the shaft that holds the saw blade.

The size of a table saw is indicated by the largest diameter of saw blade that can be used on the machine. The most common sizes are 8- and 10-inch tilt arbor saws. A 10-inch saw can crosscut a workpiece up to 3⅛ inches thick.

A variation of the large stationary table saw is the smaller benchtop model. Fig. 22-2. It is easy to carry and can be used almost anywhere. It is perfect for the beginning or small workshop.

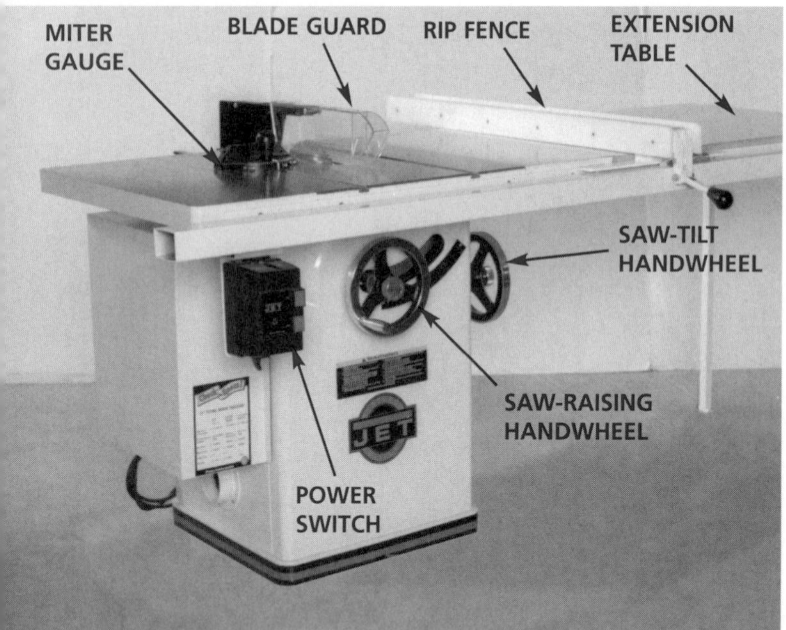

Fig. 22-1—*The table saw is used more than any other machine in the woodworker's shop. This 10-inch stationary saw is equipped with an extension table.*

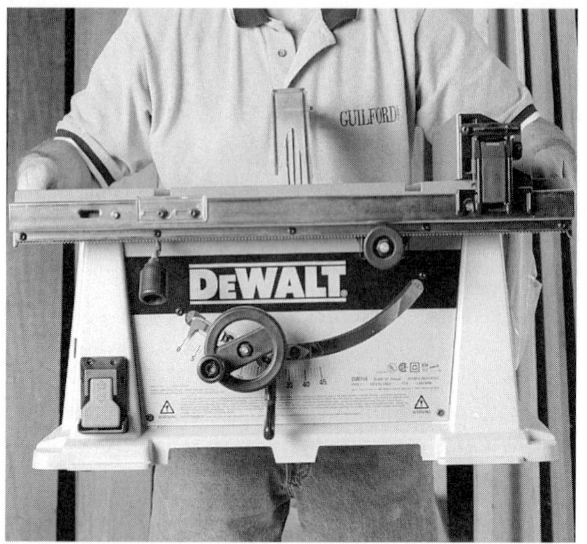

Fig. 22-2—*A benchtop saw can be easily carried to the work site. It is ideal for the small workshop.*

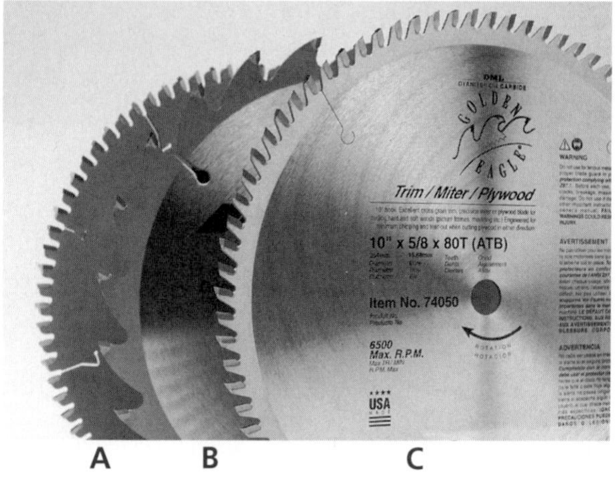

Fig. 22-3—*Common circular saw blades include: (A) Crosscut, (B) Rip, (C) Plywood.*

SELECTING A SAW BLADE

There are many kinds of table saw blades. The most common are ripsaw (riptooth), crosscut, and combination blades. Fig. 22-3. The combination blade can be used for both crosscutting and ripping. Therefore, when doing many different operations, one after another, it is a good idea to use a combination blade.

Many saw blades have **carbide tips.** Carbide is an extremely hard alloy of carbon and metal. It keeps the blade sharper for a longer period of time.

Safety First

►*Using a Table Saw*

- Wear eye protection.
- When changing saw blades or using a dado head, make sure the main switch is off. When possible, pull the plug from the receptacle to prevent accidental starting.
- Make all adjustments when the saw is at a dead stop. These adjustments include height of the saw blade, angle setting, fence adjustments, and all special setups. Always stop the saw before adjusting any setting.
- When setting up the saw for any sawing operation, check to see that the saw blade revolves freely in the correct direction, that the blade is securely fastened to the arbor, and that any clamps or knobs on the fences are tightened.
- Raise the saw blade from ⅛ to ¼ inch higher than the thickness of the stock to be cut.
- Have your instructor doublecheck all special set-ups and blade changes before the saw is started.
- Make sure that the **antikickback pawls** (fingers) behind the saw blade are always in place. These resist the tendency of the saw to throw the stock upward and toward the operator.
- Always use the saw guard. The saw guard covers the blade and the area around it, while allowing the stock to slide under it. Never operate the saw without the guard in place unless a special jig or fixture is used as a guard.
- Do not try to cut cylindrical (round) stock on the saw.
- Stand to one side of the saw blade. No one should ever stand in line with the blade, where kickback could cause serious injury.
- Hold the stock to be sawed against the fence or miter gauge. Never try to saw "freehand" (without holding the stock against the fence or miter gauge).
- Do not force stock into a saw blade faster than it will cut.
- Use a **pushstick** to push the stock past the blade if the space between the saw blade and the fence is 6 inches or less.
- Do not allow your fingers to come closer than 5 inches to the saw blade when you are cutting stock.
- *Never reach over the saw blade!* Have a helper take the stock away.
- Use a stick or board to clear away scraps close to the saw blade. Do not use your hand.
- Remove all special setups and the dado head from the saw after use.

Combination Blades

A combination saw blade could have as many as 80 teeth or as few as 18. Usually the more teeth, the smoother the cut. However, the more teeth, the longer it will take to make the cut.

CHANGING A BLADE

To change the saw blade, follow these steps:

1. Snap out the throat plate from around the saw blade.
2. Obtain a wrench that will fit the arbor nut. Check the thread on the arbor. On most table saws, the arbor has a left-hand thread and must be turned clockwise to loosen. However, some manufacturers use a right-hand thread. If so, you must turn the arbor nut counterclockwise to remove it.
3. Hold a piece of scrap wood against the blade to keep the arbor from turning as you loosen the arbor nut. Fig. 22-4. Remove the nut, the collar, and the blade.
4. Mark the arbor with a slight file cut or prick-punch mark. Always turn the arbor so the mark is "up" before putting on a new blade. Place the new blade so the trade-mark is at the top, in line with the mark on the arbor. Replace the collar and nut.

MATHEMATICS *Connection*

Rim Speed and Kickback

The rim speed of a saw blade is the rate at which the cutting edge of the blade moves. To find the rim speed of a blade, you need to know the diameter of saw blade (in inches) and the rpm of the motor. The following formula gives the rim speed in feet per minute.

$$\text{Rim speed} = \frac{3.1416 \times \text{diameter of blade (in inches)} \times \text{rpm of motor}}{12}$$

What is the rim speed in feet per minute of a 10-inch saw blade being turned by a motor operating at 3,600 rpm?

Remember, a 12-inch blade operating at 3,600 rpm is traveling at more than 120 miles per hour. Can you see why kickback can make a piece of wood a deadly missile?

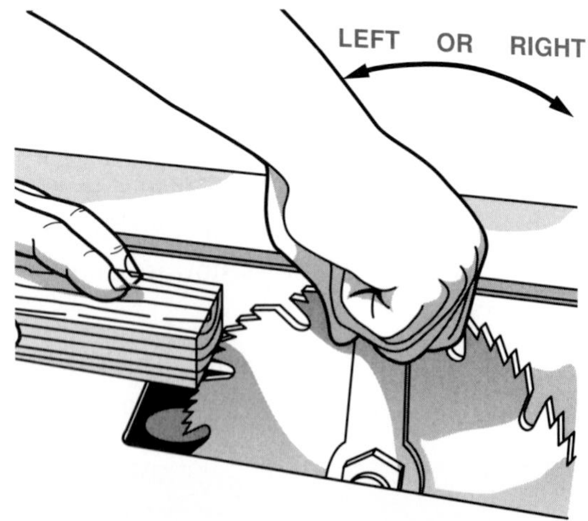

Fig. 22-4—*To remove the blade, hold a piece of wood against the blade to keep it from turning. With the proper size wrench, loosen the nut on the arbor. On most saws, the arbor has a left-hand thread and must be turned clockwise to loosen it.*

RIPPING

Before you rip stock to width on the table saw, make sure that one edge of the stock is true. If it is not, plane one edge before ripping.

Follow these steps for ripping:

1. With the power off, adjust the saw blade so that it is perpendicular with the top of the table. Fig. 22-5.
2. Raise the blade to a height of ⅛ to ¼ inch more than the thickness of the stock. Figs. 22-6 and 22-7 (page 300).

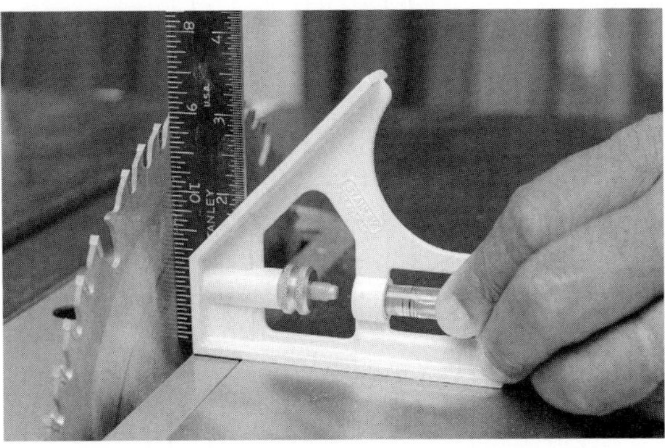

Fig. 22-5—*To get a square cut, the saw blade needs to be aligned with the table. Position the blade of the square between the saw teeth so the offset of the teeth does not affect the measurement.*

Fig. 22-6—*Adjust the saw blade to extend ⅛ to ¼ inch above the workpiece.*

Safety First

►*Ripping*

- Use a rip fence for ripping operations.
- Wear eye protection.
- Use the **splitter,** guard, and antikickback pawls when ripping. Wood cut with the grain tends to spring the saw kerf closed and bind the blade. The splitter prevents this from happening. If a splitter is not used, stop the machine and insert a wedge in the kerf as soon as the cut has passed the back of the blade.
- Stock to be ripped must have a straight, true edge and must lie flat on the table. Never cut stock on the saw if it is warped, "in wind," or has a rough or bowed edge.
- Whenever helping to "tail-off," hold the board up and allow the operator to push the stock through the saw. (To "tail-off" means to hold the already ripped pieces of stock level with the saw blade while the remainder of the stock is being ripped.) Never pull on a board being ripped.

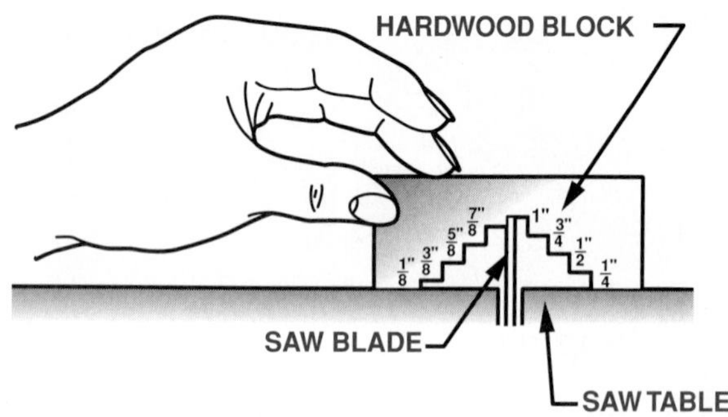

Fig. 22-7—*A blade gauge with notches accurately cut in ⅛-inch steps saves time when adjusting the blade to different heights. Raise the blade until it touches the gauge at the correct height (⅛" to ¼" higher than the thickness of the stock).*

3. Adjust the **rip fence,** which guides the workpiece straight through the saw blade, to the correct width. Three methods of setting the rip fence are shown and described in Fig. 22-8. On many machines, the width for ripping is found directly on a scale mounted on the front edge of the saw table. Always check the setting by making a short cut in a piece of scrap wood.
4. Lock the rip fence so it is tight.
5. Make sure the guard, splitter, and anti-kickback pawls are in place.
6. Turn on the power. Place the stock on the table. Stand to one side (usually the side opposite the fence), not directly behind the saw blade or the workpiece. Fig. 22-9.
7. To start the cut, apply forward pressure with one hand as you hold the stock against the fence with the other. Do not apply too much forward pressure on a small saw; this will make the saw burn or stop altogether. Continue to feed the work into the saw with even pressure.

When cutting to narrow widths, use a push-stick. As the rear edge of the stock clears the table, apply forward pressure with the push-stick until the cut is completed. Use a **feather-board** to hold the workpiece against the rip fence. Figs. 22-10 and 22-11.

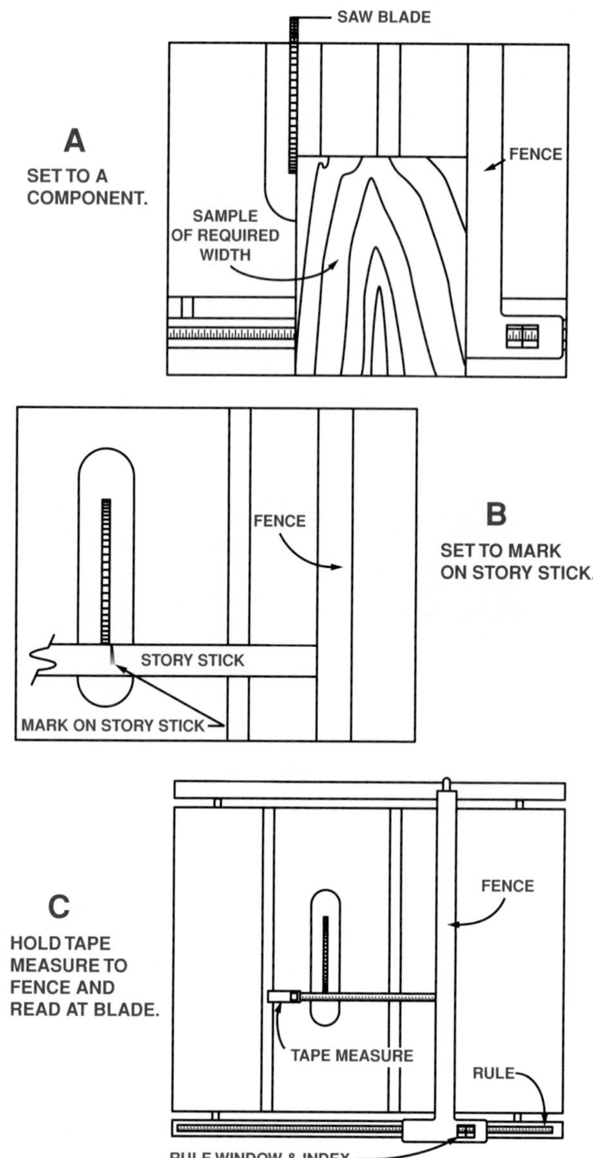

Fig. 22-8—*Setting the rip fence: (A) The most reliable method is to use a sample piece of wood that is known to be the right width. (B) Use a story stick with predetermined measurements. (C) Hold a rule or try square against the fence and measure the distance to the saw blade. (This is the least accurate method.)*

When ripping long stock, have someone help support the workpiece or use a support stand. Fig. 22-12 (page 302). If the stock is hardwood and quite thick, you may need to begin with the saw set at less than the total thickness and

Fig. 22-9—*Ripping stock to width. Position the workpiece against the rip fence and smoothly feed it into the blade. Be sure to stand to one side, not directly behind the blade or the workpiece.*

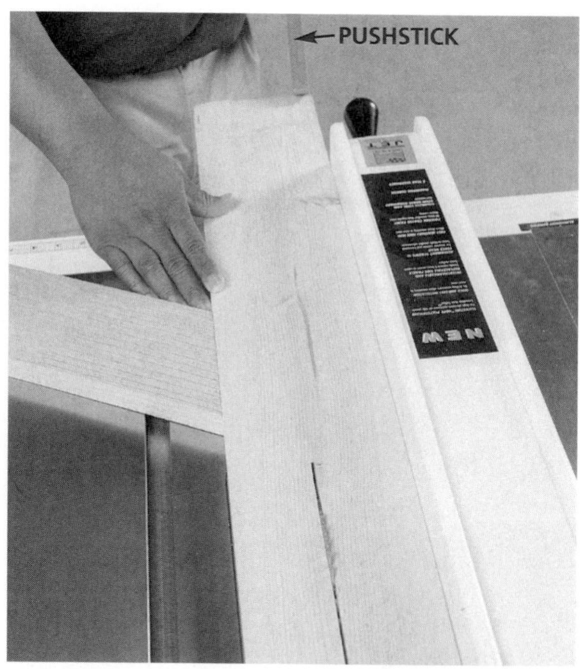

Fig. 22-10— *Use a pushstick when making a narrow cut. A featherboard clamped to the table on the infeed side of the blade will hold the workpiece against the rip fence. NOTE: The guard was omitted in order to show details. Be sure you use a guard when doing this operation.*

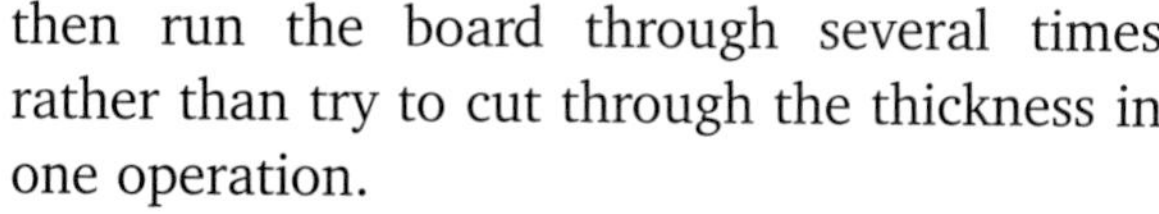

then run the board through several times rather than try to cut through the thickness in one operation.

Cutting Sheet Stock

Sheet stock, such as plywood, particleboard, and hardboard, comes in large sizes (usually 4 × 8 feet). These are hard to handle. It is often a good idea to cut the large sheets into smaller sizes with a hand saw or portable power saw. However, some table saws have extension tables that make it possible for one person to handle large sheets safely. Fig. 22-13 (page 302).

Use a fine-toothed crosscut saw blade to cut plywood. Make sure that the good side of the plywood is facing up. Use a carbide-tipped blade to cut other types of sheet stock.

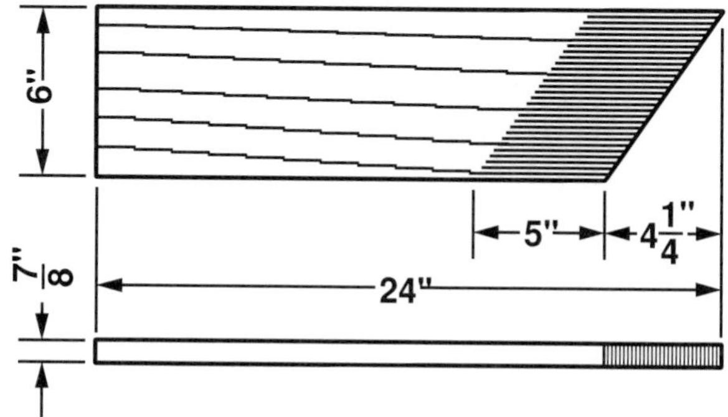

Fig. 22-11—*Measurements for making a typical featherboard. Use solid lumber.*

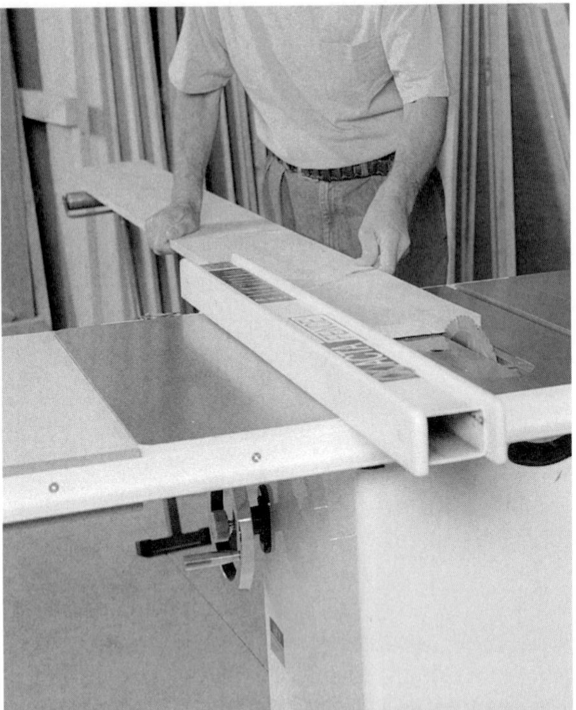

Fig. 22-12—*For long boards, use a support to hold the board level when feeding it into the saw. An outfeed support on the other side of the table is also helpful. NOTE: The guard was omitted in order to show details. Be sure you use a guard when doing this operation.*

Cutting a Taper

The best way to cut a taper is to use an adjustable taper jig. Fig. 22-14. Mark a line to indicate the taper to be cut. Then adjust the jig until the line of the taper cut is parallel with the fence. Fig. 22-15.

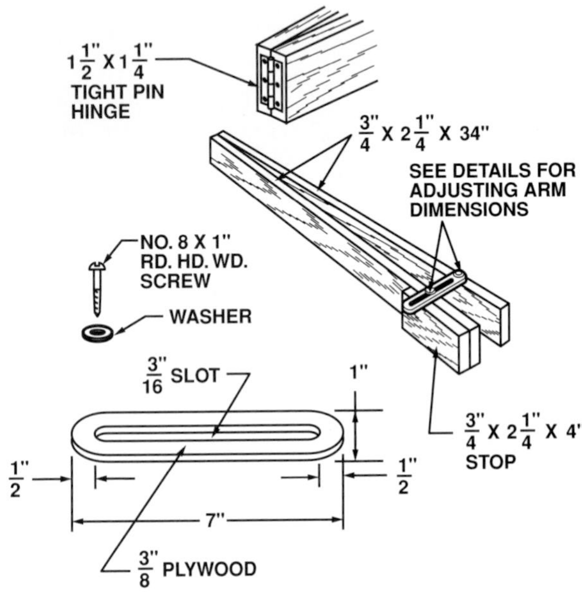

Fig. 22-14—*Measurements for a simple adjustable taper jig.*

Fig. 22-13—*Extension tables on the saw make it easier to handle sheet stock.*

Fig. 22-15—*Using an adjustable taper jig. NOTE: The guard was omitted in order to show details. Be sure you use a guard when doing this operation.*

Safety First

►*Crosscutting*

Use a miter gauge for all crosscutting operations.

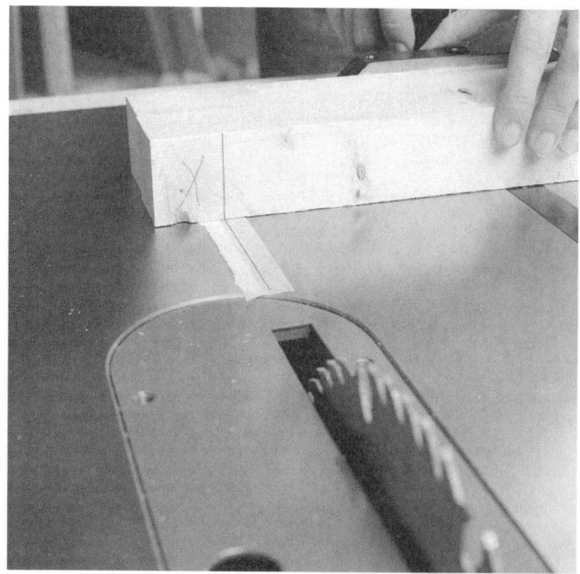

Fig. 22-16—*A line on the table and stock directly behind the saw blade will make it easier to line up the cut. NOTE: The guard was omitted in order to show details. Be sure you use a guard when doing this operation.*

CROSSCUTTING

The first step for all crosscutting operations is to put the **miter gauge** in place. The miter gauge fits into either groove of the table but is most often used in the left groove. Some operators attach a squared piece of stock the same width as the miter gauge to its face to better support the work.

To make a square cut, follow these steps:

1. Set the gauge at a 90-degree angle. Check by holding a try square against the gauge and the saw blade.
2. Mark the location of the cut clearly on the front edge or face of the stock. Fig. 22-16.
3. Set the blade to the correct height—⅛ to ¼ inch higher than the thickness of the stock.
4. Hold the stock firmly against the gauge and slide both the work and gauge along the table to complete the cut. Figs. 22-17 and 22-18 (page 304).

If you must cut several pieces to the same length, use one of the following methods:

- Clamp a small block of wood to the rip fence just in front of the saw blade. The fence,

Fig. 22-17—*Crosscutting. The miter gauge is set at a right angle to the blade. An even, forward pressure should be applied to the workpiece and gauge. NOTE: The guard was omitted in order to show details. Be sure you use a guard when doing this operation.*

WOODWORKING TIP

Reducing Tearout

Remember, the teeth on the blade of a table saw are always moving downward through the cut. To reduce *tearout* (splintering), use a good blade and make sure the good side of the piece being cut is facing up.

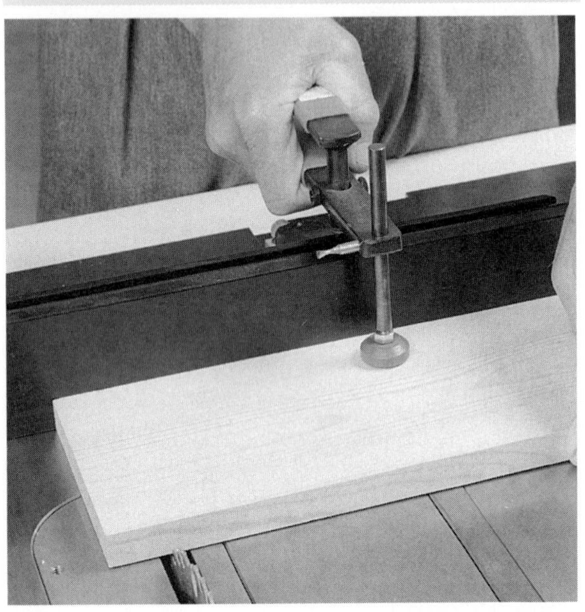

Fig. 22-18—*Some miter gauges are equipped with a clamp to hold the workpiece solidly in place. NOTE: The guard was omitted in order to show details. Be sure you use a guard when doing this operation.*

with a block attached, acts as a length guide. Fig. 22-19. *Never use the rip fence alone as a length guide.* If you do, the piece will lodge between the revolving saw blade and the rip fence. It may kick back with terrific force.

- Fasten a wood extension to the miter gauge. Then clamp a stop block to it to control the length of cut. Fig. 22-20.

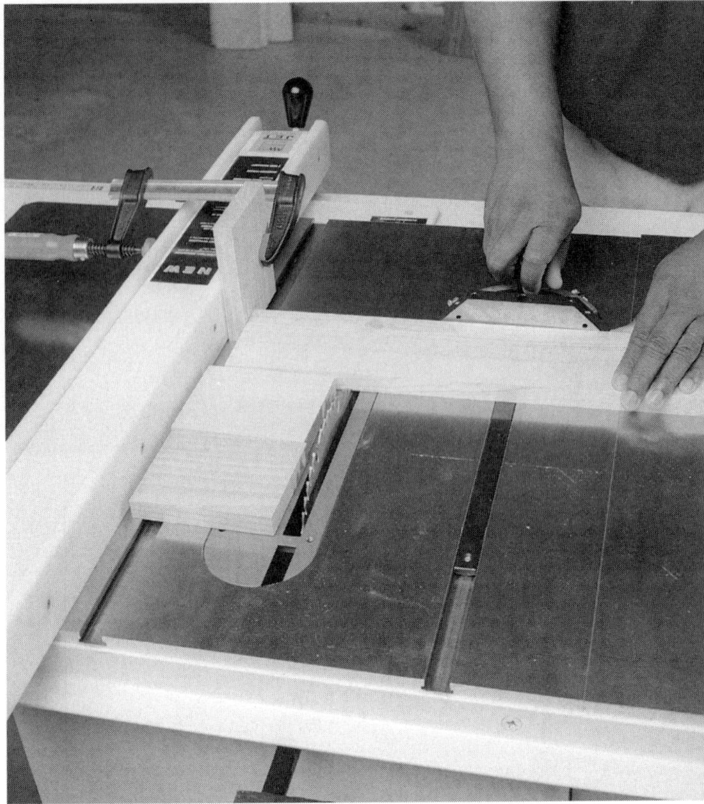

Fig. 22-19—*Attaching a stop block to the rip fence is one of the easiest methods for cutting many pieces to the same length. As the pieces are cut off, there is plenty of clearance between the saw blade and the rip fence. This will lessen the chance of kickback. NOTE: The guard was omitted in order to show details. Be sure you use a guard when doing this operation.*

Mitering

To make a miter cut, adjust the miter gauge to the correct angle and proceed as in crosscutting. Hold the stock firmly against the miter gauge because it tends to creep toward the revolving saw blade as the cut is being made. Whenever possible, use a miter gauge clamp to hold the stock securely. Figs. 22-21 and 22-22. To make a compound miter cut, set the miter gauge to the correct angle and tilt the blade. Fig. 22-23.

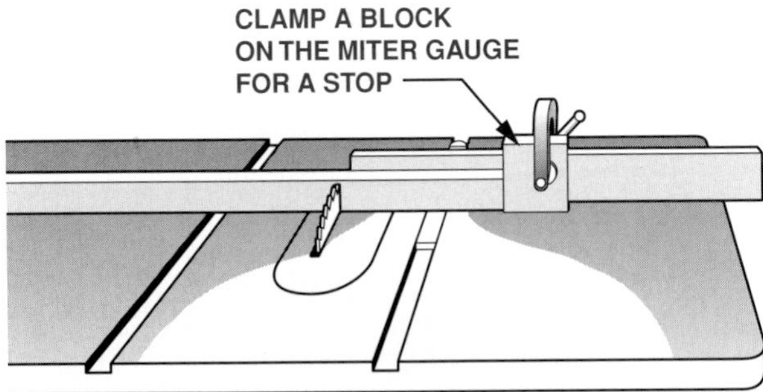

Fig. 22-20—*An extension and stop block on the miter gauge are very helpful when cutting several pieces to the same length.*

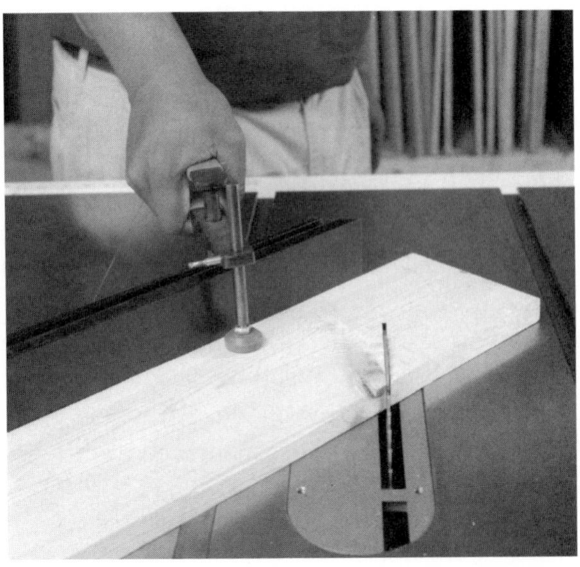

Fig. 22-22—*A miter gauge clamp is very helpful when making angled cuts. Abrasive paper on the miter gauge helps keep the workpiece from sliding. NOTE: The guard was omitted in order to show details. Be sure you use a guard when doing this operation.*

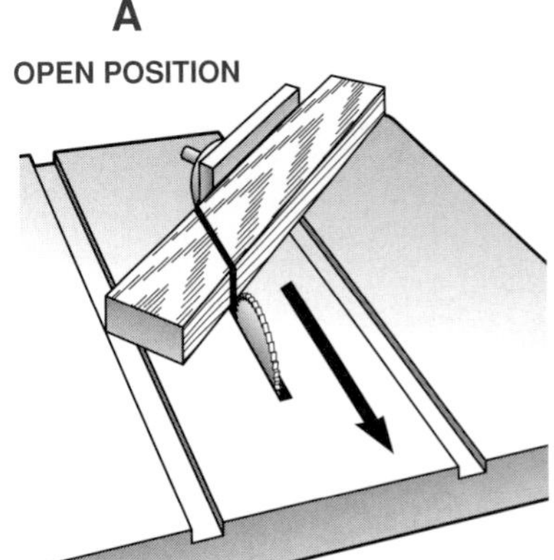

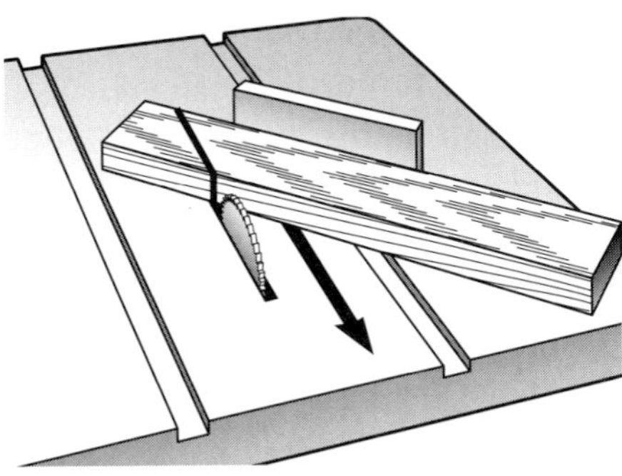

Fig. 22-21—*The miter gauge in open and closed positions. Use the closed position and a miter gauge clamp whenever possible.*

Fig. 22-23—*Cutting a compound miter. Notice that the miter gauge is turned and the arbor of the saw blade is tilted.*

CUTTING BEVELS AND CHAMFERS

To cut a bevel or chamfer when you are either ripping or crosscutting, you must tilt the saw blade to the correct angle for the cut. The gauge, which shows this angle of tilt, is on the front of the saw just below the table.

After making the adjustment, check the angle by holding a sliding T-bevel against the tabletop and saw blade. After you have the correct angle, you can proceed with ripping or crosscutting. Fig. 22-24.

CUTTING DADOES AND GROOVES

The rip, crosscut, and miter cuts divide the workpiece into two pieces. Dadoes and grooves do not separate the workpiece into two pieces; these cuts are three-sided channels in the workpiece.

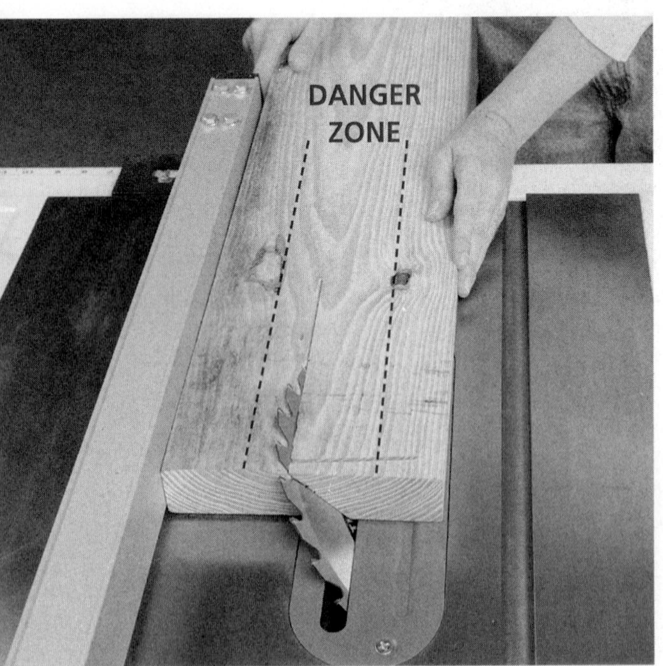

Fig. 22-24—***Be particularly careful when cutting a bevel or chamfer with the arbor tilted. The danger zone is much wider than expected.***

Fig. 22-25—***Dadoes and grooves were used in making this bookcase.***

A dado is a channel cut *across* the grain of the board. A groove runs the *same* direction as the grain of the wood. Dadoes and grooves are used in making joints in cabinets, furniture, and in a variety of other woodworking projects. Fig. 22-25.

Follow the steps below to cut a dado or a groove on the table saw.

1. Lay out the groove or dado on the stock.
2. Set the saw blade to a height equal to the depth of the channel to be cut.
3. Adjust the rip fence to allow the cut to be made just inside the layout line.
4. Hold one surface of the stock firmly against the fence and make the cut for the first side of the dado or groove.

Facts about Wood

Avoiding Waste—We are all concerned about the world problem of deforestation. The solution is very complex, but woodworkers can help. It may sound elementary, but the old saying "Measure twice and cut once" can make a difference in how much wood you use and will help eliminate a lot of wasted lumber. Plan your project very carefully. You will not only save time and money, but you will also help save our forests.

Fig. 22-27—*A typical dado head has two outside blades that are each ⅛ inch and four chipper blades: one that is 1/16 inch thick; two that are ⅛ inch thick; and one that is ¼ inch thick. With this assortment, you can cut grooves from ⅛ to 13/16 inch wide, in intervals of 1/16 inch.*

Fig. 22-26—*Make two cuts to form the two sides of the groove or dado. If necessary, make several "waste cuts." Clean out the groove with a sharp chisel.*

5. Readjust the fence and cut the second side. Fig. 22-26. If necessary, make several additional cuts in the waste stock of the dado or groove.
6. Clean out the groove with a sharp chisel.

Using a Dado Head

A **dado head** is equally adapted to cutting with or across the grain. It can be used to cut grooves or dadoes from ⅛ to 2 inches wide (in several passes). One dado blade will cut a groove or dado ⅛ inch wide. Two will make a cut ¼ inch wide. Chipper blades can be put between two dado cutters to cut a dado or groove of various widths. Fig. 22-27.

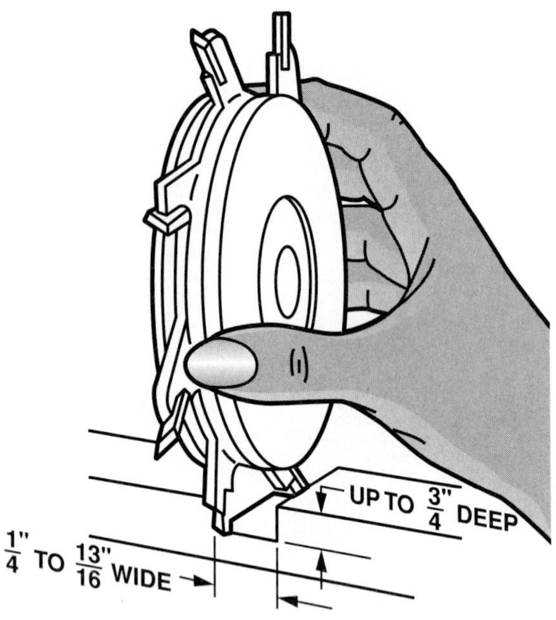

Fig. 22-28—*A 6-inch adjustable dado cutter.*

Fig. 22-29—*Because the dado head is bigger than a saw blade, the throat plate on the table saw will need to be replaced. A custom-made throat plate can be made to fit your table saw.*

The adjustable dado head shown in Fig. 22-28 will make cuts from ¼ to 13⁄16 inch wide and up to ¾ inch deep. Set the width by loosening the arbor nut and rotating the center section of the head until the width mark on this part is opposite the correct dimension.

The dado head is larger than a saw blade, so the throat plate on the table saw must have a wider opening. Fig. 22-29.

The setup for making a dado or groove is similar to the setup for ripping or crosscutting a piece of stock. With a simple fixture on the miter gauge, it is easy to make multiple dadoes, such as might be used for the sides of a bookcase. Fig. 22-30.

CUTTING TENONS

Making mortise-and-tenon joints is quite simple when the tenon is cut on the table saw. (These joints are discussed in detail in Chapter 13.) Follow the steps below to cut a tenon on the table saw.

1. Set the saw blade to a height equal to the thickness of stock to be removed from one side of the tenon.
2. Hold the stock against the miter gauge and make the shoulder cuts.
3. Set the saw blade to a height equal to the length of the tenon.

4. Select a hand-made or a commercial tenoning jig. Fig. 22-31. Clamp the stock to the tenoning jig. Position the jig and fence to cut out the cheek on the side away from the jig. Make the cut.
5. Turn the stock around and cut the other cheek without changing the location of the fence.

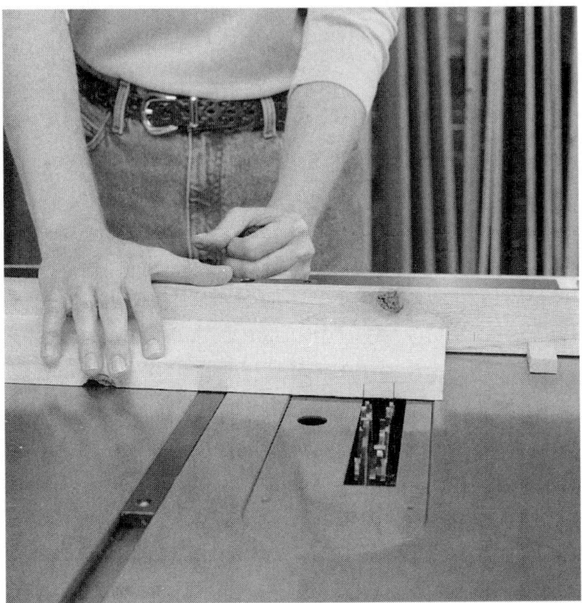

Fig. 22-30—*Cutting a series of dadoes. A small key (piece of wood) has been attached to the auxillary fence on the miter gauge so the dadoes will be equally spaced.*

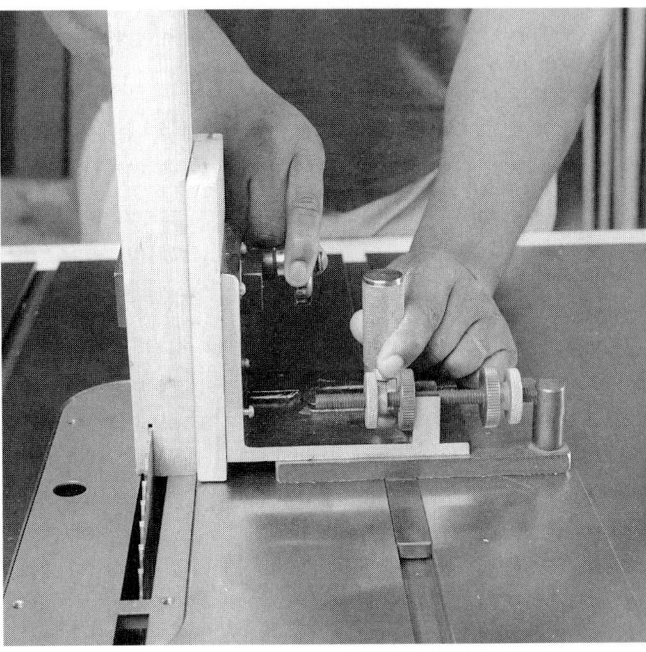

Fig. 22-31—*A commercial tenoning jig is an excellent accessory for cutting tenons and grooves.*

CHAPTER 22

Review & Applications

Chapter Summary

Major points from this chapter that you should remember include:

- The table saw is used for ripping boards to width and crosscutting them to length.
- There are a number of safety precautions to keep in mind when using the table saw.
- There are special procedures to follow when ripping narrow stock.
- A miter gauge is used for all crosscutting operations. It can be set at an angle to make miter cuts.
- The saw blade can be tilted to the desired angle to make bevel and chamfer cuts.
- There are special procedures for cutting dadoes, tapers, rabbets, and tenons on the table saw.

Review Questions

1. Describe the procedure for changing a table saw blade.
2. At what height should the saw blade be set for ripping?
3. Explain when and how to use a pushstick.
4. Explain how to cut a taper.
5. What is the first step for all crosscutting operations? Outline the basic procedure for making a straight crosscut.
6. How is a compound miter cut made?
7. How can a dado head be adjusted to cut grooves of different widths?
8. Outline the steps for cutting a tenon on the table saw.

Solving Real World Problems

In preparation for making bookshelves for his office, Tommy has installed a ¾″ stacked dado set for the first time. As he makes the first couple of passes, he finds that he has to put a lot of pressure on the plywood to hold it down on the saw. He knows that this is a potential safety problem and starts to look for a cause. As Tommy examines the sheet of plywood in which he has just cut a dado, he sees a narrow burn mark in the center of the dado along the entire length. What could be the possible cause of these problems?

CHAPTER 23

Radial-Arm Saw

Look for These Terms

- cutoff saw
- elevating crank
- overarm
- radial-arm saw
- yoke

YOU'LL BE ABLE TO:

- Describe the operation of the radial-arm saw.
- Make a straight crosscut on the radial-arm saw.
- Make miter, bevel, and dado cuts on the radial-arm saw.
- Perform a ripping operation on the radial-arm saw.

The **radial-arm saw,** sometimes called a **cutoff saw,** is ideal for crosscutting operations. While it can also be used for ripping, it is better to use the table saw for that operation. The table saw is also better suited for long crosscuts. To cut a 4′ × 8′ panel into two 4′ × 4′ pieces on a radial-arm saw, you need to make multiple cuts. On a table saw, you can do it in one cut.

Overall, however, a radial-arm saw is more versatile than a table saw. Properly equipped, the radial-arm saw can be used to dado, rip, shape, dovetail, mortise, sand, wire-brush, grind, buff, and perform many other operations.

PARTS OF THE RADIAL-ARM SAW

Extremely accurate cuts can be made on the radial-arm saw. This is because the various parts of the saw can be adjusted in order to fine-tune the machine for a precise cut.

The *saw unit* (blade and motor) can be tilted 90 degrees to the left or right. The **yoke** that holds the saw unit can be turned in a

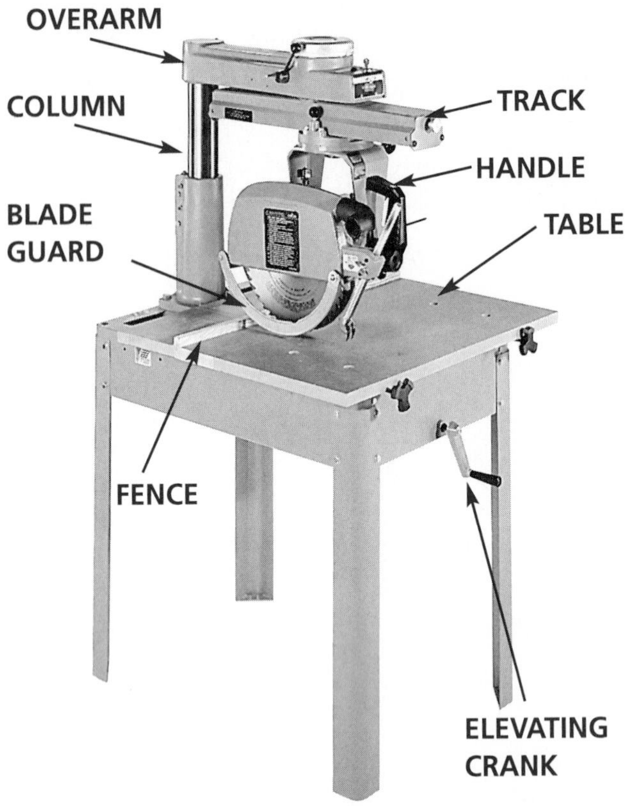

Fig. 23-1—*The parts and controls for a radial-arm saw. Each model has a slightly different arrangement. Study the manufacturer's manual for complete instructions on how to use the saw.*

360-degree circle. The yoke and saw unit can be moved back and forth along the **overarm** or track. Fig. 23-1. The overarm sits on a column and can be rotated in a complete circle around this column. By turning the overarm or track to the desired angle and locking it in place with the locking lever (located on the column or outer end of the overarm), the *angle of cut* can be adjusted. Located either directly above the column or on the front of the machine, is the **elevating crank,** which is turned to adjust the *depth of cut.*

For all crosscutting operations—including straight cutting and cutting a miter, bevel, dado, or rabbet—the stock is held firmly on the table in a stationary position, and the saw unit is moved along the overarm or track to do the cutting. Fig. 23-2. For all ripping operations, the saw unit is locked in a fixed position and the stock is moved into the revolving blade.

The size of the radial-arm saw is determined by the size of the blade. The most common size is 10 inches. The blades used on the radial-arm saw are the same as those used on the table saw.

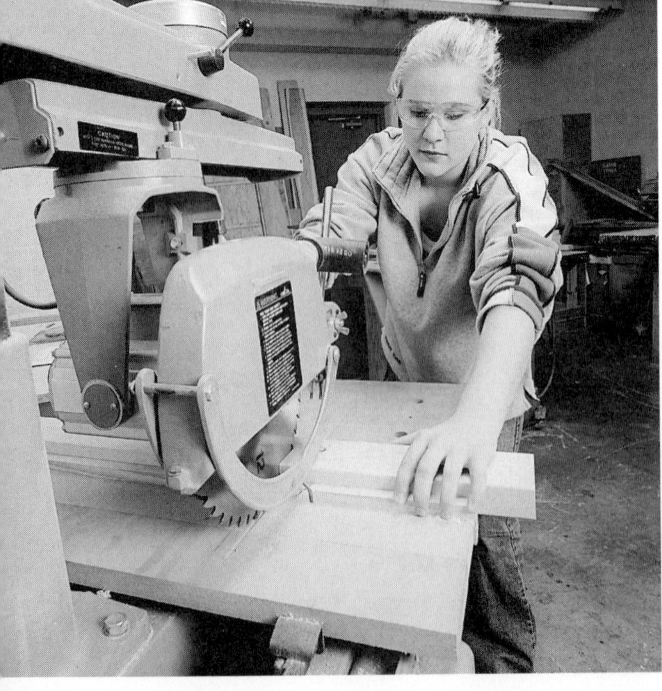

CROSSCUTTING OPERATIONS

One clear advantage the radial-arm saw has over the table saw is that all cutting is done from the top rather than from beneath the wood, so layout lines are clearly visible. Another advantage is that, when crosscutting with the radial-arm saw, the action of the blade pushes the stock down and against the fence. Fig. 23-3. The board can't move down because of the table and it can't move back because of the fence, so only light hand pressure is needed to hold the wood in place.

Fig. 23-2—*When cutting stock to length on the radial-arm saw, hold the stock in a stationary position while moving the saw back and forth on the overarm track.*

Safety First

►*Using a Radial-Arm Saw*

The radial-arm saw can be very safe to use if you follow all the safety rules; it can be very dangerous if you do not. Observe the following precautions:

- Wear eye protection.
- Make sure that the blade you install is sharp and is the correct kind for the cutting operation you will be doing.
- Mount the blade on the arbor so that the cutting edges turn toward you.
- Make sure the guard is always in place.
- See to it that all clamps are tight before starting the motor.
- Make sure the saw is at full speed before starting the cut.
- Keep your hands away from the path of the blade. Maintain a 6-inch margin of safety.
- Hold the stock firmly against the table for crosscutting operations.
- When ripping, always feed the stock into the blade so that the bottom teeth are toward you. This will be the side opposite the antikickback fingers.
- *Make all adjustments with the motor and blade at a dead stop.* After turning off the machine, do not try to stop the blade by holding a stick or similar item against it.
- Return the saw to the rear of the table after completing the cut. Never remove stock from the table until the saw is returned.

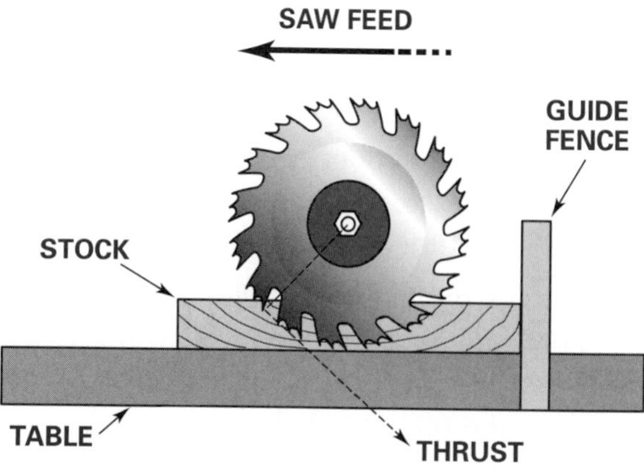

Fig. 23-3—***In crosscutting, the saw blade moves in the same direction it is rotating. The saw's thrust is downward and to the rear, thus holding the stock firmly against the guide fence.***

SCIENCE *Connection*

Carbide-Tip Blades

Like table saws, radial-arm saws can use a variety of different types of blades. As you know from Chapter 22, some blades have carbide tips. Tungsten carbide is a ceramic material made up of a 1:1 ratio of tungsten and carbon atoms. It is much harder than tool steel, so it lasts much longer.

So why don't manufacturers make the entire saw blades from tungsten carbide? Although carbide is hard, it is also brittle. When the blade hits a hard spot in the wood, the carbide is likely to crack. The tool steel is not as hard, but it is more likely to absorb enough of the shock to prevent the crack from spreading through the entire blade. By bonding a carbide tip to a blade of tool steel, manufacturers blend the best features of each material to create a sturdy, long-lasting blade.

To make a straight crosscut:

1. Make sure that the overarm or track is at right angles to the guide fence.
2. Adjust the depth of cut so that the teeth of the blade are about 1/16 inch below the surface of the wood table.
3. Set the *antikickback device* about 1/8 inch above the work surface. This will act as a safety device to keep your fingers away from the rotating blade.
4. Hold the stock firmly on the table with the good face of the board up. The cutoff line will be showing and should be in line with the saw blade.
5. Start the machine and allow it to come to full speed. Then pull the saw unit slowly, so that the blade cuts into the stock. Fig. 23-4. This will take very little effort.
6. After making the cut, return the saw to its place behind the guide fence and turn off the machine.

When several pieces of the same length are needed, a stop block can be clamped to the table. A rabbet cut at the end of the stop block will prevent sawdust from catching in the corner and interfering with the accuracy of the cuts. Fig. 23-5.

Cutting Miters and Bevels

To make a miter cut, simply adjust the overarm or track to the angle you want. Cut as you would for straight crosscutting. Fig. 23-6.

To cut a bevel, adjust the track or overarm for straight crosscutting. Then tilt the saw unit to the desired angle. Fig. 23-7. The angle is 45 degrees for most bevel or end miter cuts. To make a compound miter, adjust the overarm or track to the correct angle and then tilt the saw. Fig. 23-8.

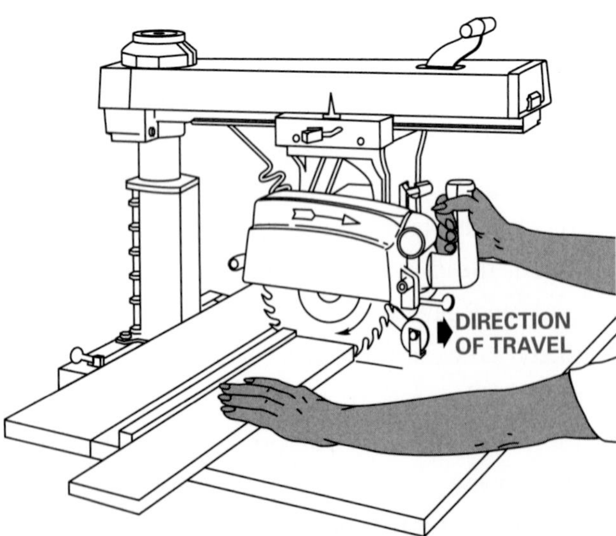

Fig. 23-4—***To do straight cutting, pull the saw smoothly through the stock.***

Fig. 23-5—***Clamp a stop block to the table to cut a number of pieces to identical length.***

Fig. 23-6—*Hold the stock firmly against the fence with one hand. Grasp the handle of the saw with the other hand and pull the saw through the cut.*

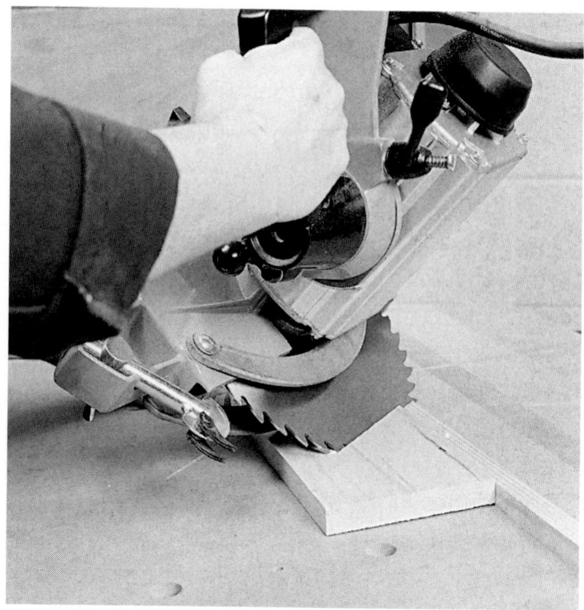

Fig. 23-8—*Making a compound miter.*

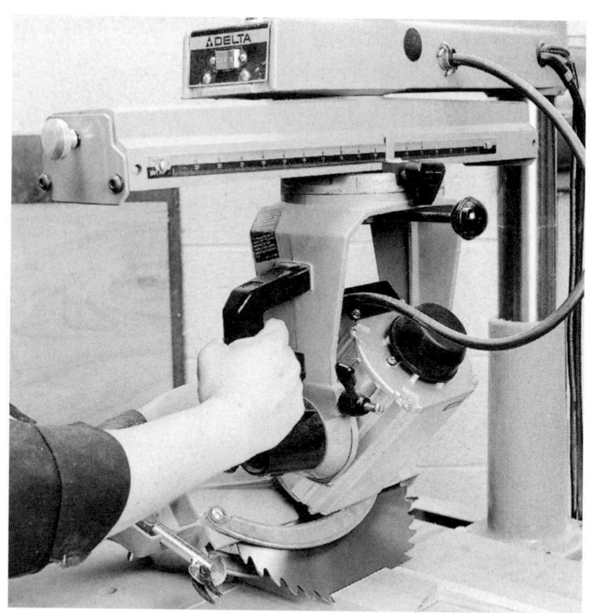

Fig. 23-7—*Making a bevel cut.*

Accurate Measurements

Cutting duplicate parts to an exact measurement on a radial-arm saw can be difficult if you are just using a tape measure. One solution is to draw ¼-inch marks on the fence on each side of the blade. Even better, buy an adhesive-backed tape measure and apply it to the fence. Then you will have a ruler to help you measure stock for the table instantly. To get the repetitive cuts, just clamp a stop on the fence at the appropriate measurement, or design a clamp that will fit over the fence. To avoid sawdust buildup, the stop should not touch the blade.

Cutting Dadoes

The radial-arm saw can cut dadoes with great accuracy. The procedure for cutting a dado is similar to the procedure for crosscutting.

Install the dado head in the same manner as the saw blade. Adjust the overarm or track to be at right angles to the fence. Lower the blade to the correct height. Start the machine and let it come to full speed. Then make the cut slowly and smoothly. Fig. 23-9. If very deep dadoes are needed, it is usually best to make them in two steps.

Sometimes it is necessary to cut matching dadoes on two separate boards. One method of doing this is with a stop block clamped to the table. Having the stop block in place as you cut dadoes in each board helps ensure that the dadoes will be cut in identical positions on both boards. If parallel dadoes are needed, a series of marks can be made or a series of stops can be fastened to the guide fence.

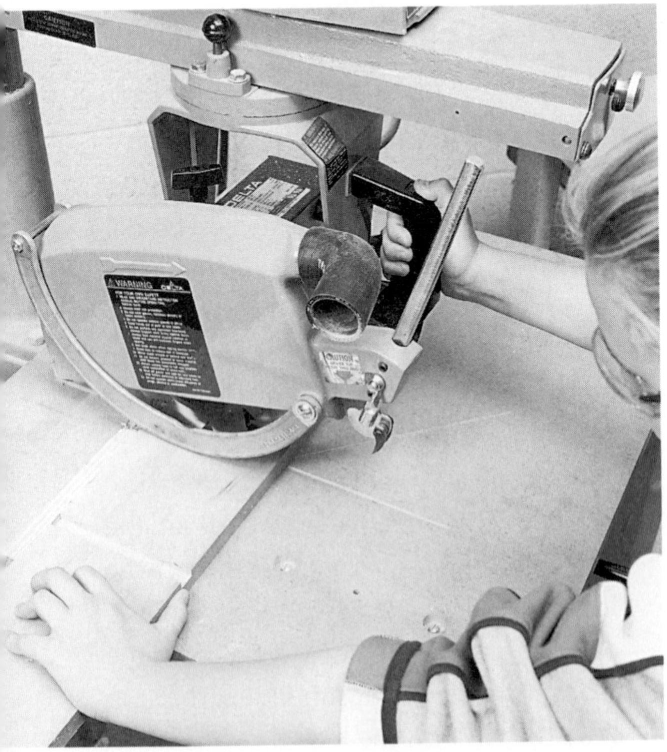

Fig. 23-9—*Cutting a plain dado.*

RIPPING OPERATIONS

As stated earlier, while the table saw is preferable for ripping operations, the radial-arm saw can be used for ripping. Follow the procedure described below.

1. Set the track or overarm at a right angle to the guide fence.
2. Turn the saw unit so that the blade is parallel to the guide fence. Then move the saw unit in or out until the correct distance between the guide fence and the blade is obtained. Lock the saw unit in position.
3. Set the depth of cut.
4. Adjust the guard so that it will be close to the work.
5. Set the antikickback device so that the fingers rest firmly on the wood surface and hold it against the table. The good face of the board should be down.
6. Check to make sure that the saw will be rotating up and toward you. Fig. 23-10.
7. Turn on the power and move the stock slowly into the saw blade. Figs. 23-11 and 23-12.

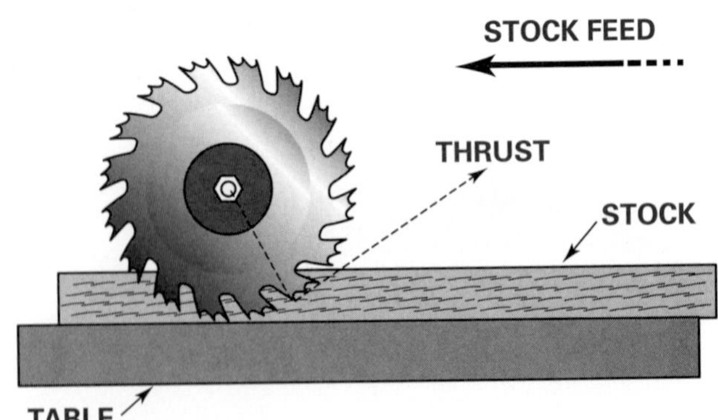

Fig. 23-10—*When ripping stock, always feed it into the rotation of the blade.*

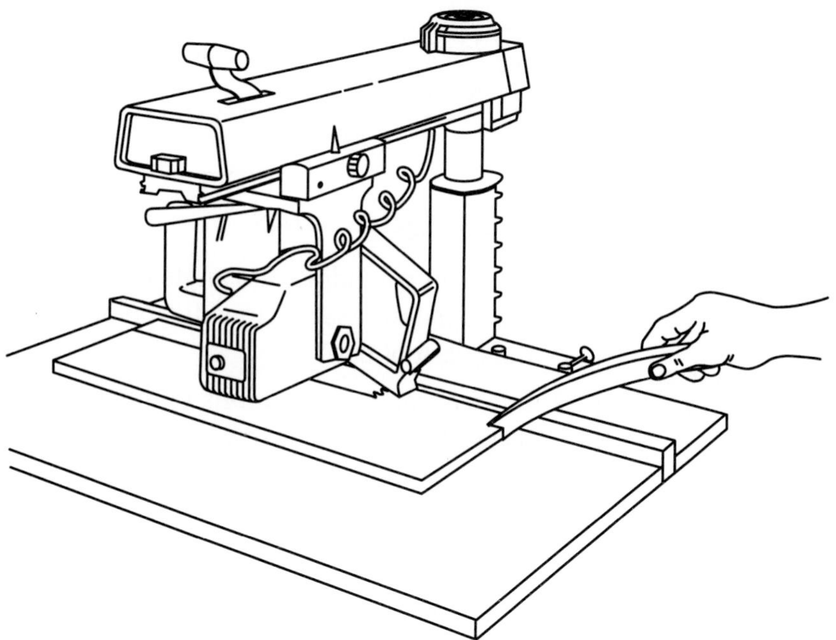

Fig. 23-11—*When ripping narrow stock, turn the saw unit so that the blade is toward the guide fence. Adjust the upper guard so that it just clears the thickness of the stock. Note that the stock is fed from the right side. Always use a pushstick to complete the cut when ripping narrow pieces.*

Fig. 23-12—*When ripping wide stock, turn the saw unit so that the blade is away from the guide fence. Adjust the upper guard so that it just clears the stock. Note that the work is fed from the left side.*

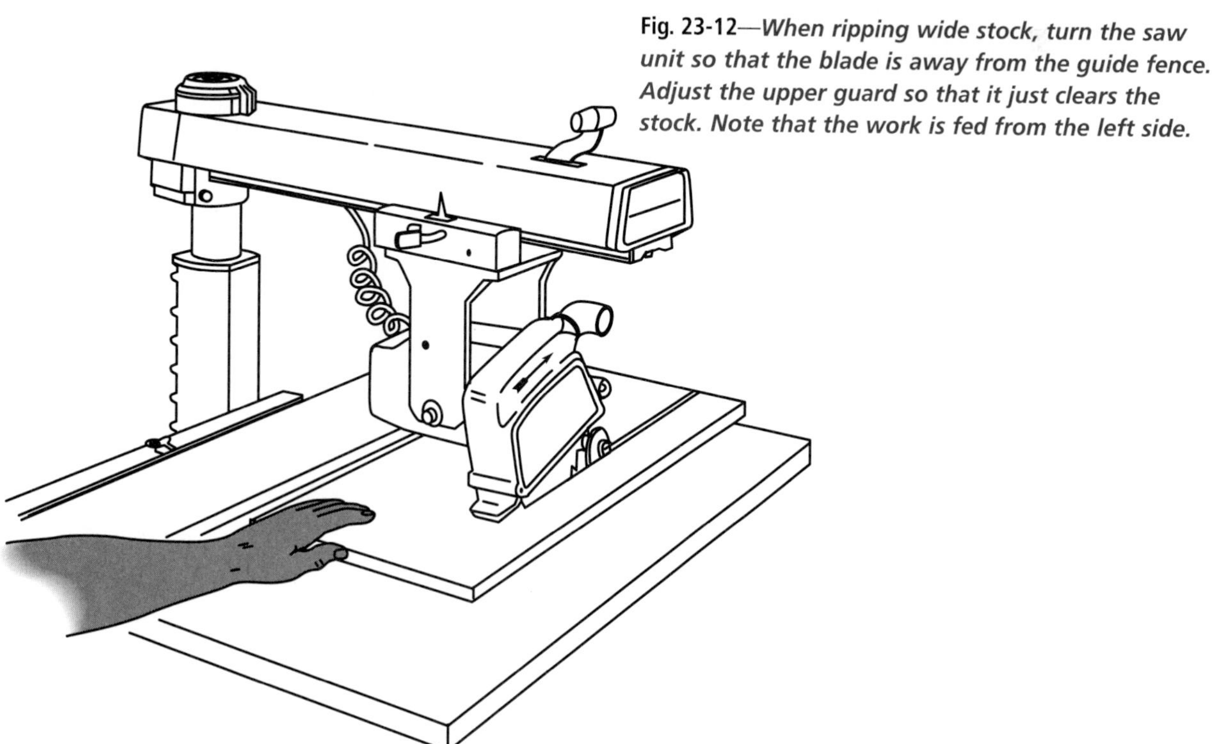

CHAPTER 23

Review & Applications

Chapter Summary

Major points in this chapter that you should remember include:

- The radial-arm saw is ideally suited for all crosscutting operations—straight, bevel, miter, and dado.
- The radial-arm saw can be used for some ripping operations, but the table saw is better suited to ripping.
- The various parts of the radial-arm saw can be adjusted so that the saw can be fine-tuned to make precise cuts.
- For crosscutting, the work is held in a stationary position and the saw unit is moved to do the cutting.
- For ripping operations, the saw unit is locked in a fixed position and the stock is moved into the blade.

Review Questions

1. How can you adjust the angle of cut on the radial-arm saw? How do you adjust the depth of cut?
2. List five safety precautions to follow when using the radial-arm saw.
3. Briefly describe the basic procedure for straight crosscutting.
4. Explain how to cut a miter on the radial-arm saw.
5. Explain how to make bevel cuts on the radial-arm saw.
6. Outline the procedure for using the radial-arm saw for ripping operations.

Solving Real World Problems

In the new home Sid is building, there is a 4-foot arch between two rooms. In order to form the arch, Sid must bend a 2 × 6 into the shape of the arch. If he special-orders a frame, the manufacturer will require at least two weeks. During that time, Sid would not be able to do anything to the house. While looking at some other bent-shape parts that were special-ordered for the house, Sid learns how some of them were made. Multiple passes were made with a saw along the back of the part. Each cut was spaced about 1/4″ from the last. Then the part was bent and glued into its new shape. How would you make the part with a radial-arm saw?

CHAPTER 24

Band Saw

Look for These Terms

- band saw
- pad sawing
- resawing

YOU'LL BE ABLE TO:

- List guidelines that must be followed in cutting with the band saw.
- Demonstrate how to cut simple and compound curves on a band saw.
- Describe how to cut circles on the band saw.
- Explain how to cut several duplicate parts at the same time on a band saw.
- Demonstrate how to change the blade on the band saw.

The **band saw** is a very versatile cutting machine. Even though it is used mostly for cutting curves, circles, and irregular shapes, it can also be used for all types of straight cutting. **Figs. 24-1** and **24-2 (page 320).**

Other advantages include the following:

- It can cut a variety of materials—wood, plastics, metal, and panel stock.
- It can make long, sweeping curves much more accurately than any handsaw.
- It has a large depth of cut that makes it perfect for resawing and cutting duplicate parts.

Fig. 24-1—*The curves on this table can be cut on a band saw.*

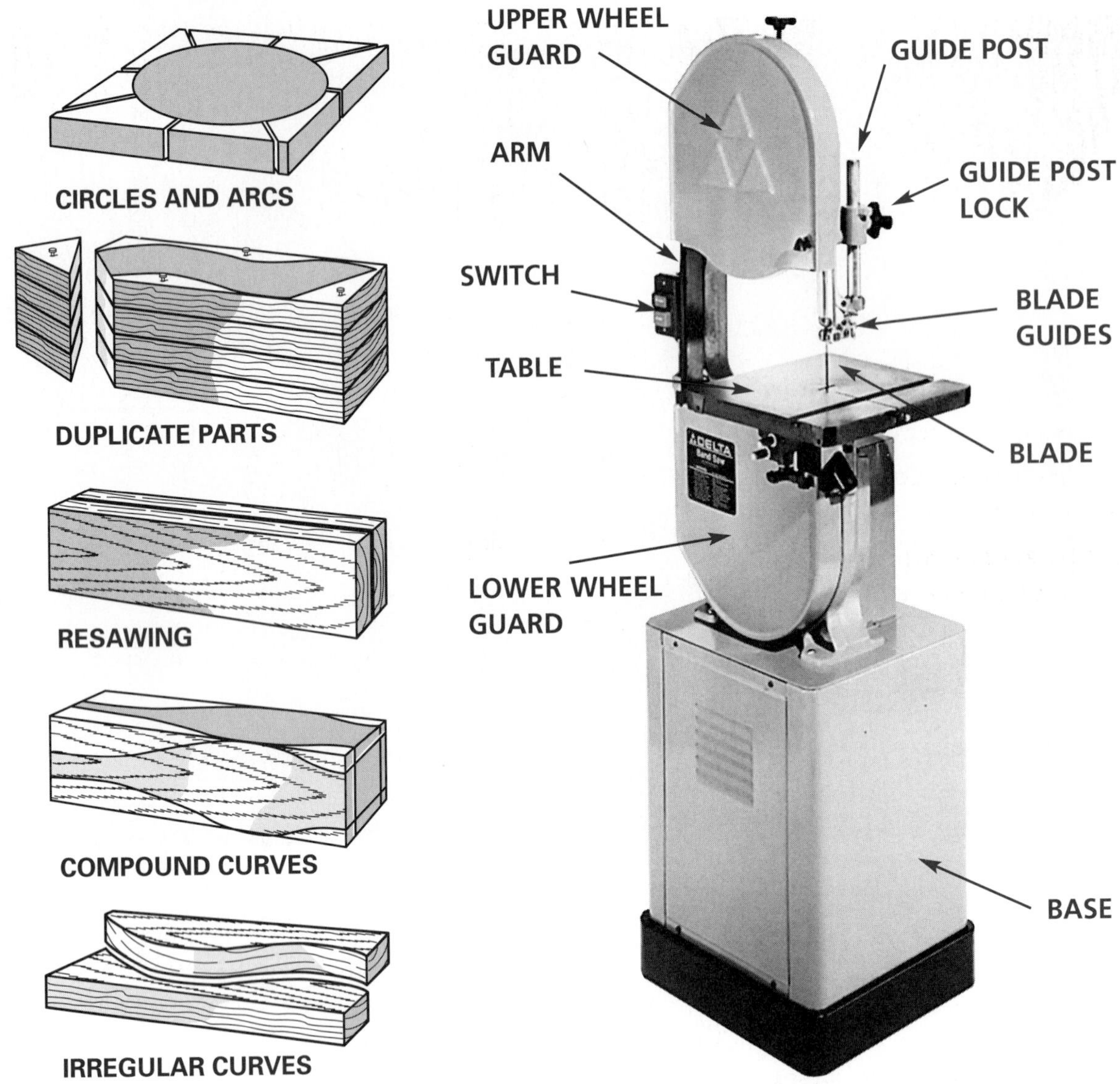

Fig. 24-2—*Common cuts that can be made on the band saw. You will learn how to make these cuts in this chapter.*

Fig. 24-3—*Parts of a band saw.*

PARTS OF A BAND SAW

The band saw has two wheels—one below the table and one above it—around which the blade (a band of steel) revolves. Figs. 24-3 and 24-4. The table can be tilted to different angles. The saw also has upper and lower wheel guards as well as guides to hold the revolving blade in line. In addition, a ripping fence and miter gauge are sometimes used.

The size of the band saw is indicated by the diameter of the wheels. The maximum workpiece thickness that can be cut is limited to the distance between the top of the table and the blade guide in its uppermost position.

Safety First

►*Using a Band Saw*

In using the band saw, always observe the following precautions:

- Wear eye protection.
- Use the correct blade. Choose the largest one with the coarsest teeth that will cut the stock cleanly.
- Before operating the saw, check the blade for proper tension and proper mounting. The teeth should point down on the downward stroke.
- Make sure that the blade is sharp and in good condition.
- Be sure the wheels turn clockwise as viewed from the front of the saw. The arrow on the motor pulley indicates the direction of rotation.
- Be sure the wheel guards are closed before turning on the machine.
- Adjust the upper guide assembly so it is ¼ inch above the stock.
- Hold the stock flat on the table.
- Cut cylindrical (round) stock by holding it in a special "V" fixture.
- Allow the saw to reach full speed before starting the cut.
- Keep your fingers at least 2 inches from the side of the blade. Never put your fingers in front of the blade.
- Feed (push) work into the band saw blade firmly and at a rate that will not overload the saw.
- Keep from twisting the blade or crowding it beyond its cutting capacity.
- If you need to back the saw blade out of a long cut, first turn off the power and allow the machine to come to a full stop.
- Clear away scraps close to the saw blade with a stick, not with your fingers.
- Make sure the band saw has stopped completely before making any adjustments.

Fig. 24-4—***A smaller tabletop band saw can handle most jobs.***

MATHEMATICS *Connection*

Coplanar Wheels

To operate correctly, the upper and lower wheels of a bandsaw must be *coplanar.* That is, all points of the wheels should lie in the same geometric plane, as shown in the illustration below. In mathematics, a *plane* is a flat (two-dimensional) surface defined by any three points in space.

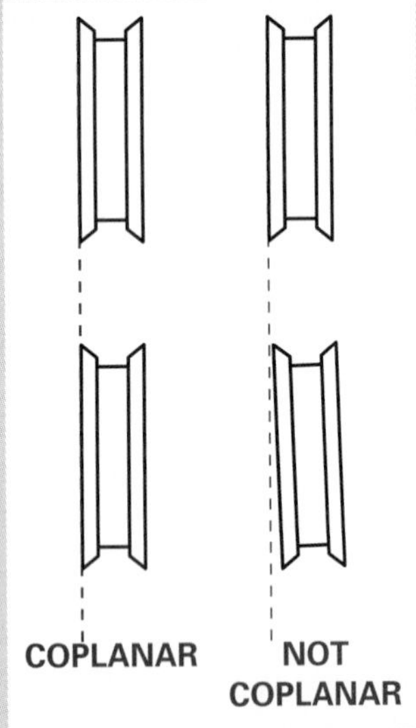

If the wheels of a band saw are slightly out of alignment so that they are not coplanar, the blade tends not to track straight. This results in inaccurate cuts. Technically, changing the blade tension can affect the alignment. In practice, tracking adjustments are seldom necessary if the wheels are properly set up.

CUTTING WITH A BAND SAW

In cutting with a band saw, follow these general guidelines:

- Watch the feed direction. Before making the cut, think through the path that the blade must make. Some pieces will swing in such a way as to hit the upper arm if your plan is not correct.
- The correct cutting position for operators is facing the blade, standing slightly to the left. Fig. 24-5.
- You must be able to follow just outside the layout line, allowing extra stock for smoothing the edges later.
- Move the stock into the blade as rapidly as it will cut. Moving it too slowly will tend to burn the wood.
- Make short cuts before long cuts. It is much easier to backtrack out of a short cut than a long one. Fig. 24-6.

Fig. 24-5—***Face the blade and stand slightly to the left.***

- Make use of turning holes. Depending on the design, a round or square hole can first be cut in the waste stock before band-sawing. Fig. 24-7.
- Break up complicated curves. Look at each job to see if a combination cut can be completed by making several simpler cuts. Fig. 24-8.
- Rough-cut complex curves. Make a simple cut through the waste stock to follow as much of the line as possible. Then cut to the layout line.
- Backtrack out of the corners of rectangular openings. First, make straight cuts down the full length of each side of the opening. Fig. 24-9. (The numbers in this illustration indicate the order of the cuts.) Backtrack slightly out of the second cut, and then cut a curve over to the other corner. For narrow rectangular openings, you may need to make "nibbling" cuts to clear away the remaining waste stock. For wider openings, the remaining waste stock can be trimmed away with one quick cut.

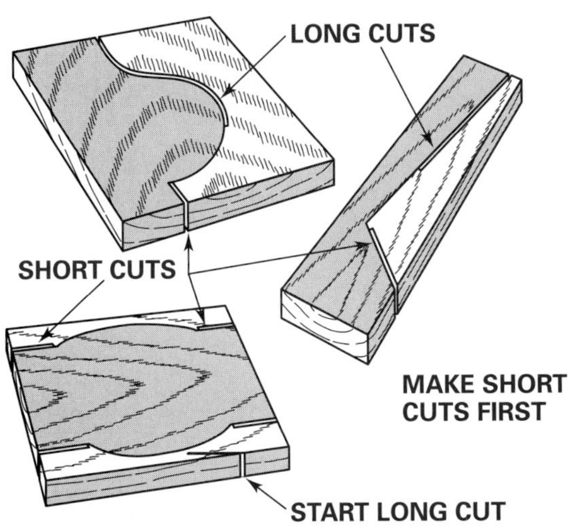

Fig. 24-6—*Make short cuts before long cuts.*

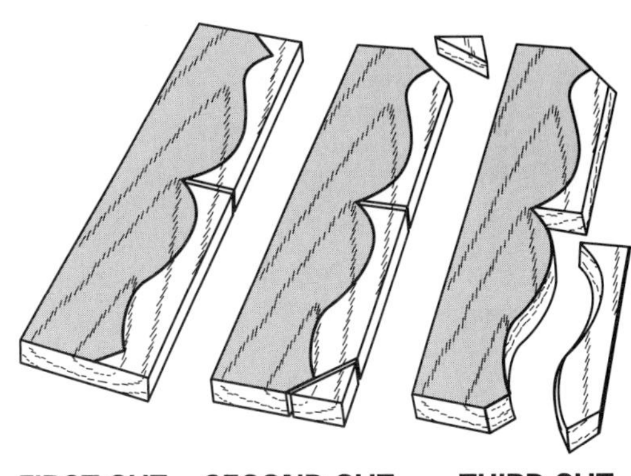

Fig. 24-8—*The correct sequence in cutting a combination curve.*

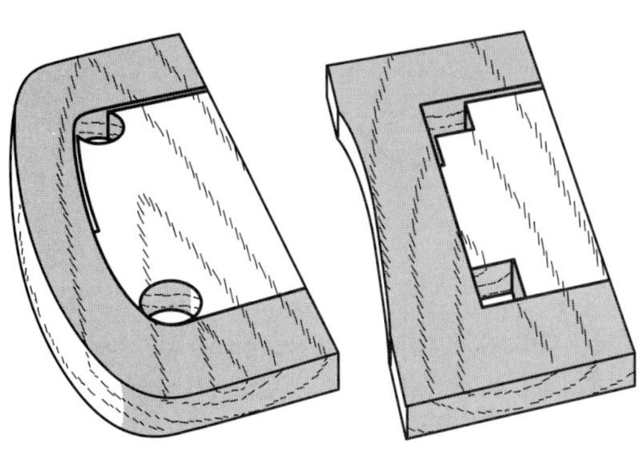

Fig. 24-7—*Use a drill or mortising tool to cut turning holes.*

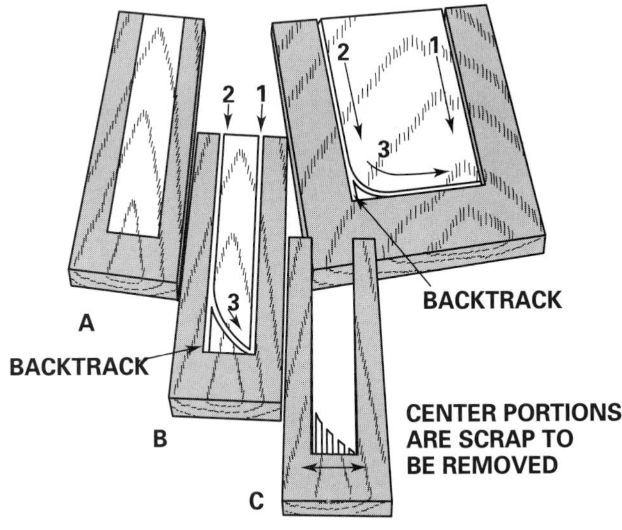

Fig. 24-9—*Sequence in cutting rectangular openings.*

CUTTING CURVES

Follow the chart in Fig. 24-10 when selecting a band saw blade for cutting curves. The width of the blade also depends on the thickness and the kind of wood to be cut.

In cutting curves, apply even, forward pressure. Carefully guide the work with your left hand to keep the cut just outside the layout line. Fig. 24-11. In cutting sharp curves, make many relief cuts from the outside edge to within 1/16 inch of the layout line. Fig. 24-12. Then, as you cut along the layout line, the waste stock will fall away freely.

Fig. 24-11—*Cutting a curve on a band saw. Carefully guide the stock to cut just outside the layout line.*

CUTTING CIRCLES

Several kinds of jigs can be used to cut circles on the band saw. The commercial circle jig has an adjustable pivot pin that the operator sets to the correct distance from the cutting edge. Fig. 24-13.

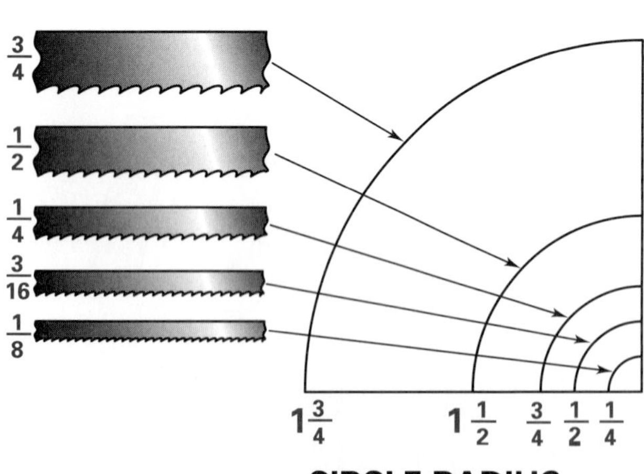

Fig. 24-10—*This chart shows how to select the right blade. For example, a ½-inch blade cannot cut a circle smaller than 1½ inches in diameter.*

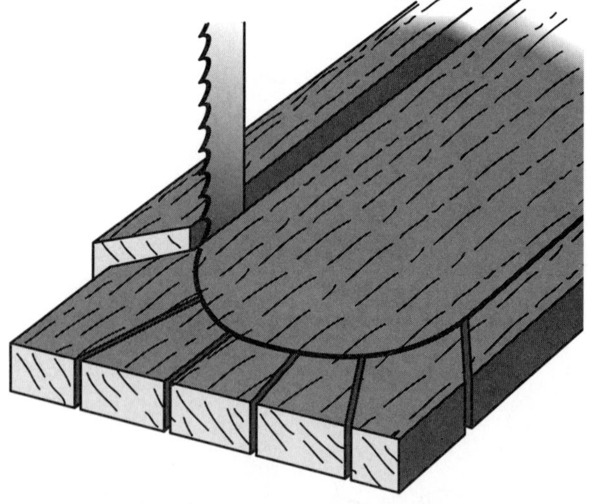

Fig. 24-12—*When cutting a sharp external curve on the band saw, make several relief cuts from the outside edge to within 1/16" of the layout line.*

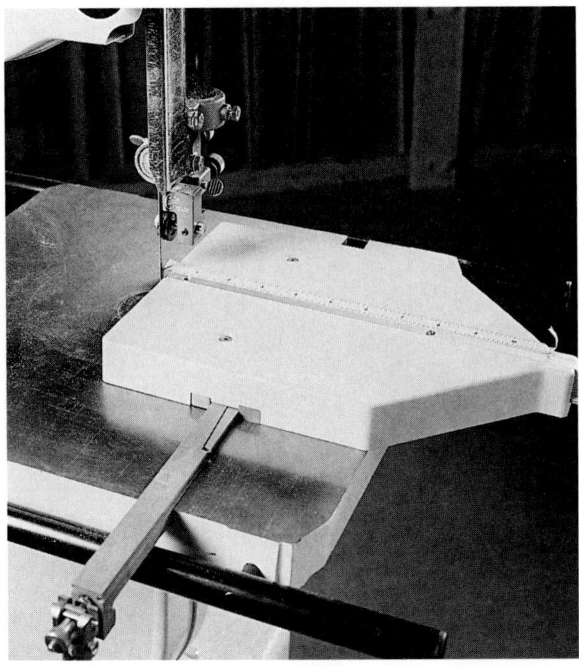
Fig. 24-13—*A commercial circle jig for cutting a true circle.*

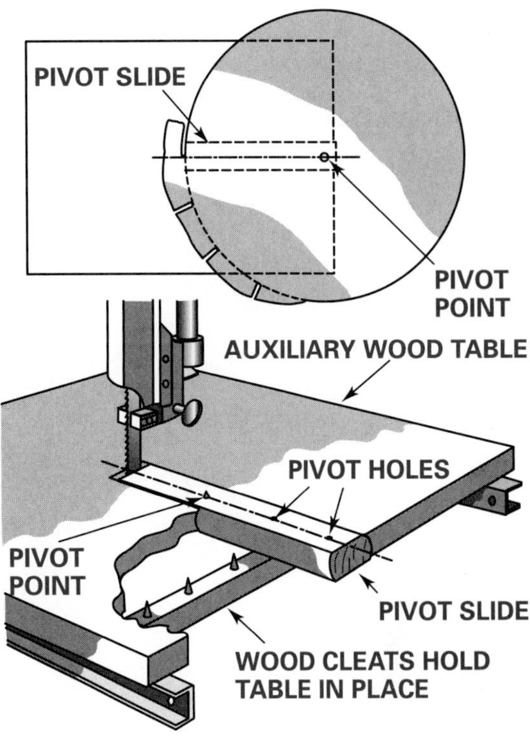

Fig. 24-14—*Details of a homemade wood jig for cutting circles.*

You can make your own circle jig from plywood. Refer to Fig. 24-14 as you follow the steps below.

1. Cut a piece of ¾-inch plywood slightly wider than the distance from the blade to the edge of the band saw table. Fasten two cleats to the underside of the piece.
2. Cut a groove or dovetail dado in the plywood at right angles to the blade, with the center of the groove at the front of the teeth.
3. To make the pivot slide, cut a small hardwood stick that will slip into the groove and be flush with the table top.
4. Place a sharp pin or screw at the end of the pivot slide. This will be the pivot point. Remember that the pivot point (pin or screw) must be at right angles to the blade and in line with the front of the blade.

Now you are ready to use your homemade jig to cut a circle. Follow these steps:

1. Move the pivot slide to adjust the pivot point to the correct radius.
2. Before placing the stock to be cut on the pivot point, cut a small notch in the edge of the stock. This notch creates a pocket for the blade to start in.
3. Center the blade in the notch and press the stock down on the pivot point.
4. Turn on the band saw. Once the saw comes up to full speed, spin the stock slowly to cut the circle. Fig. 24-15 (page 326). Don't feed the stock too fast, or the blade will wander. The blade will also wander if the pivot point is too far forward or back.

Fig. 24-15—*Using a homemade plywood jig for cutting a circle.*

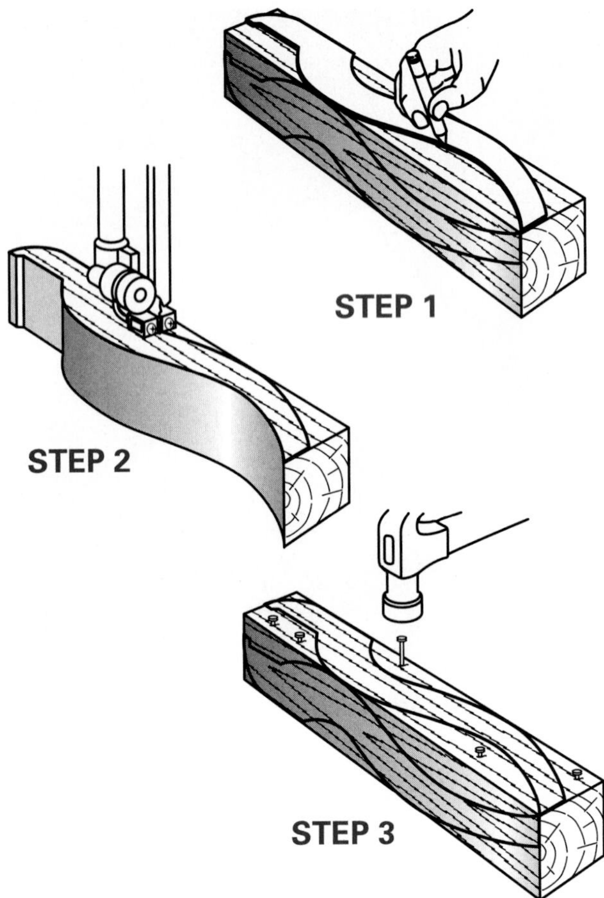

Fig. 24-16—*Steps for making a compound cut.*

COMPOUND CUTTING

Cut compound curves from two sides of the stock. Refer to Fig. 24-16 as you read the steps below.

1. Make a pattern and trace it on two adjoining faces of the workpiece.
2. Make the cuts needed to remove the waste stock from one face of the workpiece. (In Fig. 24-16, two cuts were needed.)
3. Fasten this waste stock back in its original place on the workpiece, using small nails or masking tape. (If using nails, place them so that the blade will not hit them when making the remaining cuts.)
4. Turn the stock one-quarter turn and make the needed cuts on the adjoining face. Then remove the waste stock you temporarily attached in Step 3.

RESAWING

Resawing is sawing stock to reduce its thickness. When stock is much thicker than needed, it is resawed. This can be done on the band saw.

The widest possible blade should be selected, and a fence or pivot block should be attached to the table. A layout line across the end and edge of the board is often helpful. Hold the stock firmly against the fence or pivot block and slowly feed the work against the blade. Fig. 24-17.

For very wide boards, it is better first to resaw partway from either edge on a table saw and then complete the cut on the band saw.

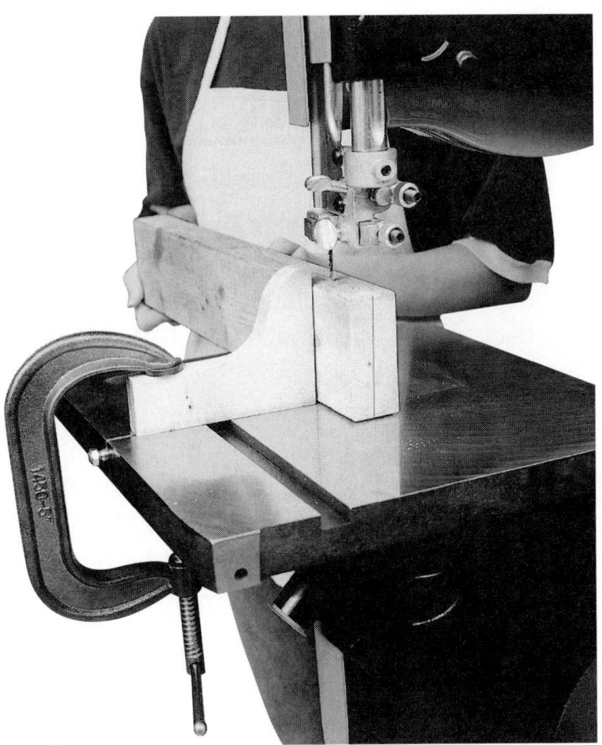

Fig. 24-17—*Resawing stock. A band saw is better for resawing than a table saw because less stock is wasted with the thinner band saw blade.*

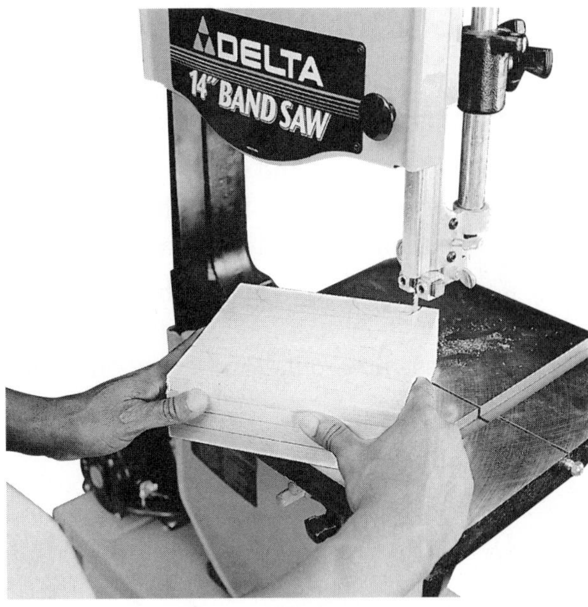

Fig. 24-18—*If cutting duplicate pieces from thin stock, you can nail the pieces together in the waste stock and cut several pieces at a time.*

PAD SAWING

Pad sawing is the cutting of several pieces at one time. On projects that require duplicate pieces, such as flat scroll work, you can nail several thin boards together and make several scrolls with one cut. Fig. 24-18. Another method would be to first cut the pattern from a thick piece of stock and then resaw it into thinner pieces. Fig. 24-19.

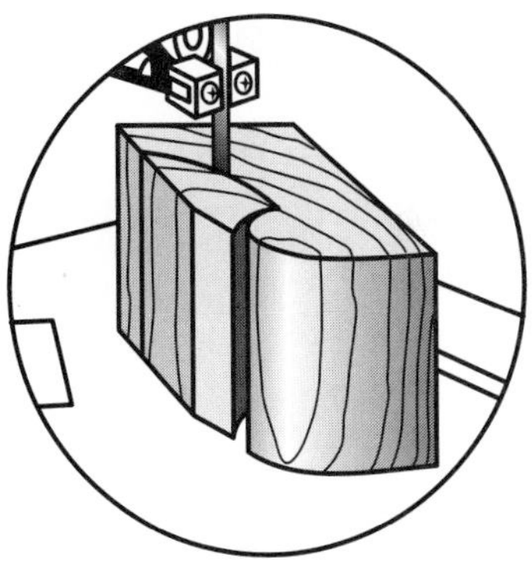

Fig. 24-19—*You can cut duplicate parts by first cutting thicker stock to shape, then resawing to the correct thickness.*

CROSSCUTTING AND RIPPING

The table of the band saw has a groove into which a miter gauge will fit. Stock can be held against this miter gauge to do accurate crosscutting or miter cutting.

If a table saw is not available, stock can be ripped to width on the band saw by fastening a fence or pivot block to the table and proceeding in the same way as with a table saw.

The table of the band saw can be tilted to do jobs such as beveling and chamfering on curves and irregular designs.

CHANGING SAW BLADES

Change saw blades as directed below:

1. Disconnect the machine.
2. Open the upper and lower guard doors. Remove the pin in the table slot, and loosen the vertical adjustment. Fig. 24-20.
3. Remove the old blade. If it has been broken, pull it off very carefully. It is "springy" and may flip around and cause an injury. Lower the upper wheel by turning the vertical adjustment. Grasp the saw blade with both hands; lift it off the upper and lower wheels. Coil it into three loops. (See the Woodworker's Tip for folding the blade.)
4. Slip the new blade onto the upper and lower wheels, with the teeth pointing downward. The blade guides should be back, out of position. Fig. 24-21. Tighten the vertical adjustment. Revolve the wheels by hand to see how the blade "rides." If the blade does not run in the center of the wheels, tilt the upper wheel with the tilt adjustment screw. The rear edge of the saw blade should be perpendicular to the table.
5. Adjust the tension of the saw with the vertical adjustment. If the saw does not have a tensioning spring, the blade should bend $\frac{1}{8}$ to $\frac{1}{4}$ inch when pushed lightly on its side.

Fig. 24-20—*Installing a band saw blade. The upper wheel has been released to permit the new blade to be slipped over the two wheels. There are adjustments on the reverse side of the upper wheel for tension and for tilting the wheel.*

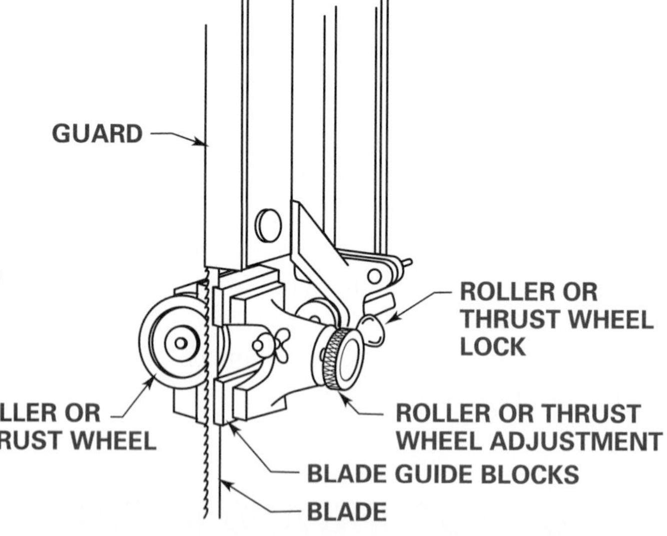

Fig. 24-21—*The roller or thrust wheel should just clear the back of the blade. The blade guide blocks should just clear the blade; they must not run on the blade. The front of the blocks should be just behind the teeth.*

WOODWORKING TIP

Folding a Band Saw Blade

Step 1: *Put on leather gloves and eye protection.* Then, hold the blade in front of you with the blade's teeth pointed away from you. Grasp the back of the blade as shown—palms facing forward, thumbs pointed away from you, and hands at about the 10 o'clock and 2 o'clock positions on the blade. Gently step on the blade with one foot just enough to hold the blade to the floor.

Step 2: Holding the blade firmly, roll your wrists inward (toward your body) until your thumbs point at each other.

Step 3: As you feel the blade's resistance weaken, gently push the folding blade to the floor while gradually lifting your foot from the blade.

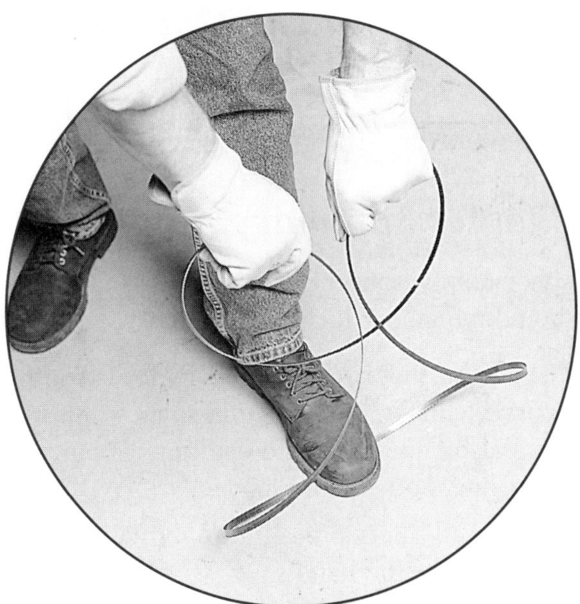

STEP 2

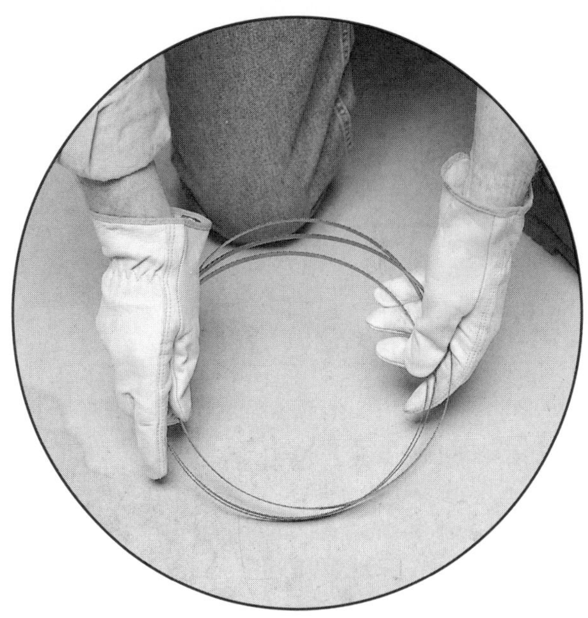

STEP 3

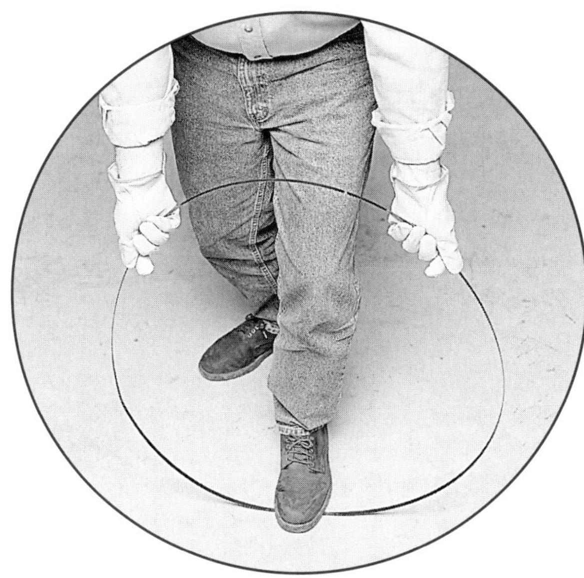

STEP 1

CHAPTER 24

Review & Applications

Chapter Summary

Major points from this chapter that you should remember include:

- The band saw is used primarily for cutting curves, circles, and irregular shapes, but it can also be used for crosscutting, ripping, beveling, and chamfering.
- Follow all safety precautions and guidelines when using the band saw.
- There are special procedures to help make cutting sharp and compound curves easier.
- A commercial or a homemade circle jig can be used to cut circles on the band saw.
- When stock is too thick, it can be resawed on the band saw to reduce its thickness.
- The band saw can be used to cut several duplicate pieces at one time.

Review Questions

1. What determines the maximum thickness that can be cut on a band saw?
2. Where should you stand when cutting with a band saw? Where should your fingers be when holding the workpiece you are cutting?
3. What is the general rule for selecting a band saw blade for cutting curves?
4. Outline the procedure for cutting sharp external curves on the band saw.
5. Explain how to cut circles on the band saw.
6. Describe two methods for cutting several duplicate parts at one time on the band saw.
7. Explain how to install a new band saw blade once the old blade has been removed from the saw.

Solving Real World Problems

As Chin adds the finishing touches to the desk he has been building, he notices that he does not have the ½″ material he needs to make two drawers. Looking over the material he has on hand, he finds an 8′ 1 × 6 and a 6′ 2 × 8. Not wanting to make another trip to the hardwood store, he decides to resaw the 2 × 8 into the material he needs. Each drawer will require a finished board ½″ thick, 6″ wide, and 4½′ long. Chin decides to cut the 2 × 8 in half with the band saw and then plane the surface to the correct thickness. The band saw blade will remove .125″ from the board, and the starting thickness measures 1½″. To cut the board exactly in the center, what should Chin set the fence distance to? If his bandsaw has a maximum cut height of 6½″, how can he cut the board?

CHAPTER 25

Sliding Compound Miter Saw

Look for These Terms

- bevel
- compound angle
- compound miter saw
- kerf board
- miter
- sliding compound miter saw

YOU'LL BE ABLE TO:

- Crosscut wood using the sliding compound miter saw.
- Correctly set the sliding compound miter saw for cutting a miter and a bevel.
- Cut a miter, bevel, and compound angle using a sliding compound miter saw.

Two of the most innovative power tools to be developed in recent times are the **compound miter saw** and the **sliding compound miter saw. Figs. 25-1** and **25-2 (page 332).** Both saws are referred to as "compound" because they can cut two angled surfaces at the same time (a **compound angle**). That is, a **miter** can be cut across the face of a board, and, at the same time, a **bevel** can be cut on the end. **Fig. 25-3.** Both saws can only be used for crosscutting. Neither can do ripping.

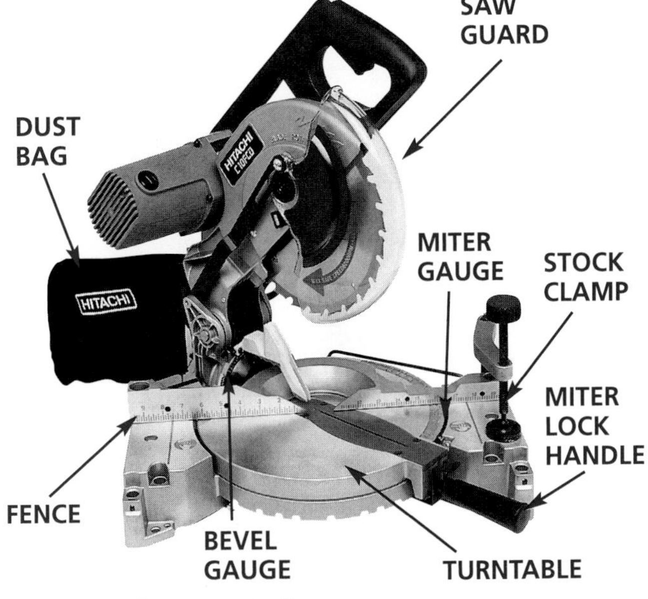

Fig. 25-1—*Compound miter saw.*

Fig. 25-2—*Sliding compound miter saw.*

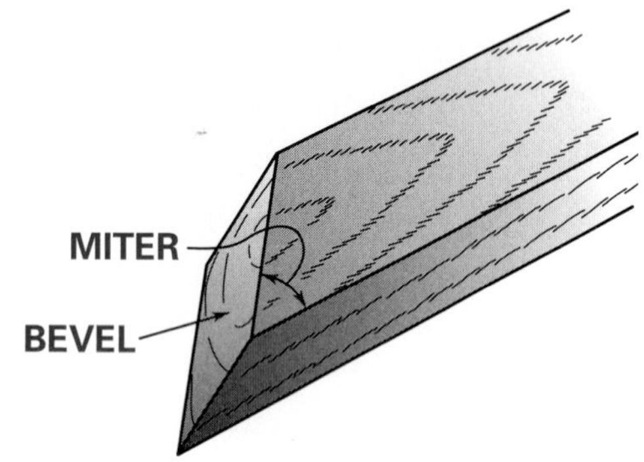

Fig. 25-3—*Compound angle.*

Fig. 25-4—*The head assembly on this saw moves on a single, bottom-mounted rail.*

The sliding compound miter saw is the newer and more versatile of the two saws. It can perform every operation that the compound miter saw can, and more, because of its sliding feature.

CUTTING ACTION

When cuts are made with the compound miter saw, the saw head is simply moved down in a *controlled* chopping motion. Thus it is often referred to as a "chop saw." When locked in position, the *sliding* compound miter saw can perform cuts in the same way. However, it can also be moved to make slide cuts.

The saw head assembly of the sliding compound miter saw is mounted on one, two, or three slide rails, or rods. These may either be top-mounted, as shown in Fig. 25-2, or bottom-mounted. Fig. 25-4.

To make a slide cut, the operator slides the saw assembly on the rails from its resting location to the front of the machine. The saw is then turned on, lowered into the cut position, and *pushed* through materials clamped to the saw's table. Fig. 25-5. The action of pushing the saw away when making a cut is an important safety factor. The force of the cutting action and the spinning saw blade are directed *away* from the operator.

PREPARING THE SLIDING COMPOUND MITER SAW FOR CUTTING

Note: Although their basic operations are the same, sliding compound miter saws vary in design from manufacturer to manufacturer and model to model. Always read and follow instructions in the owner's manual for setting up and operating a particular machine.

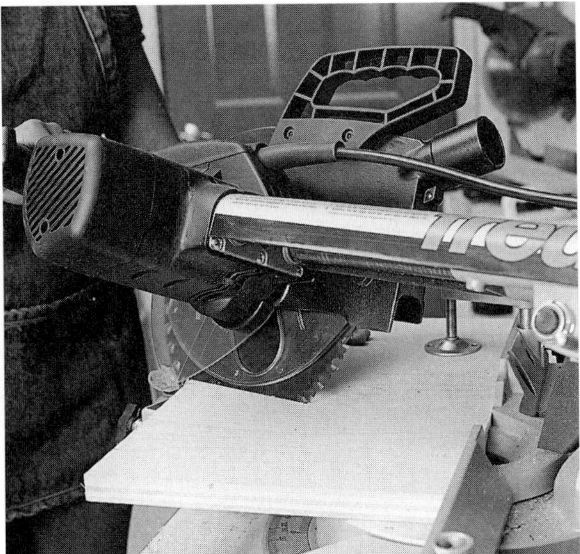

Fig. 25-5—*Because of its sliding feature, the sliding compound miter saw can cut stock that is wider and thicker than stock the compound miter saw can cut.*

MATHEMATICS *Connection*

Using the Miter Scale

When you are facing the machine, the miter scale on most machines starts at 0 degrees in the middle, with increments to 45 degrees to the left and right, as shown in the illustration. When the angle is set to 0 degrees, the saw actually cuts through the wood at a 90-degree angle. If you were to set the miter angle to 30 degrees left, at what angle would the saw actually cut?

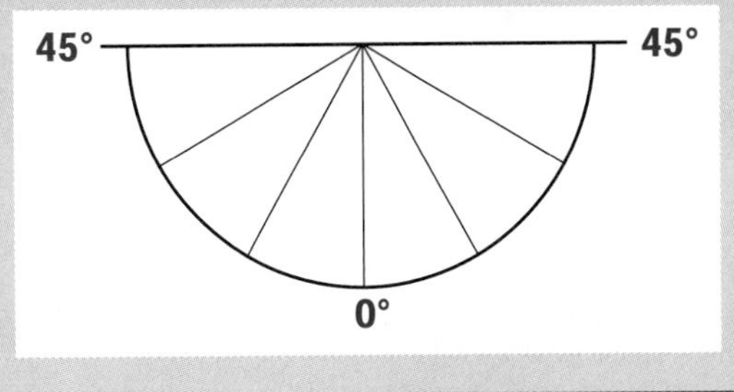

Adjusting the Settings

Miter and bevel settings may be used separately. That is, you can cut either a miter or a bevel using the sliding compound miter saw. More importantly, they can be used in conjunction with each other to cut both at the same time. This feature simplifies the task and saves both time and effort. For making a simple, straight crosscut, both scales should be set at 0 degrees.

Setting the Miter Angle

The angle of a miter to be cut is set by positioning the turntable according to a scale. The miter scale is located on the front portion of the base.

To set the miter angle, loosen the table lock mechanism. Depending on your machine, this may be done by turning a miter lock knob and depressing a lock lever.

Rotate the turntable to the desired angle. As you turn the table, the saw head is turned at the same time so that it is in position to cut the workpiece at the proper angle. The angle will be indicated with a pointer or displayed in a small window on the turntable. Fig. 25-6 (page 334). When the desired angle is indicated, release the lock lever and tighten the lock knob to hold the table and saw head in position.

You may note that there are several preset positions on the base. These are miter angles that woodworkers cut most often. The settings may vary from manufacturer to manufacturer.

Setting the Bevel Angle

The bevel angle can be set at any angle from 0 to 45 degrees. Like the miter setting, a scale, angle indicator, and locking mechanism are provided for this function. Fig. 25-7.

The locking mechanism is usually a lever located on the rear of the saw. It can be operated from the front while reading the bevel scale located near it.

Loosen the locking lever by turning it. The saw head can then be tilted by pushing it over to the left until the desired angle is indicated on the bevel scale. (*Note:* On nearly all models, the head can only be tilted in one direction.) Next, tighten the locking lever to keep the saw in the proper position for cutting the angle.

Fig. 25-6—***Move the turntable to the desired position and lock it in place.***

Kerf Boards

Located in the center of the turntable and its miter arm is a plastic or fiberboard piece called a **kerf board** or kerf insert. Fig. 25-8. It is so named because a saw cut is called a *kerf.* The blade of the saw extends into a slot in the board when making a cut.

The purpose of a kerf board is to minimize the gap between the side of the blade and the turntable. The tighter the kerf board is to the

Fig. 25-7—***The parts for setting a bevel are usually located on the rear of the machine.***

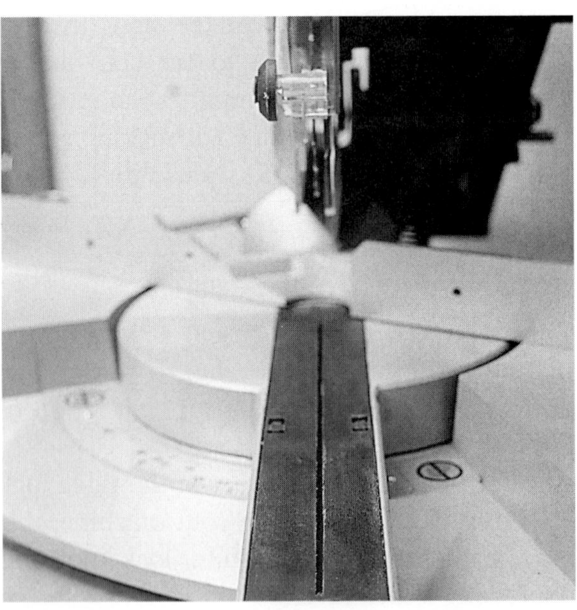

Fig. 25-8—***The kerf board should be replaced periodically.***

blade (without touching it), the less tearout will occur on the workpiece. *Tearout* is the splintering that often happens when a saw blade breaks through the bottom surface of the wood. Support under the wood, such as the kerf board, helps prevent it. Minimizing the gap also helps keep small cut-off pieces from pinching between the blade and table, possibly causing damage.

Kerf boards are removable and require periodic replacement. On some saws, kerf boards can be adjusted for different blade widths. If adjustments are not possible on your particular saw, you might need to have a different kerf board for each size blade you use.

Safety First

►*Positioning Clamps*

Position clamps so they do not interfere with machine operation. Before turning on the saw, lower the head assembly and make certain that no part of it will hit the clamp.

Clamping and Supporting Stock

Before cutting, the workpiece must be positioned firmly against the fence and clamped securely in place. There are a number of material hold-down clamps designed for use with sliding compound miter saws. The workpiece should not be held by hand.

Fig. 25-9—***Vertical hold-down clamp.***

Vertical clamps attach to the saw base. They apply force to the stock in a downward direction. Fig. 25-9. This keeps stock from lifting up into the blade, causing damage and possible injury.

Horizontal clamps apply force in against the backer fence. They are used for short, vertically oriented stock.

Before clamping, long stock must be supported at the same level as the turntable surface. If the end of the stock is allowed to hang, a lever force is created. The other end will be forced upward into the saw blade.

Accessories

One of the true strengths of a sliding compound miter saw is its capacity to make consistent repetitive cuts. Extensions for supporting materials are available for most machines. These attach to the sides of the base. By locating a set plate (adjustable stop) on an extension a measured distance from the blade edge, many like-sized parts can be cut. Fig. 25-10 (page 336).

Fig. 25-10—*Using a set plate on an extension makes it easy to cut several pieces exactly the same size.*

Fig. 25-11—*Extensions or a stand can provide support for long workpieces.*

Stands with extensions are also available. These are particularly useful when cutting long stock. Fig. 25-11.

Accessories such as crown molding stops and holder rod assemblies are very helpful for specialized jobs. Many other accessories are being developed as woodworkers discover the versatility of the sliding compound miter saw.

OPERATING A SLIDING COMPOUND MITER SAW

The sliding compound miter saw is operated in the same manner for making simple crosscuts, miter cuts, bevel cuts, and compound cuts. The settings, not the motion, determine the angles of the cuts.

Starting the Saw

All sliding compound miter saws require the operator to press a lock-off button and the power switch in order to start the saw. Fig. 25-12. This is a built-in safety system. It ensures that you cannot start the saw simply by bumping the power switch. Also, the power switch must be pressed continuously in order for the saw to stay in operation. The lock-off and power switch are commonly located on the handle of the saw head and can be operated by one hand.

Making the Cut

When making the cut, use a firm, steady motion. As the blade passes through the material being cut, you will be able to feel if you are pressing too hard or too fast. A fast cut is not a smooth cut. A cut made by pressing too hard can result in burning or scorching the wood.

As you finish the cut, do not lift the saw with the blade still rotating. Instead, release the power switch and allow the blade to coast to a full stop before lifting the head.

Safety First

►Using the Sliding Compound Miter Saw

- Always wear eye protection.
- Make sure the saw is securely bolted to a workbench or a stand designed especially to hold it.
- Always unplug the machine before inspecting it or servicing it.
- Carefully inspect the blade for cracks, missing teeth, or other damage before plugging the tool into an outlet. Replace any damaged part you find.
- On many machines, there is a slight, but distinct, movement of the saw head when the power is switched on. Hold onto the handle tightly.
- Keep all guards in place.
- Use care and common sense when operating a power saw.
- When using this saw, one hand is used to press the power switch. Keep your other hand at least 4 inches away from the blade.
- Never hold onto any workpiece by hand. It must be tightly secured by the hold-down clamps provided.
- Make sure the blade is not in contact with the workpiece before pressing the power switch.
- Always make sure the blade rotates freely before doing any work.
- Wait until the blade attains its full speed before lowering it into the material.
- When the saw is turned off, the blade will coast (continue to turn) for a time. Keep your hands away from it.
- Wait until the blade has completely stopped turning before raising the saw head and removing the workpiece.

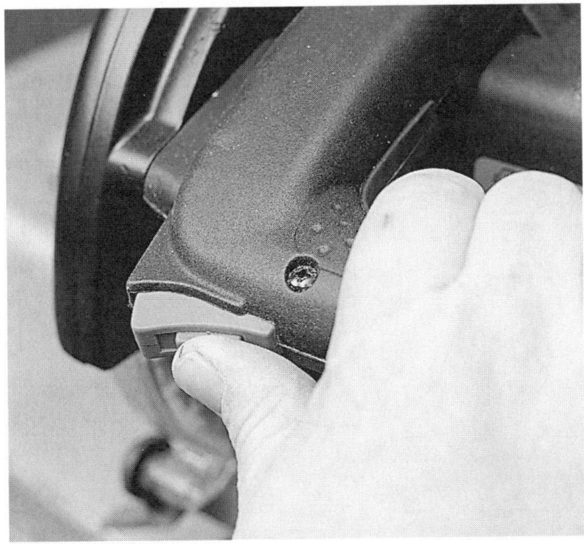

Fig. 25-12—***The lock-off button activates the power switch. The power switch must be pressed continuously to keep the saw in operation.***

Blade Thickness

Thicker blades leave wider kerfs, resulting in more waste. Therefore, woodworkers generally use the thinnest blade feasible for a given project. However, in commercial settings in which a sliding compound miter saw may be used constantly throughout the day, thicker blades are sometimes more practical. Thinner blades are more likely than thicker ones to warp with high usage. This causes a less accurate cut over time, which in turn increases waste due to rework.

ROLE IN THE SHOP

As the newest addition to the modern woodworking shop, the sliding compound miter saw is being used to perform tasks that previously were done by other power saws. In many shops, it has largely replaced the radial-arm saw. It can do many of the same operations and is safer to use. A radial-arm saw cuts as it is pulled toward the operator. See Chapter 23.

The sliding compound miter saw does have limitations, however. It cannot do rip cuts as the radial-arm and table saws can. Neither can it make long crosscuts. Also, it cannot be used to cut dadoes. The table saw is best for that. See Chapter 22.

The sliding compound miter saw is probably the best tool to come along for making relatively short crosscuts and for cutting angles quickly, easily, and, most importantly, more safely.

CHAPTER 25

Review & Applications

Chapter Summary

Major points from this chapter that you should remember include:

- Both the compound miter saw and the sliding compound miter saw can cut two angled surfaces at the same time.
- The saw head assembly of the sliding compound miter saw is moved on rails.
- When making a cut with the sliding compound miter saw, the force of the cutting action is directed away from the operator.
- Miter angles are set by moving the turntable to a position indicated on a miter scale; bevel angles are set by tilting the saw head at an angle indicated on a bevel scale.
- Before cutting, materials must be positioned against a fence and clamped securely in place.

Review Questions

1. Why are both the compound miter saw and the sliding compound miter saw referred to as "compound"?
2. How is a slide cut made using a sliding compound miter saw?
3. Describe how to set the machine for cutting a miter.
4. Describe how to set the machine for cutting a bevel.
5. What is the purpose of a kerf board?
6. How do you start a sliding compound miter saw and keep it in operation?
7. When you finish a cut, what must you do before lifting the saw head?
8. Compared to other saws, for what processes are sliding compound miter saws best?

Solving Real World Problems

Kim is getting ready to make landscaping rings around two young trees in her yard. She decides that an eight-sided ring would look nice and that each side will be 36″ long. Working with 6-foot landscaping timbers, she discovers that her circular saw does not cut deep enough to cut all the way through in one pass. Kim turns to her sliding compound miter saw, with its 4″ cut depth, to do the job. Now she must determine the angle at which to set the saw. How can she calculate the needed angle? What is it? Is there any attachment to the miter saw that would help her operate the saw more safely?

CHAPTER 26

Scroll Saw

Look for These Terms

- inlaying
- marquetry
- scroll saw

YOU'LL BE ABLE TO:

- Choose the proper scroll saw blade for the project.
- Demonstrate cutting external and internal curves and designs with the scroll saw.
- Explain how to do straight cutting on the scroll saw.
- Use the scroll saw to make simple inlay patterns.
- Describe how to install a scroll saw blade.

The **scroll saw,** sometimes called a *jigsaw,* is designed to cut sharp curves and angles on both the outside edges and the interior sections of a workpiece. **Fig. 26-1.** The saw moves up and down to do the same type of cutting that can be done by hand with a coping or compass saw. It differs from the band saw in that it can make inside cuts without cutting through any of the surrounding stock.

Fig. 26-1—*The fine scroll work on this bench is typical of the cutting done on a scroll saw.*

PARTS OF THE SCROLL SAW

The saw consists of a frame with an overarm and a base, a driving mechanism to convert rotating action into *reciprocating* (up-and-down) action, a table, a guide, and a saw blade. A tension sleeve, through which a plunger moves, is mounted in the end of the overarm. Fig. 26-2.

The size of the scroll saw indicates the saw's throat depth, which is the distance between the blade and the base of the overarm. The throat depth determines the maximum width of stock that can be cut on the saw. The throat depth varies from model to model, but most saws range between 15 and 24 inches.

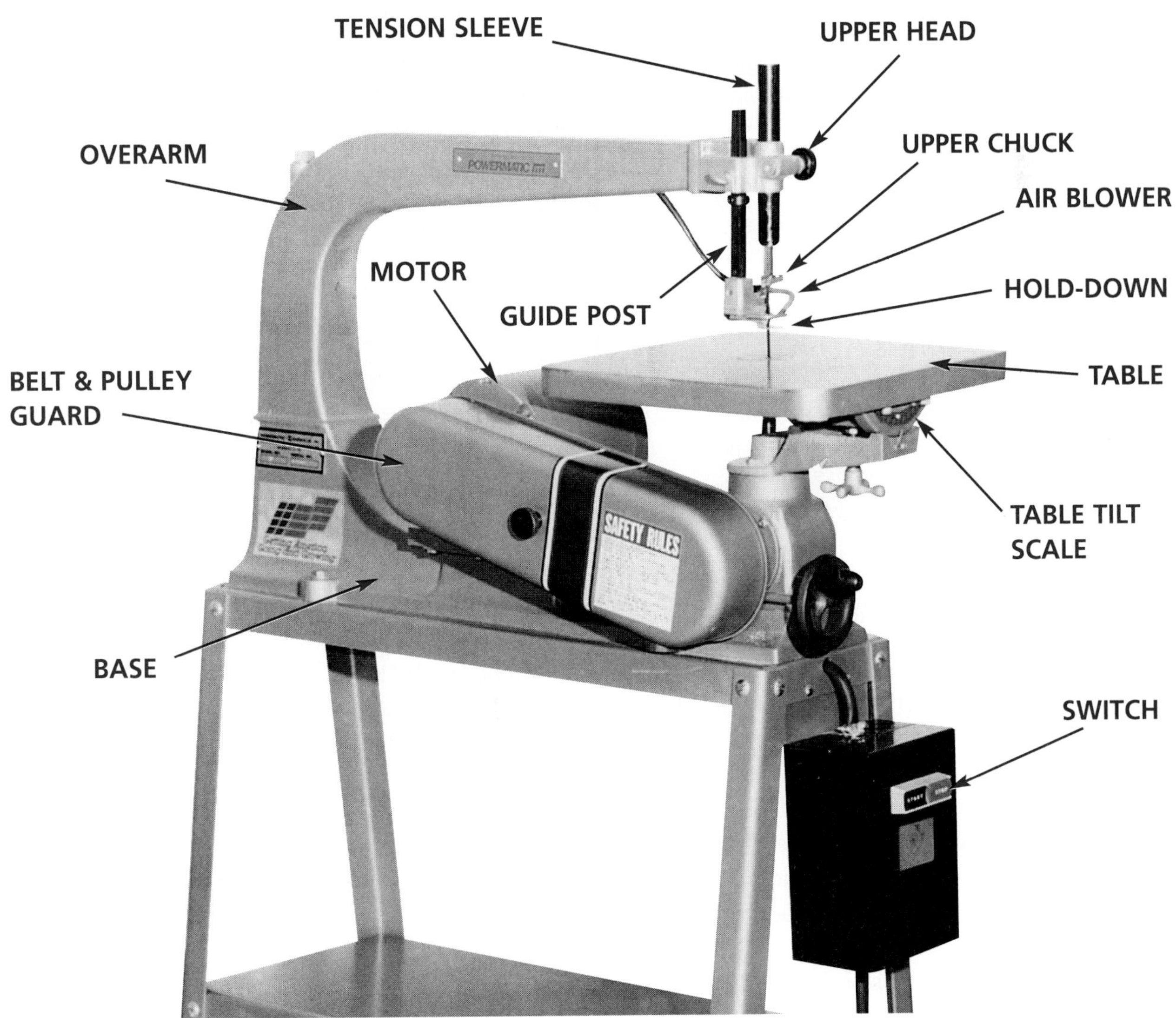

Fig. 26-2—*The parts of a scroll saw.*

Fig. 26-3—*This variable-speed scroll saw can be adjusted for speeds from 40 to 2000 cutting strokes per minute.*

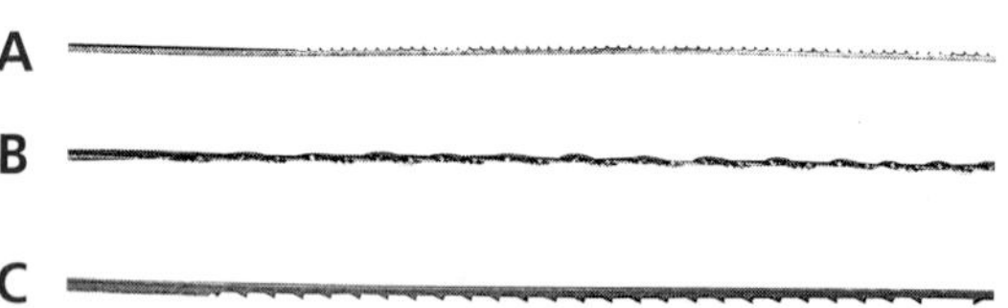

Fig. 26-4—*Scroll saw blades. (A) Metal-cutting blade. (B) Spiral blade, used for very fine cutting. (C) Standard plain-end general purpose blade.*

SCIENCE *Connection*

Converting Rotary to Reciprocal Motion

Scroll saws, like many other machines, use *eccentric cams* to convert rotary (circular) motion to reciprocating motion. An eccentric cam is circular, but its axis of rotation is off-center, as shown in the illustration. A *follower* follows the surface of the cam as it rotates. As the cam turns on its axis, the follower traces an up-and-down path, resulting in reciprocal motion.

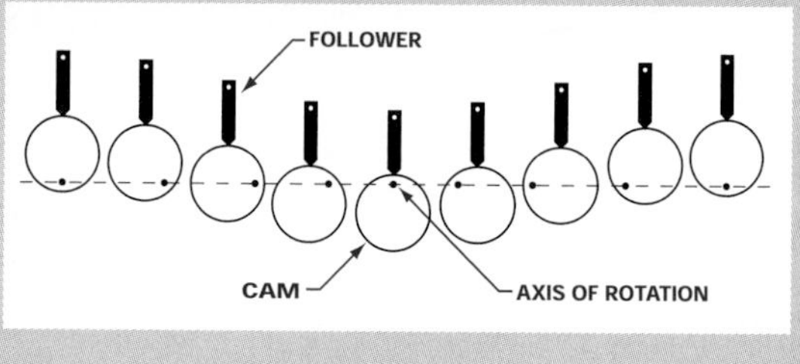

Depending on the model, the speed of the saw can be adjusted by shifting a belt to various positions or by turning the variable-speed control handle. Fig. 26-3. Most cutting is done at high speed. However, slower speeds give you more control over the cuts being made. Slower speeds also make it easier to cut metals, plastics, and veneers.

Scroll saw blades range from 1⁄16 to 1⁄4 inch wide and between 1⁄64 and 1⁄32 inch thick. A 15-TPI (teeth per inch) blade is good for general use. Fig. 26-4.

When choosing a scroll saw blade, consider both the material being cut and the complexity of the pattern. Pick the widest and thickest blade possible that will still allow you to follow the most complex details of your pattern. (A heavier blade is less likely to break or to wander from the cutting line.) If the pattern has loose curves or some straight stretches, choose a wide, thick blade. A wide, thick blade should also be used when cutting hard materials. If the pattern has sharp curves, use a narrow, thin blade.

CUTTING WITH A SCROLL SAW

Below are the basic steps for cutting with the scroll saw.

1. Adjust the guide so that the small spring tension holds the stock against the table.

Safety First

►*Using the Scroll Saw*

When operating the scroll saw, take the following precautions:

- Wear eye protection.
- Be certain the blade is properly installed—in a vertical position, with the teeth pointing down.
- Roll the belt by hand to see if there is clearance for the blade and if the tension sleeve has been properly set.
- Check the belt guard to see that it is closed and tight.
- Adjust the hold-down to the thickness of the stock being cut. This will prevent the work from moving up and down with the blade.
- Do not allow your fingers to come any closer than 2 inches from the blade when cutting stock.
- Keep your fingers to the side of the blade, never in front of it.
- Cut cylindrical (round) stock by holding it in a special "V" fixture.
- Stop the scroll saw before you make any adjustments.

WOODWORKING TIP

Transferring Patterns

To transfer a cutting pattern to a piece of wood, trace the pattern onto a piece of paper. Then use a spray adhesive to glue the paper pattern to the wood. Make sure the paper is smooth, with no bubbles or wrinkles. Cut along the lines of the pattern. Tear away the remaining paper when finished cutting.

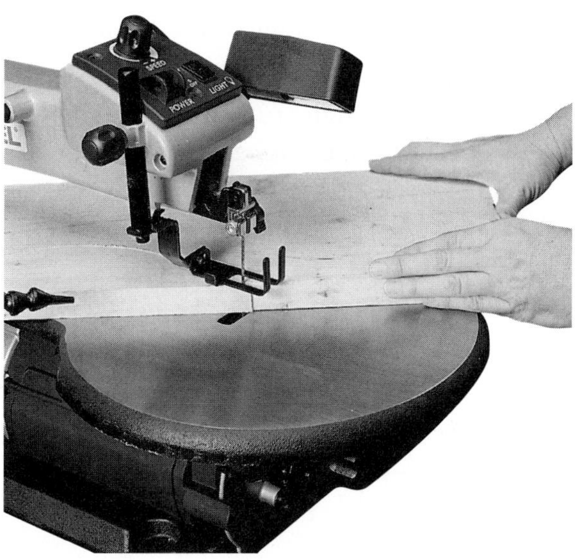

Fig. 26-5—***Guide the work with your fingers as you apply forward pressure with your thumbs. Note that the guide holds the workpiece firmly against the table.***

2. Hold the workpiece with the thumb and forefinger of both hands. Guide the work with your fingers as you apply even, forward pressure with your thumbs. Fig. 26-5. Do not force the stock into the blade.
3. Turn the stock appropriately as you are cutting the curve(s). If the stock is turned too sharply, the blade may break. If it's turned too slowly, the wood may burn.

If rather complicated cuts must be made, plan the cutting carefully before you proceed. This will help avoid interrupting the cut and backing out of the kerf. Fig. 26-6.

Cutting Internal Curves and Designs

When making intricate internal cuts, first drill a relief hole in the center of the waste stock and clear away as much of the waste stock as possible. Then make the intricate cuts. Fig. 26-7.

A jeweler's blade is often used to cut sharp angles. The steps below outline the procedure for this type of cutting.

1. Drill a relief hole in the waste stock.
2. Remove the throat plate.
3. Fasten the blade in the lower chuck. Then put the stock over the blade, with the blade going through the relief hole.
4. Fasten the other end of the blade to the upper chuck. Then replace the throat plate.
5. Adjust the guide to the correct height. Then cut from the relief hole to the layout line. Fig. 26-8.

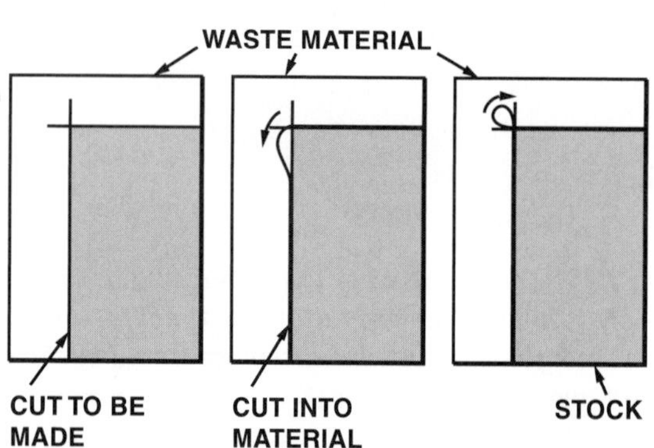

Fig. 26-6—*Two methods of cutting an exterior corner. One way is to make a slightly curved cut at the corner; this small, curved waste is then trimmed off the stock with a quick second cut. Another method is to make a complete circle in the waste stock.*

Tilt-Table Work

The table of the scroll saw can be tilted to cut a bevel on a straight, circular, or irregular design. When cutting an angle or bevel, the workpiece must always remain on the same side of the blade. If the workpiece is swung around the blade, the bevel will change direction.

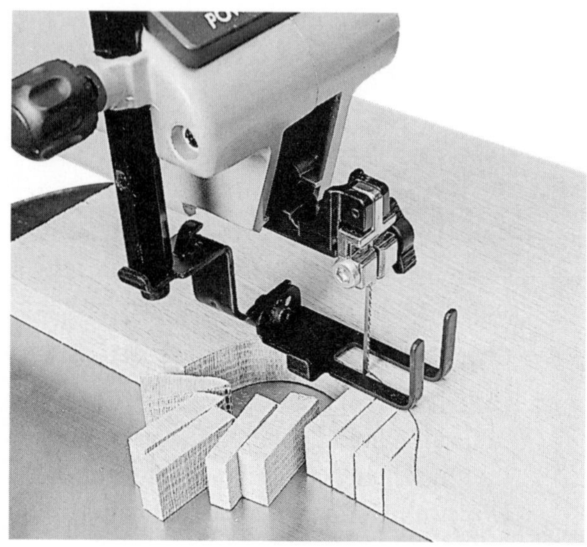

Fig. 26-7—*Make the intricate cuts after the waste material has been cleared out. Well-placed relief holes make turning easier.*

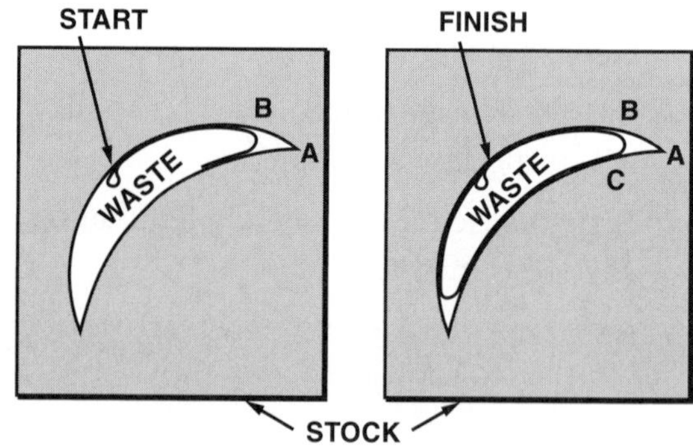

Fig. 26-8—*Cutting a sharp angle. Start from a hole drilled in the waste stock. Cut to point B, then make a curve as shown. Follow the layout line from point C to the opposite end, cut another curve, then follow the layout line again until the waste stock drops out. Finally, make the sharp corners by cutting from both B and C to A and doing the same for the lower corner.*

Straight Cutting

It is quite difficult to cut a long, straight line on the scroll saw. However, straight cuts can be made on thin stock. The best way to do this is to clamp a guide board to the table a short distance from the blade. Fig. 26-9. Choose the widest blade possible, since this will help eliminate the blade's tendency to stray off the straight line.

Making Identical Parts

If two or more identical parts must be cut on the scroll saw, one of the best methods is to make a "sandwich" of the material. Fasten the pieces together with small nails or brads in the waste stock. Then cut out your pattern. Fig. 26-10.

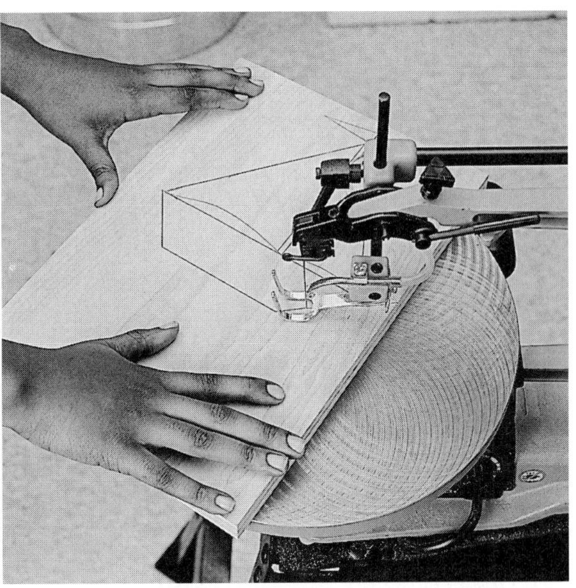

Fig. 26-10—*This project is being cut out of one piece of wood. Multiple pieces could be made by sandwiching several pieces together and nailing them in the waste stock.*

Making an Inlay or Simple Marquetry

Inlaying (or **marquetry**) is a way of forming a design by using two or more different kinds of wood that have a marked contrast in color or grain pattern. Fig. 26-11 (page 346). To make a simple inlay, follow these steps:

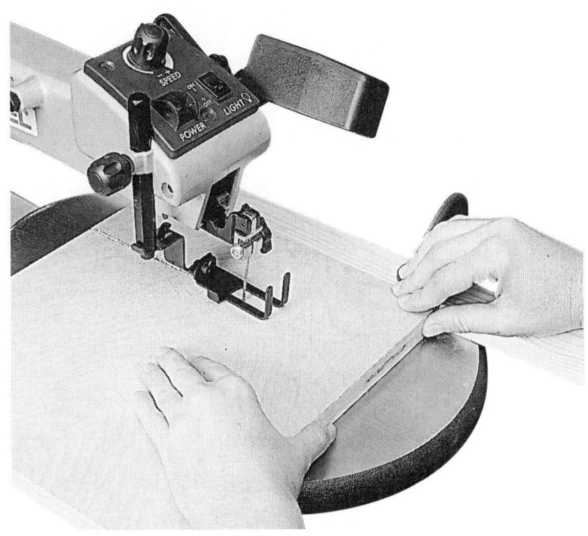

Fig. 26-9—*Using a guide board for straight cuts.*

1. Fasten the two different pieces of wood together in a pad, using small nails or brads at each corner.
2. Drill a small hole at an inside corner of the design to start the blade.
3. Tilt the table of the saw one or two degrees. Make all necessary cuts, with the work always on the same side of the blade.
4. Take the pad apart and assemble the design. When pieces with beveled edges are fitted together, there will be no space caused by a saw kerf.

INSTALLING A SCROLL SAW BLADE

Follow the steps below when installing a scroll saw blade.

1. Select the correct type and thickness of blade for the project.

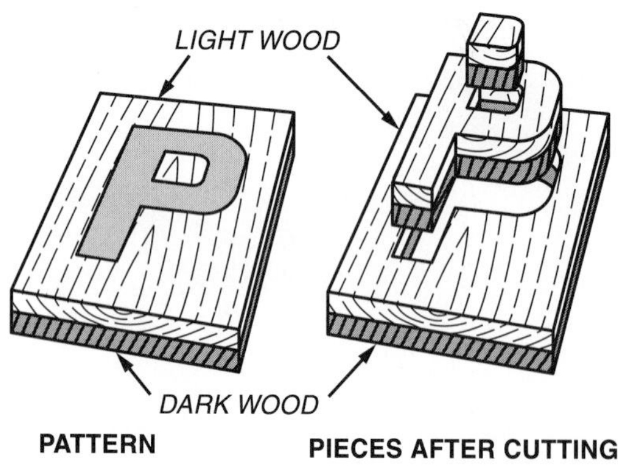

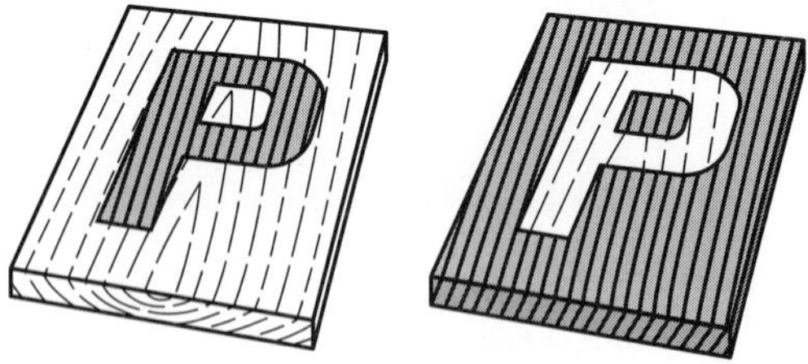

Fig. 26-11—*Making a simple inlay.*

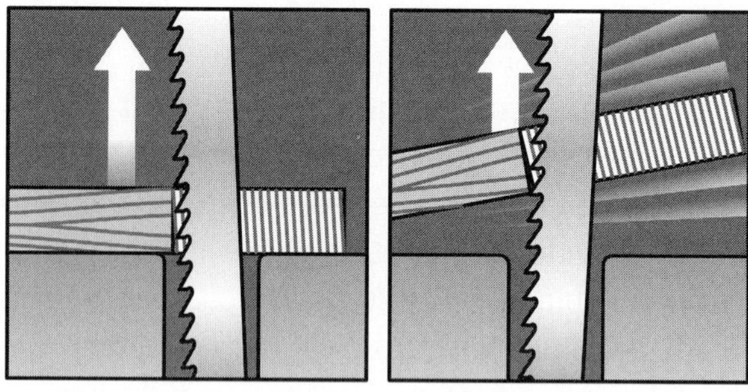

Fig. 26-12—*Install the blade with the teeth pointed down. Slant the blade toward the work to clear the stock on the upstroke. If it slants away, the teeth will catch and lift the stock.*

2. Remove the table insert. Tilt the table so that you can see the lower chuck. Place the lower end of the blade in the lower chuck with the teeth pointing down. Fig. 26-12.
3. Tighten the setscrew (or thumbscrew), making sure that the blade is held securely in the chuck. Fig. 26-13.
4. Pull down on the upper chuck and attach the other end of the blade in this chuck. If necessary, loosen the tension sleeve knob and move the sleeve up and down. Fig. 26-14. *Note:* If a saber blade is used for internal cutting, it is held only in the lower jaw. Fig. 26-15.
5. Adjust the guide post until the hold-down applies light pressure on the workpiece.
6. Adjust the roller guide to touch the back of the blade lightly. Tighten it in position.

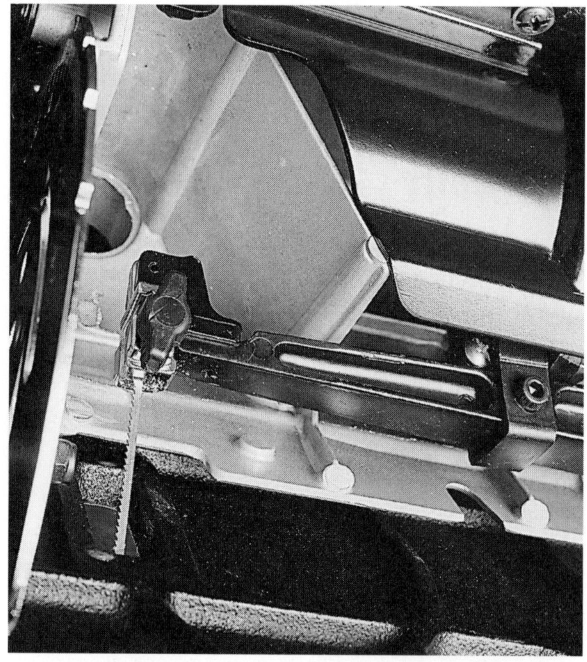

Fig. 26-13—*Clamp the blade firmly in the lower jaw, using a hexagon wrench or a thumbscrew to tighten the jaws. Do not clamp too tightly or you may shear the threads or break off the thumb control.*

7. With the power off, roll the belt by hand to turn the machine to make sure that the blade moves freely.

8. Turn the saw on. Recheck the blade movement. If the speed is too fast, readjust the variable-speed hand control. If the machine does not have such a control, turn off the power. Remove the belt guide and move the belt for a lower speed.

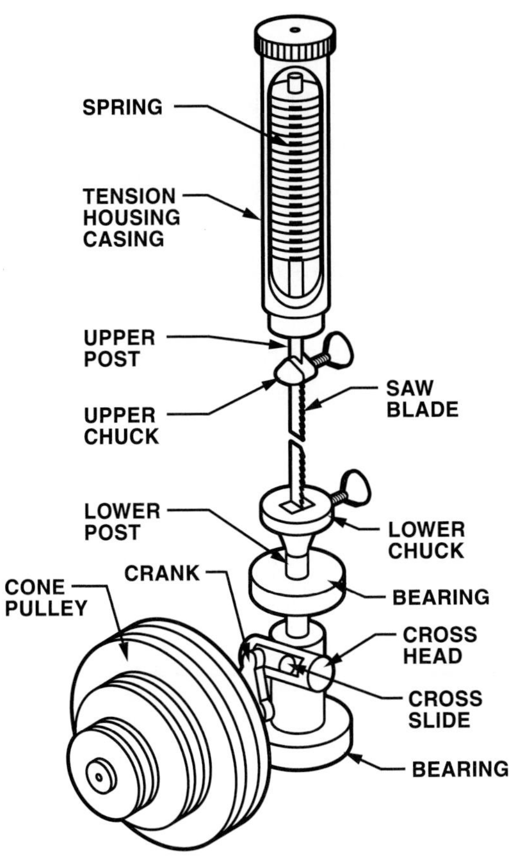

Fig. 26-14—*Secure the blade in both the lower and upper chucks. The lower chuck does the driving, while the upper chuck controls the tension.*

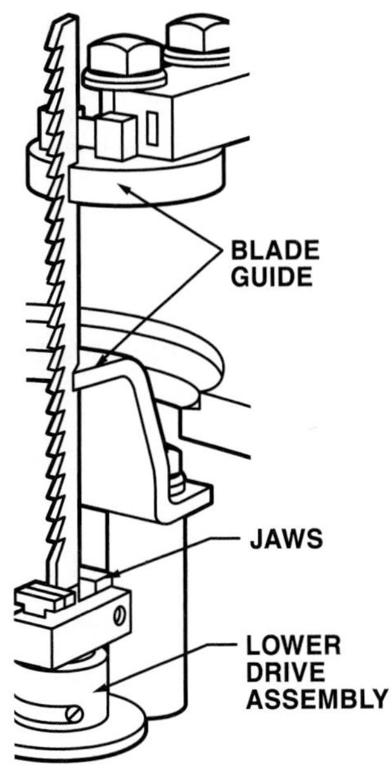

Fig. 26-15—*Fastening a saber blade in the lower chuck. The chuck has been turned a quarter turn and the blade fastened in the V-jaws. An extra support bracket has been fastened in place.*

CHAPTER 26

Review & Applications

Chapter Summary

Major points from this chapter that you should remember include:

- The scroll saw is designed to cut sharp curves and angles on both the outside edges and interior sections of a workpiece.
- You must consider the material being cut as well as the intricacy of the pattern when choosing a scroll saw blade.
- There are a number of safety precautions to keep in mind when operating the scroll saw.
- The table of the scroll saw can be tilted for cutting bevels.
- The scroll saw can be used to make identical parts and inlaid designs.
- Scroll saw blades are secured in both a lower and an upper chuck, with the blade's teeth pointing down.

Review Questions

1. What determines the maximum width of stock that can be cut on a scroll saw?
2. Describe what type of blade you would choose to cut a scallop-edged pattern consisting of a series of loose curves.
3. Explain how to make internal cuts on the scroll saw.
4. How do you do straight cutting on the scroll saw?
5. What is inlaying? Explain how to make a simple inlaid design.
6. Outline the procedure for installing a scroll saw blade.

Solving Real World Problems

Admiring the exotic fretwork he sees in a museum, Trent decides to try to make a jewelry box for which he has the plans. Sitting at the saw, he starts to follow the pattern on the wood slowly. After some time, he stops to check his work and finds that there are many areas where the wood is scorched and discolored. What are two things he can do to avoid burning the cut line so much? What if he used thinner wood?

CHAPTER 27

Drill Press

YOU'LL BE ABLE TO:

- Identify operations performed using the drill press.
- Make adjustments correctly for the operation being performed.
- Select the proper tool for the process being performed on the drill press.
- Operate a drill press correctly, observing all safety rules.

Look for These Terms

- adjustable screw
- pilot bit
- brad-point bit
- countersink bit
- drill press
- fly cutter
- Forstner bit
- hole saw
- mortising attachment
- multispur bit
- plug cutter
- sanding drum
- spade bit

The **drill press** can perform a variety of operations. **Figs. 27-1 and 27-2 (page 350).** It is used primarily for drilling holes of various diameters and depths and at various angles. Holes can be drilled accurately and with great consistency. With proper speeds, the drill press can also be used for mortising and sanding. **Fig. 27-3 (page 351).** Appropriate accessories and jigs or setups are needed to perform these processes.

The size of a drill press is expressed as twice the distance from the center of the chuck to the column. For example, if this distance is 8 inches, then the size of the drill press is 16 inches. This tells you the largest diameter stock that can be drilled through the center.

To control the rotation speed of the spindle, a drill press is equipped with either a variable speed control or step pulleys. On most machines, the speed ranges between 250 and 3,100 rpm. Those with variable speed can operate at any speed from 250 to 4,900 rpm. Lower speeds are required when using tools such as Forstner bits and circle cutters.

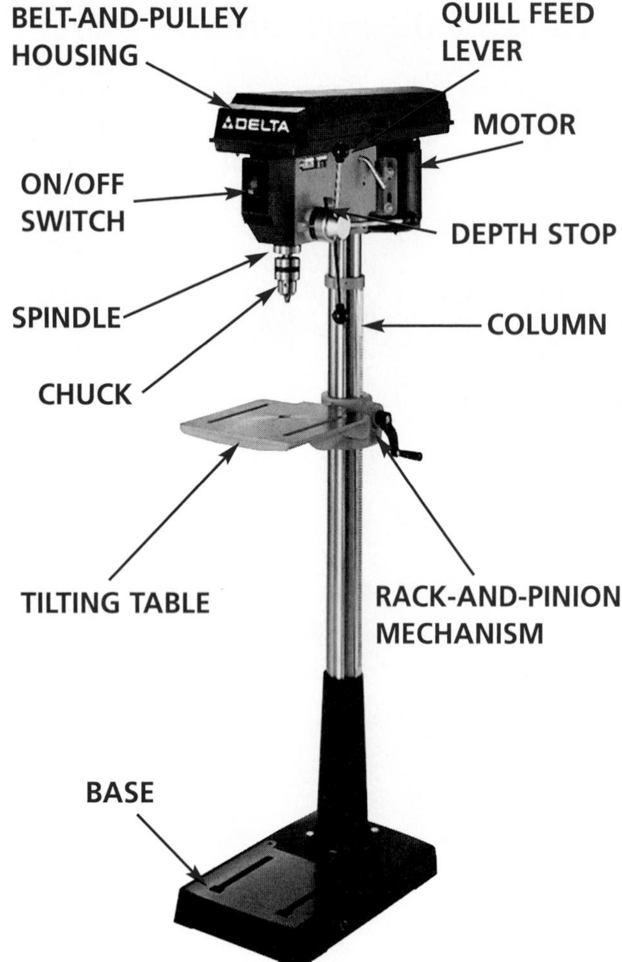

Fig. 27-1—*Parts of a drill press with step pulleys.*

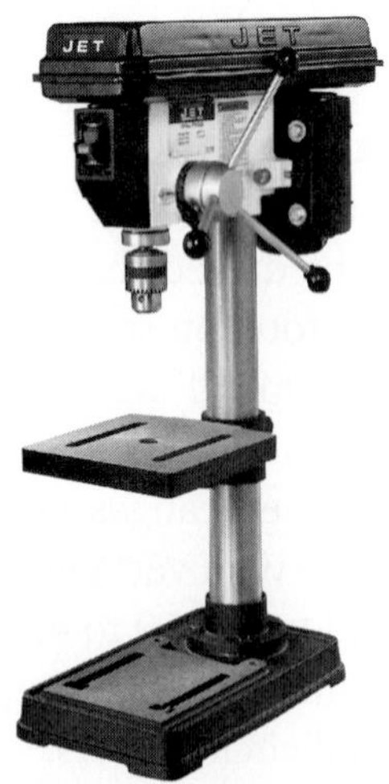

Fig. 27-2—*A benchtop drill press.*

Safety First

►*Using a Drill Press*

- Wear eye protection.
- Use a drill bit with a straight shank—never a square one.
- Check the speed. Use a slow speed for large holes and a faster speed for smaller holes.
- Clamp work securely, especially when operating the drill press at high speeds.
- Keep your hair and loose clothing away from all moving parts. Always wear eye protection. Fig. 27-4.
- Place the drill bit in the chuck straight and tighten it securely.
- Always remove the chuck key before starting the machine.
- When preparing the setup, make sure the drill bit will not mar the vise or table. Place a piece of scrap stock under the work to be drilled.
- Clamp round (cylindrical) stock in a V-block before drilling.
- Keep your fingers at least 4 inches away from the rotating tool.
- Remove chips and shavings with a brush or stick of wood—never your fingers.
- Feed the drill or bit smoothly into the workpiece. When the hole is deep, withdraw it often to clear the shavings.
- Check the setup carefully before doing special operations such as shaping or sanding.

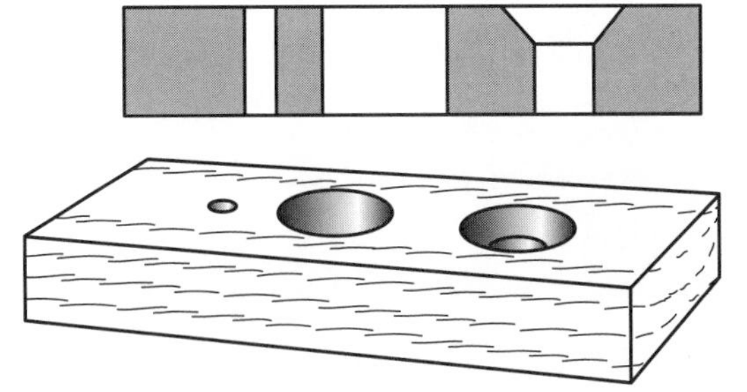

DRILLING, BORING, & COUNTERSINKING

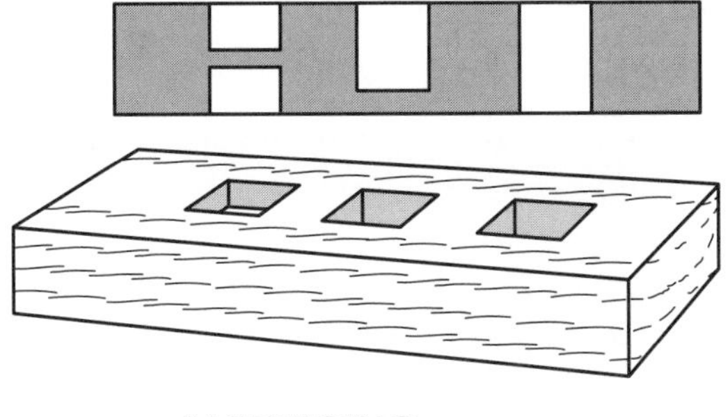

MORTISING

Fig. 27-3—*Common operations that can be performed on the drill press.*

TOOLS FOR THE DRILL PRESS

Different types of tools are used in the drill press for different purposes. Fig. 27-5 shows a few special tools. The most common drill bits are described below and are illustrated on pages 23 and 24.

- *Twist drills* are available in sizes from 1⁄16 inch to ½ inch by 64ths. These can also be used in portable electric drills. See Chapter 6.
- **Spade bits** come in diameters from ⅜ inch to 1½ inches. These are fast-cutting tools that leave a rather rough hole.
- **Brad-point bits** are ideal for drilling wood. These come in sizes from ⅛ to ½ inch by 16ths. The sharp brad point lets you place the hole exactly where you want it. The sharp cutting spurs make a clean hole in wood.
- **Multispur bits** are used to cut perfectly round, flat-bottomed holes.

Fig. 27-4—*Follow good safety practices when using the drill press.*

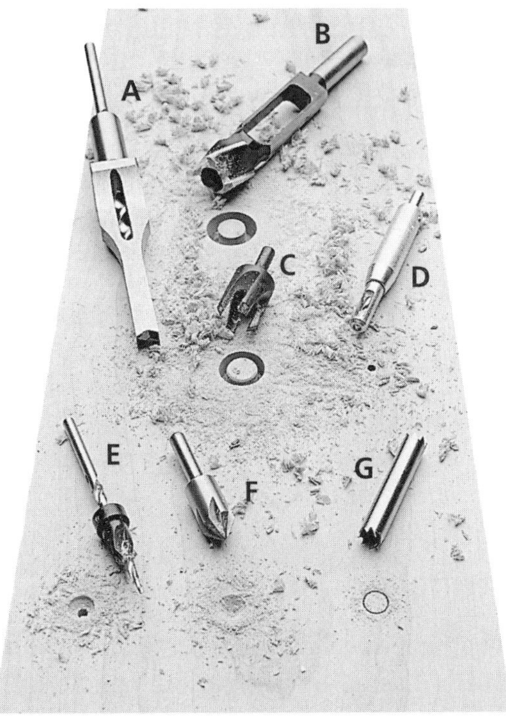

Fig. 27-5—*Tools used on the drill press. (A) Hollow chisel mortise. (B) Plug/tenon cutter. (C) Plug cutter. (D) Self-centering hinge bit. (E) Drill and counter-sinking bit. (F) Countersink bit. (G) Screw extractor.*

- **Plug cutters** are used for cutting cross-grain and end-grain plugs and dowels up to 3 inches long. They come in common sizes of 3⁄8, 1⁄2, 5⁄8, 3⁄4, and 1 inch. The smaller sizes match the diameter of dowel rods.
- **Adjustable screw pilot bits** are used when installing flathead screws. They do the same job as the drill and countersink, but they do it in one operation. See Chapter 15.
- Auger bits with straight shanks can be used for drilling relatively large holes.
- **Fly cutters,** or circle cutters, can be adjusted to cut holes from 1 inch to 4 inches.
- **Hole saws** are used for cutting large holes. They come in sizes up to 6 inches.
- **Forstner bits** are used to cut flat-bottomed holes, even through knots, end grain, and veneer.
- **Mortising attachments** used only with the drill press consist of a drill bit surrounded by a four-sided chisel. The drill bit removes the chips. Sizes range from 1⁄4 to 1 inch in diameter.
- **Countersink bits** are cone-shaped. This enlarges the top of a hole so a flathead screw can be driven in flush with the surface. Sizes range from 1⁄4 to 3⁄4 inch in diameter.

Safety First

►*Changing Speeds*

On a variable speed machine, change the speed while the machine is operating. On a step pulley machine, disconnect the power before making any changes.

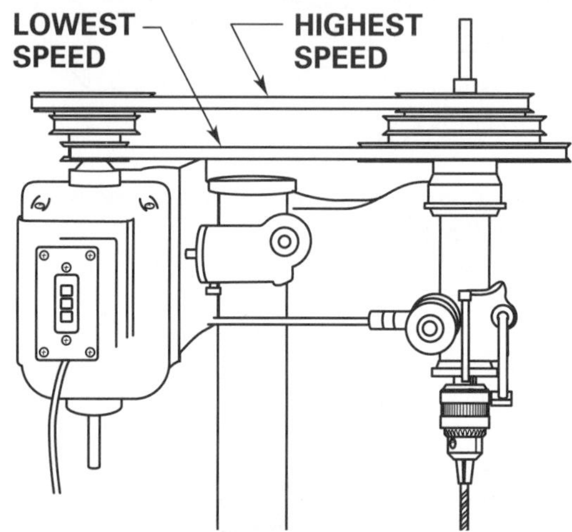

Fig. 27-6—***Positions of the belt for highest and lowest speeds. Note the relative sizes of the pulleys.***

ADJUSTMENTS

A drill press is a very flexible tool. It can be adjusted in a number of ways to perform a wide variety of tasks.

Setting the Speed

On variable speed machines, the speed of rotation is adjusted by turning the handle on the front of the speed control. Because it can be set to any speed within a range of speeds, very fine adjustments are possible.

On machines with step pulleys, speed is set by the way a belt is arranged on the pulleys. The speed is changed by moving the belt from one set of pulleys to another. Small machines have one belt and two pulleys. These can be adjusted for four or five speeds. Large machines may have two belts arranged on three pulleys, with twelve speed settings possible.

The rate of speed depends on the relative sizes of the pulleys engaged by the belt. The highest speed is obtained when the belt is on the largest pulley on the motor (*driving* pulley) and the smallest pulley on the spindle (*driven* pulley). Fig. 27-6.

Installing a Drill Bit

When installing a drill bit, first place a piece of scrap wood on the table. Use the *chuck key* to open the chuck slightly larger than the diameter of the shank of the tool. Fig. 27-7. Hold the tool in the chuck, making sure it is straight. Then tighten securely with the key.

Adjusting the Table

When adjusting the table, first release the table-locking clamp. Then raise the table so that the space between the top of the workpiece and the point of the drill bit or cutting tool is about ½ inch. Tighten the clamp.

Setting the Depth Gauge or Stop

The depth gauge or stop is on the side of the drill press. It controls how far the drill bit or cutting tool can move. You may wish to drill a hole only partway through a piece of wood. To set the gauge, draw a line on the side of the wood indicating this depth. Pull down on the handle until the point of the tool is even with this line. Then move the two adjusting nuts that limit the drilling to this point.

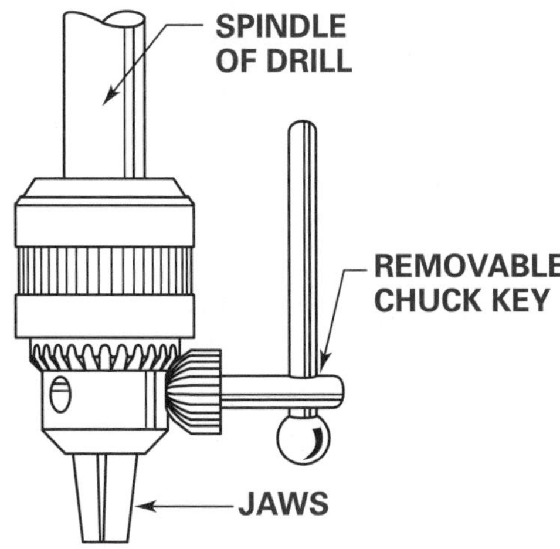

Fig. 27-7—***Make sure the tool is centered in the three jaws of the chuck. Always remove the chuck key before operating the drill.***

MATHEMATICS *Connection*

Calculating Rotation Speed

The mathematical relationship between the pulley diameters and the speed is as follows:

$$\frac{\text{speed of driven pulley}}{\text{speed of driving pulley}} = \frac{\text{diameter of driving pulley}}{\text{diameter of driven pulley}}$$

Assume that a driving pulley with a diameter of 6 inches and a speed of 1,400 rpm is connected to a driven pulley with a diameter of 9 inches. Using the formula above, the speed of the driven pulley is found as follows:

$$\frac{s}{1{,}400} = \frac{6}{9}$$

$$9 \times s = 1{,}400 \times 6$$

$$9s = 8{,}400$$

$$s = \frac{8{,}400}{9}$$

$$s = 933 \text{ rpm}$$

DRILLING PROCEDURES

Follow this procedure for drilling tasks:

1. Select the correct drill bit or cutting tool and fasten it securely in the chuck. *Caution: Always remove the chuck key before turning on the power.*
2. Make sure the proper layout has been made and that the position of the hole is well marked.

3. Make sure the drill bit is free to go through the opening in the drill-press table.
4. Place scrap wood under the workpiece. This will help prevent splintering when the drill goes through the underside.
5. Determine and set the correct cutting speed. The speed should vary with the type of bit, the size, the kind of wood, and the depth of the hole. In general, the smaller the drill bit and the softer the wood, the higher the speed.
6. Clamp the work securely. The force of the spindle in motion tends to rotate the work. Holding it by hand can be very difficult. This is particularly true when using larger drilling tools, hole saws, or similar cutting devices. Fig. 27-8.
7. Feed the tool into the material. Use good judgment when doing this. If the tool smokes, reduce the speed and the rate of feed.

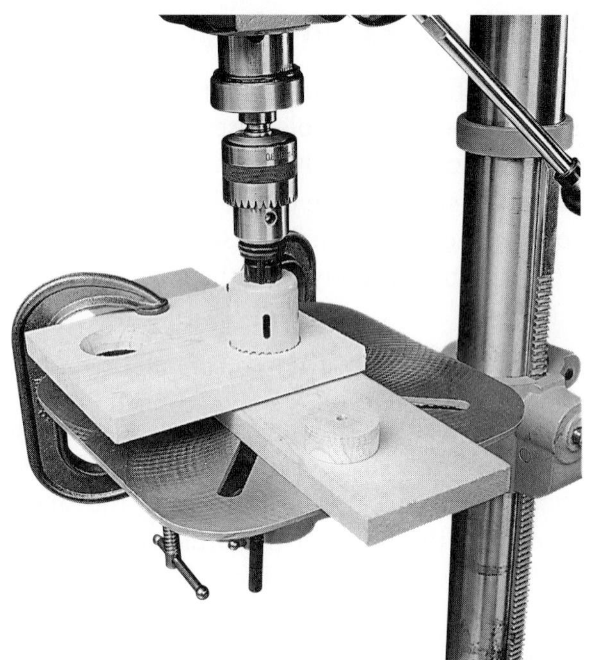

Fig. 27-8—*For safety, always clamp the workpiece to the drill press.*

Drilling Small Holes in Flat Stock

To drill a hole that is ¼ inch or smaller, use a twist drill. Follow this procedure:

1. Locate the center of the hole on the workpiece and mark it with a center punch or scratch awl.
2. Place the workpiece on the table over a piece of scrap wood.
3. Turn on the power and slowly move the point of the bit into the stock. Hold the stock firmly and apply even pressure to the handle.
4. If the stock is hardwood or the hole is deep, back up the bit once or twice to remove the chips before finishing the hole.
5. Finish drilling the hole. Always drill completely through the stock and into the scrap wood.

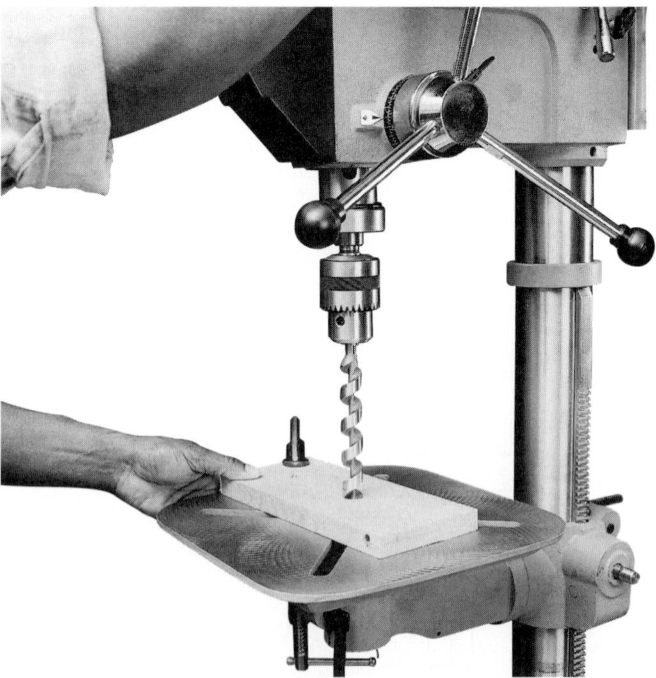

Fig. 27-9—*Drilling a hole with an auger bit. The work is clamped to the table with a C-clamp.*

Cutting Medium and Large Holes in Flat Stock

Holes from ¼ inch to 1¼ inch can be cut with a variety of drilling tools. For example, a twist drill, auger bit, Forstner bit, or spade bit could be used. Figs. 27-9 and 27-10.

There are two methods for drilling a through hole. The simplest is to place a piece of scrap wood under the hole so that the tool will cut through the stock and into the scrap piece. This keeps the underside from splintering. The second method is to cut until the point of the bit shows through the stock, then drill from the other side.

To drill a hole to a specific depth, adjust the depth stop with the *power off*. Bring the cutting tool down to the side of the work where the depth is marked. Then set the depth stop.

Holes larger than 1¼ inch are best cut with a hole saw or a circle cutter. Make sure that the work is firmly clamped, especially when using a hole saw.

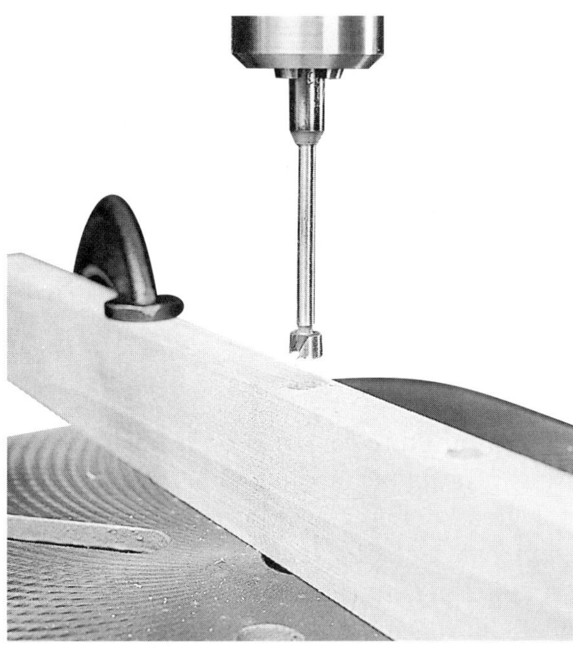

Fig. 27-10—***Using a Forstner bit with the work securely clamped to the table by C-clamp.***

Other Drilling Operations

Holes can be drilled in round stock by holding the workpiece in a V-block. Fig. 27-11. Countersinking may be necessary when installing flathead screws. Fig. 27-12. To drill holes at an angle, adjust the table right or left to the correct angle. Fig. 27-13 (page 356).

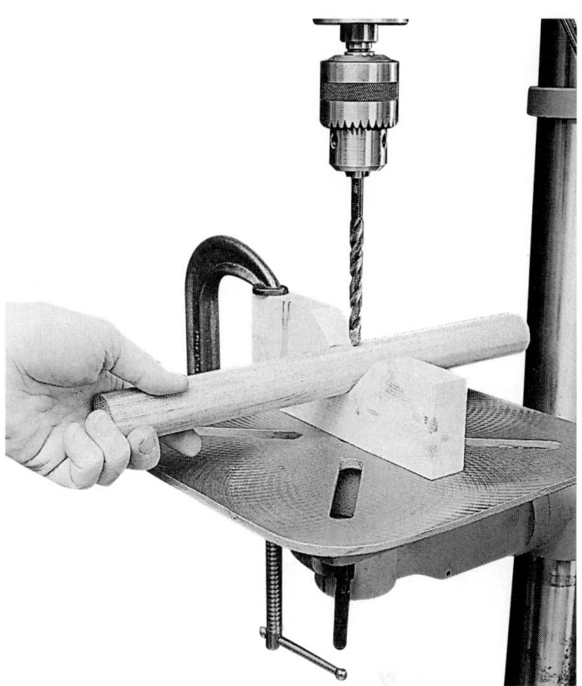

Fig. 27-11—***Use a V-block to hold dowel rods and other cylindrical stock.***

Fig. 27-12—***Using a drill press to countersink holes.***

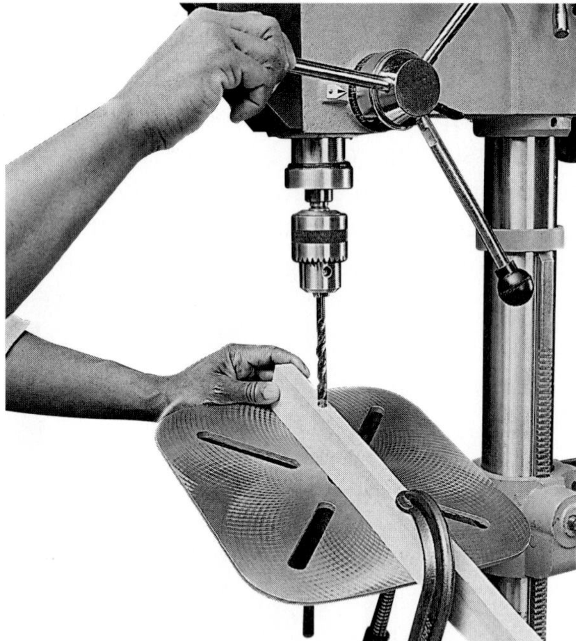

Fig. 27-13—*Drilling a hole at an angle. Mark the location with a center punch so the drill will not slide out of place as the drilling starts.*

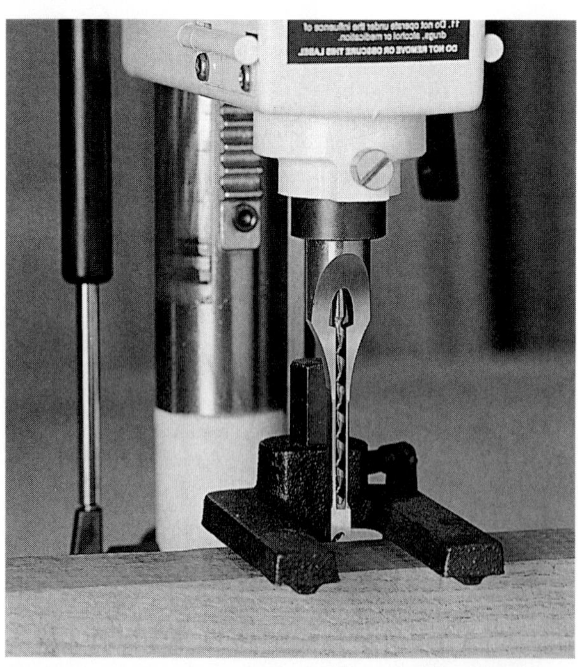

Fig. 27-14—*Using a mortising attachment. The part for holding the chisel is locked to the drill press. The auger bit is fastened in the chuck. A fence is locked to the table, and clamps are used to hold the stock in place.*

WOODWORKING TIP

Drilling an Angle

Drilling a hole at an angle—particularly an extreme one—can be tricky. Sometimes the sides of the drill bit move against the workpiece before the sharp point of the drill can begin the hole. To prevent this from happening, position a block of scrap wood on the workpiece so that it barely overlaps the drilling location. Clamp it in place. Drill through the block and into the workpiece to cut the hole cleanly.

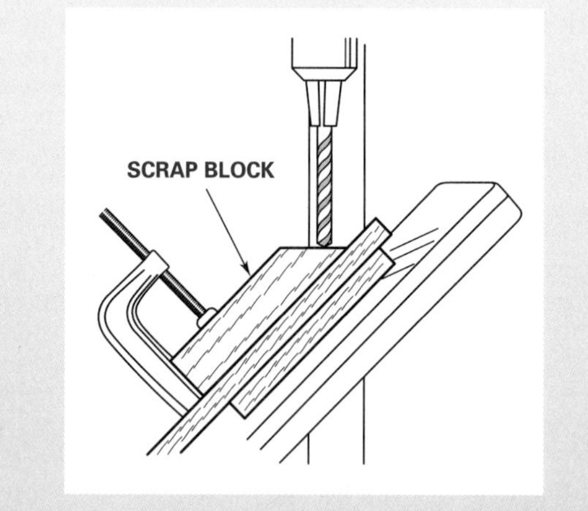

MORTISING

If much furniture construction is done, a mortising attachment should be available. This tool greatly simplifies cutting a *mortise-and-tenon joint.* See Chapter 13.

A mortising attachment (Fig. 27-14) consists of a hollow, square mortising chisel in which an auger bit revolves. The chisel itself is ground to a sharp point at each corner. These points enter the wood just after the revolving bit. They cut the square opening after the bit has removed most of the stock.

On most mortising attachments, the chisel is fastened to the drill press and a straight-shank auger bit fastened in the chuck. The chisel should be the same width as the width of the mortise to be cut. A fence should be attached to the table to guide the stock.

SANDING

A **sanding drum** can be used effectively on the drill press. Fig. 27-15. It's particularly useful for sanding curved shapes and the inside of cut-out areas, but can also be used for routine sanding tasks. Sizes of sanding drums range from ½ inch to 3 inches. Various grit sizes are available. See Chapter 7.

In order to make contact with the entire surface of the sanding drum, a jig similar to that shown in Fig. 27-15 can be made. Such a jig can be used with all sizes of drums.

Fig. 27-15—*Using a sanding drum. Align the jig so the drum is ¼ inch from the edge of the hole.*

How to Construct a Sanding Jig for a Drill Press

Follow these instructions to construct the jig shown in the illustration. Use plywood.

1. Cut two table pieces and two sides.
2. On one table piece, draw a circle 3½-inch diameter close to upper left corner.
3. Perpendicular to left edge, draw a line across the work just touching the bottom of the circle.
4. Touching the same line, draw a circle 2-inch diameter ¼ inch to the right of this circle.
5. Connect the top portions of the two circles with a straight line.
6. With the work clamped to the drill press, cut out the holes using hole saws. Cut away material between the holes.
7. Repeat the process on the other table piece, but use 1¾-inch and ⅞-inch diameter holes.
8. Glue the pieces together. In use, cutouts should be in the upper left corner on both tables.

CHAPTER 27

Review & Applications

Chapter Summary

Major points from this chapter that you should remember include:

- Drill presses use either a variable speed control or step pulleys to control the rotation speed of the spindle.
- Make sure the rotation speed is set correctly for the operation being performed.
- A number of different types of tools can be used in the drill press for different purposes.
- Adjustments made on a drill press are setting the speed, installing a drill bit or cutting tool, positioning the table, and setting the depth gauge.
- Placing scrap wood under the workpiece prevents splintering as the drill bit exits on the underside.
- Small holes are drilled at fast speeds and larger holes at slower speeds.

Review Questions

1. Name the operations that can be performed on the drill press.
2. What are the two types of speed controls used in drill presses?
3. Explain what you would do to change the speed setting on a step pulley machine (one belt, two pulleys) from the lowest speed possible to the highest speed possible.
4. Which tools might you use to cut a flat-bottomed hole?
5. List general drilling procedures.
6. What is the shape of an opening cut by a mortising attachment?
7. What device is used for sanding on a drill press?

Solving Real World Problems

Dario has to drill a series of 5 holes located 1″ from the edge of a board. The board is to be a shelf in a tool cabinet that will hold mortise chisels and bits. Each hole must be 3/4″ in diameter and positioned 2″ from the last hole. The shelf is 1″ thick and the holes must not go through the shelf. How can Dario consistently position the shelf on the drill press table? What is the best way to make sure he does not drill through the shelf?

CHAPTER 28

Router

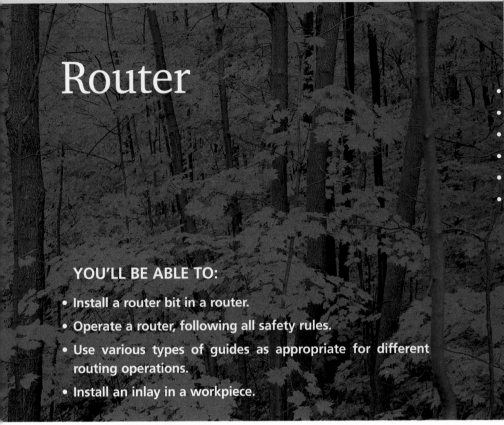

Look for These Terms

- fixed-base router
- freehand routing
- pilot end
- plunge router
- router
- router bit

YOU'LL BE ABLE TO:

- Install a router bit in a router.
- Operate a router, following all safety rules.
- Use various types of guides as appropriate for different routing operations.
- Install an inlay in a workpiece.

One of the most versatile tools for working with wood is the **router. Fig. 28-1 (page 360).** A router has a vertical shaft into which a cutting tool called a *bit* is fastened. The shaft turns very rapidly to make cuts in wood. Different types of cuts can be made depending on the type of bit used. **Fig. 28-2.** Typically the router is used to cut dadoes, rabbets, and grooves. However, it can also perform joinery, mortising, planing, and shaping processes as well.

TYPES OF ROUTERS

The two types of routers are the **fixed-base router** and the **plunge router.** The main parts of both types are the base and the motor unit.

In the fixed-base router, the base is clamped to the motor. The joined parts are moved as a single fixed unit. Edge cutting and trimming is usually done with a fixed-base router.

In a plunge router, the motor is attached to a base that has springs. The springs allow the motor unit with the router bit in it to be raised and lowered without moving the entire machine. A plunge router is typically used for cutting dadoes, mortises, and rabbets. Fig. 28-3.

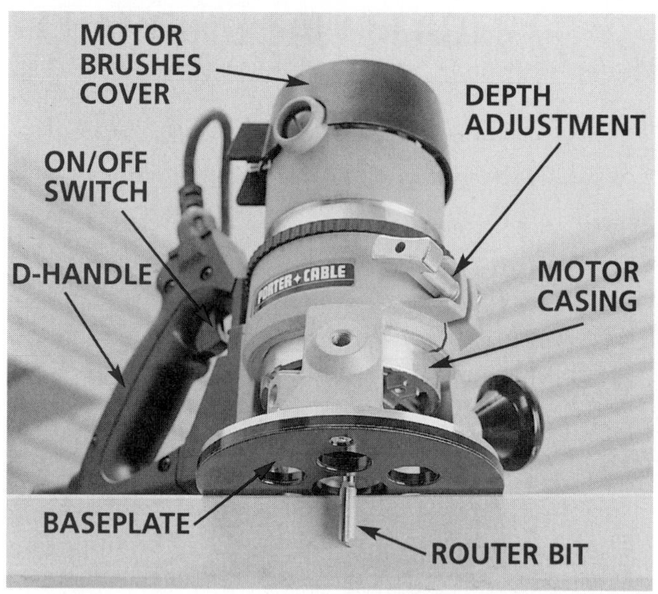

Fig. 28-1—*The main parts of a fixed-base router.*

Fig. 28-3—*A plunge router.*

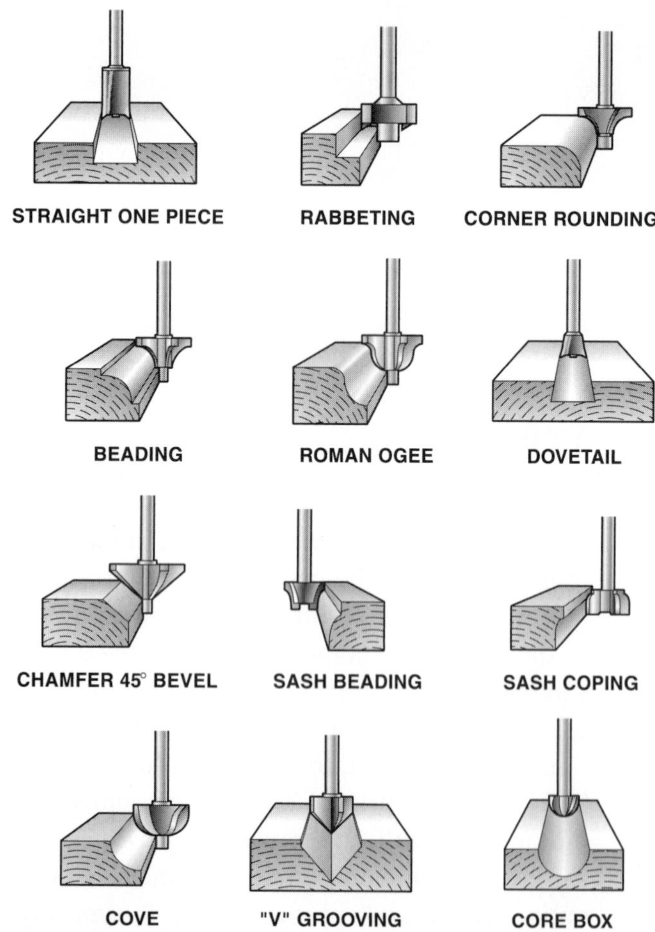

Fig. 28-2—*Examples of the types of cuts that can be made with a router.*

Fig. 28-4—*A small, lightweight router rotates at speeds up to 30,000 rpm.*

Motors used to power routers range from ¾ hp to 3 hp. Figs. 28-4 and 28-5. Routers with 1 to 2 hp work well for most tasks. Routers with low power are used for fine detailed work. High power is needed for cabinetmaking and many shaping tasks.

Fig. 28-5—***A heavy-duty, multiple-speed router is used for making deep cuts.***

MATHEMATICS *Connection*

Cutting Speed

It's amazing how many cuts a router bit makes in one second! A typical router operates at 25,000 revolutions per minute (rpm). At that rate, one cutting edge of a router bit would make 25,000 cuts per minute.

25,000 rpm ÷ 60 seconds = 417 cuts per second

How many cuts would a router bit with two cutting edges make in one second?

Safety First

►*Using a Router*

- Follow all general safety rules for power tools.
- Select the proper bit or cutter for the specific job.
- Use the proper wrenches to tighten the nuts that hold the bit or cutter in the shaft.
- Adjust the bit or cutter depth.
- Fasten the workpiece securely in a vise or with clamps.
- Check to see that the bit rotates freely and that all adjusting nuts and knobs have been tightened.
- Always wear ear and eye protection when using a router.
- Make a trial cut on a scrap piece of the same thickness before attempting the final cut.
- Keep your hands clear of the rotating bit or cutter.
- Hold the router firmly in both hands. Take special care when turning the machine on and off.
- Feed the router into the stock at an appropriate speed. Never force it faster than it will cut. Feeding too slowly will cause the bit to heat up.
- Always move the router in the proper cutting direction.
- After turning off the power, wait until the machine comes to a complete stop before setting it on its side.

ROUTER BITS

The **router bit** is the part of the router that does the cutting. Fig. 28-6. Bits most commonly used are made of high-speed steel (HSS). HSS bits with carbide tips are also used quite often.

There are two basic types of router bits: those with a **pilot end** and those without. Fig. 28-7. Bits with a pilot end are designed so that the pilot rides against the edge of the workpiece during cutting. Fig. 28-8. Therefore, at least ⅛ inch of the original edge must remain for the pilot to follow. Some pilots are solid, while others have a ball-bearing end.

Bits with pilot ends are used to shape the edges of the workpiece. For example, they can be used to cut a fancy edge on a tabletop. Bits without a pilot end are used for such operations as cutting grooves and dadoes.

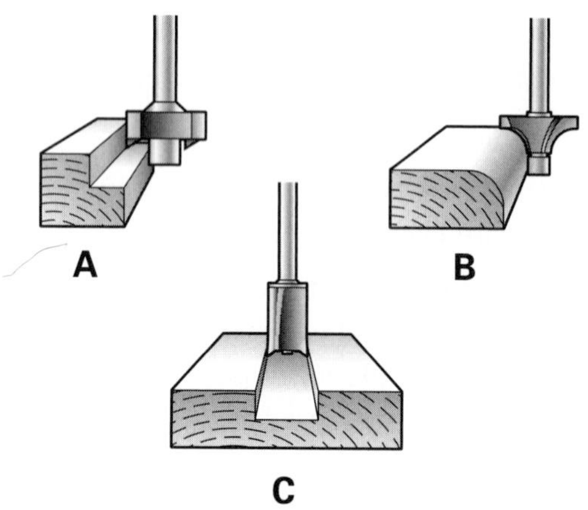

Fig. 28-6—*Three commonly used router bits. (A) Rabbet. (B) Corner rounding. (C) Straight.*

Fig. 28-8—*The pilot on the end of the cutter controls the amount of cut. It rides on the edge and does no cutting.*

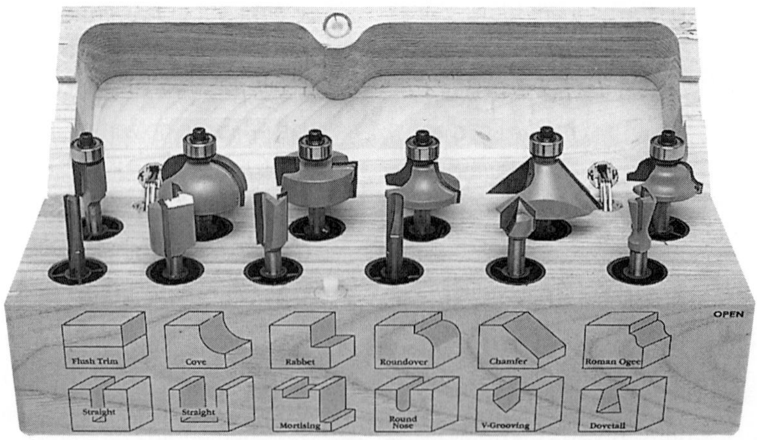

Fig. 28-7—*A set of router bits. Some of the bits have a pilot end, which rides against the edge of the workpiece during cutting.*

Installing a Router Bit

The router is turned upside down to install a router bit. The exact method of installing the bit depends on the design of the router. Some routers have a push button to lock the motor shaft. Others have two nuts, one on the shaft and one on the collet. For these, two open-end wrenches are needed. One is used to hold the nut on the shaft. The other is used to tighten the nut on the collet.

To install the router bit, insert the shank of the bit as far as possible. Then pull it out about ⅛ inch. Hold the shaft to keep it from turning and tighten the nut on the collet. Fig. 28-9.

OPERATING PROCEDURES AND TECHNIQUES

Operating a router effectively requires care and practice. Control is very important, whether the cut is done freehand or with guides. The following procedures and techniques will help you operate a router successfully and safely.

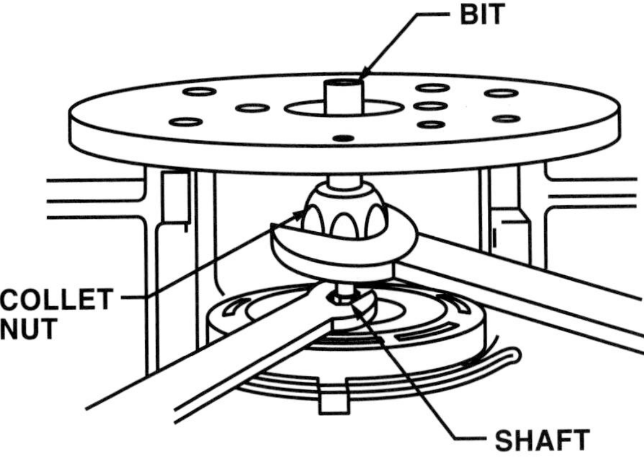

Fig. 28-9—***Installing a bit. Note that one open-end wrench keeps the motor shaft from turning while the other wrench tightens the collet nut.***

Operating a Router

Preparing to cut:

- Adjust the depth of cut while the machine is off.
- Make sure the workpiece is securely clamped down and stable enough to support the router during cutting.
- Try the router on a piece of scrap stock that is the same thickness as the finished piece.
- Hold the router firmly while switching to an "on" position.
- Control the router with both hands.

Speed considerations:

- Feed the router into the workpiece firmly and at a rate that will not overload it.
- Do not feed the router too fast or with too much cutting depth, especially with decorative bits. This will overload the motor and cause the edge of the wood to chip out. When necessary, make two or more passes. Adjust the guide or depth of cut after each pass. Fig. 28-10.
- Because the router bit rotates at a very high speed, it may heat up if the router is moved too slowly through the material. The wood will also show burn marks. Fig. 28-11 (page 364).

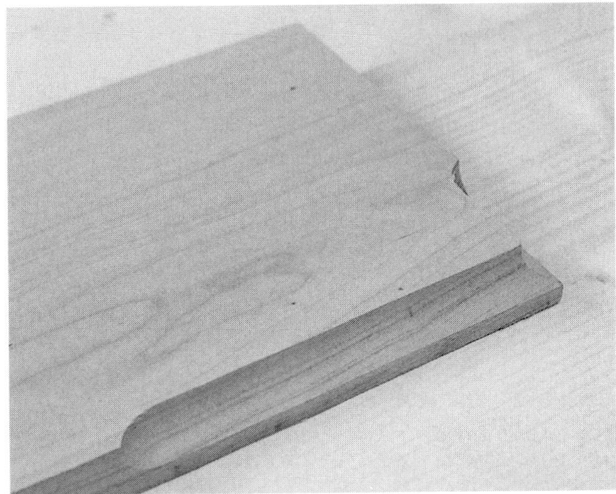

Fig. 28-10—***Two passes were required to cut this deep cove.***

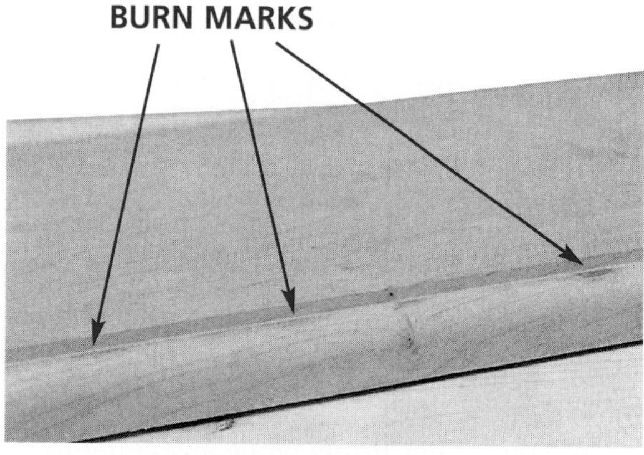

Fig. 28-11—*Feeding the router too slowly into the workpiece results in burn marks.*

- Get to know the "feel" and "sound" of the router as you practice on scrap wood. When you turn on the router, it may sound like it is operating too fast. As the cutting starts, the sound decreases. If you feed the router too fast it will begin to "moan." Also, you may smell burnt wood.

Cutting techniques:

- Since the bit rotates clockwise, cleaner cutting will result if you move the router from left to right as you face the workpiece.
- Move the router counterclockwise when cutting outside edges. Move the router clockwise when routing inside edges. Fig. 28-12.

After the cut:

- Turn "off" the power switch and wait until the machine comes to a complete stop.
- Rest the router on its side after the cut has been completed.

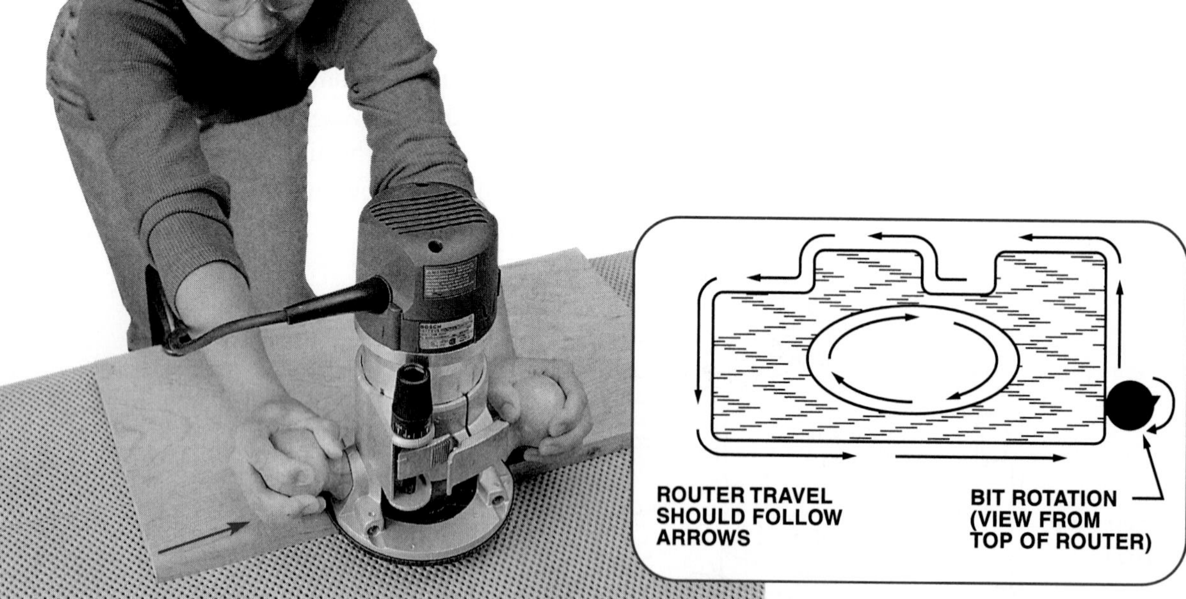

Fig. 28-12—*The correct direction of router travel for routing outside edges and inside edges.*

Freehand Routing

Freehand routing is routing that is done without guides. Fig. 28-13. A good example of this would be routing out a nameplate or numbers for an address. Inside routing is used to make these signs. Sometimes the letters or numbers are cut into the wood by the cutter bit. In other cases, the background around the letters and numbers is routed so the letters or numbers stand out. Use a narrow router bit for inside routing.

To make a sign using inside routing, first carefully trace patterns of the letters or numbers on the workpiece. Place the base of the router at a slight angle to the outline. Then lower the cutter into the wood. Carefully guide the router to form the letter or number. It will take some practice to cut sharp, smooth, even letters or numbers.

Fig. 28-13—*In freehand routing, the router is moved and controlled by the operator.*

Making and Using a T-Square Guide

A homemade T-square is a simple device for guiding the router when making straight cuts on flat workpieces. This device can be made of scrap stock. Use the dimensions shown. Make sure the edges are perfectly smooth and straight. Place the T-square on the surface being routed and clamp in place. Hold the base of the router firmly against the edge of the T-square when making straight cuts.

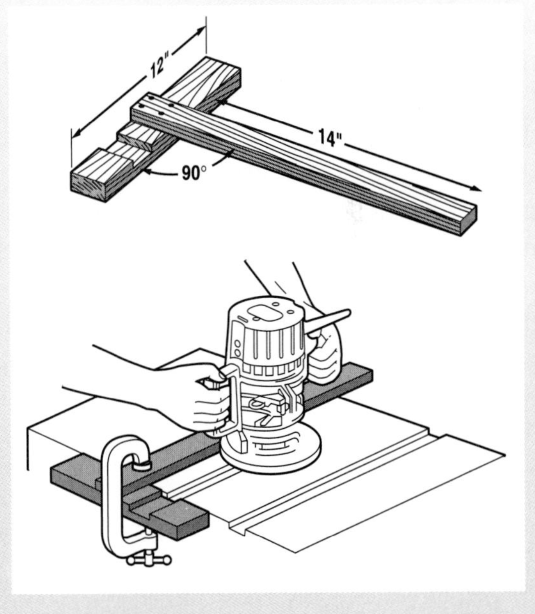

Making Straight and Rabbet Cuts

To make *straight cuts* accurately with a router, use a straightedge cutting guide. Fig. 28-14. For wide cuts, you may need to use more than one. Long guides are best. They provide greater stability and can be used for both long and short cuts. Also, repeated cuts will be more consistent.

A *rabbet* is cut with a straight bit that is adjusted to the required depth of cut. Fig. 28-15. The width of the rabbet is controlled by a guide attached to the base of the router. When making a rabbet cut completely around a piece, it is usually better to make the cuts across the end grain first and then along the grain. Doing this will help prevent chipping at the corners.

Fig. 28-15—***A plunge router being used to cut a rabbet along an edge.***

Cutting Grooves and Dadoes

To cut grooves and dadoes, use a T-square guide or a straightedge guide. Position the guide so that the groove or dado will be in the correct location. Make sure the bit is set to the proper depth. Hold the base firmly on the surface so that it will not rock. For wide dadoes or for making a cross lap joint, several passes may be necessary. Fig. 28-16. To cut a groove in a narrow piece, clamp it between two other pieces to provide support for the router base. Fig. 28-17.

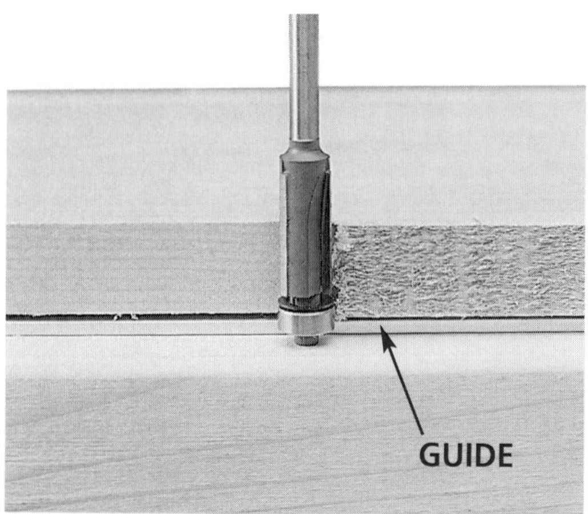

Fig. 28-14—***A straightedge guide. Here it is used when routing a rough edge.***

Fig. 28-16—***Using a guide attachment for cutting a dado.***

Making Dovetail Joints

A *dovetail joint* is a type of mortise-and-tenon joint. The tenon is flared on one side. A jig is available that will help you make attractive dovetails easily, quickly, and accurately using a router. Fig. 28-18. Dovetail joints are often used for making drawers. See Chapter 14.

Fig. 28-17—***Cutting a groove in a narrow piece. Clamping the piece between two boards will provide support for the router base.***

Using a Router Table

Routers mounted in a router table remain stationary and the workpiece is moved. Fig. 28-19 (page 368). Commercial router tables are available, or a table can be made. Depending on the job, a table-mounted router can have several advantages over a portable router. Cuts, particularly long cuts, can be made accurately and with consistency. A table-mounted router is also good for making a number of cuts and for shaping thin workpieces.

Using Templates

A *template* is a pattern used when forming or shaping a workpiece. It may be made of plywood, hardboard, plastic, other durable material. Templates are particularly useful if you need to make several pieces the same shape or if you need to repeat the same type of cut a number of times. Fig. 28-20.

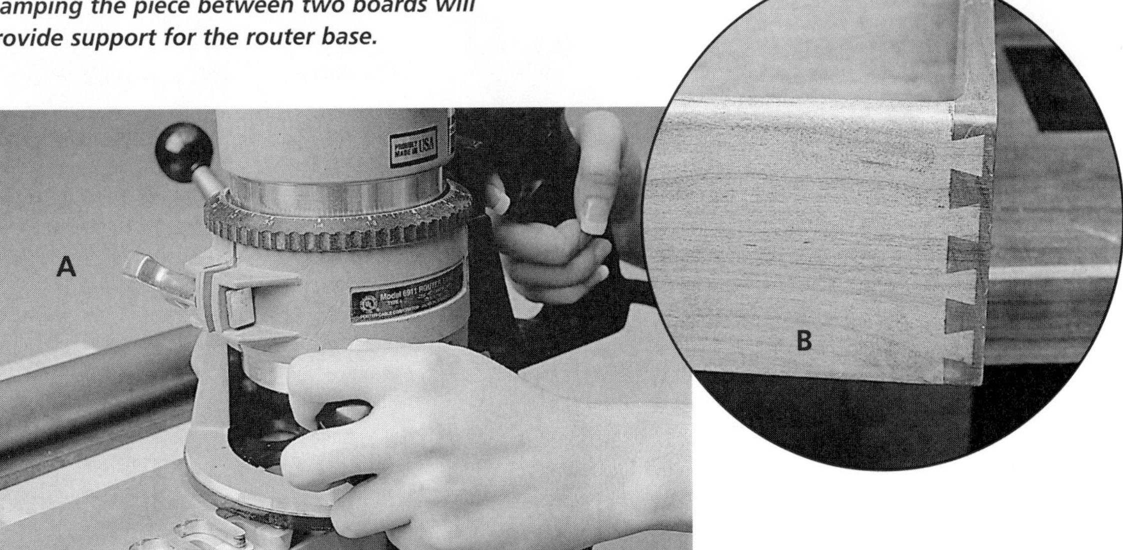

Fig. 28-18—***(A) Making a dovetail joint using a dovetail jig. The end result is well worth the time required to set up the jig. (B) A completed dovetail joint.***

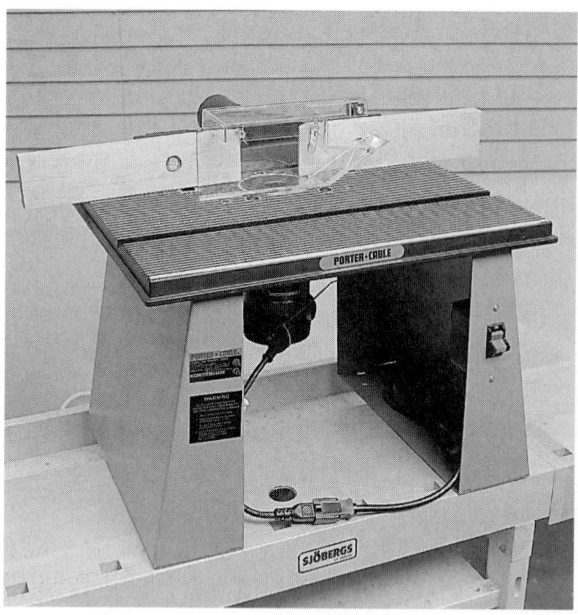

Fig. 28-19—*The router is installed upside-down in a router table. By using the fence, you can make a wide variety of cuts.*

A *template guide* may be used. The guide is a small collar-like device that is slipped over the bit and rested against the router base. It keeps the template from being damaged by the router. The finished work will be the width of the guide larger than the template. Be sure that the template is clamped securely to the workpiece so that it cannot move out of position during the routing process.

Inlaying

Inlaying is the setting of a material into a surface as a decoration. To add a strip of inlay, first cut a groove the desired distance from the edge. Use a *left-hand spiral bit* and a gauge to guide the router. Set the bit for the correct depth. This should equal the thickness of the inlaying material. The surface of the inlay must be even with the surface of the workpiece.

Fig. 28-20—*Templates are excellent for duplicating a specific pattern.*

Carefully cut a groove around the edge of the material. Fig. 28-21. The corners will be rounded and must be trimmed out with a chisel. The groove must be cut to the exact width of the inlay strip. Then cut the strip of inlay material with a *miter corner.* Fit each piece in to check the final design. Then apply glue to the back of the inlay and fasten it in the groove. Fig. 28-22. Place a piece of wood over the inlay. Clamp it until the glue is dry.

Interesting block inlays can be purchased for mounting in tabletops and other projects. Place the inlay over the location. Trace the outline with a sharp pencil. Then mark it with a sharp knife. Set a bit to the correct depth. Rout out the area in which the block inlay is to be placed. Clean out the sharp corners with a knife. Then glue the inlay in place.

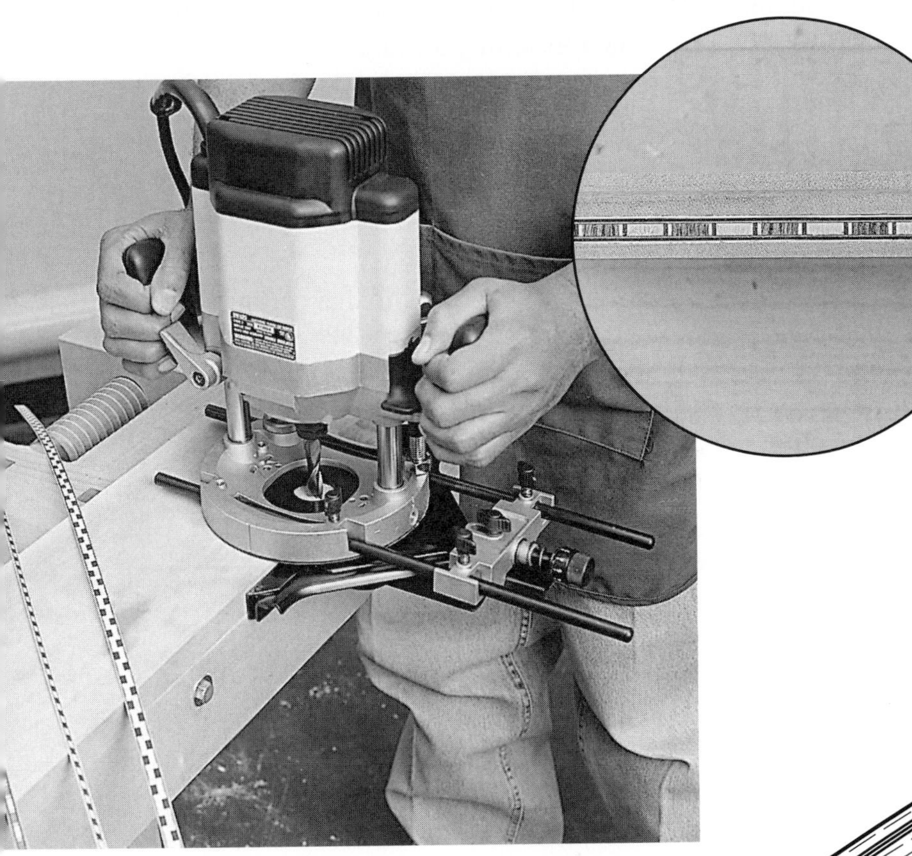

Fig. 28-21—***Using a router equipped with a small diameter bit and a guide to cut a groove for an inlay. The diameter of the bit is equal to the width of the inlay.***

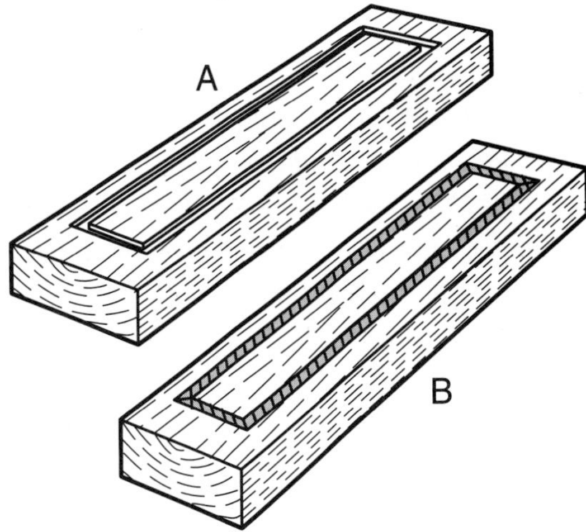

Fig. 28-22—***Inlaying. (A) Groove cut by a router. (B) Inlay glued in place.***

CHAPTER 28

Review & Applications

Chapter Summary

Major points from this chapter that you should remember include:

- A router can do not only cutting operations but joinery, mortising, planing, and shaping processes as well.
- The two types of router are fixed-base routers and plunge routers.
- Before cutting, you should make sure the correct bit has been installed properly, that it moves freely, and that all adjusting nuts and knobs have been tightened.
- During operation, a router must be carefully controlled. It should be operated at a speed and depth appropriate for the process and the material.
- Inlaying is the setting of a material into a surface as a decoration.

Review Questions

1. What safety precautions should you take before cutting with the router?
2. What are the two basic types of router bits?
3. On some routers, how are wrenches used when installing a router bit?
4. Describe the proper way to feed the router bit into the material.
5. In which direction should you move the router when routing outside edges? Inside edges?
6. Why is a long straightedge cutting guide better than a short one?
7. For what tasks are templates useful?
8. How would you install a commercial block inlay in a tabletop?

Solving Real World Problems

New cabinet doors are under construction at the Cole home. Most of the cutting and gluing phase is complete, and it's time to start making the raised panel for the doors. Mr. Cole has a router table, a variable speed router, and a nice collection of router bits. As he installs the 3½″ raised panel cutter into the router, he pauses and considers the router speed. What rpm range should he select for the router when using such a large cutter in it? Should he push the router bit all the way down into the router collet before tightening it? Explain.

CHAPTER 29

Sanders

Look for These Terms

- belt-and-disc sander
- disc sander
- narrow belt sander-grinder
- oscillating spindle sander
- pressure-sensitive adhesive (PSA)
- stationary belt sander
- stationary disc sander

YOU'LL BE ABLE TO:

- Operate both sanders in a combination belt-and-disc sander.
- Set and operate a stationary belt sander correctly, observing all safety rules.
- Change a sanding belt on a stationary belt sander.
- Operate a stationary disc sander correctly, while observing all safety rules.

Power sanders save time and effort by reducing the amount of hand- or finish-sanding needed to complete a project. They can be used to follow up every type of cutting and finishing operation. End grain, face grain, edges, curves, and irregular shapes can be sanded using stationary power sanders.

Several different kinds of power sanders are available. These include belt, disc, edge, and strip sanders. The belt and disc sanders are most common. Combination belt-and-disc sanders are available.

BELT-AND-DISC SANDER

The **belt-and-disc sander** is a useful combination sander. Fig. 29-1 (page 372). It includes both a belt sander and a disk sander in a single machine.

The *belt sander* has a sanding (abrasive) belt that moves around two rollers. One roller is motor-driven. The other turns freely. A hard, flat platen (plate) is mounted between the two rollers very near to the worktable. The belt is placed around the two rollers and against the front of the platen. When the motor runs, the driven roller rotates. This causes the belt to move around both rollers. The workpiece is placed on the worktable and pressed against the moving belt as it passes over the platen.

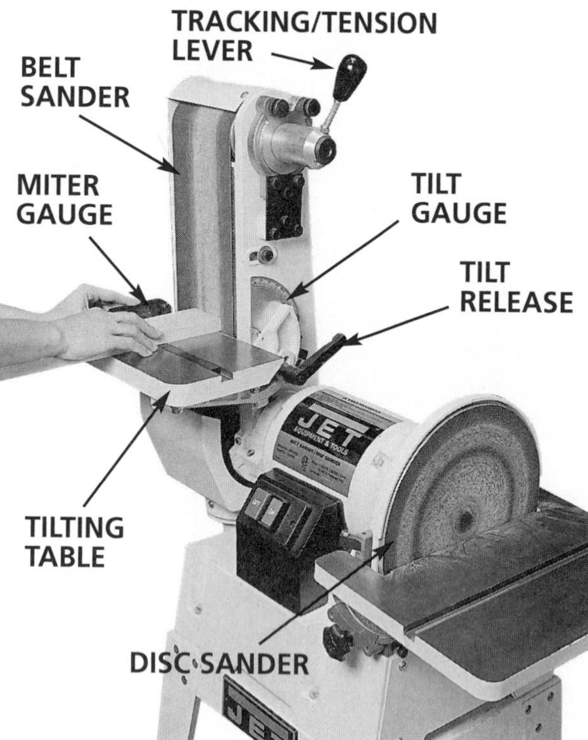

Fig. 29-1—*A combination belt-and-disc sander.*

SCIENCE *Connection*

Heat Buildup

Both belt and disc sanders operate by moving an abrasive surface against the wood surface. The movement of the abrasive against the wood generates friction. In turn, the friction causes heat to build up on the surface of both the sanding belt or disc and the wood. Excessive heat can cause burn marks on the wood. The best way to avoid burn marks is to move the wood constantly in relation to the sander. This prevents the buildup of heat by allowing the wood to cool slightly between passes.

Safety First

►*Using Power Sanders*

- Be certain the sanding belt or disc is correctly mounted. The belt must track in the center of the rollers and platen.
- Check the guards and worktable adjustments to see that they are all securely locked.
- Use the worktable, fence, and other guides to control the position of the work, whenever possible. Position the table 1⁄16 inch from the sanding belt or disc.
- Small or irregularly shaped pieces should be held in a hand clamp, or a special jig or fixture could be made.
- Do not use power sanders to shape parts when the operations could be better performed on other machines.
- Sand only clean, new wood. Do not sand work that has excess glue or finish on the surface. These materials will ruin the abrasive.
- Keep the stock in motion when sanding to prevent burning due to friction.
- Wear a dust mask and eye protection.

The **disc sander** is located adjacent to the belt sander. It is a flat, round platen on which a sanding disc is mounted. The same motor that operates the belt sander also operates the disc sander. During operation, the disc rotates. The workpiece is placed on the worktable and pressed against the downward-moving side of the rotating disc. The downward motion tends to hold the work against the table. The disc sander may have its own table, or the same table may be used for both sanders.

STATIONARY BELT SANDER

The **stationary belt sander** can be used in vertical, horizontal, or slanted positions. It is set in the desired position by loosening the hand lock and moving the entire unit. Most worktables will tilt to any angle between 0 and 45 degrees. A miter gauge can also be used on the machine. With the machine in the horizontal position, a fence can be attached to guide the work for surface sanding.

Changing the Sanding Belt

To change the sanding belt, first remove the guards. Then release the tension by turning the belt-tension knob. Remove the old belt and slip on a new one. Fig. 29-2. Apply a slight amount of tension. Then center the belt on the rollers by adjusting the idler pulley with the tracking handle. Next, increase the tension and replace the guards. Check the centering adjustment again by moving the belt by hand. Readjust when necessary. If the sander is to be used in a tilted position, the centering should be done after this adjustment.

Safety First

►*Using Stationary Belt Sanders*

- When sanding the end grain of narrow pieces on the belt sander, always support the work against the worktable.
- Feed the stock directly against the sanding belt. Never feed it in from the left or right because it may catch and rip or pull the belt off.

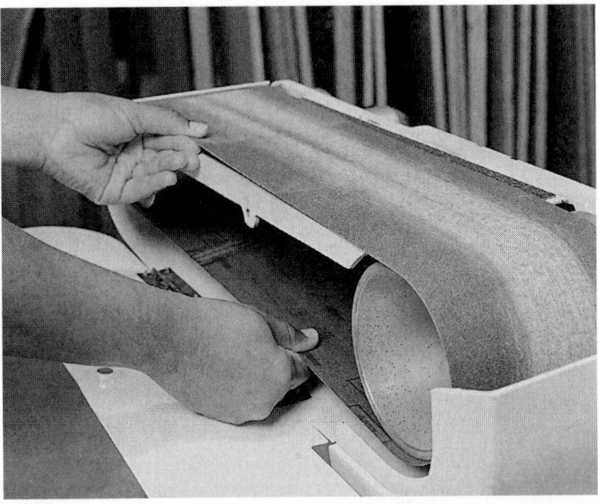

Fig. 29-2—***Changing the sanding belt on a stationary belt sander.***

Using the Stationary Belt Sander

For *surface sanding,* place the sander in a horizontal position. Fig. 29-3. The workpiece can be fed freehand across the sanding belt by steadily applying light pressure. However, for

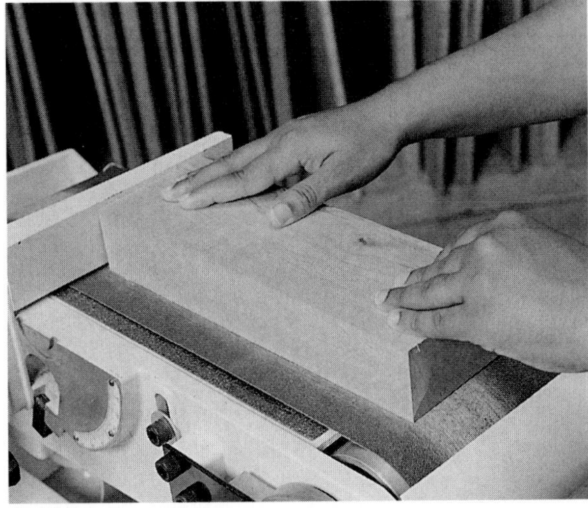

Fig. 29-3—***Surface sanding using the belt sander in a horizontal position.***

more accurate *edge sanding,* use a fence to guide the work. *Beveling* and *angle sanding* can be done by tilting the table and the sander. Fig. 29-4. Curves can be sanded on the open end of the sander when it is in a horizontal position. Fig. 29-5.

For *end-grain sanding,* the sander should be in a vertical position and the worktable should be used as a guide for squareness. Fig. 29-6. Bevels and chamfers can also be sanded in this manner by using a miter gauge as a guide.

Fig. 29-5—***Sanding a curved shape using the end of the belt sander. Note that the belt guard has been removed for this operation.***

STATIONARY DISC SANDER

The **stationary disc sander** is used mostly for sanding end grain and the edges of a workpiece. It is not recommended for sanding the face of a workpiece. It will leave sand marks.

The largest workpiece sanded on a disc sander should be a little smaller than half the diameter of the disc. This is because sanding should only be done on the downward-moving side of the rotating disc.

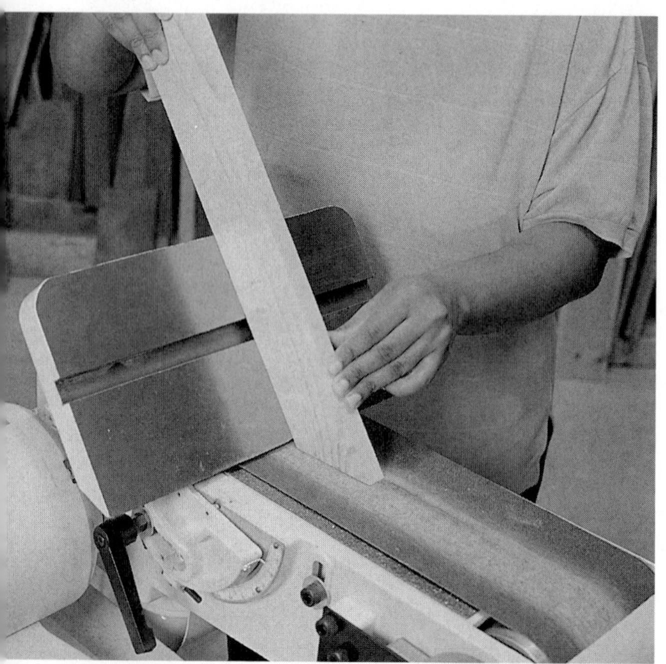

Fig. 29-4—***Sanding an angle with both the belt sander and the worktable tilted to the correct angle.***

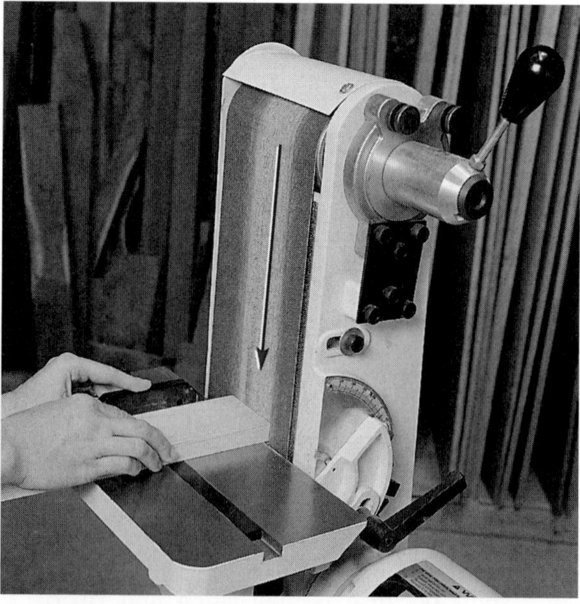

Fig. 29-6—***End-grain sanding using the belt sander in a vertical position. Note how the sanding belt moves downward toward the work, tending to hold it to the worktable.***

Safely Sanding Small Workpieces

The space between a vertical sander and the worktable makes sanding small pieces dangerous. It's difficult to hold on to the pieces and keep them steady and straight. To avoid problems, clamp a 1-inch thick piece of wood to the worktable, positioning one edge across the space. This will provide firm support for the small workpiece as you press it against the moving sanding surface.

Safety First

►*Using Stationary Disc Sanders*

- Do not operate the disc sander if the sanding paper is loose. Make sure the adhesive is holding the sanding disc tightly to the platen.
- Sand only on the side of the disc sander that is moving downward, toward the table. Check the rotation of the disc. Some rotate clockwise, others rotate counterclockwise.

Installing a Sanding Disc

Sanding discs can be purchased already cut to exact size and with a **pressure-sensitive adhesive (PSA)** back. All you have to do is to make sure the metal platen is clean. Then strip the cover paper off the abrasive and press the sanding disc onto the platen.

Using a Stationary Disc Sander

A workpiece can be held on the worktable for end-grain sanding. The table can be used in combination with a miter gauge to sand a chamfer or bevel. Fig. 29-7. It can also be used to sand the edge of a circular piece. Fig. 29-8 (page 376). Move the stock back and forth on the downward-moving side of the rotating disc. Holding it in one position may cause burning due to friction.

Most disc sanding is done freehand. Remember that the edge of the disc is moving much faster than the center. Allow for this. To sand circles or arcs, hold the work firmly on the table, push it against the sanding surface, and rotate it slowly.

Fig. 29-7—*Sanding a chamfer using a miter gauge. Note that the sanding disc is rotating clockwise.*

Fig. 29-8—*Sanding the edge of a circular workpiece using a disc sander.*

Fig. 29-9—*A narrow belt sander-grinder. The worktable can be tilted 10 degrees in and 90 degrees out.*

NARROW BELT SANDER-GRINDER

The **narrow belt sander-grinder** is a tabletop machine. It uses a narrow belt that moves around three or four pulleys. Fig. 29-9. This machine is excellent for sanding small parts and for getting into hard-to-reach places.

OSCILLATING SPINDLE SANDER

The **oscillating spindle sander** is perfect for sanding curved and irregularly shaped edges. Fig. 29-10. Depending on the sanding needs of the workpiece, drums of various sizes can be attached to the spindle of this machine. The drum projects above the worktable. It oscillates vertically (moves up and down) and spins to sand the workpiece. The table can be tilted for sanding at various angles.

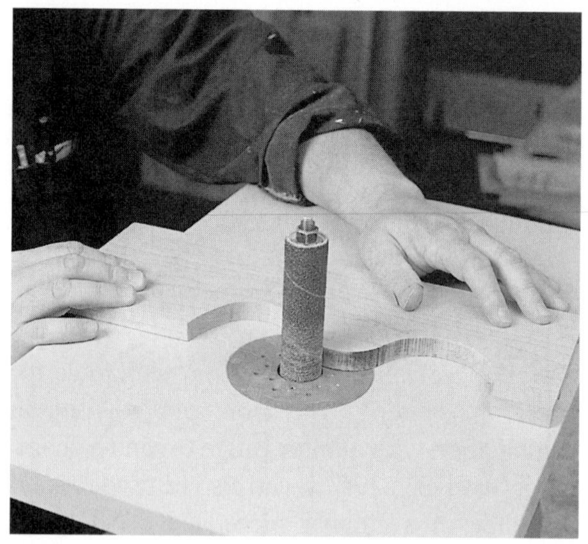

Fig. 29-10—*An oscillating spindle sander is used for sanding curved edges.*

How to Make a Sanding Jig for a Disc Sander

To aid in sanding the edges of a round workpiece, make a jig with a pivot. Use plywood. Refer to the drawing as you follow these steps:

1. Cut a square auxiliary table.
2. Cut a 1-inch groove across the grain in the center of the table.
3. Cut two strips of hardwood. The first should fit into the dado. Be sure it will slide. The second should fit the miter slot of the worktable.
4. On the first strip, install a nail 1 inch from the end that will be toward the sanding disc. The tip of the nail should protrude slightly above the surface. This is the pivot.
5. Place the workpiece (jig) on the worktable so that one end of the dado points toward the downward-moving side of the sanding disc. Position it ⅛ inch away from the sanding disc. Mark the location of the miter slot.
6. Attach the second hardwood strip on the underside of the jig where you have made the mark. When placed in the miter slot, this strip must hold the jig ⅛ inch from the sanding disc. Be sure the strip is square to the sanding disc.
7. Set the jig on the worktable, placing the wood strip in the miter slot. Clamp it in place.
8. Make a small hole in the center of the underside of the workpiece you want to sand.
9. Turn on the sander. Place the workpiece on the pivot and slide the work forward. Turn it slowly but steadily against the sanding disc.
10. Move the jig from time to time as the sanding surface tends to wear.

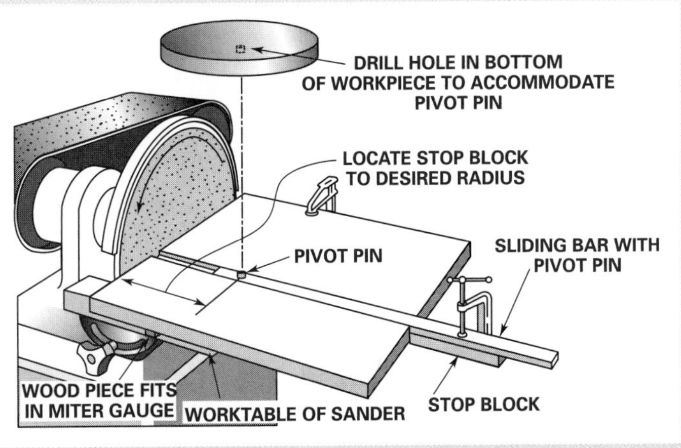

CHAPTER 29

Review & Applications

Chapter Review

Major points in this chapter that you should remember include:

- All parts of a power sander should be correctly positioned and secure before the machine is operated.
- The stationary belt sander can be used in vertical, horizontal, or slanted positions to perform different sanding operations.
- The largest workpiece sanded on a disc sander should be a little smaller than half the diameter of the disc.
- The narrow belt sander-grinder is excellent for sanding small parts.
- Curved and irregularly shaped pieces can be sanded effectively using the oscillating spindle sander.

Review Questions

1. When using a power sander, why is it important to keep stock moving against the sanding surface?
2. In what different positions can a stationary belt sander be used?
3. What is the procedure for changing a sanding belt?
4. How would you do surface sanding on a stationary belt sander?
5. On a disc sander, why is it important to check the disc rotation?
6. Describe how you would sand the edge of a circular workpiece on a disc sander.
7. For what sanding tasks would you be most likely to use a narrow belt sander-grinder?

Solving Real World Problems

Anoki has a random-orbit sander that has a 5″ hook-and-loop pad installed on it. While sanding his newly constructed coffee table, he exhausted his supply of 120-grit sanding discs. He decides that he must go to the store for more discs and for some other grits he has used up. When he places his order at the tool store, he is informed that they are out of 5″ discs but do have 6″ on hand for the same price. Can he use the 6″ discs? Does he have any other options?

CHAPTER 30

Lathe

Look for These Terms

- cup center
- faceplate turning
- gouge
- headstock
- parting tool
- roundnose tool
- skew
- spindle turning
- spur
- tailstock

YOU'LL BE ABLE TO:

- Identify common turning tools and discuss their use.
- Describe the two basic methods of turning.
- Demonstrate both rough turning and finish turning.
- Explain how to cut shoulders, Vs, beads, and coves on the lathe.
- Outline the procedure for faceplate turning.

With a wood lathe, the workpiece is shaped by a process called *turning*—a cutting operation in which the workpiece is revolved against a single-edged tool. While the lathe does have many important industrial uses, you will find that you can use it to make a variety of turned parts in the shop. **Fig. 30-1.**

Fig. 30-1—*These plant stands were turned on a lathe.*

PARTS OF THE LATHE

The wood lathe consists of a bed, a headstock, a tailstock, and a tool rest. The **headstock** is permanently fastened to the bed. The **tailstock** slides and can be locked in any position on the bed. The tool rest supports the turning tool close to the workpiece. It can be moved along the bed and adjusted to various heights and angles. Fig. 30-2.

The headstock has a hollow, tapered spindle that is threaded on both ends. For most turning operations, a **spur,** or *live* (moving) *center,* is fastened to the headstock spindle and a **cup center** is inserted in the tailstock spindle. The stock is then held between the live center on the headstock spindle and the *dead* (stationary) *center* on the tailstock spindle. This is called **spindle turning,** or turning between centers.

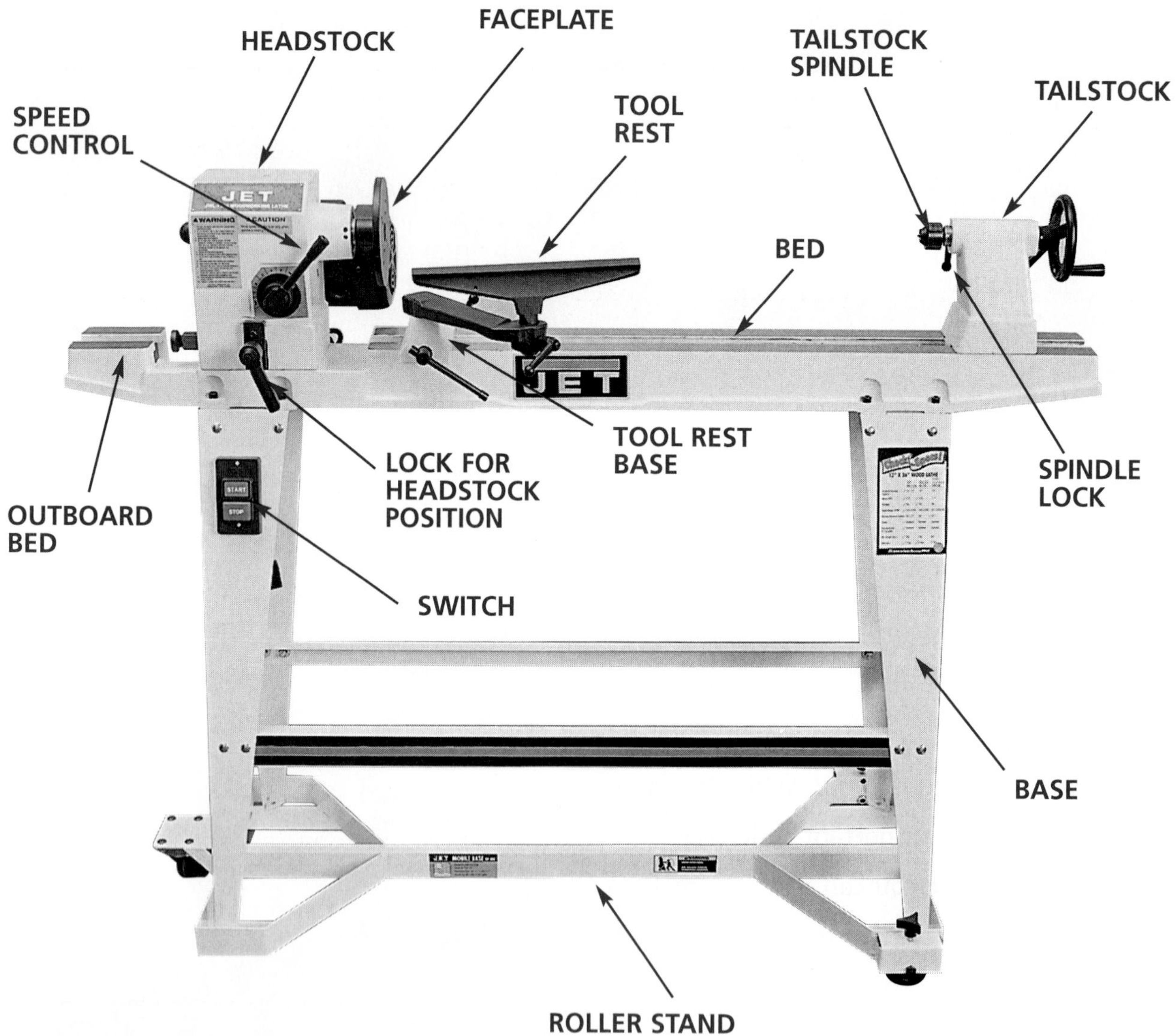

Fig. 30-2—*Parts of a lathe.*

For making bowls or other small circular objects, a metal disk called a *faceplate* may be mounted on the headstock spindle. The stock is then fastened to the faceplate with screws, and the tool rest is turned around so that it is in front of the workpiece. (The tailstock is not used.) This is called **faceplate turning.**

TURNING TOOLS

The common turning tools and their uses are shown in Table 30-A. In addition to these tools, the lathe operator needs a good bench rule, a pencil, dividers, a pair of inside and outside calipers, and a hermaphrodite caliper. Fig. 30-3 (page 382). The outside caliper is used to check the outside diameter of turned work. The inside caliper is used to make inside measurements. The *hermaphrodite caliper* is used for laying out distances from the end of stock and for locating the centers for turning.

Table 30-A. Turning Tools

Tool	Description	Use
Gouge	Curved blade with a rounded cutting edge that is beveled on the convex side	Small Gouge: For small concave cuts Large Gouge: For rough cuts, such as reducing rectangular stock to a cylindrical shape
Skew	Cutting edge is ground at an angle	Small Skew: Squaring ends; cutting shoulders, V-grooves, and beads Large Skew: Smoothing
Roundnose	Flat blade with rounded end that is beveled on one side	Cutting out concave curves
Spearpoint	Flat blade with sharp V-point end that is beveled on one side	Finishing V-grooves and beads
Parting Tool	Has two beveled sides that are angled toward each other to form a narrow cutting edge at the end of the blade	Cutting grooves; squaring off stock; cutting to specific directions

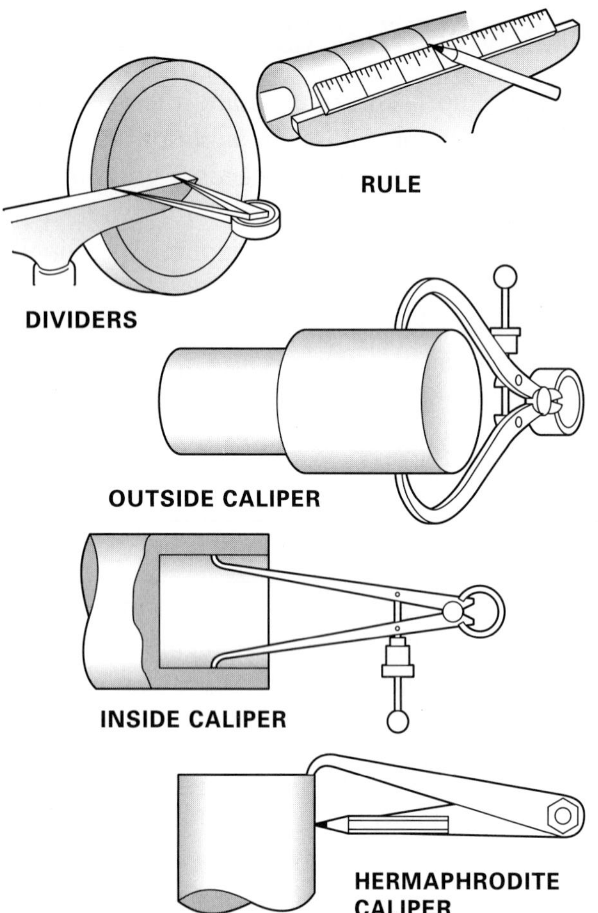

Fig. 30-3—*Some measuring tools needed for different turning operations.*

Loosening Faceplates

Sometimes a thread-on faceplate locks onto the live spindle, making it very difficult to take off the faceplate. One way to ease this problem is to slip a single sheet of waxed paper onto the threads before attaching the faceplate. This waxed paper "washer" makes it a lot easier to loosen and remove the faceplate later.

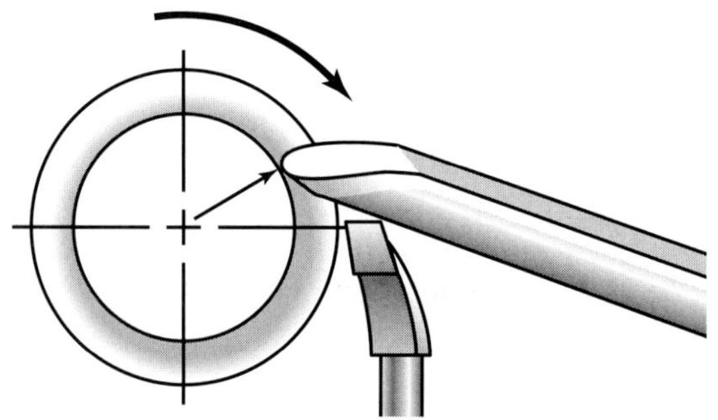

Fig. 30-4—*In cutting, the tool digs into the revolving stock to peel away small shavings.*

METHODS OF TURNING

There are two basic methods of turning wood—cutting and scraping. In *cutting,* the tool actually digs into the revolving stock, creating shavings of wood as it peels away the waste stock layer by layer. Fig. 30-4. It produces a smooth surface that requires very little sanding. This is a faster method, but it requires much skill.

In *scraping,* the tool is held at right angles to the surface of the revolving stock. Instead of peeling away thin shavings, the tool scrapes away fine particles of wood. Fig. 30-5. Scraping is easier to do than cutting and is accurate, but the surface is rougher and requires more sanding.

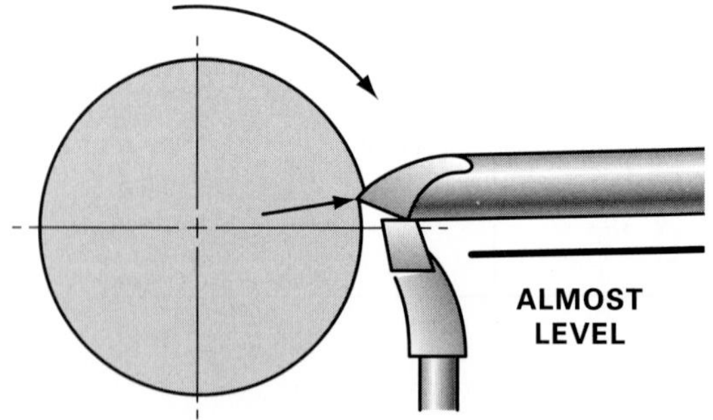

Fig. 30-5—*In scraping, the tool is held at right angles to the stock to wear away fine particles of the wood.*

Safety First

►*Using the Lathe*

- Always wear proper clothing and eye protection for any lathe operation. Remove jewelry. Tie back long hair.
- Make sure the wood is free of checks, knots, or other defects.
- All glued-up work must dry at least 24 hours before being turned on the lathe. If the glue has not set properly, the pieces may fly apart during turning.
- Lubricate the dead center (in the tailstock) with oil or beeswax.
- Securely lock the tailstock before starting the lathe.
- Check for end play (left or right movement) by rotating the stock.
- Set the tool rest as close as possible to the stock being turned, without actually touching the workpiece. If there is too much space between the tool rest and the workpiece, the turning tool may catch and be thrown from your hands.
- After centering rough stock on the lathe, turn the stock a few times by hand to make sure it will clear the tool rest.
- Keep the tool rest in the locked position. Never try to adjust the tool rest while the lathe is running.
- Start all turning operations at the lowest speed until the stock is roughed down to cylindrical (round) form. Then increase to a higher speed. *Do not turn large-diameter stock at high speed.*
- Hold all turning tools firmly with both hands.
- Maintain a firm, well-balanced stance on both feet.
- Do not exceed the recommended speed for the size stock and turning operation you are using. Excessive turning speed can cause pieces to be thrown from the lathe, which could result in serious injury.

SPINDLE TURNING

As stated earlier, spindle turning involves turning stock that is held between the live (moving) center and the dead (stationary) center.

Preparing Stock for Turning

The piece to be turned should be about one inch longer and ⅛ to ¼ inch thicker than the finished piece. Follow the steps below to prepare square stock to be turned to a cylindrical shape.

1. At each end of the stock, draw diagonals from the opposing corners to locate the center of the stock. Fig. 30-6 (page 384). For softwoods, use a scratch awl or dividers to mark the center point at each end. For hardwoods, drill a hole at each center point, then cut ⅛-inch deep saw kerfs along the diagonals you marked.

   Note: If the stock is more than 3 inches square, trim off the edges to form an octagon shape.

Fig. 30-6—*Draw two diagonal lines across opposing corners to locate the center of the stock.*

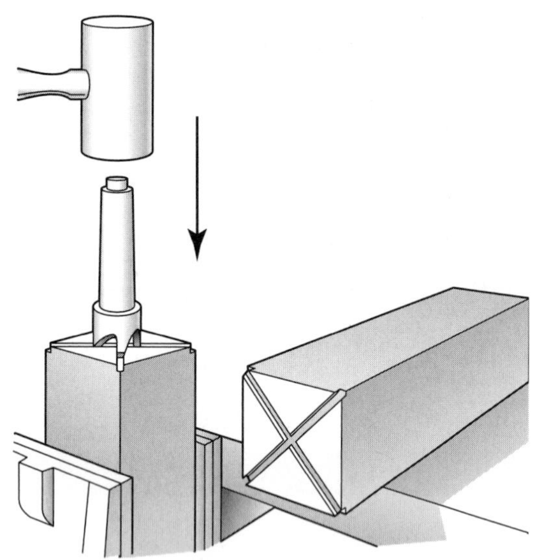

Fig. 30-7—*Use a mallet to drive the spur into the wood.*

2. Choose one end to be the headstock end. Center the spur, or live center, over this end and tap it with a mallet to drive it into the wood. (With hardwood, be sure the spurs enter the saw kerfs.) Fig. 30-7. Place the live center (spur) in the headstock spindle.
3. Bring the tailstock to within about 1½ inches of the other end of the stock. Lock the tailstock to the bed. Then turn the tailstock handle until the cup (dead) center seats firmly in the stock. Back off the tailstock slightly and rub a little oil or wax on the end of the wood to lubricate it. Then retighten the tailstock and lock it in position.
4. Adjust the tool rest so that it clears the stock by about ⅛ inch and is slightly above center. Rotate the stock by hand to see if it has enough clearance.

Setting the Lathe Speed

The speed of the lathe during spindle turning should be adjusted according to the diameter of the workpiece and the operation being performed. See Table 30-B.

Rough Turning

Rough turning is begun using a large **gouge.** (See Table 30-A.) To hold the gouge, grasp the handle firmly in one hand, keeping your forearm close to your body. Use the other hand to hold the blade lightly against the tool rest.

The blade of the gouge can be held in two ways. One way is to grasp the blade close to the cutting point, with the fingers underneath and the thumb over it; the index finger then serves as a stop against the tool rest, as shown in Fig. 30-8. The other way is to place your hand over the tool blade, with your thumb under the blade and your wrist bent at an angle to form the stop. After determining which method feels more comfortable for you, proceed as follows:

1. Refer to Table 30-B for the correct lathe turning speed.
2. Begin cutting about one-third of the way in from the tailstock. Hold the gouge tightly against the tool rest, tilting it down and toward the tailstock so that a shearing cut

Table 30-B. Speeds for Turning

Diameter of Stock	Roughing to Size (rpm)	General Cutting (rpm)	Finishing (rpm)
Under 2″	900 to 1300	2400 to 2800	3000 to 4000
2″ to 4″	600 to 1000	1800 to 2400	2400 to 3000
4″ to 6″	600 to 800	1200 to 1800	1800 to 2400
6″ to 8″	400 to 600	800 to 1200	1200 to 1800
8″ to 10″	300 to 400	600 to 800	900 to 1200
Over 10″	200 to 300	300 to 600	600 to 900

will be taken as you move the cutting edge toward the dead center. *Hold the gouge firmly;* the revolving workpiece could throw the tool out of your hand.

3. After each cut, begin about 2 inches closer to the live center. For the final cut, tip the gouge to the left and work toward the live center.
4. Stop the lathe. Check to see if the stock is nearly round. If wide, flat sides are still visible, further rounding with the gouge will be necessary.
5. Once the workpiece is round, stop the lathe and move the tool rest closer to the work.
6. Set the outside caliper to about 1⁄16 inch more than the finished diameter is supposed to be. Hold the caliper loosely in one hand near one end of the stock. In the other hand, hold a **parting tool** with its narrow edge on the tool rest. (Refer to Table 30-A.) Begin to cut a groove in the stock, holding the caliper directly over the stock in the groove being formed. Fig. 30-9 (page 386). When the correct depth is reached, the caliper will slip over the stock. Make similar cuts about every 2 inches, throughout the length of the stock.
7. The grooves you made in Step 6 are called *sizing cuts.* Once the stock between is cut away to the same depth as the sizing grooves, you know the piece will have the same diameter throughout its length. This can be done by using a square-nose chisel as a scraping tool. Hold the chisel level with the tool rest and at a right angle to the workpiece. Move the chisel back and forth across the workpiece until the grooves made by the parting tool have disappeared and the piece is smooth and of uniform diameter.

Fig. 30-8—*One method of holding a gouge. NOTE: Use the safety shield when actually doing this work.*

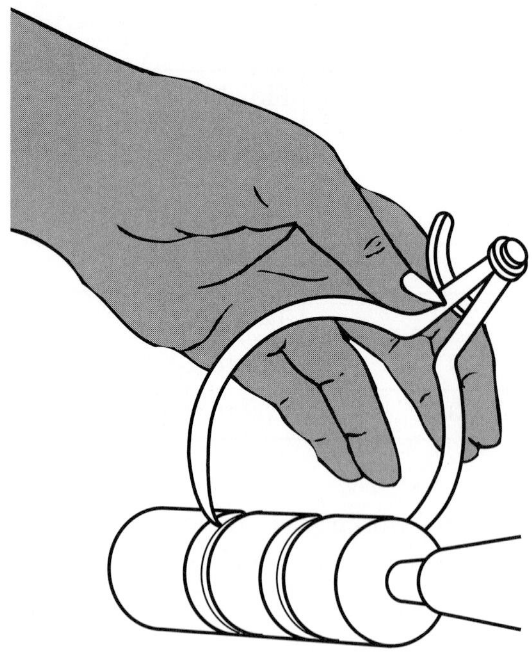

Fig. 30-9—*Hold the caliper in the groove being formed by the parting tool. NOTE: Do not apply pressure to the caliper—the caliper may spring over the revolving stock and cause an accident.*

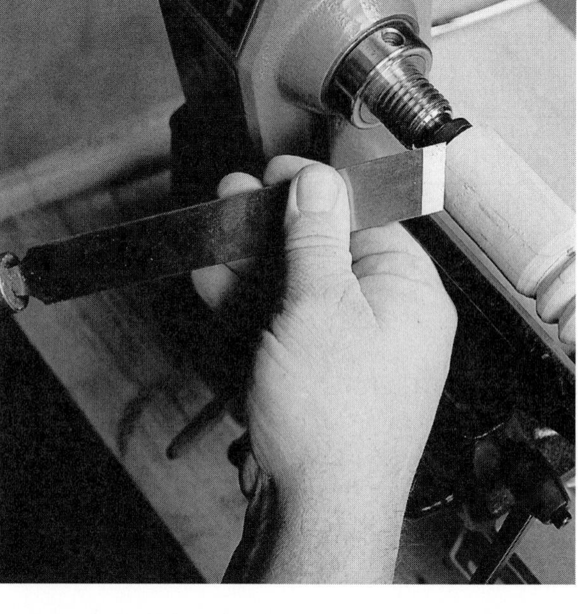

Fig. 30-10—*Using the skew as a cutting tool for finish turning. Note that the skew touches the cylinder about halfway between the toe and the heel of the cutting edge. NOTE: Use the safety shield when actually doing this work.*

Finish Turning

Finish turning on a cylinder-shaped piece is usually done with a **skew,** which has a tapered cutting edge. (Refer to Table 30-A.) The uppermost point of the skew's cutting edge is called the *toe,* and the lower point is called the *heel.*

Finish turning can be done either by the cutting or the scraping method. No matter which method you use, begin at the center of the workpiece and work toward each end to keep the tool from catching and splitting the wood.

After referring to Table 30-B to determine the correct speed, follow the steps below to finish-turn using the cutting method.

1. Place the skew on its side with its cutting edge well above the workpiece. Holding the side of the skew firmly against the tool rest, slowly draw the tool down, turning it at a slight angle until the center of the cutting edge comes in contact with the workpiece. Fig. 30-10.
2. Lift the handle slightly and force the cutting edge into the wood. Make a shearing cut, working from the center of the stock toward the dead center.
3. Reverse the tool and repeat the process, working from the center of the stock toward the live center.

You can also scrape the workpiece to finish size using a larger skew or a flat-nose tool. With the lathe set at a high speed, hold the center of the cutting edge parallel to the workpiece. Fig. 30-11. Force the tool into the stock until the scraping begins. Work from side to side, removing only a small amount of material at a time.

On long cylinders and tapers, you can smooth down the work accurately with a small block plane. Be sure to adjust the plane to make a light cut so that it makes a fine, continuous shaving. Using the tool rest to support the plane, hold the plane at an angle of about

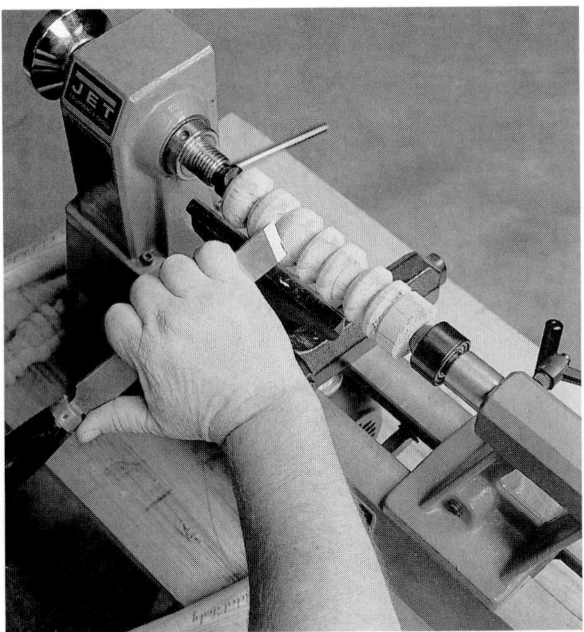

Fig. 30-11—*Using the skew as a scraping tool to finish. NOTE: Use the safety shield when actually doing this work.*

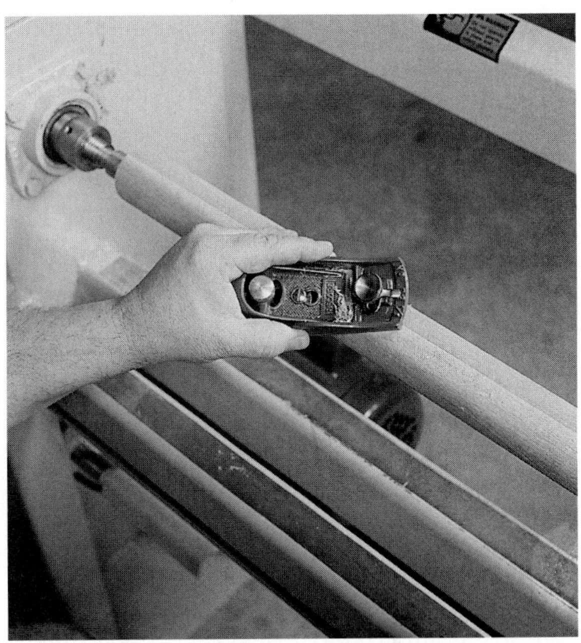

Fig. 30-12—*Using a block plane to smooth a cylinder. NOTE: Use the safety shield when actually doing this work.*

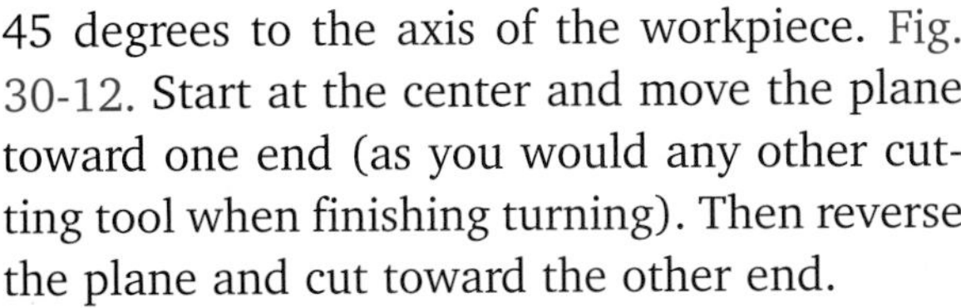

45 degrees to the axis of the workpiece. Fig. 30-12. Start at the center and move the plane toward one end (as you would any other cutting tool when finishing turning). Then reverse the plane and cut toward the other end.

The workpiece can be made even smoother with abrasive paper. Grasp a strip of abrasive paper at both ends, and hold the middle against the workpiece.

Squaring Off the Ends

Square off the ends of the workpiece by following the steps below.

1. Square the tailstock end of the workpiece with a parting tool or a skew. Hold the tool so that its side is flat on the tool rest. Fig. 30-13.
2. From this squared end, measure the proper length of the workpiece. Use a pencil to mark the finish size.

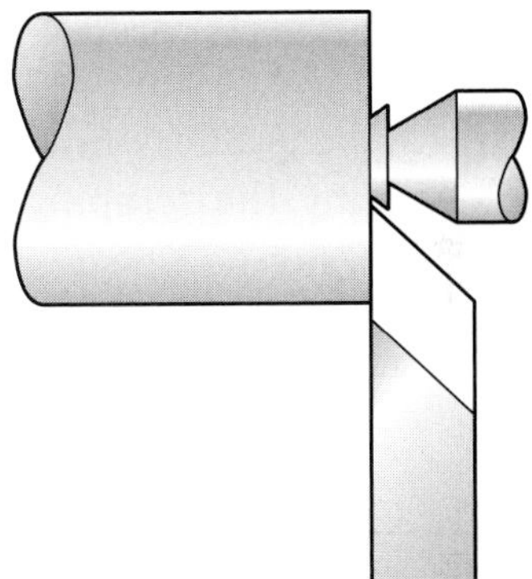

Fig. 30-13—*Hold the skew with its side flat on the tool rest to square the tailstock end of the revolving workpiece.*

3. Place the toe of the skew at the point marked in Step 2. Holding the bevel of the skew parallel to the end being cut, force the toe into the wood. Remove a $\frac{1}{32}$-inch deep shaving with each cut, until only $\frac{1}{4}$ inch of stock remains. What remains will be cut off after the workpiece is removed from the lathe. Fig. 30-14 (page 388).

As an alternative to using a skew to square off the headstock end of the workpiece, you can use a parting tool. Follow Steps 1 and 2 as described. Then force a parting tool into the revolving stock, about ⅛ inch beyond the measured length. Cut down with the parting tool until only ¼ inch of stock remains. It will be cut off later.

Cutting a Shoulder

The procedure for cutting a shoulder is similar to that for squaring off the end.

1. Use a parting tool to cut a groove. Cut down until the diameter is slightly larger than the desired smaller size. Fig. 30-15.
2. Use a small gouge to remove most of the stock from the area of the smaller diameter.
3. Use the toe of the skew to cut the vertical part of the shoulder.
4. Cut the horizontal part of the shoulder with the heel of the skew in a manner similar to finish turning.

You can also cut a shoulder with a skew or a flat-nose tool, using the scraping method. Fig. 30-16.

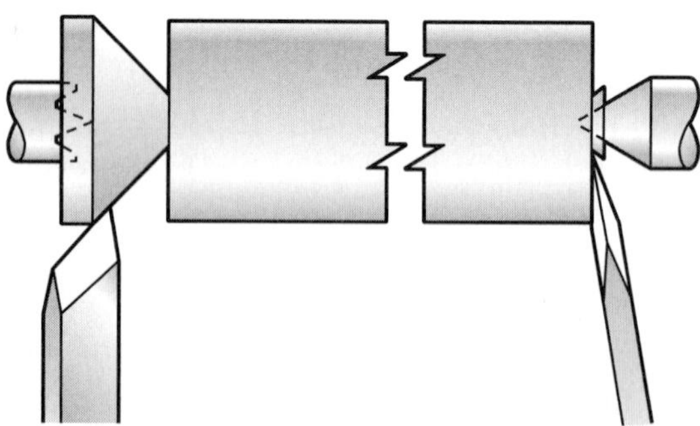

Fig. 30-14—*Squaring the ends of stock with a skew. As the cut grows deeper when you are squaring the headstock end of the workpiece (shown at left), you will need to provide clearance for the skew by making a series of taper cuts to form a half-V. The small remaining piece will be cut off by hand later.*

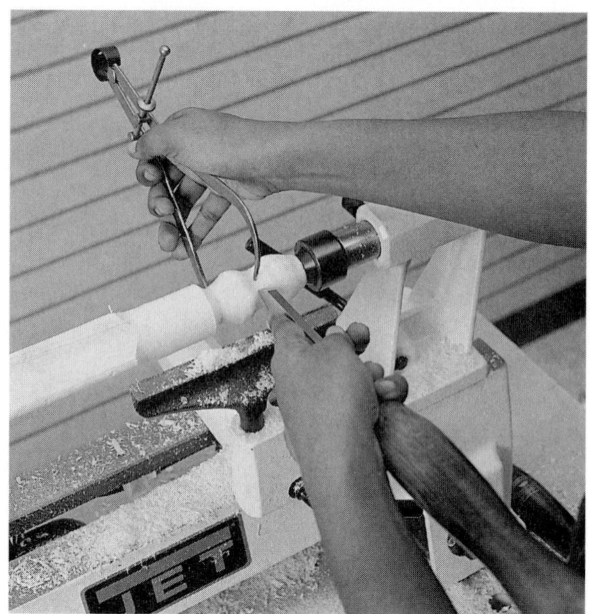

Fig. 30-15—*Use a caliper to check the smaller diameter created by the groove being cut with the parting tool. Stop cutting once the diameter is slightly larger than the size required for the shoulder. NOTE: Use the safety shield when actually doing this work.*

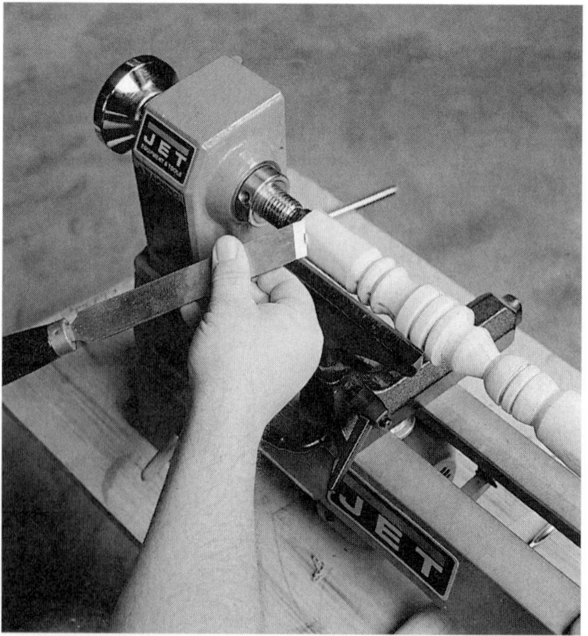

Fig. 30-16—*Using a skew and the scraping method to cut a shoulder. NOTE: Use the safety shield when actually doing this work.*

Cutting a Tapered Surface

Follow the steps below to cut a tapered surface.

1. Turn the stock to the largest diameter.
2. Use a parting tool to cut a groove to mark the smallest diameter. Also cut several grooves to mark intermediate diameters (smaller than the largest diameter but larger than the smallest) to serve as guides for turning the taper.
3. Rough out the taper with a gouge.
4. Finish-turn the taper with a skew, using the heel to do most of the cutting. Fig. 30-17.

Cutting Vs

A small skew can be used to cut Vs. Holding the skew on edge, force the heel of the tool into the point on the stock where you want the center of the V to be. Then work in at an angle, as shown in Fig. 30-18 to cut one side of the V. Continue to the correct depth. Then turn the skew in the opposite direction to finish the V on the other side.

MATHEMATICS *Connection*

Figuring Tapers

Suppose you want to create tapered legs for a table. Each leg is to have a 3-inch diameter at the top and a 1-inch diameter at the bottom. The legs will be 20 inches long. What is the diameter of each leg exactly 10 inches from the bottom of the leg?

Cutting Beads

Cutting accurate beads is rather difficult. The procedure is described below.

1. Mark the position of the bead with lines to indicate both sides of the bead as well as its center. Use the toe of a skew to cut a V-shaped groove along each side guide line.

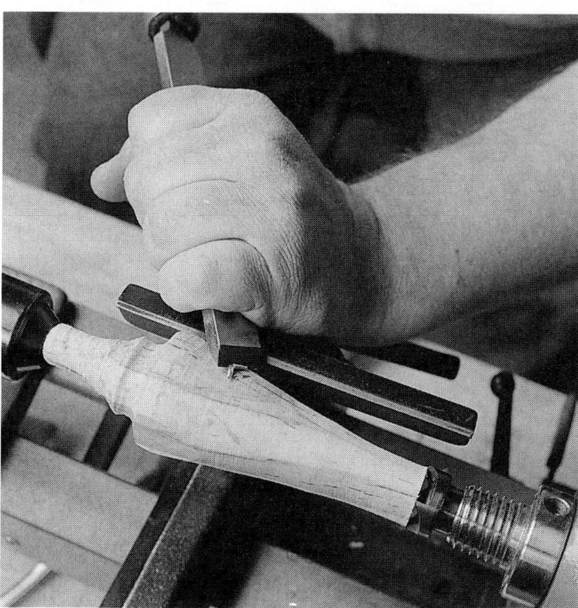

Fig. 30-17—***Using the skew to finish-turn a taper. NOTE: Use the safety shield when actually doing this work.***

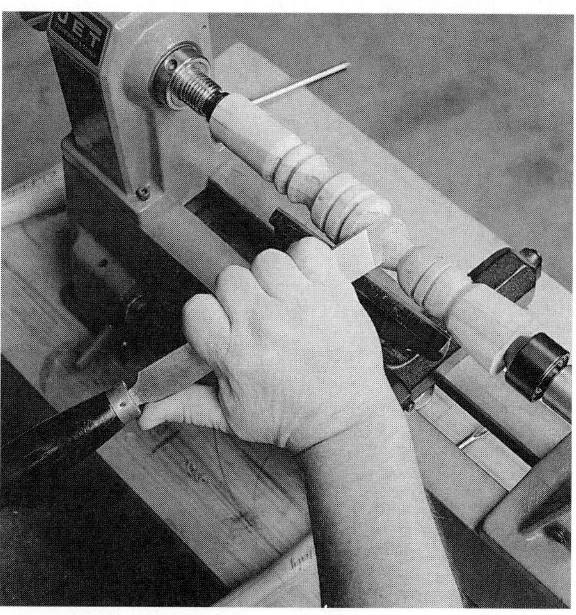

Fig. 30-18—***Cutting Vs with the heel of the skew. NOTE: Use the safety shield when actually doing this work.***

2. First cut the left side of the bead. Begin to cut high on the stock. Hold the skew on its side on the tool rest at the center guide line, with the bevel (*not* the cutting edge) touching the center guide line. Position the heel of the skew (which will do most of the cutting) to the left of and lower than the toe. Fig. 30-19A. Angle the handle slightly downward.
3. To begin the cut, raise and twist the handle slightly, until the heel just shears the wood. Roll the heel of the blade toward the left groove while, at the same time, raising the handle and pushing the blade forward into the wood. End cutting with the skew's cutting edge vertical in the groove, with its heel down. Fig. 30-19B.
4. Cut the right side of the bead in the same manner, but start with the heel pointing to the right.

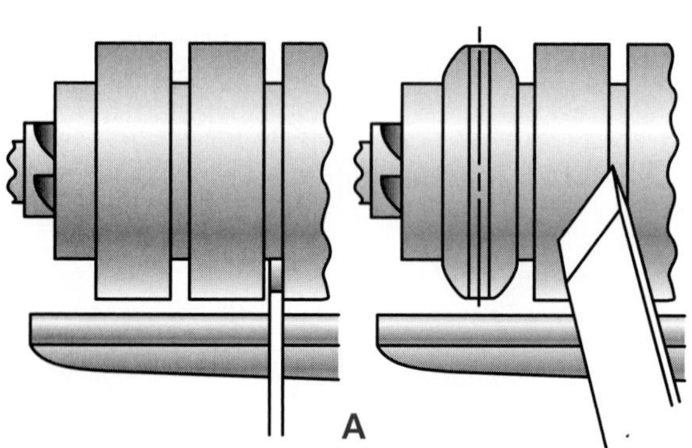

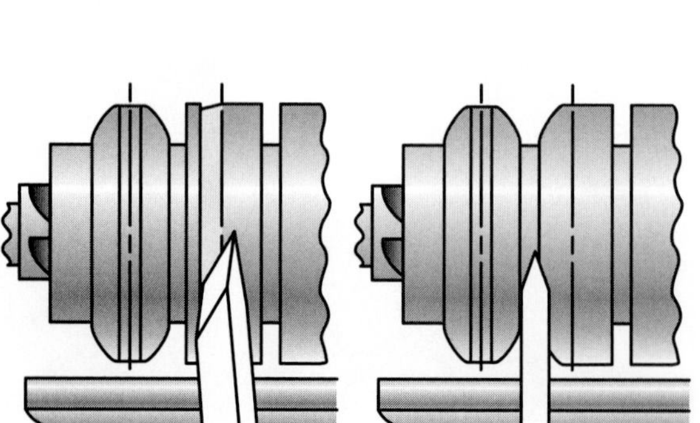

Fig. 30-19—*Cutting a bead.*

Turning Concave Surfaces

Concave surfaces (*coves*), such as those shown in Fig. 30-20, can be turned either by scraping with a **roundnose tool** or by cutting with a small gouge. (Refer to Table 30-A.)

The simplest way to cut a concave surface is to force a roundnose tool into the marked centerpoint of the cove. Using the tool rest as a fulcrum, work the handle back and forth to form the concave surface.

If you use a small gouge, tip it on edge. Begin the cut near what is to be the center of the cove and work toward the sides, rolling the gouge from side to side as pressure is applied. Continue to take shearing cuts, first from one side and then the other, until the concave surface is formed. Fig. 30-21.

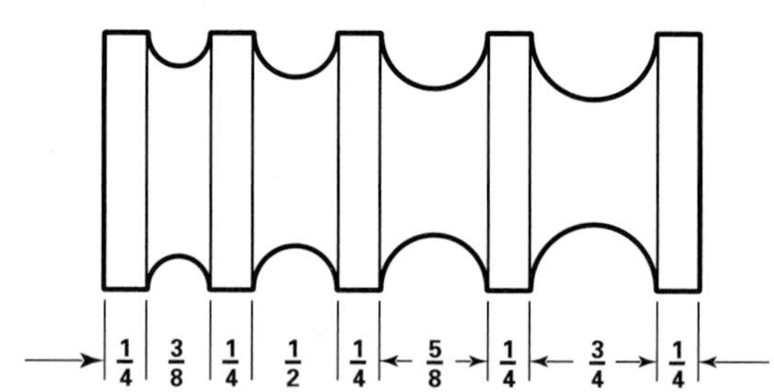

Fig. 30-20—*An example of a cove-cut surface.*

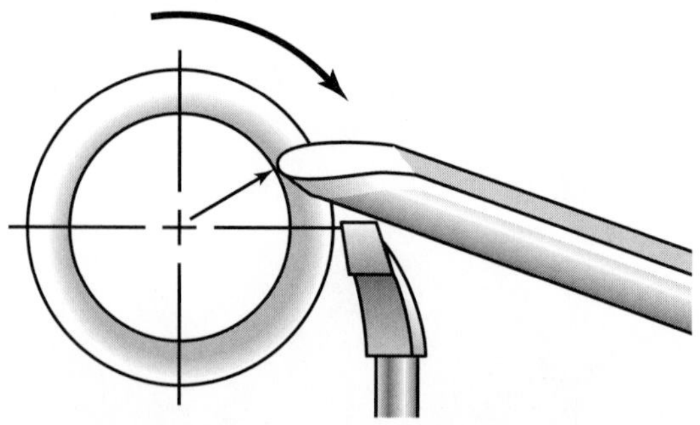

Fig. 30-21—*Holding the gouge on edge, take shearing cuts, first from one side and then the other, to form the concave surface.*

Turning Convex Designs

Most turned pieces are a combination of straight turning, beads, Vs, and long concave or convex surfaces.

For convex work, the usual procedure is to turn the piece to the largest diameter to be finished. Then, with a parting tool, mark points along the stock where extra material is to be removed. In many cases, the parting tool is used at several points to show where stock is to be removed and to what depth. Then the gouge, skew, and roundnose tool are used to form the design.

Duplicator Attachments

Many types of duplicator attachments can be used to make duplicate parts with relative ease. The *template,* or pattern, used to make the duplicate(s) can be either the original turning or a flat template that has been cut on the scroll saw. Fig. 30-22.

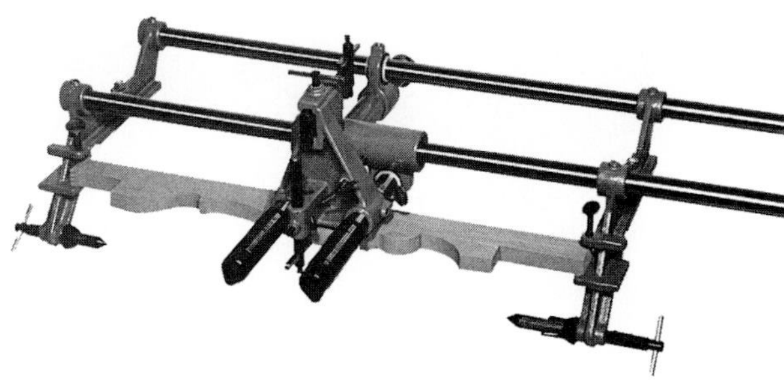

Fig. 30-22—*The flat pattern in the lower part of this duplicator attachment will serve as a guide for making duplicate pieces.*

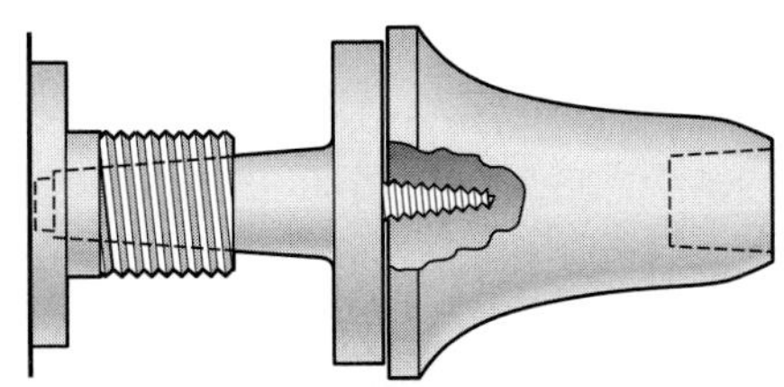

Fig. 30-23—*The screw center, used for small stock, screws directly into the workpiece.*

FACEPLATE TURNING

Bowls, trays, and many other small circular objects can be turned on a faceplate. To do the turning, the stock is fastened to a faceplate, which is fastened to the headstock spindle; the piece is shaped by the scraping method.

The two most common types of faceplates are the *screw center,* which is used on pieces no larger than 4 inches in diameter (Fig. 30-23), and the *standard faceplate,* which is used for larger work. The standard faceplate has holes through which screws are inserted and then fastened into the workpiece. The cutting tools most commonly used for faceplate turning include the roundnose, spear, and gouge.

To make a bowl, first use a bandsaw or scroll saw to cut the stock to a circular shape that is slightly larger than the finish size of the project. If the back of the stock will be damaged greatly by the screws that will hold it to the faceplate, protect it with a piece of scrap stock. Cut a piece of scrap stock at least an inch in thickness and about the same size as the desired finish size for the base of the bowl. Glue the two pieces together with a piece of brown wrapping paper between them so that they will separate easily later. Fig. 30-24. Then follow the steps on page 392.

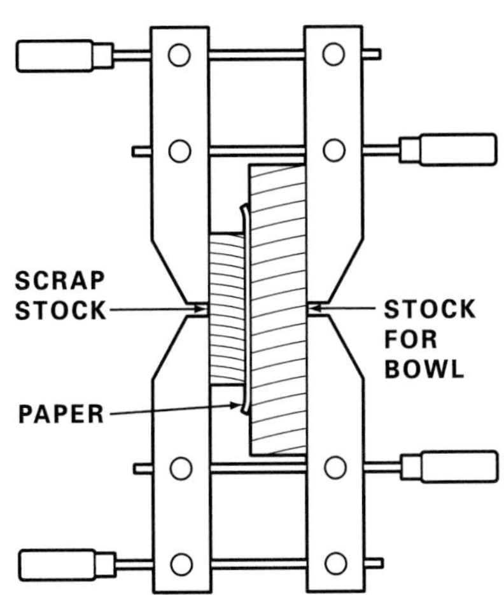

Fig. 30-24—*Gluing up stock in preparation for turning a bowl.*

1. Fasten the stock to the faceplate. Make sure that the screws are not so long that they will mar the bottom of your bowl.
2. Remove the live center and fasten the faceplate on the headstock spindle. Fig. 30-25.
3. Adjust the tool rest so that it is parallel to the outside of the round stock, about ¼ inch away from the stock.
4. Set the lathe to a slow speed and use a roundnose tool to true the outside edge of the stock. Fig. 30-26.

   Note: At this point, you may want to make a cardboard template that will match the interior and exterior shape of the bowl.
5. Turn the tool rest parallel to the face of the bowl and begin to shape the inside with a gouge or a roundnose tool. Fig. 30-27. *Never try to cut across the entire diameter*—once you pass the center, the tool will move up and away from a tool holder.

Fig. 30-26—***Truing the outside edge. The faceplate can be attached to the "outboard" side on some lathes, as shown here. NOTE: Use the safety shield when actually doing this work.***

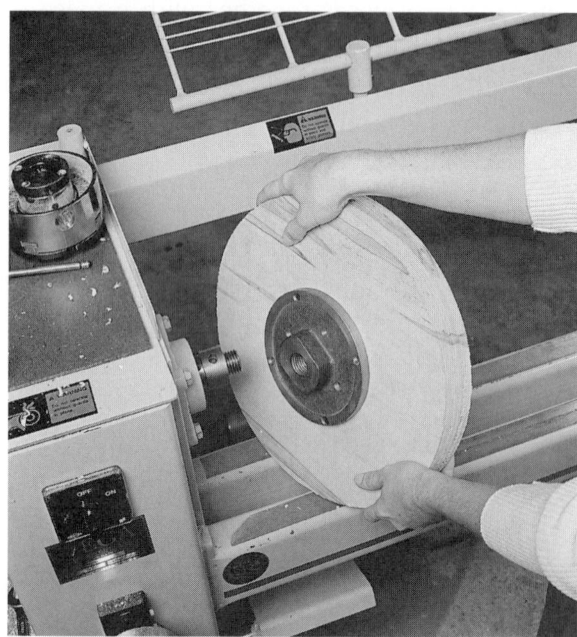

Fig. 30-25—***The stock has been fastened to the faceplate with short wood screws. The faceplate is now being attached to the headstock spindle.***

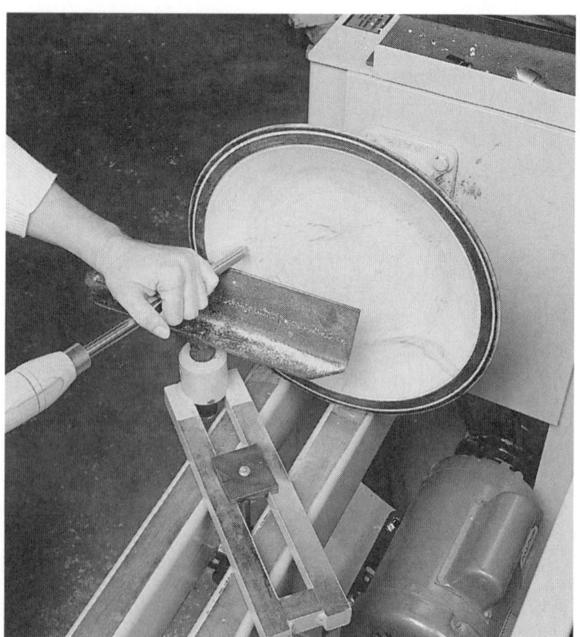

Fig. 30-27—***Shaping the inside of the bowl. NOTE: Use the safety shield when actually doing this work.***

6. Once the inside is shaped, sand it with 2/0 or 1/0 abrasive paper. Apply a finish as described in the next section.
7. Move the tool rest around and shape the bowl's exterior. Then sand this surface and apply a finish.

After completing the turning, cut or split away the scrap stock from the finished product with a sharp chisel. If the base of the workpiece also must be formed, use a recessed wood chuck to hold the stock while shaping the base. Fig. 30-28.

Safety First

►Sanding and Polishing

When sanding or polishing on a lathe, remove the tool rest to prevent accidents.

Applying a Finish

There are several ways to apply a finish as a bowl or similar piece revolves on the lathe. Fig. 30-29. A simple method is to apply paste wax to a folded cloth and hold the cloth against the revolving stock. Repeat this application about a dozen times.

To apply a French-type oil polish, fold cotton or linen cloth into a pad. Apply about one teaspoon of white shellac to the pad. Then add several drops of linseed or mineral oil. Hold the pad over the spinning work, moving it back and forth. As the pad begins to get dry, apply shellac and oil to keep it moist. Apply several coats until the wood has a mirror-like finish.

Still another method of finishing a bowl is to apply several coats of clear lacquer before removing the bowl from the faceplate. Apply a little rubbing oil and rottenstone to the cloth, then polish.

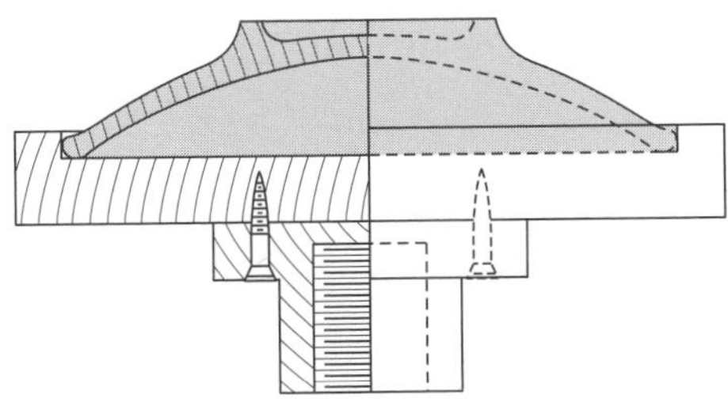

Fig. 30-28—***When it is necessary to turn both the front and back of a piece, use a recessed chuck to hold the stock for turning the back.***

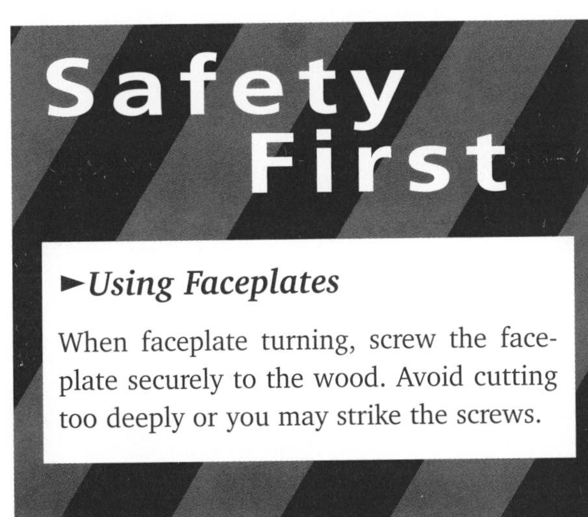

Safety First
►Using Faceplates
When faceplate turning, screw the faceplate securely to the wood. Avoid cutting too deeply or you may strike the screws.

Fig. 30-29—***The finished projects. CAUTION: For projects such as these salad bowls, be sure to use a finish that is safe for human consumption.***

CHAPTER 30

Review & Applications

Chapter Summary

Major points from this chapter that you should remember include:

- Turning is an operation in which the workpiece is shaped by revolving it against a single-edged tool. The workpiece may be revolved between centers or on a faceplate.
- The two basic methods of turning are cutting and scraping.
- There are a number of safety precautions to keep in mind when using the lathe.
- In spindle turning, a workpiece must first be rough turned and then finish turned. Then the ends must be squared off.
- Duplicator attachments for the lathe make it possible to turn duplicate pieces with ease.
- Small bowls, trays, and other circular pieces can be turned on a faceplate. A finish can be applied while revolving the workpiece.

Review Questions

1. Name and describe five common turning tools and tell the types of cuts that can be made by each.
2. Describe the two basic methods of turning wood on a lathe.
3. At what point on the stock should you begin rough turning? After this first cut, where would you begin your next cut?
4. Describe the procedure for finish turning using the scraping method.
5. Explain how to cut a tapered surface.
6. Describe the simplest way to cut a cove on the lathe.
7. Explain how to turn the interior of a bowl.

Solving Real World Problems

Mr. Haley is making a new bed for his son. His son wants a four-poster bed with 4′ posts, and he has given his father a picture of the bed he would like to have. Mr. Haley's lathe has a maximum stock capacity of 32″. How can he make the posts needed for the bed?

CHAPTER 31

Preparing for Finishing

Look for These Terms

- bleaching
- filler
- knots
- linseed oil
- mill marks
- pumice
- rottenstone
- solvent
- steel wool
- tack rag

YOU'LL BE ABLE TO:

- Choose the proper finish for a project.
- Correct common defects found in wood projects.
- Describe the use of various finishing supplies.
- Choose and care for brushes.
- Outline the basic steps in applying a fine finish.

There are only two reasons for finishing wood—to protect it and to add to its beauty. The beauty of any fine wood project can be greatly enhanced by a proper finish that has been correctly applied. **Fig. 31-1.**

Fig. 31-1—*Much of the beauty of a project lies in its fine finish.*

The most important rule to remember when preparing the project for finishing as well as throughout the finishing process is "Don't get in a hurry." You put a lot of time and work into constructing your project, so take the time needed to finish it properly—even if it takes as long as it did to build it in the first place.

TYPES OF FINISHES

All wood finishes can be grouped into three general types:

- *Transparent coatings* are coatings that can be seen through. Examples include shellac, varnish, and lacquer.
- *Penetrating finishes* actually soak into the wood rather than just coat it. They may be applied plain or mixed with color. Examples include linseed oil, Danish oil, and tung oil.
- *Opaque finishes* cover the surface completely. They cannot be seen through. Examples include paint and enamel.

The project's design, the type of wood used, and the project's intended use are all important considerations when choosing a finish. Do you want to bring out the grain, enhance it, or cover it up? Do you want the finish to be glossy, soft and satiny, or bright and colorful? Do you want the piece to blend in well with or complement surrounding furnishings, or perhaps to offer a sharp contrast? Will the piece be exposed to heat, moisture, spills, or heavy use? Ease of application and cleanup, as well as drying time, are also considerations when choosing a finish.

SCIENCE *Connection*

Environment-Friendly Finishes

Consider the environment when choosing a finish. In response to stricter air emission standards established by the Environmental Protection Agency (EPA), manufacturers of finishing materials have developed a complete line of water-based products, including wood putty, stains, filler, and clear topcoat finishes. Standard finishes are made with solvents that emit a great deal of pollutants into the air. The water-based products substitute water for a large portion of these polluting solvents. They contain about 50 to 60 percent less solvent than standard finishes. In addition, soap and water rather than solvents are used for post-finishing clean-up when water-based finishes are used.

PREPARING THE SURFACE

Choosing what finish to use is important, but that is really the easiest part. Preparing the surface correctly is crucial to achieving a fine finish.

After assembling the project, check it carefully to make sure it is smooth and free from imperfections. A finish never covers up dried glue, ridges, dents, chips, gouges, or scratches. Instead, these flaws and defects will show more clearly after the finish is applied.

Removing Excess Glue

One of the worst contributors to a bad finish is glue that has squeezed out. Glue will not take stain. Sometimes the glue doesn't even show up until after you have stained the wood or applied the final finish, so you need to make sure that every fleck of glue has been removed.

Wipe away as much glue as possible while it is still wet. Glue that has squeezed out around joints (or been accidentally spilled and gone unnoticed) can be removed with a putty knife, scraper, or chisel once it has thickened. Fig. 31-2. Glue that has hardened around a joint can usually be removed with a sharp chisel. However, hardened glue is difficult to remove, and you risk damaging the wood during the removal process.

If casein glue has been used, you may need to apply bleach to the glue spots. Certain kinds of casein glue will darken the wood, especially porous woods such as oak and mahogany.

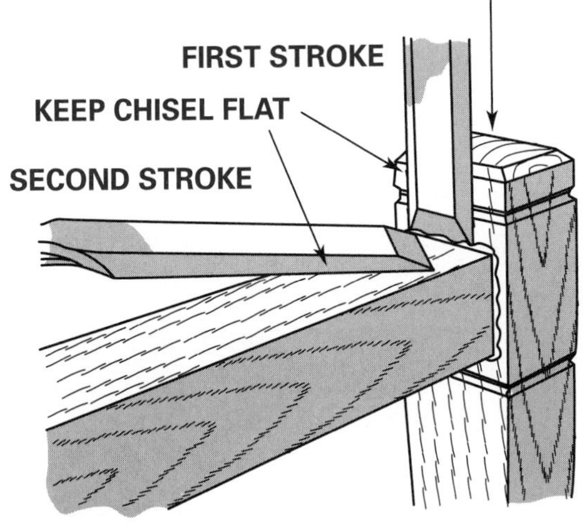

Fig. 31-2—*If the glue is on a flat surface, use a putty knife or scraper.*

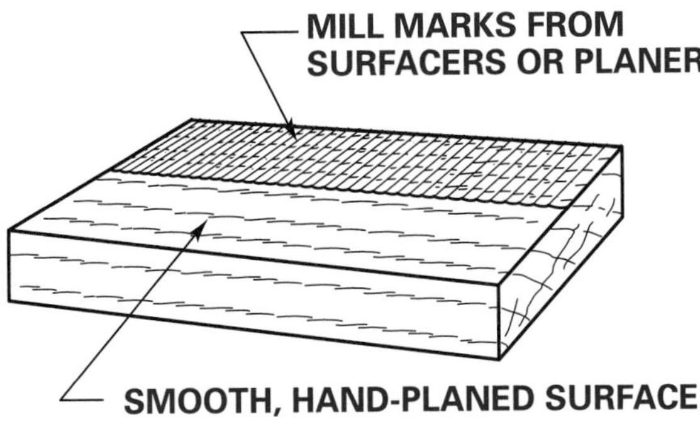

Fig. 31-3—*Uniform mill are marks caused by the revolving cutterhead of the planer.*

Removing Mill Marks

Mill marks are uniform ridges that run across the width of planed lumber. They are caused by the revolving cutterhead of the planer (or jointer). Fig. 31-3.

Mill marks are hard to see on unfinished wood, but they will be visible after the finish is applied so they must be removed. Hold the bare wood at a very low angle to your eyes so that you can see the marks. Then make several light passes with a sharp plane or scrape the surface to remove the ridges.

Correcting Defects

The three most common defects in wood are shallow *dents* (the wood cells have been crushed), *chips* (wood is missing), and *gouges* or *scratches* (the wood cells are severed). Look at the wood surfaces from different angles and under very strong light to find these defects. If it appears there is a defect, highlight it by wiping the area with a cloth dipped in grain or denatured alcohol.

A shallow dent can often be raised by using steam to swell the wood cells. Place a wet cloth over the dent. Then press with a household iron set on "High." Steam will penetrate the dent and cause it to swell and rise. Fig. 31-4 (page 398). Be sure to keep the iron moving so that the wood is not scorched. Allow the surface to dry thoroughly before sanding.

For deep defects, such as chips and gouges, first clean the area to remove loose or crushed wood fibers. To repair a chip, try to find the splinter and reglue it in place. If the splinter cannot be found, or if the defect is a gouge mark or a deep dent that cannot be raised, the damaged area will need to be filled.

A variety of commercial fillers are available to repair damaged wood. One example is a *wood patch*, which is made from wood fibers and takes stain easily. Many repair fillers are available in natural as well as in oak, cherry, walnut, and other colors. Follow package directions for proper use and safety.

Another type of defect in boards is a natural defect—knots. **Knots** are the hard lumps that are formed at the point that branches begin to extend out from the trunk and limbs of the tree. When the tree is sawn into boards, we have cross sections of these knots in the boards. These cross sections of tree knots are also referred to as *knots* in describing boards. Knots must be covered with a knot sealer such as white shellac before any finish, including paint, can be applied. Fig. 31-5. The sealer keeps the resin in the knot from bleeding into the finish.

Final Sanding

It is especially important to sand the wood as smooth as possible before applying any finish. Open-coat aluminum oxide paper is generally preferred for sanding furniture.

Sand with progressively finer grits, starting with medium or fine (100 to 180) grit. (Stay away from coarser grits; anytime you use grits of 80 and below, you are visibly scratching the surface.) Finish off with a very fine (220 to 280) grit.

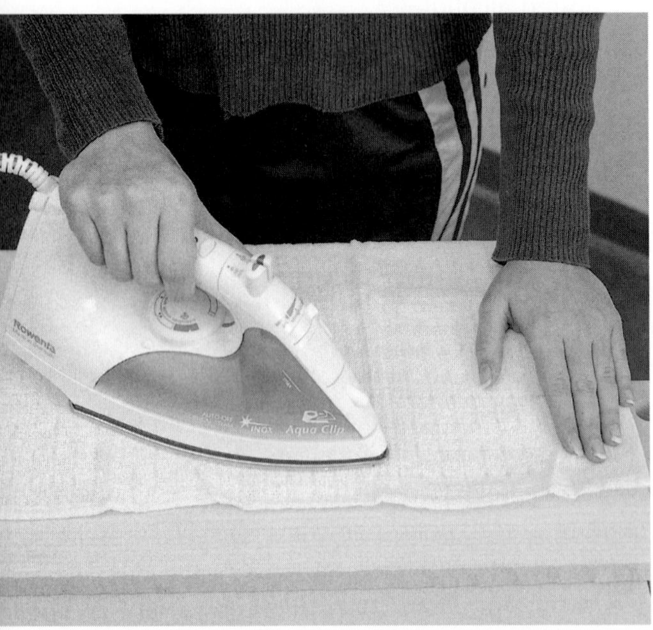

Fig. 31-4—*To remove a shallow dent, place a damp cloth over the dent. Then press the dent with a hot iron to force steam into the dent, causing it to swell and rise.*

Homemade Repair Filler

You can make your own filler material to repair damaged wood or to fill nail and screw holes. Simply mix equal parts of wood sanding dust and powdered glue. Then add enough water to form a thick paste. Apply enough filler to fill the defect until it is a little higher than the surrounding area. Allow the repair to dry and harden thoroughly. Then sand until smooth and level.

Fig. 31-5—*Be sure to seal knots by applying a coat of white shellac.*

When sanding, remember to rub *with* the grain. Sanding across the grain will cause scratches that will show through the finish. On flat surfaces, use a sanding block to prevent unevenness and round edges. For curved or irregular surfaces, tear the paper into strips and use it without a block. Pads of medium to fine steel wool are excellent for smoothing difficult areas. Fig. 31-6.

After sanding, remove fine particles and dust by wiping the wood with a **tack rag.** This is a cloth that has been treated to attract and hold dust. Fig. 31-7.

Fig. 31-6—***Steel wool can be used to smooth carved surfaces and other irregular areas before finishing.***

Fig. 31-7—***A tack rag, several grades of sandpaper, and a sanding block are needed for final sanding.***

FINISHING SUPPLIES

A number of different tools and materials—other than the finish itself, of course—are used during the finishing process. These include abrasives, brushes, solvents, and waxes and oils.

Abrasives

As stated earlier, before finishing is begun, final sanding is done with aluminum oxide paper of progressively finer grits, beginning with medium to fine (100 to 180) grit and ending with very fine (220 to 280). Other abrasive materials are used during the actual finishing process.

Number 4/0 (150) garnet or aluminum oxide finishing paper is used for sanding after staining, after applying the first coat of shellac, and before applying the filler coat. Number 6/0 (220) garnet or aluminum oxide is used for final smoothing after coats of shellac or other finish. These materials may be used dry or with oil.

Making a Tack Rag

Tack rags can be purchased, but it is easy to make your own. Sprinkle a piece of lint-free cotton cloth with varnish that has been diluted with about 25 percent turpentine. Fold the moistened cloth tightly, then wring it out until it is almost dry. Store the tack rag in a fireproof container when not in use.

Waterproof (wet-or-dry) abrasive papers from 240 to 400 grit are used with water for hand sanding between lacquer coats or for rubbing enamel or lacquer.

Steel wool is used in place of sandpaper in some finishing operations. Natural steel wool, made of thin metal shavings, is used if oil-based stains and/or finishes are going to be applied. Synthetic steel wool is used if water-based stain and/or finishes are going to be applied. Both natural and synthetic steel wool can be purchased in grits ranging from 0000 (extra fine) to 3 (coarse).

Pumice is a white powder made from volcanic material. It is combined with water or oil to rub down the finish. Pumice is available in several grades. The most common grades used for wood finishing are FF and FFF.

Rottenstone is a reddish-brown or grayish-black iron oxide material that comes from shale. Much finer than pumice, it is used with water or oil to produce a smoother finish after the surface has been rubbed with pumice.

Brushes

Both bristle brushes and foam polybrushes are used for finishing. *Natural bristle* brushes, made with hog hair, were once considered the only type of brush to use with oil-base paints, varnishes, and lacquers. However, today's *synthetic bristle* brushes, made of nylon or polyester, are generally considered fine to use with water-soluble finishes as well as with most oil-base finishes. These brushes are usually less expensive than natural bristle brushes. The parts of bristle brushes are shown in Fig. 31-8.

Foam polybrushes, made of plastic foam tapered on both sides to a sharp point, can be used to apply most finishes. Fig. 31-9. These brushes cost so little that they can be discarded after use.

Choose your brush carefully. Brush quality determines ease of application as well as the quality of the finished job. A good brush holds more finishing material, controls dripping and spattering, and applies the finish more smoothly to minimize brush marks. To assure that you are buying a quality brush, check the following factors:

- *Flagged bristles* have split ends that help load the brush with more finish, while allowing the finish to flow more smoothly. Cheaper brushes will have less flagging or none at all. Fig. 31-10.

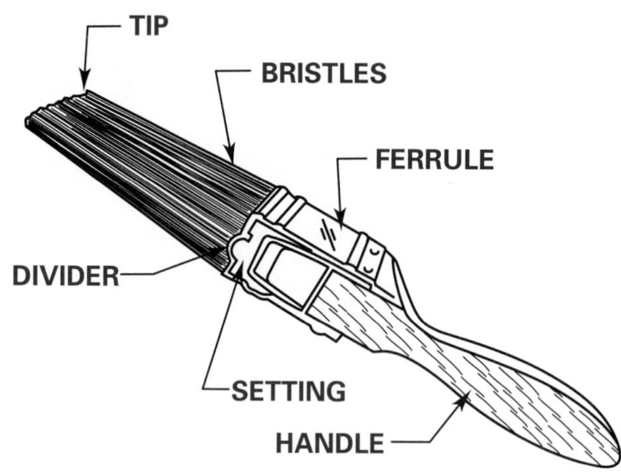

Fig. 31-8—*Parts of a bristle brush.*

Fig. 31-9—*Inexpensive foam polybrushes do a nice job with most finishes.*

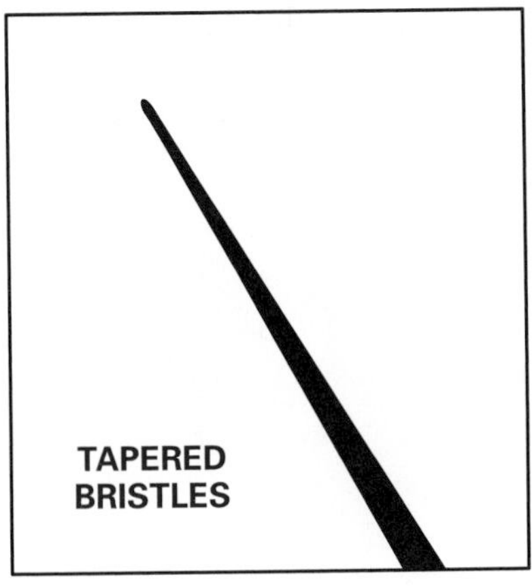

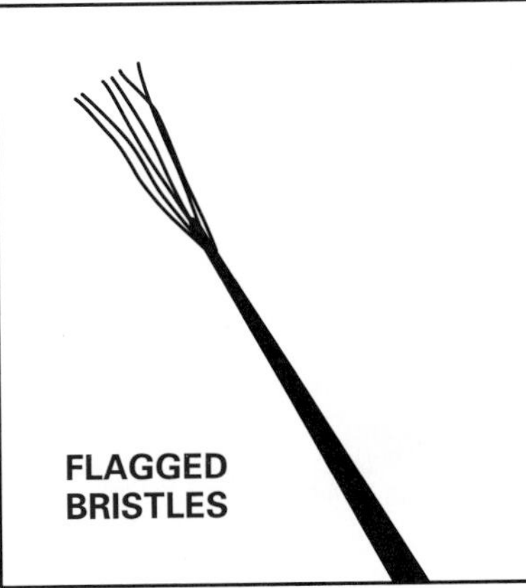

Fig. 31-10—*Tapered bristles are thicker at the base than at the tip. Flagged bristles have split ends.*

- *Tapered bristles* help the finish release easily and flow onto the surface smoothly. Check that the base of each bristle is thicker than the tip. (Refer again to Fig. 31-10.) This helps give the brush tip a fine edge for more even and accurate work.
- *Fullness.* As you press the bristles against your hand, they should feel full and springy. If the divider in the brush setting is too large, the bristles will feel skimpy and limp and there will be a large hollow space in the center of the brush.
- *Bristle length* should vary. As you run your hand over the bristles, some shorter ones should pop up first. This indicates a variety of bristle lengths for better loading and smoother release of the finish.
- A *strong setting* is important for bristle retention and maximum brush life. Bristles should be firmly bonded into the setting with epoxy glue. Nails should be used only to hold the ferrule to the handle.

Size and shape are also important factors when choosing a brush. An angular-cut brush helps do clean, neat work on narrow trim. Choice of brush width is determined by the amount of open or flat area to be finished. Table 31-A can be used as a guide for choosing a brush of the right size.

Table 31-A. Choosing Brush Size

Brush Width	Application
1″ to 1½″	Touch-ups and little jobs, such as toys, furniture legs, and hard-to-reach corners
2″ to 3″	Trim work, such as sashes, frames, molding, or other flat surfaces
3½″ to 4″	Large, flat surfaces, such as floors, walls, or ceilings

Using and Caring for Brushes

The following are some general suggestions for using and caring for brushes:

- Dislodge any loose bristles from a new brush by twirling the brush handle between your palms as shown in Fig. 31-11 (page 402).
- Dip only about one-third the bristle length into the finishing material. To remove any excess finish, gently tap the brush against the side of the inside rim of the can. Fig. 31-12. Never scrape the brush against the rim.
- Always hold the brush at a slight angle to the work surface.
- Never apply finish with the side of the brush.

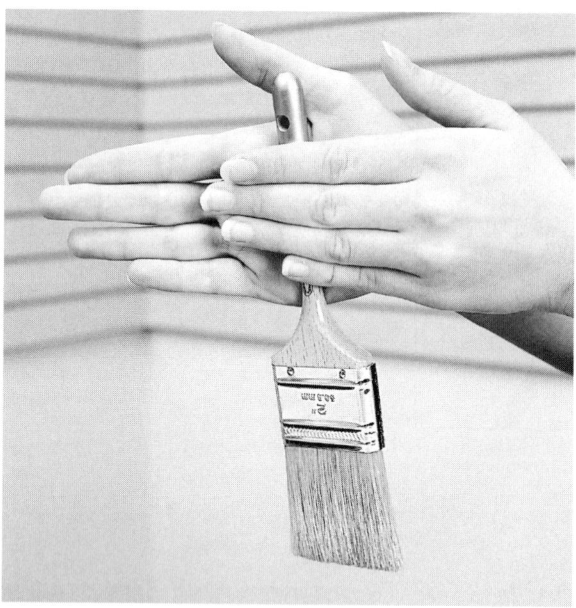

Fig. 31-11—*Twirl the brush between your palms to get rid of any loose bristles.*

Fig. 31-12—*Dip only about one-third of the bristle length into the finish. Then gently tap off any excess along the rim of the can. Never rub the bristles along the edge of the can.*

- Never use a wide brush to finish small, round surfaces such as dowel rods.
- Clean brushes immediately after they have been used. The solvents used to clean brushes are generally the same as those recommended to thin down finishes. (See Table 31-B.) Work the cleaner thoroughly into the heel of the brush. With your fingers, open the bristles to clean out the waste material. Wipe the brush dry. Then suspend the brush to allow it to air dry. Fig. 31-13. *Never allow the brush to stand on its bristle end.*
- If a brush is to be stored longer than overnight, wash it in good commercial cleaning solvent or a good grade of detergent mixed in water. Wash the brush thoroughly. Comb the bristles with a metal comb. Rinse well, shake out the excess water, and allow to dry. Then wrap the brush in heavy waxed paper. Fig. 31-14.

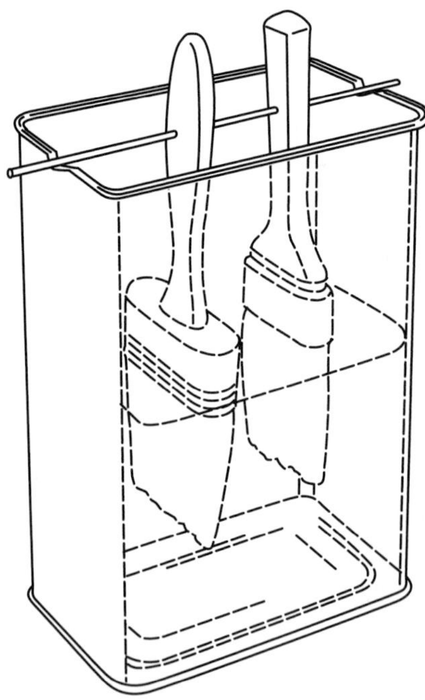

Fig. 31-13—*One method of suspending brushes to make sure that the bristles do not become bent. (Note: The front of the can in this picture has been cut away for display purposes only.)*

Solvents

Solvents are liquids that dissolve another substance. Solvents are used to thin a finishing material (when necessary) and to clean brushes after use. Table 31-B lists some common solvents, along with the finishes with which they are used.

Waxes and Oils

Waxes can be either liquid or paste. Both are made from a base of beeswax, paraffin, carnauba wax, and turpentine. Often used as a protective overcoat for other finishes, wax gives wood a durable sheen that resists everyday dirt and scuffs.

Wax can also be used by itself as a finish on most woods. Wax is applied by hand rubbing along the grain, using a soft cloth. Waxes do not darken the wood as do most oils. However, wax can yellow over time, so a waxed surface needs to be renewed once a year.

Fig. 31-14—*For long-term storage, wrap the dried brush in waxed paper.*

Table 31-B. Solvents

Solvent	Finish
Mineral spirits	Oil-based paints and stains; varnish
Turpentine	Same uses as mineral spirits, but has extremely strong odor
Denatured alcohol	Shellac
Lacquer thinner	Lacquers

Safety First

►*Using and Storing Solvents*

- Solvents emit dangerous fumes. Use only in a well-ventilated area.
- Solvents are extremely flammable. Keep all solvents away from sources of heat, sparks, and fires.
- Store solvents in their original containers. If this is not possible, be sure the new container is clearly labeled.
- Be sure to read and obey the labels on each type of solvent and finish.

Safety First

►*Finishing Safety*

Keep safety uppermost in your mind when working with finishing materials. All finishing materials have some degree of danger associated with them. Be sure to observe the following precautions:

Personal Safety

- Wear approved eye protection.
- Cover your clothing with a suitable apron.
- Wear a respirator when you are spraying finishing materials.
- Avoid breathing fumes from toxic materials.
- Wear rubber or vinyl gloves to minimize risk of skin irritations when using a cloth or pad to apply solvents, bleaches, stains, and finishes and when cleaning brushes.
- Wash your hands thoroughly after using any finishing materials.

Materials Safety

- Read labels on containers carefully before use. Follow directions *exactly* as stated on the container.
- Keep all solvents away from sources of heat, fire, or sparks.
- Be sure to close all cans and containers tightly after use.
- Whenever possible, store all thinners and solvents in their original containers. Make sure they are clearly labeled.
- Dispose of all used rags and disposable gloves in a fireproof container.
- Dispose of all leftover combustible materials immediately or store them in approved containers for later disposal.
- Store aerosol containers in a cool place.
- Keep all finishing materials and solutions in a fireproof cabinet.

Finishing Room Safety

- Perform all finishing tasks in a separate, well-ventilated area that is specifically designed for finishing.
- Make sure the proper types of fire extinguishers are available in the room.
- For all spraying tasks, use a properly installed, well-ventilated spray booth that is kept clean and well maintained.
- Keep the entire finishing area clean and free from spills.
- Never leave opened containers of finishing materials unattended.
- Never use tools or machines that can cause sparks or fire in the finishing area.

Linseed oil is sometimes mixed into oil-based paints. More commonly, however, it is used by itself to finish certain types of furniture. When used as a finish, the oil is applied with a rag and then rubbed into the surface. Several coats are applied in this manner. After the oil is dry, a paste wax is applied and rubbed to a high polish.

Mineral oil, also known as *paraffin oil,* is a clear, nontoxic oil that comes from a petroleum base. It can be used as an oil finish on projects that will hold food (although a commercial salad bowl finish is more commonly used for such projects). A medium-heavy mineral oil called *rubbing oil* is used with pumice or other abrasives as a lubricant for rubbing a dried film of finishing material.

STEPS IN FINISHING

Although the materials may vary, all finishing is done in about the same way. There are several basic steps to follow. Not all are necessary, however, for finishing every piece of wood. Choose from the following only the steps needed for the kind of wood used and for the finish you want.

1. *Bleaching.* **Bleaching** removes the color from wood. It is done to lighten or even out the color of unfinished wood. Very light, natural wood finishes are popular for contemporary furniture. Bleaching is not necessary if a darker finish is desired.
2. *Staining.* Staining is done to enhance the grain and to achieve the color you want.
3. *Sealing.* Sealing is done to seal the stain and thus prevent it from bleeding into the topcoat. It enriches the stain and stiffens the fibers of the wood.
4. *Filling.* **Fillers** are needed to close the pores of *open-grained* (porous and semiporous) woods. Woods such as oak, mahogany, and walnut contain large cells, making these woods very porous. When the lumber is cut and planed, the cells are ruptured, revealing tiny troughs that run lengthwise. These troughs must be filled with a paste filler to obtain a smooth finish. Woods such as birch, maple, and gum have smaller cells. These semiporous woods require a thinner liquid filler or no filler at all. Nonporous (*close-grained*) woods such as pine, cedar, and redwood do not require a filler.

   Note: If a filler is used, a sealer should be applied again over the filler.
5. *Applying the topcoat.* The topcoat may be shellac, varnish, lacquer, penetrating oil, paint, or enamel. Usually, two or more coats are needed and the surface is sanded between each coat.
6. *Rubbing, buffing, and waxing.* After a clear topcoat such as varnish or linseed oil is dry, the surface may be rubbed, polished, and waxed to produce the desired sheen and to provide additional protective coating.

CHAPTER 31

Review & Applications

Chapter Summary

Major points from this chapter that you should remember include:

- There are three basic types of finishes—transparent coatings, penetrating finishes, and opaque finishes.
- Mill marks and all traces of glue must be removed before any stain or finish can be applied correctly.
- Before applying any finish, damage must be repaired, knots must be sealed, and the wood must be sanded as smooth as possible.
- Sandpaper, steel wool, pumice, solvents, and various oils and waxes are among the supplies needed for finishing.
- There are a number of safety precautions to keep in mind when using finishing materials.

Review Questions

1. What determines the type of finish to use?
2. What are mill marks? How can they be removed before the project is finished?
3. Explain how to repair a shallow dent.
4. How must knots be treated before finishing?
5. Identify and describe the features to look for when choosing a quality brush.
6. Imagine that you used a brush today to apply varnish and will be using it again tomorrow to apply a second coat. Briefly describe how you would clean and store the brush.
7. What is the purpose of fillers? Give three examples of woods that require fillers and three examples that don't need fillers.

Solving Real World Problems

Lois is building a blanket chest that will stand at the foot of her bed. The bed is a very lightly colored red oak, and she has selected the wood for the chest with this in mind. She only plans on using a clear finish, so any blemish or defect in the wood will show. Lois attended a furniture builder's seminar at a woodworking show to learn more about finishing such a piece. There she learned that the key to a great finish is preparation. "Plane, sand, scrape," was the instructor's motto. Lois has planed the surfaces and is now ready to sand, but she has forgotten what grit to start with. The surface is already smooth. Can she just scrape?

CHAPTER 32

Applying Stains and Clear Finishes

YOU'LL BE ABLE TO:

- Use an oil-based or water-based stain to stain wood.
- Apply a wood sealer.
- Know when a filler is needed and how to apply it.
- Apply clear surface finishes.
- Choose from and apply a variety of penetrating finishes.

Look for These Terms

- alkyd resin
- Danish oil
- lacquer
- penetrating oil stain
- pigment stain
- polyurethane
- spar
- staining
- tung oil
- wash coat

Once prefinishing operations—removing any glue, repairing imperfections, and final sanding—are complete, it's time to get down to the business of finishing. **Fig. 32-1.** As you proceed, be sure to keep that all-important rule in mind—"Don't get in a hurry."

Fig. 32-1—*The beautiful finish on this table was obtained by properly applying stain, sealer, filler, and a clear topcoat finish.*

STAINING

Staining is changing the color of wood without changing its texture. This is done by applying transparent or semitransparent liquids made from dyes, pigments, and/or chemicals.

Types of Stains

Stains may be oil-based or water-based. The two common kinds of oil-based stains are pigment and penetrating stains.

Pigment stains are made by adding color pigments to boiled linseed oil and turpentine. These stains can be purchased ready-mixed in a variety of colors. You can also mix your own. Table 32-A indicates which pigment(s) to mix with the oil/turpentine base to achieve the desired color.

Penetrating oil stains are made by mixing aniline dyes in oil. These stains can also be purchased in many different colors. Different shades can be made by mixing stains of the same type or by adding tinting colors.

Water-based stain can be purchased ready-mixed, or you can make your own by mixing powdered aniline dye with hot water. Water-based stain has several advantages over oil-based stains. It is cheaper, has a more even color, dries more quickly, and is less likely to fade. Water-based stain is also preferable to oil-based stain when a clear lacquer topcoat is going to be applied because it has less tendency to "bleed" into the lacquer. The chief disadvantage of water-based stain is that it raises the grain of the wood.

Applying the Stain

Whether you are using oil stain or water stain, you should always test the stain to make sure it will produce the desired color. Make your test on a piece of scrap wood left over from making the project or on an inconspicuous part of the project, such as the back of a drawer. Fig. 32-2. For a true idea of the final color, apply the total finish—sealer, filler (if needed), and topcoat (varnish, shellac, oil, or lacquer)—over the stain after it has dried.

The same basic procedure is used to apply both oil- and water-based stains. Table 32-B summarizes the steps to follow.

Table 32-A. Making Pigment Stains

Desired Color	Pigment(s)
White	Zinc oxide ground in oil
Golden oak	White zinc tinted with yellow ochre and raw sienna
Medium oak	Raw sienna and burnt sienna
Light brown	Vandyke brown
Dark brown	Vandyke brown and drop black
Walnut	Half Vandyke brown and half burnt umber
Black	Drop black

Fig. 32-2—*Test the color of the stain on an area that will not show when the piece is finished or on a piece of scrap left over from making the project.*

Table 32-B. Applying Stains

Oil-Based Stains	Water-Based Stains
1. Wipe all surfaces with a tack rag to remove any dust. 2. Put on rubber or vinyl gloves. 3. Apply a thin coat of linseed oil to the end grain so that it won't soak up too much stain and appear darker than the rest of the piece. Fig. 32-3. 4. Stir the stain thoroughly. 5. Dip about ⅓ of a good-quality brush into the stain. Wipe off excess along the rim of the container. Begin at the center of the surface. Using light strokes, work out toward the edges, following the grain. With each new brushful, begin on the unfinished surface and stroke toward the stained surface, covering the surface evenly. 6. Allow the freshly stained area to dry for a few minutes. When it has lost its wet appearance, use a clean, lint-free cloth folded into a pad to wipe lightly *in the direction of the grain.* Fig. 32-4. 7. Repeat Steps 3 through 6 until the entire project has been stained. 8. Allow the stain to dry for 12 to 24 hours before continuing.	1. Sponge the surface of the wood lightly with water to help prevent the stain from raising the grain. Do not saturate the surface. 2. Allow the surface to dry. Then sand with 2/0 sandpaper and use a tack rag to remove any dust particles. 3. Put on rubber or vinyl gloves. 4. Sponge or brush the end grain with water to prevent it from absorbing too much stain. 5. Apply water stain in the same general way as oil stain. Refer to Steps 4 through 7 under "Oil-Based Stains." 6. Allow the stain to dry 12 to 24 hours before continuing with the finishing operation.

Fig. 32-3—***End grain tends to soak up stain at a faster rate. Brush or wipe linseed oil onto the end grain before applying any oil-based stain. For water-based stains, apply water to the end grain.***

Fig. 32-4—***Using a clean, lint-free cloth folded into a pad, wipe lightly, following the grain of the wood.***

Applying a Sealer

After the stain has dried for 12 to 24 hours, a sealer should be applied over the stain before a clear finish is applied. The sealer prevents the stain from bleeding into the clear topcoat, which causes a cloudy finish.

Safety First

►*Staining*

- Always work in a well-ventilated area.
- Wear rubber or vinyl gloves.

Follow all other rules of good safety when using stains, solvents, and other finishing materials. Review the Finishing Safety rules in Chapter 31.

Fig. 32-5—*The type of filler needed depends on the type of wood that is used.*

A **wash coat** of one part shellac to seven parts of denatured alcohol is a commonly used sealer over most stains. If a lacquer finish will be applied, a lacquer sealer should be used in place of the shellac wash coat. Frequently, the best and easiest sealer is just a thinned-down wash coat of whatever you plan to use as your final finish.

Brush on a uniform coat of the sealer. After the sealer has dried, sand the surface lightly with 6/0 sandpaper. Then use a tack rag to remove the dust particles.

Applying a Filler

If your project is constructed of open-grained or semiporous woods, you will need to apply a filler over the sealer to seal the pores of the wood before applying a clear surface finish. A thick paste filler is needed to seal the pores of open-grained woods such as oak. A thinner liquid filler is needed for semiporous woods such as maple. Fig. 32-5. Close-grained woods such as pine need no filler. See Table 32-C.

Paste filler comes in a natural color or in colors to match wood stains. The natural color can be tinted with oil colors or stain pigments to match the color of the stain or the wood, but it is better to buy the shade you desire.

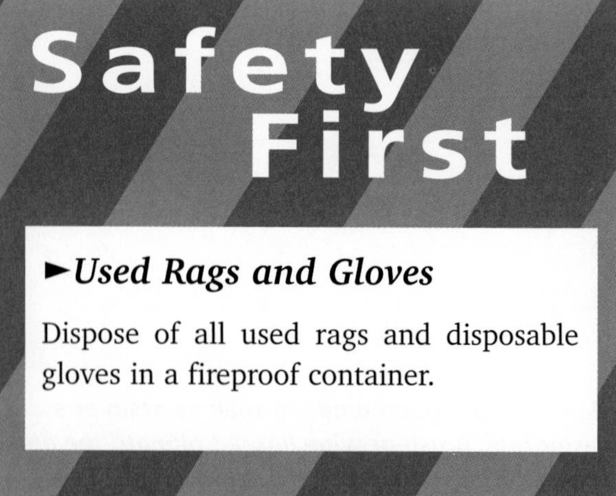

Safety First

►*Used Rags and Gloves*

Dispose of all used rags and disposable gloves in a fireproof container.

Table 32-C. Fillers

Thick Paste Filler (open-grained)	Liquid Filler (semiporous)	No Filler Needed (close-grained)
Ash	Beech	Basswood
Chestnut	Birch	Cedar
Elm	Cherry	Fir
Hickory	Gum	Pine
Lauan	Maple	Poplar
Oak		Redwood
Walnut		Spruce

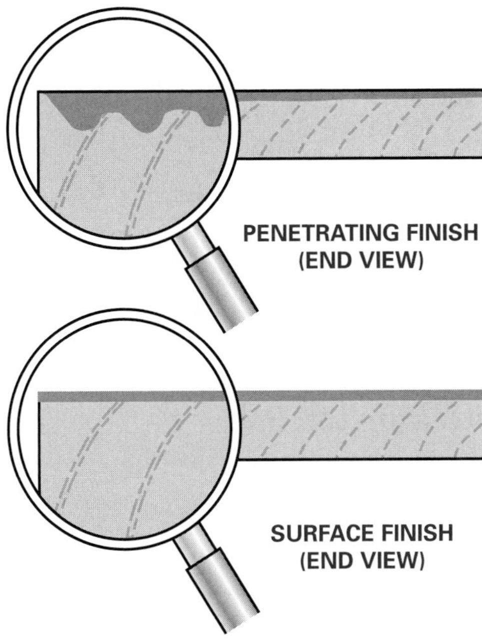

Fig. 32-6—*These two drawings show the difference between a penetrating finish and a surface finish.*

Liquid filler can be made by thinning paste filler with mineral spirits until it is very thin, or you can buy it ready-mixed. Follow the directions on the container carefully to apply the filler, being sure to follow all safety rules. Allow the filler to dry 24 hours. Then apply another coat of sealer over the filler, and allow it to dry before applying a clear topcoat finish.

APPLYING CLEAR FINISHES

There are two basic types of clear topcoat finishes: those that penetrate into the wood and those that stay on the surface of the wood. Fig. 32-6.

Surface finishes include shellac, varnish, and lacquer. Penetrating finishes include natural oils, such as linseed oil, as well as oils fortified with resins, such as Danish oil.

Safety First

►*Finishing*

Follow all rules of good safety when using finishing materials. Review the rules for personal, materials, and finishing room safety outlined in Chapter 31. Before you begin to work with any finishing product, be sure to read safety information on the product label.

Shellac

Shellac produces a strong, smooth, and durable finish. It dries so quickly that it is dust-proof within minutes, and other coats can be applied within hours. It is a good finish for all *interior* projects except tabletops, where water or alcohol will discolor it.

Because of its orange color, natural shellac gives light woods an unattractive, yellow-orange tint, so bleached shellac, called *white shellac,* is more satisfactory for general use. The steps below outline the general procedure for applying shellac.

1. Put on rubber or vinyl gloves and wipe the wood surface clean with a lint-free cloth that has been dipped in denatured alcohol.
2. Follow the directions on the label of the product you are using to apply three coats of shellac. Begin at the center of a flat surface or near the top of a vertical surface and work out toward the edges. Take light, long strokes. Overlap the strokes slightly to keep ridges and streaks at a minimum. Fig. 32-7. Be careful along the edges to keep the shellac from piling up and running. Between coats, allow the surface to dry and then go over it with fine steel wool or 5/0 garnet paper. *Rub with the grain of the wood.* Wipe the surface clean with a tack rag.
3. Clean the brush with denatured alcohol as soon as you have finished applying each coat.
4. To get a very even, smooth surface, mix some ground pumice in oil, and rub down the surface with a felt pad. After this, a still smoother surface can be obtained by mixing rottenstone with oil and rubbing it in. Then clean the surface and allow it to dry.
5. Apply a coat of paste wax. Place the wax on a soft, lint-free cloth and, using circular motion, rub it all over the surface. Apply just enough wax to make a thin film. Allow the wax to dry several minutes. (Check the product label for drying time.) Then rub briskly with a clean, soft cloth or a buffer pad.

Checking Shellac

Shellac deteriorates with age. To make sure shellac is not too old, test a small amount on a piece of scrap. If the shellac takes a long time to dry or remains tacky, don't use it—get a fresh supply.

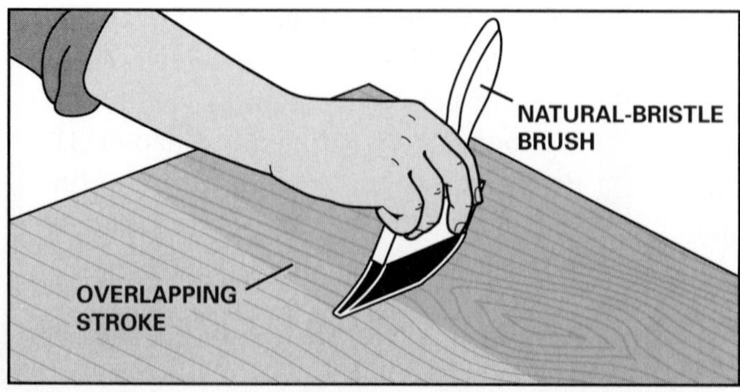

Fig. 32-7—*Use long, smooth strokes. Slightly overlap each stroke to keep ridges and streaks to a minimum.*

Varnishes

Varnishes provide a tough, protective topcoat and brighten the color of the stain (or the natural wood color if no stain is used). Varnishes are available in gloss, semigloss, and satin finishes. The number of coats determines the depth and smoothness of the finish.

There are three basic types of varnish—**alkyd resin, spar,** and **polyurethane.** Water-based varieties are available in addition to the more traditional solvent-based varnishes.

Applying Varnish

Varnishes should be applied in a dust-free, well-ventilated area. Varnish should not be applied on cold and damp or hot and humid days. Ideally, the temperature should be between 70 and 80 degrees Fahrenheit (21-27 degrees Celsius). The *general* steps for applying varnish are outlined below. Always read and follow the directions on the label of the product you are using.

1. Wipe the project thoroughly with a tack rag to remove all dust or particles.
2. For the first coat, thin the varnish with mineral spirits. Then mix 3 parts varnish with one part mineral spirits.
3. Use a 2- to 3-inch wide, long-bristled brush or similar-sized polyfoam brush to apply the varnish. Dip about one-third of the bristle-length into varnish. Be careful not to overload the brush. *Do not wipe the brush along the rim of the container*—doing so can create air bubbles in the varnish. Apply the varnish with long, easy strokes. Brush first with the grain and then across the grain. Fig 32-8. When the brush begins to "dry," brush out the varnish with the grain, using only the tip of the brush. Brush from the center toward the outside edges. As you near the edges, have very little varnish on the brush. This will keep the varnish from running over the edges and from piling up along the corners.

A

B

C

Fig. 32-8—*(A) Apply varnish with long smooth strokes, brushing with the wood grain. (B) Make several quick brush strokes across the grain to level out the varnish. (C) "Tip off" along the grain by using very light strokes with the tips of the bristles.*

4. Allow the first coat to dry (overnight for standard varnishes or 2 to 4 hours for water-based ones). Be sure all tackiness is gone. Then, using No. 6/0 aluminum oxide sandpaper, lightly rub the surface *with the grain.* Use just enough pressure to dull the sheen and produce a uniform, smooth finish. When sanding is complete, wipe the entire surface with a tack rag to remove all dust particles.

   Note: Very fine steel wool can be used instead of sandpaper when sanding between coats of solvent-based varnishes. Do not use natural steel wool when using water-based varnishes—doing so will leave particles that can discolor the finish.

5. Apply second and third coats of varnish as described in Step 3, but do not thin the varnish. Once the second coat has dried thoroughly, sand the surface lightly with super-fine (#400) waterproof sandpaper. Use a tack rag to remove all dust particles before applying a third coat.

Final Rubbing

No matter how carefully it is applied, the final coat of varnish may have a few dust specks or tiny air bubbles that mar its finished appearance. Fig. 32-9. If a high-gloss furniture varnish was used, it may also have too much gloss. Correct these conditions by a final rubbing and polishing with special abrasives *after* the varnish has been allowed to harden for several days.

Mix a paste of powdered pumice and oil, such as machine oil. *Do not use linseed or drying oils.* Then rub with a folded pad of lint-free cloth or a felt pad. Rub with straight strokes, parallel with the grain. Use only moderate pressure. Continue until the surface feels perfectly smooth. Then wipe off all remaining powder and oil with a clean, dry cloth.

To restore the high luster that some prefer, a second rubbing or polishing is required with a still finer powdered abrasive. Mix a second paste of powdered rottenstone and crude oil. Again, rub in parallel strokes following the grain. After polishing the surface to the luster desired, wipe off all paste and oil. Do this by rubbing repeatedly with clean cloth until the surface is so dry that it "squeaks" when rubbed.

After the varnish and final rubbing(s) have dried thoroughly, apply a good paste wax (as described on page 412, under "Shellac").

> *Note:* Some manufacturers recommend that wax *not* be applied over their products. Be sure to check the directions on the product label.

Varnish Stains

A *varnish stain* will give the desired color and finish in one coat. Varnish stain is useful when time cannot be given to applying many coats of regular varnish. Varnish stains can be purchased in different colors. These stains are generally applied in the same way as varnish. Drying time varies from 2 to 8 hours.

Fig. 32-9—*Air bubbles can mar the finish.*

Safety First

►*Spraying*

- When using aerosol sprays, do the spraying in a well-ventilated room or outdoors in an area protected from wind.
- When using spraying machines, work in a properly installed spray booth that is kept clean and well maintained.
- Wear a mask or respirator, goggles, rubber or vinyl gloves, and clothing protection (e.g., apron or lab coat) when spraying.

Lacquer

Lacquer is a finish composed of nitrocellulose, resins, and solvents. It dries quickly and produces a hard finish.

If a lacquer finish is to be applied over stain, the stain should be water-based rather than oil-based. Water-based stain has less tendency to bleed into the lacquer. A lacquer sealer should be applied over the stain before applying the lacquer finish.

Most commercial lacquer finishes are applied by spraying, but the brush finish is usually done in most small shops. Table 32-D summarizes the steps for applying lacquer with a brush and from an aerosol container.

Table 32-D. Applying Lacquer

Brush Lacquer	Aerosol Lacquer
1. Open a can of clear brushing lacquer and stir it well. Lacquer usually does not have to be thinned. If it does, use lacquer thinner of the same brand.	1. Place the item to be sprayed on a bench that has been covered with newspapers.
2. Select a brush with soft bristles, such as camel's hair, or a polyfoam brush. Dip the brush about one-third of the way into the lacquer, but do not wipe it on the side of the container. Load the brush heavily. Apply with long, rapid strokes, overlapping the sides of each stroke. Do not attempt to brush the lacquer in as you would varnish—remember lacquer dries very quickly.	2. Hold the aerosol can so that the valve is the recommended distance from the project throughout the spraying process. (This distance can vary from brand to brand, so be sure to check the product label.) Start at the front (side nearest you) and spray back and forth, moving toward the rear. Overlap each stroke. If sags or runs develop, you are covering the surface too quickly.
3. Allow the lacquer to dry about 2 hours. Then rub the surface lightly with 6/0 sandpaper and use a tack rag to remove dust particles.	3. Turn the project a quarter of a turn and spray again. Fig. 32-10 (page 416). Spray several light coats rather than one heavy one.
4. Apply the second and third coats in the same way. After the third coat is dry (1 to 4 hours), rub and polish the surface.	4. When finished, turn the can upside down. Press the spray head for a few seconds to clean out the line. If the spray head becomes clogged, remove it. Then use a pin or needle to clean out the small opening and a knife to clean out the small slot at the bottom.

You can also use a spray gun to apply lacquer. If you are using a small, portable spray gun like the one shown in Fig. 32-11, you may want to thin the lacquer by mixing a solution of one part lacquer to one part lacquer thinner. Then follow these steps:

1. Clean the surface with a tack rag.
2. Check the gun to make sure that it is clean.
3. Put on your respirator mask, goggles, gloves, and clothing protection.
4. Fill the spray container with lacquer (and thinner, if needed).
5. Test the spray gun on a piece of scrap stock or paper. It should spray with a fine, even mist. Also check the distance between the gun and the surface. Fig. 32-12. Usually 6 to 8 inches is a good distance.
6. Move your arm—not your wrist—as you spray, keeping the gun at right angles to the surface at all points along the stroke. Fig. 32-13. Begin the stroke before pulling the trigger, and release the trigger before ending the stroke. Overlap the strokes about 50 percent, using alternate left and right strokes.

   Note: When spraying multi-sided surfaces such as a chest or cabinet, spray the corner lines created by the adjoining surfaces first. Aim the gun directly at the line at which the two surfaces meet so that each side receives an equal amount of finish. Fig. 32-14 (page 418).
7. Spray on four or five coats of thin lacquer.
8. When spraying curved surfaces, use the same spraying techniques as for other surfaces. Fig. 32-15.
9. After you have finished, clean all equipment with lacquer thinner.

Fig. 32-10—***Start at the front and spray horizontally with overlapping strokes as you move toward the rear, as shown by numbers 1 through 5. Then turn the project a quarter of a turn and spray horizontally again, as shown in numbers 6 through 10.***

Fig. 32-11—***A portable spraying unit and compressor is satisfactory for small projects.***

SCIENCE *Connection*

HVLP Spraying

High velocity, lower pressure (HVLP) spraying is one way to reduce the pollutants that go into the air. With conventional spraying, a large amount of the finishing material goes into the air. With HVLP spraying, 65 percent or more is applied directly to the wood surface. A special spray gun is needed to use this method.

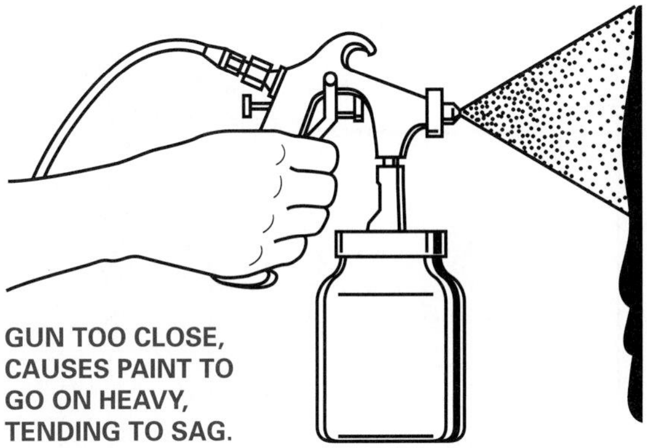

GUN TOO CLOSE, CAUSES PAINT TO GO ON HEAVY, TENDING TO SAG.

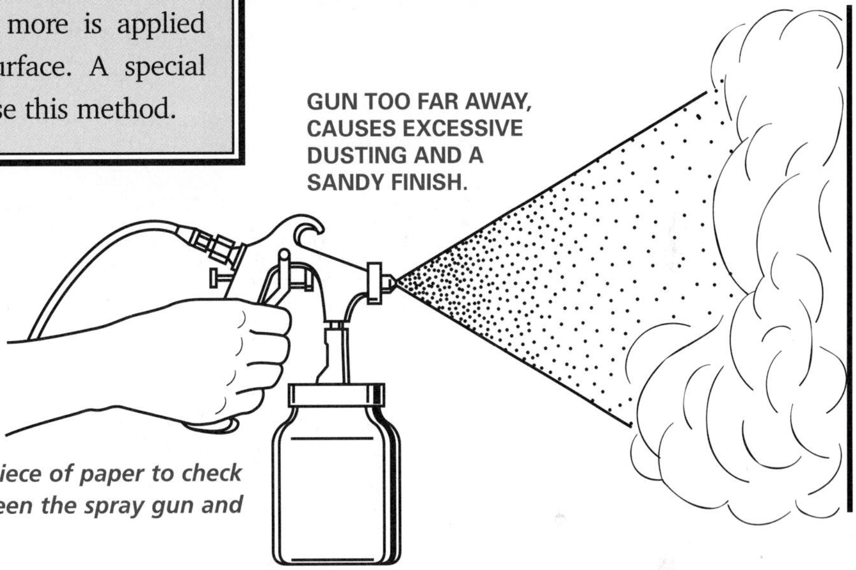

GUN TOO FAR AWAY, CAUSES EXCESSIVE DUSTING AND A SANDY FINISH.

Fig. 32-12—*Make a test on a piece of paper to check for the correct distance between the spray gun and the surface.*

MOVE GUN IN A STRAIGHT LINE
KEEP WRIST FLEXIBLE
RIGHT
ARCING GIVES AN UNEVEN COATING
WRIST IS TOO STIFF
WRONG

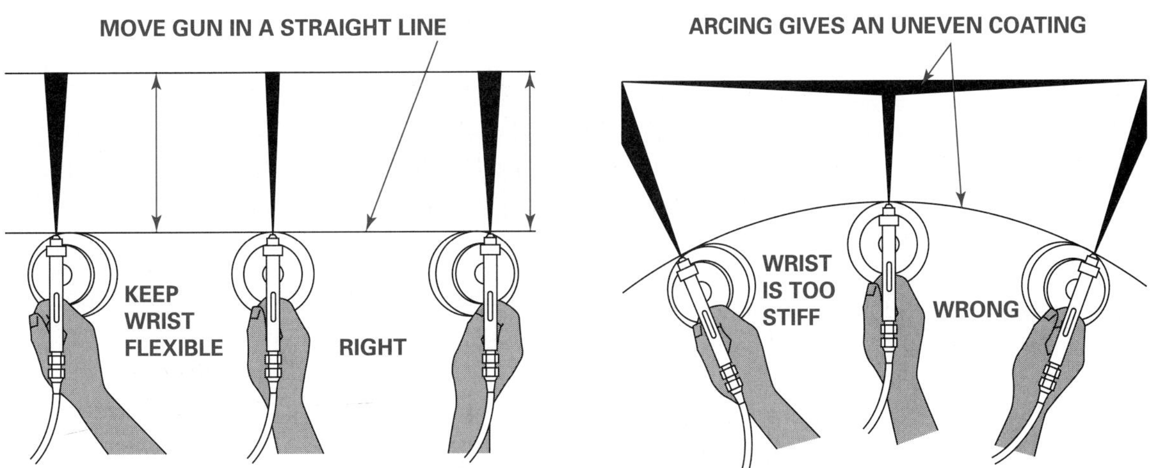

Fig. 32-13—*Move the spray gun in a straight line, never in an arc. Keep your wrist straight.*

Fig. 32-14—*Aim directly at the corner line created by the adjoining surfaces so that each side receives an equal amount of finish.*

Safety First

►*Using Wipe-on Oil Finishes*

- Use all finishing materials in a well-ventilated area.
- Wear vinyl or rubber gloves when applying wipe-on finishes.
- Dispose of all used rags and disposable gloves in a fireproof container.

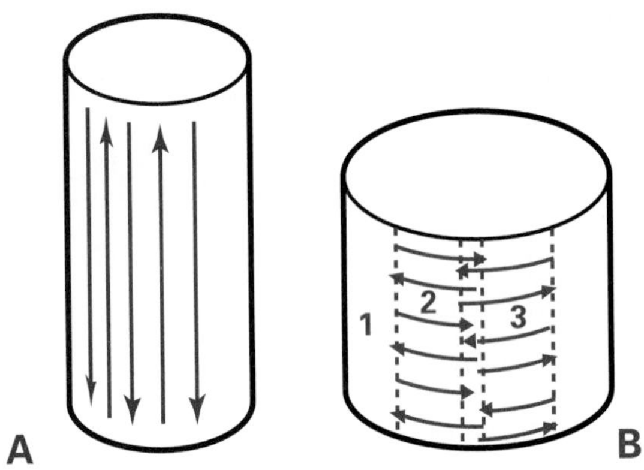

Fig. 32-15—*(A) Use vertical strokes for small-diameter curved surfaces. (B) Use horizontal strokes for large-diameter surfaces.*

Oil and Wipe-On Finishes

Oil finishes and penetrating resin-oil finishes such as Danish oil add a finish *in the wood,* not on the surface. These oil finishes emphasize the grain and leave a flat, natural appearance. They darken slightly with age and take on a rich glow. Most can be applied with a small soft cloth, brush, sponge, or pad.

- *Linseed Oil.* Boiled linseed oil can be used as a final finish on bare or stained wood—no sealer is required. It tends to darken wood and may turn sticky in damp weather. Though not highly water-resistant, this finish will withstand hot dishes. It is also less likely to show scratches than a varnish finish.
- *Tung Oil.* **Tung oil** is obtained by pressing the nuts from the tung tree, which grows in China. It penetrates deep into the wood and forms a long-lasting moisture barrier. *Several* hand-rubbed coats, applied in much the same way as boiled linseed oil, are required. As with linseed oil, the more coats that are applied the more attractive and durable the finish.
- *Danish Oil.* **Danish oil** is a blend of oils and resins that penetrates, seals, and preserves wood surfaces. This finish is available in natural, which has no pigments to color the wood, as well as in oak, cherry, and several shades of walnut. Danish oil requires only two coats.
- *Salad Bowl Finish.* Vegetable oil can be used to finish kitchenware such as salad bowls and cutting boards. However, a vegetable oil

finish tends to become rancid over time, which affects the taste and safety of food. Instead, use a commercial *salad bowl finish,* made of tung oil and alkyd resins. This odorless, taste-free finish has been approved by the U.S. Food and Drug Administration (FDA) as being completely foodsafe. Salad bowl finishes are also excellent for finishing wooden toys. Fig. 32-16.

- *Deft® Finish. Deft®* is a semigloss, clear, interior wood finish made of tung oil with urethane (plastic) for greater durability. It is easy to use, does not show brush marks, and will not darken over time. Deft® seals and finishes the wood in as few as three quick-drying coats. The first coat seals the wood. The second adds depth. The third coat usually results in a smooth, fine furniture finish.

Filling Open-Grained Woods

When using wipe-on finishes on open-grained woods such as oak and walnut, you will get a much smoother finish if the pores of the wood are filled. Here is a simple homemade remedy that can be used for all types of wipe-on finishes.

Apply the first coat of finish. While it is still wet, sand the surface with 400-grit waterproof abrasive paper. This will produce a slurry of wood dust and finish that will fill the pores of the wood. Allow this slurry to stay on the surface about 20 minutes. Then wipe off the excess slurry *with* the grain of the wood. Some of the slurry will remain in the wood's pores. Allow the filled surface to dry overnight before applying further coats of wipe-on finish.

Fig. 32-16—*A salad bowl finish was used to complete this project.*

CHAPTER 32

Review & Applications

Chapter Summary

Major points from this chapter that you should remember include:

- Staining is done for a variety of reasons.
- Stains may be oil- or water-based. Both types are applied in a similar manner, but there are differences in surface preparation and drying times.
- A sealer may need to be applied over stain.
- Fillers should be applied to close the pores of open-grained and semiporous woods.
- The two basic types of clear finishes are surface finishes and penetrating finishes. Surface finishes include shellac, varnish, and lacquer. Penetrating finishes include linseed oil, tung oil, Danish oil, salad bowl finish, and Deft®.

Review Questions

1. Describe the procedure for wiping stain.
2. What treatment should be given to end grain when using oil-based stain? When using water-based stain? Why is this special treatment needed?
3. What is the purpose of a sealer? Briefly describe the procedure for sealing.
4. What type of filler is needed before a clear finish can be applied to a project made of oak? To one of maple? To one of pine?
5. Explain how to apply a coat of varnish.
6. Describe how to apply aerosol lacquer.
7. Compare tung oil and Danish oil finishes.

Solving Real World Problems

Sudip has finished assembling his new dining room chairs. All the pine surfaces have been finely sanded and vacuumed off. He uses a tack cloth to ensure that all dust particles are removed from the chairs. Keeping in mind that the chairs are made of pine, what should be his next step if he is going to stain the wood? How many coats of sanding sealer should he apply before the finish?

CHAPTER 33

Applying Paints and Enamels

Look for These Terms

- enamel
- gloss
- latex
- primer

YOU'LL BE ABLE TO:

- Choose an appropriate paint or enamel finish for a project.
- Prepare bare wood surfaces for painting.

Paints and enamels are *opaque finishes*—they cannot be seen through. Available in a wide range of colors—literally thousands of colors if they are custom-mixed at the store—paints and enamels are used to cover less expensive woods and to add beauty and durability to a project. **Fig. 33-1**.

Paints and enamels may be water-based **latex** (also called *latex acrylic*) or oil-based. Latex paints and enamels dry much more quickly than oil-based ones and offer quick, easy, soap-and-water cleanup. (Oil-based paints and enamels require solvents such as mineral spirits for cleanups.) In addition, latex finishes don't usually have as strong an odor as oil-based ones and, because latex paints dry more quickly, the paint odor clears away more quickly.

Paints are available in flat, satin, semigloss, and gloss finishes. (**Gloss** refers to the amount of surface brightness. For example, flat paint has no gloss; satin paint has a soft luster.) **Enamels** are high-gloss paints that provide a slick, hard surface that is easy to clean.

Fig. 33-1—***There's a paint for every job, big or small, in just about any color imaginable.***

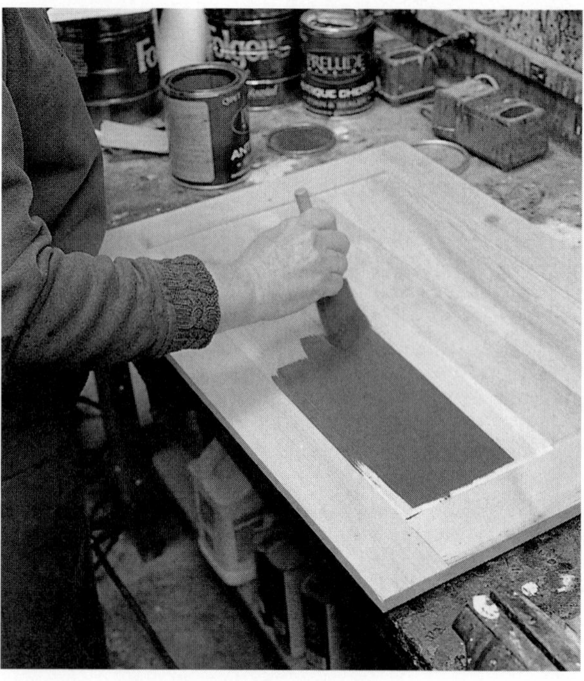

Fig. 33-2—***If painting over bare wood, a primer is always needed to seal the wood.***

Both paint and enamel are available as brush-on or aerosol finishes. This chapter discusses brush-on applications. Aerosol paints and enamels are applied in much the same manner as aerosol lacquers. You can read about aerosol and spray-on applications in Chapter 32.

PREPARING TO PAINT

Prepare the surface by raising or filling any dents or gouges, sealing any knots, and final sanding. (See the "Preparing the Surface" section of Chapter 31, pages 396-399.)

Apply a **primer** to bare wood surfaces. An undercoat of primer seals the wood and prepares it to take the paint or enamel finish. Fig. 33-2. *Be sure that the primer is compatible with the type of paint (oil-based or latex) you plan to use.*

Safety First

►*Painting*

- Always work in a well-ventilated area when painting.
- Review the rules for personal, materials, and finishing room safety in Chapter 31 before beginning a painting project.
- Thoroughly read the safety precautions listed on the label of the product you are using.
- Always wear eye protection when working with volatile chemicals such as paints and solvents.

APPLYING PAINT OR ENAMEL BY BRUSH

Below are the basic steps for applying paint and enamel with a brush. Drying times and directions for application can vary from brand to brand, so be sure to carefully read the directions on the label of the product you will be using.

1. Make sure the surfaces to be painted are clean and free of all dust particles.
2. Stir the paint until smooth and well blended. Fig. 33-3.

   Note: If the paint or enamel needs to be thinned, use the thinner recommended on the product label.
3. Dip about one-third of the bristle length of the brush into the paint. When the bristles are well filled, remove them from the paint and tap lightly against the inside edge of the rim to remove excess paint—the bristles should be filled but not dripping with paint. Fig. 33-4. Use long, even strokes to brush a generous amount of paint onto the surface. Then feather out the paint, using lighter brush strokes in the direction of the grain. The surface should be well covered, but take care not to apply too much paint or runs will develop in the finish.

   Note: For large areas, you can apply the paint or enamel with a small roller, then feather it out with light brush strokes. Fig. 33-5 (page 424).

SCIENCE *Connection*

Paint Pigments

In good-quality paints, the primary *pigment* (tint or color) is titanium oxide (TiO_2). Good-quality paints contain a much higher concentration of TiO_2 than lower-quality paints. Lower-quality paints often substitute less expensive pigments. Lower-quality pigments often result in a grayish tinge to the paint color.

Fig. 33-3—*Be sure the paint has been stirred well before using.*

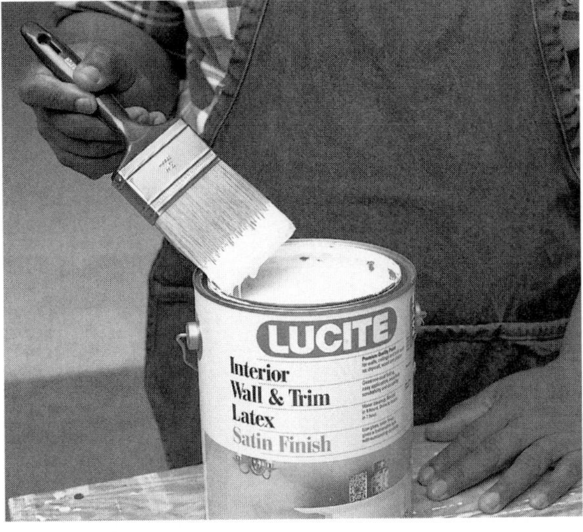

Fig. 33-4—*Dip about one-third of the bristle length into the paint. When the brush is wet-filled, remove it and tap lightly against the inside rim of the can.*

4. Allow the surface to dry thoroughly (at least 24 hours). Sand smooth with 320- or 400-grit sandpaper. Fig. 33-6. Then clean the surface with a cloth to remove any dust.
5. Apply the second coat in the same manner as the first, but do not sand the final coat.

Be sure you clean brushes and any other materials used for painting immediately after each use. Those used with latex paint or enamel should be washed thoroughly in warm, soapy water, then rinsed thoroughly in cool, clear water. Fig. 33-7. If oil-based paint or enamel was used, use a solvent such as mineral spirits for cleanup. Hang brushes to dry. *Never allow the brush to stand on its bristle end.* (See Chapter 31 for a full discussion of the care and use of brushes.)

Fig. 33-5—*For large areas, you can use a small roller to apply the paint. Then use light brush strokes to feather it out.*

Fig. 33-6—*Which of these three would you use for final sanding?*

PAINTING FURNITURE

The discussions below outline special techniques for painting different types of furniture. A basic principle to keep in mind when painting any type of furniture is to always paint the hard-to-reach parts—legs, undersides, cross spindles, backs, etc.—first.

Before beginning to paint furniture, remove all knobs and handles. Paint these separately, using a small brush. Fig. 33-8.

Tables

Begin by turning the table upside down and painting the bottom of the tabletop and its edges. Then paint the legs, painting the inner sides first. Finish off the legs with smooth, lengthwise strokes. (When painting round or turned legs or braces, brush all the way around them; do not finish with lengthwise strokes.)

Fig. 33-7—*Equipment used with latex paint can be easily cleaned with soap and water.*

Fig. 33-8—*You can place screws through the bottom of heavy cardboard, Styrofoam®, or scrap wood to hold the knobs as you paint them.*

Turn the table right side up. Paint all the top edges and frame sides. Then paint the tabletop, brushing across the narrowest dimension. Follow with long, light strokes the length of the tabletop, using only the tip of the brush to achieve a smooth, even finish.

Chairs

Turn the chair upside down. Place the seat of the chair on a box or a table. Paint all bottom surfaces—seat bottom, legs, and cross braces. Then stand the chair upright and paint all top surfaces, finishing with the seat. Fig. 33-9.

Chests and Cabinets

First, remove any drawers from the chest or cabinet and put them to one side. Begin by painting the moldings (if there are any) that surround the panels of the cabinetwork. Then paint the panels, picking up any runs or sags at the corners. Paint the frame and then the top of the cabinet or chest, using a cross-brushing technique.

Fig. 33-9—*A combination of paint and stain was used to finish this project.*

To paint the drawers, first paint each side. Start at the front and work back about 6 inches. Next, paint all exposed edges and the front panel. (If painting the inside of the drawer, do that *before* painting the exposed edges and the front.) Stack all painted drawers to dry with the bottom edges in an upright position.

Eliminating Touch-ups

When painting two or more colors on the same project, or when painting any surface next to a surface of a different color, use masking tape for color separation. This will help eliminate time-consuming touch-ups. The tape is readily peeled off after you are done painting.

CHAPTER 33

Review & Applications

Chapter Summary

Major points from this chapter that you should remember include:

- Paints and enamels may be water-based latex or oil-based.
- Paints are available in flat, satin, semigloss, and gloss finishes.
- Enamels are high-gloss paints that provide a slick, easy-to-clean surface.
- A primer should always be applied to bare wood before painting.
- Generally, two finish coats of paint or enamel are applied, with a light sanding between the coats.
- Always paint hard-to-reach areas of furniture first.
- There are special techniques for painting tables, chairs, chests, and cabinets.

Review Questions

1. What are some advantages of latex paint?
2. What is the benefit of enamel paints?
3. When should a primer be used? What is the purpose of a primer?
4. Describe the basic technique for brushing paint onto a flat surface.
5. How long should you let the first coat of paint dry before applying the second coat? What, if any, surface preparation should be done before applying a second coat?
6. Give a step-by-step description of how to paint a table.
7. Outline the steps for painting a chair.

Solving Real World Problems

Steve has an unfinished armoire that he wants to paint. The armoire is constructed of pine and has been sanded with 220g sandpaper. After priming with an oil-based primer, Steve decides to paint it with an old 4″ nylon paintbrush. The paint that Steve has chosen is an oil-based paint that has a very thick texture. What can Steve use to thin the paint so that it can be brushed? Will his nylon brush do a good job, or is there another that would work better?

CHAPTER 34

Preparing for Construction

Look for These Terms

- abstract
- building codes
- deed
- floor plan
- manufactured housing
- materials list
- specifications
- survey

YOU'LL BE ABLE TO:

- Explain the difference between conventional framing and manufacturing methods when building a home.
- Name factors to be considered when selecting a lot.
- Identify ways in which information on constructing a house is communicated.
- List views shown in construction drawings.

When looking for housing, people have three main choices: renting, buying an existing home, or building. If the choice is to build, then an understanding of the construction process, building techniques, and building materials is very important.

A great deal of planning must precede the actual construction of a home. Some of the planning relates to finances and some to legal and building code issues. However, most relates to the design and construction of the house itself. Fig. 34-1 (page 428).

CONSTRUCTION SYSTEMS

Houses can be built using conventional construction methods, or they can be manufactured in a factory. Those that are manufactured may be prefabricated as entire structures, parts of structures, or modular parts to be put together according to the customer's requirements.

Fig. 34-1—*Before this attractive home could be constructed, much planning had to be done.*

Conventional Framing

The majority of single-family homes in the United States and Canada are built using a system of wood-frame construction called *conventional framing,* or stick framing. Individual pieces of wood framing, from wall studs to roof rafters, are delivered to the jobsite in bundles and then assembled piece by piece into a house. Fig. 34-2. See Chapter 36.

Conventional framing is somewhat more time-consuming than manufacturing methods. It also calls for great skill on the part of individual tradespeople, such as carpenters. However, it is highly versatile, and it can be used to construct nearly any shape or size of building up to three stories tall.

Manufactured Housing

Manufactured, or industrialized, **housing** is built wholly or in part on factory assembly lines. This method of construction is very efficient. Fig. 34-3. The construction process is reduced to a series of relatively simple steps.

Wood vs. Structural Steel—Wood, when exposed to fire temperatures, burns and is converted to charcoal. Steel does not burn under similar conditions. When used in heavy timber construction, though, wood has a great advantage over unprotected steel. In a severe fire, the outside surfaces of thick beams become charred. However, the core of the wood remains at a low temperature. Because of its low heat conductivity, the center remains uncharred and intact. It retains most of its strength. Under the same fire conditions, steel quickly becomes heated throughout because of its good heat conduction. It loses much of its rigidity and load-bearing capacity. For this reason, steel members are commonly required to be enclosed in concrete or some other protective material.

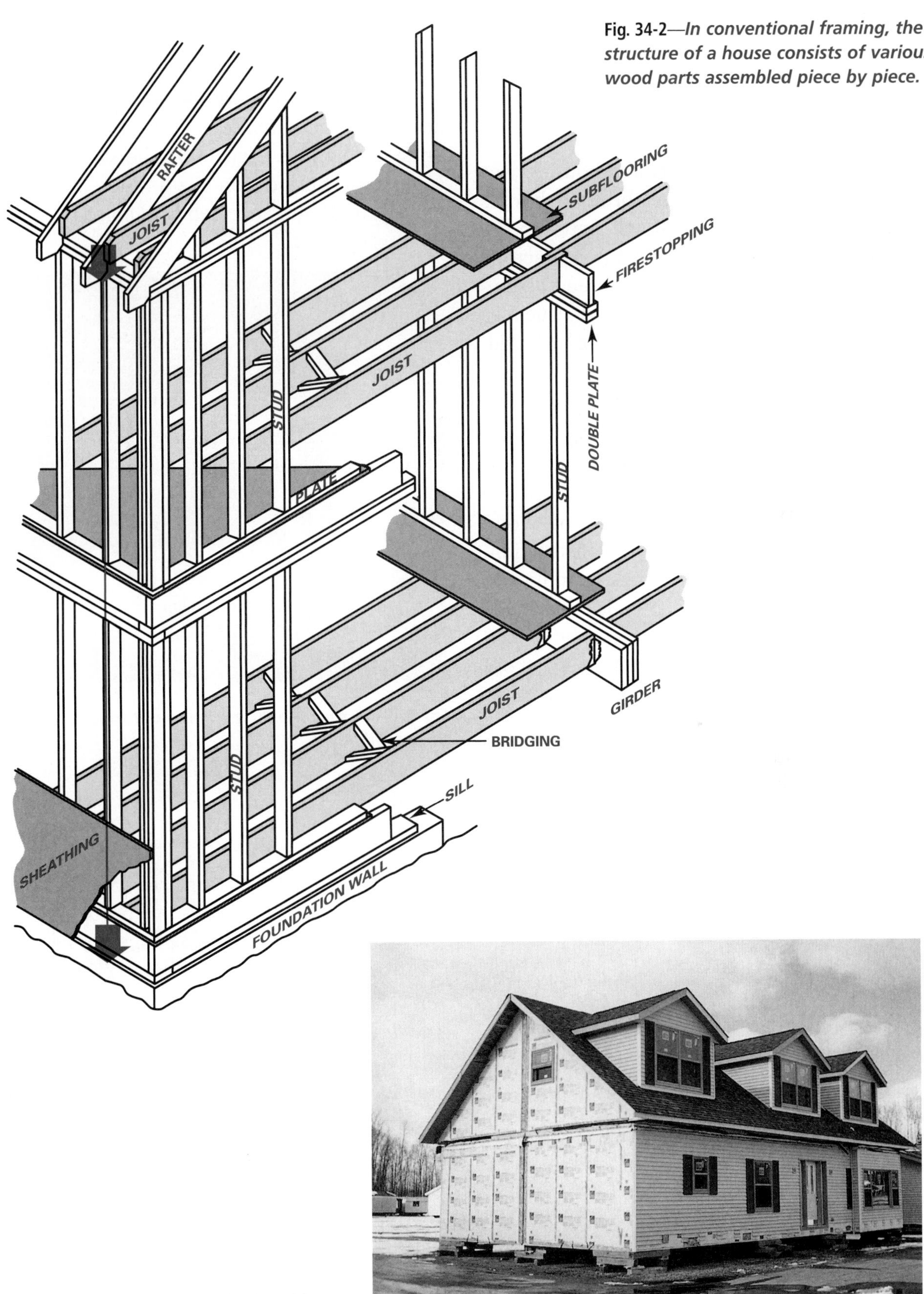

Fig. 34-2—*In conventional framing, the structure of a house consists of various wood parts assembled piece by piece.*

Fig. 34-3—*A manufactured home.*

Thus, it requires fewer carpenters and other skilled workers than does the conventional framing method. Few materials are wasted. Delays caused by weather are minimized. In some cases, there are also considerable cost savings.

An early type of manufactured housing was the *prefabricated house.* Prefabricated houses consisted of only the shell. All parts of the house were precut and prefit at the factory. They were then moved to the site and assembled. None of the interior systems, such as wiring, plumbing, and heating, were included.

Prefabricated Parts

Prefabricated parts are standard-sized parts made in a factory. Trusses were the first prefabricated parts to be used commonly. These are triangular-shaped assemblies of wood members often used in roof construction. Trusses can be produced in large quantities in manufacturing plants. The same idea has been used in developing prehung doors, prefinished paneling, and many other items that speed on-site home building. Fig. 34-4.

Sectional Houses

Today, not only parts but entire sections of houses can be factory-built. As its name suggests, a *sectional house* is built in sections on an assembly line. Fig. 34-5. The sections are completely finished on the inside. They are moved to the site and then assembled.

Fig. 34-5—*A sectional home being put together on an assembly line.*

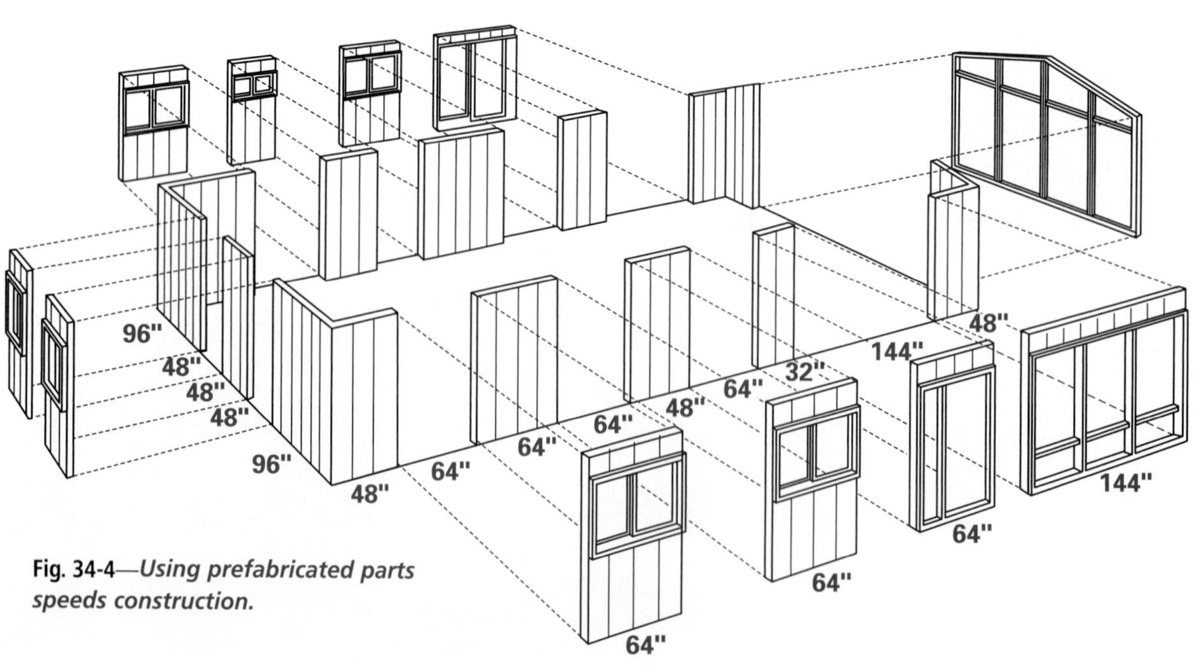

Fig. 34-4—*Using prefabricated parts speeds construction.*

Modular Construction

The most completely industrialized form of housing consists of *modules* (modular units). With the boxlike modular system, completely finished units of structures are built, shipped to sites, and may be assembled with other units. Fig. 34-6. For a small house, a single module may be used. For larger houses, several modules might be required. Modules can also be used in building apartment houses, motels, and commercial buildings.

PLANNING CONSTRUCTION

Since the home is the largest single investment for most individuals and families, it is important to protect it through careful planning. Many decisions must be made regarding the location, design, and actual construction of the house. Fig. 34-7. Financial, legal, and building code issues must be considered.

The construction process is much the same for multifamily housing as for single homes. Multifamily housing includes apartments, condominiums, cooperatives, town houses, and similar types of housing. Fig. 34-8 (page 432).

Selecting and Purchasing a Lot

Many factors are involved in selecting and purchasing a lot. Points to consider include:

- *Convenience.* Is the lot easy to get to? Are jobs, schools, community services, and recreational facilities nearby? Will the lot be served by municipal water and sewer systems or by a well and a septic system? If water and gas lines, sewers, and streets are not yet installed, will there be an assessment for them later?

Fig. 34-6—*A module being moved on the site.*

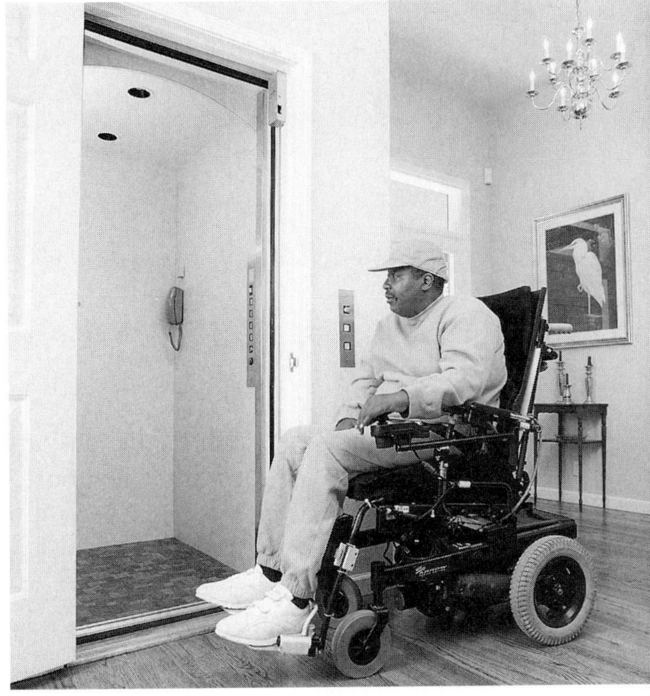

Fig. 34-7—*Take the needs of family members into consideration when you consider possible designs for a house.*

Fig. 34-8—*Planning for a multifamily home is very similar to planning for a single home.*

Fig. 34-9—*The natural terrain and vegetation are things to consider when selecting a site for a new home.*

- *Lot shape, contour, vegetation.* Is the lot wide enough and deep enough for the house? How easy will the lot be to build on? Will it be difficult to reach with all the materials and equipment needed for construction? Are there trees and other vegetation? Fig. 34-9.
- *Future prospects.* Is the neighborhood likely to remain relatively stable? Will the nature of the area change as the city grows?

Legal Documents

Real estate, or real property, consists of land and the buildings on it. At least three legal documents are needed to buy such property.

- A **survey** identifies boundaries and makes certain that the property meets building needs.
- An **abstract** traces the ownership of the property through legal documents. This is usually done by a company that specializes in abstracts.

- A **deed** is a document that provides evidence of ownership. It is the legal means by which ownership is transferred.

Real estate agents and lawyers can help the purchaser obtain these documents.

Selecting House Plans

Several methods can be used to obtain house plans for the lot. One popular way is to purchase them from a company that specializes in designing stock plans. These are standard plans that can fit many different lots. Design companies usually have plan books that show *floor plans* as well as drawings of how the finished house will look. Fig. 34-10. A **floor plan** is a drawing that shows sizes and locations of rooms, windows, doors, and many other features. (Floor plans are discussed in more detail later in this chapter.)

Once a suitable house plan has been chosen, complete information for building the house can be purchased from the company. The information is communicated in three ways. *Working drawings* are a set of drawings that provide information needed to construct the house. **Materials lists** identify all materials needed. **Specifications** are written details on construction that have not been described elsewhere.

Purchasing stock plans is the least expensive way to obtain a house plan. House plans can also be obtained from local builders who have a limited design service. However, if the house is to be completely and individually designed, an architect or building designer is usually required. The fee for design services may be based on a percentage of the total building cost, or it may be a flat fee.

> **MATHEMATICS** *Connection*
>
> ***Cost per Square Foot***
>
> To get a rough idea of how much the house should cost, find out the average cost per square foot of residential building in your area. Then multiply this figure by the number of square feet in the plans.
>
> For example, if the average cost in your area is $72 per square foot, and your house will be 2,100 square feet, the house will cost roughly $151,200. How much would a 2,400-square-foot house cost to build at $85 per square foot?

Fig. 34-10—*An artist's drawing of what a planned house will look like when completed.*

Financial Arrangements

After deciding on the house and the plans, the next step is to ask contractors to bid on the cost of construction. The cost will vary greatly depending on features such as fireplaces, built-ins, and special wood trim that may be included in the home, as well as on the quality of materials. Fig. 34-11.

After a contractor has been chosen, it is usually necessary to obtain financing from a bank or savings and loan company. Once financing has been arranged, construction contracts can be signed. It is then the responsibility of the contractor and/or architect to make sure that the building goes as planned.

Fig. 34-11—*The quality of lumber selected for a construction project will affect its cost.*

Building Codes

Building codes are laws that establish *minimum* standards of quality and safety in housing. Each code is actually a collection of regulations that governs the details of construction. A builder can construct a house that surpasses the code requirements, but he or she cannot construct a house that falls short of the code requirements.

Inspections and Permits

In areas that are covered by building codes (some rural areas are not), it is necessary to get a *permit* to begin construction of a house. This is done by submitting a full set of working drawings to the local building department (usually a branch of the city or county government). A plans inspector at the building department examines the plans to ensure that they meet the local building codes. If they do, the builder is issued a permit to begin construction. If they do not, the plans must be revised. The building permit must be kept on the building site at all times.

The building department usually inspects the building site periodically during construction. When the house is complete, one last inspection is made. If there are no problems, a *certificate of occupancy* is issued. This certifies that the house is ready to be lived in.

CONSTRUCTION DRAWINGS

The ability to read and understand building plans is basic to all construction. The ideas of the designer or architect are conveyed in sketches and working drawings.

Elements of Drawing

A drawing consists of lines, dimensions, and symbols. *Lines* show the shape of house and include many details of construction. *Dimensions* are numbers that tell the sizes of each part as well as overall sizes. The builder or an employee uses these dimensions in making the materials list. *Symbols* are used to represent things that would be impractical to show by drawing, such as doors, windows, electrical circuits, and plumbing and heating equipment. Fig. 34-12.

Some drawings also contain *notes* or written information to explain something not otherwise shown. Frequently in these notes, abbreviations are given for common words.

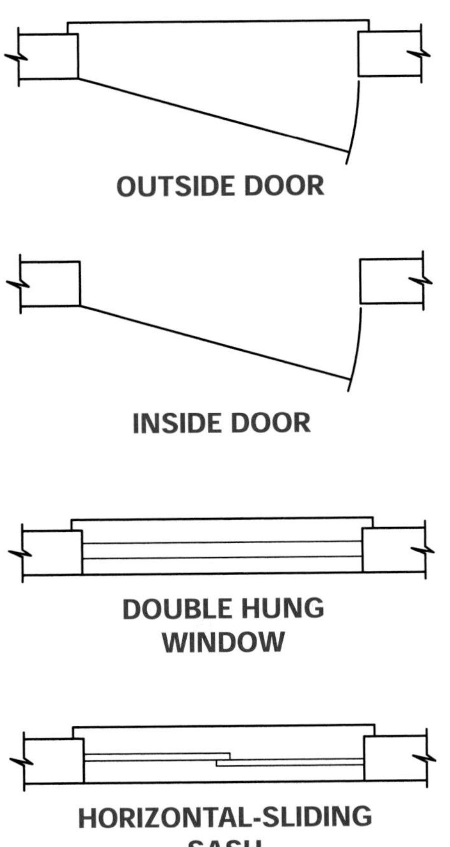

Fig. 34-12—*Drawing symbols for doors and windows.*

Views in Construction Drawings

Construction drawings include both technical working drawings and picture-like presentation drawings. The views of a house are presented in different types of drawings, depending on the type of information that must be conveyed.

Plans

A *plan* is a top view. Several types of plans are used for specific purposes.

- A *plot plan* shows the building site with boundaries, contours, existing roads, utilities, and other physical details such as trees and buildings.
- A *foundation plan* is a top view of the footing and foundation walls, showing their precise size and location. All openings in foundation walls are shown, as well as details of grading, waterproofing, and backfilling.
- *Floor plans* are cross-section views of a building. The horizontal cutting plane is placed so that it shows all door and window openings. Fig. 34-13 (page 436). A floor plan shows:
  - Outside shape of the home.
  - Arrangement, sizes, and shapes of rooms.
  - Types of materials to be used.
  - Thickness of walls and partitions.
  - Types, sizes, and locations of all of the doors and windows.

Elevations

An *elevation* is a picture-like view of a building that shows exterior materials and the height of windows, doors, and rooms. It may also show the ground level surrounding the structure, nearby vegetation, and elements such as decks and walkways.

Framing Plans

Framing plans show the size, number, and location of the structural members that make up the building framework. Separate framing plans may be drawn for the floors, walls, and roof. The *floor framing plan* specifies the sizes and spacing of joists, girders, and columns used to support the floor. *Wall framing plans* show the location and method of framing openings and ceiling heights so that studs and posts can be cut. *Roof framing plans* show the construction of the rafters used to span the building and support the roof.

Sectional Views

Sectional views, or sections, show how a structure looks when cut vertically by a plane. They provide important information about height, materials, fastening and support systems, and concealed features. Since they show details, they are drawn to large scale.

Details

Details are large-scale drawings showing the builders of a structure how its various parts are to be connected and placed. The construction at doors, windows, and stairs is customarily shown in detail drawings. Such drawings are also used whenever the information provided in plans, elevations, and sections is not clear enough for the tradespeople on the job.

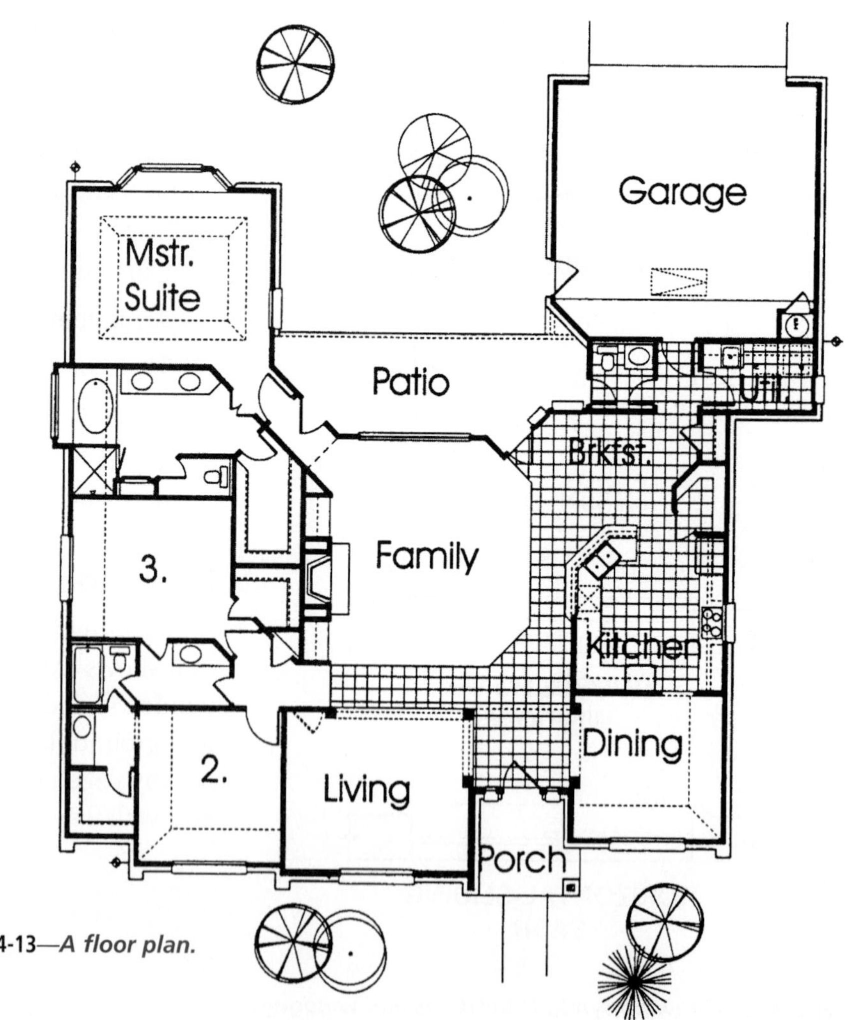

Fig. 34-13—*A floor plan.*

CHAPTER 34

Review & Applications

Chapter Summary

Major points from this chapter that you should remember include:

- Houses can be built using conventional construction methods, or they can be manufactured in a factory.
- Construction planning includes decisions about the location, design, and actual construction of the house.
- Legal documents needed when purchasing a lot include a survey, abstract, and deed.
- Specifications are written details on construction that are not given on working drawings and materials lists.
- Construction drawings include both technical working drawings and picture-like presentation drawings.

Review Questions

1. How are conventional framing methods different from manufacturing methods in building a house?
2. Name and describe the three types of manufactured houses.
3. Name at least three factors that must be considered when selecting a lot.
4. When purchasing a lot, what three legal documents need to be obtained?
5. What are three sources of house plans?
6. In what ways is complete information on building a house communicated?
7. List the views most commonly included in a set of construction drawings.
8. Which plan shows physical details of the building lot?

Solving Real World Problems

Blake and Noria are preparing to build a new home. They have found a design that they like and have purchased a set of plans from the design company. They have even discussed its construction with a building contractor. Their contractor has given them a quote for building the new home, and they are ready to start arranging financing. They discover that they will need a full set of plans for the bank where they apply for the loan, for the inspection department, and for each contractor involved with the project. How many sets of prints will they need to order, and for whom?

CHAPTER 35

Sitework and Foundations

YOU'LL BE ABLE TO:

- Briefly describe the process of locating a building on a site.
- Describe the components of a foundation.
- Name advantages of concrete block foundations versus poured concrete.
- Discuss the purpose of sill plates.
- Explain the difference between concrete and grout or mortar.

Look for These Terms

- anchor bolt
- batter boards
- builder's level
- concrete
- footing
- foundation
- plot plan
- sill plate
- sitework
- transit level

Laying out and building the foundation of a house is a critical part in the construction process. The **foundation** anchors the house to the land, supports the weight of the house, and provides a level base for the floor framing. Before the foundation can be built, its future location must be accurately identified. This is done in a **plot plan. Fig. 35-1.**

LOCATING A BUILDING ON A SITE

Before construction on a house or other building can begin, the site must be prepared. Site preparation, or **sitework**, includes:

- Surveying the site to determine the exact boundaries of the property.
- Clearing the land.
- Laying out the building on the site.

The exact location of the building is affected by building and zoning codes as well as natural features. Codes often specify the closest distance a structure can come to property lines, its distance from the street, and other factors.

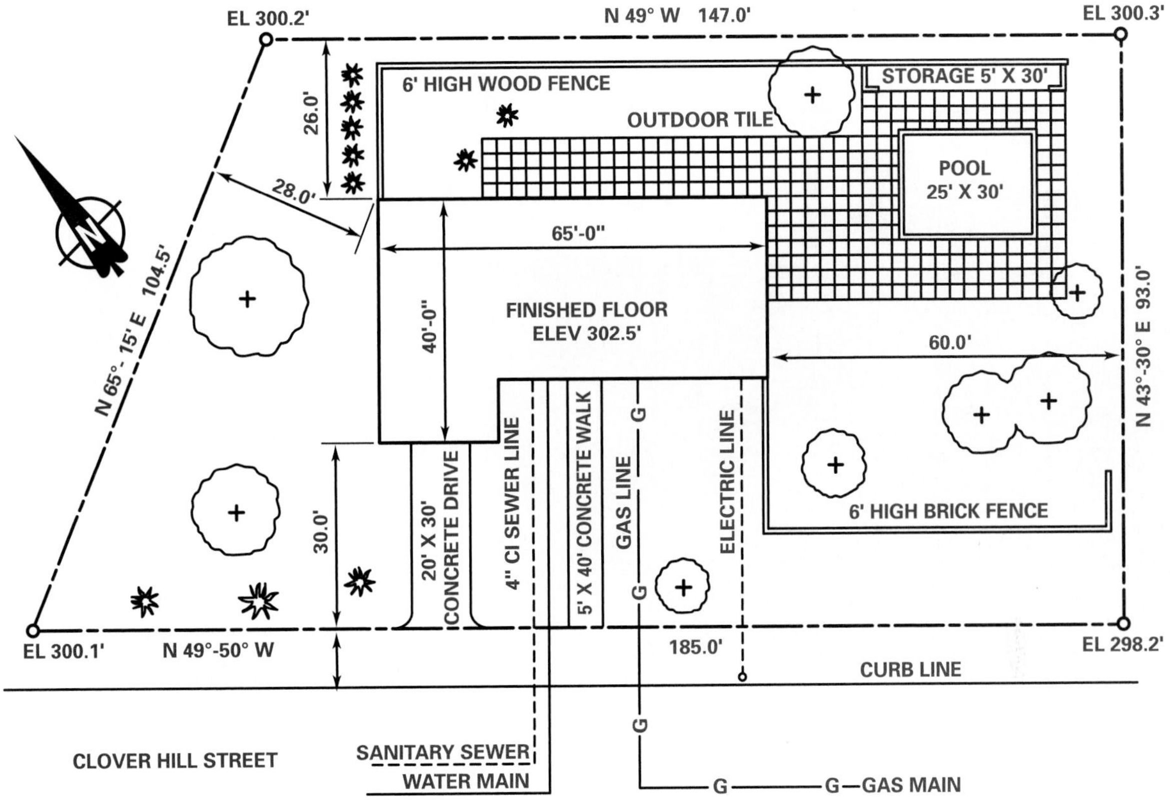

Fig. 35-1—*The plot plan accurately defines where the building is located on the building lot.*

There are two basic methods of determining the location of a building on the property. One method is to take measurements from an established reference line, such as a street or a well-marked property line. A more accurate method is to use a level. Fig. 35-2.

Using a Level

Two types of levels are commonly used in building construction. The **builder's level** (sometimes called an *optical level*) has a telescope fixed in a horizontal position. Fig. 35-3 (page 440). A **transit level** (transit) is similar to a builder's level, but its telescope can be moved up and down as well as sideways. Fig. 35-4.

To lay out a building using a level, you must start with a basic reference point. This point is usually called the *bench mark*. The bench mark is established by surveyors. It is often a small marker set in the ground at a designated location.

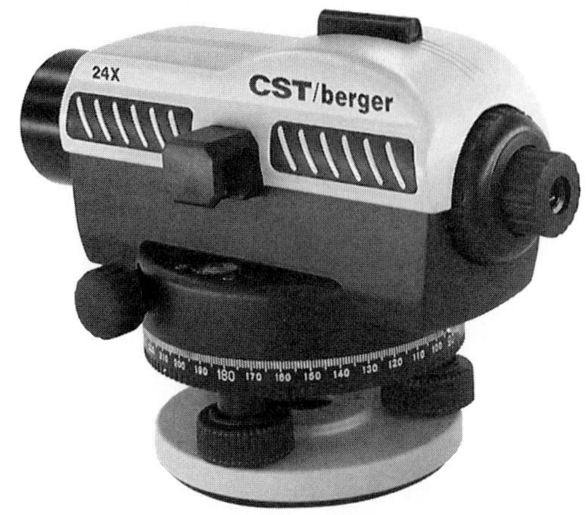

Fig. 35-2—*A builder's level allows a surveyor to take very accurate measurements.*

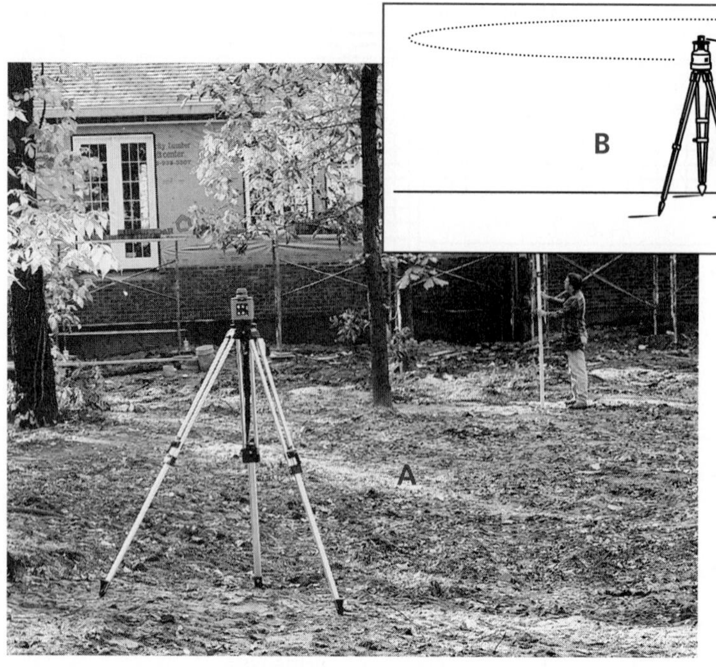

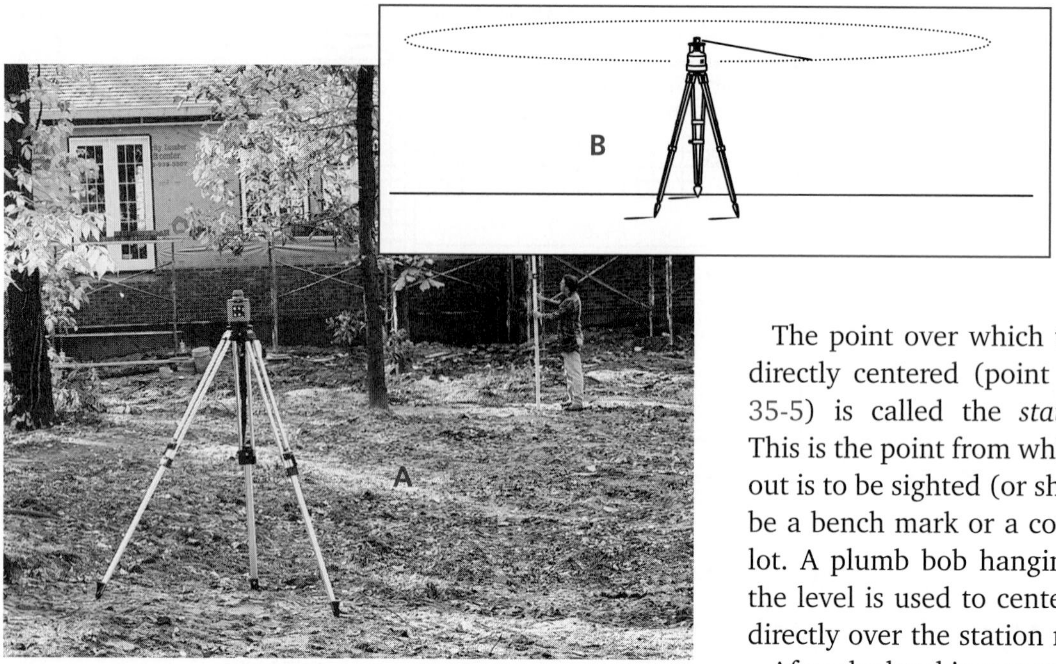

Fig. 35-3— ***(A) The telescope of a builder's level is fixed in a horizontal position. (B) It projects a horizontal reference plane across the job site.***

The point over which the level is directly centered (point "A" in Fig. 35-5) is called the *station mark.* This is the point from which the layout is to be sighted (or shot). It may be a bench mark or a corner of the lot. A plumb bob hanging beneath the level is used to center the level directly over the station mark.

After the level is centered over the station mark, another point (point "B" in Fig. 35-5) is located by sighting through the level and then is marked with a stake. Then the surveyor swings the head of the level 90 degrees, and sites and marks a third point (point "C" in Fig. 35-5). When connected by straight lines, points CAB form a right angle. Even a complex house can be positioned precisely on a job site by repeatedly plotting right angles.

Fig. 35-4—***The telescope on a transit level can be adjusted to different angles.***

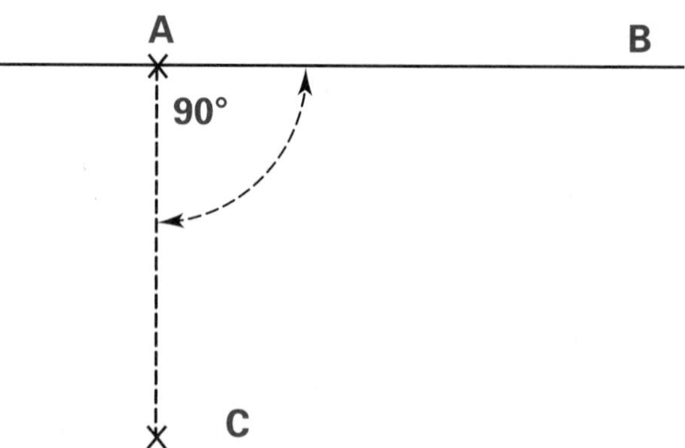

Fig. 35-5—***Laying out a right angle with a transit or a level.***

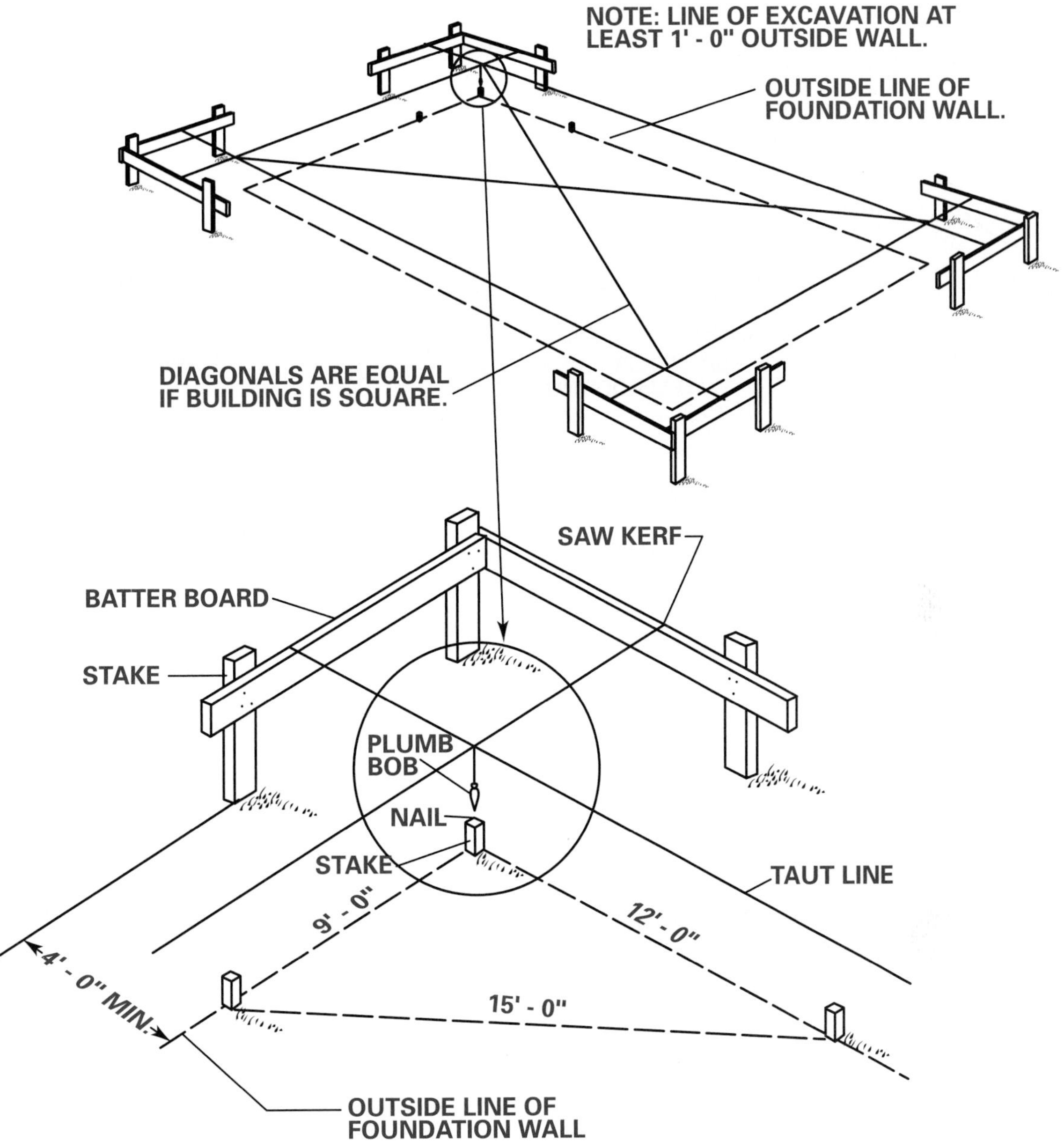

Fig. 35-6—*Using batter boards to establish the outline of a building. In some cases, the top edge of the batter board represents the height of the foundation wall.*

Laying Out the Building

After the corners of the house have been identified, the outline of the building can be laid out on the site. One way to do this is to place horizontal boards on small posts a minimum of 4 feet outside of where the corners of the building will be. String is stretched between these **batter boards** to mark the outside line of the foundation walls. This provides one method of locating and marking the outline of the building. Fig. 35-6.

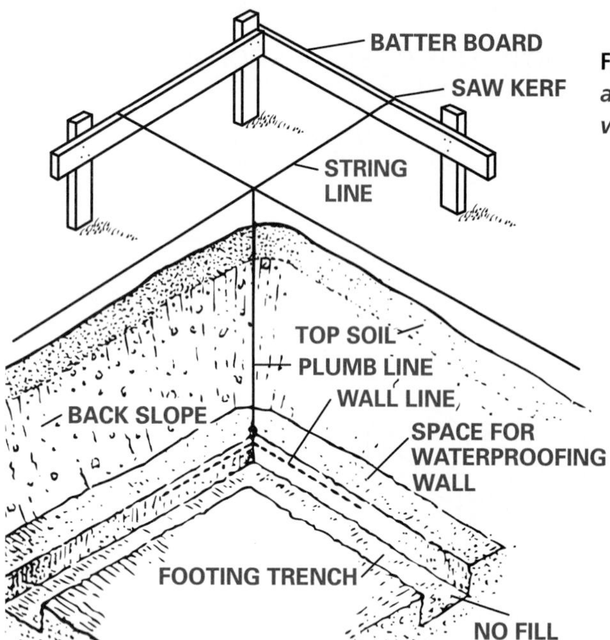

Fig. 35-7—*A plumb line is dropped from strings attached to the batter boards to locate the foundation wall accurately.*

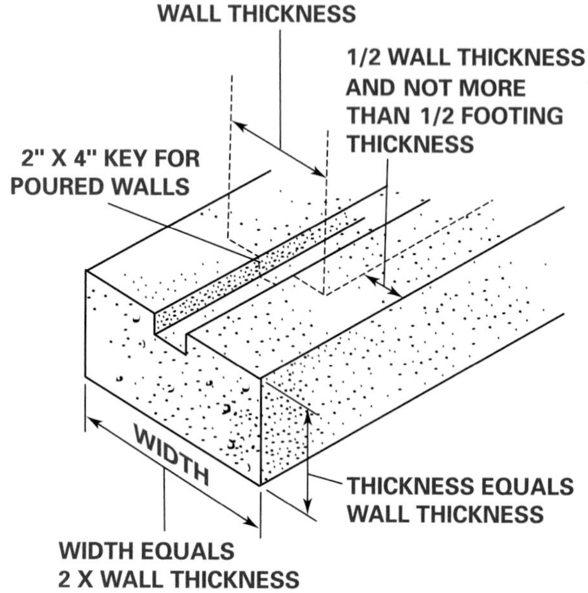

WALL THICKNESS
1/2 WALL THICKNESS AND NOT MORE THAN 1/2 FOOTING THICKNESS
2" X 4" KEY FOR POURED WALLS
WIDTH
THICKNESS EQUALS WALL THICKNESS
WIDTH EQUALS 2 X WALL THICKNESS

Fig. 35-8—*Footings must be at least 6 inches thick. In normal soils, the footing should be the same thickness as the wall it supports. It should project beyond each side of the wall by one-half the thickness of the wall.*

MATHEMATICS *Connection*

Pythagorean Theorum

Look again at the inset (closeup) for Fig. 35-6. Notice the triangle shown in dashed lines, with sides labeled 9′-0″, 12′-0″, and 15′-0″. Building contractors often use the 3-4-5, or Pythagorean, system of proving that a corner is "square." A square angle is one that measures exactly 90 degrees.

Although the Babylonians knew about this relationship more than 1000 years before the Greek Pythagorus formalized it, he was probably the first to state it in mathematical terms. (Pythagorus proved that the sum of the squares of the two sides of a right triangle equals the square of the hypotenuse—the side opposite the right angle). This theorum has been very useful in construction and in various other applications for the last 2000 years.

THE FOUNDATION

The foundation supports the rest of a building. Fig. 35-7. It may be made up of foundation walls on footings, or it may consist of a concrete slab resting on the ground. Foundations are usually installed by contractors who specialize in such work.

Foundation walls form an enclosure for basements or crawl spaces. They carry wall, floor, roof, and other building loads.

Foundation walls rest on an enlarged base called a **footing.** The footing provides a larger bearing surface against the soil beneath the building. Fig. 35-8. Footings are important in preventing settling or cracks in the wall.

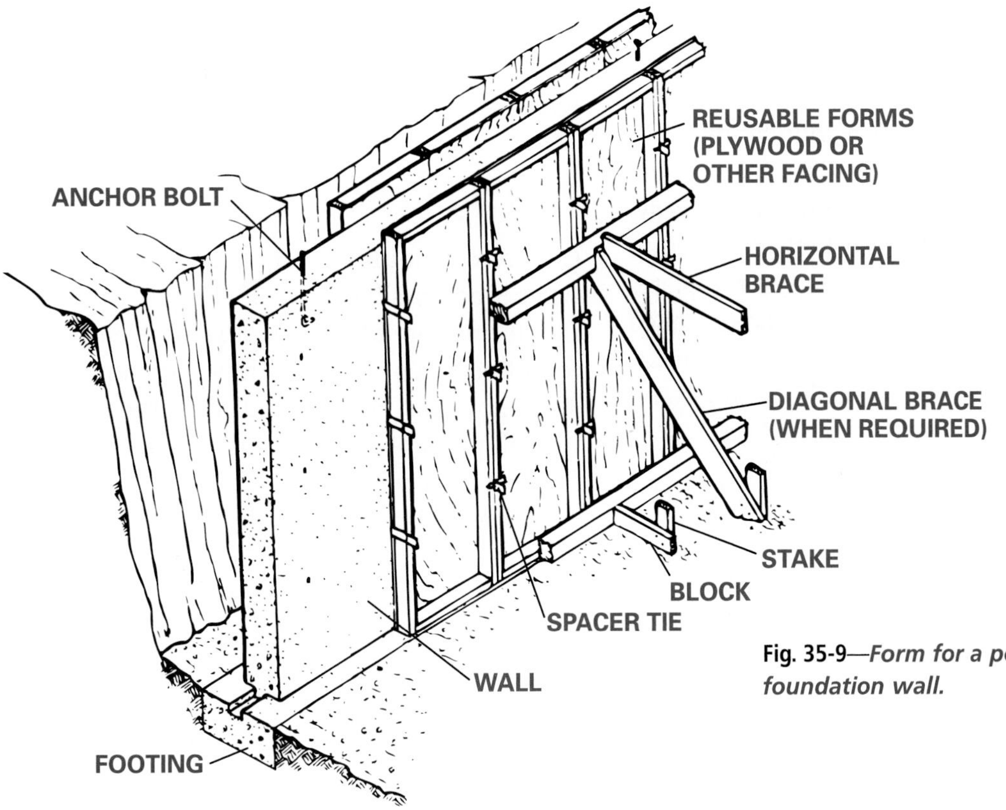

REUSABLE FORMS
(PLYWOOD OR
OTHER FACING)
ANCHOR BOLT
HORIZONTAL
BRACE
DIAGONAL BRACE
(WHEN REQUIRED)
STAKE
BLOCK
SPACER TIE
WALL
FOOTING

Fig. 35-9—*Form for a poured-concrete foundation wall.*

4" SOLID
CAP BLOCK
ANCHOR BOLT
KEY
WINDOW
FRAME
CONCRETE
BLOCK
COMMON
BOND
CEMENT MORTAR
COATING
WATERPROOF
COATING
COVE
FOOTING DRAIN
FOOTING

Fig. 35-10—*Concrete block foundation walls.*

Some foundations are made from a single mass of concrete. These slab foundations have no walls. The foundation is a concrete slab that rests directly on the ground.

Types of Foundation Walls

The two types of foundation walls most commonly used for houses are poured concrete and concrete block. Concrete block construction is popular for several reasons. It does not require formwork, and it is relatively inexpensive. Also, work on a block foundation can proceed as time permits; a concrete foundation should be poured all at once. Figs. 35-9 and 35-10.

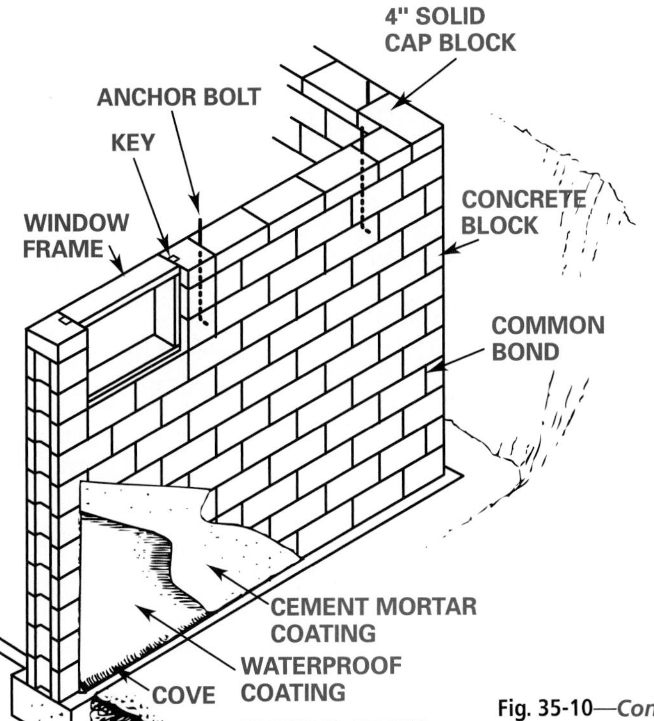

Sill Plates

In typical house construction, a **sill plate** creates the connection between foundation walls and the wood framing above. The plate should be pressure-treated wood or some other decay-resistant wood. It must be securely fastened to the foundation. Most builders fasten the plates with ½-inch diameter L-shaped bolts called **anchor bolts.** In high-wind and storm areas, well-anchored plates are particularly important. Fig. 35-11.

CONCRETE

Concrete is one of the most important construction materials. It is made by mixing cement, fine aggregate (usually sand), coarse aggregate (usually gravel or crushed stone), and water in the proper proportions. The product is not concrete unless all four of these ingredients are present. A mixture of cement, sand, and water—without the coarse aggregate—is called *mortar* or *grout.*

Pouring Concrete

Concrete should be poured continuously whenever possible. Fig. 35-12. This prevents the formation of *cold joints,* or weak areas found between pours made at different times. To remove air pockets and force the concrete into all parts of the forms, the concrete should be spaded or vibrated.

Concrete *flatwork* consists of flat areas of poured concrete, usually 5 inches or less in thickness. Examples of flatwork include concrete slab foundations, basement floors, driveways, and sidewalks.

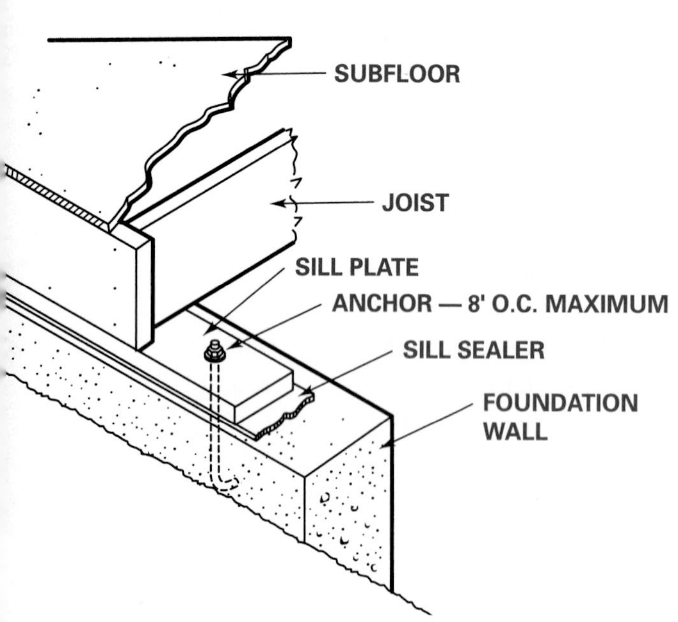

Fig. 35-11—*Anchor bolts embedded in the foundation wall are used to secure the floor frame.*

Fig. 35-12—*Pouring concrete for a new house.*

Finishing

After the concrete has been placed (poured) and thoroughly vibrated and spaded, it must be *finished*. Figs. 35-13 and 35-14. The finish depends on the intended use. A good, non-slippery finish is best for sidewalks, and a coarse, scored surface is best for driveways. A smooth-troweled finish is best for porches and basement floors.

Fig. 35-13—*After a slab is poured, it is finished to give it a surface appropriate for its final use.*

Facts about Wood

Trees, Soil, and Water—Trees, soil, and water are the three greatest renewable natural resources. They depend on each other for their existence. If there were no trees, the uncontrollable runoff of rain or melted snow would wash away valuable topsoil and cause serious flooding downstream. Although trees will not help clear up a polluted lake or stream, they do prevent additional pollution. Soil and timber conservation practices make it possible for the land to absorb and store water. This helps prevent erosion and floods.

Fig. 35-14—*Large areas of concrete are usually finished with special equipment such as this power trowel.*

CHAPTER 35

Review & Applications

Chapter Summary

Major points from this chapter that you should remember include:

- Sitework includes surveying the site, clearing the land, and laying out the building.
- Builder's levels and transits are instruments used to accurately locate a building on a site.
- Foundations may consist of foundation walls and footings or a concrete slab.
- Foundation walls may be of poured concrete or concrete block construction.
- Concrete is a mixture of cement, fine and coarse aggregate, and water.
- Mortar, or grout, is a mixture of cement, fine aggregate (sand), and water.
- Concrete should be poured all at one time to prevent the formation of weak areas.

Review Questions

1. What factors affect the exact location of a house on a building site?
2. Briefly describe the process for using a level.
3. What is the difference between a builder's level and a transit level?
4. Why are footings important?
5. What advantages do concrete block foundations have over poured concrete?
6. In what way is concrete different from grout or mortar?
7. Explain how sill plates are fastened to a foundation.

Solving Real World Problems

While doing research about building a new home, Tim and Robin found that in different areas of the country the requirements for foundations are different. For the Midwest (where they live), a 4-foot-deep concrete foundation is normal for a home built on a cement slab. However, in parts of Florida, a 12-foot foundation will suffice. Given that a 4-foot foundation is required anyway, Tim and Robin decide to dig a little deeper and build a basement. They decide to talk with a contractor about what materials and options they have, so they contact Luis for help. What materials can Luis suggest to them for basement walls? Is one material better than another?

CHAPTER 36

Framing and Enclosing the Structure

YOU'LL BE ABLE TO:

- Describe the two main categories of framing for buildings framed of lumber.
- Describe floor framing in a conventionally framed house.
- Explain the composition of a framed wall.
- Explain the purpose of ceiling joists.
- List advantages of both conventional and trussed roof construction.

Look for These Terms

- bridging
- conventional framing
- conventional roof construction
- floor joist
- plate
- post-and-beam framing
- sheathing
- stud
- subfloor
- trussed roof construction

Once a building's foundation is in place, its frame can be constructed. The floor, walls, ceilings, and roof are framed to provide support for the loads the building is expected to bear.

TYPES OF WOOD FRAMING

Buildings framed of lumber fall into one of two main categories. Fig. 36-1 (page 448).

- **Conventional framing,** often called *stick framing,* consists of closely-spaced small members assembled into a rigid structure.
- **Post-and-beam framing** consists of heavier members that are more widely spaced. Timber framing is one variation of post-and-beam framing.

Conventional Framing

Conventional framing consists of joists, studs, and rafters. These structural members are joined so they act together and share the loads in supporting the structure. These members are spaced 12 inches to 24 inches apart. When assembled and covered, they form complete floor, wall, and roof surfaces.

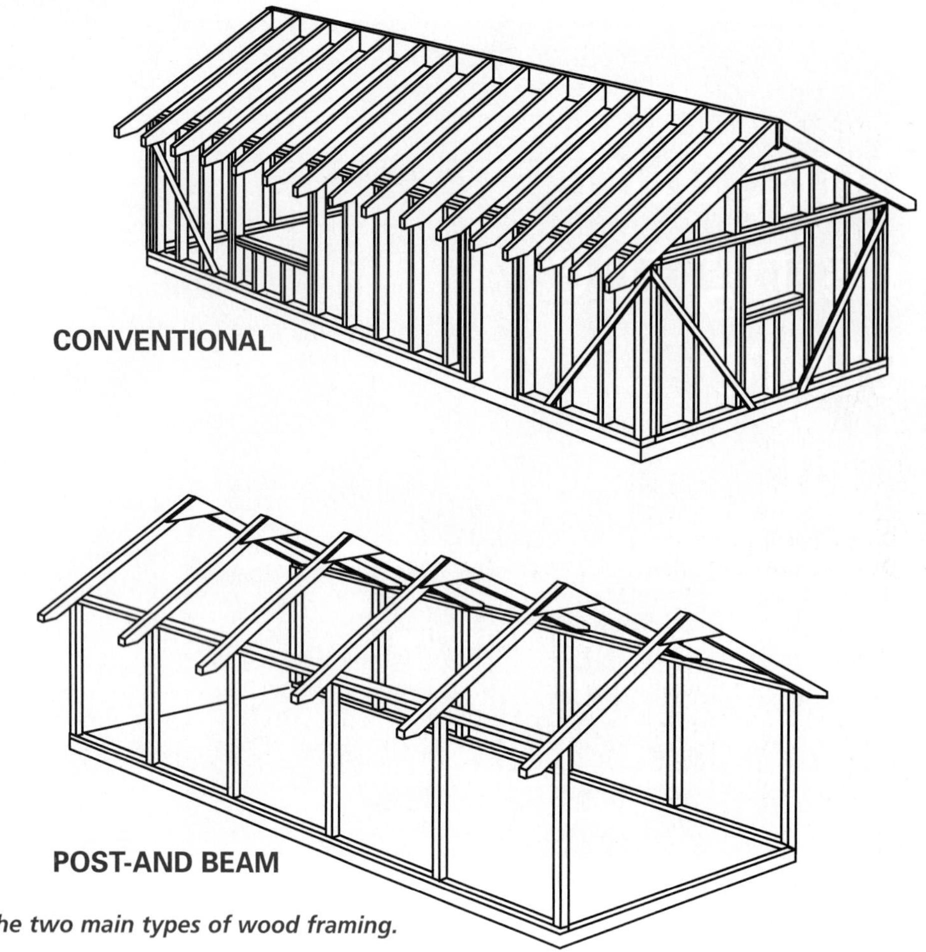

Fig. 36-1—*The two main types of wood framing.*

The most common type of conventional framing is called *platform framing.* Fig. 36-2. In this type of construction, the floors are complete platforms that rest atop the walls. The platform extends to the outside edges of the building and provides a base upon which outside and inside walls are built. Platform construction is suitable for houses up to three stories tall.

Post-and-Beam Framing

In post-and-beam framing, subfloors and roofs are supported by a series of beams spaced up to 8 feet apart. The beams are supported by posts, or piers. Compared to platform framing, post-and-beam framing calls for fewer, but larger, pieces spaced farther apart. Fig. 36-3.

FLOOR FRAMING

After the foundation is in place, the first part of a conventionally framed house to be built is the floor system. Floor framing consists of posts, girders, sill plates, joists, and subflooring. All of these members are connected with nails (and sometimes with construction adhesives) to support the rest of the house.

Fig. 36-2—*Platform framing.*

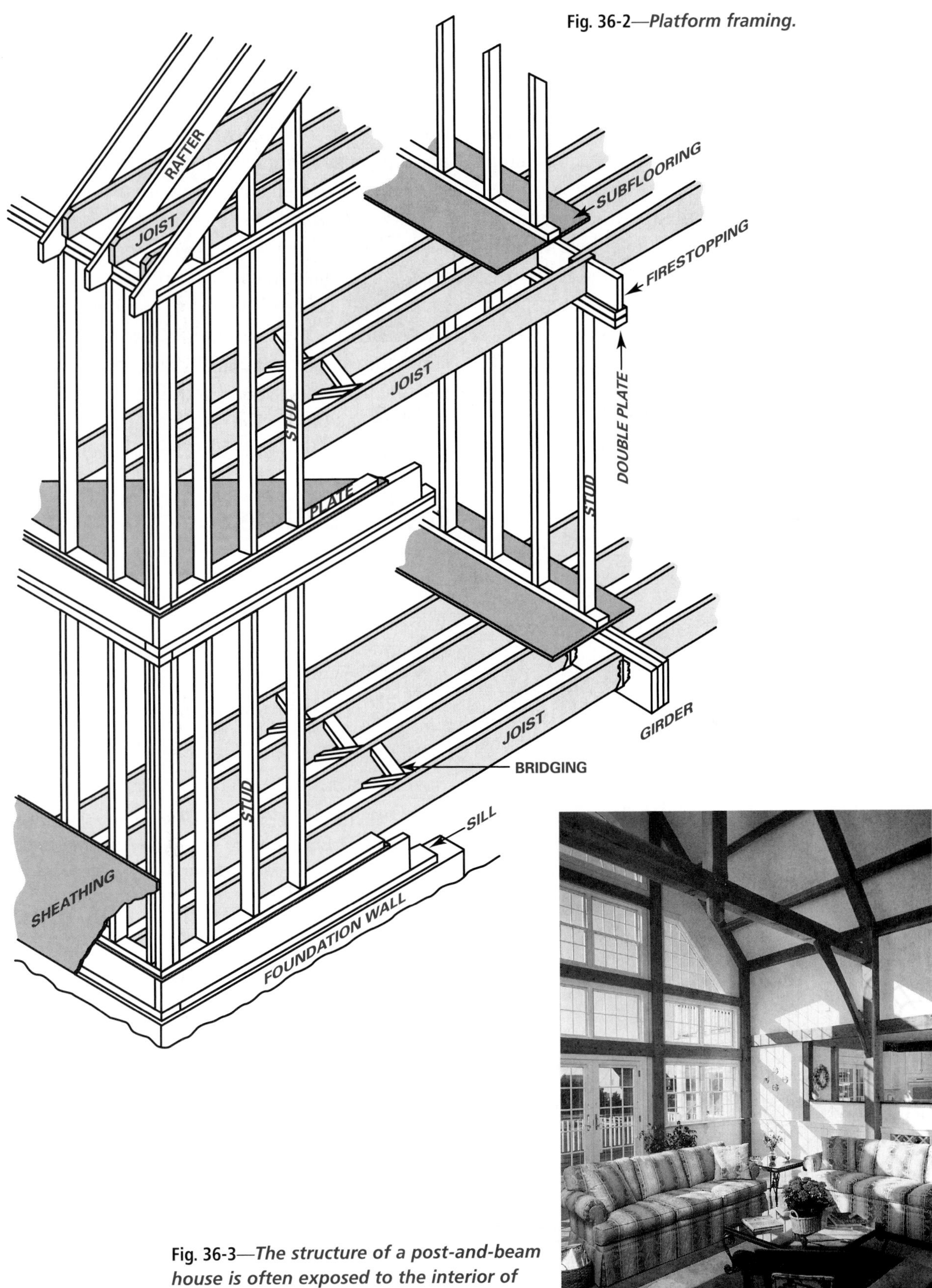

Fig. 36-3—*The structure of a post-and-beam house is often exposed to the interior of the house.*

Sill Plates and Band Joists

An assembly called a *box sill* is usually used to frame a platform floor system. It consists of a pressure-treated sill plate (sometimes called a *mudsill*) that is bolted to the foundation, and a *band,* or *rim,* joist that is nailed to the sill plate. The plate provides a secure link to the foundation. It also provides an even, level bearing surface for the floor joists. Fig. 36-4. Sills are anchored to the foundation with ½-inch bolts (anchor bolts) spaced about 6 to 8 feet apart.

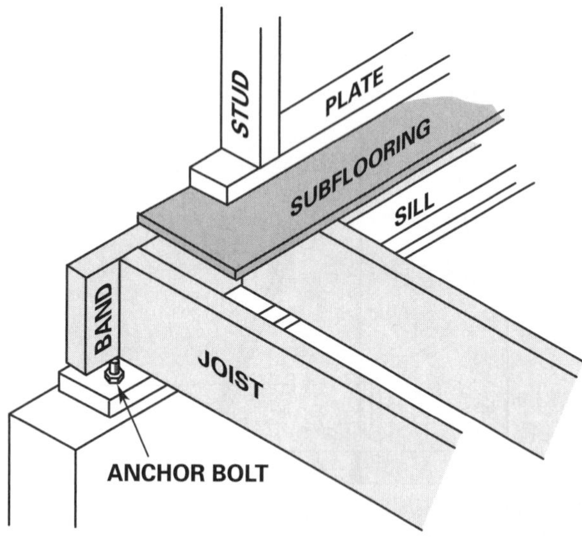

Fig. 36-4—***First-floor framing at the exterior wall in platform-frame construction.***

Floor Joists

Floor joists are the horizontal structural members that support a floor. The joists are spaced and sized to support expected loads, but they are typically 16 inches or 24 inches on center (the distance measured from one board centerline to the next).

When joists are placed over a long span, they have a tendency to sway from side to side. To help solve this problem, a bracing method called **bridging** is commonly used. Fig. 36-2 shows *diagonal bridging.* This method of bridging is very effective and requires less material than using solid lumber for bridging. Diagonal bridging may be made of wood or metal.

Posts and Girders

Floor joists generally cannot span the entire width of a house. They must be supported in mid-span by one or more large beams called *girders.* The ends of a girder typically rest on the foundation walls. The girder may gain additional support from wood or steel posts. Fig. 36-5.

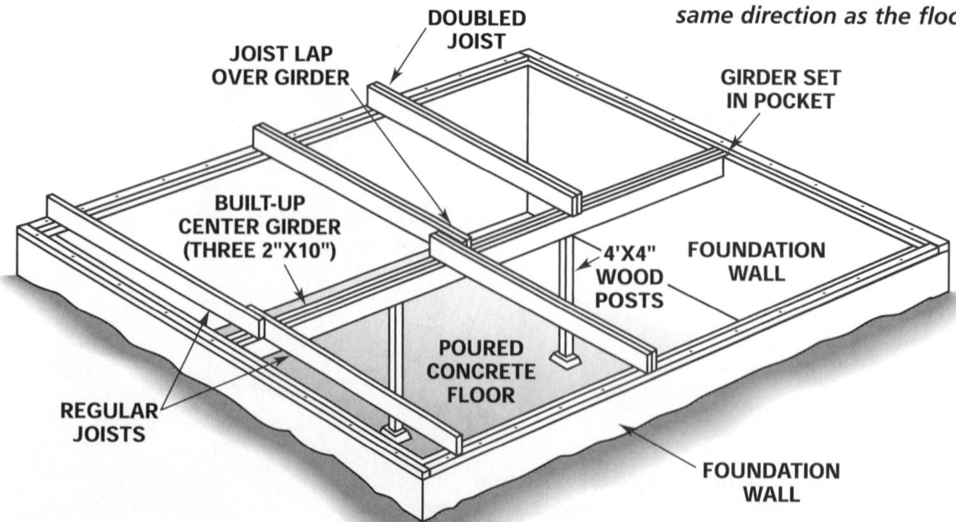

Fig. 36-5—***Partially framed floor. Note that the joists lap over the girder, Doubled joist are used under partitions that will run in the same direction as the floor joists.***

Girders may be made of wood or steel. Steel I-beam girders can span greater distances. Fig. 36-6. They do not shrink as wood does, but they are very heavy and can be hard to set into place. Wood girders are lighter and easier to install. They are easily connected to a support post. However, wood girders require more posts for support, which can obstruct floor space in the basement. Whatever the material, the top of a girder is usually level with tops of sill plates on foundation walls.

Other Floor Framing Materials

Materials used for floor framing have changed a lot in the past few years. The biggest changes affect joists and subflooring. Nominal 2-inch lumber is still common for floor joists; however, *floor trusses,* wood *I-joists,* and laminated-veneer lumber are used more and more often. Fig. 36-7. These materials offer the advantages of light weight, consistent strength, and long lengths.

Subflooring

A **subfloor** is a wood floor attached directly to the floor joists, under the finished floor. Many carpenters use construction adhesive and nails to ensure tight, squeak-free subflooring. Subflooring lends bracing strength to the building. It provides a solid base for the finish floor. In addition, it provides a safe working surface for building the house.

WALL FRAMING

A building has two types of walls: exterior and interior. Exterior walls are usually *load-bearing walls.* This means they help support the weight of the structure. Interior walls divide the building into different rooms. Some interior walls help bear the load; others do not. Interior walls may also be called *partitions.*

Fig. 36-6—*A steel girder and support post.*

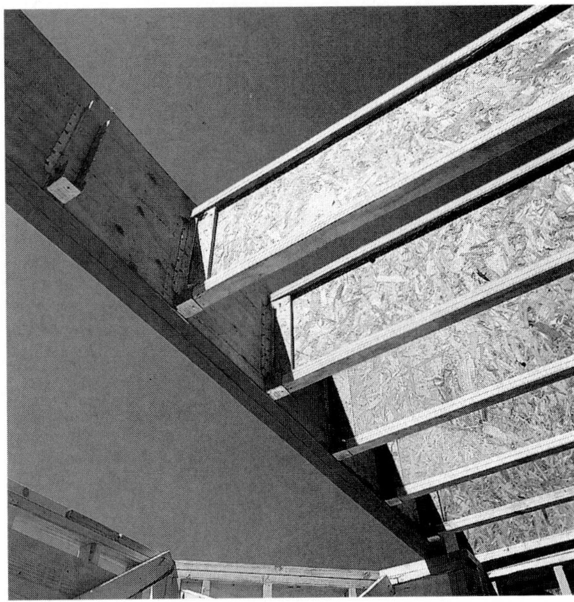

Fig. 36-7— *Wood I-joists are often used where predictable stiffness and long lengths are required. This I-joist is being supported by a metal joist hanger.*

Structural Members

Frame walls, both exterior and interior, are made up of two types of structural members: horizontal and vertical. The horizontal members are called **plates.** (At the tops and bottoms of windows and doorways, they're called *headers.*) The vertical members are called **studs.**

Sole Plates and Top Plates

All framed walls rest on a horizontal structural member called a *sole plate,* or *bottom plate.* Fig. 36-8. The sole plate is nailed into the floor.

At the upper end of a wall is the *top plate.* The top plate ties studs together at the top of the wall. It serves as a connecting link between the wall and the roof framing. The top plate is usually doubled to distribute the weight of joists that don't rest exactly above a stud.

Studs

Studs are the vertical structural members placed between wall plates. In one- and two-story buildings, studs are usually 2 × 4 or 2 × 6 lumber. In conventional construction, the studs are 2 × 4s spaced 16 inches on center. In cold climates, 2 × 6s are often used to frame outside walls because they allow more insulation to be fit into the walls.

Before a wall is assembled, the stud locations are marked out on the plates. Then, each stud is secured by nailing into it through the plates. An alternate fastening technique is to toenail (nail at an angle) each stud into the plate.

In a wall, allowances must be made for framing doors and windows. These allowances, when added to the size of the finished openings, make up what is called a *rough opening.* The rough opening is the distance between the *trimmer studs.* (Refer again to Fig. 36-8.) Most building plans provide the rough opening size for framing.

Assembling Walls

In the most common method of assembling walls, all the lumber is precut to length by carpenters. The pieces are then placed at the area of the subfloor where they will be nailed together. Once nailed together, the wall is measured to be sure it's square. It's then raised into place and nailed down. Fig. 36-9. Sometimes **sheathing,** the layer placed between a building's frame and outside surface, and siding are put on the wall while it is still on the subfloor.

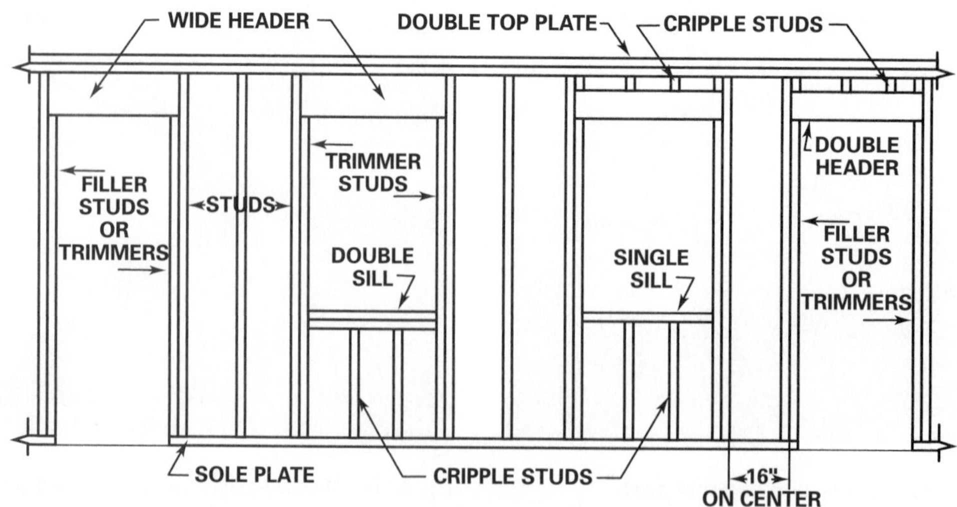

Fig. 36-8—*Parts of a framed wall section. Note that filler studs are also called jack studs.*

Fig. 36-9—*Raising a long stud wall.*

Fig. 36-10—*A house wrap is applied between the sheathing and the siding to help reduce air infiltration.*

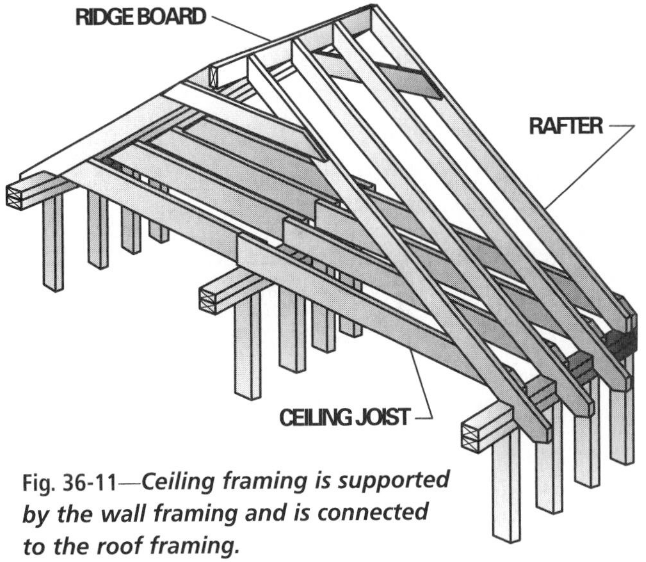

Fig. 36-11—*Ceiling framing is supported by the wall framing and is connected to the roof framing.*

Once all the outside walls have been tilted up into place and braced, the inside walls are framed. Inside walls are framed in the same manner as outside walls.

House Wrap

Shortly after the walls have been sheathed and erected, *house wrap* is stapled to the sheathing. House wrap reduces air infiltration and protects the sheathing from any moisture that might get behind the siding.

A house wrap has several advantages over traditional felt building papers. Because of its light weight, it can be installed using rolls 9 feet wide. This helps speed installation. Also, a house wrap is difficult to rip; therefore, it's less likely to be damaged during installation. Fig. 36-10.

CEILING FRAMING

Ceiling joists create a link between the outside and inside walls. The ceiling joists also serve as floor joists for an attic or a second story. In conventional construction, the ceiling joists are laid out, cut, and fastened one piece at a time to the building. Fig. 36-11. They are supported by beams, girders, or bearing walls.

Facts about Wood

The Log Cabin—The log cabin was a simple shelter with several advantages. It could be built quickly from what often was the most available resource—wood. One of the great strengths of the log cabin was that it could be built using only a few tools. An axe would be needed, and probably an auger. In selecting trees, the builder took some care to ensure that the logs would be roughly of the same diameter. After being felled, the trees were cut to the needed length and trimmed of their branches. The ends were notched to interlock with each other.

The walls were built by stacking one log on top of another. The spaces between the logs were chinked, or filled with mud or moss. The roof was made of several small logs, which served as rafters, and a ridgepole, which ran the length of the cabin. The roof framework was covered with roughly hewn flat wooden slabs. The roof was perhaps the least weather-resistant part of the log cabin. In fact, the lyrics of many pioneer songs mention the problems of a leaking cabin roof.

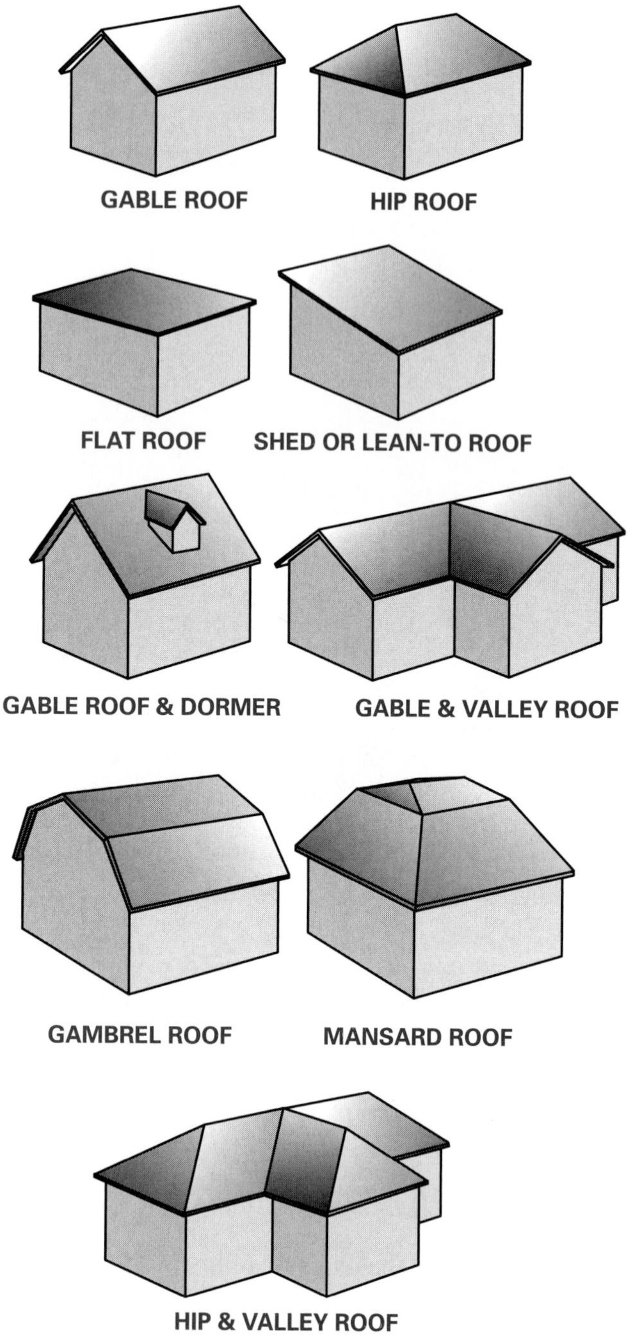

Fig. 36-12—*Common roof styles.*

ROOF FRAMING

To many carpenters, roof framing is the greatest test of carpentry skill. Roof construction must be strong in order to withstand snow and wind loads. It is also the most complex carpentry found in the frame of a typical house. Further, because the shape of a roof is such a key part of the look of a house, it demands extra effort during construction. Various roof styles are used to create different architectural effects. Fig. 36-12. A carpenter must understand and be able to frame these many styles.

Types of Roof Framing

The two types of frames for building a roof are conventional and truss frames. In **conventional roof construction,** joists and *rafters*—the framing pieces that support the roof itself—are cut to length at the jobsite and assembled one by one into a roof. Fig. 36-13.

In **trussed roof construction,** the lumber is prefabricated in a factory into large, triangular frames called *trusses.* The trusses combine rafters and joists into one unit. These units are then delivered to the jobsite and hoisted into place. Fig. 36-14.

Fig. 36-13—*A conventionally framed roof is installed piece by piece. This is a gable roof.*

Fig. 36-14—*Fastening roof trusses in place.*

MATHEMATICS *Connection*

Rise and Run

The angle of a roof is not defined in terms of degrees. Instead, it is described as a relationship between the roof's "run" and its "rise."

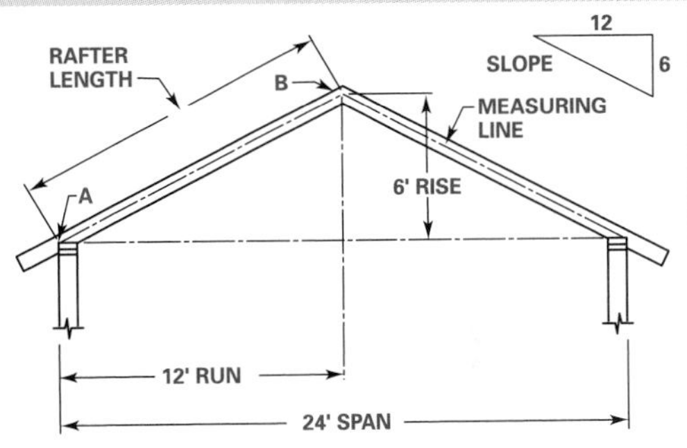

For a symmetrical roof, the *run* is exactly half the width of the house. The *rise* is the vertical distance from the top of the wall framing to a theoretical midpoint on the ridge. All roof framing calculations are based on this geometry. On building plans, this relationship is often shown as a triangle, listed with numbers representing the run and rise for a roof. For example, a roof that gets 12 inches higher for every 12 inches of run would be referred to as a "12/12" (pronounced "twelve-twelve") roof. A 5/12 roof would be less steep; a 13/12 roof would be more steep.

A roof truss is a prime example of efficient use of wood products. Trusses can be erected quickly, so the house can be enclosed in a short time. They are usually designed to span from one outside wall to the other; no inside bearing walls are required.

A conventional roof takes longer to install than a trussed roof; however, conventional roof construction affords more flexibility, allowing any type of roof to be constructed. Also, it lets carpenters make design changes at the job site to serve client needs. Fig. 36-15.

Roof Sheathing

The final step in roof framing is to nail sheathing to the rafters. Roof sheathing, like wall sheathing and subflooring, is a structural element. It provides a nailing base for the finish roof covering. It gives rigidity and strength to the roof framing. Panel sheathing such as plywood can be installed quickly over large areas. It also provides a smooth, solid base with a minimum of joints. Fig. 36-16.

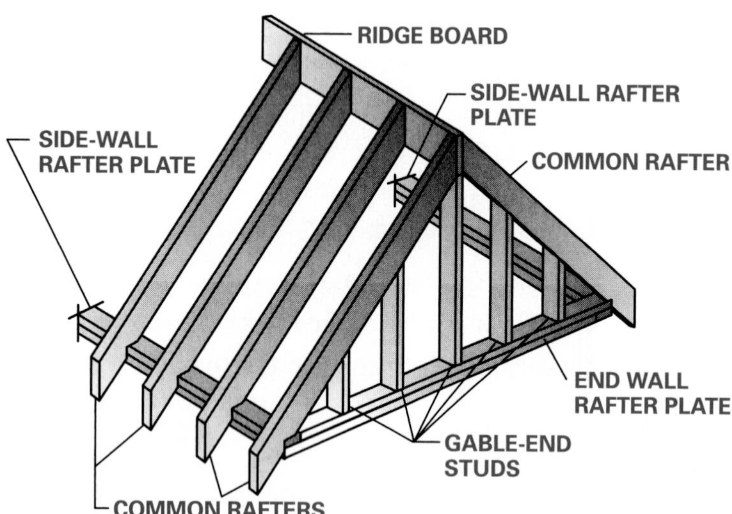

Fig. 36-15—*Framing a gable-end roof.*

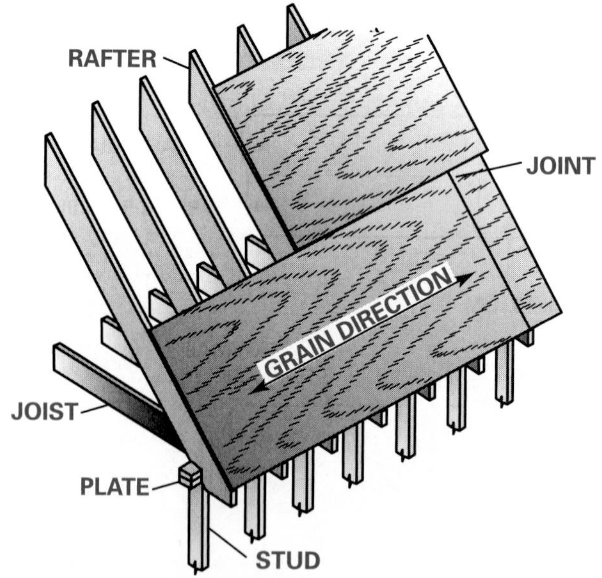

Fig. 36-16—*The grain of plywood sheathing should be at right angles to the supporting members.*

CHAPTER 36

Review & Applications

Chapter Summary

Major points from this chapter that you should remember include:

- Conventional framing consists of small, closely-spaced members built into a rigid wall and floor structure.
- Post-and-beam framing consists of heavy structural members that are widely spaced.
- Floor framing of a conventionally framed house consists of posts, girders, sill plates, joists, and subflooring.
- The horizontal structural members of a wall are called *plates*; the vertical members are *studs*.
- Ceiling joists serve as floor joists for an attic or a second story.
- The two types of frames for building a roof are conventional and trussed frames.

Review Questions

1. What are the two main categories of framing for buildings framed of lumber?
2. Describe the floor framing in a conventionally framed house.
3. Explain the composition of a framed wall.
4. What is the purpose of house wrap?
5. What is the purpose of ceiling joists?
6. What is a roof truss?
7. List advantages of both conventional and trussed roof construction.

Solving Real World Problems

Zack's new home is progressing well, and the foundation work is finished. The general contractor stops by Zack's present home to let him know that the lumber order will be dropped off in two days. Almost as a side comment, the contractor asks if Zack has decided what he wants for wall sheathing. Zack says he is not sure what would be best and asks for a recommendation. What do you think the contractor will offer as options? Which option would be the best?

CHAPTER 37

Completing the Exterior

YOU'LL BE ABLE TO:

- List materials used as roofing.
- Name three types of windows.
- List materials that can be used as siding.
- Explain why outside wood surfaces should be painted.

Look for These Terms

- doorframe
- doorjamb
- downspout
- gutter
- shingle
- siding
- skylight
- underlayment

The exterior of a structure consists of roofing, siding, windows, and exterior doors. These go together to form a protective "shell," sealing out the weather and providing security. **Fig. 37-1.** In addition, materials selected for the roofing and siding and the type, number, and placement of windows and doors are very important. They greatly affect the overall appearance of the home as well as the ease of maintenance. Generally, the roofing is installed first, then windows and exterior doors, and then the siding. After these installations are complete, paint is applied as needed.

ROOFING

The choice of roofing materials and the method of application are influenced by cost, roof slope, expected service life of the roofing, wind resistance, fire resistance, and local climate. Due to the large amount of exposed surface, appearance is also important.

Materials used for pitched (slanted) roofs include wood, asphalt, or fiberglass shingles, cement or clay tiles, and slate. Steel sheet materials are also used in some parts of the country.

On flat or low-pitched roofs, built-up roofing is typically used. This consists of layers of roofing felt covered with asphalt and a top layer of gravel. Synthetic rubber or plastic sheets are sometimes used instead. These may or may not be covered with a layer of gravel.

Fig. 37-1—*Each part of the exterior of this house contributes to the attractiveness of the overall design.*

Underlayment

Underlayment is a material such as roll roofing that is placed under finish coverings to provide a smooth, even surface. Underlayment is often installed before the exterior roofing materials are applied. Its use is normally required with asphalt and slate shingles and tile roofing. Fig. 37-2. Other methods may be used with wood shingles. Fig. 37-3.

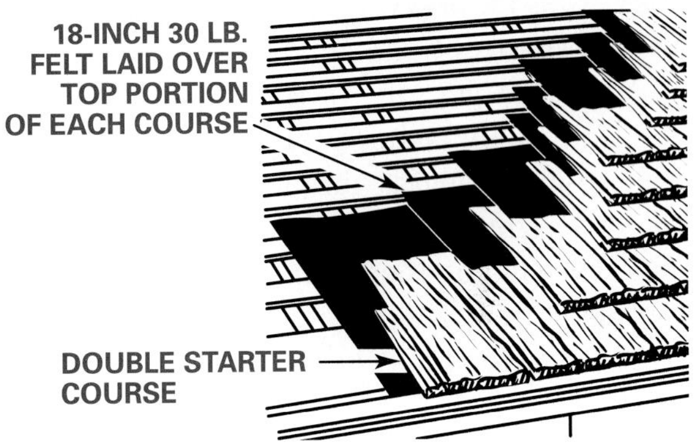

Fig. 37-3—*Wood shakes (shingles) may be installed over spaced sheathing with roofing felt placed between courses.*

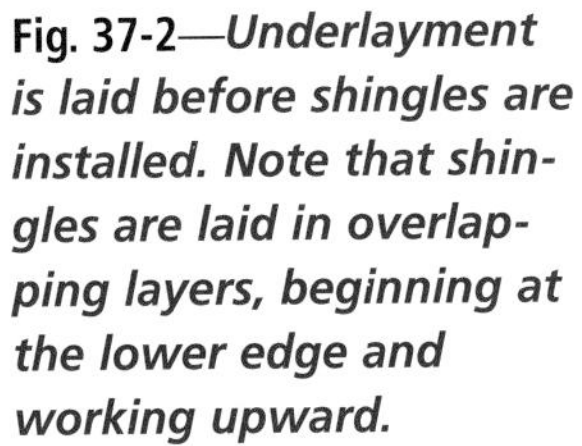

Fig. 37-2—*Underlayment is laid before shingles are installed. Note that shingles are laid in overlapping layers, beginning at the lower edge and working upward.*

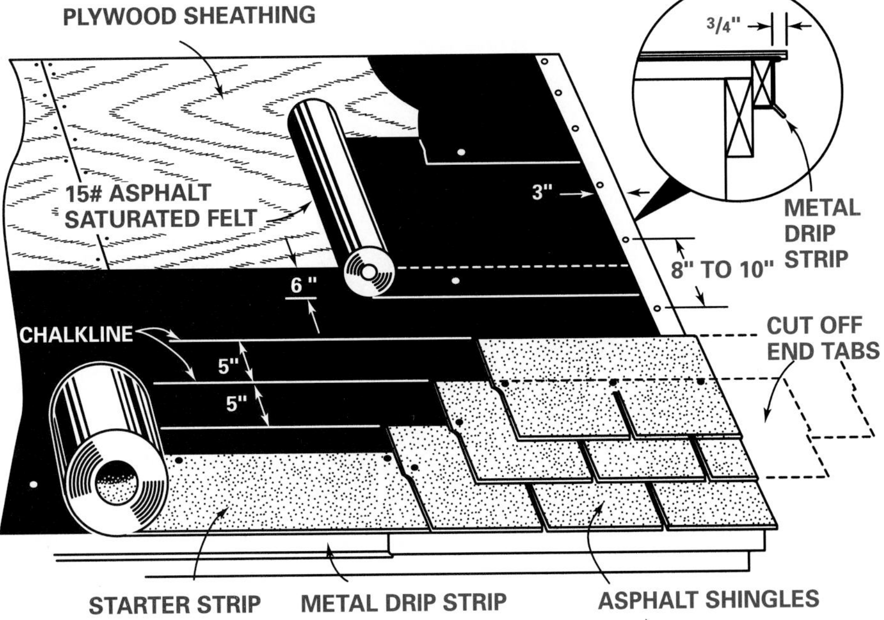

Shingles

Shingles are thin pieces of building material laid in overlapping courses (rows). They are used to cover the roof or exterior walls of a house.

Asphalt and fiberglass roof shingles come in several forms. The most common is the strip shingle, sometimes called a 3-tab shingle. These shingles are nailed to the roof deck, and they have asphaltic sealing strips that help to hold the shingle in place under windy conditions. Nails for applying asphalt roofing should be corrosion-resistant. Fig. 37-4.

Wood shingles are a traditional favorite. They are manufactured in 24-inch, 18-inch, and 16-inch lengths and in several grades. Wood shingles are not recommended on roofs that have shallow slopes. In areas of the country where fire hazards prevail, roofing products other than wood may be required.

Gutters and Downspouts

Gutters are troughs installed at the lower edge of the roof to catch water and carry it to downspouts. **Downspouts** are pipes that carry the water down and away from the foundation or into a storm sewer. Most gutters and downspouts are made of aluminum formed into prefabricated sections.

Gutters are sometimes installed along with the roofing. More often, both gutters and downspouts are installed after the siding is in place and the painting has been done. Fig. 37-5. They may be attached to the house with a variety of metal clips and hangers.

Fig. 37-4—*Using a pneumatic nailer to install shingles.*

Roof Ventilation

Attics and roof spaces must be ventilated. That is, a means must be provided for air to flow from under the roof. Adequate ventilation prevents a buildup of moisture and/or heat.

In cold weather, the warm air seeping up from the interior cools against the cold surface areas under the roof. If not allowed to escape, the moisture in the air will condense on the cold surfaces. It can soak insulation and encourage rot in wood framing. In hot weather, ventilation of attic and roof spaces allows hot air to escape. This keeps the temperature lower in the rest of the house.

One way of providing ventilation for attic and roof spaces is to install louvered vents in openings in end walls near the peak (highest edge) of the roof. Fig. 37-6. Many of these

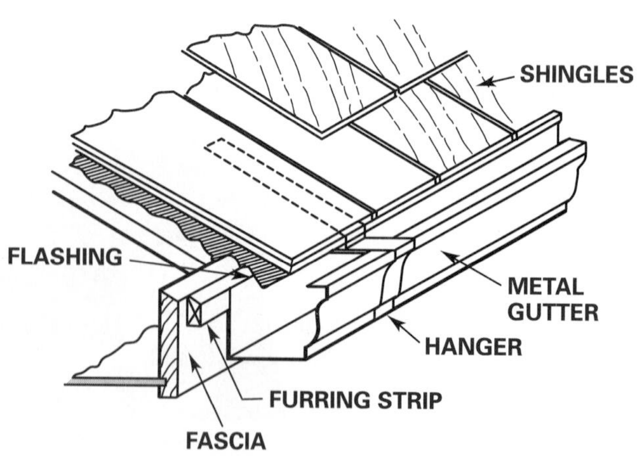

Fig. 37-5—*Gutter installed on a house.*

vents are made of metal. Others may be made of wood or vinyl to fit the house design more closely. To increase the flow of air through the attic and roof spaces, vents are often installed in the soffits (areas under the overhang of the roof). Fig. 37-7.

WINDOWS AND DOORS

Windows allow light and air to enter the building. Doors are a means of access and help provide privacy and security. Windows and doors are also important to the architectural design of a house. Fig. 37-8.

Fig. 37-6—*Outlet vents come in a variety of shapes and sizes.*

Fig. 37-8—*The windows and exterior doors are attractive features in this house design.*

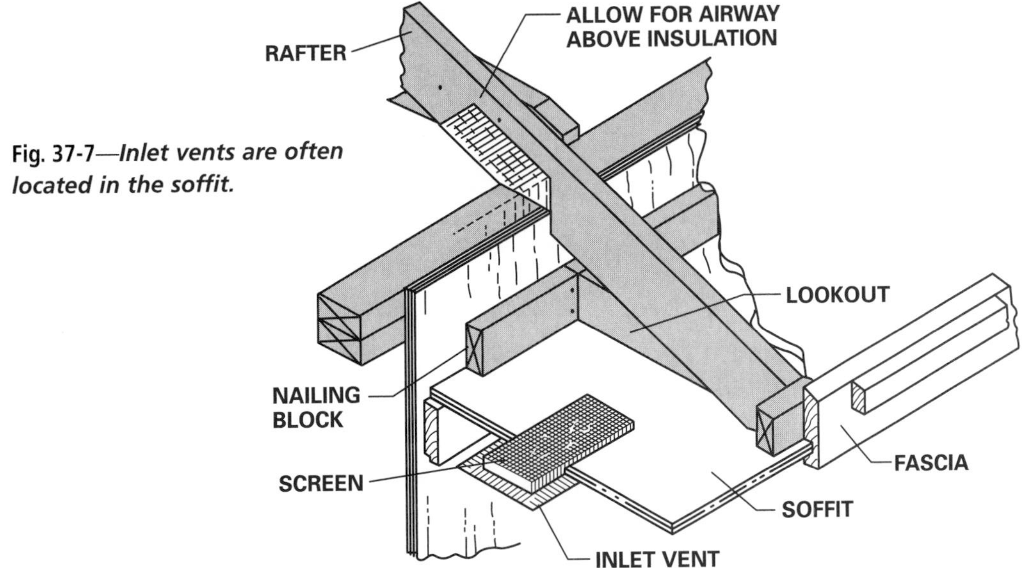

Fig. 37-7—*Inlet vents are often located in the soffit.*

Windows

Windows are usually assembled at the factory, complete with sash, screens, weather stripping, frame, and exterior casing. Fig. 37-9. Wood is the most common structural component of a typical window. It is used for the sash as well as the window frame. All wood components should be treated with a water-repellent preservative at the factory to provide protection before and after they are placed in the walls.

Some windows have extruded vinyl or aluminum components instead of wood. Windows with structural vinyl sashes and frames are easy to maintain and have good thermal (heat-related) performance.

Another type of window is often called a *clad window.* Fig. 37-10. It utilizes wood components that have been covered with rigid vinyl. The resulting product combines the strength and high energy efficiency of wood with the maintenance-free performance of vinyl.

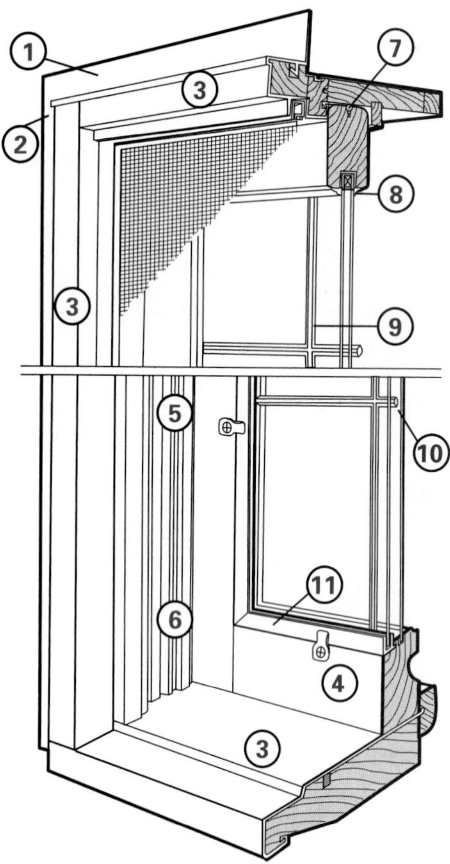

Parts of an assembled double-hung window: 1. Head flashing. 2. Flange. 3. Casing. 4. Sash. 5. Counterbalancing unit. 6. Tracks. 7. Weather stripping. 8. Glazing. 9. Grill (installed on the inside when insulated glass is used). 10. Grill (installed between the glass when storm panels are used). 11. Storm panel.

Fig. 37-9—*Parts of a double-hung window.*

Types of Windows

The three most common types of windows are double-hung, casement, and fixed. Fig. 37-11. Not every room will need the same size and type of window. For example, in bedrooms, light and ventilation are necessities. However, privacy and wall space for furniture are also important factors. A row of narrow operating windows placed high on two walls could be used.

Fig. 37-10—*A vinyl-clad window.*

SCIENCE Connection

Energy-Efficient Windows

Double-pane windows (windows that have two pieces of glass with an air space between them) are roughly twice as energy efficient as single-pane windows. The air between the panes slows the transfer of heat. However, double-pane windows can be made even more efficient by filling the air space with an inert gas such as argon or krypton. These gases are both safe and invisible. In addition, they also decrease problems with moisture condensation between the panes. Krypton is more effective than argon, but it is rarely used because it is also more expensive. Argon adds little or nothing to the cost of the window, but it increases efficiency by 15 to 20 percent. Some window manufacturers even offer it as a standard feature.

Fig. 37-11—***Types of windows. (A) Double-hung and fixed. (B) Casement.***

Energy Efficiency

Window manufacturers have greatly improved the energy efficiency of windows. This is important because 20 to 30 percent of the heat lost from an average house is lost through the windows. Heat loss can happen due to air leaking around the window or by heat being radiated through the glass. In hot climates, cool indoor air can be lost in the same ways.

Skylights

Skylights are transparent or translucent components installed on either pitched or flat roofs to provide light. Some also provide ventilation.

There are many styles of skylights. They may be either of two types: ventilating or fixed. *Ventilating* skylights swing open on hinges. Fig. 37-12 (page 464). They can allow heated air to escape the house in hot weather. They can also funnel cooling breezes into the house. *Fixed* (non-ventilating) skylights cannot be opened. They are generally less expensive and easier to install.

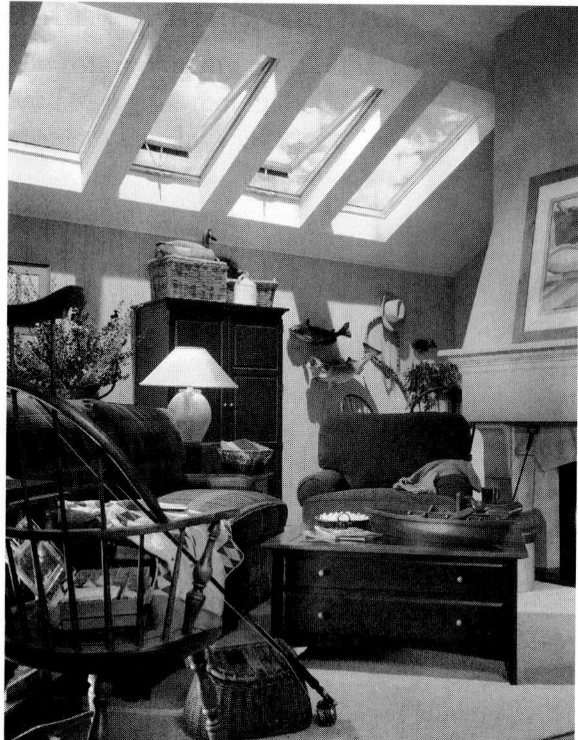

Fig. 37-12—*A ventilating skylight.*

Facts about Wood

Trees and Noise Pollution—During the past thirty years, the noise level in the average urban community has increased more than eight times. Trees tend to absorb some of the noise. Proper placement of trees and shrubs can reduce noise level up to 50 percent.

Exterior Doors

Exterior doors are made from wood or metal. They are available in many styles. Fig. 37-13. Metal doors usually have cores of rigid insulation. This greatly reduces the amount of heat lost through the door. Wood doors often feature panel construction. See Chapter 14.

Doorframes

A **doorframe** is an assembly that supports the door and enables it to be connected to the house framing. It consists of the doorjamb, sill, interior trim, exterior trim, and other molding, depending on the architectural design of the building. Usually, all parts of the doorframe are made from wood. Fig. 37-14.

The **doorjamb** is the part of the frame which fits inside the rough opening. Wood has been the traditional material for jambs, but steel and aluminum are also used.

Fig. 37-13—*Door styles can vary a great deal.*

SIDING

Siding is the exterior wall covering of a house. A homebuilder today can select from a wide variety of siding materials. These include solid wood, plywood, steel, aluminum, and vinyl. In addition, brick veneer is often specified where durability is of primary concern. Wood shingles and shakes are widely used as sidings. They are available in various lengths and grades.

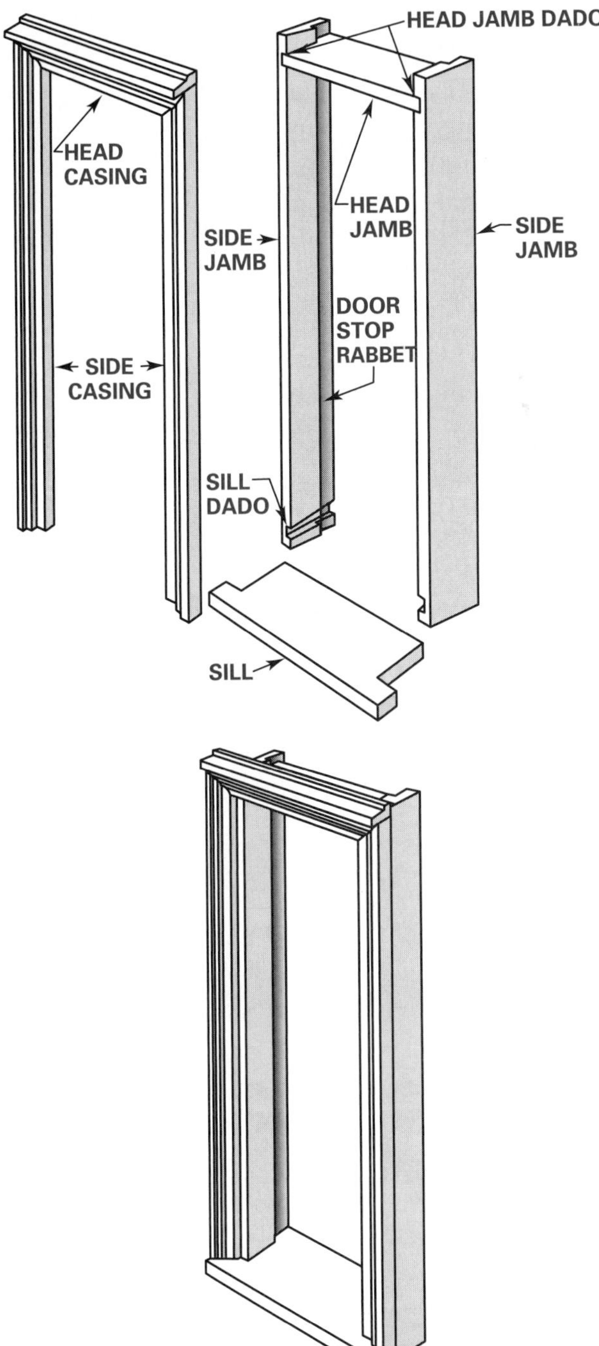

Fig. 37-14—*Parts of a doorframe.*

Masonry Veneer

Brick and stone are masonry materials. A veneer (layer) of these materials is often used for part or all of the siding over wood frame walls.

Brick is the most popular masonry material used for siding. Fig. 37-15. Its initial cost is relatively high. However, it requires little maintenance and lasts a long time. It can also fit into a variety of architectural styles.

PAINTING

Paint is used extensively to protect and decorate exterior surfaces. The siding on many homes must be painted. Trim and window frames and sashes are typically painted.

Properly applied, paint can prolong the life of a home and improve its appearance. Outside surfaces, particularly wood, require paints that will offer protection against weathering and moisture. Paint has other functions as well. In summer, white and light-colored paints reflect heat away from the house.

Painting is one of the last jobs to be completed when a new house is being built. It is done after the roofing and siding are complete and the windows and doors are in place.

Fig. 37-15—*Brick is a common masonry siding material.*

CHAPTER 37

Review & Applications

Chapter Summary

Major points from this chapter that you should remember include:

- The exterior of a structure consists of roofing, siding, windows, and doors.
- Factors considered when choosing roofing materials include cost, appearance, and the expected service life of the roofing.
- Most gutters and downspouts are aluminum.
- Ventilation of attics and roof areas prevents a buildup of moisture and heat.
- Windows and doors are important to the architectural design of a house.
- Various types of siding can be used to cover the exterior wall of a house.
- Properly applied, paint can prolong the life of a home and improve its appearance.

Review Questions

1. In what order are the exterior parts of a house installed?
2. What materials can be used for covering pitched roofs?
3. What is underlayment?
4. If you were installing shingles for a roof, where would you lay the first course?
5. What is one method of providing ventilation for an attic?
6. Name three types of windows.
7. What is the difference between a ventilating and a fixed skylight?
8. What part of the doorframe fits inside the rough opening?
9. Why should wood surfaces be painted?

Solving Real World Problems

Lars and Yelta have elected to finish the exterior of their new home in such a way that they will have zero maintenance. This has required them to use vinyl-clad doors and windows. Although they could have used aluminum, they decided that the durability of vinyl was best for their young family. Now they have to choose the material for siding their new home. They could use brick, aluminum, or vinyl. List the advantages of each and make a suggestion based on what you know about these materials.

CHAPTER 38

Completing the Interior

YOU'LL BE ABLE TO:

- Name the stages in which plumbing, electrical, and heating systems are installed, and give one example of a task performed during each stage for each system.
- Explain why insulation is important in a structure.
- Describe how drywall is installed.
- Give examples of finishes applied to walls and floors.
- Name the principal structural parts of a stairway.

Look for These Terms

- alkyd paint
- baluster
- drywall
- guardrail
- handrail
- heat pump
- insulation
- riser
- stringer
- tread
- vapor barrier

Work on completing the interior begins when the exterior shell of a house is covered and the interior is protected from the weather. **Fig. 38-1.** Special contractors begin the work of installing plumbing, electrical, and heating and air conditioning systems.

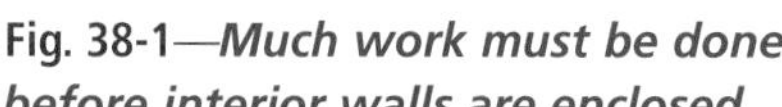

Fig. 38-1—***Much work must be done before interior walls are enclosed.***

Construction tasks included in completing the interior are:

- Enclosing walls.
- Building special features such as stairs.
- Laying floor coverings.
- Installing cabinets.
- Hanging interior doors.
- Attaching trim.
- Applying finishes to walls and ceilings.

INSTALLING SYSTEMS

Plumbing, electrical, and heating and air conditioning systems are installed in two stages. *Rough-in* work is done before insulation is installed. *Finish* work on systems is done later.

The carpenter must work closely with special contractors. Openings for pipes, wires, and heating system features are cut through structural members. Care must be taken to make sure the members are not weakened.

Plumbing System

The plumbing system brings fresh water into the house (water supply system) and removes waste water and wastes (waste disposal system). Fig. 38-2. The piping system for water and drainage is installed by a plumber. Plumbers must possess a license and have a wide variety of skills. They deal with a great number of materials and products. Plumbers must also be familiar with plumbing codes that regulate how systems should be installed.

A *service main* is piping used to bring water into the house from a public water system. A water meter is located just outside or inside the basement wall to record the amount of water used. If a public system is not available, the fresh-water pipes are joined to a water pump that supplies the water from an underground well. In either case, the water is under pressure. The pressure causes the water to flow evenly through the house as needed for drinking, cooking, washing, and bathing.

Soil pipe carries waste water and other wastes out of the house to a public sewage system or to a private septic tank. Waste systems are not under pressure. All pipes are installed to slope downward. Wastes flow because of gravity.

Piping is run during the rough-in stage of work. Fixtures and hardware are installed during the finish stage.

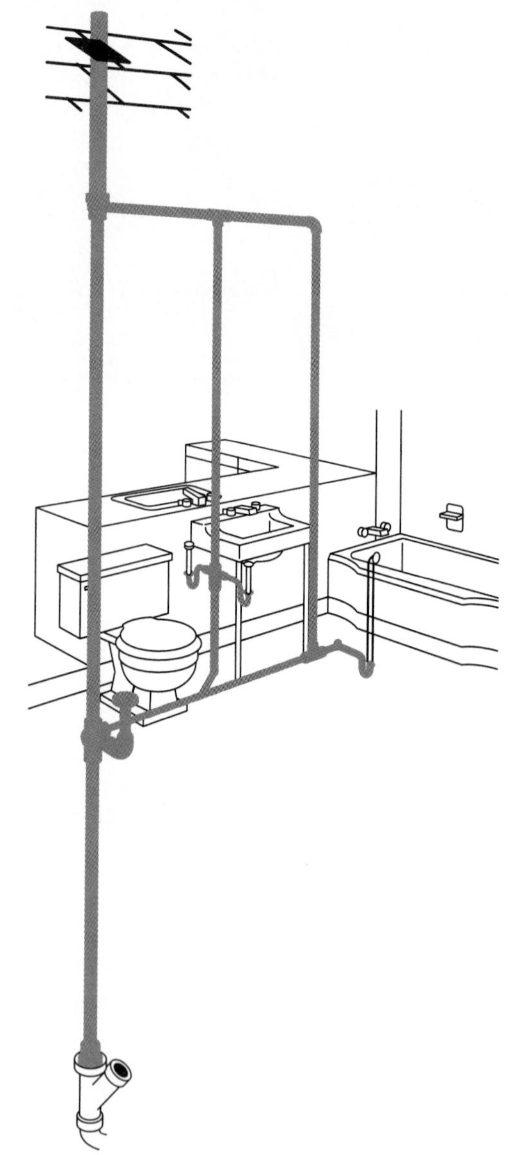

Fig. 38-2—***Rough-in plumbing for a waste disposal system.***

Electrical System

The job of wiring a home must be performed by a licensed electrician. The materials and methods used to wire a house are controlled by local and national codes.

Rough-in work includes installing a meter and distribution panel and running the wiring throughout the house. The meter measures how much electricity is used. The distribution panel provides the way of distributing electricity as needed to various parts of the house. It also contains safety devices called *circuit breakers.*

The finish work is done after wall surfaces are in place. It includes installing switches, outlets, lighting fixtures, and other devices.

Heating and Air-Conditioning System

Heating and air-conditioning needs vary from region to region. Several types of heating systems are available. The most popular is forced hot-air heating. Fig. 38-3. This type of system consists of a furnace, ducts, and registers. A blower in the furnace circulates warm air to various rooms through supply ducts and registers. Return air registers and ducts carry the cooled room air back to the furnace. There it is reheated and recirculated. Hot air systems are commonly fueled with oil, electricity, or natural gas.

In the forced hot-air system, ductwork is installed during the rough-in stage. Register grills and temperature controls are installed during the finish stage.

Some forced hot-air systems can be adapted for cooling by the addition of cooling coils. Combination heating and cooling systems may be installed. Both of these allow the same ducts to be used for both heating and cooling.

A **heat pump** is a device that can heat or cool the air in a house. It is connected to standard duct systems. Heat pumps are widely used in the southern regions of the United States. Like forced hot-air systems, they are often modified to provide both cooling and heating for a home.

Fig. 38-3—*A forced hot-air heating system.*

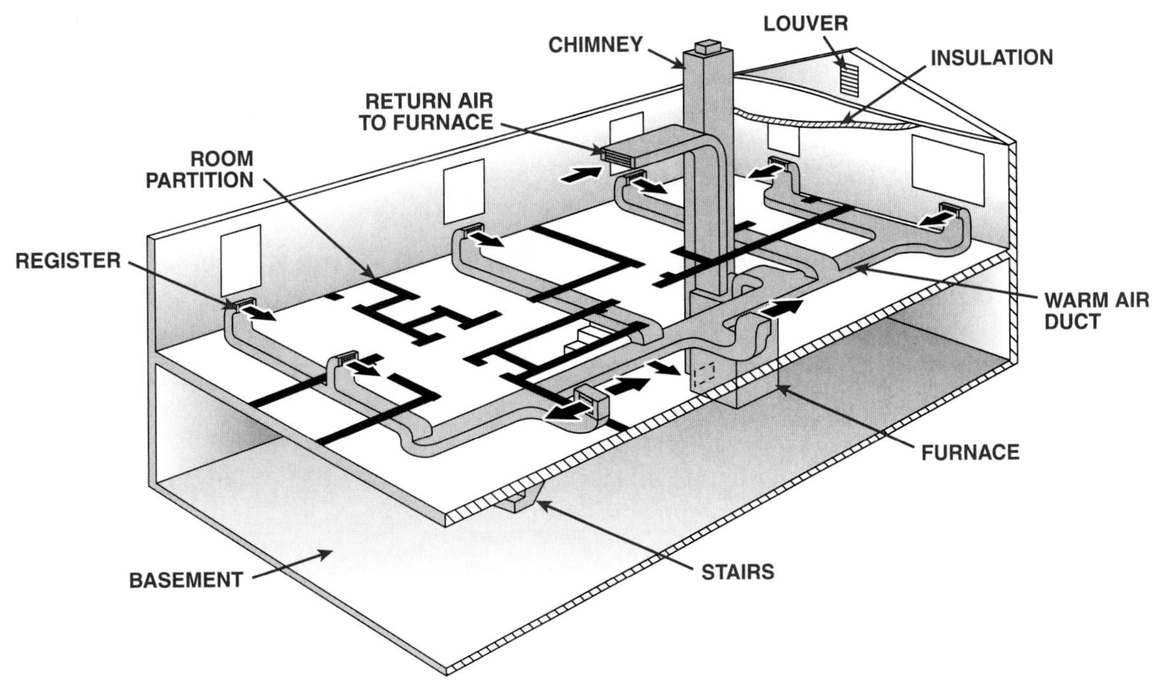

INSULATION

Insulation is a material used in construction that slows down the transmission of heat. Properly installed, it keeps homes warmer in winter and cooler in summer. The most common insulation is fiberglass.

All walls, ceilings, roofs, and floors that separate heated from unheated spaces should be insulated. Fig. 38-4. One side of insulation is covered by a **vapor barrier.** This is a material that prevents moisture from getting into interior walls, floors, or ceilings. The vapor barrier should face toward the inside of the house when insulation is installed.

INTERIOR WALLS AND CEILINGS

After the walls have been insulated, and the rough-in of plumbing, electrical, and heating systems is complete, the wall and ceiling surfaces can be installed. **Drywall** is a sheet material made of gypsum filler faced with paper. It is the most used interior wall and ceiling finish in residential construction.

Sheets of drywall are normally 4 feet wide and 8 feet long. However, they are available in lengths up to 16 feet. Standard thicknesses are ½ inch and ⅝ inch. Special moisture-resistant drywall should be used in bathrooms. Fire-resistant drywall is also available.

Drywall can be applied quickly and efficiently. It can be nailed or screwed to studs and ceiling joists. Fig. 38-5. These must be in alignment to provide a smooth, even surface.

The joints between drywall panels require special treatment. First, joint compound, a plaster-like material, is applied. Then perforated tape is pressed into it, and additional coats of compound are applied. This is then sanded to make the surface smooth and level with the wall surface. Fig. 38-6.

MATHEMATICS *Connection*

Calculating R-Values

Insulation is rated according to a system developed by the U.S. Federal Trade Commission. This rating, called *R-value,* is based on *R-factor,* a material's ability to stop or slow heat transfer. To determine the R-value of insulation, multiply its stated R-factor by its thickness. For example, a roll of 5-inch insulation batting that has an R-factor of 3.8 per inch has an R-value of $5 \times 3.8 = 19$.

Drywall is very versatile. It can serve as a *substrate* (base material) to which finish materials such as wallpaper or tile can be applied. It can also stand alone with a painted finish.

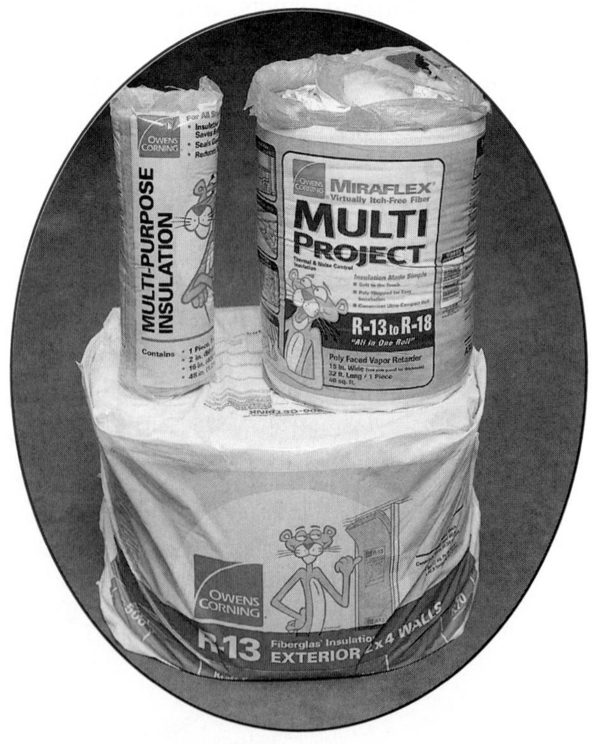

Fig. 38-4—*Fiberglass insulation.*

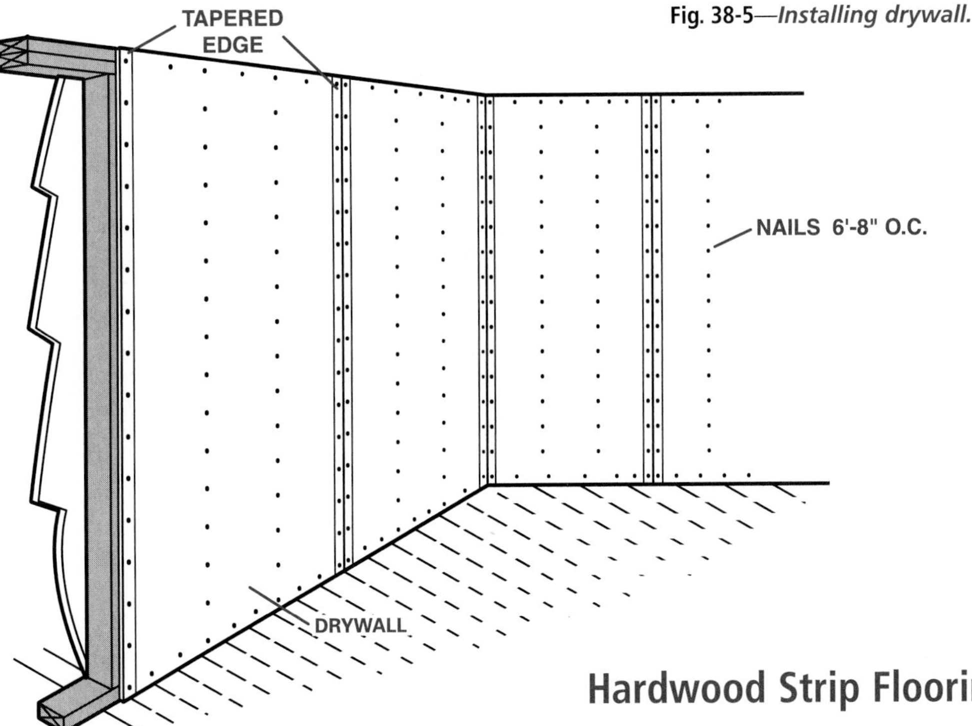

Fig. 38-5—*Installing drywall.*

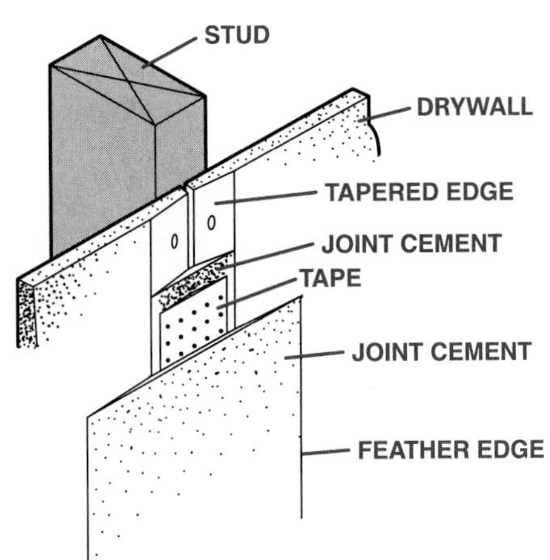

Fig. 38-6—*Drywall joint treatment.*

WOOD AND VINYL FLOORING

Finish flooring is the final wearing surface applied to a floor. Many materials are used as finish flooring. Each one has properties suited to a particular usage. Durability and ease of cleaning are essential in all cases. The installation of finish flooring should be among the last construction operations in a house.

Hardwood Strip Flooring

Strip flooring is the most widely used type of hardwood flooring. As the name implies, it consists of flooring pieces cut in narrow strips. Fig. 38-7. The most common hardwoods for strip flooring are oak, maple, beech, birch, and pecan. Oak, the most plentiful, is also the most popular. Strip flooring can be laid over a plywood subfloor.

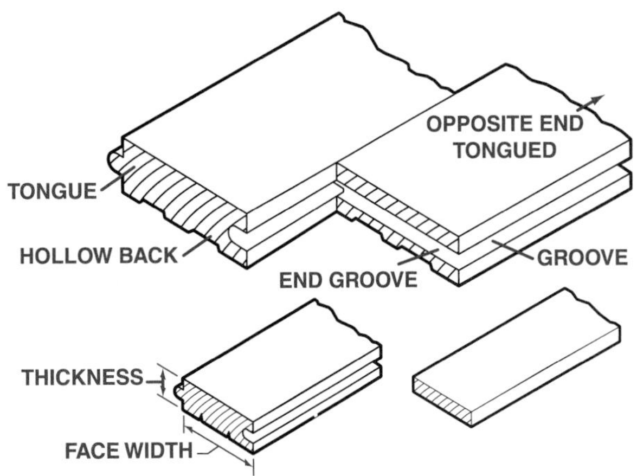

Fig. 38-7—*Most hardwood strip flooring is tongued and grooved at the factory so that each piece joins the next one snugly when installed.*

Facts about Wood

Engineered Wood Flooring—Engineered flooring is a fairly recent development in wood flooring materials. Technically, it is considered a type of plywood. Like other plywoods, it is made from several layers of wood bonded together. However, it is available in strips and planks like solid wood flooring. In general, engineered flooring is more dimensionally stable than solid wood, and it comes prefinished. It is available in a variety of woods, thicknesses, and widths.

Fig. 38-8—*Vinyl flooring is durable enough to withstand hard use.*

Vinyl Flooring

Vinyl flooring is durable, stain-resistant, and relatively inexpensive. It is often used in kitchens, bathrooms, and recreation rooms. Fig. 38-8.

Vinyl flooring can be purchased in sheet or tile form. It is installed over ¼-inch plywood underlayment. A thick adhesive called *mastic* is used to attach the flooring to the underlayment.

CERAMIC TILE

Ceramic tile has become an increasingly useful and popular interior finish material. It can be used throughout the house to provide durable, colorful, and easy-to-clean surfaces. Ceramic tile is used primarily in kitchens and bathrooms. Fig. 38-9. However, it can also be used in mudrooms, entry halls, and even exterior patios.

Tile is adhered with mastics, mortars, or epoxy adhesives. Once the tiles are in place and the adhesive has cured, the spaces between the tiles are filled in with *grout*. Grout is a form of mortar that prevents moisture and dirt from getting between the tiles.

Fig. 38-9—*Ceramic tile is waterproof and easy to clean.*

STAIRS

Building a stairway is generally considered to be the highest test of carpentry as a craft. There are two general types of stairs: principal and service. *Principal stairs* are designed to provide ease and comfort. They are often made a feature of house design. Fig. 38-10. Service stairs lead to the basement or attic. They are usually somewhat steeper and constructed of less expensive materials. Details for building stairs are provided in house plans.

All stairs have **treads** on which the stair user steps and **stringers** that support the treads. *Closed* stairs also include **risers.** These are the vertical members between the treads. Fig. 38-11. If the stairway will be carpeted, the treads and risers may be made of plywood. Otherwise they might be made of oak.

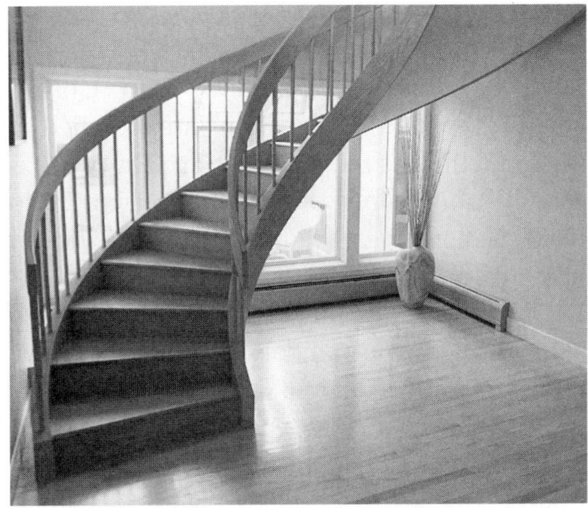

Fig. 38-10—*A stairway may be an important design feature in a house.*

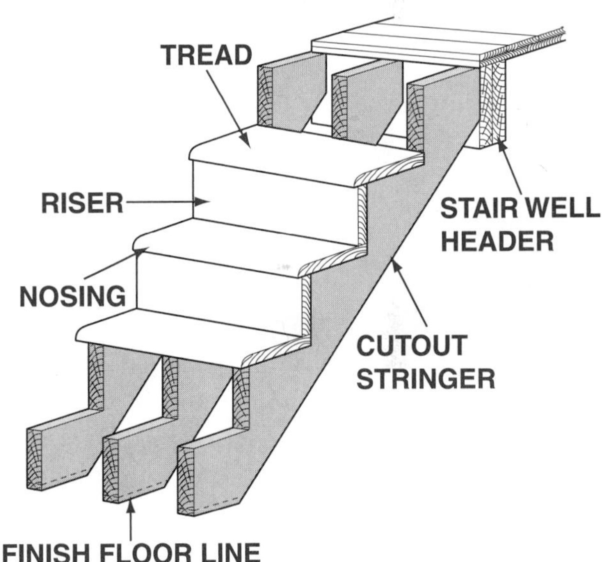

Fig. 38-11—*Parts of a stair.*

Designing a Stairway

Building codes tightly regulate the dimensions and design of stairs for safety reasons. They focus on the following features in particular:

- Stair width.
- Headroom.
- Relationship between the height of the riser and the width of the tread.
- Location of handrails and guardrails.
- Spacing of balusters.

Along with safety-related issues, stair design must address comfort. The relationship between the height of the riser and the width of the tread determines the ease with which the stairs may be ascended or descended. If the combination of run and rise is too great, the steps are tiring. There is a strain on leg muscles and heart. If the combination is too short, toes may kick the riser at each step.

One of the most important features of a good stairway is a properly designed and installed handrail. The **handrail** is the top part of the railing that people grasp when using stairs. A solid, easily grasped handrail can prevent a stairway user from suffering serious injury in the event of a fall. On landings where the railing is horizontal, the top part is called a **guardrail.**

Balusters are the vertical members that support the handrail and guardrail. Their purpose is to prevent anyone, particularly children, from slipping under the railings and falling off the stairs.

CABINETS

Kitchen cabinets, wardrobes, linen closets, china cases, and similar items are installed at the same time as the interior trim. This work is usually done after the finish flooring is in place. Cabinets are available in a wide variety of styles and woods. Fig. 38-12.

Base and wall cabinets should be designed and installed to a standard height and depth. Limits for counter height range from 30 inches to 38 inches. However, the standard height is 36 inches. Wall cabinets may be anywhere from 12 inches to 33 inches high. The usual height is 30 inches. Wall cabinets may be hung free or under a 12-inch *soffit*.

Fig. 38-12—***Well-designed cabinets can be attractive as well as useful.***

Bathroom cabinets differ from kitchen cabinets slightly in size. They are usually 30 inches high and 21 inches deep. The drawer opening is usually 4 inches deep, rather than 5 inches as in the kitchen cabinet.

Countertops for kitchen cabinets, bathroom vanities, built-in desks, and room dividers are often covered with plastic laminate or ceramic tile. Preformed countertops can be bought in slabs or custom made.

INTERIOR DOORS AND TRIM

Careful trim carpentry is one of the hallmarks of high-quality building construction. The installation of interior doors is typically included in the trim carpenter's duties. Decorative wall *moldings* and *casing* (framing) around doors and windows conceal gaps that occur where different materials meet. These products can also have a strong influence on the style of a room.

Interior Doors

The two general interior door types are the flush door and the panel door. Fig. 38-13. The *flush door* is usually made up of a hollow core of light framework faced with thin plywood or hardboard. Plywood-faced flush doors are readily available in birch, oak, and mahogany, which are suited to a natural finish. The *panel door* consists of solid wood stiles (vertical side members), rails (crosspieces), and panels.

Various other types of doors can be found in a typical house.

- *Louvered doors* are commonly used for closets because they provide some ventilation.
- *Sliding doors* are designed for closets and storage walls because they can cover unusually large openings.
- *Bifold doors* may be used to enclose a closet area, storage wall, or laundry area.

Fig. 38-13—*Types of interior doors.*

PANEL DOOR

PLYWOOD-FACED FLUSH DOOR

POCKET DOOR

BIFOLD DOOR

- *Folding doors* may be used as a room divider or to close off a laundry area, closet, or storage wall.
- *Pocket doors* are convenient where there is little or no space for a door to swing outward because they slide into a cavity in wall.

Interior Trim

Trim can be used in many places in the home. It is most often used around doors, windows, and along walls and ceilings. Fig. 38-14. The trim can be traditional or contemporary. Standard patterns and shapes are readily available from local lumber dealers. However, special trim and trim accessories are also available.

Fig. 38-14—*Trim gives a room a finished appearance.*

PAINTING

Paints are used to decorate and protect surfaces. Fig. 38-15. Interior paints serve these purposes:

- Making surfaces easy to clean.
- Improving wear resistance.
- Sealing surfaces from moisture.
- Decorating.

The two basic types of paint are latex and alkyd. *Latex paints* are water-based paints. These are easy to work with because tools clean readily in soap and water. **Alkyd paints** are oil-based paints. These require the use of a *solvent,* such as turpentine, for cleanup.

Alkyd and latex paints come in various finishes. Usually, flat finish paints are used for walls. Semigloss is used for trim, and enamel (glossy) is used wherever maximum protection is required.

Most drywall surfaces are finished with two coats of flat paint. However, kitchen and bathroom walls are exposed to high humidity and receive more wear. These are best finished with one application of an undercoating and two coats of semigloss enamel. This type of finish wears well, is easy to clean, and resists moisture.

Fig. 38-15—*Paint is the most popular interior finishing material.*

CHAPTER 38

Review & Applications

Chapter Summary

Major points from this chapter that you should remember include:

- Plumbing, electrical, and heating systems are installed in two stages: rough-in and finish.
- Insulation is a material that slows down the transmission of heat.
- Drywall is the most used interior wall and ceiling finish material in houses.
- Stair design must address safety issues and comfort of use.
- Cabinets are installed at the same time as interior trim.
- Flush and panel doors are general types of interior doors.
- One purpose that paint serves is to make surfaces easy to clean.

Review Questions

1. In what two stages are plumbing, electrical, and heating systems installed? Name at least one task done during each stage for each system.
2. Why is insulation important in a structure?
3. Describe how drywall is installed.
4. Name two types of flooring.
5. In what rooms is ceramic tile used most?
6. What are three major structural parts of a closed stairway?
7. What features of a stairway are regulated by codes?
8. List three types of doors that might be used for closets.
9. What is the difference between latex and alkyd paints?

Solving Real World Problems

The drywall is hung and the walls are painted in Georgia's new home. Since Georgia worked for a contractor while she was in school, she has experience installing trim. By making arrangements with the contractor to do this part herself, she saved some money and got a chance to be a part of the home-building process.

Georgia soon discovers that although she has the skill to do the work, she lacks some of the tools she needs to cut and install the trim work. What tools might she need?

Remodeling and Renovation

Look for These Terms

- cost effective
- house inspector
- remodeling
- renovation
- suspended ceiling
- wood lath

YOU'LL BE ABLE TO:

- Provide reasons for and examples of remodeling.
- Explain the importance of inspection before remodeling.
- Outline the procedure for planning and designing a room addition.
- Locate wall studs.
- Explain what must be done when walls must be removed.
- Describe special considerations for remodeling basements.

Many people purchase a house that already exists, rather than build a new one. Buying an existing house allows a family to move in quickly.

An older house, however, does not always meet all the needs of its new owners and often requires **remodeling** to make it more suitable for them. It may need to be enlarged, or the buyer may want to make changes that will add modern conveniences or improve the appearance of the house. Fig. 39-1. The house may need repairs or need improvements to make it more energy efficient. For many of these same reasons, a homeowner who has lived in a house for a long time may decide to remodel it.

Fig. 39-1—*Installing new exterior siding can give an older house a bright, new look as well as help improve its energy efficiency.*

Fig. 39-2—*This house, typical of the 1950s, will be remodeled without adding floor space.*

Fig. 39-3—*The same house after remodeling. It now has a stunning, contemporary exterior.*

Another reason for wanting to make changes in an older house is that the house has some historical or architectural importance, and the buyer or owner wants to restore it to the way it originally looked. This is called **renovation.** This chapter focuses mainly on remodeling. However, renovation frequently involves many of the same processes used in remodeling.

TYPES OF REMODELING

The types of improvements to be made depend on the condition of the house and the needs of the owner. There are two general types of remodeling. One involves making changes to the exterior or interior without adding to the size of the house. Figs. 39-2 and 39-3. The other involves making additions, such as one or more rooms, that add living space.

Remodeling that falls into the first category is generally less expensive. Often, much of it can be done by the homeowner. Remodeling that adds living space often calls for the services of a builder, and the process is similar in many respects to building a new house.

A few typical remodeling projects are described below.

- *Replacing windows.* Old windows may be replaced with new ones to make the house more comfortable and energy efficient. The number, size, or arrangement of windows can be changed to make the house more attractive. Fig. 39-4.

Fig. 39-4—*This insulated window group replaced a narrow, rectangular window of single-pane glass. The new windows add a decorative look, provide more light, and are more energy efficient.*

- *Adding closets and storage areas.* If a house does not have enough storage space, the homeowner may want to add some.
- *Remodeling a kitchen or bath.* In older homes, kitchen and bathroom remodeling is often necessary—and often costly. Sinks, tubs, faucets, cabinets, and countertops are frequently replaced. Plumbing and electrical wiring may need to be upgraded. Islands or peninsulas may be included in kitchen remodeling to increase workspace and convenience. Fig. 39-5.
- *Changing the layout.* Large rooms can be made out of smaller ones by removing walls, or a large room may be divided into smaller rooms by adding a wall. Doors can be relocated to change traffic patterns. Fig. 39-6.

Fig. 39-5—*An island can add workspace and convenience to a kitchen.*

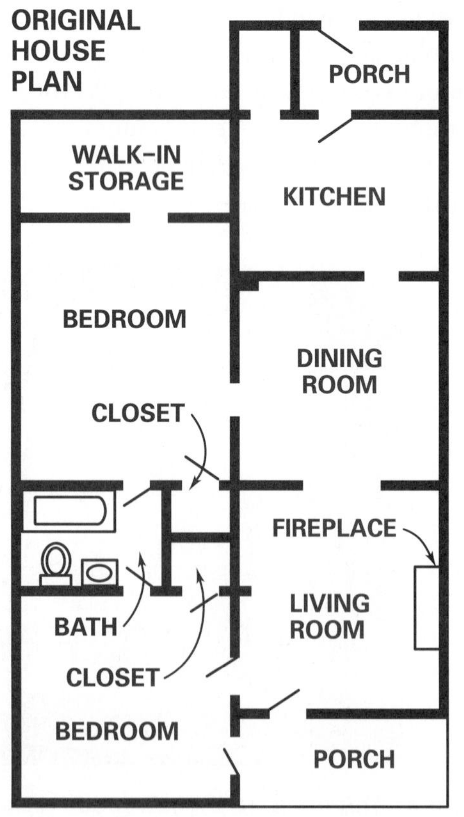

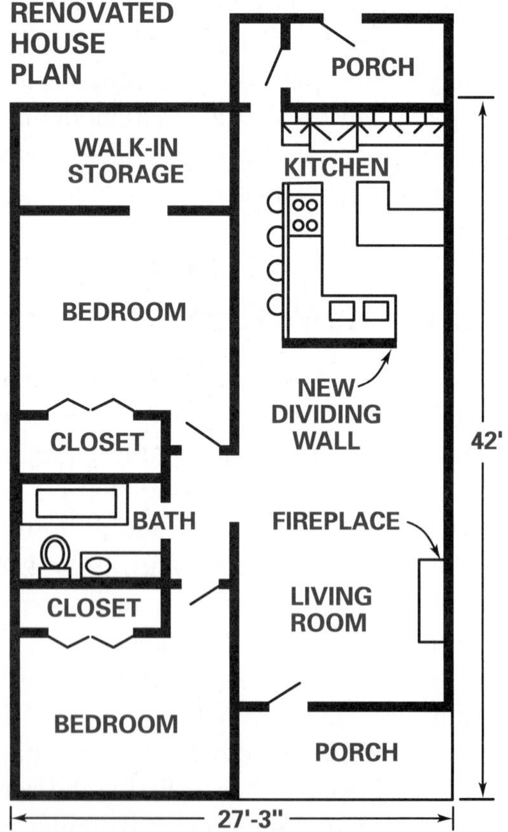

Fig. 39-6—*A house's floor plan can be changed to meet a new owner's needs.*

- *Converting existing space.* An attic, basement, or attached garage may provide space for an additional room. For a room in the attic, it may be necessary to add a dormer.
- *Adding a porch, sunroom, or deck.* Not only does this improve the appearance of a house, it can also increase living space. Fig. 39-7.

INSPECTION

Some houses need extensive and expensive repairs to put them in good shape. The costs of adding or structurally altering one or more rooms can vary greatly, depending on how the house was originally constructed. Therefore, before the decision to remodel is made, the house should be inspected to determine what must be done and how much it would cost to do it. The homeowner or prospective buyer needs to make sure that remodeling would be **cost effective**—that the benefits brought about by the improvements justify the costs of making these changes.

In the case of a home that is being considered for purchase, the inspection should be done by a qualified professional called a **house inspector.** Unlike a building inspector, who examines a house during construction to ensure that the work meets local building codes, a house inspector checks the general condition of a house to find existing or potential problems. Fig. 39-8 (page 482).

If you are a homeowner considering remodeling, you may be able to conduct much of the inspection yourself. Be sure to wear old clothes so you can check any areas that might be affected by the remodeling, including the attic and crawl spaces.

Equip yourself with a flashlight, pocketknife, pen, notebook, and tape measure. Make a thorough review of the structure. For example, if you are considering removing a wall, check the location of utilities such as heating, wiring, and plumbing. You could run into a problem that would require moving those lines before remodeling could proceed.

As you inspect the house, remember that old homes are often not very energy efficient. Look for any flaws that waste heat or air conditioning and be sure to correct them during the remodeling. Fig. 39-9 (page 483).

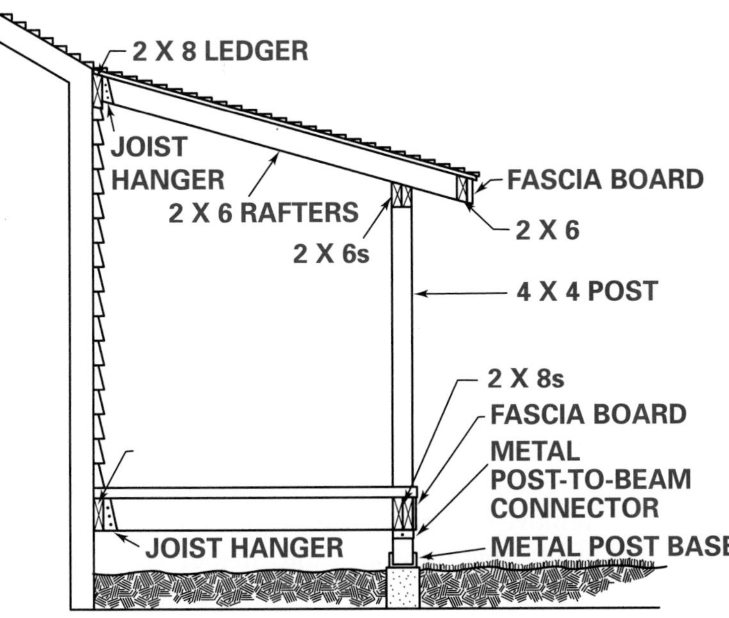

Fig. 39-7—*The existing roof of this house was extended to form the roof for a new sunroom with an attached deck.*

HOUSE EXTERIOR

I. Site Plan

A. House Orientation
   1. North-South
   2. East-West
B. Trees and Shrubs
   1. Provide windbreak
   2. Provide shade
   3. Allow south sunlight
C. House Location
   1. On hill
   2. Protected by trees
   3. In open plain

II. House Appearance

A. Yard
   1. Clean, well-kept
   2. In need of landscaping
   3. Unkept, messy
B. House
   1. Well-maintained appearance
   2. Needs paint
   3. Needs roof
   4. Broken windows

III. Electrical

A. Service from pole
   1. 2-wire 110V/120V
   2. 3-wire 220V/240V

IV. Siding

A. Type
   1. Wood
   2. Stucco
   3. Masonry
   4. Aluminum or vinyl
   5. Other
B. Condition
   1. Satisfactory
   2. Repairs needed

V. Roof

A. Type
   1. Asphalt shingles
   2. Metal
   3. Rolled asphalt
   4. Other
B. Condition
   1. Satisfactory
   2. Repairs needed
C. Chimney
   1. Satisfactory condition
   2. Repairs needed
D. Flashing
   1. Satisfactory condition
   2. Repairs needed
E. Eaves and Downspouts
   1. Type
   2. Condition

VI. Exterior Foundation

A. Type
   1. Stone
   2. Stone capped with mortar
   3. Cement block
   4. Other
B. Condition
   1. Satisfactory condition
   2. Repairs needed

HOUSE INTERIOR

I. Basement

A. Floor
   1. Stone
   2. Cement
   3. Dirt
B. Walls
   1. Stone
   2. Block
   3. Cement
C. Condition
   1. Wet or damp
   2. Dry - no sign of moisture
D. Joists, Sillplate and Header
   1. Good condition, straight joists, no cracks
   2. Rotting wood, insect infestation, sagging wood
E. Heating System
   1. Type
   2. Appearance
   3. Original or new
   4. Type of fuel

II. Insulation

A. Type
B. Location (how many?)
   1. Wall
   2. Ceiling
   3. Floor

III. Electrical

A. Service Entrance
   1. 60 amperes
   2. 100 amperes
   3. 200 amperes
   4. Fuse box
   5. Circuit breakers
B. Branch Circuits (how many?)
   1. Area lighting circuits
   2. Appliance circuits
   3. Electric range/dryer/ hot water heater circuit

IV. Plumbing

A. Sewage System
   1. City sewer
   2. Septic tank
   3. Other
B. Septic Tank (if appropriate)
   1. Concrete
   2. Metal
   3. Size (500, 800, 1000 gal.)
C. Supply lines
   1. City water
   2. Well (dug or drilled)
   3. Flow rate
D. Water Lines
   1. Copper
   2. Galvanized iron
   3. PVC plastic
   4. Other
E. Hot Water Tank
   1. Gas
   2. Electric
   3. Fuel oil
   4. Age and condition
   5. Size
F. Well Pump
   1. Type
   2. Condition

V. Room Layout

A. Traffic Flow
   1. Satisfactory
   2. Changes needed
B. Kitchen
   1. Modern/convenient
   2. Changes needed
C. Interior Walls
   1. Type (plaster, paneled, drywall)
   2. Condition

Fig. 39-8—*A general checklist for inspecting the exterior and interior of a house.*

Another factor that can affect the cost of the remodeling is the type of finish walls in the house. Older houses often have plaster walls, and **wood lath** was once used as a base for plaster. These slender strips of split or sawn wood were nailed over the studs to cover the wall cavities completely. Plaster was then applied over the lath. This technique is seldom used today, but remodelers frequently encounter wood lath under plaster as they cut into old walls. It takes more time and effort to remove or repair plaster walls than it does drywall. This translates into higher costs if you are paying someone to do the work.

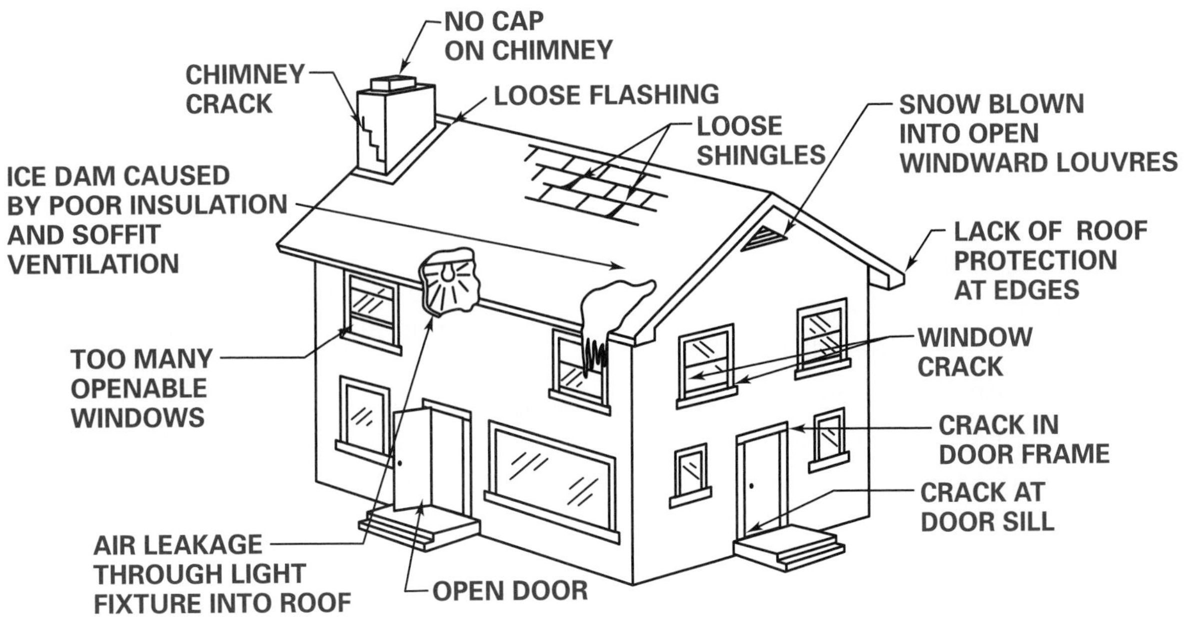

Fig. 39-9—*Common flaws that reduce heating and cooling efficiency.*

PLANNING AND DESIGNING

Once the house has been inspected and deemed suitable for remodeling, it's time to design and plan the project. If the remodeling project involves structural changes, such as changing the layout of the rooms or adding a room, sketch a floor plan for the room(s). Consider the layout of the existing house when planning a new room. In general, family rooms should be placed by the kitchen or dining area. Bedrooms should be off by themselves, not close to the family room. Avoid awkward floor plans, such as a bedroom that opens onto a kitchen or dining area. (If available, a set of plans for the existing home can be very helpful.) Indicate the location of doors, windows, closets, etc.

If the remodeling project involves adding on to the house, you will also need to design the roof line. The roof line should be influenced by the current roof line of the house. The shape and pitch of the roof should either duplicate the present roof or harmonize with it. Fig. 39-10.

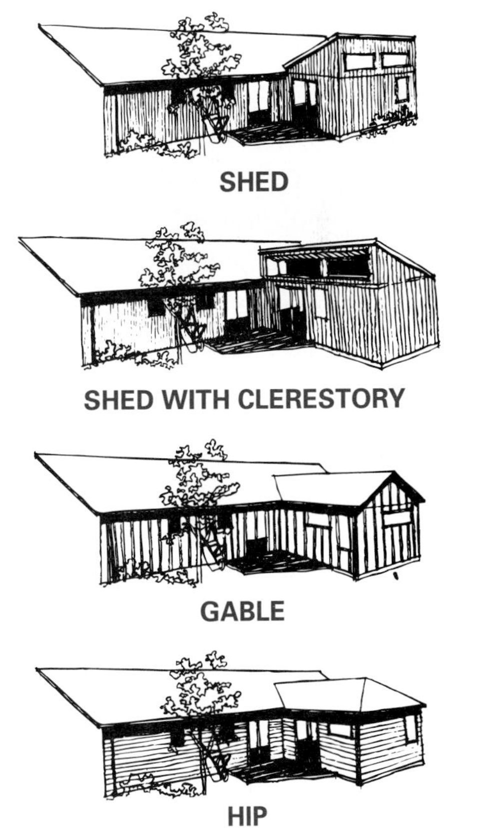

Fig. 39-10—*The roof line of any addition should be the same as or complement the existing roof.*

Salvaging Materials

Be sure to consider whether the old materials can be reused. If any original material, such as the trim, can be used again, then demolition must be done with special care. This is especially true when renovating a structure. In removing trim around a window or door, for example, start at one corner to pry gently and lift the trim away from the wall. When the trim is loose, use a soft-faced mallet to tap gently on the trim—not the nails—as you hold it in place in its original position. This will usually expose the heads of the finishing nails, which can then be removed one at a time with a claw hammer or heavy pliers. This technique should prevent the trim from cracking as you pull it away from the frame so that it can be reused.

Be sure to check building codes and zoning laws to make sure your plans will meet all requirements. *Building codes* specify the types of materials and the methods that can and cannot be used. *Zoning laws* specify how close the addition can be to the property line and may include building height limitations. Finally, using the completed design, develop a bill of materials and apply for the necessary building permits.

MAKING STRUCTURAL CHANGES

Before structural remodeling can begin, some demolition must take place. *Demolition* is the removal of old building materials, one piece at a time, so that rebuilding can begin. The general procedure is first to remove all hardware, then the trim, then the millwork (doors, windows, flooring, etc.), and finally, the wall covering and structural parts. If utilities must be removed and replaced, skilled electricians and/or plumbers should become involved.

Locating Wall Studs

Many remodeling projects require you to know where wall studs are. This is the case, for example, when a new door must be added in an existing wall. Whenever possible, the door opening should begin at a stud so that not more than two studs will have to be removed to install a single door.

You can locate the studs in several different ways. The simplest method is to use a magnetic or electronic stud finder. You can also check carefully for taped wall joints or nails in the trim to locate the studs. Still another method is to measure 16 inches (24 inches for some walls) from the corner of the room and tap along the wall with the *handle* of a hammer until you hear a solid sound.

Removing Partition Walls

Often, rooms are not the right size and it may be necessary to remove one or more partitions. This is not difficult if the partition is a nonbearing wall and if plumbing, electrical, and heating utilities are not involved.

All outside walls and usually the center interior partition, parallel to the long side of the structure, are load-bearing walls. To determine whether a wall is load-bearing, check the direction of the floor and ceiling joists.

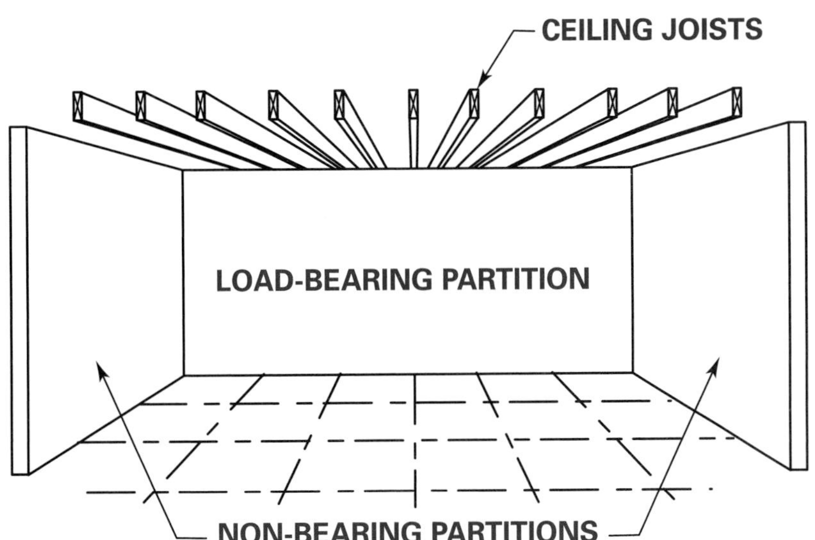

Fig. 39-11—*Load-bearing and nonbearing partitions. A second floor may place a load on any partition.*

Partitions parallel to the joists are usually non-bearing walls. Fig. 39-11. However, the wall may support a second-floor load, so that too must be checked. In most house construction, when the second floor joists are perpendicular to the partition, the joists require support, making the partition below load-bearing.

When all or part of a load-bearing wall must be removed, the wall must be *shored up* (temporarily supported) before it can be removed. If it is an outside wall, as in the case of removing the wall to add a new room, shoring is needed only on the inside. If the bearing wall is on the inside, shoring is needed on both sides of the wall.

The correct way to remove a wall depends on how the wall was assembled. If the studs are nailed through the top plate and toenailed through the sole plate, use a pry bar or nail remover to take out the lower nails. Then pull the stud out from the top plate. However, if a reciprocating saw with a metal cutting blade is available, cut the studs just below the top plate and just above the sole plate. Then remove the major portion of the studs, which can be reused.

WOODWORKING TIP

Tool Rental

Remodeling or renovation may require tools that you do not have on hand. However, jobs that require these tools may not be necessary again for many years. Rather than buying the tools for one-time use, consider renting them. Many rental centers have a good selection of hand and power tools.

After the studs are removed, pry loose the sole plate and remove the top plate. If the ends are tied to the side walls, cut them flush with the wall. After the wall is removed, the ceiling, walls, and floor will require repair where the partition intersected them.

REMODELING BASEMENTS

Remodeling a basement to extend living space is one of the most common remodeling projects. However, by their very nature, basements pose unique remodeling problems.

Walls

Wood paneling is often used in remodeling projects—especially in basements—because it quickly covers imperfections in other wall surfaces. The basic types of paneling are plywood, hardboard, and solid wood strip paneling. All of these can be painted or stained. Some paneling is even prefinished. Fig. 39-12. Although wood paneling is usually installed vertically, it can also be installed horizontally or at an angle to add interest.

Plywood paneling is available in a variety of textures and patterns, including saw-textured, relief grain, embossed, and grooved. It is also available in a variety of veneers, including domestic and tropical hardwoods.

Ceilings

When basement spaces are remodeled, the ceiling area is often problematic. This is because water piping, electrical wires, and heating/cooling ducts are usually fastened to the underside of the floor joists above. To avoid having to relocate these items to create a finished ceiling in the basement, a suspended ceiling can be installed.

Suspended ceilings consist of panels held in place by a grid system at a desired distance from the existing ceiling structure. The panels are usually made of fiberglass or plastic and are 24 inches by 48 inches or smaller. The grid system that supports these panels includes main runners, cross-tees, and wall molding.

The cross-tees are installed at right angles to the main runners. Fig. 39-13. Suspended ceiling components come in a variety of styles and patterns.

Fig. 39-12—*Prefinished paneling.*

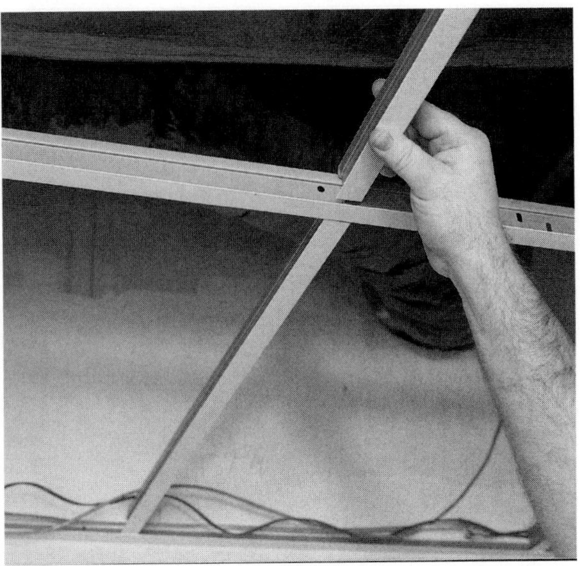

Fig. 39-13—*At left, the cross-tees are being installed between the main runners. At right, the ceiling panels have been laid in the grid system.*

MATHEMATICS Connection

Estimating Materials for Suspended Ceilings

You are planning to install a suspended ceiling in a basement that measures 32′ × 16′. The ceiling tiles you have chosen measure 2′ × 4′. You plan to place the main runners down the length of the room, spacing them 4′ on center. The cross-tees will run 2′ on center. Main runners are available in 12′ lengths, and the cross-tees are 4′ long. How many ceiling tiles will you need? How many main runners and cross-tees? (Assume that the main runners can be spliced together to create the appropriate length.) *Hint:* It may help you to draw the room outline on a piece of graph paper.

A suspended ceiling can be installed easily and quickly. It conveniently covers up bare joists, exposed pipes, and wiring. Access to unsightly valves, switches, and controls hidden by the suspended ceiling is no problem because the panel in question can merely be slid to one side.

Flooring

Another decision when remodeling a basement is what to do about the concrete floor. Because of the possibility of moisture migrating through the material, extra care must be taken when choosing and installing finish floors.

If carpeting or vinyl sheet flooring is to be used, check to make sure that it can be used *below grade* (below ground level). Also make sure that any adhesives or other materials are designed for below-grade installation.

If hardwood flooring is to be installed, a vapor barrier must be installed directly over the concrete floor. Fig. 39-14. This ensures that moisture is kept away from the wood. Figure 39-15 shows another type of vapor barrier installation. After the vapor barrier has been installed, lay a subfloor of ¾-inch exterior plywood. Stagger the end joints by 4 feet. Fasten the plywood to the slab with concrete nails. To allow for expansion, you should leave a ⅜-inch gap between panels and a ¾-inch gap at the walls. Then install the hardwood flooring over the subfloor. Fig. 39-16.

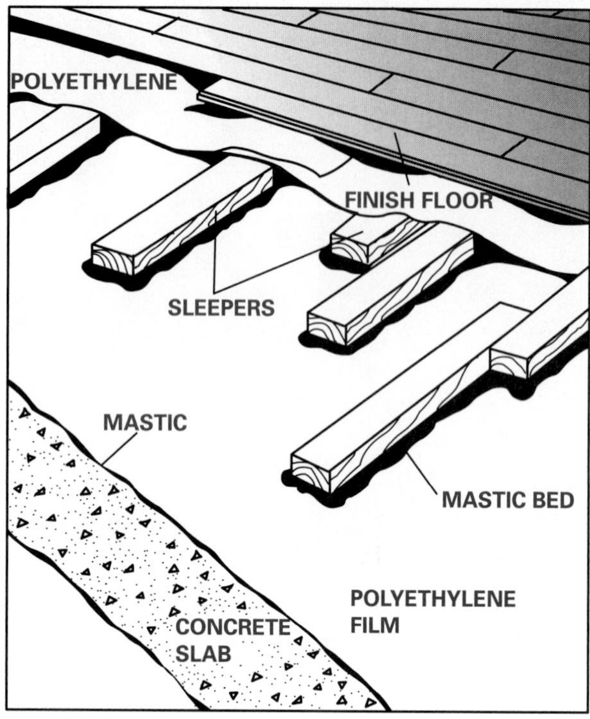

Fig. 39-15—*Another type of vapor barrier—in this case, polyethylene.*

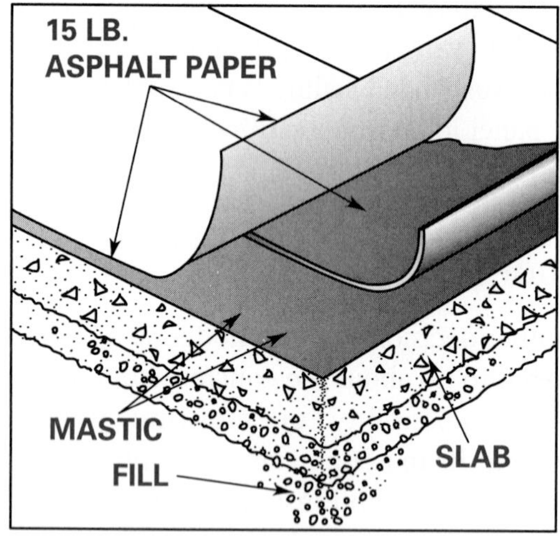

Fig. 39-14—*A vapor barrier must be placed over a concrete slab before a hardwood floor can be installed.*

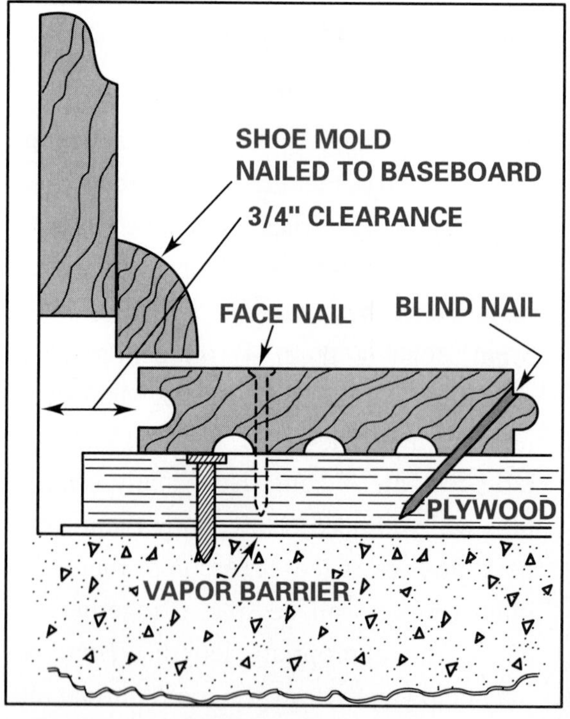

Fig. 39-16—*Hardwood flooring should be placed over a ¾-inch plywood subfloor.*

CHAPTER 39

Review & Applications

Chapter Summary

Major points from this chapter that you should remember include:

- There are many different reasons for and types of remodeling.
- A house should be inspected before remodeling to determine the cost and feasibility of making the improvements.
- Any remodeling project should be designed and planned carefully.
- Many remodeling projects require that you locate wall studs.
- Special care must be taken when removing a load-bearing wall.
- Basement remodeling presents unique problems that require special planning.

Review Questions

1. Why might someone want to remodel a house or room?
2. What is the main difference between remodeling and renovation?
3. Give three examples of remodeling projects and describe what might be done in each.
4. Why should a house be inspected before deciding to remodel? Give examples of things that should be checked.
5. Why should building and zoning codes be checked before adding a room?
6. Explain how to locate wall studs.
7. Why are suspended ceilings especially suited to remodeling a basement?

Solving Real World Problems

Fran looks around her bedroom and decides that the old wallpaper just has to go. First she attacks a seam in the paper with a putty knife, hoping that she can get it to peel off with just some light exercise. Quickly she discovers that the glue is too strong and the paper just tears away. Next she tries to sand the paper off with a small sander. Again she has no success. Mildly frustrated with her inability to remove the paper, she tries soaking the paper with water to loosen the glue. This works a little better, but it is very time-consuming. What other method or tool could Fran try?

CHAPTER 40

Maintenance and Repair

Look for These Terms

- caulk
- damper
- hood
- sealant
- spark arrester
- tuckpointing

YOU'LL BE ABLE TO:

- Describe an organized approach to inspecting a house.
- Describe common repair and maintenance tasks for houses.
- Choose and apply an appropriate caulk or sealant.
- Describe typical fireplace and chimney construction.
- Explain how to identify and prevent decay and insect damage in wood members of a house.

Though most of the materials and systems used in a house are durable, all of them eventually require some maintenance. Properly maintained, a wood-framed house can last indefinitely. **Fig. 40-1.** If maintenance is ignored, however, the house gradually deteriorates, losing much of its value. Such a house is less safe, less comfortable, and requires more energy to heat and cool than a properly maintained house. In addition, periodic maintenance reduces the need for expensive, time-consuming repairs.

Even houses that have been maintained properly sometimes require repairs, however. A roof may be damaged by falling tree limbs or by weather extremes, for example. Indoors, a stair baluster might be damaged while furniture is moved upstairs, or plaster walls might crack due to settlement of the house foundation. Modest repairs can be done by the homeowner. More extensive work often requires the services of a professional contractor.

INSPECTING A HOUSE

To develop a regular program of maintenance, get in the habit of inspecting your house inside and out at least once each year. This will help

Fig. 40-1—*A typical house is a complex structure that must be maintained properly. Joints between materials or assemblies, such as between the upstairs deck and the house, call for particular care.*

you locate and repair small problems before they turn into bigger problems. The heating system, plumbing, and electrical wiring should be looked at closely by someone who knows these systems. Problems with these systems often develop in older houses and can be costly to repair if neglected.

Many household problems are caused by water or water vapor. Some of the most important maintenance steps, therefore, relate to preventing water from penetrating materials. Fig. 40-2. Paint and caulk are the most important maintenance products in the battle against water.

Every few years it may be worthwhile having other professional help in making your inspection. A commercial termite inspector, for example, will often uncover damage that the untrained person may overlook. If there has been extensive damage due to termites, marine borers, or dry rot, repairs to the house may be very costly. Spotting such problems early can save money.

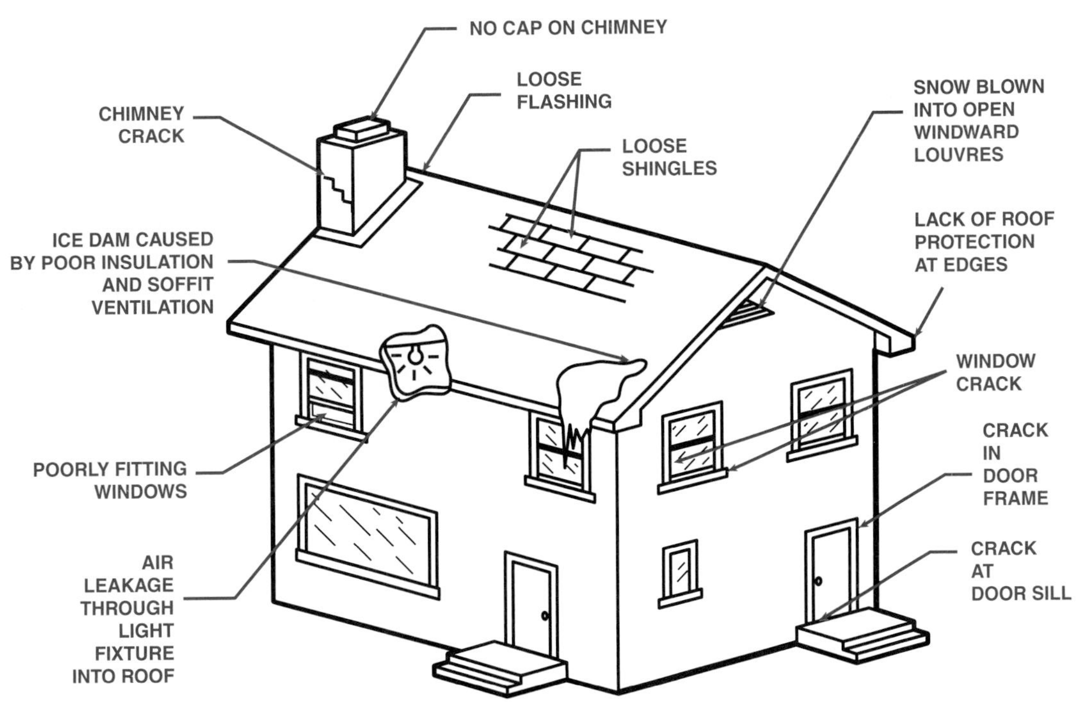

Fig. 40-2—*Moisture can enter a house through many avenues. Older homes, especially, may need work to prevent water damage.*

Exterior Walls

Begin your inspection on the outside of the house. First, check the exterior walls for any obvious problems.

Paint

Peeling paint or bare wood in random areas around a house indicates a moisture problem coming from inside the house. Paint peeling away from another layer of paint indicates that different kinds of paints have been used on the same surface. In either case, the old paint must be completely removed or new siding must be attached.

Trim

Check the wood trim around the eaves and soffit to make sure there is no dry rot due to ice damage, leaky gutters, insects, or mildew. Also check the base of the walls to make sure that soil is no closer than 8 inches to any wood.

Fig. 40-3—*Tuckpointing damaged mortar joints. New mortar is pressed into the joints. Later, it is tooled to match surrounding joints.*

Mortar Joints

Over time, the mortar joints in a brick veneer wall may need repair. This may be due to mechanical damage, the use of poor-quality mortar in the original wall, or simply age. To repair the joints, the loose and damaged mortar should first be removed from the joints. New mortar can then be pushed into the joints. This is called **tuckpointing.** The repaired joints may then be tooled to match surrounding joints. Fig. 40-3.

Roof

Check the condition of the roof and its components. Pay attention to the following items:

- If the roof ridge is not straight, make a note to check the basement walls, support columns, and joists. Foundation settling in an older home often shows up in a swayback or leaning roof ridge. Fig. 40-4.
- Roof shingles must lie flat and intact. Old roof shingles often curl and lose their granular surface. They can also curl under and break.

Safety First

►*Inspecting a Roof*

It is usually not necessary to climb onto a roof to inspect it. Such work can be hazardous to the untrained homeowner. Instead, make as many inspections as possible from the ground, using binoculars to investigate potential problems.

Fig. 40-4—*Problems with a house roof may be due to foundation problems caused by excessive moisture.*

- Sometimes too many asphalt shingles have been placed over each other. There should not be more than two layers of shingles on a roof.
- Check the roof for large amounts of black tar. This indicates that there have been roof leaks. Sometimes these roof leaks cause the roof sheathing to rot.
- Check the chimney for loose mortar or broken bricks.
- Make sure the metal flashing fits well and is in good condition.

Windows and Doors

Check to see that doors and windows throughout the house are fit snugly. Storm doors and windows are advisable in most climates. Poorly fitted doors and windows, and those without proper weatherstripping, are a major cause of energy loss. Be sure to check the window operation by raising and lowering double-hung windows and by cranking open casement windows. Check each window for cracked, loose, or missing glazing putty. This is a common flaw that leads to considerable water damage in a window.

Basement

Check the basement or crawl space, floors, and walls for damp spots. This should be done at least once each season. Mustiness indicates that moisture may be finding its way into the living space. Frequent dampness in a basement can damage the furnace and other equipment.

Cracks on the inside of a foundation wall often indicate that the wall is being pushed in from the outside. Check the structural members for termites and dry rot, particularly near foundation walls.

Interior Walls and Ceilings

Examine interior walls and ceilings carefully for the following:

- Stains that might indicate the presence of a plumbing or roof leak. Many older homes have plastered walls that should be checked for cracks.
- Damage to gypsum board. Newer homes typically have gypsum board wall surfaces; check particularly where they may have been damaged by doorknobs and furniture.

• “Nail pops” (nails that are working loose from the framing and pushing through the wall surface). Fig. 40-5.

REPAIRING EXTERIOR WALLS

Protection of exterior walls against moisture from both inside and outside sources is very important. Painting surfaces is the most common way to protect them, but paint requires regular maintenance to be effective. Poor maintenance, improper paint application, and poor construction techniques can cause many problems. Described here are some of the most common paint problems that you should look for during periodic house inspections.

Cracking and Alligatoring

Cracks that look similar to an alligator’s skin are informally known as “alligatoring.” Fig. 40-6. This problem usually results from one of two errors. The previous paint film may have been applied in several heavy coats without sufficient drying time between coats, or the undercoat may not have been compatible with the finish coat.

Cracked or alligatored paint can be corrected by following these steps:

1. Sand the surface smooth.
2. Apply one coat of undercoat and one top coat of a recommended house paint according to label directions.

Blistering and Localized Peeling

Blistering or localized peeling can occur when moisture that was trapped in siding is drawn from the wood by the sun’s heat and pushes the paint from the surface. Fig. 40-7. Peeling can be corrected by following these steps:

1. Locate and eliminate sources of moisture. Is the area near a bathroom or kitchen? Is there seepage or leakage from eaves, roofs, or plumbing?
2. Scrape off the old paint. Either scrape down to the wood on the entire board, or scrape off the old paint from the peeling area and for about 12 inches around the area.
3. Sand the surface to fresh wood and spot-prime with a recommended undercoat.
4. Seal all seams, holes, and cracks against moisture with caulk.
5. Apply a top coat of recommended house paint according to label directions.

Fig. 40-5—*A framing member that has not been properly squared with the plate increases the possibility of puncturing the gypsum board with the nailhead. Also, the stud may twist as it dries out, causing the board to be loosely nailed. This often results in a “nail pop.”*

Fig. 40-6—*Cracking and alligatoring.*

Fig. 40-7—*Peeling and blistering result when the sun draws moisture under the paint toward the surface.*

Fig. 40-8—*Flaking.*

Flaking

Flaking results when siding alternately swells and shrinks as the moisture from behind the siding is absorbed and then evaporates. The brittle film of paint cracks under the strain and pulls away from the wood. Fig. 40-8. To correct this problem, follow the steps outlined in the previous section, "Blistering and Localized Peeling."

Mildew

Mildew thrives in high humidity and high temperature. Fig. 40-9. If left on the surface and painted over, it will grow through the new coat of paint. A problem with mildew can be corrected by following these steps:

1. Scrub the entire surface with a solution of ⅓ cup of trisodium phosphate (TSP) and ½ cup of household bleach mixed with one gallon of warm water.
2. Apply one coat of wood undercoat. *Note:* Mildew-resistant additive may be added to an undercoat if mildew conditions are severe and an oil-base top coat is used. The additive should not be put in the finish coat.
3. Apply one top coat of mildew-resistant latex house paint.

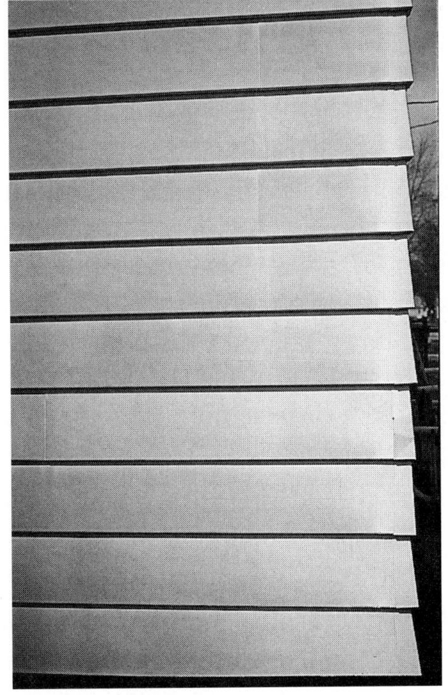

Fig. 40-9—*Mildew.*

Safety First

►*Hazardous Chemicals*

Trisodium phosphate and bleach (sodium hypochlorite) are strong chemicals that require precautions for safe use.

- Wear eye protection. These chemicals can react violently with other strong chemicals, causing corrosive spatters.
- Avoid contact with the skin. Wear appropriate gloves when working with these chemicals.
- Use trisodium phosphate and bleach only in well-ventilated areas to avoid inhaling a high concentration.

Top Coat Peeling

This problem is usually found on overhanging horizontal surfaces and other areas protected from weather. Fig. 40-10. It is due to the poor adhesion of a previous coat of paint from buildup of "salt" deposits that are not washed away by rain. To correct top coat peeling, follow these steps:

1. Sand the surface thoroughly to remove all peeling paint.
2. Wash the sanded surface with a solution of ⅓ cup trisodium phosphate to one gallon of water. Rinse well and allow to dry.
3. Apply two coats of house paint undercoat, or one undercoat and one top coat of a recommended house paint according to label directions.

Fig. 40-10—***Top coat peeling.***

SCIENCE *Connection*

Mildew Taxonomy

Taxonomy is the scientific classification of living (and extinct) organisms. The highest level of classification is the kingdom. You have probably heard of the two best-known kingdoms: the Plant (*Plantae*) and Animal (*Animalia*) kingdoms.

The substance we call "mildew" is not one, but an entire group of organisms that belong to the biological kingdom called *Fungi* or *Mycota.* In addition to mildew, the Fungi kingdom includes molds, mushrooms, yeasts, and similar organisms.

The organisms in this kingdom were originally included with the Plant kingdom. However, they do not have an organized plant structure, and they do not have chlorophyll—two important characteristics of plants. They have therefore been reclassified into a separate kingdom.

REPAIRING INTERIOR WALLS

Interior walls are usually smoother and more finely finished than exterior walls. Therefore, interior walls often require more preparation and care than exterior walls. Follow these steps to repair an interior wall:

1. Start with a clean surface. A thorough dusting of the surfaces to be painted is usually enough. However, kitchen walls or badly soiled or glossy surfaces should be washed to remove dirt and grease and to dull the surface. Wash from the bottom up with an abrasive cleanser or a solution of trisodium phosphate. Rinse with clean, warm water.
2. Fill small cracks. Fine cracks in walls or nail holes in wood trim can be filled with a ready-mix spackling compound. Press the paste into the cracks with a spatula or putty knife. Fig. 40-11. Force material into tiny cracks or corners with your fingers. When dry, the patched areas can be sanded smooth.
3. Use patching plaster to fill large cracks in a plaster wall. On cracks 1⁄16 inch wide or larger, undercut the crack to an inverted V-shape to "lock" the repair material in place. Fig. 40-12. Wet the edges of old plaster so that the new plaster will bond. Mix the patching plaster according to directions and fill the cracks. Remove the excess and smooth the surface with a putty knife.

Safety First

►*Removing Lead Paint*

If you suspect that lead-based paint has been used in the house, *do not* use a heat gun to remove it. Take special precautions to minimize contact with paint dust and chips. Lead paint was outlawed for residential use in 1978 because of concerns about its effect on health.

Fig. 40-11—*Putty all cracks. Be sure to leave a little buildup on the surface to allow for shrinkage when drying. Sand to a smooth surface.*

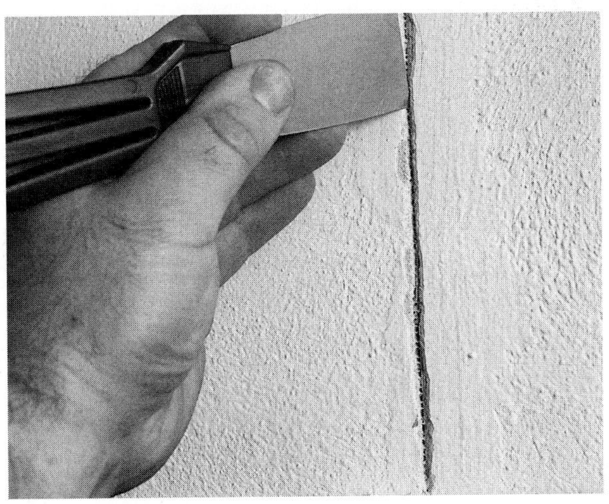

Fig. 40-12—*A large crack should be undercut with the putty knife to ensure good holding power for the fill material.*

4. When filling large holes or broken areas, clean out the old plaster or drywall and tack a piece of wire screen or hardware cloth to the lath to anchor the patch. Fig. 40-13. Fill the hole and smooth the plaster level with the surface. A second filling may be necessary to make up for shrinkage. Large cracks or holes should be repaired a day or two before painting to allow for adequate drying time.
5. When the patched area is dry and smooth, spot-prime it with the same paint that you will use as a finish coat or with an appropriate primer. Fig. 40-14. To determine whether you should use a primer or some of the same paint as the finish coat, check the directions on the label of the container.
6. If old paint has peeled, the loose flakes can be removed by scraping. Fig. 40-15. Paint can also be removed with a heat gun. This is an electrical tool that produces hot air at temperatures that range from 250 degrees F to more than 1,000 degrees F. The heated air softens the paint so that it can easily be removed. Fig. 40-16.
7. When the loose paint has been removed, sand the remaining surface smooth. Spot-prime the bare surfaces before painting. Vulnerable areas such as window sills will benefit from an extra overall coat of primer. Consult label directions for an appropriate primer.

Fig. 40-14—*Spot-prime patched areas before you apply the finish coat.*

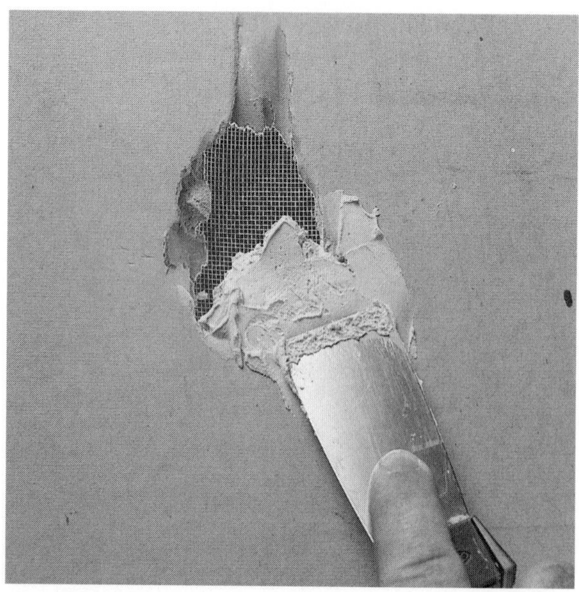

Fig. 40-13—*For large holes, tack a piece of screen or hardware cloth in place to support the plaster fill material.*

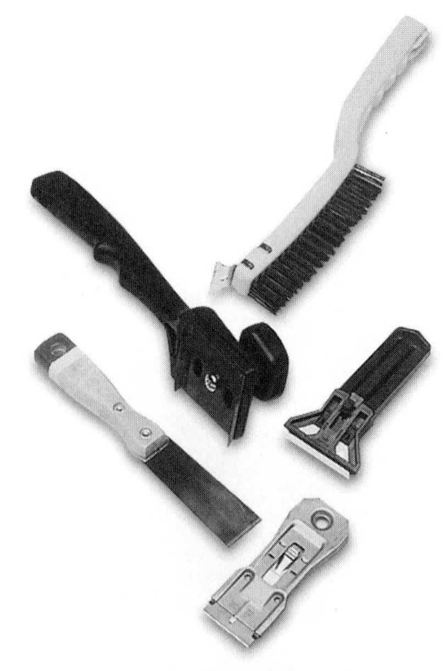

Fig. 40-15—*There are many sizes and shapes of scrapers for cleaning surfaces before finishing.*

Fig. 40-16—*An electric heat gun can be used to remove old paint.*

CAULKS AND SEALANTS

Every joint between two materials is a potential entry point for water. But another important reason to seal joints is to improve energy efficiency. About one-fifth of our total national use of energy is for climate conditioning of residences, including houses and apartments.

If it is cold outside, heat is lost by *exfiltration* (leakage of warm air to the outside through cracks in windows, walls, and doors) and *infiltration* (leakage of cold outside air into the house). Fig. 40-17.

WOODWORKING TIP

Preparing to Paint Interior Walls

Good preparation will make the painting job much easier and faster. Consider these steps:

- Protect the floors and furniture.
- Remove pictures, accessories, and small pieces of furniture from the room.
- Remove hardware and switch plates from the walls. It is easier to take these items off before you start than to try to paint around them, and a more professional job is the result. Switch plates, door knobs, lock plates, and handles can be removed easily with a screwdriver.

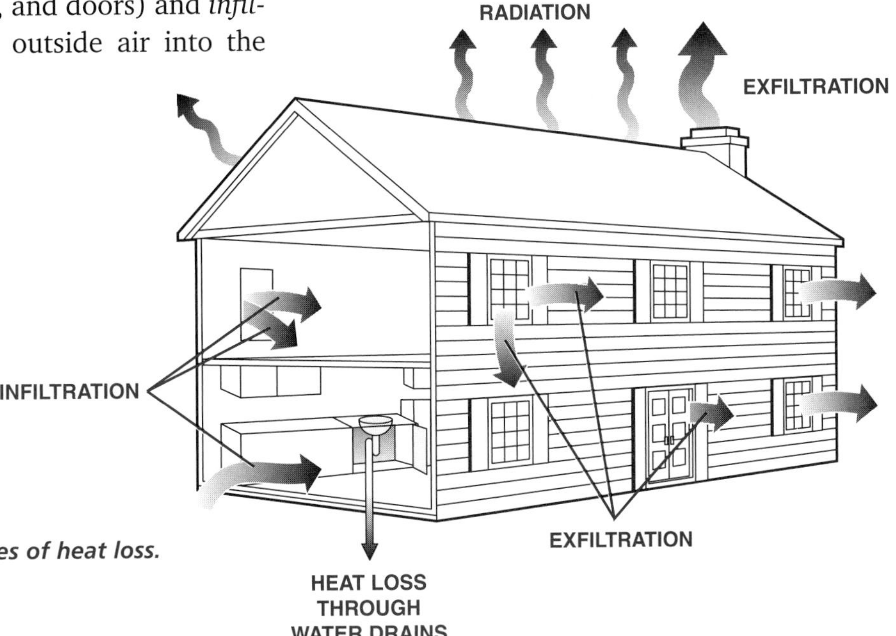

Fig. 40-17—*Types of heat loss.*

One of the easiest and least expensive ways to reduce energy consumption and prevent air infiltration is to seal the house by caulk-ing. **Caulks** are inexpensive and easy to apply. Joints should be caulked where air or water leakage can be expected, such as around the casings of windows and doors. Fig. 40-18.

If considerable movement of a joint is expected, a **sealant** is the best material to use. Sealants are similar to caulks, but they are generally more flexible and more expensive. Because of their flexibility, sealants are often applied to the joint between two dissimilar materials. The joint between stone and wood trim is an example of where a sealant would be appropriate. Caulks and sealants come in various forms. Most commonly, they come in a sealed cartridge for use in a caulking gun. Fig. 40-19.

Caulks or sealants should also be applied in the following places:

- Around an outside faucet.
- Between the basement and floor framing.
- Between the drip cap and siding.
- On corners formed by siding.
- Where pipes and wires penetrate the ceiling.
- Below an unheated attic or chimney.
- Where chimney and masonry meet siding.

Any area that has been caulked or sealed should be inspected periodically to ensure the integrity of the joint.

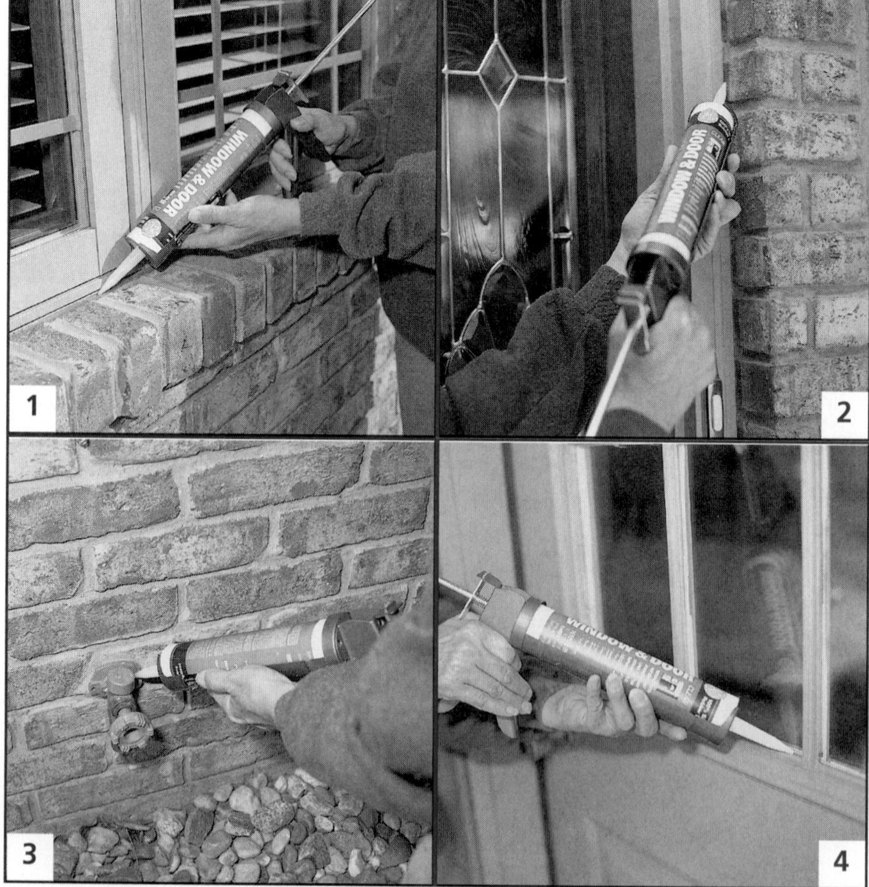

Fig. 40-18—*Caulks or sealants should be applied: (1) Around the window where the frame meets brick, siding, or drywall. (2) Along the top and sides of the door where the frame meets brick, siding, or drywall. (3) Around water faucets. (4) Around window panes and frames, using glazing compound where glass meets frames.*

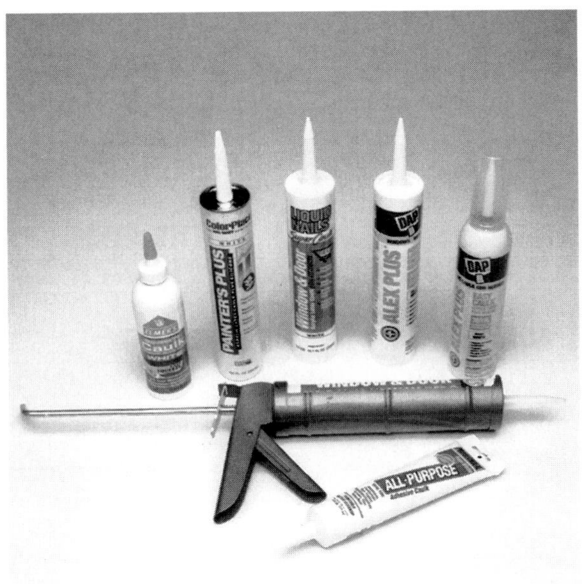

Fig. 40-19—*Caulks and sealants usually come in disposable tubes.*

Choosing a Caulk or Sealant

It is important to select the correct kind of caulk or sealant for each purpose. Table 40-A. Then be sure to follow the instructions of the manufacturer. Also note that the durability of these materials varies depending on exposure to weather.

Applying Caulks and Sealants

For caulks and sealants to work properly, they must be applied correctly. The materials to be caulked should be clean, dry, and free of dust, grease, or other residue. Concrete and other masonry should be dry and thoroughly cured.

When using a caulking gun, hold it at a 45-degree angle to the surface and squeeze with steady pressure. Keep the rear of the caulking gun slightly slanted in the direction you are moving. Slowly draw it along so that the caulk not only fills the crack, but also overlaps the edges. Move the gun with a pulling motion. To get a smooth bead, fill a seam in one stroke.

Table 40-A. Choosing a Sealant

	Caulk/Sealant	Approximate Lifespan Used Outdoors (Years)	Flexability	Adhesion	Compatible Paints	Other Characteristics
	Oil-Base	1-2	Fair	Good	• Latex • Solvent-Base	• Least expensive
	Butyl Rubber	5-15	Good	Good	• Latex • Some Solvent-Base	• Never hardens completely
	Styrene Butadiene Rubber	3-10	Fair to Good	Fair	• Latex	• Adheres to damp surfaces • Adheres to treated lumber
Latex	Polyvinyl Acetate	1-3	Fair	Fair to Good	• Latex • Solvent-Base	• Adheres to damp surfaces • Can be used as a light-duty adhesive
	Vinyl Acrylic	5-20	Good	Good to Very Good	• Latex • Solvent-Base	• Adheres to damp surfaces • Sometimes includes a fungicide
	Solvent Acrylic	5-20	Good	Very Good	• Latex • Solvent-Base	• Not recommended for indoor use
	Kraton	5-15	Good	Good	• Latex • Solvent-Base	• Adheres to asphalt
	Polysulfide	10-25	Very Good	Good	• Latex • Solvent-Base	• Resists chemicals
	Polyurethane	5-15	Very Good	Fair to Good	• Latex • Solvent-Base	• Often used in commercial construction
	Silicone	20 or more	Outstanding	Fair to Good	• Alkyd	• Sometimes includes a fungicide

Safety First

►*Tooling Caulks and Sealants*

You may have heard of people tooling a joint with a wet finger, but this *not* a good idea. Many caulks and sealants contain toxic chemicals. Also, some caulks and sealants can be difficult to remove from the skin. Instead, tool the joint with a plastic spoon or similar device.

The temperature of the material to be caulked should be at least 45 degrees F. Lower temperatures prevent the caulk from adhering properly. The best time to apply caulk to an exterior joint is in the spring or fall. At those times the width of the joint will be halfway between its seasonal extremes.

Most caulks and sealants should be *tooled* after application. This means that they should be pushed into the joint and smoothed over.

FIREPLACES AND CHIMNEYS

Fireplaces and chimneys require regular maintenance not only to maintain good working order, but also for safety. Fig. 40-20. All fireplaces and fuel-burning equipment such as woodstoves, gas water heaters, and furnaces require some type of chimney. The chimney should be cleaned regularly to remove substances such as *creosote* that collect on the inside of the chimney. *Chimney cleaning is a job that should be left to professionals.*

Fig. 40-20—***Fireplaces and chimneys must be maintained regularly to prevent fire hazards.***

Masonry chimneys are sometimes built without flue lining to reduce cost, but those with lined flues are safer and more efficient. When the flue is not lined, mortar and bricks directly exposed to the action of flue gases disintegrate. This disintegration, plus that caused by temperature changes, can open cracks in the masonry. This increases the fire hazard.

Brick chimneys that extend through the roof may sway enough in heavy winds to open up mortar joints at the roof line. Openings to the flue at that point are dangerous because sparks from the flue may start fires in the woodwork or roofing.

Chimneys must be flashed and counterflashed to make the junction with the roof watertight. Fig. 40-21. Corrosion-resistant metal, such as copper, aluminum, or zinc, should be used for flashing. Flashing should be inspected periodically.

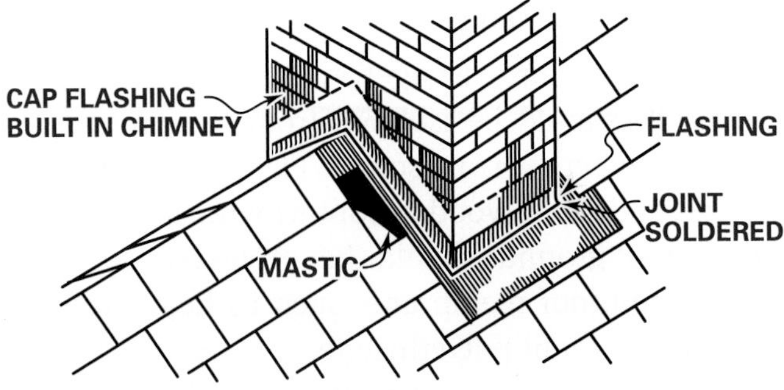

Fig. 40-21—*Flashing at a chimney located on a ridge.*

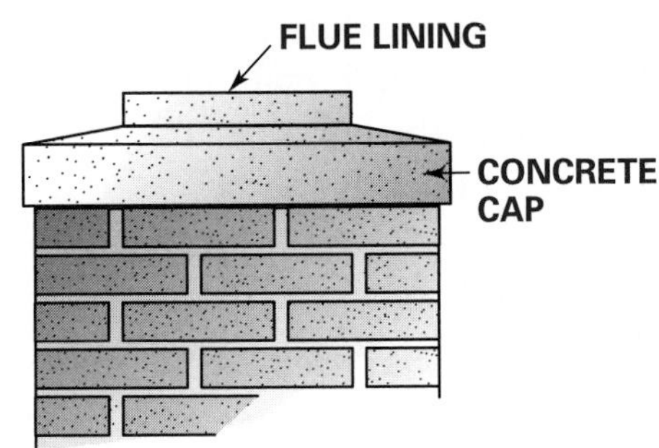

Fig. 40-22—*A chimney cap.*

To prevent moisture from entering between the brick and flue lining, a concrete cap is usually poured over the top course of brick. Precast or stone caps are also used. The mortar is finished with a slope to drain water away from the top of the chimney. Fig. 40-22. Over time, the cap often cracks, admitting moisture into the chimney area. When inspecting the cap for damage, also check the chimney hood and (if present) the spark arrester. **Hoods** are used to keep rain out of chimneys and to prevent downdraft. **Spark arresters** may be required when chimneys are on or near combustible roofs, woodland, lumber, or other combustible material. They should be made of rust-resistant screen and should be securely fastened to the top of the chimney.

A fireplace **damper** consists of a cast iron frame with a hinged lid that opens or closes to vary the throat opening. A properly installed damper allows the flue to be closed to prevent loss of heat from the room when there is no fire in the fireplace. In the summer, the flue should be closed to prevent loss of cool air from the air-conditioning system, and to prevent insects from entering the house through the chimney. Use a flashlight to inspect the damper each year to ensure that it seals correctly.

PROTECTION AGAINST DECAY AND INSECT DAMAGE

Wood has proved itself through the years to be an excellent building material. Damage from decay and termites has been small in proportion to the total value of wood in homes, but it has always been a troublesome problem for many homeowners.

Wood used under conditions where it will always be dry, or even where it becomes wet briefly and dries rapidly, will not decay. However, all wood and wood products are subject to decay if kept wet for long periods under temperature conditions favorable to the growth of decay organisms. Most of the wood used in a house is not subjected to such conditions. There are places where water can work into the structure, but such places can be protected by proper design and construction as well as by use of caulks and sealants.

Decay

Wood decay is caused by certain fungi that can process wood for food. These fungi require air, warmth, food, and moisture for growth. Early stages of decay caused by these fungi may be accompanied by a discoloration of the wood. Advanced decay is easily recognized because the wood has by then undergone definite changes in properties and appearance. In advanced stages of building decay, the affected wood is often brown and crumbly or rather white and spongy. These changes may not be apparent on the surface. The loss of sound wood inside, however, is often indicated by sunken areas on the surface or by a hollow sound when the wood is tapped with a hammer. Where the surrounding atmosphere is very damp, the decay fungus may grow out on the surface, appearing as white or brownish growths in patches or strands.

Fungi grow most rapidly at temperatures of about 70 degrees to 85 degrees F. Elevated temperatures such as those used in kiln-drying lumber kill fungi. Low temperatures, even far below zero, merely cause them to remain dormant.

Wood-destroying fungi cannot grow in wood with a moisture content of 20 percent or less. Moisture content greater than this is unlikely in wood that is sheltered against rain, condensation, and fog. Decay can be stopped permanently by simply taking measures to dry out the infected wood and keep it dry.

Design Details

Steps and stair carriages, posts, wallplates, and sills should be isolated from the ground with concrete or masonry. Fig. 40-23. Sill plates and other wood in contact with concrete near the ground should be protected by a moisture-proof membrane, such as heavy roll roofing or 6-mil polyethylene.

Surfaces such as steps, porches, door and window frames, roofs, and other projections should be sloped to promote runoff of water. Metal flashing should be used around chimneys, windows, doors, and other places where water might seep in. Gutters and downspouts should be placed and maintained to divert water away from the building. They should be cleaned at least once each year.

Water Vapor from the Soil

Crawl spaces of houses built on poorly drained sites may be subjected to high humidity. During the winter, when the sills and

Fig. 40-23—*Exterior wood steps should be at least 6 inches off the ground. They should be supported on a concrete footing.*

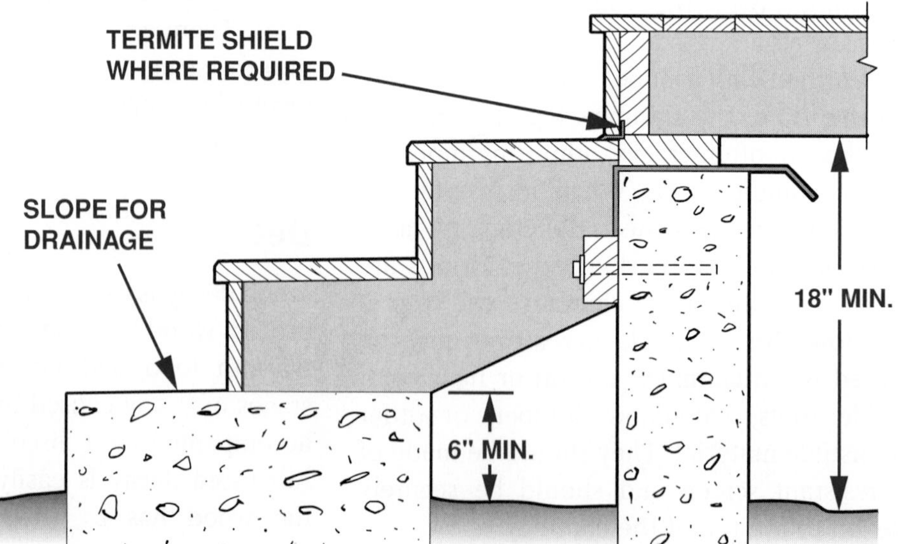

outer joists are cold, moisture condenses on them. In time, the wood absorbs so much moisture that it is susceptible to attack by fungi. Unless this moisture dries out before temperatures favorable for fungus growth are reached, considerable decay may result. However, this decay may progress so slowly that no weakening of the wood becomes apparent for years. Placing a 6-mil sheet of polyethylene over the soil to keep the vapor from getting into the crawl space prevents such decay. The barrier should be inspected every few years to ensure that it remains in place and undamaged.

Insects

Wood is subject to attack by termites and some other insects. There are more than fifty species of termites in the United States. Fig. 40-24. Termites can be grouped into two main classes: subterranean and dry-wood. Fig. 40-25. Subterranean termites account for about 95% of all termite damage.

Other insects include carpenter ants and powderpost beetles. Carpenter ants nest in the ground as well as in dead trees, firewood, and houses. They do not eat wood. Fig. 40-26 (page 506). Instead, the damage they cause to a house comes from the irregular tunnels they create in wood. These tunnels serve as their nest.

Fig. 40-25—***A termite. Termites are the most destructive insect pest in the United States.***

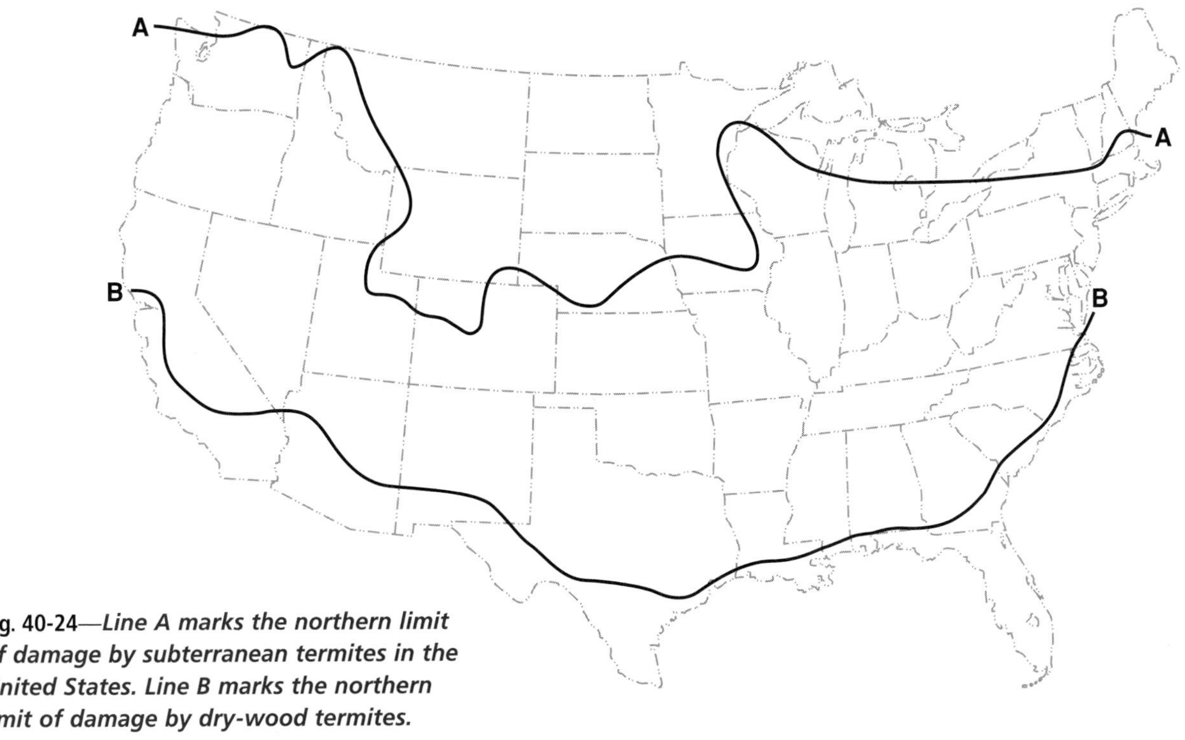

Fig. 40-24—***Line A marks the northern limit of damage by subterranean termites in the United States. Line B marks the northern limit of damage by dry-wood termites.***

The powderpost beetle is the most common beetle to infest wood. Fig. 40-27. It is second only to termites in the amount of damage caused. Powderpost beetles attack hardwoods only, particularly ash, oak, mahogany, hickory, maple, and walnut. Beetles commonly enter the house via wood that is already infested. This can include firewood, roughsawn timbers, and barn wood. One common sign of a powderpost beetle infestation is a tiny pile of fine, flour-like powder. This material is pushed out of infested wood as the beetles emerge. The greatest beetle activity is in wood with a moisture content between 10 and 20 percent.

There are some simple, general ways to guard against all wood-destroying insects. Remove all woody debris, such as stumps and discarded form boards, from the soil around the house. Firewood piles should be kept well away from the house, and wood should never be stacked indoors. Steps should also be taken to keep the soil under the house as dry as possible. No wood member of the house structure should be in contact with the soil.

Fig. 40-26—***Carpenter ant. These insects do not eat wood, but they do burrow through it to create nests.***

Fig. 40-27—***Powderpost beetle.***

CHAPTER 40

Review & Applications

Chapter Summary

Major points from this chapter that you should remember include:

- Houses should be inspected inside and out on a regular basis.
- Preventing water and water vapor from penetrating materials is a major part of home maintenance.
- Caulks and sealants help seal a house against moisture.
- Fireplaces and chimneys require regular maintenance to keep them safe and in good working order.
- Decay of wood is due to wood-eating fungi; insects such as termites, carpenter ants, and powderpost beetles can cause further damage.

Review Questions

1. What exterior items should you check during regular house inspections?
2. What interior items should you check during regular house inspections?
3. What is the most common way to protect exterior walls against moisture?
4. Briefly explain how to correct a problem with mildew.
5. What is the difference between a caulk and a sealant?
6. Describe the purpose of the damper in a fireplace and explain how to inspect it.
7. What causes wood to decay?
8. List three types of insects that can cause damage to a house.

Solving Real World Problems

It is Solana's first winter in her house. When she bought the large, Victorian house last summer, she liked the large rooms and the many windows. Now she notices it is especially cold around the windows. When it is windy outside, the curtains will actually move. The heating system runs a lot and Solana's heating bills are very high, yet the house is still cold. The only solution she has for keeping warm is to wear extra clothing and to cover up with a blanket when she watches her favorite TV shows. Solana likes the house and she wants to continue living there, but she has to do something about those heating bills. What can Solana do to make her house more comfortable in winter and keep the heating bills at a more reasonable level?

Construction Appendix

JOB SCHEDULING

To keep the building progressing at a smooth rate, the general contractor has the responsibility of scheduling the jobs. Whenever it is necessary to have a subcontractor come in, the general contractor must arrange the time. Careful scheduling can minimize the frustrations of undue delays caused by subcontractors whose work needs to be done before other progress can be made.

The following is a list of steps in house construction. It is the general contractor's responsibility to see that these steps are carried out.

1. **Survey.** The job site is surveyed and the abstract brought up to date so that application for title insurance can be made.
2. **Permit.** A building permit is obtained from proper authorities so that work can begin.
3. **Excavation.** The excavator brings in power equipment and strips the topsoil away, piling it in one corner of the lot for future use. If the building will have a basement, it is excavated at this time.
4. **Temporary power.** The electrical company is contacted to arrange for a power pole to be set in place on the building site and a hookup made. The electricity is needed for operating power tools.
5. **Temporary water.** On some job sites the plumber makes the temporary water hookup, which must be coordinated with the city utilities. Sometimes a power-driven water pump for a well is used. In existing neighborhoods, water can be obtained from a neighbor. In this case, the permanent hookup for water to the building is not made until the foundation walls have been installed.
6. **Foundation.** Footings and foundation walls are installed by the concrete and masonry subcontractor.
7. **Soil treatment.** In areas that require it, a termite control specialist will treat the soil at the base of the foundation wall and the footing.
8. **Plumbing.** Plumbing contractor installs pipelines for plumbing in the subsoil.
9. **Backfill.** At this point the general contractor must make certain decisions concerning how the foundation walls are to be braced to permit backfilling. Backfilling cannot be done unless the foundation walls are braced or the building is framed to provide plenty of weight to the foundation. Even when the walls and roof have been framed, the foundation must be braced to some extent before backfilling. Whether to backfill immediately after the foundation is in or to wait until after the framing is completed is the general contractor's choice.

If backfilling is delayed until framing is completed, the workers have the inconvenience of working around a foundation with a large excavation. This is unsafe and can lead to injuries because the carpenters normally have to use planking to carry materials over the excavation into the building. On the other hand, if backfilling is done before construction, extra time and costs are involved in bracing the foundation securely so that it does not cave in. Before any backfilling can be completed, the exterior walls of the foundation must be moisture-proofed.

10. **Framing.** The carpenters can now do the floor, wall, and roof framing.

11. **Masonry.** Chimneys and fireplaces are built by the masonry contractor after the rough framing is completed.

12. **Mechanicals.** At this point in construction, a variety of activities may be carried out simultaneously, or at least in rapid succession. These include plumbing, heating, and electrical work. All of these mechanical subcontractors must work in two stages; namely, the "rough-in" and the "finish" work.

    For example, when the rough framing is complete, the electrician comes in to do the rough wiring, including installing the outlet boxes and feeding all the wires through the framing. This is the rough-in portion of the work. Later after the interior walls are all completed, the electrician comes back to install the switches, outlets, and fixtures. This is also true of the plumber and the heating and air-conditioning workers. (The plumber will install bathtubs with the rough-in, since tubs are a built-in feature of the house.)

13. **Windows and doors.** While the mechanical subcontractors are doing the rough-in work, the carpenters install exterior doors, windows, and special framing.

14. **Insulation.** After all rough-in work is done, insulation is installed in the walls and as necessary in the ceiling.

15. **Interior and exterior finishes.** Most interiors of homes are finished either with gypsum drywall or with plaster. If plastering is specified, this should be done immediately. At the same time, carpenters can work on the exterior of the building installing siding, exterior trim, and the garage door.

    Normally, plaster is applied in two stages, the rough coat and the finish coat. Plenty of time must be allowed for drying between stages and after the finish coat. Often a week or ten days must be allowed before proceeding with any interior work. If gypsum drywall is installed, the drying period is much shorter since the only wet application is taping the joints and covering nailheads.

16. **Slabs.** If the house has a basement, the concrete floor is poured after the rough plumbing is installed. However, this is done before the finish interior work. Concrete, too, must dry out thoroughly. The garage floor is put in anytime after the backfill is completed. Often this is done at the same time as the basement floor. The concrete is delivered to the site by special trucks. Concrete driveways and sidewalks, however, are installed at the very last stages of the building construction, after finish grading.

17. **Finish carpentry.** At this stage the carpenters are ready to do the interior finishing, provided the plaster and concrete are thoroughly dry. If lumber is delivered and stored in a house where there is high humidity due to wet plaster and concrete, the wood absorbs the moisture and swells. Later it will dry out and show large cracks.

    In completing the interior, the first job is to install underlayment, then flooring, wood paneling, and finally the finish flooring. The interior frames, doors, and cabinets are next installed. Finally the interior moldings are applied, including the base, shoe, ceiling, and window trim.

18. **Exterior painting.** As carpenters are working on the inside of the house, painters can be finishing the exterior. The ideal arrangement is for the painters to work closely behind the carpenters on the exterior so that the wood is properly sealed. If exterior trim has been preprimed at the factory, this schedule is not so critical.

19. **Grading.** While the carpenters are completing the interior of the house, the exterior grading can be done and all flat concrete work such as sidewalks and driveways can be installed.
20. **Landscaping.** The final step in completing the exterior of the house is the landscaping.
21. **Interior painting.** When the carpenters have completed the interior of the house, the painting can be done.
22. **Tile and floor coverings.** After the paint is dry, all tile and floor coverings except hardwood and carpet are installed.
23. **Finish electrical.** At this point the electricians can return to install switches, outlets, and fixtures.
24. **Finish plumbing.** The plumbing contractor can now install the plumbing fixtures.
25. **Wood flooring.** One of the last jobs on the interior of the house is to finish the wood flooring. Many homes are completely covered by carpeting and require no floor finishing. Carpets are often laid directly over a good plywood or particleboard base. However, if hardwood floors are used, floor sanding should be done after the interior painting to remove any paint drops or spillage. The actual finishing is done as one of the last jobs so that traffic will not raise dust when the floor finish is drying. Hardwood flooring can be purchased prefinished, which greatly simplifies this part of the job.
26. **Carpeting.** After the wood floors are finished, carpeting is installed.
27. **Cleanup.** the general contractor is responsible for the final cleanup. A good contractor will make sure that the windows are washed and all waste materials removed from the job site.
28. **Punch list.** After the entire house has been completed, the general contractor walks through the house with the new owner. This is a chance for the owner to make sure everything has been done to his or her satisfaction. Often the owner will spot such things as scuffed paint, cracked woodwork, or light fixtures that don't work properly. The contractor then makes a list, called a "punch list." This lists all the items that must be completed before the house is acceptable to the owner.

Table of Lumber Abbreviations

AD—Air-dried
ADF—After deduction freight
ALS—American Lumber Standards
AVG—Average
AW&L—All widths and lengths
BD—Board
BD FT—Board feet
BDL—Bundle
BEV—Bevel
BH—Boxed heart
B/L, BL—Bill of lading
BM—Board measure; bench mark
B&S—Beams and stringers
BSND—Bright sapwood no defect
BTR—Better
CB—Center beaded
CF—Cost and freight
CIF—Cost, insurance and freight
CIFE—Cost, insurance, freight, exchange
C/L—Carload
CLG—Ceiling
CLR—Clear
CM—Center matched
CS—Caulking seam
CSG—Casing
CV—Center V
DET—Double end trimmed
DF—Douglas fir
DF-L—Douglas fir-larch
DIM—Dimension
DKG—Decking
D&M—Dressed and matched
D/S, DS—Drop siding
E—Edge or modulus of elasticity
EB1S—Edge bead one side
EB2S—Edge bead two sides
E&CB2S—Edge and center beads two sides
E&CV1S—Edge and center vee one side
E&CV2S—Edge and center vee two sides
EE—Eased edged
EG—Edge (vertical) grain
EM—End matched
ES—Englemann spruce
EV1S—Edge vee one side
EV2S—Edge vee two sides
f—Allowable fiber stress in bending (also Fb)
FAS—Firsts and seconds
FG—Flat or slash grain
FLG—Flooring
FOB—Free on board (Named point)
FOHC—Free of heart center
FRT—Freight
Ft—Foot
FT.BM—Feet board measure (also FBM)
FT.SM—Feet surface measure
H.B.—Hollow back
HEM—Hemlock
H&M—Hit and miss
H or M—Hit or miss
IC—Incense cedar
IN—Inch or inches
IND—Industrial
IWP—Idaho white pine
J&P—Joists and planks
JTD—Jointed
KD—Kiln-dried
L—Larch
LBR—Lumber
LCL—Less than carload
LF—Light framing
LFVC—Loaded full visible capacity
LGR—Longer
LGTH—Length
LIN—Lineal
LNG—Lining
LP—Lodgepole pine
M—Thousand
M. BM—Thousand (ft.) board measure
MC—Moisture content
MG—Mixed grain
MLDG—Molding
MOE—Modulus of elasticity or “E”
MOR—Modulus of rupture
MSR—Machine stress rated
NBM—Net board measure
N1E—Nose one edge
PAD—Partly air dried
PARA—Paragraph
PART—Partition
PAT—Pattern
PET—Precision end trimmed
PP—Ponderosa pine
P&T—Posts and timbers
RC—Red cedar
RDM—Random
REG—Regular
RGH—Rough
R/L, RL—Random lengths
R/S—Resawn
R/W, or RW—Random widths
R/W, R/L, RWL—Random width, Random length
SB1S—Single bead one side
SDG—Siding
SEL—Select
SG—Slash or flat grain
S/L, or SL—Shiplap
S&E—Side and edge
S1E—Surfaced one edge
S2E—Surfaced two edges
S1S—Surfaced one side
S2S—Surfaced two sides
S4S—Surfaced four sides
S1S&CM—Surfaced one side and center matched
S2S&CM—Surfaced two sides and center matched
S4S&CS—Surfaced four sides and caulking seam
S1S1E—Surfaced one side, one edge
S1S2E—Surfaced one side, two edges
S2S1E—Surfaced two sides, one edge
SM—Surface measure
SP—Sugar pine
SQ—Square
STD. M—Standard measure
STK—Stock
STPG—Stepping
STR—Structural
TBR—Timber
T&G—Tongued and grooved
VG—Vertical (edge) grain
WDR—Wider
WF—White fir
WT—Weight
WTH—Width
WRC—Western red cedar
WWPA—Western Wood Products Association

SYMBOLS
″—Inch or inches
′—Foot or feet
×—By, as 4 × 4
4/4, 5/4, 6/4, etc.—nominal thickness expressed in fractions

ESTIMATING CONCRETE

Concrete is measured and sold by the cubic yard (27 cubic feet). To calculate the amount of concrete required for a job, figure the volume of the forms in cubic feet (thickness × width × length) and divide by 27 to obtain the total number of cubic yards. For example, to pour a 4″ thick driveway which is 20′ wide and 40′ long, calculate the cubic footage by multiplying the thickness, 4″ (or ⅓ of a foot) times 20′ (width) times 40′ (length). This multiplies out to 266 ⅔ or 267 cubic feet. Converting this to cubic yards, 267⁄27 = 9 24⁄27 or 10 cubic yards of concrete are needed to complete the driveway.

Table B. Estimating Table for Concrete Slabs

Thickness of Slab	Cubic Feet of Concrete per Square Foot	Square Feet from One Cubic Yard
2″	0.167	162
3″	0.25	108
4″	0.333	81
5″	0.417	65
6″	0.50	54

Table A. Concrete Estimating

Thickness In.	Sq. ft.	Thickness In.	Sq. Ft.	Thickness In.	Sq. ft.
1	324	4¾	68	8½	38
1¼	259	5	65	8¾	37
1½	216	5¼	62	9	36
1¾	185	5½	59	9¼	35
2	162	5¾	56	9½	34
2¼	144	6	54	9¾	33
2½	130	6¼	52	10	32.5
2¾	118	6½	50	10¼	31.5
3	108	6¾	48	10½	31
3¼	100	7	46	10¾	30
3½	93	7¼	45	11	29.5
3¾	86	7½	43	11¼	29
4	81	7¾	42	11½	28
4¼	76	8	40	11¾	27.5
4½	72	8¼	39	12	27

THIS TABLE INDICATES THE AREA IN SQUARE FEET THAT 1 CUBIC YARD OF CONCRETE WILL FILL FOR A VARIETY OF THICKNESSES. FOR EXAMPLE, 1 CUBIC YARD OF CONCRETE WILL FILL A FORM AREA OF 40 SQUARE FEET FOR A WALL 8″ THICK.

Better Homes and Gardens® WOOD®

Stay-Put Tool Holders

Pegboard provides an ideal way to store your tools. One of the biggest problems with it is how to get the hanger "pegs" to stay put. Here are a few suggestions, along with two mini-projects that will help you organize your pegboard tool rack.

- For the pencil holder, accurately measure the perforated holes for placement of the dowels.
- The tabletop (desktop) fasteners can be purchased at a woodworking store or through a woodworking catalog.
- The "recharging station" can be customized to your drill and charger. Some chargers adapt better to drilled holes than to dowels. The "footprint" of the charger can be transferred to the workpiece by lightly pressing it into an inkpad and then setting it on the board.
- When installing the mounting screws in the pegboard, just snug them up; overtightening will easily strip out the holes in the pegboard.

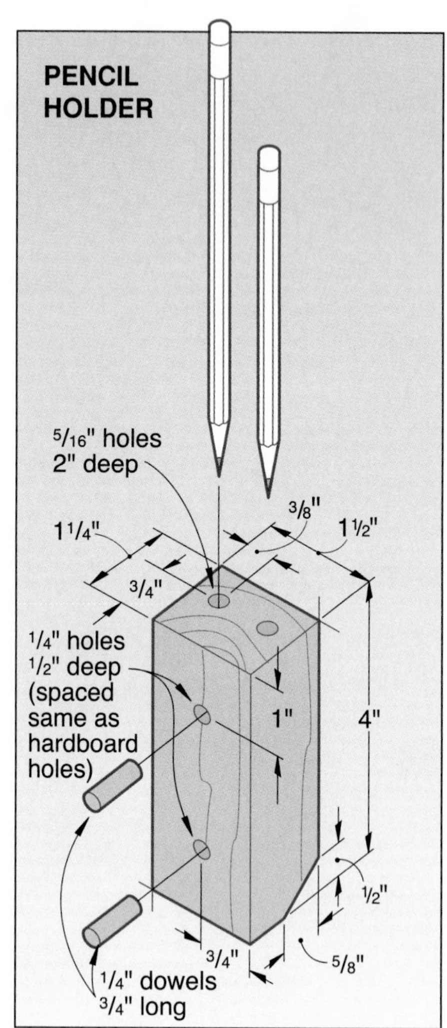

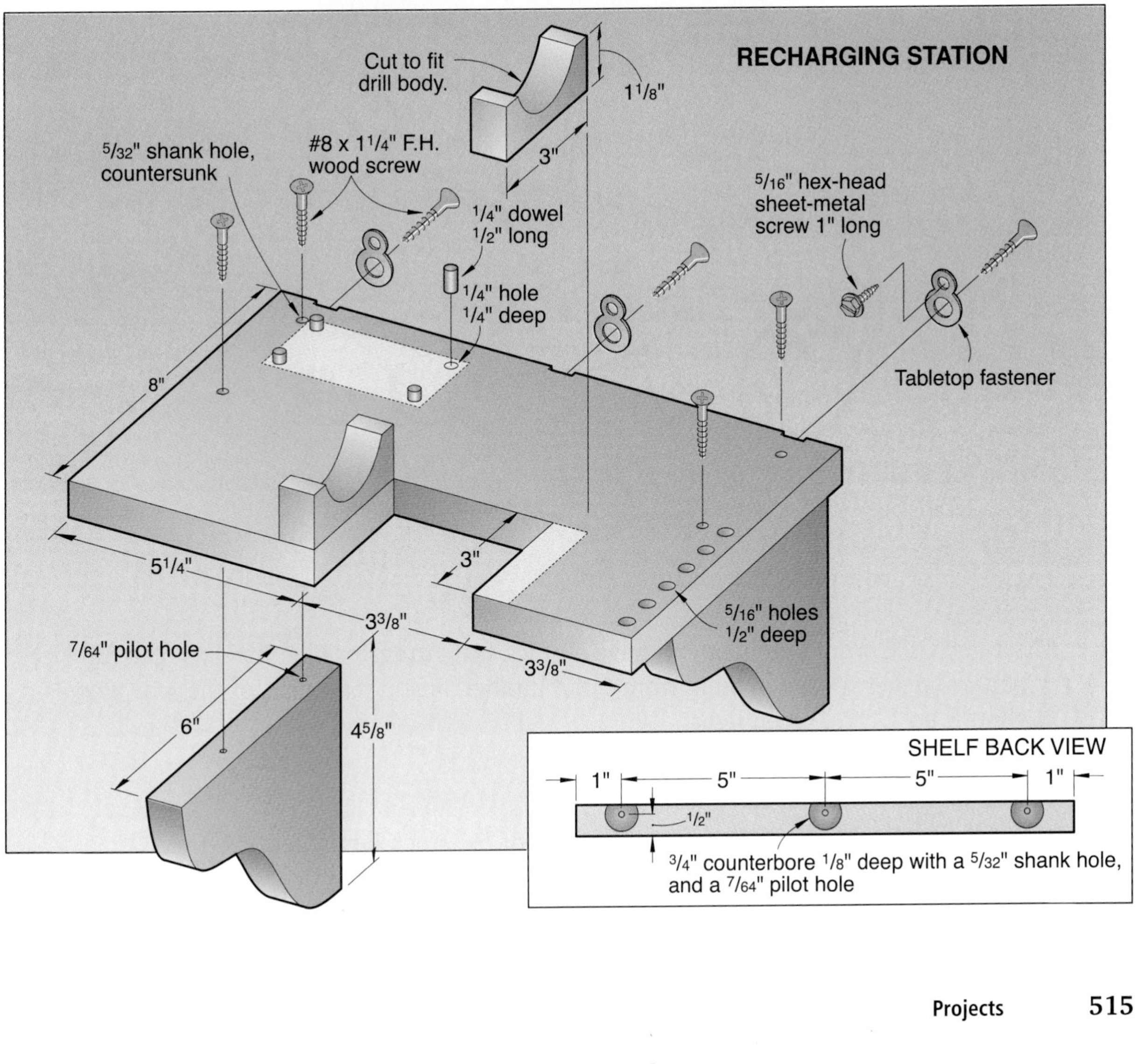

RECHARGING STATION

Cut to fit drill body.

1 1/8"

3"

5/32" shank hole, countersunk

#8 x 1 1/4" F.H. wood screw

1/4" dowel 1/2" long

1/4" hole 1/4" deep

5/16" hex-head sheet-metal screw 1" long

Tabletop fastener

8"

5 1/4"

3"

3 3/8"

3 3/8"

5/16" holes 1/2" deep

7/64" pilot hole

6"

4 5/8"

SHELF BACK VIEW

1" 5" 5" 1"

1/2"

3/4" counterbore 1/8" deep with a 5/32" shank hole, and a 7/64" pilot hole

Rubber Band-Powered Dragster

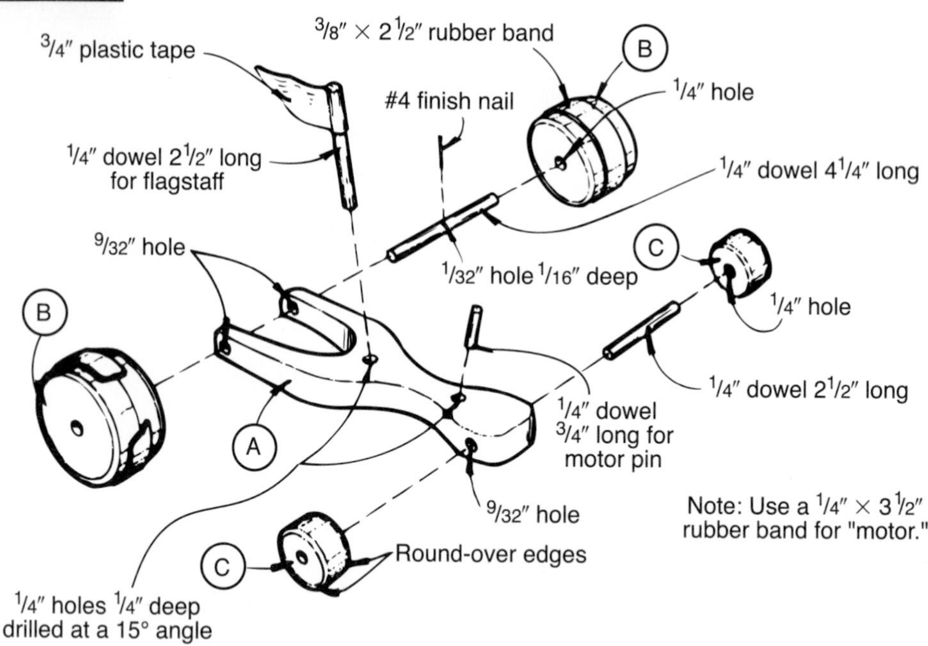

Most any child would be happy to have a quick little dragster like this one. The full-size pattern makes it easy to build from scrap lumber. Just a few turns of the rear wheels wind it up for the green light.

Suggestions:

- Use carbon paper to transfer the pattern.
- The rear “slicks” should be cut from two 3″ pieces of ½″ stock laminated together.
- Cut the slicks out with a 2″ hole saw.
- Cut the front wheels from ½″ stock with a 1″ hole saw.
- Grind the tip of a #4 finish nail until it measures ⅝″ long.

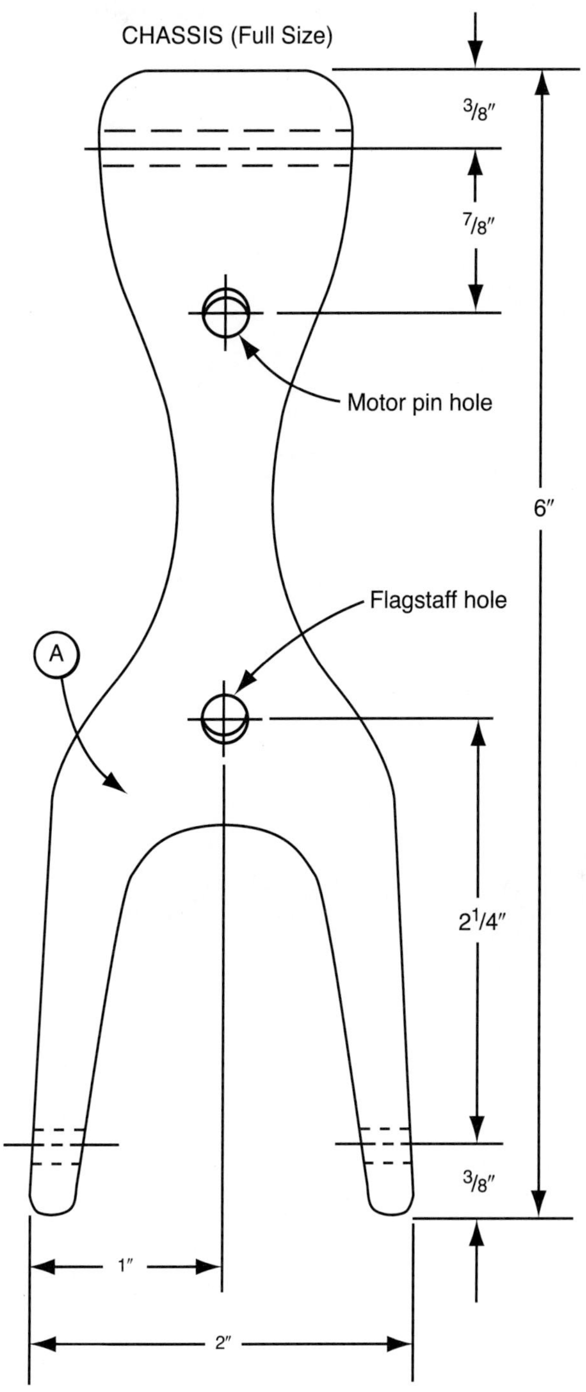
CHASSIS (Full Size)
3/8″
7/8″
Motor pin hole
6″
Flagstaff hole
A
2 1/4″
3/8″
1″
2″

A Building-Block Castle

Better Homes and Gardens® WOOD® MAGAZINE

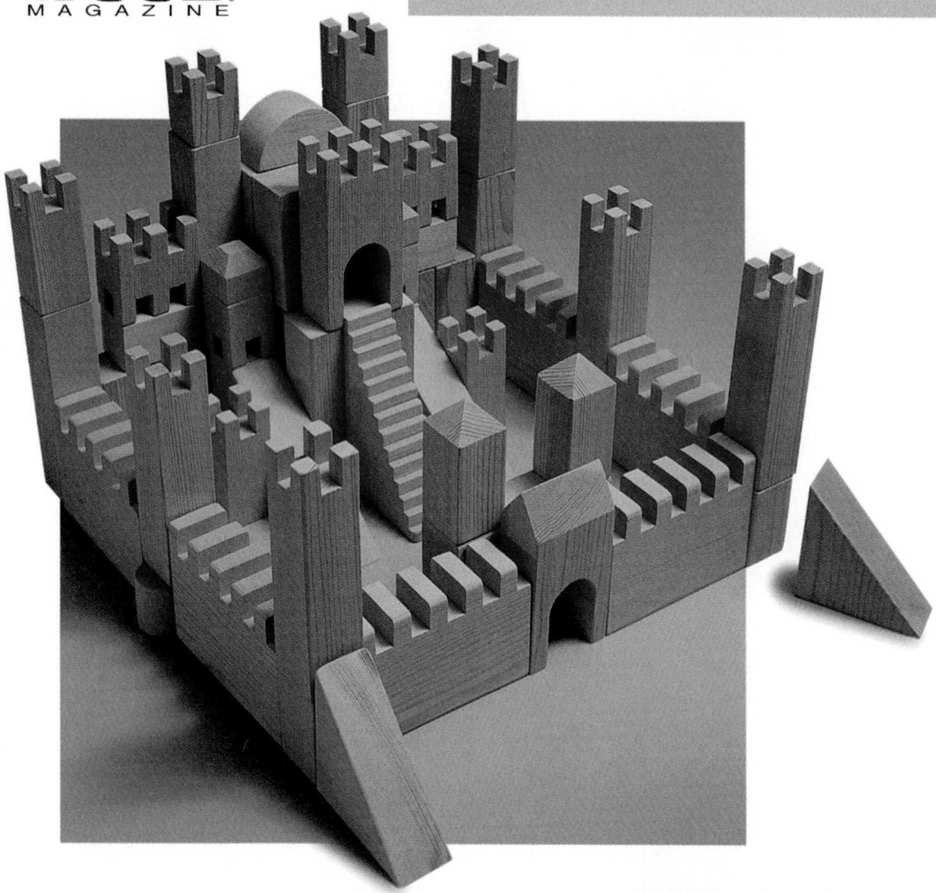

Children will spend a lot of time playing with this set of building blocks. It's a great holiday or birthday gift!

Suggestions for building the blocks:

- All the pieces can be cut from 2 × 4 and 2 × 6 clear-heart redwood, pine, or fir.
- Save the cutout when you band-saw the arc in parts C and F.
- Cut a piece of 1½″ stock to 3½″ × 3½″. Mark a diagonal from one corner to the other, and band-saw the block into two triangles for parts D. With a larger block, do the same for parts E.

Clamp the castle piece to an auxiliary fence fastened to your miter gauge when cutting the dadoes.

Angle-cut the ends of a 12″ length of 1½″ square stock to form the pointed ends on each part H. For safety, we started with an extra-long piece.

Determine the location, and clamp a stop block to your miter-gauge fence to position and support the castle pieces when machining the centered dado.

- Use a ½″ dado for the castle tops.
- Parts R and S use the same steps. They should be cut on the band saw.
- Sand all pieces and slightly round-over all corners.
- If you apply a finish, be sure it is safe for children.

Bill of Materials

Part	Finished Size			Material	Qty.
	T	W	L		
A	1½″	3½″	5″	redwood	4
B	1½″	3½″	3½″	redwood	4
C	1½″	3½″	5″	redwood	1
D	1½″	3½″	3½″	redwood	2
E	1½″	3½″	5¼″	redwood	2
F	1½″	2½″	5″	redwood	1
G	1½″	3½″	5″	redwood	6
H	1½″	2½″	5″	redwood	1
I	1½″	2½″	2½″	redwood	2
J	1½″	1½″	6″	redwood	4
K*	1½″	1½″	3½″	redwood	4
L*	1½″	1½″	2″	redwood	4
M	1½″	1½″	6″	redwood	4
N	1½″	1½″	3½″	redwood	6
O	1½″	1½″	2½″	redwood	4
P*	1½″	1½″	3½″	redwood	2
Q*	1½″	1½″	1¼″	redwood	2
R*	1½″	3½″	3½″	redwood	1
S*	1½″	3½″	5¼″	redwood	1

*Initially cut parts marked with an * oversized. Then, trim each to finished size according to the how-to instructions.

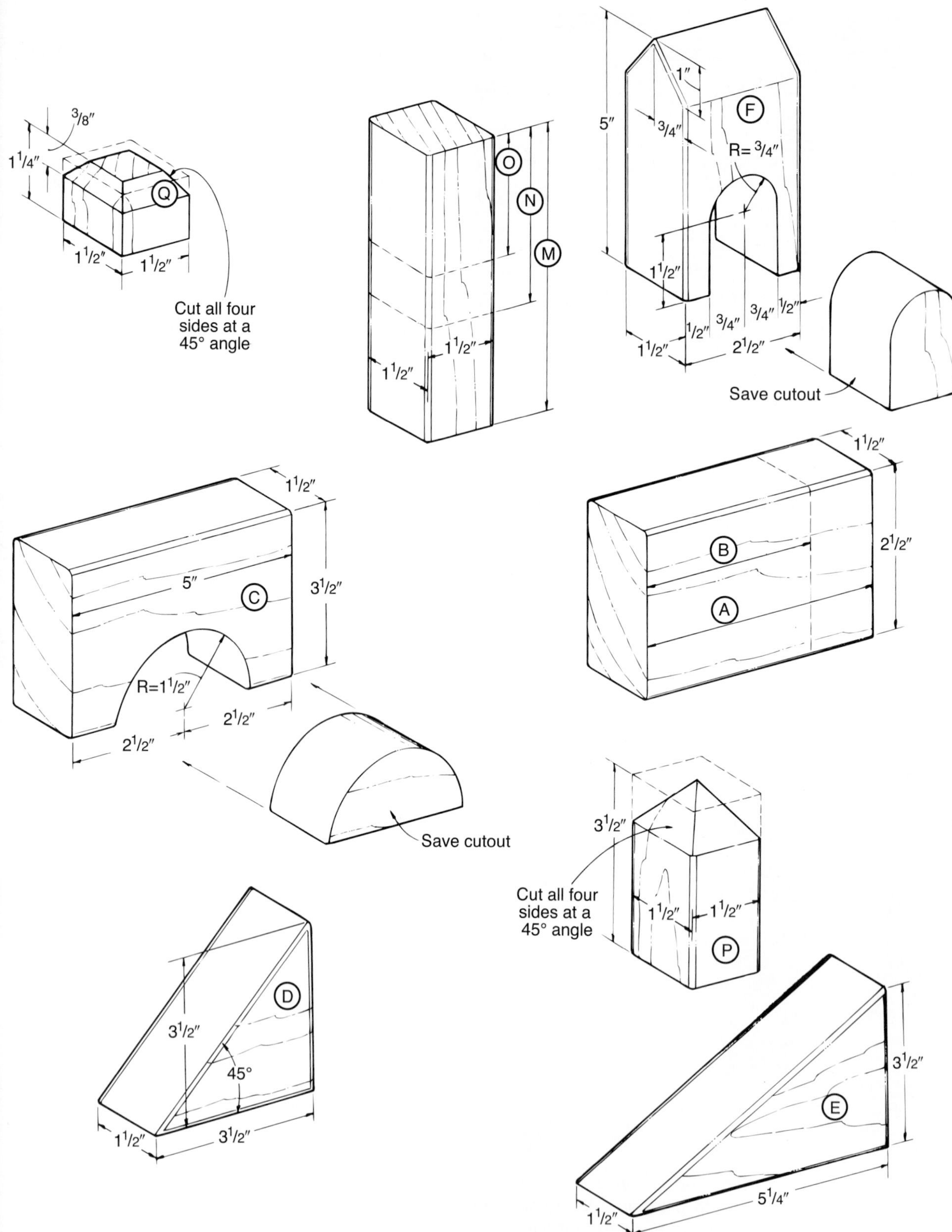
3/8″
1 1/4″
Q
1 1/2″
1 1/2″
Cut all four sides at a 45° angle
O
N
M
1 1/2″
1 1/2″
5″
1″
3/4″
F
R= 3/4″
1 1/2″
1/2″
3/4″
3/4″
1/2″
1 1/2″
2 1/2″
Save cutout
1 1/2″
3 1/2″
5″
C
R=1 1/2″
2 1/2″
2 1/2″
Save cutout
1 1/2″
B
A
2 1/2″
3 1/2″
Cut all four sides at a 45° angle
1 1/2″
1 1/2″
P
D
3 1/2″
45°
1 1/2″
3 1/2″
3 1/2″
E
1 1/2″
5 1/4″

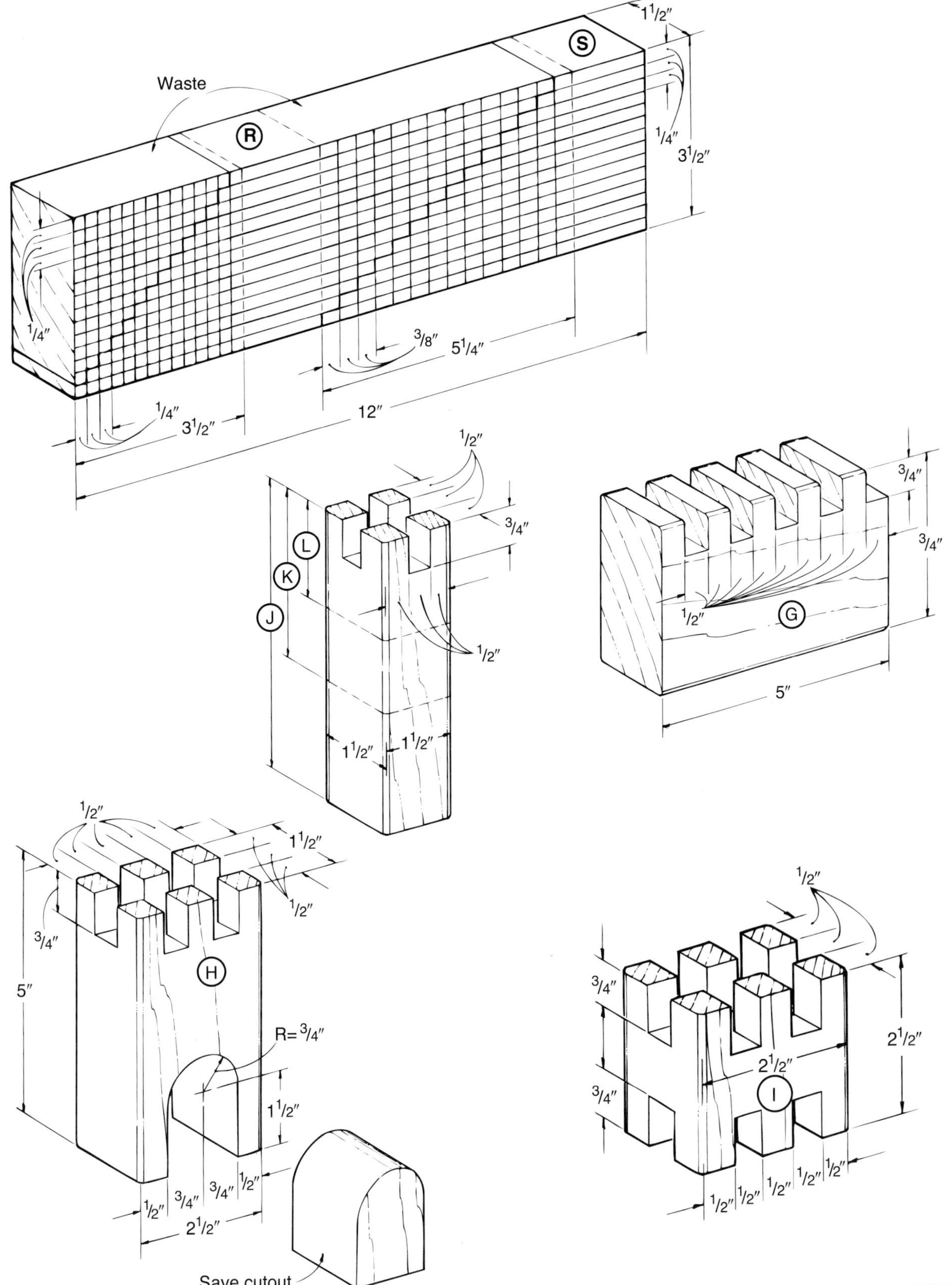

Waste
R
S
1 1/2″
1/4″
3 1/2″
1/4″
3/8″
5 1/4″
12″
1/4″
3 1/2″
1/2″
3/4″
L
K
J
1/2″
1 1/2″
1 1/2″
3/4″
3/4″
1/2″
G
5″
1/2″
1 1/2″
1/2″
3/4″
5″
H
R= 3/4″
1 1/2″
1/2″
3/4″
3/4″
1/2″
2 1/2″
Save cutout
1/2″
3/4″
2 1/2″
2 1/2″
I
3/4″
1/2″
1/2″
1/2″
1/2″
1/2″

Photo Stand

A prized photo calls for a special frame. Here's a handsome one of oak that fills the bill. What's more, it's quick and easy to build.

Suggestions for building the photo stand:

- To form the tenon on the bottom of each upright, use a 3⁄8″ dado on the table saw. Be sure to use a scrapwood auxiliary fence.
- As shown in the illustration, use a fence on your drill press to position and drill the holes in the top of the frame.
- Fasten the uprights together with double-faced tape. Band-saw the tapered edge slightly outside the line. The taper goes from 1¼″ at the base to ¾″ at the top of part B.
- Add a 1⁄16″ chamfer to the edges of the uprights.
- Do not glue the dowels into the uprights.

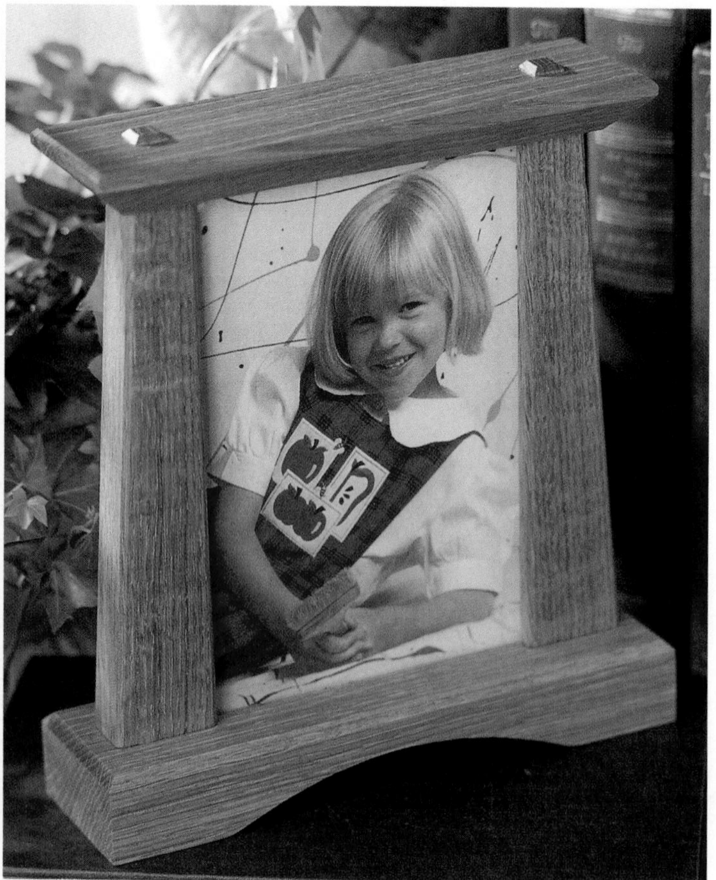

Bill of Materials

Part	Finished Size			Matl.	Qty.
	T	W	L		
A base	1″	2″	7 1/4″	O	1
B upright	3/4″	1 1/4″	6 3/16″	O	2
C cap	5/8″	1 7/8″	6 3/4″	O	2

Materials Key: O-white oak

Supplies: 1/4″ dowel, glazing (either single-strength glass or 0.1″ acrylic sheet).

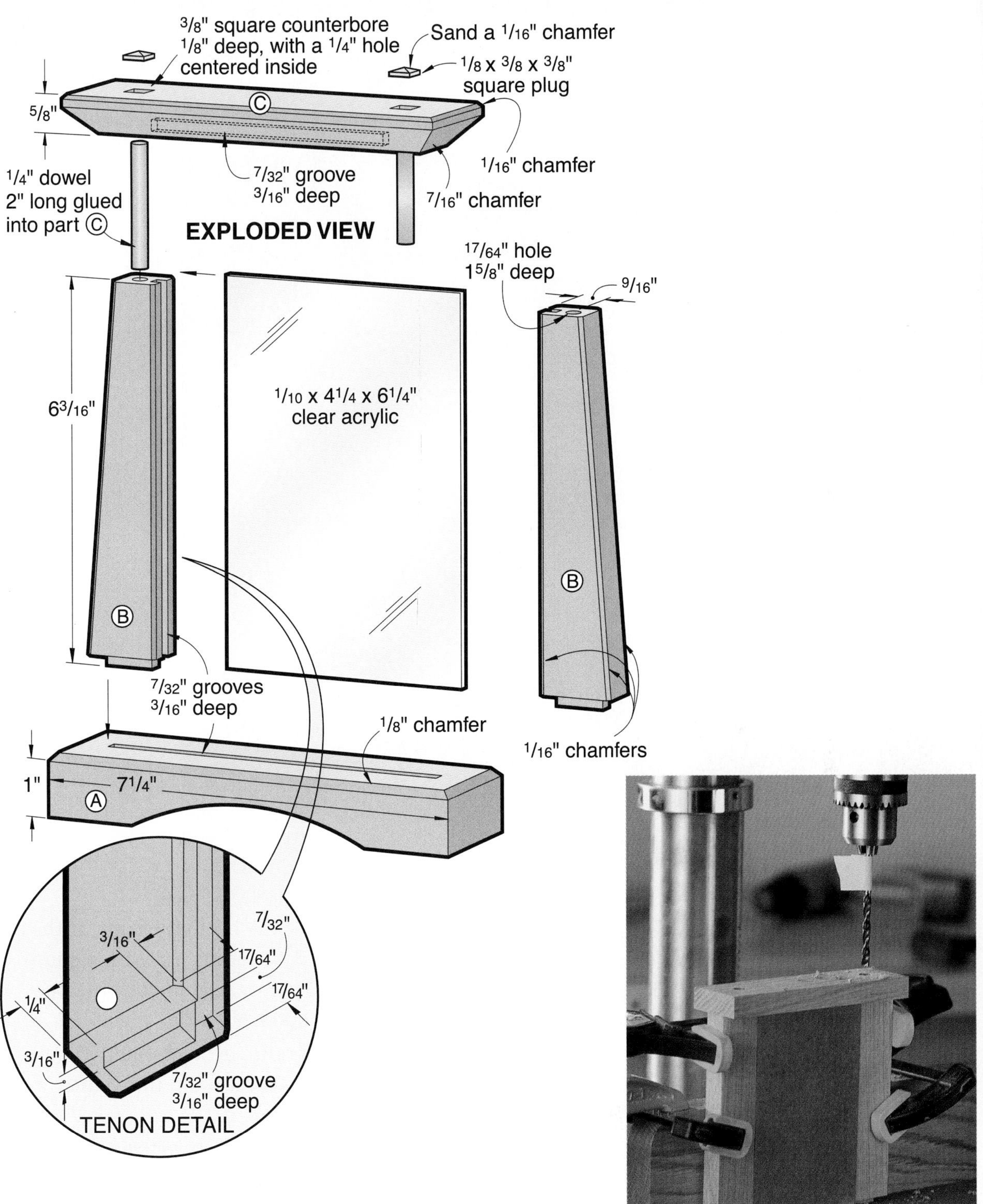

Drill the holes through the cap into the uprights with a drill press. Tape on the twist drill serves as a depth indicator.

Sturdy Shaker Stool

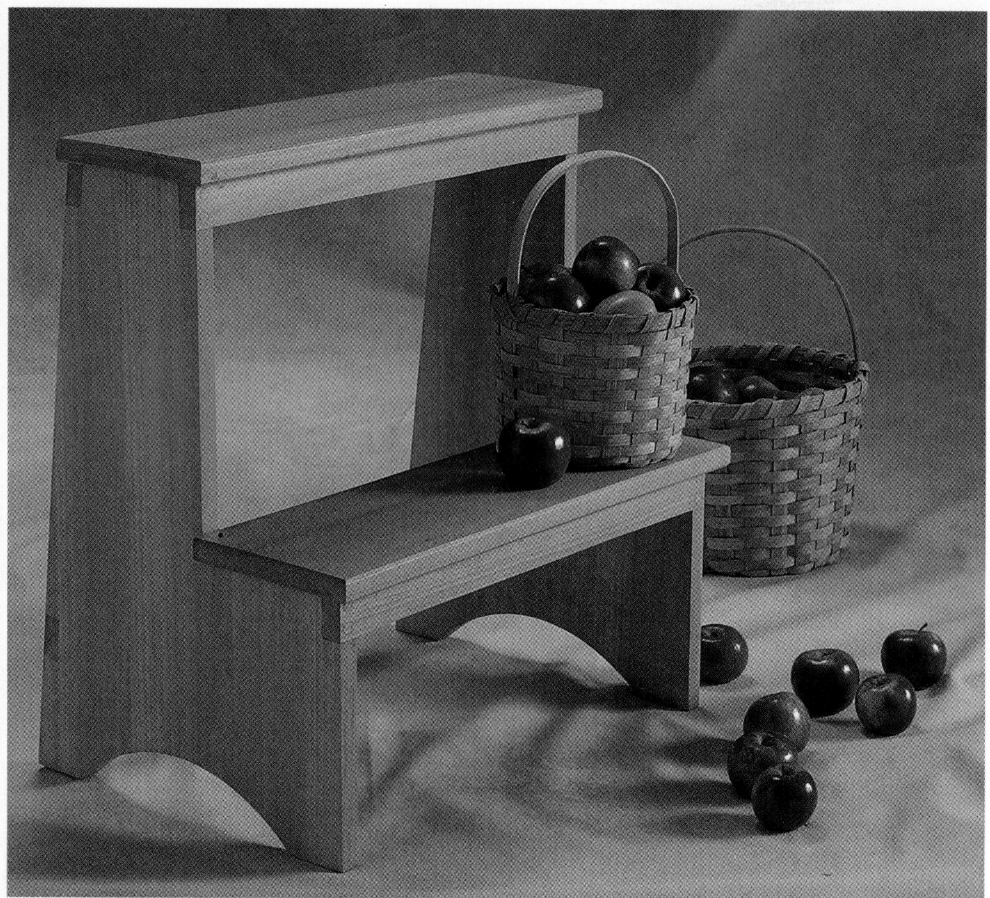

It looks like a hand-me-down from a 19th century schoolhouse, but this simple, timeless design will be equally at home in your home. You'll appreciate how stable it feels underfoot, thanks to the angled sides that shift the top step's center of gravity a bit forward.

Suggestions for building the Shaker stool:

- From 3/4″ × 5½″ stock (poplar is a good choice), crosscut six pieces 22″ long to glue together for the sides.
- Using a 1 × 8 at the bottom of one of the sides, plot the 7¼″ radius arc with a shop-made trammel point. Also lay out the back angle and the notches on this piece.
- Using double-face tape, stack the two panels together face-to-face and make the appropriate cuts.
- To cut the inside tip notch, set the saw blade to cut 1½″ deep. Place the fence to cut a ¾″-wide notch. Make several passes to cut out the notch.

CUTTING DIAGRAM

C
C
B
3/4 x 7 1/4 x 72" Poplar
A
A
A
3/4 x 5 1/2 x 72" Poplar (2 pieces)

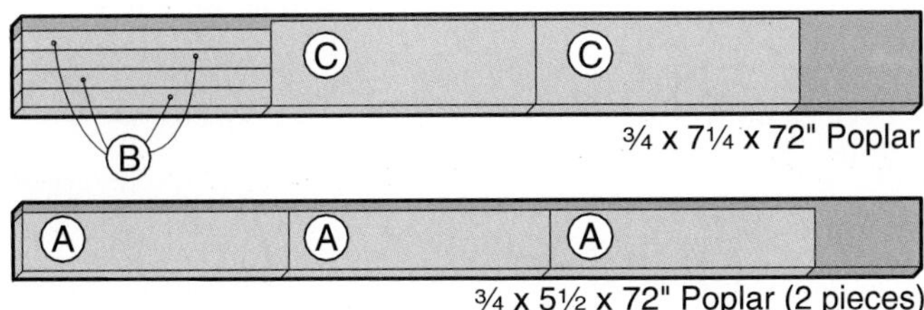

EXPLODED VIEW

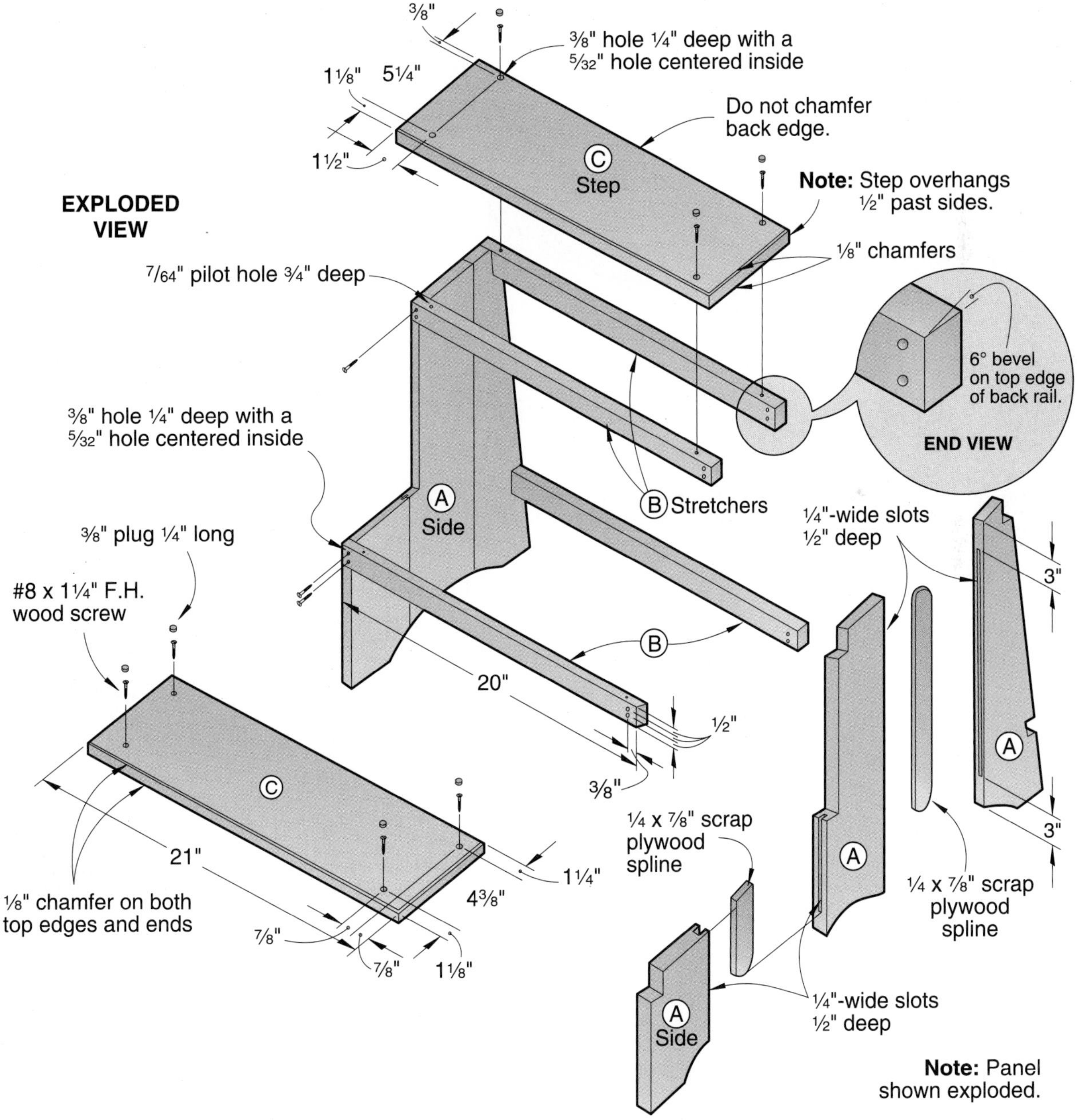

3/8"
3/8" hole 1/4" deep with a 5/32" hole centered inside
1 1/8"
5 1/4"
1 1/2"
Do not chamfer back edge.
C Step
Note: Step overhangs 1/2" past sides.
1/8" chamfers
7/64" pilot hole 3/4" deep
6° bevel on top edge of back rail.
END VIEW
3/8" hole 1/4" deep with a 5/32" hole centered inside
A Side
B Stretchers
3/8" plug 1/4" long
#8 x 1 1/4" F.H. wood screw
B
20"
1/2"
3/8"
1/4"-wide slots 1/2" deep
3"
3"
A
A
1/4 x 7/8" scrap plywood spline
1/4 x 7/8" scrap plywood spline
C
21"
1 1/4"
4 3/8"
1/8" chamfer on both top edges and ends
7/8"
7/8"
1 1/8"
1/4"-wide slots 1/2" deep
A Side
Note: Panel shown exploded.

Laying Out the Bottom Arc

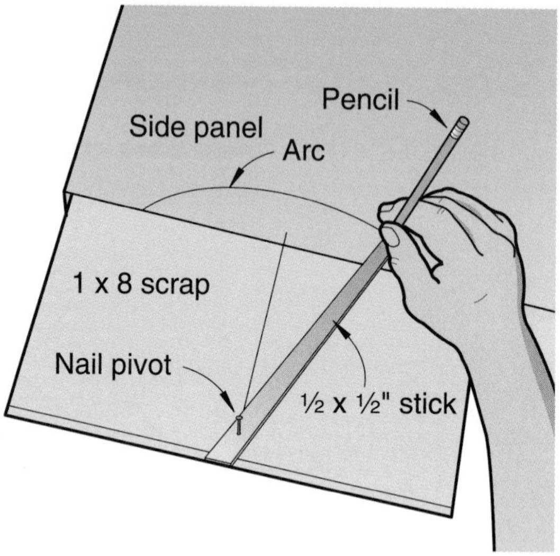

Pencil
Side panel
Arc
1 x 8 scrap
Nail pivot
½ x ½" stick

Cutting the Stretcher Notches

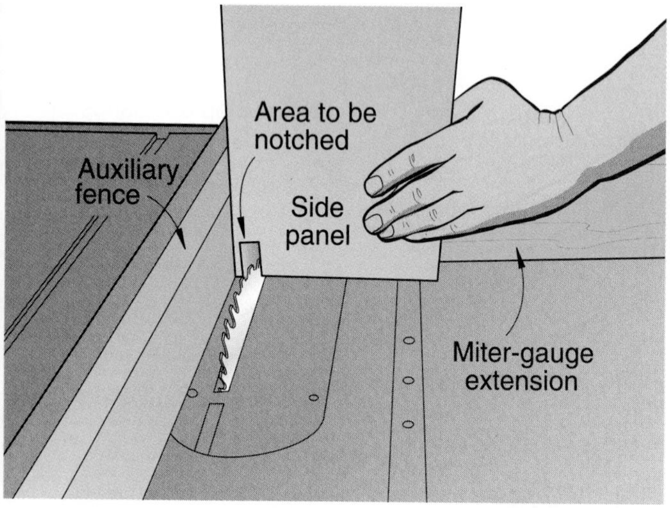

Area to be notched
Auxiliary fence
Side panel
Miter-gauge extension

Ripping the Stair Cutout

Fence
Stopblock
Step cutout waste

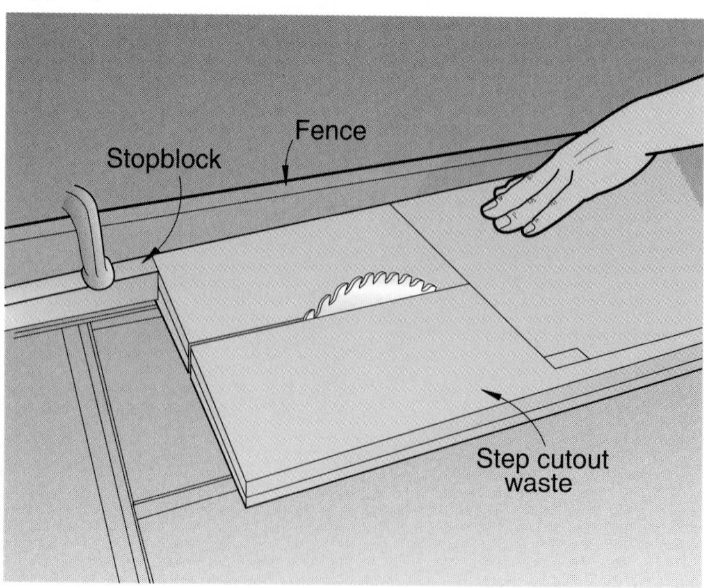

Bill of Materials					
Part	Finished Size			Matl.	Qty.
	T	W	L		
A* side	3/4″	14 1/2″	21″	O	2
B stretcher	3/4″	1 1/2″	20″	O	4
C step	3/4″	6 3/4″	21″	O	2

*Initially cut part oversize. Please read all instructions before cutting the parts.

Material Key: P—poplar.

Supplies: #8 × 1/4″ flathead wood screws, finish.

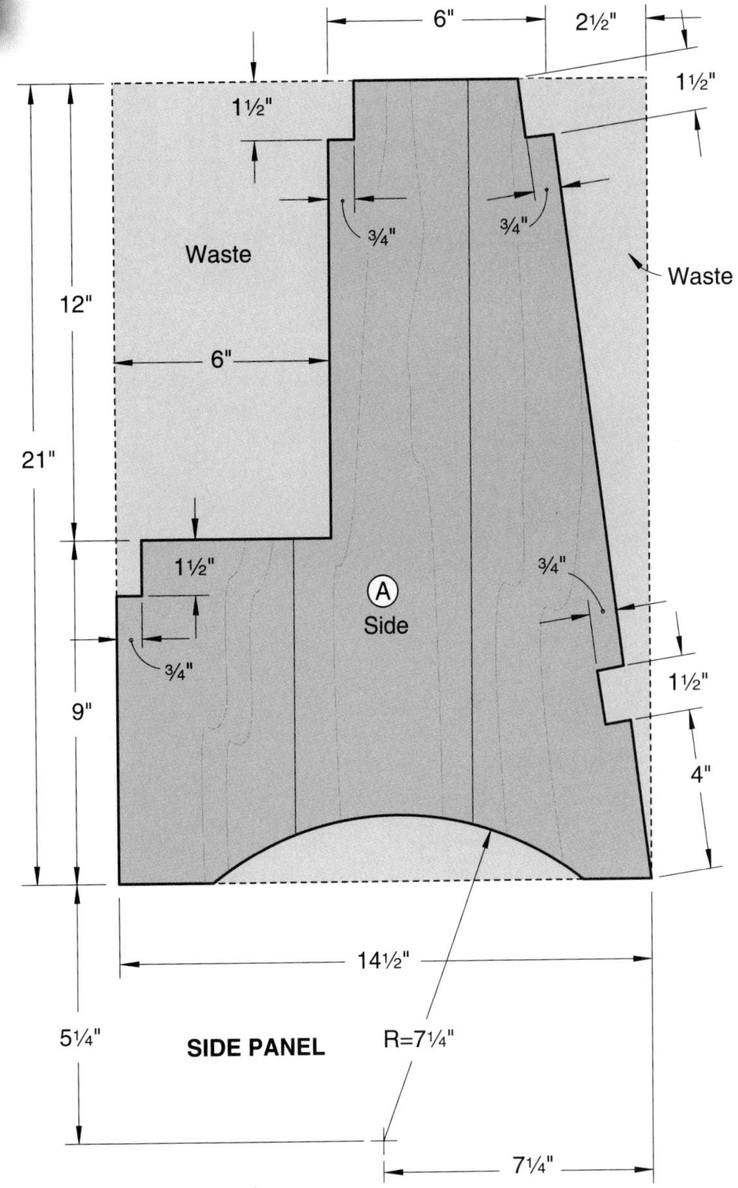

PROJECT 6

Rack-a-Tier

Better Homes and Gardens
WOOD
MAGAZINE

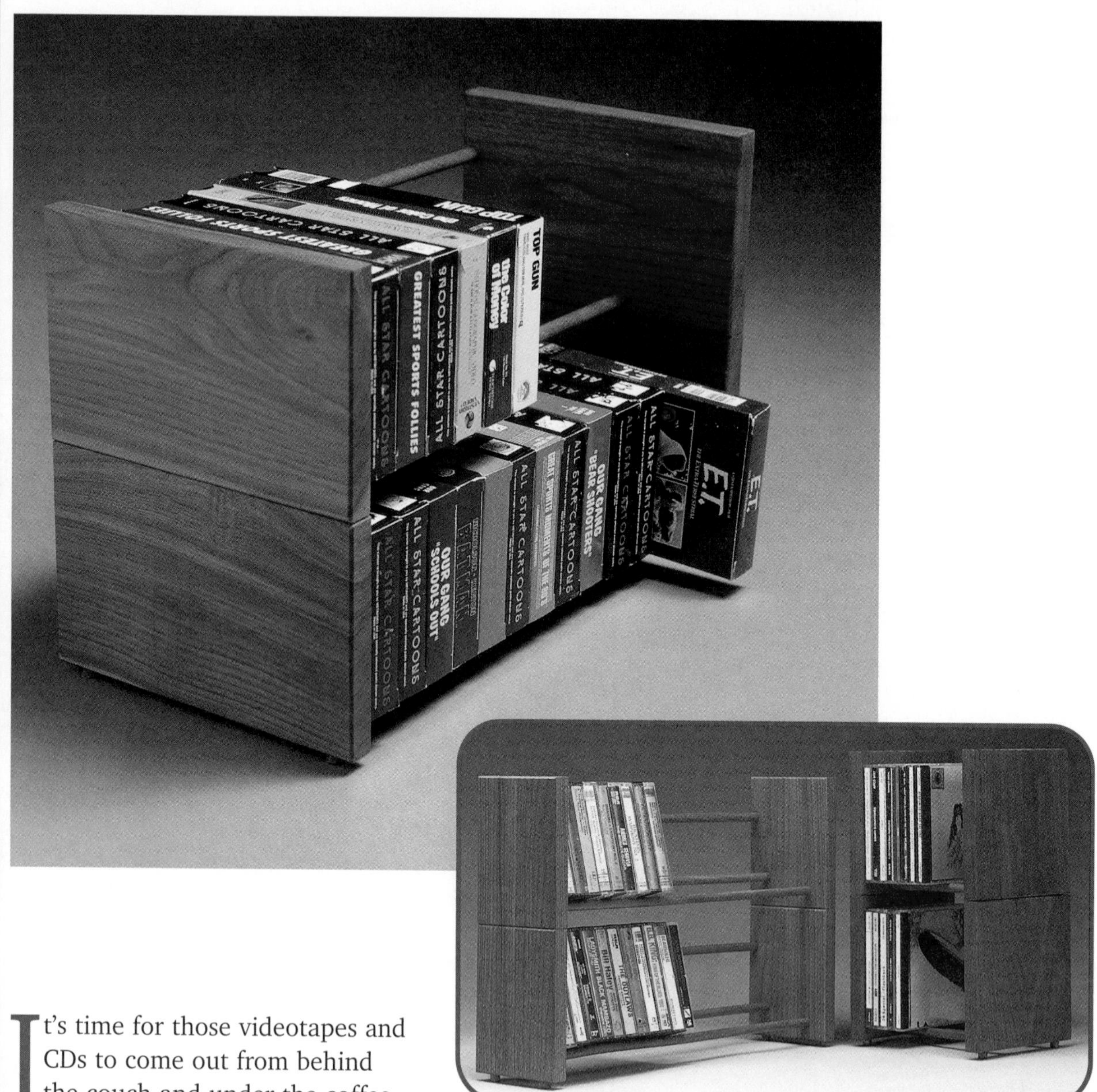

It's time for those videotapes and CDs to come out from behind the couch and under the coffee table. Here's a simple solution to keep your collection organized and tastefully displayed. Stack one rack on top of another to increase your storage space as your collection grows. Or, if you're just starting out, build a single organizer now and construct more as you need them.

Suggestions for building the rack:

- Note the grain directions as shown in the illustrations.
- One way to ensure that the dowel holes match on the ends is to drill the holes in one endpiece, then use dowel centers to transfer the hole locations to the other endpiece.
- Glue the dowels between the endpieces. To avoid wobble in the finished rack, clamp the endpieces to a flat surface so the bottom edges are level with each other.

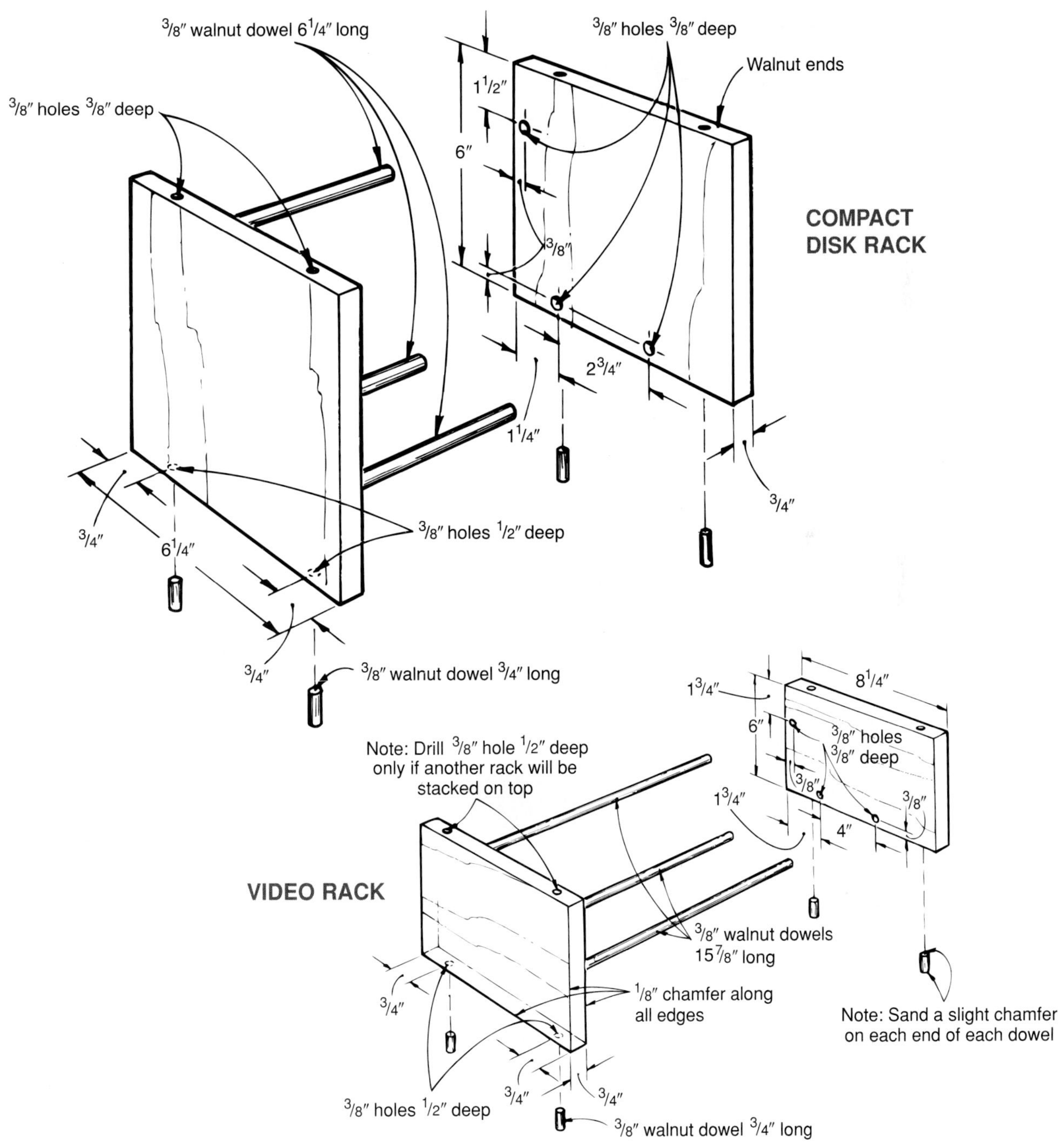

Collector's Display Case

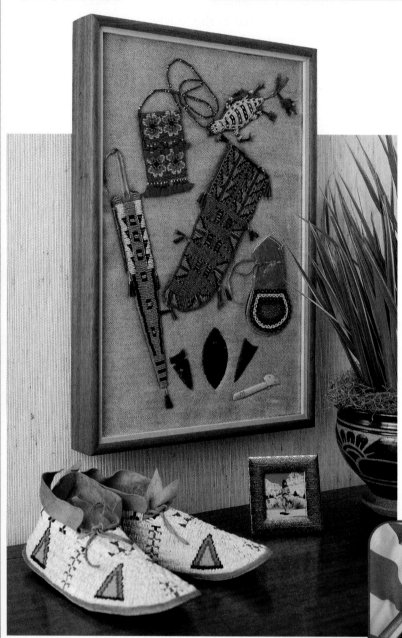

Display your favorite collectibles in this easy-to-build display case. You can use the dimensions shown or customize them to your collection size. The machining and construction of the tabletop and wall-hung display cases are nearly identical, other than the overall size of the projects.

Suggestions for building the display case:

- Use a dado blade and auxiliary fence for the rabbet and dado cuts in the frame.
- Check your miter cuts on scrap wood to be sure they are exactly 45°.
- To minimize splintering the glass stops when mitering, wrap masking tape around them to support the fibers while you make the cut.
- Apply the fabric to the foam with a spray adhesive.

Use end and top guides to guide and stop the router when routing the keyhole slots in the plywood back.

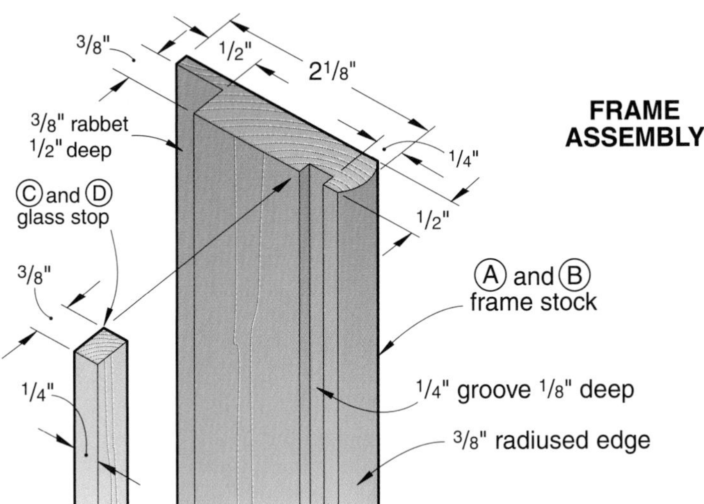

MACHINING THE FRAME MOLDING

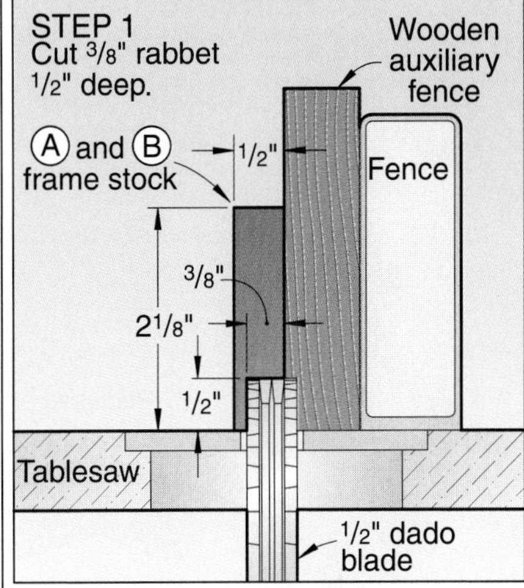

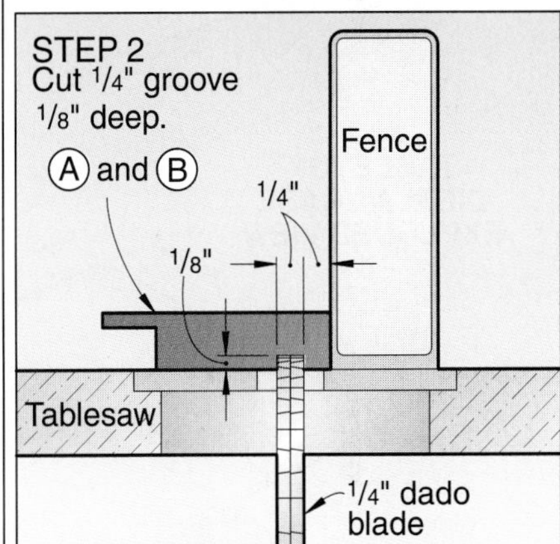

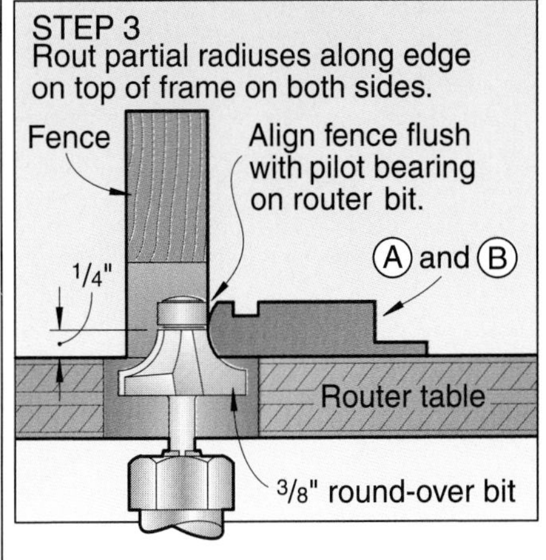

Display object
Glass
C and D
1/2"
Silicone sealant
2 1/8"
1/2"
E Plywood back
Foam
Screw-on rubber foot
7/8"
A and B
SECTION VIEW DETAIL (END VIEW)

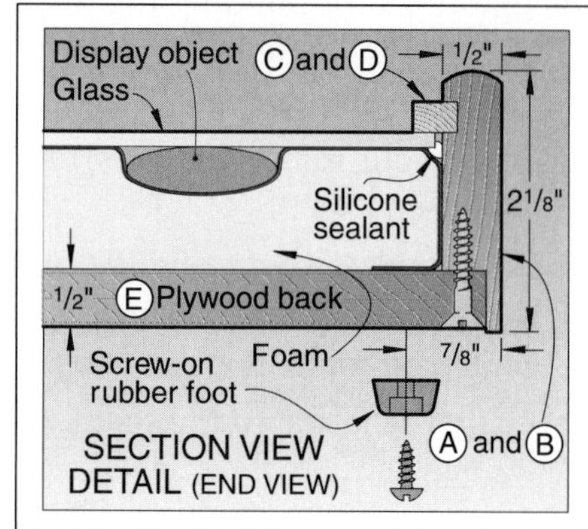

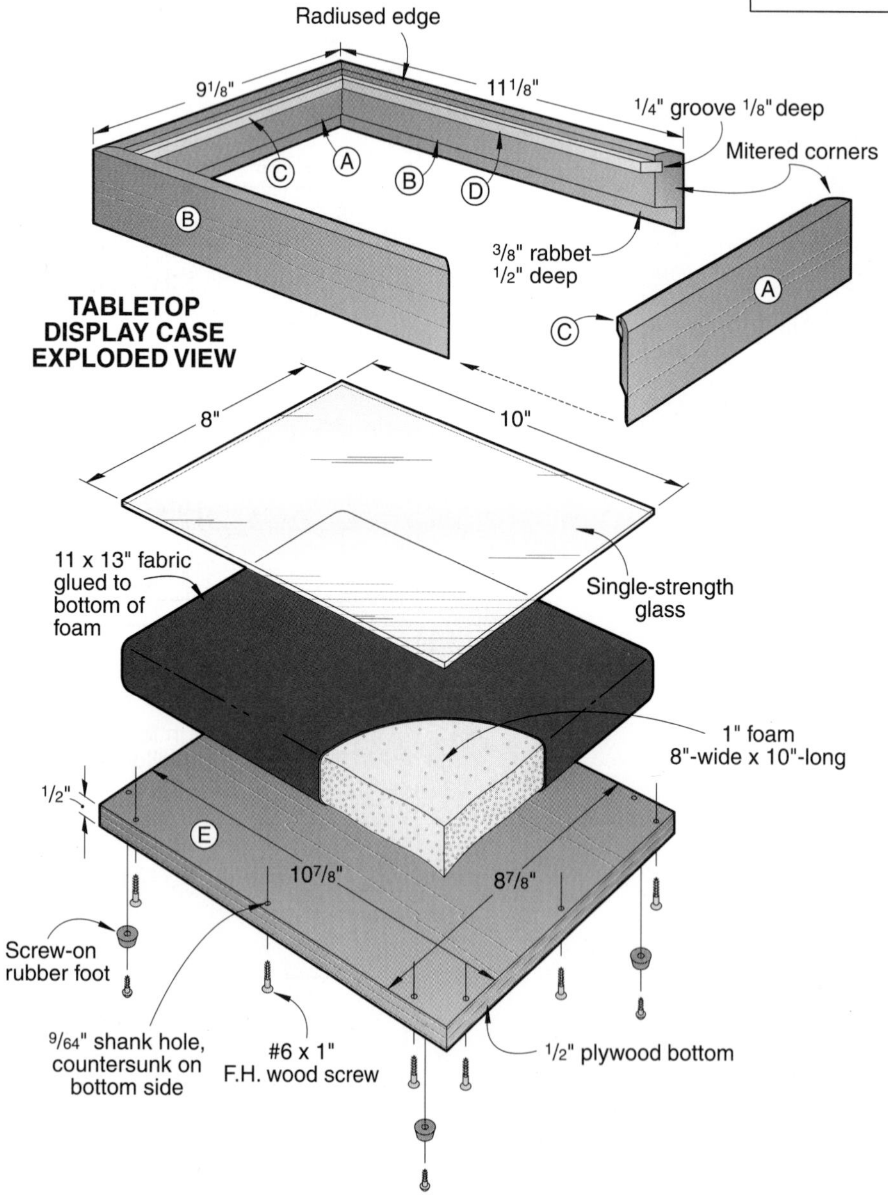

Radiused edge
9 1/8"
11 1/8"
1/4" groove 1/8" deep
Mitered corners
C
A
B
D
B
3/8" rabbet 1/2" deep
A
C
TABLETOP DISPLAY CASE EXPLODED VIEW
8"
10"
11 x 13" fabric glued to bottom of foam
Single-strength glass
1" foam 8"-wide x 10"-long
1/2"
E
10 7/8"
8 7/8"
Screw-on rubber foot
9/64" shank hole, countersunk on bottom side
#6 x 1" F.H. wood screw
1/2" plywood bottom

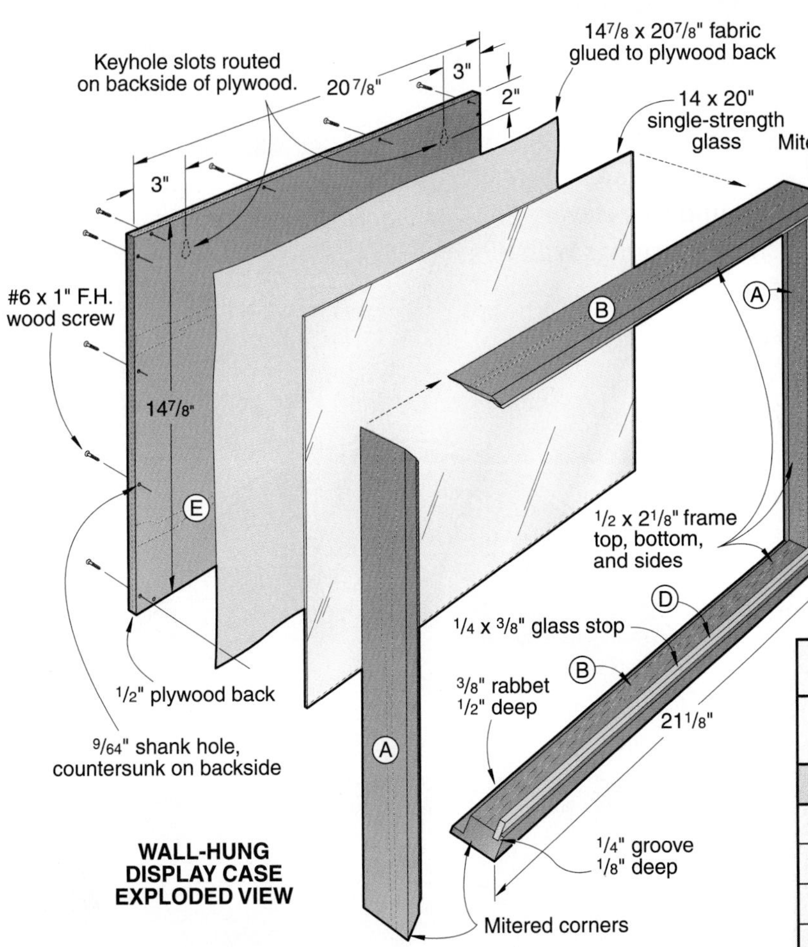

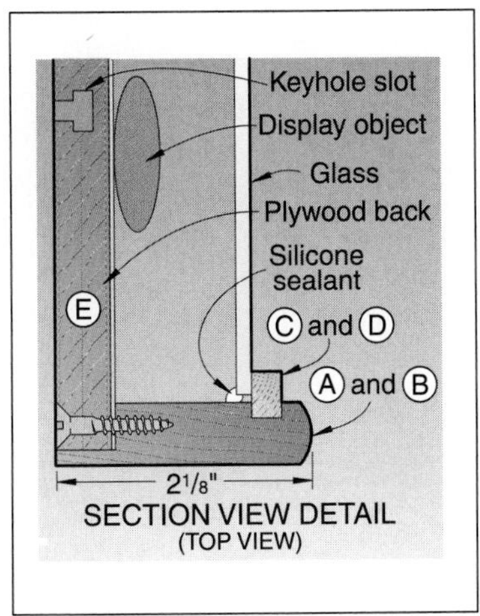

Bill of Materials

Part	Finished Size T	Finished Size W	Finished Size L	Matl.	Qty.
TABLETOP DISPLAY CASE					
A* ends	1/2"	2 1/8"	9 1/8"	W	2
B* sides	1/2"	2 1/8"	11 1/8"	W	2
C* end stops	1/4"	3/8"	8 3/8"	M	2
D* side stops	1/4"	3/8"	10 3/8"	M	2
E back	1/2"	8 7/8"	10 7/8"	BP	1
WALL-HUNG DISPLAY CASE					
A* ends	1/2"	2 1/8"	15 1/8"	W	2
B* sides	1/2"	2 1/8"	21 1/8"	W	2
C* end stops	1/4"	3/8"	14 3/8"	M	2
D* side stops	1/4"	3/8"	20 3/8"	M	2
E back	1/2"	14 7/8"	20 7/8"	BP	1

*Initially cut parts marked with an * oversized. Trim to finished size according to the instructions.

Materials Key: W–walnut, M–maple, BP–birch plywood

Supplies: 1/8" single-strength glass, clear silicone sealant, 1"-thick poly foam, #6 × 1" flathead wood screws, screw-on or self-adhesive rubber feet for tabletop version, fabric, clear finish.

Bird Feeder in the Round

Expect a bevy of birds to gather around this clever feeder, which combines a few simple turnings with a section from a plastic 2-liter bottle.

Suggestions for building this bird feeder:

- Use rough cedar or redwood for a rustic look.
- When truing the discs, particularly the large one, run the lathe at 500-800 rpm.
- Increase lathe speed to 1,200-1,500 rpm for the profile turning.
- Using the full-size pattern, draw two seed openings on masking tape on opposite sides of the plastic bottle. You can wrap the masking tape all the way around the bottle flush with the bottom edge.
- Be sure the auxiliary faceplate is a true 8″ diameter.
- Drill the center hole on the auxiliary faceplate while it's on the lathe.
- Epoxy the ⅜″ bolt into the auxiliary faceplate.
- Turn the finials between centers in the traditional way.

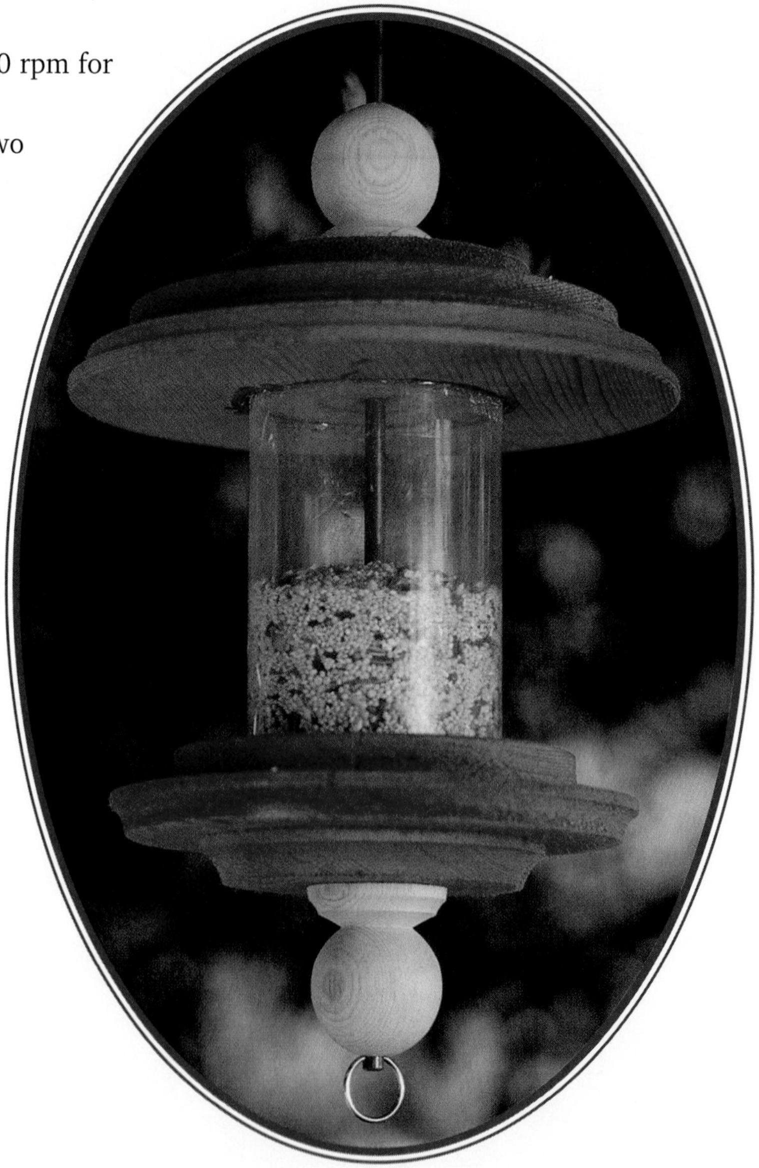

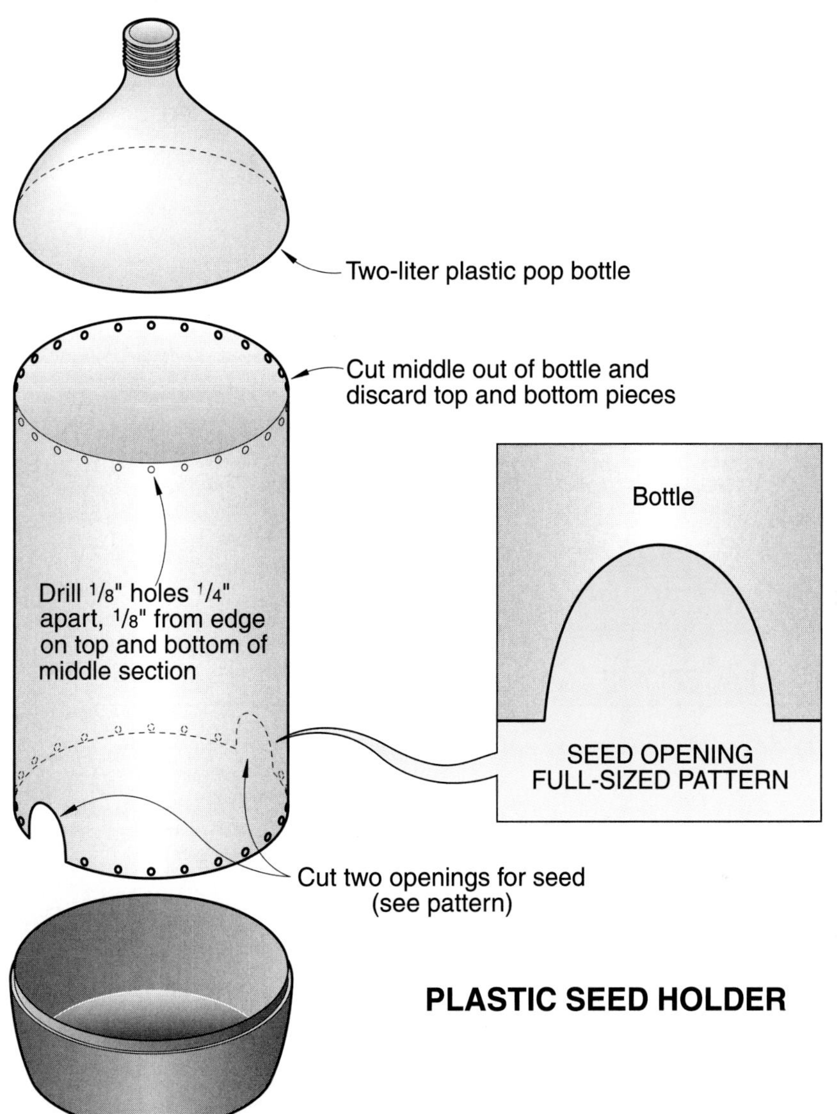
Two-liter plastic pop bottle
Cut middle out of bottle and discard top and bottom pieces
Bottle
Drill 1/8" holes 1/4" apart, 1/8" from edge on top and bottom of middle section
SEED OPENING FULL-SIZED PATTERN
Cut two openings for seed (see pattern)

PLASTIC SEED HOLDER

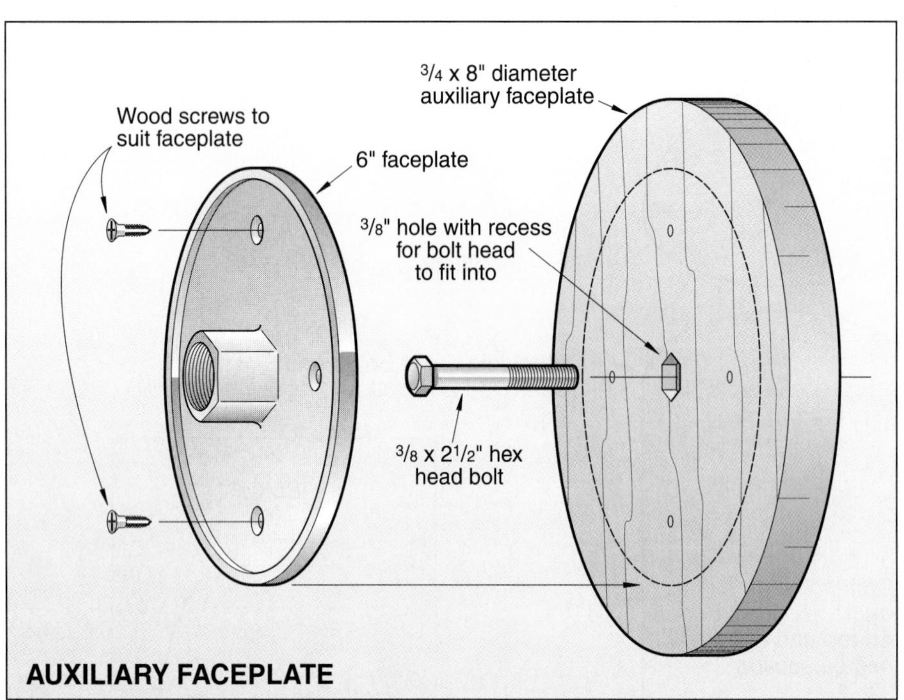
3/4 x 8" diameter auxiliary faceplate
Wood screws to suit faceplate
6" faceplate
3/8" hole with recess for bolt head to fit into
3/8 x 2 1/2" hex head bolt
AUXILIARY FACEPLATE

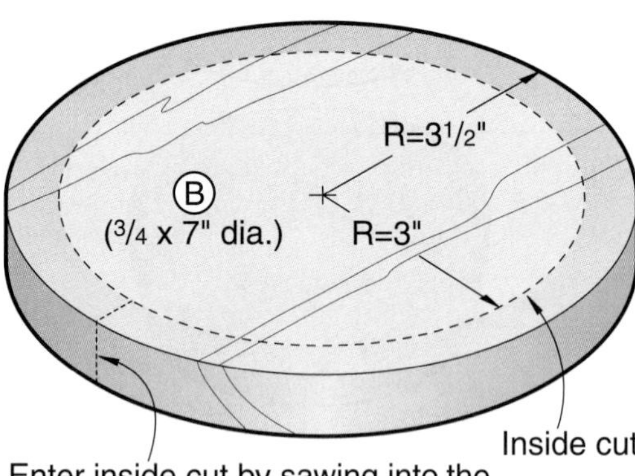
R=3 1/2"
B
(3/4 x 7" dia.)
R=3"
Inside cut
Enter inside cut by sawing into the stock with the grain

CUTTING THE RING

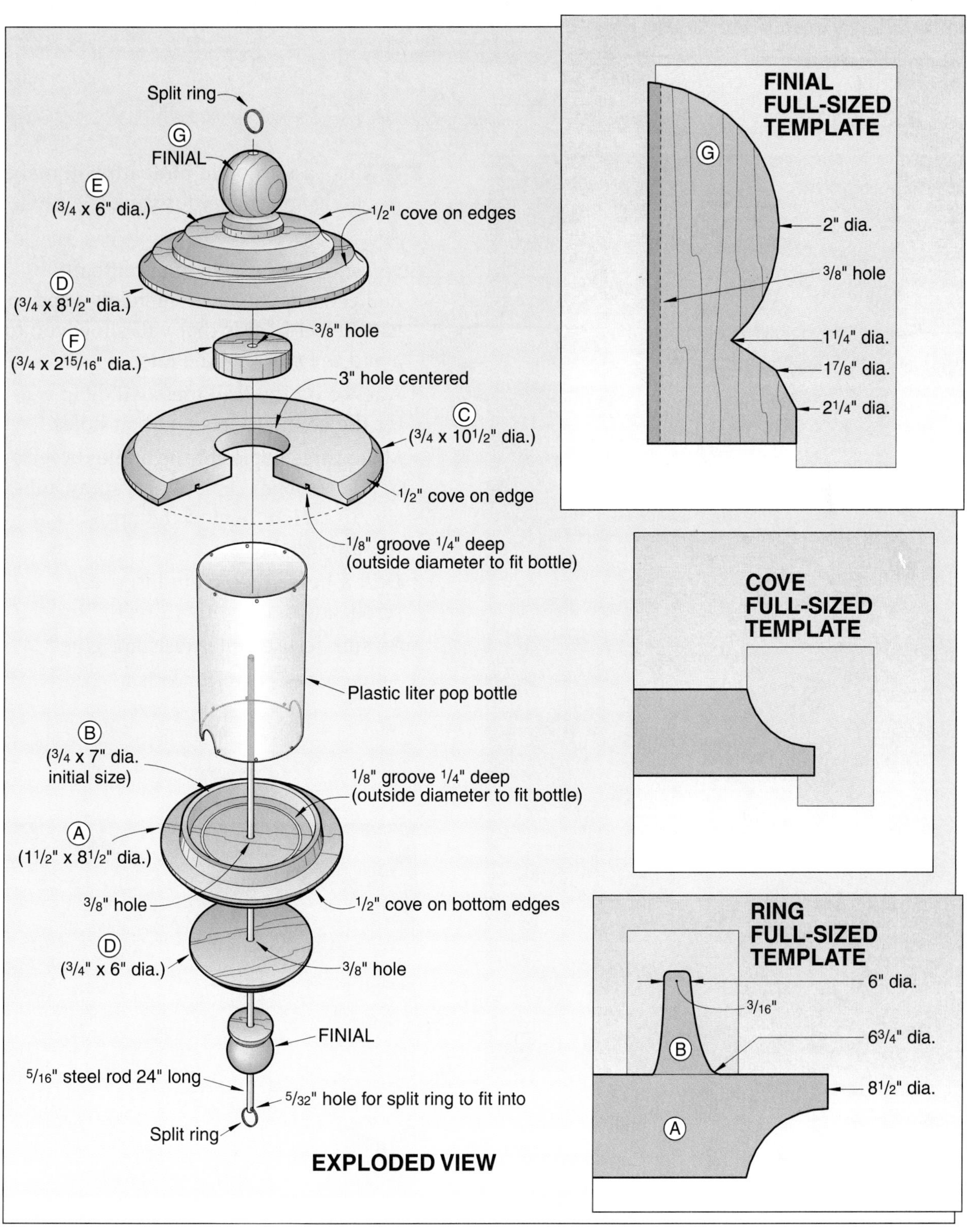
Split ring
G
FINIAL
E
(3/4 x 6" dia.)
1/2" cove on edges
D
(3/4 x 8 1/2" dia.)
F
(3/4 x 2 15/16" dia.)
3/8" hole
3" hole centered
C
(3/4 x 10 1/2" dia.)
1/2" cove on edge
1/8" groove 1/4" deep
(outside diameter to fit bottle)
Plastic liter pop bottle
B
(3/4 x 7" dia.
initial size)
1/8" groove 1/4" deep
(outside diameter to fit bottle)
A
(1 1/2" x 8 1/2" dia.)
3/8" hole
1/2" cove on bottom edges
D
(3/4" x 6" dia.)
3/8" hole
FINIAL
5/16" steel rod 24" long
5/32" hole for split ring to fit into
Split ring
EXPLODED VIEW
FINIAL
FULL-SIZED
TEMPLATE
G
2" dia.
3/8" hole
1 1/4" dia.
1 7/8" dia.
2 1/4" dia.
COVE
FULL-SIZED
TEMPLATE
RING
FULL-SIZED
TEMPLATE
6" dia.
3/16"
B
6 3/4" dia.
8 1/2" dia.
A

Bathing Beauty Birdbath

This easy-to-build birdbath will make a big splash in your yard.

Suggestions for making the birdbath:

- Red cedar is the recommended wood to use. Select boards that will allow you to avoid the worst of the knots.
- A fence on the drill press will help center the counterbores and bolt holes.
- Make three copies of the apple design and use rubber cement or a spray adhesive to attach them to the uprights in the proper places.
- Drill a hole in which to insert a #7 scroll saw blade with 12 teeth per inch.
- Be sure to use water-resistant glue.
- Apply a clear exterior finish.

To saw the spline slots in the wide uprights, saw blade is tilted 30° from vertical. Position the rip fence to locate the slot correctly.

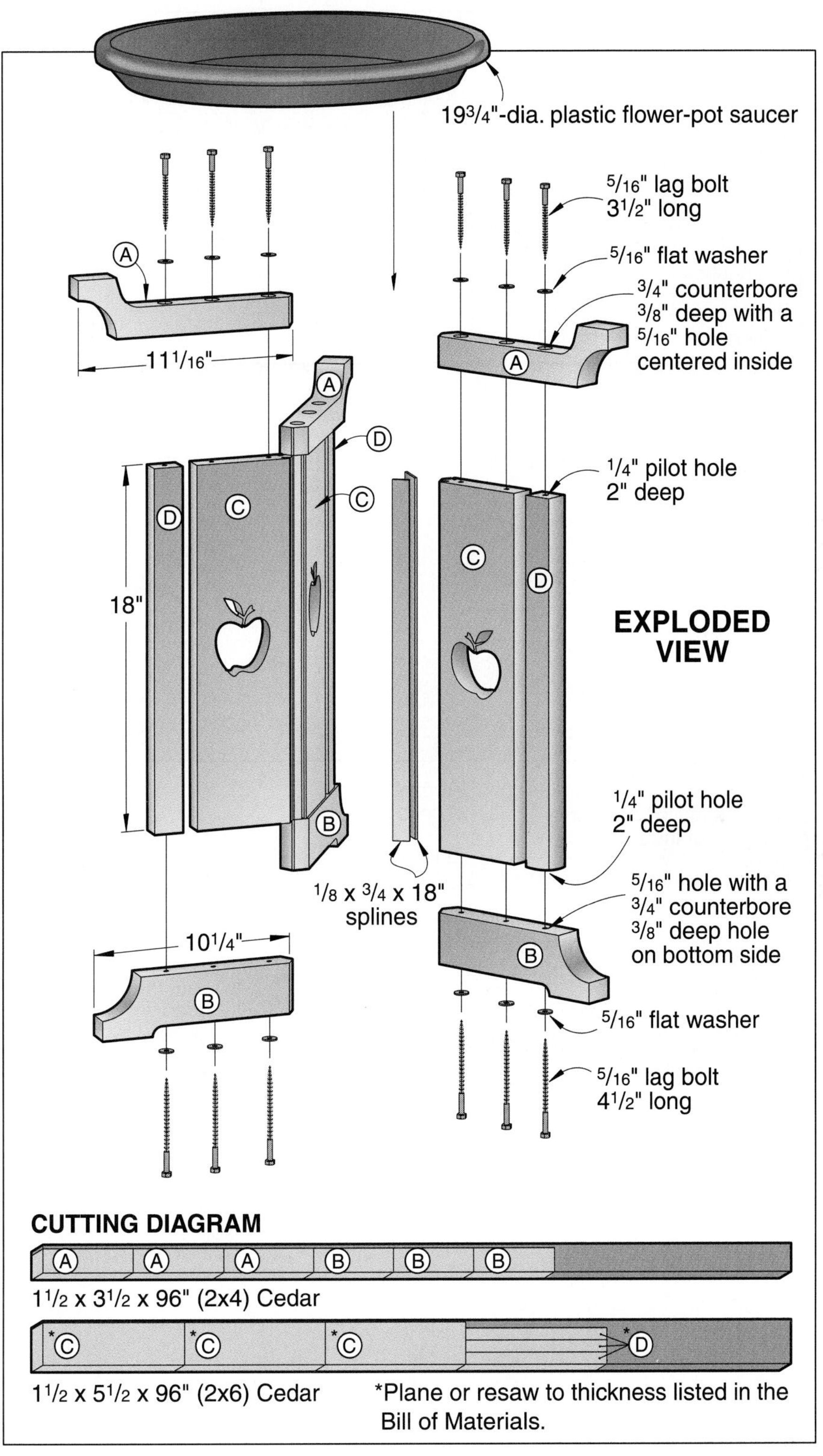

19³/4"-dia. plastic flower-pot saucer
5/16" lag bolt 3 1/2" long
5/16" flat washer
3/4" counterbore 3/8" deep with a 5/16" hole centered inside
A
11 1/16"
A
D
C
1/4" pilot hole 2" deep
18"
D
C
C
D
EXPLODED VIEW
1/4" pilot hole 2" deep
B
1/8 x 3/4 x 18" splines
5/16" hole with a 3/4" counterbore 3/8" deep hole on bottom side
10 1/4"
B
B
5/16" flat washer
5/16" lag bolt 4 1/2" long
CUTTING DIAGRAM
A A A B B B
1 1/2 x 3 1/2 x 96" (2x4) Cedar
*C *C *C *D
1 1/2 x 5 1/2 x 96" (2x6) Cedar
*Plane or resaw to thickness listed in the Bill of Materials.

Bill of Materials						
Part	Finished Size			Matl.	Qty.	**Material Key:** C–western red cedar **Supplies:** Woodworker's water-resistant glue; 5/16″ × 4 1/2″ lag bolts; 5/16″ flat washers; clear lacquer; red and green exterior enamel; clear exterior finish; silicone adhesive; 19 3/4″-dia. plastic flower-pot saucer.
	T	W	L			
A bowl support	1 1/2″	3″	11 1/16″	C	3	
B foot	1 1/2″	3″	10 1/4″	C	3	
C wide upright	1 1/4″	5 1/4″	18″	C	3	
D thin upright	1 1/4″	1 1/2″	18″	C	3	

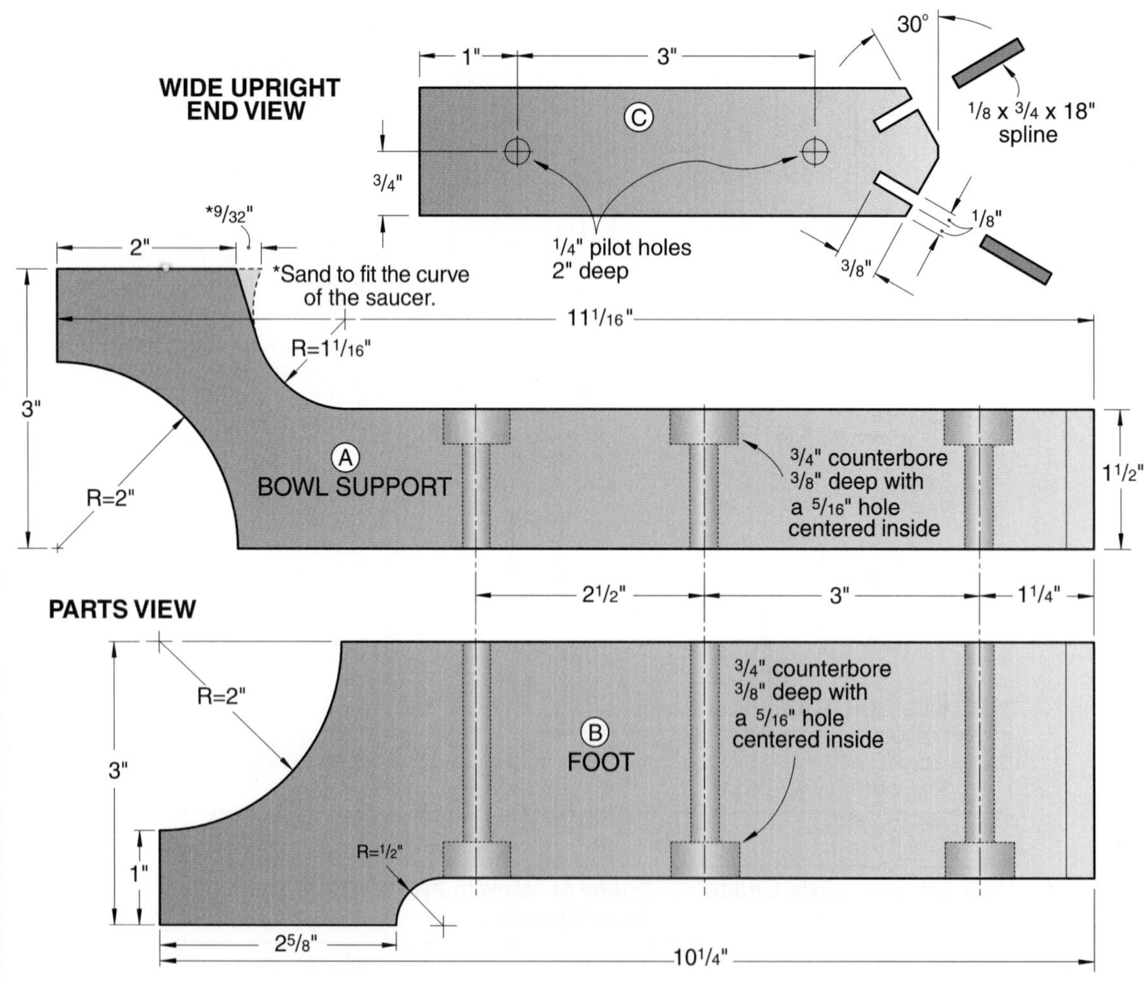

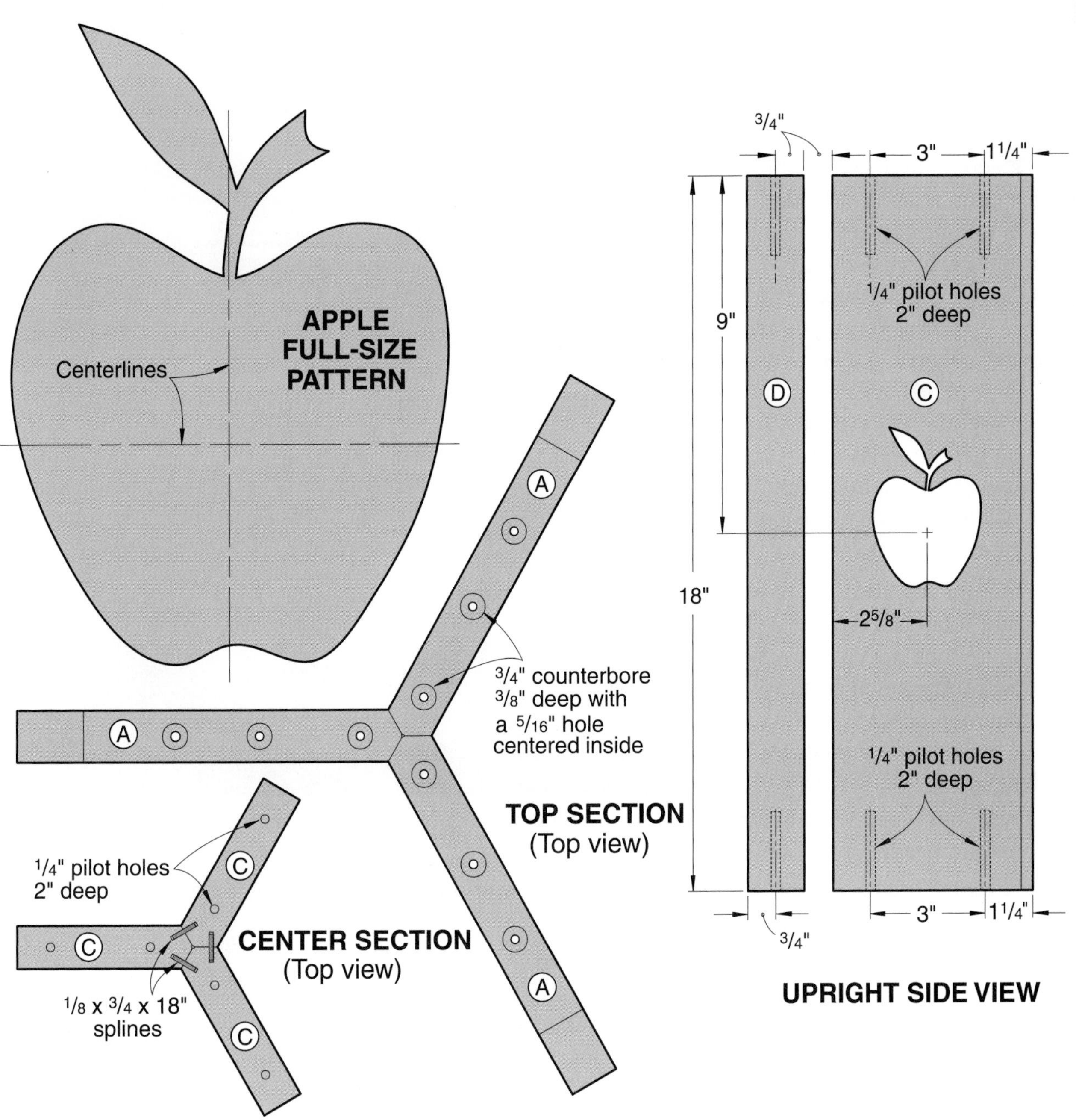
APPLE FULL-SIZE PATTERN
Centerlines
TOP SECTION
(Top view)
A
3/4" counterbore 3/8" deep with a 5/16" hole centered inside
CENTER SECTION
(Top view)
C
1/4" pilot holes 2" deep
1/8 x 3/4 x 18" splines
UPRIGHT SIDE VIEW
D
3/4"
3"
1 1/4"
9"
18"
2 5/8"
1/4" pilot holes 2" deep

Better Homes and Gardens®
WOOD®
MAGAZINE

Drafting/Reference Center

Keep reference material close at hand with this two-project setup. Simply swing the table up when you need to take a few notes or consult reference material. When you are not using it, collapse the brackets and the table folds back against the wall. The cabinet, with its drop-down door, provides necessary storage for books, magazines, tool catalogs, and other shop-related papers.

Suggestions for building the drafting/reference center:

- Because they are not included, take special care in creating a plan of procedure, cutting list, and bill of materials for this project.
- Check your woodworking catalogs or local woodworking store for the special hardware needed for this project.
- Take extra care to be sure the dadoes on the opposing sides align with each other.

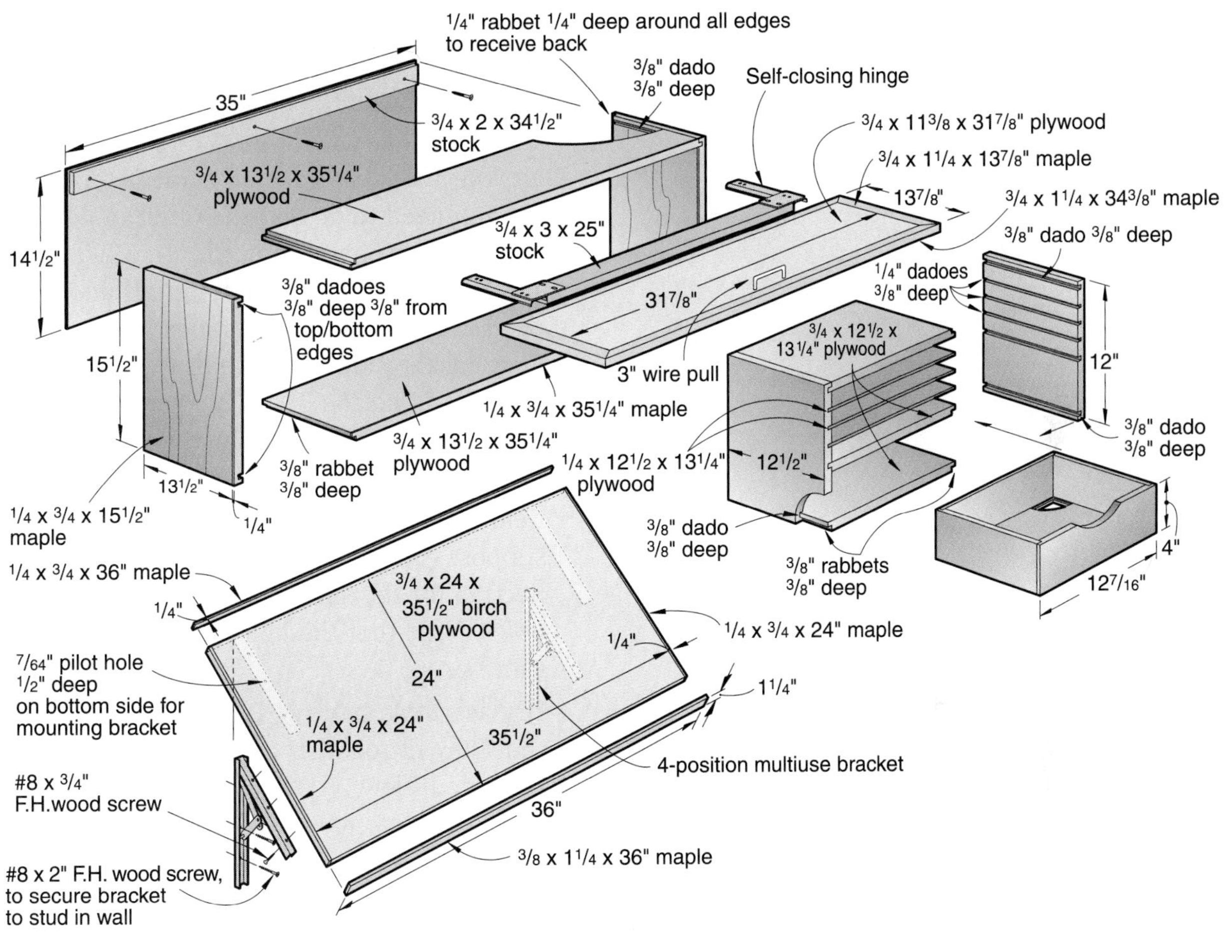
1/4" rabbet 1/4" deep around all edges to receive back
35"
3/8" dado
3/8" deep
Self-closing hinge
3/4 x 2 x 34 1/2" stock
3/4 x 11 3/8 x 31 7/8" plywood
3/4 x 1 1/4 x 13 7/8" maple
3/4 x 13 1/2 x 35 1/4" plywood
13 7/8"
3/4 x 1 1/4 x 34 3/8" maple
3/8" dado 3/8" deep
14 1/2"
3/4 x 3 x 25" stock
1/4" dadoes
3/8" deep
3/8" dadoes
3/8" deep 3/8" from
top/bottom
edges
31 7/8"
3/4 x 12 1/2 x
13 1/4" plywood
15 1/2"
3" wire pull
12"
1/4 x 3/4 x 35 1/4" maple
3/8" dado
3/8" deep
3/4 x 13 1/2 x 35 1/4" plywood
3/8" rabbet
3/8" deep
1/4 x 12 1/2 x 13 1/4" plywood
12 1/2"
13 1/2"
1/4 x 3/4 x 15 1/2" maple
1/4"
3/8" dado
3/8" deep
4"
3/8" rabbets
3/8" deep
12 7/16"
1/4 x 3/4 x 36" maple
1/4"
3/4 x 24 x
35 1/2" birch
plywood
1/4"
1/4 x 3/4 x 24" maple
7/64" pilot hole
1/2" deep
on bottom side for
mounting bracket
24"
1 1/4"
1/4 x 3/4 x 24"
maple
35 1/2"
4-position multiuse bracket
#8 x 3/4"
F.H.wood screw
36"
3/8 x 1 1/4 x 36" maple
#8 x 2" F.H. wood screw,
to secure bracket
to stud in wall

Country-Style Lamp

This country-style lamp includes handy drawers. It's perfect to set beside your favorite chair or to use as a bedside lamp.

Plan your work before you begin. Determine what you need to do and the order in which you will do it (plan of procedure). Make sure you have all the right materials and tools. Here are a few suggestions:

- First cut the sides (A), the drawer supports (B), and the back (C). Refer to the drawings and the bill of materials for the exact sizes.
- Before drilling the holes in the drawer support (B), look closely at the Side View Section for the size and placement of the holes. Note that the holes are not all the same size.
- Dry-assemble parts A, B, and C to see that they fit together as shown in the Exploded View.
- When routing the edges to the top (D), drawer fronts (F), and base trim (E), take special care routing the end grain to avoid tearing out the sides.
- When preparing your lamp for finishing, make the extra effort to do it right. The extra time will be worth it.
- Be extra careful when wiring the lamp. Have someone check your work to assure it is done right.

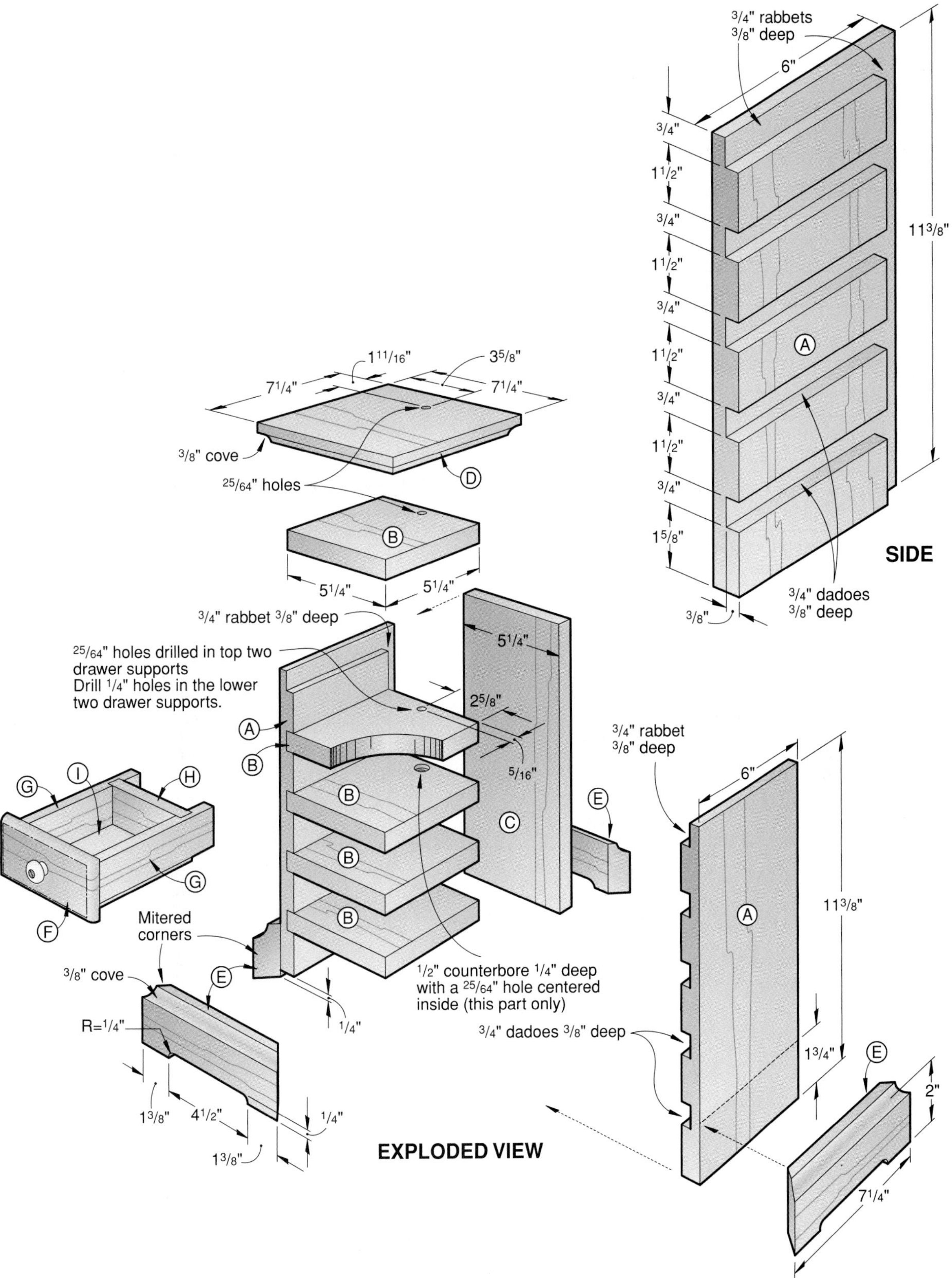
3/4" rabbets
3/8" deep
6"
3/4"
1 1/2"
3/4"
1 1/2"
3/4"
1 1/2"
3/4"
1 1/2"
3/4"
1 5/8"
11 3/8"
A
SIDE
3/4" dadoes
3/8" deep
3/8"
1 11/16"
3 5/8"
7 1/4"
7 1/4"
3/8" cove
25/64" holes
D
B
5 1/4"
5 1/4"
3/4" rabbet 3/8" deep
25/64" holes drilled in top two drawer supports
Drill 1/4" holes in the lower two drawer supports.
5 1/4"
2 5/8"
5/16"
C
E
3/4" rabbet
3/8" deep
6"
11 3/8"
Mitered corners
3/8" cove
R=1/4"
1 3/8"
4 1/2"
1 3/8"
1/4"
1/4"
1/2" counterbore 1/4" deep with a 25/64" hole centered inside (this part only)
3/4" dadoes 3/8" deep
1 3/4"
2"
7 1/4"
G
I
H
F
EXPLODED VIEW

Bill of Materials

Part	Finished Size			Matl.	Qty.
	T	W	L		
A side	3/4″	6″	11 3/8″	O	2
B drawer support	3/4″	5 1/4″	5 1/4″	O	5
C back	3/4″	5 1/4″	11 3/8″	O	1
D top	3/4″	7 1/4″	7 1/4″	O	1
E* base trim	5/8″	2″	7 1/4″	O	4
F drawer front	1/2″	1 15/16″	4 15/16″	O	4
G drawer side	1/2″	1 5/16″	5 1/8″	O	8
H drawer back	1/2″	1 5/16″	3 15/16″	O	4
I drawer bottom	1/8″	4 7/16″	4 3/8″	HB	4

*Initially cut these parts oversized. Then, trim to finished size according to the how-to instructions.

Material Key: O–oak, HB–hardboard

Supplies: Offset lamp pipe with nuts, socket with push-button switch, cord set, lamp harp, drawer knobs, lamp shade, woodworker's glue, #17 × 3/4″ and 7/8″ brads, salem maple stain, and polyurethane finish.

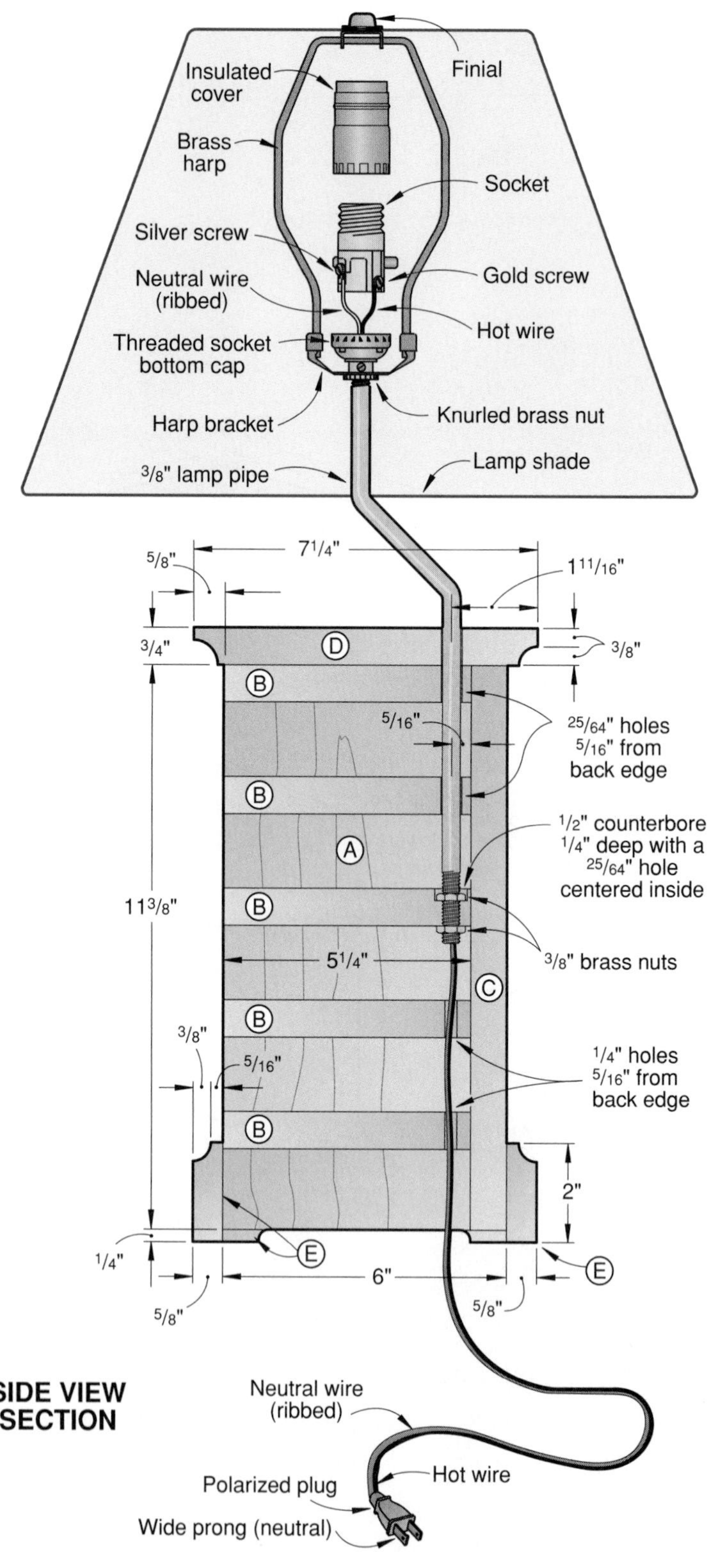

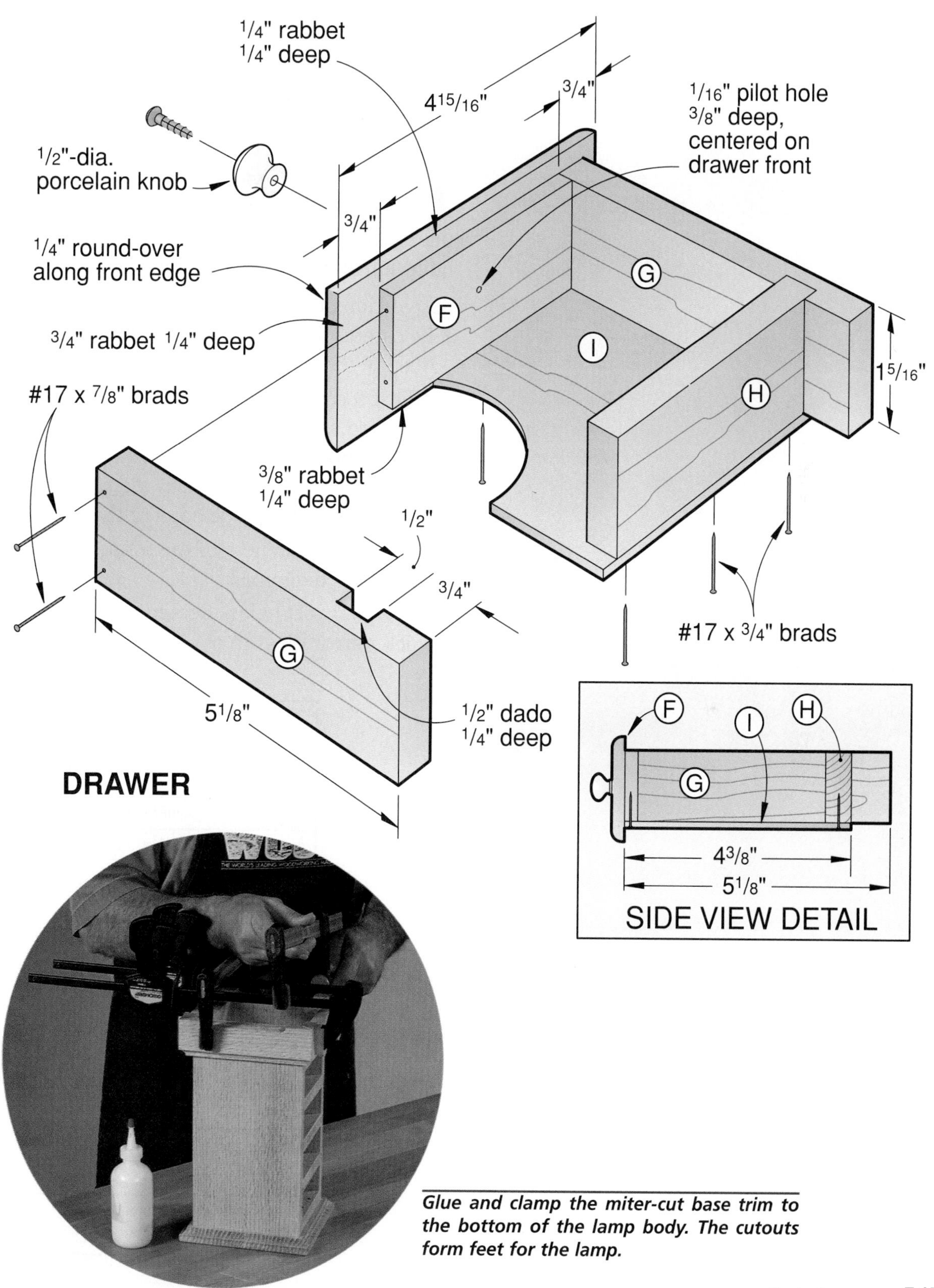

Glue and clamp the miter-cut base trim to the bottom of the lamp body. The cutouts form feet for the lamp.

A Mirror That Reflects the Past

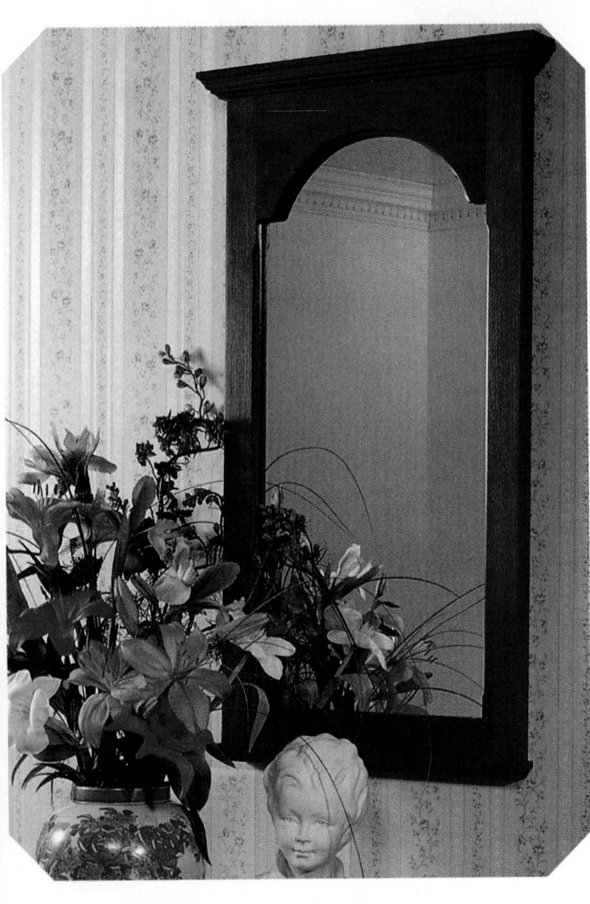

For a touch of Americana, you can't beat this Federal style hall mirror. The Federal Style became popular just after the Revolutionary War. This is an easy-to-build mirror in a style that reflects the past.

Suggestions for building the Federal mirror:

- Be careful that you don't drill the holes too deep in the top rail.
- Adjust a compass or a pair of trammel points to a 5⅜″ radius and locate the arc for the top rail. The center will be off the workpiece midway between the stiles and 1³⁄16″ below the inside edge of the top rail.
- Check for squareness of the frame by measuring the diagonals.
- Have your local glass dealer cut the ⅛″-thick mirror glass to a 12½″ × 25¼″ size.

Bill of Materials

Part	Finished Size T	W	L	Matl.	Qty.
MIRROR					
A stile	1/2″	1 3/4″	28 3/4″	W	2
B top rail	1/2″	6 1/2″	12″	W	1
C bottom rail	1/2″	1 3/4″	12″	W	1
D filler	5/8″	1″	28 3/4″	W	2
E trim	3/8″	1 5/8″	16 1/2″	W	1
F top cap	3/4″	2 1/8″	17 1/2″	W	1
G bottom cap	3/4″	1 3/8″	16″	W	1
H stop	1/2″	5/8″	24 3/4″	W	2
I stop	1/2″	5/8″	13 3/8″	W	2
J back	1/8″	12 1/2″	25 1/4″	HB	1

Material Key: HB–hardboard, W–walnut

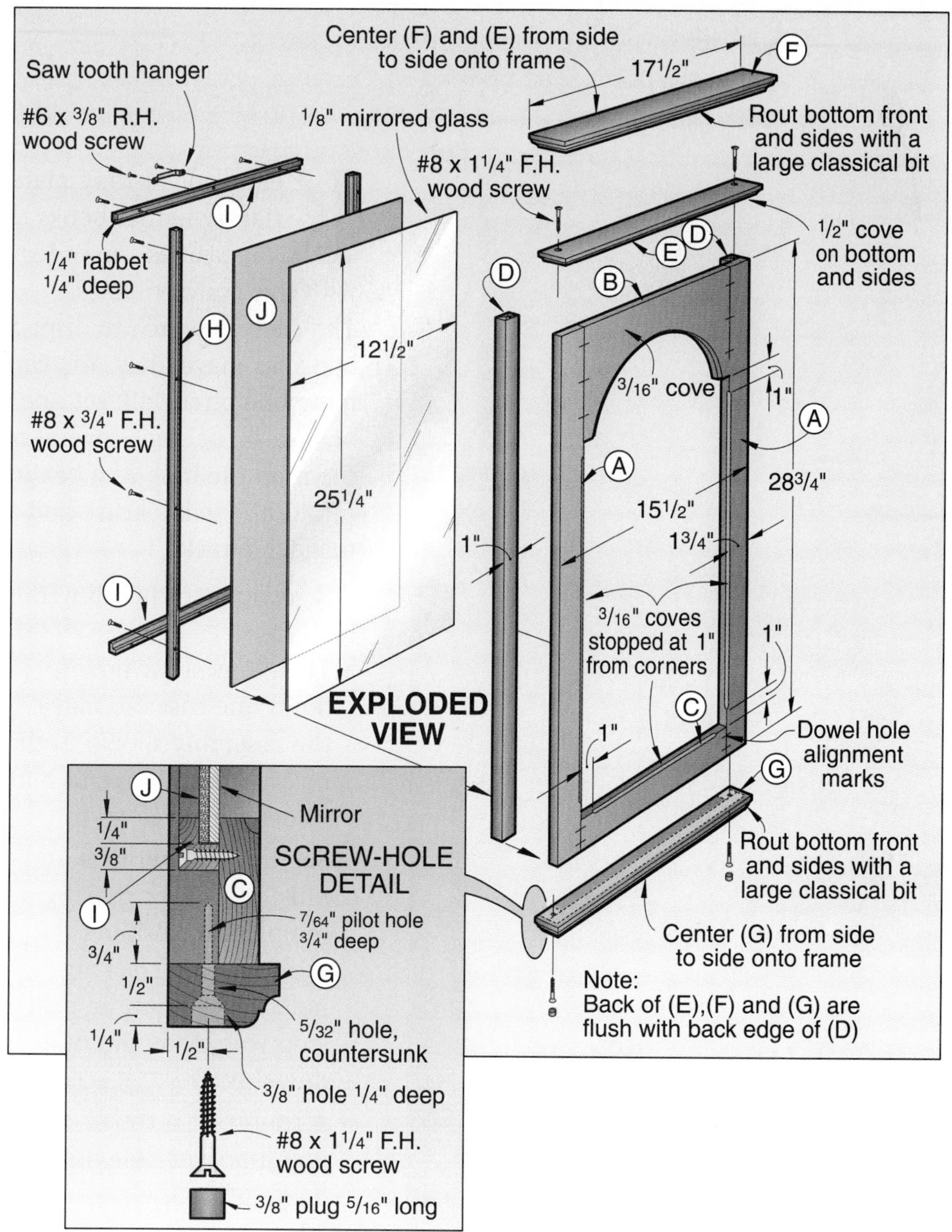
Center (F) and (E) from side to side onto frame
17 1/2"
F
Saw tooth hanger
#6 x 3/8" R.H. wood screw
1/8" mirrored glass
Rout bottom front and sides with a large classical bit
#8 x 1 1/4" F.H. wood screw
I
1/4" rabbet 1/4" deep
E
D
1/2" cove on bottom and sides
D
B
J
H
12 1/2"
3/16" cove
1"
A
#8 x 3/4" F.H. wood screw
A
28 3/4"
25 1/4"
15 1/2"
1"
1 3/4"
I
3/16" coves stopped at 1" from corners
1"
C
EXPLODED VIEW
1"
Dowel hole alignment marks
G
J
Mirror
1/4"
3/8"
SCREW-HOLE DETAIL
C
I
7/64" pilot hole 3/4" deep
3/4"
1/2"
G
Rout bottom front and sides with a large classical bit
Center (G) from side to side onto frame
Note:
Back of (E),(F) and (G) are flush with back edge of (D)
1/4"
1/2"
5/32" hole, countersunk
3/8" hole 1/4" deep
#8 x 1 1/4" F.H. wood screw
3/8" plug 5/16" long

PROJECT 13

Reading Rack

Keep your reading material—whether books or magazines—in order with this convertible case. The shelves switch to display books or magazines. You can hang one on a wall, set one on a desk, or stack several of them on the base as a beautiful all-in-one bookcase and magazine rack.

Suggestions for building this bookcase:

- The plans show how to build one case. To make the assembly shown in the picture, you'll have to build three of these.
- See the photographs and illustrations for making a template for drilling the shelf and the stacking holes. Note that the drill stop was made out of a dowel rod.
- A router on a router table and a pocket-hole jig will be helpful.
- If you have limited time, build just one case and make it a wall-hung unit.

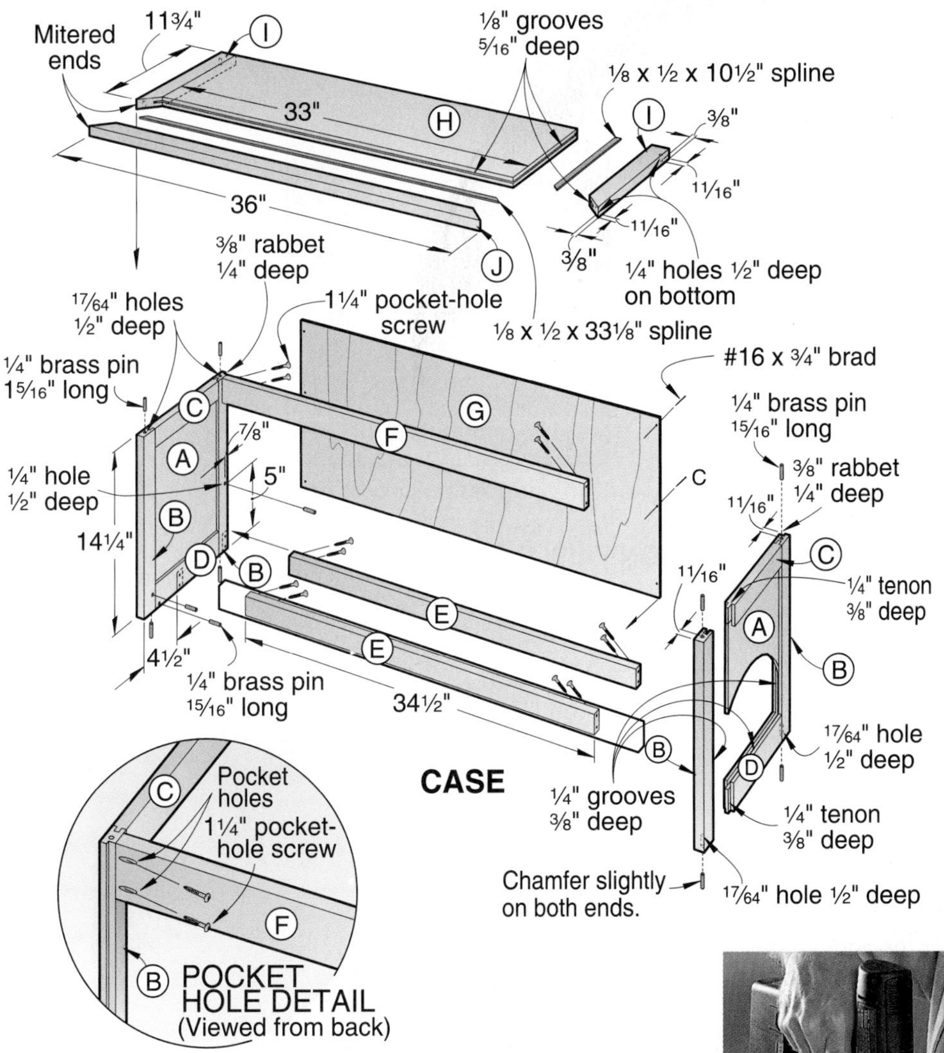

Mitered ends
11 3/4"
I
33"
H
1/8" grooves 5/16" deep
1/8 x 1/2 x 10 1/2" spline
I
3/8"
11/16"
11/16"
3/8"
1/4" holes 1/2" deep on bottom
36"
J
3/8" rabbet 1/4" deep
17/64" holes 1/2" deep
1 1/4" pocket-hole screw
1/8 x 1/2 x 33 1/8" spline
#16 x 3/4" brad
1/4" brass pin 1 5/16" long
C
G
F
A
7/8"
1/4" hole 1/2" deep
5"
B
14 1/4"
D
B
E
E
4 1/2"
1/4" brass pin 15/16" long
34 1/2"
C
1/4" brass pin 15/16" long
3/8" rabbet 1/4" deep
11/16"
C
11/16"
1/4" tenon 3/8" deep
A
B
17/64" hole 1/2" deep
D
B
1/4" grooves 3/8" deep
1/4" tenon 3/8" deep
Chamfer slightly on both ends.
17/64" hole 1/2" deep

CASE

C
Pocket holes
1 1/4" pocket-hole screw
F
B
POCKET HOLE DETAIL
(Viewed from back)

A hardboard template accurately locates the holes for the shelf-support pins in the case sides.

A drilling fixture locates indexing holes for stacking in the top and bottom of the case sides. The dowel on the bit serves as a depth stop.

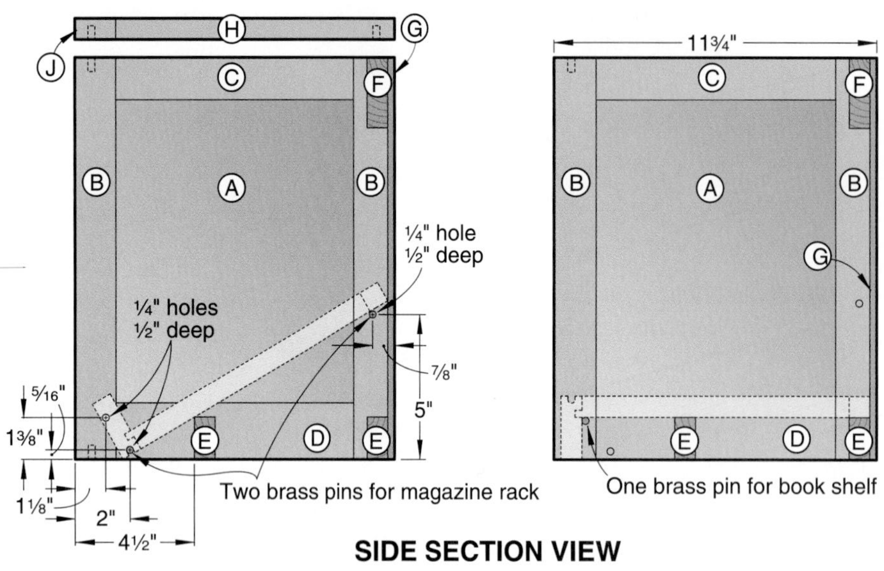

SIDE SECTION VIEW

Drill pocket holes in the ends of the crossbars for the screws that hold the case together.

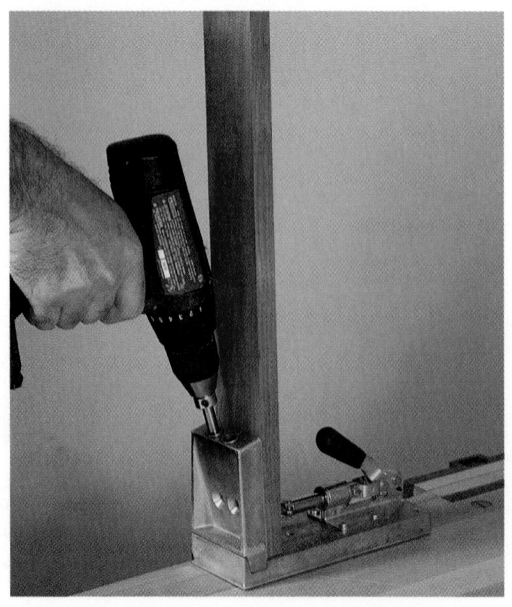

With one leg cut off, the fixture used to drill indexing holes in the case sides will locate the mating holes correctly on the top and base.

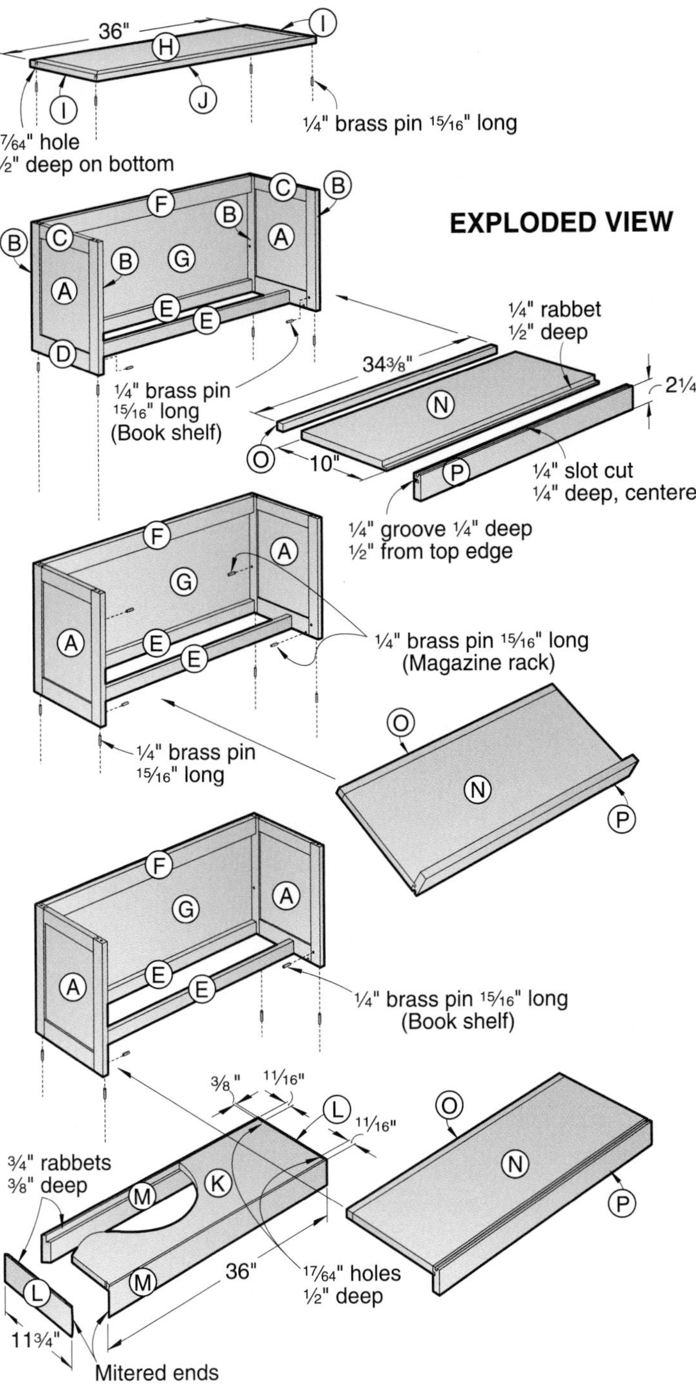

Bill of Materials

Part	Finished Size T	W	L	Matl.	Qty.
A side panel	1/4″	9 7/16″	11 7/16″	EW	2
B stile	3/4″	1 1/2″	14 1/4″	W	4
C top rail	3/4″	1 1/2″	9 1/2″	W	2
D bottom rail	3/4″	2″	9 1/2″	W	2
E crossbar	3/4″	1 1/2″	34 1/2″	W	2
F top crossbar	3/4″	2 1/2″	34 1/2″	W	1
G back	1/4″	35 1/4″	14 1/4″	WP	1
H top	3/4″	10 1/4″	33″	WP	1
I* side band	3/4″	1 1/2″	11 3/4″	W	2
J* front band	3/4″	1 1/2″	36″	W	1
K base center	3/4″	11″	35 1/4″	WP	1
L* side band	3/4″	3″	11 3/4″	W	2
M* edge band	3/4″	3″	36″	W	2
N shelf	3/4″	10″	34 3/8″	WP	1
O back band	3/4″	3/4″	34 3/8″	W	1
P front ledge	3/4″	2 1/4″	34 3/8″	W	1

*Cut these parts slightly longer initially, then trim to finished length in accordance with instructions.

Materials Key: W–walnut, WP–walnut plywood, EW–edge-glued walnut.

Supplies: dark woodworker's glue, 1/4″ brass rod, 1 1/4″ pocket-hole screws, #16 × 3/4″ brads, finish.

Doghouse

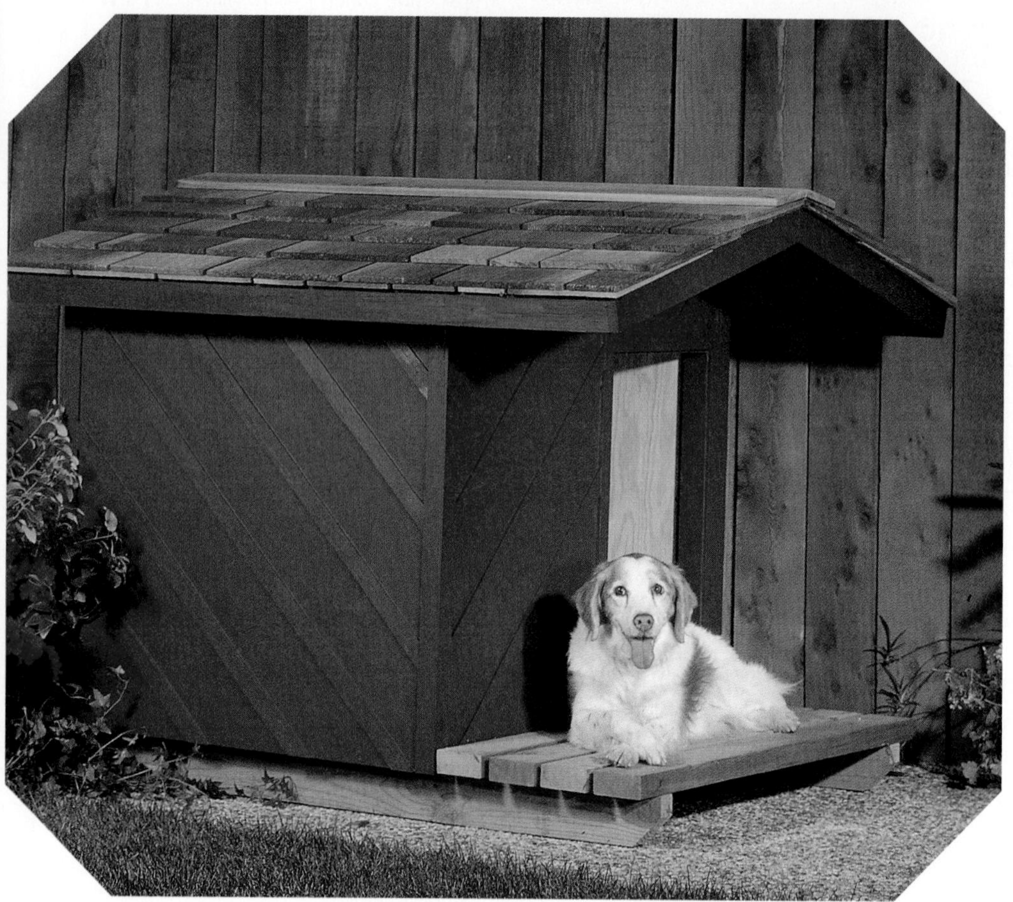

A smart year-round house designed for man's best friend. This doghouse features a design that considers summer's heat as well as winter's cold. A removable center panel provides summer or winter "quarters."

Suggestions for building the doghouse:

- This doghouse is designed for a medium-size dog. To determine the size for your dog with the center panel in place for winter conditions, use the following guidelines:
  - length: 1½ times the length of your dog
  - width: ⅔ the length of your dog
  - height: 1⅕ times the height of your dog (measured to the top of the dog's head)
- Study the drawings very carefully and adjust them accordingly for your dog.
- Be sure to use wood materials, hardware, and finishes that are appropriate for outdoor use.

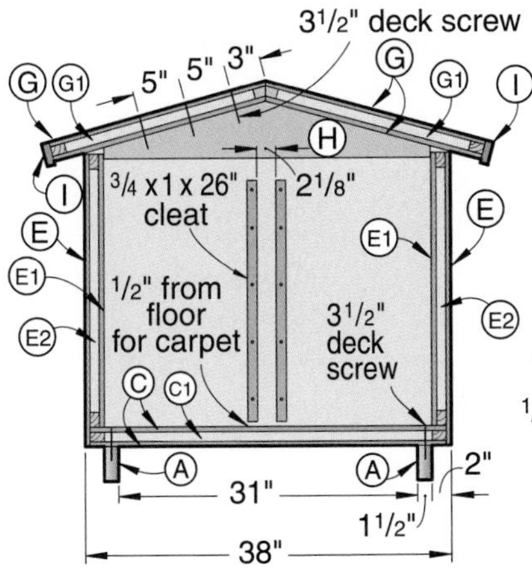
3 1/2" deck screw
G
G1
5"
5"
3"
G
G1
I
H
I
3/4 x 1 x 26" cleat
2 1/8"
E
E1
E2
E
E1
E2
1/2" from floor for carpet
3 1/2" deck screw
C
C1
A
A
2"
31"
1 1/2"
38"
FRONT SECTION VIEW

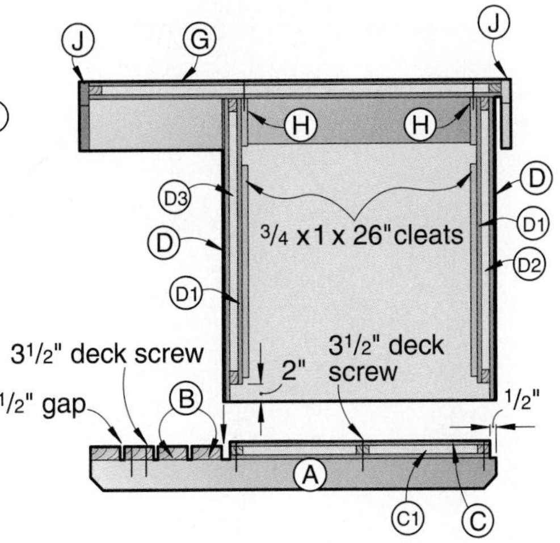
J
G
J
H
H
D3
D
D1
D
D1
D2
3/4 x 1 x 26" cleats
3 1/2" deck screw
3 1/2" deck screw
2"
1/2" gap
B
1/2"
A
C1
C
SIDE SECTION VIEW

EXPLODED VIEW

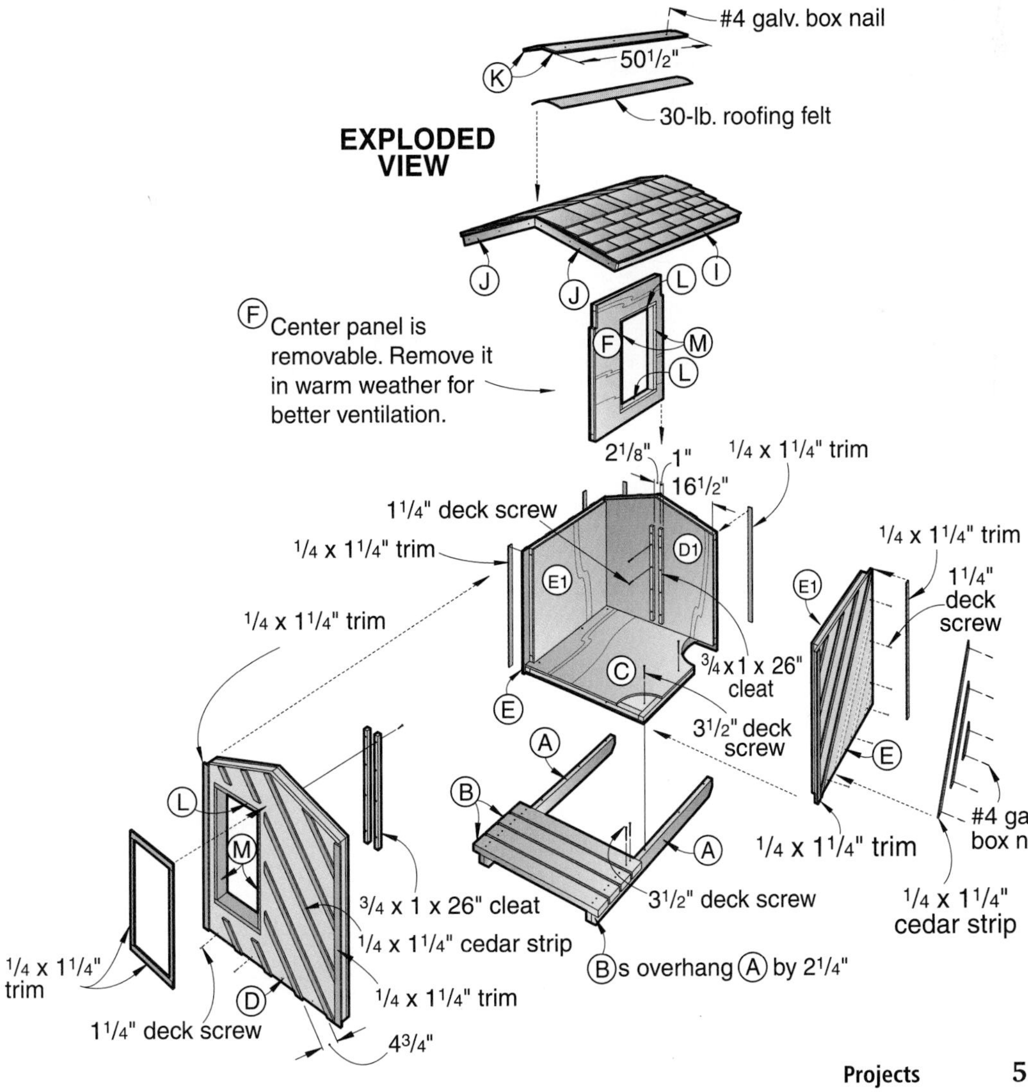
#4 galv. box nail
K
50 1/2"
30-lb. roofing felt
J
J
I
L
F
Center panel is removable. Remove it in warm weather for better ventilation.
F
M
L
2 1/8"
1"
16 1/2"
1/4 x 1 1/4" trim
1 1/4" deck screw
1/4 x 1 1/4" trim
D1
E1
1/4 x 1 1/4" trim
1/4 x 1 1/4" trim
E1
1 1/4" deck screw
3/4 x 1 x 26" cleat
C
E
3 1/2" deck screw
A
E
L
B
M
A
1/4 x 1 1/4" trim
#4 galv. box nail
3 1/2" deck screw
3/4 x 1 x 26" cleat
1/4 x 1 1/4" cedar strip
1/4 x 1 1/4" cedar strip
1/4 x 1 1/4" trim
B s overhang A by 2 1/4"
D
1/4 x 1 1/4" trim
1 1/4" deck screw
4 3/4"

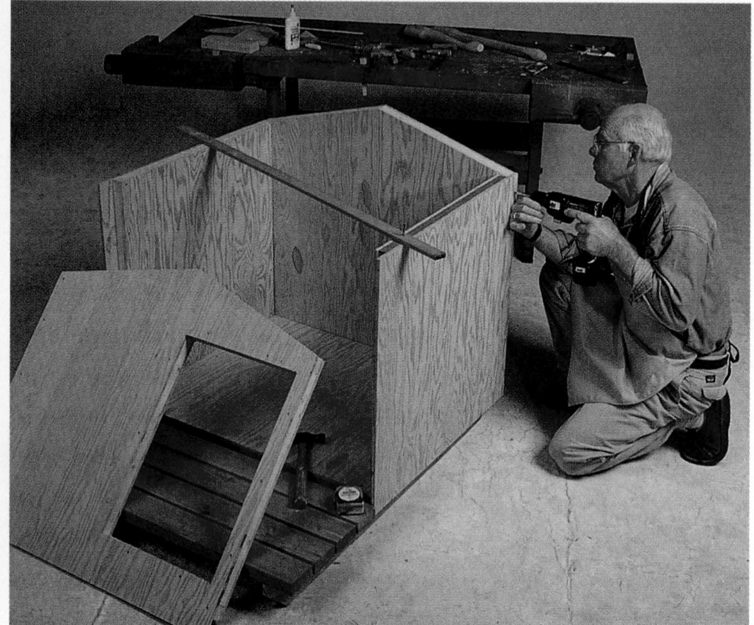

Screw the side panels to the floor and back panels. A scrap strip helps hold the side panels upright and square.

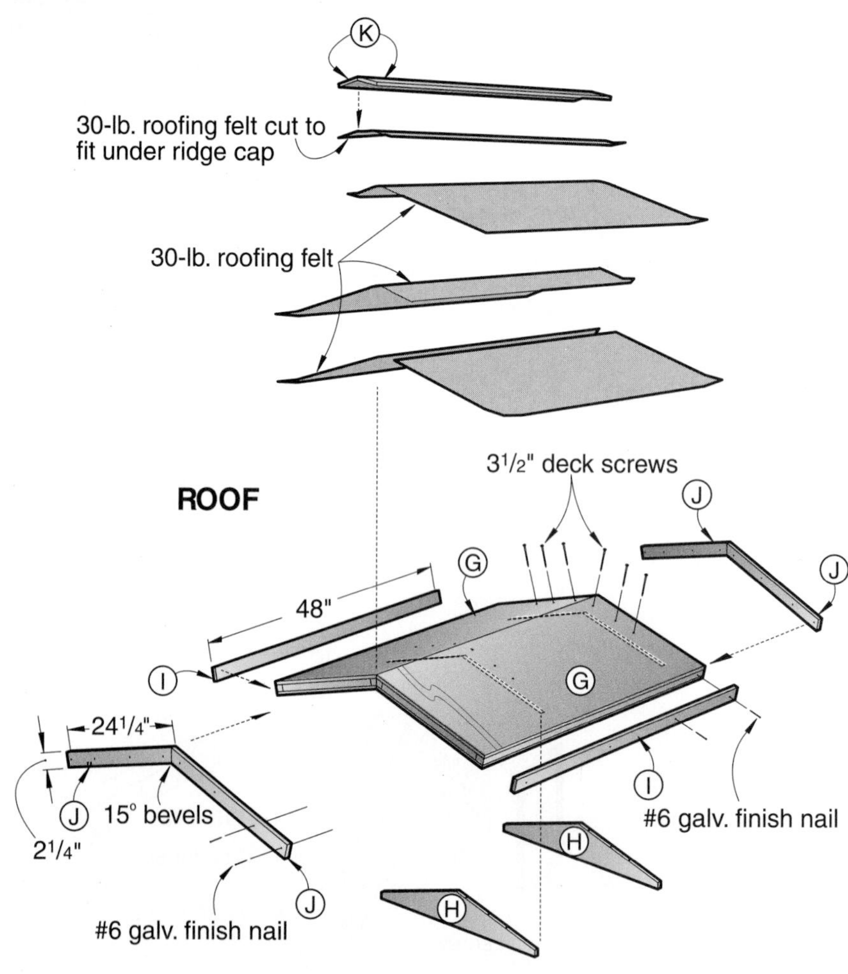

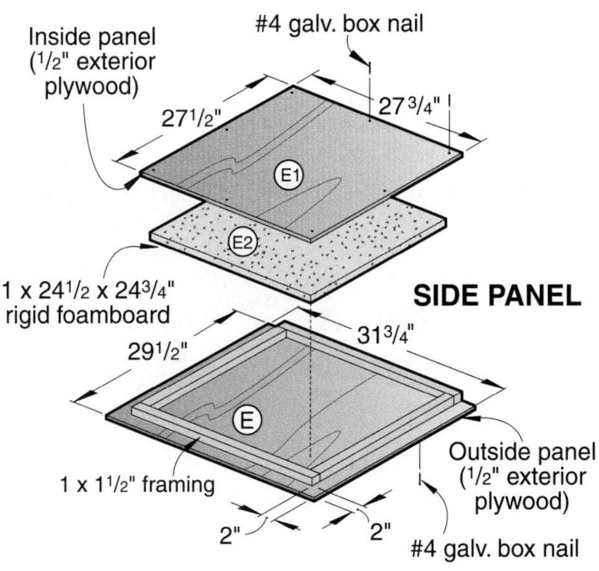

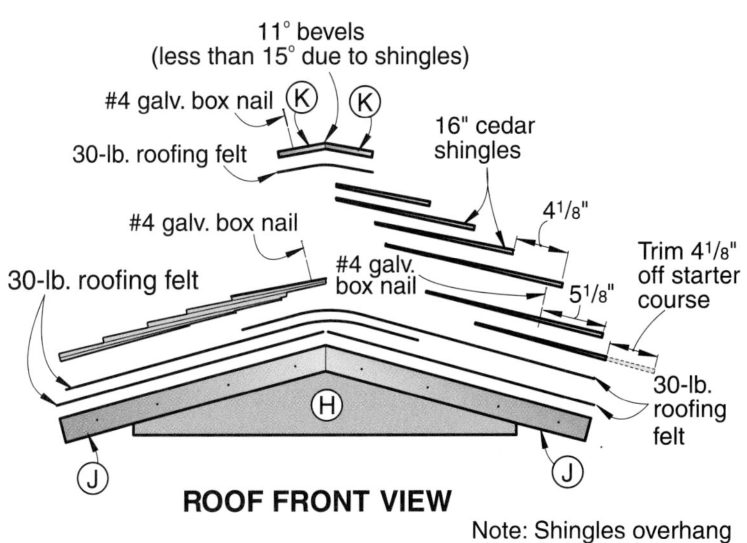

Note: Shingles overhang all edges by 1/2".

Bill of Materials

Part	Finished Size T	W	L	Matl.	Qty.
DECK					
A runners	1 1/2″	3 1/2″	47 3/4″	C	2
B decking	1 1/2″	3 1/2″	38 1/2″	C	4
FLOOR					
C panels	1/2″	37″	30 3/4″	EP	2
C1 insulation	1″	13 1/8″	34″	RF	2
FRONT AND BACK					
D outside panel	1/2″	37″	34 1/2″	EP	2
D1 inside panel	1/2″	37″	32 1/2″	EP	2
D2 insulation	1″	29 1/2″	34″	RF	1
D3 insulation	1″	17 1/2″	29 1/2″	RF	1
SIDES					
E outside panels	1/2″	31 3/4″	29 1/2″	EP	2
E1 inside panels	1/2″	27 3/4″	27 1/2″	EP	2
E2 insulation	1″	24 1/2″	24 3/4″	RF	2
PARTITION					
F panels	1/2″	32″	27 5/8″	EP	2
F1 insulation	1″	24 5/8″	29″	RF	1
ROOF					
G top & btm. panels	1/2″	48″	23 1/2″	EP	4
G1 insulation	1″	20 1/2″	45″	RF	2
H gussets	3/4″	5 1/2″	33 7/8″	C	2
I fascia	3/4″	2 1/4″	48″	C	2
J fascia	3/4″	2 1/4″	24 1/4″	C	4
K ridge cap	1/2″	4 1/4″	50 1/2″	C	2
ROOF					
L top & bottom	3/4″	2″	15″	C	4
M sides	3/4″	2″	20 3/4″	C	4

Materials Key: C–cedar, EP–exterior plywood, RF–1″-thick rigid foamboard.

Supplies: 1 1/4″ and 3 1/2″ galvanized deck screws; #4 galvanized box nails; #6 galvanized finish nails; construction adhesive; 30-lb. roofing felt; one bundle of 16″ cedar shingles; indoor/outdoor carpet; exterior stain or primer and paint.

Patio Table and Bench

Enjoy picnics on your porch or patio with this beautiful table and matching benches. They are fairly easy to build if you take your time and study the drawings carefully. The table shown here was built with cedar, but redwood could be used.

Suggestions:

- Begin by cutting the pieces for the legs (A, B, C, D). Use the bill of materials or the "Base Half Assembly" drawing for the exact dimensions.
- Plan your cutting schedule carefully to use your material efficiently.
- Make a template for the feet. Note that the notch on one (A) is opposite of the other (A).
- Dry-clamp all the pieces for the legs and use a spacer block measuring $1\frac{5}{8} \times 1\frac{5}{8} \times 18''$ to separate the inner uprights.
- With a Forstner bit drill the umbrella hole from the face side.
- Be sure to use a waterproof glue.
- Be careful not to drive the screwhead too deep into the soft cedar wood.
- Take the extra time to sand the project to prepare it for a nice wood-tone finish.
- Very soon after sanding brush on a couple coats of CWF-UV penetrating sealer.

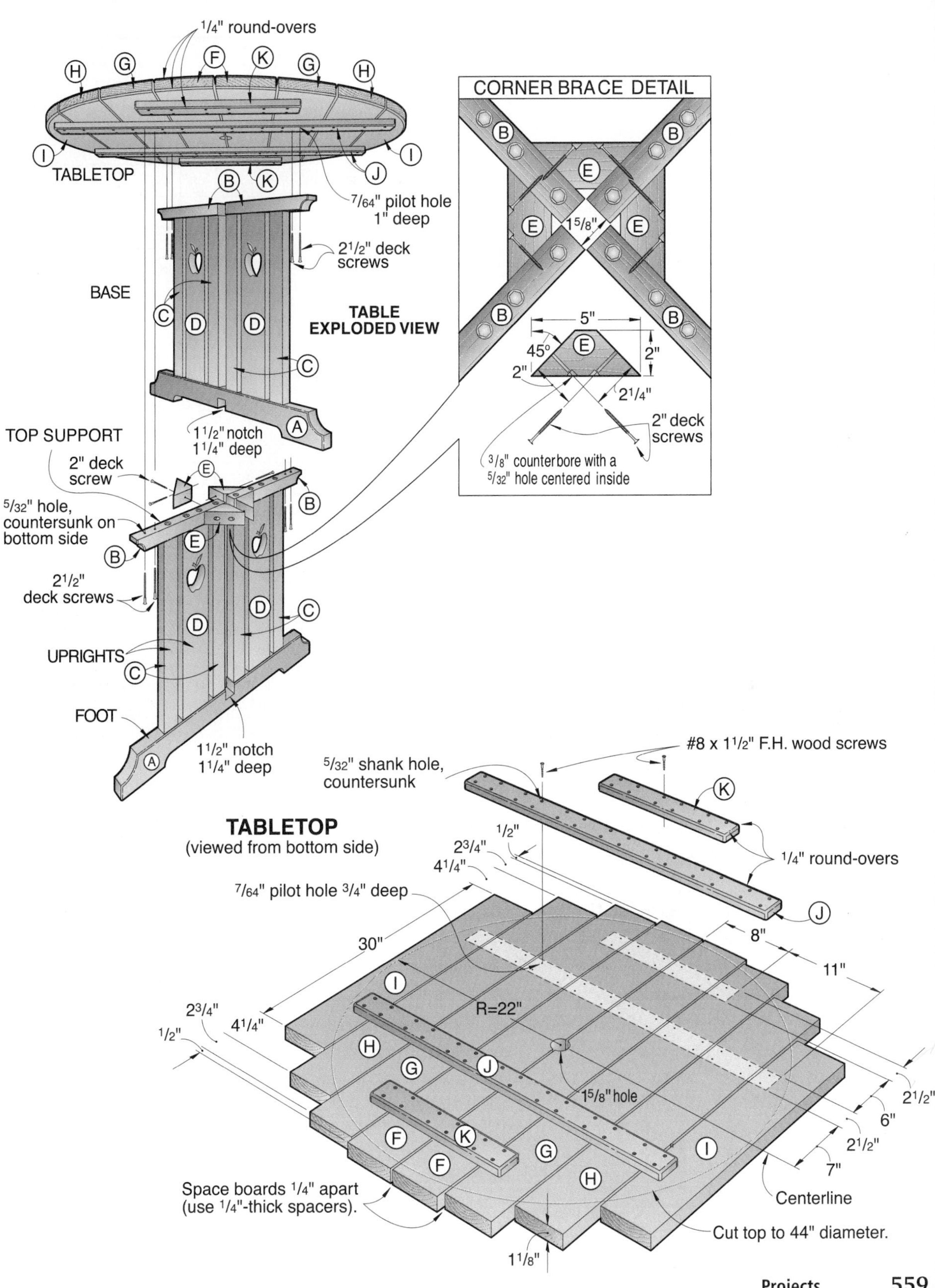

1/4" round-overs
H
G
F
K
G
H
I
I
TABLETOP
B
K
J
7/64" pilot hole 1" deep
2 1/2" deck screws
BASE
TABLE EXPLODED VIEW
C
D
D
C
A
1 1/2" notch 1 1/4" deep
TOP SUPPORT
2" deck screw
E
B
5/32" hole, countersunk on bottom side
B
E
2 1/2" deck screws
D
C
D
UPRIGHTS
C
FOOT
A
1 1/2" notch 1 1/4" deep
CORNER BRACE DETAIL
B
B
E
E
1 5/8"
E
B
B
5"
45°
E
2"
2"
2 1/4"
2" deck screws
3/8" counterbore with a 5/32" hole centered inside
#8 x 1 1/2" F.H. wood screws
5/32" shank hole, countersunk
K
TABLETOP
(viewed from bottom side)
1/2"
2 3/4"
4 1/4"
1/4" round-overs
7/64" pilot hole 3/4" deep
J
30"
8"
11"
I
R=22"
2 3/4"
4 1/4"
1/2"
H
G
J
1 5/8" hole
2 1/2"
6"
2 1/2"
7"
F
K
F
G
I
H
Centerline
Space boards 1/4" apart (use 1/4"-thick spacers).
Cut top to 44" diameter.
1 1/8"

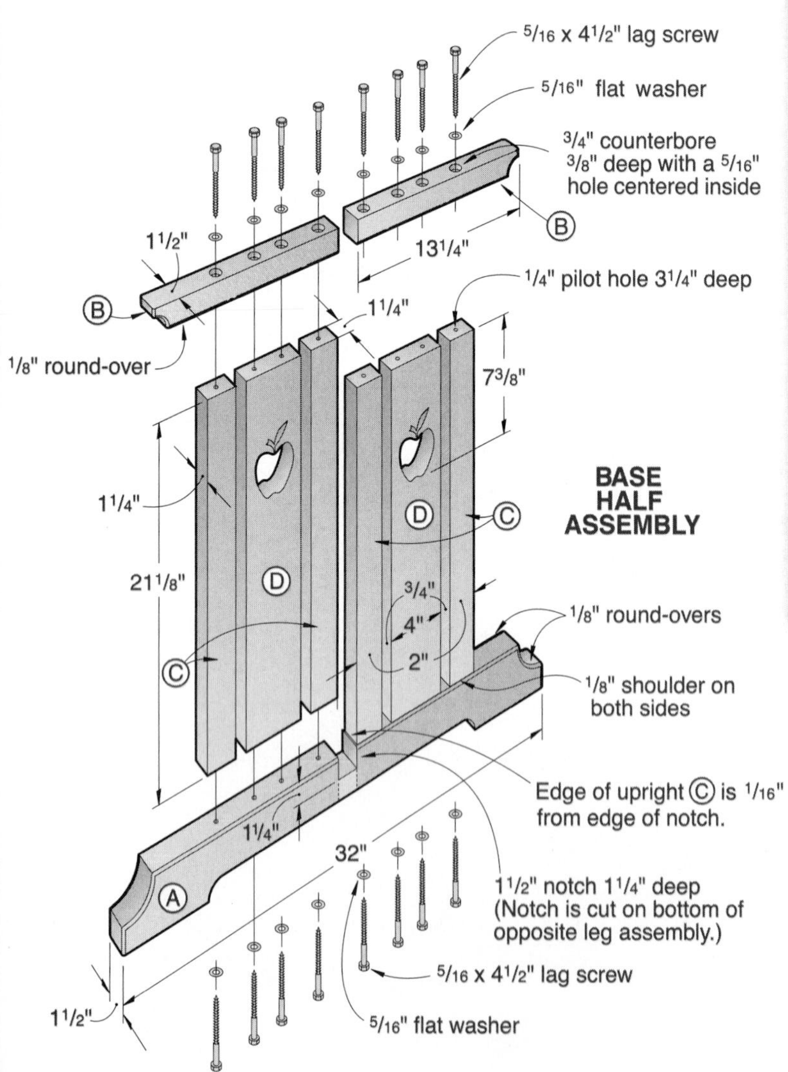

Bill of Materials

Part	Finished Size			Matl.	Qty.
	T	W	L		
TABLE BASE					
A feet	1 1/2″	3 1/2″	32″	C	2
B top supports	1 1/2″	1 1/2″	13 1/4″	C	4
C uprights	1 1/2″	2″	21 1/8″	C	8
D uprights w/cutout	1 1/4″	4″	21 1/8″	C	4
E corner braces	1 1/2″	2″	4 1/2″	C	4
TABLETOP					
F* top	1 1/8″	5 3/8″	45″	C	2
G* top	1 1/8″	5 3/8″	44″	C	2
H* top	1 1/8″	5 3/8″	38 1/2″	C	2
I* top	1 1/8″	5 3/8″	30″	C	2
J top cleats	3/4″	2 1/2″	38″	C	2
K top cleats	3/4″	2 1/2″	16″	C	2

*Blank size, cut to shape according to the how-to instructions

Material Key: C–cedar

Supplies: 2″ and 2 1/2″ deck or stainless steel screws, 32–5/16 × 4 1/2″ lag screws with 5/16″ flat washers, #8 × 1 1/2″ flathead wood screws, red and green high-gloss exterior enamel paint for the apple cutout, clear exterior finish.

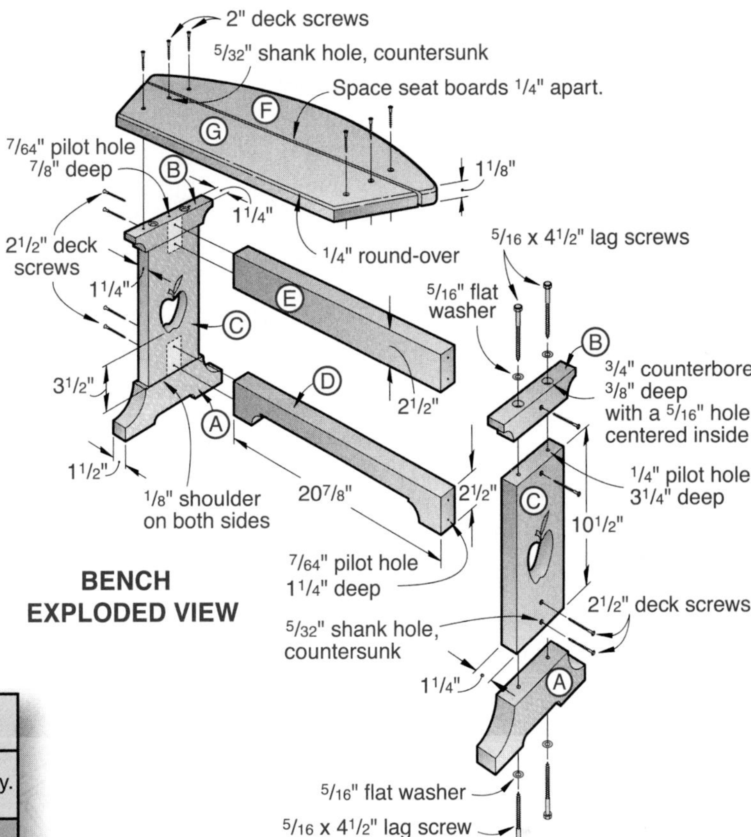

Bill of Materials

Part	Finished Size T	Finished Size W	Finished Size L	Matl.	Qty.
FOUR BENCHES					
A feet	1 1/2″	3″	10″	C	8
B top supports	1 1/4″	1 1/2″	8″	C	8
C uprights	1 1/4″	5 1/4″	10 1/2″	C	8
D bottom crsmbrs.	1 1/2″	2 1/2″	20 7/8″	C	4
E top crsmbrs.	1 1/2″	2 1/2″	20 7/8″	C	4
F* seat boards	1 1/8″	5 3/8″	32″	C	4
G* seat boards	1 1/8″	5 3/8″	32″	C	4

*Blank size, cut to shape according to the how-to instructions

Material Key: C–cedar

Supplies: 2″ and 2 1/2″ deck screws, 32–5/16 × 4 1/2″ lag screws with 5/16″ flat washers, red and green high-gloss exterior enamel paint for the apple cutout, clear exterior finish.

GLOSSARY

The chapters in which terms are important are identified by numbers in parentheses following the definitions.

A

abstract a legal document that traces the ownership of a piece of property (real estate). (Ch. 34)

adhesion (bonding) Combining or uniting force that develops between adhesive and wood. (Ch. 16)

adhesive A substance used to hold other materials together. (Ch. 16)

adjustable screw pilot bit Drilling tool used when installing flathead screws. It acts as a combination drill and countersink. (Ch. 27)

alkyd paint Oil-based paint. Requires the use of a solvent for cleanup. (Ch. 38)

alkyd resin A solvent-based type of varnish. (Ch. 32)

American Screw Wire Gauge Identification system which indicates the screw shank diameter with numbers from 0 to 24. (Ch. 15)

anchor bolt Fastener used to attach the sill plate to the foundation. (Ch. 35)

antikickback pawls Finger-like protective devices behind the blade of a table saw that resist the tendency of the saw to throw the stock upward and toward the operator. (Ch. 22)

arbor The shaft that holds the saw blade on a table saw. (Ch. 22)

auger bit Drilling tool with a point and an upwardly spiraled surface around the body of the bit which carries cut materials up and out of the hole as it turns. Used in braces, power drills, and drill presses. (Ch. 6, 27)

automatic drill (push drill) Hand tool that turns a bit when the operator pushes down on the handle. (Ch. 6)

B

backing board A true, smooth board at least ¾ inch thick, placed over thin stock being surfaced in a planer. (Ch. 20)

backsaw Handsaw that has a very thin blade with fine teeth. Used to make fine cuts both across and with the grain. (Ch. 4)

baluster Vertical member of a stair railing that helps support the handrail or guardrail. (Ch. 38)

band clamp (web clamp) A nylon strap that tightens around projects. Used when gluing up multisided projects. (Ch. 16)

band saw Versatile cutting machine with a blade that is a steel band which revolves around an upper and a lower wheel. Used mostly for cutting curves, circles, and irregular shapes. (Ch. 24)

bar clamp (cabinet clamp) Device used when gluing up large surfaces edge to edge and for clamping parts together during assembly. (Ch. 16)

batter boards Pairs of horizontal boards nailed to posts set at the corner locations of the house to be built. String is stretched between them from corner to corner to mark the outside line of the foundation walls. (Ch. 35)

belt sander Portable power tool which has a replaceable abrasive belt that is turned around two rollers. Used to perform rough sanding tasks, such as removing waste wood, and fine finishing tasks. (Ch. 7) *See also* **stationary belt sander.** (Ch. 29)

belt-and-disc sander Machine that is a combination of two types of power sanders, mounted adjacent to each other and operated by the same motor. They may have separate worktables or the same table may be used for both. (Ch. 29)

bevel or **bevel cut** Angled cut, other than a right angle, made along the edge or end of the stock. (Ch. 4, 21, 25)

bill of materials Complete list of materials, fasteners, and accessories needed for the project. (Ch. 3)

biscuit Football-shaped piece of wood installed in a joint to strengthen it. (Ch. 8)

biscuit joint Wood joint strengthened by a biscuit. (Ch. 8)

bit Small cutting tool used with a brace, drill, or router. (Ch. 28)

bleaching Process done to lighten or even out the color of unfinished wood. (Ch. 31)

blind dado joint Joint formed by cutting a dado partway across a board and notching the piece that fits into the dado so that the joint does not show from the front. (Ch. 10)

blind mortise Hole for receiving a tenon that is not cut all the way through the piece. The end of the tenon cannot be seen when the parts are assembled. (Ch. 13)

block plane Hand plane used for planing end grain. (Ch. 7)

board foot A unit of measurement, equal to a 1" thick board, 12" wide and 12" long. (Ch. 3)

box joint *See* **finger-lap joint.** (Ch. 11)

box nail Small diameter flathead nail used mostly for light carpentry and the construction of packing cases. (Ch. 5)

brace A hand tool used to hold and drive an auger bit to drill holes. (Ch. 6)

brad point (feed screw) Spirally threaded tip on the cutting end of an auger bit that helps draw the bit into the wood. (Ch. 6)

brad-point bit Drilling tool with a spirally threaded tip and sharp cutting spurs. (Ch. 27)

bridging Bracing method used with joists. (Ch. 36)

builder's level (optical level) Instrument used when laying out a building site for construction. It has a telescope that is fixed in a horizontal position. (Ch. 35)

building code Law that establishes a minimum standard of quality or safety in housing. (Ch. 34)

butt chisel Straight-edged cutting tool with a short blade, used to shape joints and cut recesses for hinges. (Ch. 7)

butt joint Type of joint in which the edge, end, or face surface of one piece of wood is joined with the edge, end, or face surface of another piece. (Ch. 8)

C

cabinet clamp *See* **bar clamp.** (Ch. 16)

cabinet hardware Metal items such as hinges, pulls, and catches that make a project with drawers and doors more usable. (Ch. 17)

caliper Measuring tool used to measure cylindrical and flat stock. (Ch. 4)

carbide tip Hard, sharp end on each tooth of some saw blades and router bits. (Ch. 22)

cardiopulmonary resuscitation (CPR) Rescue technique used to restore the breathing and heartbeat of accident victims. (Ch. 2)

casework Type of furniture construction that resembles a box turned on its end or edge, such as a bookcase. (Ch. 14)

casing nail A heavy nail with a small head that can be set below the surface. (Ch. 5)

catch A device used for holding a cabinet door closed. (Ch. 17)

caulk An inexpensive, easy-to-apply sealer used where air or water leakage can be expected in a structure, such as around windows. (Ch. 40)

C-clamp Device that resembles the letter "C," used for clamping small workpieces. (Ch. 16)

center square Measuring tool used to locate the center of a circle and to check 90-degree measurements. May also be used as a protractor. (Ch. 4)

chamfer An angled surface that results when a sharp edge on a piece of stock is cut off at a slight bevel. (Ch. 21)

chip breaker Part of a planer that presses firmly on the top of wood to prevent the grain from tearing out. (Ch. 20)

chisel A straight-edge cutting tool used to shape and trim wood. (Ch. 7)

chuck Device on a drill for holding twist drills and bits. (Ch. 6)

circular saw Portable power saw used for both crosscutting and ripping. It can be moved freehand or used with guides. (Ch. 4)

clamp Device used to hold glued materials together while adhesive sets and to hold workpieces in place during processing. (Ch. 16)

claw hammer Hand tool used in nailing. The claw is used to remove nails from wood. (Ch. 5)

clear cutting Logging method in which all trees in a given area are removed. (Ch. 1)

cleat Piece of wood clamped over workpieces being glued edge to edge. It holds the work in place and keeps the surface true and free of warpage. (Ch. 16)

clip (snipe) Undesirable but avoidable small, concave cut made by a planer at the end of stock. (Ch. 20)

closed-coat sandpaper Abrasive paper completely covered with grains of abrasives. Used mostly for working with hardwoods. (Ch. 7)

combination square Versatile measuring tool that can be used as a square, marker, level, rule, or gauge. (Ch. 4)

common nail Flathead nail used in building construction for rough carpentry. (Ch. 5)

common yard lumber Grade of softwood lumber suitable for rough carpentry. (Ch. 1)

compass saw Handsaw with a narrow, pointed blade, used for sawing curves or irregular shapes. (Ch. 4)

compound angle A beveled miter cut. (Ch. 25)

compound miter saw Power tool that can cut two angled surfaces at the same time. (Ch. 25)

computer-aided drafting (CAD) Software used in creating drawings and product plans on a computer. (Ch. 1)

computer-aided manufacturing (CAM) System in which a CAD drawing is sent directly to a computerized machine that then makes the part. (Ch. 1)

computer numerical control (CNC) Manufacturing system that involves the use of computers and a numerical code to control the machines used to make parts for products. (Ch. 1)

concrete Construction material made by mixing cement, fine aggregate, coarse aggregate, and water in the proper proportions. (Ch. 35)

coniferous Describes cone-bearing trees. (Ch. 1)

contact cement A ready-mixed, rubber-type bonding agent that is coated on two surfaces to be joined and then allowed to dry. Immediate and permanent bonding between the surfaces takes place on contact. (Ch. 18)

conventional framing (stick framing) consists of closely-spaced, relatively small individual wood members assembled piece by piece into a rigid structure. (Ch. 34, 36)

conventional roof construction Roof framing method in which joists and rafters are cut to length at the job site and assembled one by one. (Ch. 36)

coping saw Handsaw with a U-shaped frame and a replaceable blade with ripsaw-like teeth, used for cutting internal and external shapes on thin wood. (Ch. 4)

cost effective When the benefits brought about by improvements made justify the costs of making the changes. (Ch. 39)

countersink A bit or drill used to cut a recess in a surface for setting the head of a screw flush with or below the surface. (Ch. 15)

countersink bit Cone-shaped drilling tool that enlarges the top of a hole so a flathead screw can be driven in flush with the surface. (Ch. 27)

countersinking A way of enlarging the top portion of a hole to a cone shape so that the head of a flathead screw will be flush with the surface of the wood. (Ch. 15)

crosscut Cut made across the wood grain to cut stock to length. (Ch. 4)

cross-lap joint Joint formed by removing equal amounts of material from the area of contact on two pieces to be joined, and fitting the pieces together so the surfaces are flush. (Ch. 11)

cup center Part of a wood lathe that is inserted in the tailstock spindle. (Ch. 30)

customary system The system of measure used most in the United States. (Ch. 4)

cutoff saw *See* **radial-arm saw.** (Ch. 23)

cutterhead Part of a planer that surfaces a board to the desired thickness. (Ch. 20)

D

dado A channel or a groove cut across the grain of wood. (Ch. 10)

dado head A blade assembly for a table saw used to cut dadoes and grooves. It usually consists of two outside blades with a dado cutter (chipper) in between. (Ch. 22)

dado joint Type of joint formed by cutting a dado across one board to receive the end of another board. (Ch. 10)

damper Device installed within a chimney that consists of a cast iron frame with a hinged lid that opens or closes to vary the throat opening. (Ch. 40)

Danish oil A clear finish that penetrates, seals, and preserves wood surfaces. (Ch. 32)

deciduous Describes trees that shed their leaves annually. (Ch. 1)

deed A document that provides evidence of ownership and is the legal means by which ownership is transferred. (Ch. 34)

depth stops Devices used to control the depth of a hole being drilled. (Ch. 6)

design The outline, shape, or plan of something. (Ch. 3)

detail sander Portable power sander used for sanding spots larger sanders can't reach, such as inside corners. (Ch. 7)

dimension Actual size measurement used when laying out and building a project. (Ch. 3)

disc sander A flat, round platen (plate) on which a sanding disc is mounted. As it rotates, the workpiece is pressed against the downward-moving side. (Ch. 7) *See also* **stationary disc sander.** (Ch. 29)

doorframe Assembly that supports a door and enables it to be connected to the house framing. (Ch. 37)

doorjamb The part of the doorframe which fits inside the rough opening. (Ch. 37)

double spread Gluing method in which glue is applied to both surfaces being joined. (Ch. 16)

dovetail joint A strong, interlocking joint formed by fitting a dovetail-shaped tenon into a matching socket. Often used in drawer construction. (Ch. 14)

dovetail saw Handsaw with a narrow blade and fine teeth, used for extremely accurate work. (Ch. 4)

dowel Wood or plastic pin placed in matching holes in the two pieces of a joint to make it stronger. (Ch. 8)

dowel center Small metal pin used for matching and marking the location of holes on the two parts of a joint. (Ch. 8)

doweling jigs Drilling accessories attached to the edge of stock to identify hole locations and guide the drill to cut straight, perpendicular dowel holes. (Ch. 6, 8)

dowel joint A joint reinforced with dowels. (Ch. 8)

dowel rod Standard stock, ⅛ inch to 1 inch in diameter and usually 3 feet long, that is cut to make dowels for use in jointing. (Ch. 8)

dowel screw A wood screw threaded on each end. (Ch. 15)

downspout Pipe that carries water down from a roof and away from the foundation or into a storm sewer. (Ch. 37)

drawer guide Device on which a drawer slides. (Ch. 14)

dressed lumber (surfaced lumber) Lumber that has been put through a planer. (Ch. 1)

drill press Machine used primarily for drilling holes of various diameters and depths and at various angles. With appropriate jigs or setups, it can also be used for mortising and sanding. (Ch. 27)

drywall A sheet material made of gypsum filler faced with paper. It is the most used interior wall and ceiling finish in residential construction. (Ch. 38)

E

edge butt joint *See* **edge joint.** (Ch. 8)

edge half-lap joint Joint formed by cutting deep, narrow dadoes in the edges of two pieces of stock and fitting them together with the edges flush. (Ch. 11)

edge joint Type of joint in which edges of stock are glued together to form a single piece with a larger face surface. (Ch. 8, 16)

edging clamp Clamping device designed to hold moldings, veneer, and laminates to the edge of a workpiece. (Ch. 16)

elevating crank The part of a radial-arm saw which is turned to adjust the depth of cut. (Ch. 23)

enamel High-gloss paint that provides a slick, hard, easy-to-clean surface. (Ch. 33)

end butt joint Joint formed by connecting the end of one piece to the face surface, edge, or end of the second piece. (Ch. 8)

end grain The closely packed tips of cut wood fibers revealed when stock is cut across the grain. (Ch. 7)

end-to-end half-lap joint Joint formed by removing half the thickness from two equally thick pieces of wood and overlapping the cut ends so the surfaces are flush. (Ch. 11)

engineered wood Strong, long-lasting manufactured materials made from sawdust or small wood pieces and plastics. (Ch. 1)

entrepreneur Person who owns and is responsible for his or her own business. (Ch. 1)

Environmental Protection Agency (EPA) Government agency that sets limits on the amount of hazardous waste permitted to accumulate at a work site and for how long. (Ch. 2)

epoxy cement A two-part adhesive that can be used with most materials. (Ch. 16)

exploded view Drawing that shows the project as "taken apart." It gives the dimensions of each part and shows how the parts go together. (Ch. 3)

F

face planing (surfacing) Planing the surfaces to true up stock. (Ch. 21)

faceplate turning Lathe operation in which stock is mounted and rotated on a metal disk (faceplate) for shaping. (Ch. 30)

FAS Label meaning "firsts and seconds" placed on the top grade of hardwood. (Ch. 1)

featherboard A piece of lumber with a series of saw kerfs on one end, used to hold narrow stock against the rip fence when making a rip cut with a table saw. (Ch. 22)

filler Paste applied to open-grained woods to close the pores. (Ch. 31)

finger-lap joint (box joint) Strong, attractive joint in which series of fingers and notches are meshed together. (Ch. 11)

finish flooring The final wearing surface applied to a floor. (Ch. 38)

finishing nail The finest of all nails. Used for fine cabinet and construction work. (Ch. 5)

fire safety plan A map of a building showing at least two escape routes in case of fire. (Ch. 2)

fixed-base router Type of router in which the base is clamped to the motor and the joined parts are moved as a single unit. Used for edge cutting and trimming. (Ch. 28)

flat cutting method Method of cutting a log into narrow strips of veneer. The log is moved back and forth across a blade and cut into slices. (Ch. 19)

flitch Ordered bundle of veneer slices cut from a single log. (Ch. 19)

floor joist One of a series of horizontal structural members which support a floor. (Ch. 36)

floor plan An architectural working drawing that shows sizes and locations of rooms, windows, doors, and many other features. (Ch. 34)

flush door Door that fits inside the frame, flush with the surface. (Ch. 14)

flush drawer A drawer that fits flush with the frame opening. (Ch. 14)

fly cutter (circle cutter) Adjustable drilling tool used to cut holes from 1 to 4 inches in diameter. (Ch. 27)

folding rule Measuring tool that is 8 inches long when folded but can be unfolded to various lengths up to 6 feet. (Ch. 4)

footing Base on which the foundation of a building rests. (Ch. 35)

fore plane Hand plane used for planing large surfaces and edges. (Ch. 7)

Forstner bit A drilling tool used to cut flat-bottomed holes. It has a circular cutting edge with no screw tip extending below it. (Ch. 27)

foundation The part of a house that anchors it to the land, supports its weight, and provides a level base for floor framing. (Ch. 35)

freehand routing Routing that is done without guides. (Ch. 28)

full-lap joint Joint used when pieces with different thicknesses are joined. The thinner piece is fit into a cut that has been made in the thicker piece. (Ch. 11)

function The purpose of a product. (Ch. 3)

G

gain A recess cut into a door or frame in which a leaf of a hinge is installed. (Ch. 17)

girder Large wood or steel structural member which provides support for a wood frame. (Ch. 36)

gloss Amount of surface brightness of finishes. (Ch. 33)

gouge A chisel with a curved blade. (Ch. 30)

grit The size of the grains on abrasive paper. (Ch. 7)

ground-fault circuit interrupter (GFCI) Protective device that detects electrical leaks to ground and breaks the circuit. (Ch. 2)

guardrail The top part of a stair railing on a landing or other area where the railing is horizontal. (Ch. 38)

gutter Trough installed at the lower edge of the roof to catch water and carry it to a downspout. (Ch. 37)

H

half-lap joint Joint formed by removing half the thickness of two equally thick pieces of wood and fitting them together so the surfaces are flush. (Ch. 11)

hand drill Tool used with twist drills to drill small holes. It is operated by turning a crank. (Ch. 6)

hand plane Cutting tool used to shape and smooth stock. (Ch. 7)

handrail The top part of the railing that people grasp when using stairs. (Ch. 38)

hand scraper A flat, rectangular blade used for smoothing wood surfaces prior to sanding. (Ch. 7)

hand screws Wooden parallel clamps used mainly when gluing stock face to face. (Ch. 16)

hardboard Manufactured wood material made by refining wood chips into fibers that are formed into panels under heat and pressure. (Ch. 1)

hardware Parts (usually metal) needed to complete a project, make a project usable, or provide structural support within the project. (Ch. 17)

hardwood Wood cut from deciduous trees, such as maple and oak. (Ch. 1)

hazards Dangers. (Ch. 2)

headstock Part of a wood lathe that is permanently fastened to the bed. (Ch. 30)

heat pump A device that can heat or cool the air in a house. It is connected to standard duct systems. (Ch. 38)

hinge A piece of hardware used as a joint. It allows one of two joined parts to move. (Ch. 17)

hole saw Drilling tool used for cutting large holes. It comes in sizes up to 6 inches. (Ch. 27)

hood A covering with open ends placed over a chimney to keep rain out and prevent downdraft. (Ch. 40)

house inspector Qualified professional person who checks the general condition of a house in order to find existing or potential problems. (Ch. 39)

I

infeed roll Part of a planer that grips the stock and moves it toward the cutterhead. (Ch. 20)

infeed table Part of a jointer or planer that supports the work before it is cut. It is raised or lowered to adjust the depth of the cut. (Ch. 21)

inlaying (marquetry) A way of forming a design by using two or more different kinds of wood that have a marked contrast in color or grain pattern. (Ch. 26) Also, the setting of a material into a surface as a decoration. (Ch. 28)

insulation A material installed in walls, ceilings, roofs, and floors that slows down the transmission of heat. (Ch. 38)

J

jack plane Planing tool used for general planing. (Ch. 7)

jigsaw (saber saw) Portable power saw used for making both curved and straight cuts in hard-to-reach places. (Ch. 4)

jointer A machine used to true up stock. The operator can straighten, smooth, square up, and size boards to be accurately formed. (Ch. 21)

jointer plane Planing tool used for planing edges straight when making joints on long pieces of stock. (Ch. 7)

jointing Process of making joints. (Ch. 7) When using a jointer, it refers to smoothing and straightening an edge to make it square with the face surface. (Ch. 21)

K

kerf Cut made by a saw. (Ch. 4)

kerf board (kerf insert) Slotted piece set in the turntable below the blade of a sliding compound miter saw. It minimizes the gap between the side of the blade and the table. (Ch. 25)

kerf splitter *See* **splitter.** (Ch. 4)

kickback A sudden, violent thrust upward and back that stock can make under certain circumstances during processing. Precautions and care must be taken to avoid it. (Ch. 21)

knot Hard lump in lumber formed at the point where a branch begins to extend out from a tree. (Ch. 31)

L

lacquer A clear finishing material that dries quickly and produces a hard finish. (Ch. 32)

laminate A material made up of several different layers that are firmly united together. (Ch. 18)

lamination The process of building up the thickness or width of material by gluing and clamping together several layers. (Ch. 16)

lap joint Type of joint formed by laying (lapping) one piece of wood over another and fastening the two pieces together in the contact area. (Ch. 11)

latex paint Water-based paint. Tools clean readily in soap and water. (Ch. 33, 38)

layout Measuring and marking stock to size and shape. (Ch. 3)

level Tool used to check a horizontal surface to see if it is level or a vertical surface to see if it is plumb. (Ch. 4)

linseed oil Natural oil used to finish certain types of furniture. Sometimes mixed into oil-based paints. (Ch. 31)

lip door Door that has a rabbet cut around the inside edge, and the rabbeted edge (lip) overlaps the frame. (Ch. 14)

lip drawer A drawer that has a rabbet cut on the inside edge of the front piece, and the rabbeted edge (lip) overlaps the frame. (Ch. 14)

M

manufactured housing (industrialized housing) House built wholly or in part on factory assembly lines. (Ch. 34)

marquetry *See* **inlaying.** (Ch. 26)

matching Arranging pieces of veneer to create special decorative effects with grain patterns. (Ch. 19)

Material Safety Data Sheets (MSDS) Comprehensive information required by OSHA about hazardous materials workers must handle. Includes health hazards, precautions to take, and conditions to avoid when using the material. (Ch. 2)

materials list Complete listing of all materials needed for a construction project. Includes sizes, amounts, and other information. (Ch. 34)

mill marks Uniform ridges on planed lumber that must be smoothed before finishing. (Ch. 31)

miter or **miter cut** Angled cut across the face of stock. (Ch. 4, 25)

miter box A device used to guide a handsaw to make a cut through a piece of wood at a set angle from 30 to 90 degrees. (Ch. 12)

miter gauge Sliding device used to push stock and maintain a straight line or preselected angle. (Ch. 22)

miter joint An angle joint that hides the end grain of both pieces. (Ch. 12)

module (modular unit) Complete unit of a structure that is built, shipped to a site, and assembled with other units. (Ch. 34)

molding A narrow strip of wood shaped to a uniform curved profile throughout its length. Used to conceal a joint or to ornament furniture or a room interior. (Ch. 1)

mortise A rectangular hole cut in wood to receive a tenon that is the same size and shape as the hole. (Ch. 13)

mortise-and-tenon joint A very strong joint formed by fitting a projecting piece of wood (tenon) into a hole of the same dimensions (mortise) cut in another piece. Often used in fine furniture. (Ch. 13)

mortise chisel Sturdy cutting tool with a thick blade, used for clearing out a mortise to fit a tenon. (Ch. 7, 13)

mortising attachment Drill press tool that consists of a drill bit surrounded by a four-sided chisel. (Ch. 27)

multispur bit Drilling tool used in a drill press to cut perfectly round, flat-bottomed holes. Its semicircular leading edge has saw teeth, and a single cutter extends from the edge to the center. (Ch. 27)

N

nailer Air-powered tool used for driving and setting nails. (Ch. 5)

nail set A short metal punch with a cup-shaped head, used to drive the head of a nail below the surface of wood. (Ch. 5)

narrow belt sander-grinder A tabletop machine that moves a narrow abrasive belt around three or four pulleys. Used for sanding small parts and for getting into hard-to-reach places. (Ch. 29)

National Institute for Occupational Safety and Health (NIOSH) Federal agency responsible for conducting research and making recommendations for the prevention of work-related illnesses and injuries. (Ch. 2)

O

Occupational Safety and Health Administration (OSHA) Federal agency that sets and enforces standards for safety in the workplace. (Ch. 2)

offset screwdriver Hand tool used to install or remove screws located in tight places where a standard screwdriver cannot be used. (Ch. 15)

opaque finish Type of finish that covers a surface completely and cannot be seen through, such as paint or enamel. (Ch. 31, 33)

open-coat sandpaper Abrasive paper on which grains of abrasives are spaced far apart. Used mostly for working with softwoods. (Ch. 7)

orbital sander Portable power sander that uses a circular motion. (Ch. 7)

oscillating spindle sander A machine that moves a sanding drum up and down and spins it at the same time. The workpiece is pressed against the moving drum. Useful for sanding curved and irregularly shaped edges. (Ch. 29)

outfeed roll Part of a planer that helps move stock out of the back of the machine. (Ch. 20)

outfeed tables Part of a jointer or planer that supports the work after it has been cut. (Ch. 21)

overarm The part of a radial-arm saw along which the saw unit is moved. It sits on a column and can be rotated in a complete circle around the column. (Ch. 23)

P

pad sawing The cutting of several pieces at one time with a band saw. (Ch. 24)

panel stock (sheet materials) Wood, such as plywood, that has been processed and formed into panels (or sheets). (Ch. 1)

paring chisel Lightweight cutting tool used for final trimming and fitting. (Ch. 7)

parting tool Lathe tool used to cut grooves and to cut away stock during faceplate turning. (Ch. 30)

pawl Movable jaw on a bar clamp. (Ch. 16)

penetrating finish Type of finish that soaks into wood rather than just coating it. (Ch. 31)

penetrating oil stain Aniline dye mixed in oil that soaks into wood. (Ch. 32)

penny number (d) Number that relates to nail length. (Ch. 5)

Phillips-head screwdriver Hand tool made for driving crossed-head screws. (Ch. 15)

pictorial drawing Drawing or sketch that shows an object the way it looks in use. (Ch. 3)

pigment stain Color pigments added to boiled linseed oil and turpentine. (Ch. 32)

pilot end Part of a router bit that rides against the edge of a workpiece to guide and limit cutting. (Ch. 28)

plain screwdriver Regular screwdriver used to install slotted-head screws. (Ch. 15)

plane iron Adjustable blade in a plane that does the cutting. (Ch. 7)

planer (surfacer) Machine designed to surface boards to thickness and to smooth rough-cut lumber. (Ch. 20)

plastic laminate Thin, hard manufactured material, used as a long-wearing, scratch-resistant top for such items as tables, cabinets, and desks. (Ch. 18)

plate Horizontal member of a frame wall. (Ch. 36)

plot plan Drawing that shows the building site, including boundaries and physical details as well as the planned location of the house. (Ch. 34, 35)

plug cutter Drilling tool for cutting cross-grain and end-grain plugs and dowels up to 3 inches long. (Ch. 27)

plumb Straight up and down. (Ch. 4)

plunge router Type of router which has the motor attached to a base that has springs. The cutting unit can be raised and lowered without moving the base. Used for cutting dadoes, mortises, and rabbets. (Ch. 28)

plywood Sheet material made by gluing thin layers (plies) of wood together. (Ch. 1)

pocket hole jigs Devices used as an aid for drilling holes that will be used when fastening two wood parts together, such as the rails to the underside of the table. (Ch. 6)

points The tips of saw teeth. The more points, the finer the cut. (Ch. 4)

polyurethane varnish A basic type of varnish. (Ch. 32)

post-and-beam framing Wood framing system that consists of widely spaced, relatively heavy members. (Ch. 36)

prefabricated house Type of house in which all parts of the shell were precut and prefit at a factory. (Ch. 34)

prefabricated part Standard-sized parts made in a factory. (Ch. 34)

pressure bar Part of a planer that holds stock against the table. (Ch. 20)

pressure-sensitive adhesive (PSA) Adhesive that sticks to a surface when pressure is applied. Used on the back of abrasive sanding discs for easy installation. (Ch. 29)

primer An undercoat that seals wood and prepares it to take a paint or enamel finish. (Ch. 33)

problem-solving process System used to develop solutions to the problems. (Ch. 1)

proportion Size relationship of parts or features. (Ch. 3)

pumice Powdered volcanic material combined with water or oil to rub down a finish. (Ch. 31)

pushblock or **pushstick** A wood block or stick used to move stock through a cutting machine. When using a jointer, it is held with one hand as the other hand holds the front of the stock down. (Ch. 21) When using a table saw, a push-stick is used to push the stock past the blade if the space between the saw blade and the fence is 6 inches or less. (Ch. 22)

push drill *See* **automatic drill.** (Ch. 6)

R

rabbet An L-shaped cut along the end or edge of a board. (Ch. 9, 14)

rabbet-and-dado joint Joint formed by fitting a rabbet with a tongue into a dado. Used when additional strength and stiffness are needed. (Ch. 10)

rabbet joint Type of joint formed by fitting the end or edge of one piece into a rabbet cut at the end or edge of another piece. (Ch. 9)

radial-arm saw (cutoff saw) Cutting machine used primarily for crosscutting operations. The saw unit is moved along an overarm or track to make cuts. (Ch. 23)

rail Horizontal part of a frame. (Ch. 14)

random-orbit sander A variation of the orbital sander that uses both a circular motion and a side-to-side motion at the same time. (Ch. 7)

real estate (real property) Land and the buildings on it. (Ch. 34)

reciprocating saw Portable power saw with a back-and-forth motion, used mainly for rough cutting. (Ch. 4)

remodeling Altering the structure of an older home to make it more suitable for residents. (Ch. 39)

renovation Restoring a home to the way it originally looked. (Ch. 39)

repair plate Metal piece installed across a joint to add strength and provide structural support in a project. (Ch. 17)

resawing Sawing stock to reduce its thickness. (Ch. 24)

rip cut Cut made with the grain to cut stock to width. (Ch. 4)

rip fence Part on a table saw used to guide a workpiece straight through the saw blade. (Ch. 22)

ripping bar A heavy metal bar with a nail slot on the curved end and a chisel-shape on the straight end. Used for pulling nails and for prying things apart. (Ch. 5)

ripping fence Guide used when making a rip cut with a circular saw near the edge of the stock. (Ch. 4)

ripsaw Handsaw for cutting with the grain. Used to cut stock to width. (Ch. 4)

riser Vertical member between the treads of a stair. (Ch. 38)

rotary cutting method Method of cutting a log into a continuous sheet of veneer. The log is turned lengthwise against a blade. (Ch. 19)

rottenstone Fine abrasive material that comes from shale. It is used with water or oil after pumice to produce a smoother finish. (Ch. 31)

roundnose tool Flat turning chisel with a rounded cutting edge. (Ch. 30)

router Versatile power tool which uses different types of bits to cut dadoes, rabbets, and grooves and to perform joinery, mortising, planing, and shaping processes. (Ch. 28)

router bit Tool fastened into a router that does the cutting. (Ch. 28)

rule Measuring tool marked off in equal units (graduations). Used to measure lengths. (Ch. 4)

S

sanding Process of smoothing wood by rubbing it with an abrasive. (Ch. 7)

sanding drum Cylindrical sanding attachment for a drill press or an oscillating spindle sander. (Ch. 27, 29)

scale Proportion used in drawing. A smaller measurement can represent a larger measurement or vice versa. (Ch. 3)

scratch awl A slender, metal, pointed tool with a wooden handle, used for marking and punching the location of holes to be drilled. (Ch. 4)

screwdriver bit Bit used in a power drill to drive screws. (Ch. 15)

screw-mate counterbore Bit which does all the operations performed by the screw-mate drill and countersink, as well as drilling plug holes for wooden plugs. (Ch. 15)

screw-mate drill and countersink Bit designed to drill to correct depth, do countersinking, make correct shank clearance, and drill pilot holes for flathead screws. (Ch. 15)

scroll saw (jigsaw) Machine with an up-and-down cutting action, designed to cut sharp curves and angles on both the outside edges and the interior sections of a workpiece. (Ch. 26)

sealant A material similar to caulk, but generally more flexible and expensive. (Ch. 40)

sectional house A single house is built in sections on an assembly line. (Ch. 34)

selective cutting Logging method in which only trees of a certain diameter and species are harvested. (Ch. 1)

select yard lumber Grade of lumber that has a good appearance and takes different finishes such as stain and paint. (Ch. 1)

sheathing Layer of sheet material placed between a building's frame and outside surface. (Ch. 36)

shelf-drilling jig Guide used when drilling a series of evenly spaced holes, such as for shelf pins for adjustable bookshelves. (Ch. 6)

shelf standard A slotted wood or metal strip used with brackets or clips to hold shelves in bookcase construction. (Ch. 14)

shingle One of a number of thin pieces of building material laid in overlapping courses (rows) to cover the roof or exterior walls of a house. (Ch. 37)

siding The exterior wall covering of a house. (Ch. 37)

sill plate Wood member bolted horizontally to the top of the foundation wall. It creates the connection between the foundation and the wood framing above it. (Ch. 35)

SI Metric System A decimal system of measure. (*SI* stands for "System International.") (Ch. 4)

single spread Gluing method in which glue is applied to only one of the two surfaces to be joined. (Ch. 16)

sitework Site preparation. Tasks include surveying, clearing the land, and laying out the building. (Ch. 35)

skew Lathe tool which has a tapered cutting edge. (Ch. 30)

skylight Transparent or translucent component installed on either a pitched or flat roof to provide light and, sometimes, ventilation. (Ch. 37)

sliding compound miter saw Power tool that can cut two angled surfaces at the same time. It can make slide cuts as well as cutting down and through. (Ch. 25)

sliding door Door that is moved horizontally to open. (Ch. 14)

sliding T-bevel Measuring tool that has an adjustable blade in a handle. Used for laying out all angles other than right angles. (Ch. 4)

smooth plane Planing tool used for general planing and finishing work. (Ch. 7)

snipe *See* **clip.** (Ch. 20)

softwood Wood from coniferous trees, such as pine and redwood. (Ch. 1)

solvent Liquid that dissolves another substance. Used to thin finishing materials and to clean brushes after use. (Ch. 31)

spade bit Fast-cutting drilling tool used for cutting general-purpose holes. (Ch. 27)

spark arrester Wire mesh placed over a chimney to reduce the discharge of sparks. (Ch. 40)

spar varnish A basic type of varnish.(Ch. 32)

specifications Written details on construction not given elsewhere, such as in drawings or materials list. (Ch. 34)

spindle turning Turning between centers on a lathe. (Ch. 30)

splitter (kerf splitter) Device or part that prevents wood being cut with the grain from the springing saw kerf closed and binding the blade. (Ch. 4, 22)

spur (live center) Part of a wood lathe fastened to the headstock spindle. (Ch. 30)

square Tool used for checking the squareness (right angles) of stock, cuts, or joined pieces. (Ch. 4)

staining Applying transparent or semitransparent liquids made from dyes, pigments, and/or chemicals to change the color of wood without changing its texture. (Ch. 32)

standard stock Lumber cut to a widely used size and shape. (Ch. 3)

stationary belt sander Machine that has an abrasive belt that revolves on two pulleys. It can be used in vertical, horizontal, or slanted positions. (Ch. 29)

stationary disc sander Machine that has an abrasive disc attached to a metal platen which rotates. Workpieces are moved back and forth against the downward moving side of the disc. (Ch. 29)

steel wool Abrasive material made of thin metal shavings, used in place of sandpaper in some finishing operations. (Ch. 31)

stick framing *See* **conventional framing.** (Ch. 34, 36)

stile A vertical side piece in the frame of a paneled door. (Ch. 14)

stock-cutting list List in which the thickness, width, and length of the stock listed in the bill of materials are increased to allow for cutting and other operations. (Ch. 3)

straightedge Any tool or object, such as a board, along which a straight line can be drawn. (Ch. 4)

stringer Part of a stair that supports the treads. (Ch. 38)

structural hardware Metal items, such as repair plates and corner blocks, that are used to strengthen joints and hold unseen parts together in a project. (Ch. 17)

structural lumber Classification of lumber used in construction. Grades are based upon the strength of the pieces. (Ch. 1)

stud Vertical member of a frame wall. (Ch. 36)

subfloor Wood floor attached directly to the floor joists, under the finished floor. (Ch. 36)

substrate Wood underlayment to which plastic laminate or veneer is applied. (Ch. 18, 19)

surface-lap joint Joint formed when pieces of wood are joined without additional processing. (Ch. 11)

surfacer *See* **planer.** (Ch. 20)

survey Measurements that identify boundaries of a property. (Ch.34)

suspended ceiling Ceiling panels held in place by a metal grid system at a desired distance from the existing ceiling structure. (Ch. 39)

T

table saw Cutting machine with a fixed, horizontal table and an adjustable blade. Used for ripping and crosscutting boards. (Ch. 22)

tack rag A cloth treated to attract and hold dust. (Ch. 31)

tailstock Part of a wood lathe that slides along and can be locked in any position on the bed. (Ch. 30)

tang The noncutting end of an auger bit that is fastened into a brace. (Ch. 6)

tearout The splintering that often happens when a saw blade breaks through the bottom surface of the wood. (Ch. 22, 25)

template A pattern used when forming or shaping a workpiece. (Ch. 28)

tenon A projecting piece of wood sized and shaped to fit into a mortise cut in another piece of wood. (Ch. 13)

through mortise A hole for receiving a tenon that is cut all the way through a piece of wood. (Ch. 13)

toenailing Driving nails at an angle to fasten the end of one piece of wood to the side of another piece. (Ch. 5)

transit level (transit) Instrument similar to a builder's level, but its telescope can be moved up and down as well as sideways. Used when laying out a building site for construction. (Ch. 35)

transparent coating Finish that can be seen through, such as shellac, varnish, or lacquer. (Ch. 31)

tread Horizontal part of a stair on which the stair user steps. (Ch. 38)

trial assembly A temporary assembly of all parts of a project, made before gluing to check for proper fit. (Ch. 16)

trussed roof construction Roof framing method in which factory-made trusses are used. (Ch. 36)

tuckpointing Technique used to repair mortar joints. The loose and damaged mortar is removed and new mortar is pushed in. (Ch. 40)

tung oil A clear, penetrating finish which dries to a low sheen that can be rubbed to a gloss finish. (Ch. 32)

turning A cutting operation on a lathe in which the workpiece is revolved against a single-edged tool. (Ch. 30)

twist drill Drilling tool that has a deep spiraling groove running from tip to shank. (Ch. 6, 27)

U

underlayment A material placed under finish coverings to provide a smooth, even surface. (Ch. 37)

upholster Apply padding and cover with fabric. (Ch. 14)

utility knife Sharp tool used for accurate marking and some cutting tasks. (Ch. 4)

V

vapor barrier A material that prevents moisture from getting into interior walls, floors, or ceilings. (Ch. 38)

variable-speed (drill) Feature of a drill that allows the speed to be adjusted by the amount of pressure applied to the trigger. (Ch. 6)

veneer A thin layer of wood of good quality that adds beauty and character to furniture and other wood projects. (Ch. 16, 19)

veneering The process of applying a thin layer of fine wood, called *veneer*, to the surface of a wood of lesser quality. (Ch. 19)

W

warp Any variation from a true, or plane, surface on a board. Includes crook, bow, cut, wind (twist), or any combination of these. (Ch. 7)

wash coat Sealer applied after stain and before a clear finish. (Ch. 32)

ways Guiding surfaces on a jointer. (Ch. 21)

white liquid-resin glue A general all-purpose glue that works well for small woodworking projects. (Ch. 16)

wood lath Slender strips of split or sawn wood that were used as a base for plaster. (Ch. 39)

wood screw A fastener with a groove twisting around part of its length. It cuts its own threads into the piece it is fastening. (Ch. 15)

working drawings Drawings that provide dimensions and other information needed to construct a project. (Ch. 3, 34)

Y

yard lumber Classification of lumber cut for a wide variety of uses. It may be common or select grade. (Ch. 1)

yellow liquid-resin glue An all-purpose glue much like white liquid-resin glue, but with greater moisture-resistance and tack and better sandability. (Ch. 16)

yoke The part of a radial-arm saw that holds the saw unit. (Ch. 23)

INDEX

C

D

I

J

K

N

O

P

S

T

U

V

W

Y

Z

CREDITS

Cover Design:
Greg Nettles, DesignNet

Cover Image:
Kevin May Corporation

Interior Design:
Greg Nettles and Brent Parrish, DesignNet

Projects:
Better Homes & Gardens® Wood® Magazine

Technical Writer:
Marlene Weigel

Photography Assistance:
The following individuals and companies helped with tools, materials, and building of projects and offered the use of their facilities for photo shoots.

Amerhart
Amish Outlet, Inc.
C&H Repair Plus Supply Co.
Colonial Hardwoods
Dave Drum
Frames Plus
Garrett Wade
Lowe's of Peoria
Naked Furniture
Sutherland Hardware
The Recipe Box
WILSONART® International

Photo Credits:
Alyeska Cedar Works, 461
Andersen Windows Inc., 461, 462, 463, 479
Arnold & Brown, 81
Art Mac Dillo's – Gary Skillestad, 88, 90, 91, 92, 93, 94, 95, 96, 97, 99, 100, 101, 104, 105, 106, 107, 108, 109, 110, 113, 114, 115, 116, 117, 118, 122, 124, 125, 127, 128, 129, 133, 135, 136, 137, 139, 140, 142, 143, 144, 145, 146, 148, 149, 150, 152, 153, 157, 158, 159, 160, 162, 163, 164, 165, 166, 167, 168, 169, 173, 175, 176, 177, 179, 180, 182, 185, 187, 188, 189, 194, 195, 196, 198, 203, 204, 205, 206, 208, 211, 214, 215, 217, 218, 219, 220, 221, 222, 240, 242, 244, 245, 246, 247, 248, 249, 250, 271, 273, 275, 279, 282, 283, 287, 288, 289, 291, 292, 313, 314, 316, 317, 320, 323, 324, 325, 326, 327, 328, 332, 351, 352, 353, 356, 377, 397, 398, 400, 401, 402, 429, 430, 439, 440, 441, 442, 443, 444, 448, 449, 450, 452, 453, 454, 456, 459, 460, 461, 462, 465, 473, 480, 481, 483, 485, 488
Bassett Furniture, 202, 218
Roger Bean, 455
Black & Decker, 499
Creative Architect & Planners - Scott B. Roberts, Architect / Garland Independent School District, 428, 433, 436
Carved by Ron Ramsey, 464
CTS Corp. / Berger, 439, 440
D.S. Nelson Co. Inc., 473
Davis Frame Company, 449
Deck House Inc., 467
Delta International Machinery Corp., 10, 286, 287, 312, 320, 350
Dewalt, 332
Dynamic Laser Application, 70
Curt Fisher, 6, 12, 214, 219, 431, 474, 476
David Frazier Photography, 430, 502
Ann Garvin, 445
Alan Goldstein, 480
Peter Hemsley (UK) Woodturner, 65, 87
Jet Equipment & Tools Inc., 350

Keystone Steel & Wire, 61
Louisiana – Pacific Corp., 451
Kevin May Corporation, 4, 7, 8, 9, 16, 17, 18, 19, 20, 21, 22, 23, 24, 25, 26, 27, 28, 29, 30, 31, 32, 33, 34, 35, 36, 37, 38, 39, 40, 41, 42, 43, 44, 45, 46, 47, 48, 49, 50, 51, 52, 53, 54, 55, 56, 57, 58, 59, 60, 64, 76, 78, 80, 81, 82, 83, 90, 91, 103, 106, 107, 108, 109, 110, 111, 112, 113, 114, 115, 116, 119, 120, 121, 122, 125, 126, 127, 128, 129, 130, 132, 133, 134, 135, 136, 137, 138, 140, 143, 144, 145, 147, 149, 150, 151, 152, 153, 154, 157, 159, 160, 161, 163, 164, 169, 170, 171, 172, 173, 176, 177, 178, 179, 181, 182, 186, 188, 191, 193, 194, 195, 197, 198, 199, 200, 205, 207, 210, 211, 212, 213, 215, 216, 217, 221, 225, 226, 228, 230, 231, 232, 235, 239, 240, 241, 242, 243, 244, 245, 246, 252, 255, 256, 257, 259, 262, 263, 264, 265, 267, 268, 272, 273, 280, 281, 282, 283, 284, 288, 290, 292, 293, 296, 299, 301, 302, 303, 304, 305, 306, 307, 308, 309, 312, 314, 315, 316, 322, 324, 325, 326, 327, 328, 329, 331, 332, 333, 334, 335, 336, 337, 340, 343, 344, 345, 346, 351, 354, 355, 356, 357, 360, 361, 362, 363, 364, 365, 366, 367, 368, 369, 372, 373, 374, 375, 376, 379, 380, 385, 386, 387, 388, 389, 392, 393, 395, 399, 400, 402, 403, 407, 408, 409, 410, 413, 419, 422, 423, 424, 425, 432, 434, 444, 451, 459, 460, 461, 465, 470, 472, 475, 478, 485, 486, 487, 491, 492, 495, 496, 497, 498, 500, 501, 513
Makita U.S.A. Inc., 8, 279
Midway Homes Inc., 429, 431
Jon P. Muzzarelli, 453, 455, 472
Porter-Cable Corporation, 11, 360
Powermatic Inc., 279, 341
Prographics / Estwing, 149
S4S Industries, 475
Sears, 321, 342, 376, 416
Speight Studio Architects Inc. – D. Wayne Speight, 479
Jeff Stoecker, 50, 58, 62, 67, 68, 77, 85, 224, 225, 226, 227, 228, 230, 231, 232, 233, 253, 254, 255, 256, 258, 259, 260, 264, 266, 268, 298, 300, 301, 302, 305, 308, 344, 346, 347, 360, 362, 363, 364, 365, 367, 369, 382, 384, 386, 387, 388, 390, 391, 393, 398, 411, 412, 414, 416, 417, 418, 468, 469, 471, 491, 493, 494, 499, 503, 504, 505
Teledyne Post, 91
Thomasville, 89, 306, 319
Vacuum Pressing Systems Inc. / Darryl Keil, 89, 270, 271, 274
Photos Courtesy of Van Waters & Rogers Inc., 505, 506
Velux – America Inc., 78, 464
Wacker, 445
Better Homes & Gardens® Wood® Magazine
King Au, 544, 547, 558
Bob Calmer, 516
Kim Downing, 516, 517, 520, 521, 529, 531, 532, 533, 535, 536, 537, 545, 546, 547, 549, 559, 560, 561
John Hetherington, 5, 13, 518, 519, 522, 523, 524, 534, 538, 544, 547, 548, 550, 551, 552, 554, 556
Bill Hopkins, 515, 528, 538, 542
Lorna Johnson, 523, 526, 527, 539, 540, 541, 545, 546, 547, 551, 552, 553, 559, 560, 561
Roxanne LeMoine, 514, 515, 523, 539, 540, 541, 543, 551, 549, 552, 553, 555, 556, 557
Steve Struse, 522, 523
Bill Zaun, 516, 517, 520, 521, 529